Lecture Notes in Computer Science 4527

José Mira José R. Álvarez (Eds.)

Bio-inspired Modeling of Cognitive Tasks

Second International Work-Conference on the Interplay Between Natural and Artificial Computation, IWINAC 2007
La Manga del Mar Menor, Spain, June 18-21, 2007
Proceedings, Part I

Springer

Volume Editors

José Mira
José R. Álvarez
Universidad Nacional de Educación a Distancia
E.T.S. de Ingeniería Informática
Departamento de Inteligencia Artificial
Juan del Rosal, 16, 28040 Madrid, Spain
E-mail: {jmira, jras}@dia.uned.es

Library of Congress Control Number: 2007928350

CR Subject Classification (1998): F.1, F.2, I.2, G.2, I.4, I.5, J.3, J.4, J.1

LNCS Sublibrary: SL 1 – Theoretical Computer Science and General Issues

ISSN 0302-9743
ISBN-10 3-540-73052-4 Springer Berlin Heidelberg New York
ISBN-13 978-3-540-73052-1 Springer Berlin Heidelberg New York

Springer is a part of Springer Science+Business Media

springer.com

Printed in Germany

Typesetting: Camera-ready by author, data conversion by Scientific Publishing Services, Chennai, India
Printed on acid-free paper SPIN: 12076161 06/3180 5 4 3 2 1 0

Preface

The Semantic Gap

There is a set of recurrent problems in AI and neuroscience which have restricted their progress from the foundation times of cybernetics and bionics. These problems have to do with the enormous semantic leap that exists between the ontology of physical signals and that of meanings. Between physiology and cognition. Between natural language and computer hardware. We encounter this gap when we want to formulate computationally the cognitive processes associated with reasoning, planning and the control of action and, in fact, all the phenomenology associated with thought and language.

All "bio-inspired" and "interplay" movement between the natural and artificial, into which our workshop (IWINAC) fits, faces this same problem every two years. We know how to model and reproduce those biological processes that are associated with measurable physical magnitudes and, consequently, we know how to design and build robots that imitate the corresponding behaviors. On the other hand, we have enormous difficulties in understanding, modeling, formalizing and implementing all the phenomenology associated with the cognition field. We do not know the language of thought. We mask our ignorance of conscience with the term emergentism.

This very problem recurs in AI. We know how to process images, but we do not know how to represent the process for interpreting the meaning of behaviors that appear in a sequence of images computationally, for example. We know how to plan a robot's path, but we do not know how to model and build robots with conscience and intentions. When the scientific community can link signals and neuronal mechanisms with "cognitive magnitudes" causally we will have resolved at the same time the serious problems of bio-inspired engineering and AI. In other words, we will know how to synthesize "general intelligence in machines."

To attempt to solve this problem, for some time now we have defended the need to distinguish between own-domain descriptions of each level and those of the external observer domain. We also believe that it is necessary to stress conceptual and formal developments more. We are not sure that we have a reasonable theory of the brain or the appropriate mathematics to formalize cognition. Neither do we know how to escape classical physics to look for more appropriate paradigms.

The difficulty of building bridges over the semantic gap justifies the difficulties encountered up to now. We have been looking for some light at the end of the tunnel for many years and this has been the underlying spirit and intention of the organization of IWINAC 2007. In the various chapters of these two books of proceedings, the works of the invited speakers, Professors Monserrat and Paun, and the 126 works selected by the Scientific Committee, after the refereeing process, are included. In the first volume, entitled "*Bio-inspired Modeling of Cognitive Tasks*," we include all the contributions that are closer to the theoretical, conceptual and methodological aspects linking AI and knowledge engineering with neurophysiology, clinics and cognition. The second volume entitled "*Nature-Inspired Problem-Solving Methods in Knowledge Engineering*" contains all the contributions connected with biologically inspired methods and techniques for solving AI and knowledge engineering problems in different application domains.

An event of the nature of IWINAC 2007 cannot be organized without the collaboration of a group of institutions and people who we would like to thank now, starting with our university, *UNED*, and its Associate Center in Cartagena. The collaboration of the Universitat Politécnica de Cartagena and the Universitat de Murcia has been crucial, as has the enthusiastic and efficient work of José Manuel Ferrández and the rest of the Local Committee. In addition to our universities, we received financial support from the Spanish Ministerio de Educación y Ciencia, the Fundación SENECA-Agencia Regional de Ciencia y Tecnología de la Comunidad de Murcia, *DISTRON s.l.* and the Excelentísimo Ayuntamiento de Cartagena. Finally, we would also like to thank the authors for their interest in our call and the effort in preparing the papers, condition *sine qua non* for these proceedings, and to all the Scientific and Organizing Committees, in particular, the members of these committees who have acted as effective and efficient referees and as promoters and managers of pre-organized sessions on autonomous and relevant topics under the IWINAC global scope.

My debt of gratitude with José Ramón Alvarez and Félix de la Paz goes, as always, further than the limits of a preface. And the same is true concerning Springer and Alfred Hofmann and their collaborators Anna Kramer and Erika Siebert-Cole, for the continuous receptivity and collaboration in all our editorial joint ventures on the interplay between neuroscience and computation.

June 2007 José Mira

Organization

General Chairman

José Mira, UNED (Spain)

Organizing Committee

José Ramón Álvarez Sánchez, UNED (Spain)
Félix de la Paz López, UNED (Spain)

Local Organizing Committee

José Manuel Ferrández, Univ. Politécnica de Cartagena (Spain).
Roque L. Marín Morales, Univ. de Murcia (Spain).
Ramón Ruiz Merino, Univ. Politécnica de Cartagena (Spain).
Gonzalo Rubio Irigoyen, UNED (Spain).
Gines Doménech Asensi, Univ. Politécnica de Cartagena (Spain).
Vicente Garcerán Hernández, Univ. Politécnica de Cartagena (Spain).
Javier Garrigós Guerrero, Univ. Politécnica de Cartagena (Spain).
Javier Toledo Moreo, Univ. Politécnica de Cartagena (Spain).
José Javier Martínez Álvarez, Univ. Politécnica de Cartagena (Spain).

Invited Speakers

Gheorge Paun, Univ. de Sevilla (Spain)
Javier Monserrat, Univ. Autónoma de Madrid (Spain)
Álvaro Pascual-Leone, Harvard Medical School (USA)

Field Editors

Emilia I. Barakova, Eindhoven University of Technology (The Netherlands)
Eris Chinellato, Universitat Jaume-I (Spain)
Javier de Lope, Universitat Politécnica de Madrid (Spain)
Pedro J. García-Laencina, Universitat Politécnica de Cartagena (Spain)
Dario Maravall, Universitat Politécnica de Madrid (Spain)
José Manuel Molina López, Univ. Carlos III de Madrid (Spain)
Juan Morales Sánchez, Universitat Politécnica de Cartagena (Spain)
Miguel Angel Patricio Guisado, Universitat Carlos III de Madrid (Spain)
Mariano Rincón Zamorano, UNED (Spain)
Camino Rodríguez Vela, Universitat de Oviedo (Spain)

José Luis Sancho-Gómez, Universitat Politécnica de Cartagena (Spain)
Jesús Serrano, Universitat Politécnica de Cartagena (Spain)
Ramiro Varela Arias, Universitat de Oviedo (Spain)

Scientific Committee (Referees)

Ajith Abraham, Chung Ang University (South Korea)
Andy Adamatzky, University of the West of England (UK)
Michael Affenzeller, Upper Austrian University of Applied Sciences (Austria)
Igor Aleksander, Imperial College of Science Technology and Medicine (UK)
Amparo Alonso Betanzos, Universitate A Coruña (Spain)
José Ramón Álvarez Sánchez, UNED (Spain)
Shun-ichi Amari, RIKEN (Japan)
Razvan Andonie, Central Washington University (USA)
Davide Anguita, University of Genoa (Italy)
Margarita Bachiller Mayoral, UNED (Spain)
Antonio Bahamonde, Universitat de Oviedo (Spain)
Emilia I. Barakova, Eindhoven University of Technology (The Netherlands)
Alvaro Barreiro, Univ. A Coruña (Spain)
Josh Bongard, University of Vermont (USA)
Fiemke Both, Vrije Universiteit Amsterdam (The Netherlands)
François Brémond, INRIA (France)
Enrique J. Carmona Suárez, UNED (Spain)
Joaquín Cerdá Boluda, Univ. Politécnica de Valencia (Spain)
Enric Cervera Mateu, Universitat Jaume I (Spain)
Antonio Chella, Università degli Studi di Palermo (Italy)
Eris Chinellato, Universitat Jaume I (Spain)
Emilio S. Corchado, Universitat de Burgos (Spain)
Carlos Cotta, University of Málaga (Spain)
Erzsébet Csuhaj-Varjú, Hungarian Academy of Sciences (Hungary)
José Manuel Cuadra Troncoso, UNED (Spain)
Félix de la Paz López, UNED (Spain)
Ana E. Delgado García, UNED (Spain)
Javier de Lope, Universitat Politécnica de Madrid (Spain)
Ginés Doménech Asensi, Universitat Politécnica de Cartagena (Spain)
Jose Dorronsoro, Universitat Autónoma de Madrid (Spain)
Gérard Dreyfus, ESCPI (France)
Richard Duro, Universitate da Coruña (Spain)
Juan Pedro Febles Rodriguez, Centro Nacional de Bioinformática (Cuba)
Eduardo Fernández, University Miguel Hernandez (Spain)
Antonio Fernández-Caballero, Universitat de Castilla-La Mancha (Spain)
Jose Manuel Ferrández, Univ. Politécnica de Cartagena (Spain)
Kunihiko Fukushima, Kansai University (Japan)
Jose A. Gámez, Universitat de Castilla-La Mancha (Spain)
Vicente Garceran Hernández, Universitat Politécnica de Cartagena (Spain)

Pedro J. García-Laencina, Universitat Politécnica de Cartagena (Spain)
Juan Antonio García Madruga, UNED (Spain)
Francisco J. Garrigos Guerrero, Universitat Politécnica de Cartagena (Spain)
Tamás (Tom) D. Gedeon, The Australian National University (Australia)
Charlotte Gerritsen, Vrije Universiteit Amsterdam (The Netherlands)
Marian Gherghe, University of Sheffield (UK)
Pedro Gómez Vilda, Universitat Politécnica de Madrid (Spain)
Carlos G. Puntonet, Universitat de Granada (Spain)
Manuel Graña Romay, Universitat Pais Vasco (Spain)
Francisco Guil-Reyes, Universitat de Almería (Spain)
John Hallam, University of Southern Denmark (Denmark)
Juan Carlos Herrero, (Spain)
César Hervás Martínez, Universitat de Córdoba (Spain)
Tom Heskes, Radboud University Nijmegen (The Netherlands)
Fernando Jimenez Barrionuevo, Universitat de Murcia (Spain)
Jose M. Juarez, Universitat de Murcia (Spain)
Joost N. Kok, Leiden University (The Netherlands)
Yasuo Kuniyoshi, Univ. of Tokyo (Japan)
Petr Lánsky, Academy of Sciences of Czech Republic (Czech Republic)
Hod Lipson, Cornell University (USA)
Maria Longobardi, Università di Napoli Federico II (Italy)
Maria Teresa López Bonal, Universitat de Castilla-La Mancha (Spain)
Ramon López de Mántaras, CSIC (Spain)
Tino Lourens, Philips Medical Systems (The Netherlands)
Max Lungarella, University of Tokyo (Japan)
Manuel Luque Gallego, UNED (Spain)
Francisco Maciá Pérez, Universitat de Alicante (Spain)
george Maistros, The University of Edinburgh (UK)
Vincenzo Manca, University of Verona (Italy)
Riccardo Manzotti, IULM University (Italy)
Dario Maravall, Universitat Politécnica de Madrid (Spain)
Roque Marín, Universitat de Murcia (Spain)
Jose Javier Martinez Álvarez, Universitat Politécnica de Cartagena (Spain)
Rafael Martínez Tomás, UNED (Spain)
Jesus Medina Moreno, Univ. de Málaga (Spain)
Jose del R. Millan, IDIAP (Switzerland)
José Mira, UNED (Spain)
Victor Mitrana, Universitat Rovira i Virgili (Spain)
José Manuel Molina López, Univ. Carlos III de Madrid (Spain)
Juan Morales Sánchez, Universitat Politécnica de Cartagena (Spain)
Federico Morán, Universitat Complutense de Madrid (Spain)
Arminda Moreno Díaz, Universitat Politécnica de Madrid (Spain)
Ana Belén Moreno Díaz, Universitat Rey Juan Carlos (Spain)
Isabel Navarrete Sánchez, Universitat de Murcia (Spain)
Nadia Nedjah, State University of Rio de Janeiro (Brazil)

Taishin Y. Nishida, Toyama Prefectural University (Japan)
Richard A. Normann, University of Utah (USA)
Manuel Ojeda-Aciego, Univ. de Málaga (Spain)
Lucas Paletta, Joanneum Research (Austria)
José T. Palma Méndez, University of Murcia (Spain)
Miguel Angel Patricio Guisado, Universitat Carlos III de Madrid (Spain)
Gheorghe Paun, Universitat de Sevilla (Spain)
Juan Pazos Sierra, Universitat Politécnica de Madrid (Spain)
Mario J. Pérez Jiménez, Universitat de Sevilla (Spain)
José Manuel Pérez-Lorenzo, Universitat de Jaén (Spain)
Jose M. Puerta, University of Castilla-La Mancha (Spain)
Alexis Quesada Arencibia, Universitat de Las Palmas de Gran Canaria (Spain)
Günther R. Raidl, Vienna University of Technology (Austria)
Luigi M. Ricciardi, Università di Napoli Federico II (Italy)
Mariano Rincón Zamorano, UNED (Spain)
Victoria Rodellar, Universitat Politécnica de Madrid (Spain)
Camino Rodríguez Vela, Universitat de Oviedo (Spain)
Daniel Ruiz Fernández, Univ. de Alicante (Spain)
Ramón Ruiz Merino, Universitat Politécnica de Cartagena (Spain)
Ángel Sánchez Calle, Universitat Rey Juan Carlos (Spain)
Eduardo Sánchez Vila, Universitat de Santiago de Compostela (Spain)
José Luis Sancho-Gómez, Universitat Politécnica de Cartagena (Spain)
Gabriella Sanniti di Baja, CNR (Italy)
José Santos Reyes, Universitate da Coruña (Spain)
Shunsuke Sato, Aino University (Japan)
Andreas Schierwagen, Universität Leipzig (Germany)
Guido Sciavicco, Universitat de Murcia (Spain)
Radu Serban, Vrije Universiteit Amsterdam (The Netherlands)
Jesús Serrano, Universitat Politécnica de Cartagena (Spain)
Jordi Solé i Casals, Universitat de Vic (Spain)
Antonio Soriano Payá, Univ. de Alicante (Spain)
Mª. Jesus Taboada, Univ. Santiago de Compostela (Spain)
Settimo Termini, Università di Palermo (Italy)
Javier Toledo Moreo, Universitat Politécnica de Cartagena (Spain)
Jan Treur, Vrije Universiteit Amsterdam (The Netherlands)
Ramiro Varela Arias, Universitat de Oviedo (Spain)
Marley Vellasco, Pontifical Catholic University of Rio de Janeiro (Brazil)
Lipo Wang, Nanyang Technological University (Singapore)
Stefan Wermter, University of Sunderland (UK)
J. Gerard Wolff, Cognition Research (UK)
Hujun Yin, University of Manchester (UK)

Table of Contents

Neural Networks and Quantum Neurology: Speculative Heuristic Towards the Architecture of Psychism

Javier Monserrat

Universidad Autónoma de Madrid

Abstract. A new line of investigation known as quantum neurology has been born in recent years. One of its objectives is to accomplish a better explanation of psychism. It basically explains the unity of consciousness, its holistic character, and the indeterminism of its responses. How is this "phenomenological *explicandum*" explained in classical neurological architecture? After commenting on the properties of classical architecture, we focus on the proposal of Edelman, since we consider it as probably one of the better proposals explaining psychism. The discussion of Edelman's proposal, from the viewpoint of the problem about the "physical support" of psychism in classical physics, allows us to evaluate the strengths of his proposal, as well as the remaining insufficiencies in his explanation. The "heuristic" way of quantum neurology offers a new approach to the "phenomenological *explicandum*" that does not contradict, but completes classical architecture. The discussion regarding the Hameroff-Penrose hypothesis allows us to propose that the psycho-bio-physical ontology would have an architecture with three levels (or sub-architectures) and two (or three) interface systems among them. This hypothetical architecture permits us to reflect on the production of ontologies, architectures, and functional logics (real and artificial). In any case, the new quantum neurology would suggest new formulations of the psycho-bio-physical ontology by means of the graph theory (classical neurology) and of topology (quantum neurology).

1 Introduction

These reflections are being proposed based on my professional and personal interests, namely, epistemology, cognitive psychology and vision science.

When speaking about neural networks, I do not refer primarily to systems of artificial neurons that permit parallel distributed processing within the framework of connectionism as a man-made architecture, but to biological networks of living neurons which make up our brains. These networks, according to authors, are also referred to as patterns, cannons, structures, engrams or neural maps.

What we could call classic neurology, according to our understanding, would be, on one hand, the understanding of these neural networks: their diversification, their modularization, their branching structure of interconnected and interactive systems as, among other things, functioning units; and, on the other hand, the

J. Mira and J.R. Álvarez (Eds.): IWINAC 2007, Part I, LNCS 4527, pp. 1–20, 2007.

knowledge of their correlational order (is it causal?) with these events which are phenomenologically called *qualia* and which constitute the essential elements of our physical lives.

Along with this classic neurology, today there is being developed what could be called *quantum neurology.* In this new field there is a bitter argument among those who reject it passionately, those who respect it with interest, and those who support it, persuaded that it is opening paths to new knowledge which will enrich the scientific explanation of the human and animal psyche. In any case, using our own evaluation, we only refer to quantum neurology as a scientific heuristic, as there does not seem to be enough scientific evidence to confirm it (given the provisional character of all scientific confirmations).

It could be said that the main characteristic of this new quantum neurology is this: to suppose that quantum phenomena – or, more exactly, phenomena that are quantum coherent – could occur in the internal biological tissue of neurons (but not only) and that it would be possible to attribute to them the "physical support" or real ontology, which produces the emergence of sensation or germinal "senticence" and, after complex evolutionary transformations, the emergence of sensation-perception-consciousness which is the architectural base of the animal and human psyche. If this were the case, neural networks would then constitute the skeleton in which the structure of "quantum niches" that give way to sensation-consciousness would live. This neural skeleton could work as a connector, both in ascendant and descendent ways, between "quantum niches" and the internal and external environment of organisms.

The first idea we present is that quantum neurology has opened up a new form of heuristic speculation for science, which is pertinent for two reasons: 1) because it is scientifically and methodologically well constructed when compared with alternative theories that are open to experimental and empirical criticism, and 2) because it points out a suggestive explanation for psychic phenomenon that science cannot ignore and must understand but up until now has not sufficiently explained within the alternatives offered by the framework of classical neurology. It is obvious that all this must be clarified and we hope to do so in this presentation.

Moreover, and this is the second idea we will defend, we consider that, if quantum neurology is discovering the real architecture of the physical ontology that provides organisms the sensation of themselves and of their external environment, then this architecture is also capable of suggesting models for either: a) an interaction with the real physical ontology, or b) a design and construction of artificial physical architectures which are oriented to some specific goal, or c) an abstract conception of the formalisms which could support the suppositions presented in points a) and b). We will clarify this in what follows.

2 What Scientific Argumentation Demands

We consider that these two ideas can be seen as the logical result of a well-constructed argument, not as a logical necessity, but as a means to diffuse logical probability, which we consider not only justified, but necessary in the process of "doing" science.

Science should really respond to the rules established by epistemological argumentation. We understand that the theory that gives an empirical base to science today is not positivism, but a theory closer to Popper's ideas or even to the conceptions developed after Popper: scientific facts are an "interpretation" made by the human receptive system. Moreover, the human "phenomenal" experience itself is also part of the *explicandum* of the human sciences. Science must produce knowledge based on "empirical evidence". This is not trivial, because it means that science has an epistemological moral obligation to explain all facts (to find the real causes that produce them). In other words: it is not "scientific" to ignore or discriminate some facts in order to explain them.

Because of this, the science that deals with humans does not begin with theories, but with facts that demand an explanation. It is facts that should orient the (heuristic) search for explanations (or theories), and facts are the last appeal to judge the suitability of proposed explanations (or theories). What we referred to previously as classical neurology is a theory that is supported by a dense weave of interconnected empirical evidence; it belongs to the field of theories. Quantum neurology is also a heuristic theory, although it is less mature and less accepted than classical neurology. This means that it is being constructed right now. It assumes and integrates quantum neurology and all its empirical evidence, but also opens up new horizons for the explanation of humans and builds new theoretical frames that could orient new designs of empirical research which might be able to confirm it.

Classical neurology has been able to explain many facts; obviously no one, not even we, will dare to question this. But what we are trying to defend is that some empirical (phenomenological) facts, which are important and unquestionable and cannot be ignored but explained by science, do not seem to be explained adequately by classical neurology: many authors have proposed different arguments which defend that classical neurology is not really able to explain these facts (although other authors, for example, Edelman, do think that classical neurology does explain them).

Quantum neurology was born precisely as a theoretical heuristic whose main purpose was to explain those facts which seemed not to have an explanation in the context of classical neurology. Being the only alternative that tries to explain these facts, the methodological and epistemological requirements of science demand the promotion of a heuristic reflection towards new proposals that could explain these anomalous facts (we should remember Kuhn). The "heuristic" search for new theories does not mean that we should accept them before they are ready; but the rich proliferation of theories is essential for science (Feyerabend) as it allows internal discussion, the promotion of different lines of experimentation and the election of the best theory. For these reasons we think that the promotion of quantum neurology today is a necessity of science which is epistemologically well grounded.

Quantum neurology is therefore building a heuristic theory whose affirmations have a logical, diffuse probability, but which needs a more solid structure of empirical evidence to be able to consolidate itself. However, as a provisional theoretical construct (in a heuristic search) it is sufficiently supported by the methodical exigencies of science: in science basic theorization is at least as necessary as the search for empirical evidence. In many cases it would not even be possible to know which kinds of evidence should be searched for without theorization. So our urge to build this theoretical alternative really responds to an epistemological logic which is sufficiently justified today.

3 Neurological Explicandum and Phenomenology

So, what are the problematic facts we are referring to? In principle, these facts are a part of the *explicandum* of the human sciences. All sciences are based on phenomena or facts of immediate experience which "should be explained" by the knowledge of the causes that produce them; these causes are the scientific *explicans*. In the case of the human sciences, the *explicandum* is made up by the totality of our phenomenological experience, that is, all our immediate, personal experience, and also all our consensual and socially - or intersubjectively - lived experience, all what it means to be "a human being".

The scientific discipline that describes phenomenological experience is phenomenology. Phenomenology itself can be debated and should be contended. It is, in fact, a very complex discipline: different authors and schools, from fields like philosophy, psychology or even neurology, have proposed different basic ways for the phenomenological analysis of human beings.

We could think, for example, about the immediate experience of our own cognitive activity: doing science as an activity has been described differently in the scientific epistemological theories of positivism, Popper's or post-popperianism (at the bottom, many theories of science are simply functional descriptions of how we act when doing science). Language is also a fact that can be described. Therefore, knowledge, science, language, all are the *explicandum* that must be explained scientifically by the knowledge of their causal systems or *explicans*. Even "reason" is also a functional experience (which connects cognitive and linguistic contents) which appears as a phenomenon and which should also be explained by science from its causal system.

It is obvious that in this presentation we cannot undertake a deep analysis and discussion of phenomenology. But we want to refer to three very important phenomenological features or contents of our human experience (reducible to two), which are part of the basic *explicandum* of the human sciences and to which we will refer later. Science cannot ignore them nor can it avoid explaining them. They are as follows:

1. The unitary character of *consciousness*. Our consciousness is noticed as a system which integrates in a unity the different sensory modalities (vision, audition, proprioception ...) which are projected to the psychical subject that coordinates them and sets responses.

2. The *holistic* character of consciousness. Our experience of consciousness is wide open: we feel the openness of the external space through vision, the unitary extension of our sensations through our own body, or the wide unity of our internal experience when we close our eyes and follow the stream of our thoughts.
3. The *indeterminate nature* of the responses of a conscious subject. Subjects notice themselves as open to a multitude of possible responses and consider themselves as the cause of these responses. Responses may be driven by programmed automatisms, but subjects strategically exercise their control and feel that their lives are played out without an absolute determinism, with free indeterminate options (which does not mean absolutely unconditioned). This phenomenology of our own experience of indetermination (free will) creates the basic persuasion that gives sense to our personal and social life.

It is evident that these three features are not exclusive to the human domain. According to the modern views of comparative psychology, ethology, biology and evolutionary neurology, we can make a scientific inference based on the fact that these three features - the unity of holistic consciousness, and the flexibility and indetermination of adaptive responses - are present in higher animals in varying degrees and with their corresponding characteristics. It is clear that animal indetermination is not comparable to the free will of man, opened by the exercise of reason; but it is certainly an evolutionary prologue. For that reason, the features that we select as a reference in our presentation are common, in this sense, to both the animal and human domains.

4 Psychic Architecture in Classical Neurology

All of the essential lines of psychic architecture are already known. Further on in the presentation we will refer to Edelman, but we consider it convenient to stop here to present a brief synthesis of classical neurology.

We will begin with visual images. A pattern of light, codified by its differential reflection in the external world, is processed by the optics of the eye before being eliminated in the layer of photoreceptors in the retina. If the point differences of the image were codified in the light patterns, they should produce a trans-codification in the retina: it passes from a photonic code to a neural code. The electro-chemical signals, via ordinary synaptic communication, transmit the image to the brain. The signals arrive at the superior colliculus, the oldest visual nucleus in evolutionary terms, and then to more modern centers such as the LGN (lateral geniculate nucleus). From there the signals travel to zones V_1, V_2, and V_3 of the visual cortex. These zones connect with nearby zones like V_4 and V_5 and more widely with the brain by way of the "where route" (towards the superior parietal lobe) and by way of the "what route" (towards the parietal lobe). The correct activation of the neural engram, pattern, or canon of a specific image creates the psychic effect of "seeing". In this active system (from the retina to the cortex), every one of the parts plays a special role in producing the image.

The visual system produces the activation of a complex neural pattern which produces the psychic effect of "seeing" the image with the wide range of *qualia* it has.

The image is constructed, then, in the module that processes images (for example, the image of a lion). But this image is also connected with the temporal lobe, in which its cognitive interpretation occurs (what is a lion); visual agnosia allows us to determine that it is possible to have an image without a cognitive interpretation (to see a lion without knowing what it is). The idea of a lion is connected too with the semantic and phonetic areas of the brain that process language, so we can shout: "a lion!". In a similar way, there are connections with the limbic system (the amygdala), so an emotional reaction of fear occurs because of the lion. Likewise other modules are activated in turn, especially in the prefrontal and frontal areas, and subsequently a plan of action is defined to confront the situation [25, chap. IV] [28].

We could say that in the psychological subject all the *qualia* produced in the different modules of the neural system come together in parallel (the subject sees the image, hears sounds, feels his body in a holistic manner, notices the emotional effect...) and, therefore, give impulse to the actions that constitute the subject's response (the subject produces language, controls his own movements, prepares a plan of action, builds his own thoughts...). This psychological subject, which is also present in the animal domain, has emerged little by little in the process of evolution through the process of neurally mapping the body in the brain (as has been explained particularly by Antonio R. Damasio [2, part III]).

The activation of engrams that produces the *qualia* that the psychological subject feels is only the tip of the *iceberg* of the neural system. There is an incredibly dense weave of engrams which do not produce *qualia,* but which still determine all organic regulations (already built in the evolutionary oldest parts of the brain) and multiple automatisms which support conscious behavior (for example, motor or linguistic automatisms). With these arguments we are not trying to discuss the problem of determination and free will in the field of neurology. We just want to point out that, in normal subjects, this psychological and neural architecture is not a closed or static system, but a very flexible one which can reach surprising degrees of plasticity. For example, when some parts of the body are missing, the brain can simulate them (i. e., phantom limbs); but when it is some part of the brain that is missing, the brain can reorganize itself in amazing ways to still "deal" with the stimulation that comes from the body (i. e., reorganization of motor or linguistic areas after a brain lesion). *Classic architecture*has its own characteristics. We would point out seven of them:

1. It is stable, but also oscillating. It is not a neural network of retropropagation which can be controlled from some other system. Afferent stimulations (which arrive in the sensitive brain and move to connected areas) produce interactive structures (engrams) in a classic, unitary system in which they are registered or "facilitated" (Hull), becoming then available for later reactivation. These structures are stable, but not in a rigid but oscillatory way (as we can see in the fuzziness of our memories).

2. These structures (engrams) are co-participative connections: the same neurons and the same branches of synaptic connections (each neuron can have thousands of synapses) can co-participate in multiple different engrams.
3. These networks of connections expand in a classical three-dimensional macroscopic space which responds to the shape of the brain.
4. The networks are connected and activated in parallel inside the same three-dimensional spatial topology, i. e., when seeing an image in real time, the subject simultaneously notices that vague reminiscences flow into his memory, feels his own body and follows a line of thought.
5. These networks are built following the logic of a well-arranged topology: this ordering allows for an ordered interaction of, for example, the engrams which are activated and de-activated when I explore a piece of knowledge which was registered in my mind (in the frontal and prefrontal areas but interacting with other cerebral modules). This gives rise to what William James called the stream of consciousness, whether the engrams are images or thoughts. This ordering is both intra-modular (i.e., an ordered record of folders with images or sounds) and inter-modular (i.e., a knowledge system which, after the activation of the frontal areas, connects in real time with the images activated in parallel in the visual registry module).
6. These networks are dynamic. This means that, although the neural records are stable, as we said before, they are being transformed continuously, as we postulated that the transformation should occur, for example, in the continuous stream of visual or auditory images, in the unconscious occurrence of engrams which control language, motility or the stream of thoughts.
7. These networks are plastic in the sense that functions allow the construction or improvisation of the architecture itself, with the properties that we discussed before: the brain can re-organize itself when either some substantial part is missing or after a brain lesion occurs. Therefore, we could say that the classical neural architecture is self-generating: a germinal architecture which is not yet developed allows its own functions to adequately generate the complex architecture that we observe in the mature system.

5 Gerald M. Edelman and the Sufficiency of the Classical Neuronal Architecture

In this article, when we speculate about sufficiency, we ask whether classical neurology offers a satisfactory explanation of the phenomenological *explicandum.* If we limit ourselves to previously chosen phenomenological features, we wonder if classical neurology can explain a) the unity of consciousness, b) the holistic experience of consciousness, and c) the indeterminate nature of the responses of the conscious subject. Many neurologists have, of course, taken for granted the classical explanation without noticing any problems. Others have observed, at least to some degree, that the classical view can be problematic and have tried to offer a convincing answer. According to our point of view, Gerald M. Edelman has developed the most profound system and the most well-constructed arguments.

Therefore, referring to him can serve as a way to reflect on the sufficiency of classical neurology.

In fact, Edelman understands that neurology should explain phenomenological experience. In "*The Universe of Consciousness*" his phenomenology (which we will not detail here) presents two essential features described as "continuous unity" and "infinite variety", which are, to our mind, a light version of Edelman's ideas on holistic unity and indeterminate free will. "Continuous unity" refers to the unitary sensation of the body and of all the psychological modalities (sensations, emotions, etc...) and their convergence in the conscious subject. "Infinite variety" refers to the modality of human or animal actions caused from consciousness to the unknown (against the "instructional" determinism of computers). So, where then does Edelman's explanation take us? [9, part I].

Edelman's explicative system is based on neural Darwinism and the theory of neural group selection (TNGS [5] [9, chap. 7]). Many physical and biological processes are explained assuming that Nature has produced an enormous amount of different states which enable the evolutionary selection of those traits that are better adapted. Human brains are thought to be built by neural darwinistic selection as well, therefore constructing a quasi-infinite variety of possibilities for engrams. What has been selected are neural groups (not individual neurons but groups of them) and the connections which form the most adaptive maps or engrams. The neural architecture, as Edelman conceives it, agrees with the descriptive characteristics of the classical architecture that we presented before. Edelman has contributed mainly to the explanation of the emergence of psychological and cognitive activity through his analysis of the representative processes in the mechanisms of memory. But now I would like to pay attention to his concept of "dynamic nucleus" because it will be the basis for an explanation of the phenomenological features that we mentioned before.

The dynamic nucleus hypothesis is an explanation of how the brain functions. It is the final consequence of TNGS and the conception of the nervous system as a diversified specialized group that produces by means of neural darwinism a unitary psychological activity that is diversified and specific. Think about our psychological experience: Our conscious self coordinates proprioceptive, visual, auditory, tactile and kinesthetic experiences in a single moment as if they were a remembered present of complex auto-images, dense systems of awareness, thought, registered imagination, emotional states, etc., that flow over into the present. All these guide the direction of behavior and coordinate our motor functions, although they vary and are redirected following changes in stimuli and the use of the ability to choose, degenerate and generate an infinite array of new possibilities [9, chap. 12].

How is such complexity possible? Edelman responds with the dynamic nucleus hypothesis: In real time, in the hundreds of milliseconds that constitute collective activations occurring over and over again (are generated and degenerated) and mapped from diverse modules that contain the neural bases for all the different psychological activities, everything flows together in the psychological subject as a single system because of complex activation and de-activation buses that are coordinated by multi-directional re-entries [7, pp. 64–90] [9, chap. 10]. These complex relationships of re-entry among modules are the neural correlates that support conscious activity, both as a continuous unity and the way it can be informed (modular diversity and registered content).

Therefore, what science should now explain is continuous unity and infinite variety. As we have pointed out earlier, the dynamic nucleus hypothesis must justify two properties of the mind: integration and re-entry (which form the basis for continuous unity) and differential complexity (which form the basis of differentiation and infinite variety). Edelman believes that his dynamic nucleus hypothesis, as a synthesis of macroscopic neurology (of neurons and synaptic networks), explains how the different maps unitarily flow together in real time and how the complexity (i.e., the huge population) of the maps permits a selection that is controlled by the subject in the context of the environment. The mind is thus unitary as a parallel system, i.e., it is "selective". For Edelman, this is the same as saying it is indeterminate, not instructional. In this sense, neural darwinism, because of its selectivity, would be based both on indetermination and on animal and human behavior [31].

6 The Physical Support Problem of Psychism

Does Edelman's hypothesis explain the phenomenological *explicandum* that we started with? It seems clear that, in part, it does contribute something, at least when explaining it. The activation of parallel engrams and their references (also parallel) that are produced by the subject make, without a doubt, the unitary and holistic experience of consciousness intelligible. The continuous selection among a multitude of engrams in their optimal state would also permit the understanding of variability and of the indeterminate unpredictability of consciousness. Nevertheless, the problem should be analyzed in the light of our own ideas (as understood by the physical sciences) about "physical support," which proceed from our understanding of psycho-biophysics. From this point of view, we can discuss whether Edelman's version of the classical explanation of phenomenological experience is sufficient.

The scientific expectation, as we have said before, is monistic. The biological world has been evolutionarily produced because of a preexisting ontology of "physical support". In turn, the psychological world has also been produced due to the same preexisting ontology of "physical support". How do we know that this "physical support" depends on physical science? At this point, we have to make an important observation: Not all of physics can make psychological experience a product of evolution, basing it on the ontology of the physical world; this is understandable. For example: If the physical world was made up of what the Greeks called "atoms", invisible and closed-in on themselves (as imagined by certain mechanisms of the 19TH and even 20TH centuries), it would be impossible to explain not only how the experience of sensation-perception-consciousness is produced, but also unity, holism and psychic indetermination. A clear idea of science could therefore take us from reductionism to dualism.

However, current ideas about matter no longer follow the atomic model of the Greeks. The primal matter of the *big bang* is radiation, which extends in physical fields. Particles are "folded radiation" that gradually forms what we call "matter" or physical objects. There is, in certain conditions, a conversion of matter into energy or radiation, and vice-versa. Matter "unfolds" and converts into energy; the energy in radiation can "fold up" into matter. The wave-corpuscle (or particle-field) duality is one of the principles of quantum mechanics. The physical world has as many "field" properties as "corpuscle" properties. The ultimate idea, however, is that the ontology of real things is an "energy field", given that particles (and physical bodies) are made up of a folding or alteration of the base energy in that field (which has received diverse names throughout the history of physics and that today remains related to the concept of a quantum vacuum).

We should remember that physics now differentiates between two types of particles or matter. First, there is bosonic matter which is formed by a certain type of particle that has the property of unfolding more easily in fields of unitary vibration. In this way, the mass of bosonic particles, for example, the photon, lose their individuality when they enter into a state of unitary vibration that is extended in a field constituting a state recognised as quantum coherence. The wave function is symmetric and it is considered that this depends on these properties. The first description of these states of coherence were Bose-Einstein condensates. Today, in modern physics, a multitude of quantum coherence states have been described within the most strict experimental conditions.

Second, there is fermionic matter. These are particles whose wave function is asymmetrical, so that their vibrations have difficulties entering into coherence with other particles. They persistently maintain their individuality, not fusing with other particles and remaining in a state of unitary indifferentiation. Electrons and protons, essential constituents of atoms, for example, unite and form material structures according to the 4 natural forces: gravity, electromagnetic force, strong nuclear force and weak nuclear force. Nevertheless, every particle maintains its individuality. Every electron in an atom, for example, has its orbit, which, when completed, makes the electron vibrate in its orbital space. According to quantum principles, we cannot know exactly where the electron is. The location in space depends on the "collapse of the wave function" of said particular electron; the collapse is produced, for example, by the experimental intervention of an observer. Because of the fact that the energy of the *big bang* caused the folding of this type of fermionic particles, the classical macroscopic world exists: stellar bodies, planets, living things and man. Their folding accounts for differentiation and the possibility of a multitude of unfathomable things, like the survival of living beings with stable bodies and standing firmly on the surface of the earth.

This enables us to have an idea regarding how causality happened in the classical macroscopic world, i.e., the world organized in terms of fermions. One can think of two stones crashing into each other and breaking, or of a watch whose gears transmit motion. These physical entities, stones or metallic pieces, remain as closed and differentiated units. If we go down to the quantum level of microscopic fermionic entities, we can see that in molecules and macromolecules, every atom and every particle continue to have the same identity. Actions and cause-effect series are, in this world, associations and dissociations of independent particles, atoms and molecules by means of ionic unions and covalences abiding by the four previously mentioned forces. Shared orbits of electrons can be formed in covalent links, but they are very localised and probably do not nullify the independence of the electrons. However, what is interesting to note here is a consequence: Causal interactions do not break the enclosure and differentiation of the component elements in the classical macroscopic world made up of fermions. In other words, holistic fields do not appear in the world of physical fields.

Furthermore, these causal systems are partly deterministic: The conditions that blindly produce a bond or dissociation follow the laws of physics and chemistry. On the other hand, however, these systems can give rise to indeterminate states: We will not know the precise effect of a state that is produced among a multitude of possible states. We attribute the effect to a chance that is unpredictable certain. This happens in the physics of chaotic systems and in biology, for example, in cytoplasmic biochemistry that gives rise to Darwinian selection. The fermionic evolution (mechanical-classical) of the universe has produced states or loops of indetermination; but what is finally produced in these indeterminate environs is caused by cause-effect series that are blind and deterministic.

We now return to the question that was asked before: Does the architecture of Edelman, as an excellent theory of classical neurology, explain phenomenological experience? The first thing we should notice is that classical neurology constructs its explanations based on a microscopic fermionic world. In discrete events occurring among neurons of the network, which is our brain, deterministic cause-effect series are transmitted (along with the previously mentioned chaotic reservation) that do not create fields nor break the differentiation of entities in each neuron or in other structural entities (macromolecules, molecules, atoms, particles etc.) conforming with their fermionic nature.

As a consequence: (1) The "unity of consciousness" is partially explained, as with Edelman, by the parallel convergence of all the engrams that project their effect in the psychic subject-coordinator, but this unity is made up of differentiated and isolated parts; it is like the unity seen in the complex mechanism of a watch. (2) The "holistic unity of consciousness" does not seem to be explained as a function of an adequate "physical support" for the same reasons: In vision, for example, an image transmitted by a photonic code in light becomes disintegrated in the brain, and it is not possible to understand what the integration field observed in the phenomenological experience could consists of. (3) The explanation concerning the "indetermination of the conscious subject" can be accomplished in part through the mechanics of chaos and darwinistic biological selection within a fermionic classical macroscopic framework. However, phenomenological experience contains something more that is not explained: Animals choose responses based on the telenomic logic of their systems, and man, in addition, chooses responses based on rational and emotional thinking (it is not a pure chaotic indetermination chosen by chance because of darwinistic selection).

7 Quantum Neurology in Search of a "Physical Support" of Psychism

Edelman makes the observation that "sensation" cannot be explained by science [8, pp. 116–117 and 138–139]. We cannot know why matter is susceptible to producing sensation. We agree completely with this observation. Questions of the type like "Why does matter produce sensation instead of not producing it?" or questions also like the classic question of Leibnitz, reformulated by other philosophers, "Why does something exist instead of nothing?" are questions that do not have a response. We begin from the fact that something exists, and that we should attribute the ontological property of producing consciousness to the primordial substrate that caused the universe. The pieces of empirical evidence from the process of evolution oblige us to do so within the monistic expectation of science. The problem of science, then, is not so much whether matter produces consciousness or not (something which is a fact), but to understand how the ontology of matter can explain its phenomenological properties.

Can a discontinuous world - with some entities isolated from others, corpuscular or "fermionic" in the previous sense - explain the unity of consciousness and its holistic contents (sensitive integration of fields of reality as in vision or in proprioception)? Can the causality produced by deterministic cause-effect series from interactions among entities made of fermionic matter explain certain variations in the indeterministic flexibility in animal behavior and human freedom? Everything is debatable, but many certainly think that it cannot be explained.

Where to find, then, an adequate "physical support" to ground in a sufficient manner the phenomenological properties of psychism? Current psycho-physics is moving towards the field aspects of physical reality, already known for many years now. It was almost an inevitable option to think that the solution, or at least a more convincing manner of explaining that comes closer to the phenomenological *explicandum*, could be found by searching in physical fields and among the properties of matter described in quantum mechanics. In 19^{TH}century classical mechanics, physical reality was corpuscular matter and radiation. Quantum mechanics unified these two aspects in the corpuscle-wave duality, with particles as "folded radiation" (as we said before). Every matter, bosonic and fermionic, is "radiation" in its ontological core. Bosonic matter tends to be diluted easily in everything unified, in vibrating fields, losing the individuality of its particles in states of "quantum coherence." But although fermionic matter firmly maintains its individuality, it can also produce states of coherence, as has been verified in extreme experimental situations. Note that fermionic matter also pertains to the quantum world. In other words, knowledge about quantum mechanics (e.g., the electron in its orbit is a vibrating wave) is applied to it.

What do we understand, then, by "quantum neurology"? A more general definition could be the following: It is the search for and investigation about the quantum properties of the most primitive matter in order to relate them to the neuronal system in view of establishing the appropriate "physical support" to explain the phenomenological properties of psychism. To this end, authors like Henry Stapp, Herbert Fröhlich, Stuart Hameroff, Roger Penrose, Albert F. Popp, among others, have contributed ideas. These authors have contributed ideas and proposals, but they do not exhaust quantum neurology. Their contributions can be more or less certain, and above all, debatable, always setting aside experimental and empirical evidence. It is very possible that the truly prolific ideas for quantum neurology have not yet been proposed, and that crucial empirical contrasts perhaps have not yet been designed.

In this presentation, we cannot deal with the exposition of the ideas of these authors. But we take for granted that their ideas are already known. Since we are now going to refer to them, we will at least recall the basic outline of the Penrose-Hameroff hypothesis [33, chap. VII] [19], now the center of discussions. Very briefly synthesized, the hypothesis consists in arguing that some structures of the cellular cytoskeleton, microtubules, distributed widely in the entire neuron, could possess the appropriate physico-biological characteristics, so that the phenomenon of quantum coherence could occur in them. Vibrating states in quantum coherence would have a wave function that would be in "quantum superposition" (being in multiple states at the same time and not being in any state). But in certain moments, a "wave-function collapse" of the system would be produced. The Hameroff-Penrose hypothesis would postulate that states of consciousness (and all the *qualia* that accompany them, e.g., in a visual image) would result from the entrance into quantum coherence of vast quantities of microtubules of different neurons and brain modules due to the effects of action-at-a-distance or non-local causation, already known in quantum mechanics from the imaginary experiment of Einstein, Podolsky and Rosen in 1935 (EPR effects).

The Hameroff-Penrose hypothesis, then, opens new avenues to explain the phenomenological properties of psychism. Quantum coherence states due to action-at-a-distance (EPR effects) would be the most appropriate "physical support" to explain the unity of consciousness and field sensations (proprioception and vision); to produce "sensation" would be a field property of matter, as long as there would be a "psychical subject" capable of "sensing it." The indetermination-freedom of behavior would have its physical support in indeterminate quantum states and in the property of superposition. The subject could induce the collapse of the wave function in a flexible manner that would allow the descending control of the mecano-classical mechanisms of movement.

Let us suppose that the Hameroff-Penrose hypothesis were correct, and let us think about the consequences it could have for vision science. In principle, we would consider the neuronal engram of an image, when activated, as producing the collapse of the wave function in a subsystem of microtubules belonging to that engram. The sensation of the visual image would be the psychic effect (phenomenological) of the system interaction because of action-at-a-distance (EPR effects) of the state of quantum coherence of those microtubules. The pattern of the image would be given outside, objectively in the world, and would consist in the pattern of light that reaches both retinas. Since images are continually different in optical flow, one would have to think that the subsystem of microtubules involved would be varying in a continuous manner.

This makes us realize that the fundamental explicative problem would consist in knowing the mechanisms or series of interactions that begin from the determinant pattern of light (the external physical world that "imposes" the content of an image) up to the collapse of some or other systems of microtubules. It would be a bottom-up process. In this process, a quantum-classical interface should mediate, since the transmission of an image is made by means of classical neural engrams (fermionic) that should induce precise effects in the states of quantum superposition of the microtubules within each of the neurons activated in an image, producing quantum coherence at a distance among microtubules as EPR effect. The practical totality of these interface processes are not known to us. The Hameroff-Penrose hypothesis and many other things that are being investigated today concerning the biochemistry of neurons (e.g., the proposals concerning how to understand the functions of tubulin dimers on the walls of microtubules, or the manner of producing quantum coherence, or the function of the so-called "hydrophobic pocket," or clatrins, etc.) are only initial proposals that should be given a relative value, and, needless to say, are debatable. We will not go into them.

In the same manner, but inversely (that is, top-down), the conscious psychical subject would be the result of a "subject engram" and of a special system of associated microtubules. Evolution should have designed a descending mechanism (top-down), so that the decisions (variable, flexible, indeterminate) of the subject would control action (motor system) or the flow of the same thoughts (mind). Superposition and quantum indeterminism would allow us to understand how the subject could induce the collapse of the wave function of some or other microtubules, and how from there would be generated a quantum-classical interface that, supported by motor automatisms, would end in the final production of movement [37].

8 Psycho-bio-physical Ontology from the Perspective of Quantum Neurology

How living beings are really constructed, together with the nature of their components, constitutes their *ontology.* If the superior factor is the mind, we would be speaking of the ontology of the mind. It is a physical ontology, because it is made of a "physical world." It is a biological ontology, because it is a "physical world" organized as biological or living matter. It is a psychic ontology, because in the mind are produced psychic effects (sensation-perception-consciousness-subject) that interact (bottom-up and top-down) with the biological and the physical. In contrast, we see that, in agreement with all the available empirical evidence, current computers have a different ontology that is purely physical (neither biological nor psychic).

This psycho-bio-physical ontology has an *architecture.* In turn, "architecture" is defined as the structural form of the physical construction of that psycho-bio-physical ontology. Depending on the preceding analysis, and within the supposed hypothesis of a quantum neurology, we could say that this architecture has three levels (or three sub-architectures) and two (or three) systems of interface among them: (1) The *physical architecture* of reference, since the mind is united systemically to the physical world (e.g., united to the electromagnetic fields of light for vision). (2) The *classical architecture,* constituted by the nervous system or neuronal system connected to the global physical structure by the system of senses, internal and external. The architecture of engrams, patterns, canons or neural networks connects stimulations to automatic (without producing *qualia*) and conscious (psychic life and the sensation of *qualia*) response loops. In this architecture, physical-biological-neuronal processes happen in a differentiated world of macroscopic, fermionic, objects, in which cause-effect series are transmitted among independent entities. Previously, in this same presentation, we analyzed more extensively the properties of classical architecture. (3) The *quantum architecture,* in which living organisms would have to construct "biological niches" that made possible the presence of matter in quantum states that were the support for sensations and for their holistic and indeterminate dimensions. Quantum coherence, superposition of states, and action-at-a-distance or non-local causation (EPR effects) would be the foundations of this architecture. Bosonic and fermionic matter could be involved in this architecture, since fermionic matter (although it produces individual differentiation) has a quantum nature (e.g., electron) and it has been verified that it can also enter into states of quantum coherence. (4) The *classical-quantum interface* would be the totality of ascending, bottom-up, mechanisms, because of which the world imposes the selection of activated microtubules (e.g., in visual image). (5) The *quantum-classical interface,* because of which the conscious subject is capable of generating a descending, top-down, cause-effect series; of controlling the mecano-classical, fermionic structures; and of breaking biological determinism by introducing continually the factor of psychic unpredictability (freedom). (6) Furthermore, one could add a *physico-biological interface* of lesser importance (e.g., the connection of light pattern to the retina through interface with the eyeball optic), which we omit so as not to prolong this presentation.

Functionality (operativity) of the psycho-bio-physical ontology. Every ontology has an architecture that, *eo ipso* (by itself), involves a certain mode of functioning that excludes other modes. The same is true for the psycho-bio-physical ontology of living beings and of man. a) It allows a functionality founded on a deterministic causality proper to the mecano-classical world that produced all the automatisms of the system. b) But it also allows a new functionality, generated from sensation-perception-consciousness-subject states, that is supported by quantum coherence states. c) One actually deals with an integrated functionality in which what is automatic is coordinated with, and at the service of, a holistic functionality that is terminally directed from consciousness.

Operative logic of the psycho-bio-physical functionality. Some ontologies with their own functional systems can be presented as systems that operate with certain logical systems. This applies to the psycho-bio-physical ontology, since it has been formed evolutively in order to assimilate and to operate adaptively on the order of the natural world. Sensation, perception, consciousness, subjectivity, attention, memory, cognition, language, learning, thought, etc., have emerged by evolution to operate this natural order.

Phenomenological access to the logic of neural networks of the operative system. The Aristotelian logic itself was a first description of how the mind logically functions; the first space-time mathematics (arithmetic and geometry) was also a first description. A generalized phenomenological analysis of how our mind works (cognitive psychology) allows us to infer the probable manner of constitution of the logical networks of engrams of the neural system in its special modules and in the intermodular coordination of brain activity as a whole. Thus, for example, visual images are registered and organized in "folders" that have a logical order, allowing orderly access to them. Another example: When we study a certain university subject, we produce in our frontal and pre-frontal zones an enormous quantity of ordered engrams, permitting access from one to another (connected, in turn, to other brain modules, e.g., vision), that, when activated, produce an orderly flow of reasoning. This logic is possible because the architecture of the psycho-bio-physical ontology grounds it. But we still do not totally know today the codes of the space-time order of those neural networks and the rules of their interconnection. Deciphering the code of that physical order would be a discovery as important as, or even greater than, the discovery of the spatial ordering code of the DNA due to the work of Watson and Crick.

9 Ontologies, Architectures, and Artificial Functional Logics

By the word "artificial," we refer here to their production by man in a real physical or imaginary (abstract) manner. We begin with some observations about functional logics.

Functional logics, formal systems, and simulation. The natural mind already carried out some useful functional logics and mathematics in the discourse about life in the environment and about calculation. But the human mind, inspired by the structural and space-time form of the world, has come up with formal systems that assume the natural operations of calculation and permit many other new, more complex, superior, and useful operations. We have in mind mathematical analysis itself (potentiation, logarithmation, derivatives, integral calculus, theory of functions, etc.). But we have in mind not only all the systems conceived by modern mathematics, but also artificial formal systems that allow the amplification of natural logical functioning not only for calculation but also for life in general. Contrary to Penrose, I think the human mind can conceive formal systems that simulate and exceed the functioning of the natural logic of the mind (it has already been done abstractly both in mathematical formalization and in logical formalization, e.g., in axiomatic systems). But the problem would be not so much in the abstract conception of formal systems that integrally simulate the functional logic of the mind, as in the integral knowledge of the natural logic of the mind that we should simulate. At least, one could always design partial systems of simulation.

Ontologies and artificial architectures. Abstract and formal complex systems created by the human mind have been able to operate (that is, have been able to "function") in the human mind. Engineers, with paper and pencil, have been resolving numerous mathematical calculations. But the human mind has been capable of conceiving and constructing new ontologies, with their own architectures, that allow receiving information and "operating" on it (processing it, working on it) through the application of abstract formal systems created by the same mind. Two ontologies are created today. First, the brilliant conception of Turing's universal machine that, in an algorithmic, serial, and computational manner, has allowed extraordinarily useful applications of all types, and will continue to allow them for many years. The second ontology would be that of the parallel distributed processing (PDP) connectionist computer.

Turing's machine will be very useful, but it is undoubtedly different ontologically, functionally, and architecturally from the human mind. In effect, it has no deposits of 1's and 0's; it has no CPU; nor is it algorithmic, etc. PDP systems are more similar to the ontology of the mind (this is what they intend), but much is still needed. The mind is not a neuronal network engaged in propagation that produces outputs analyzed from another system, and that permits the control of values by retro-propagation for the next propagation. In order for PDP systems to approximate what is neuronal, they would need, at least, one of the three architectures (quantum architecture) and the two systems of interface mentioned. Furthermore, if we do a one-by-one analysis of the seven points that we previously emphasized as characteristics of classical architecture, we will also see how the so-called "artificial neuronal networks" are still far from "real neuronal systems."

Thus, neither the serial-algorithmic ontologies nor the PDP ontologies have properties allowing us to argue that they are 1) ontological, 2) functional (since every ontology presupposes some possibilities and determinant functional-operative exigencies), and 3) architectonically comparable to the human mind. Nevertheless, the human mind has serial aspects (e.g., in cataloging images, in thought, in language, in motricity) and parallel aspects that can be understood from the perspective of a "weak metaphor" by applying the model of a computer, be it serial or connectionist. Ontologies and architectures, serial and connectionist computers, have thus been created that can serve us to "operate" logico-formal systems, which are created in order to simulate the human mind. This simulation, as we have said before, will be possible and credible, but it will not be perfect, nor will it presuppose ontological or functional identity with the real animal-human mind.

It is possible to continue searching for new ontologies and new architectures. Microphysical physical states susceptible to two states (0-1) and capable of registering and recuperating information are sought; it is, for example, the Qubits' road to quantum computation. Physical engineering related to the field properties of the quantum world (for example, teleportation and quantum cryptography) is also possible. This engineering could progressively be applied to design systems in which artificial holistic fields are "sensed" by an "artificial subject." But then, rather than the creation of "computers," it probably would be more appropriate to speak of the creation of "artificial life." Penrose has referred to it recently.

10 Formalization Towards New Ontologies and Architectures

Although it may be difficult to think today about the creation of ontologies and architectures similar to the natural mind (which we actually still do not know well), it is possible to create approximations that are ever more useful. PDP connectionism has already been a useful approximation. Nevertheless, what formal systems could serve as instrument to shape these new ontologies? We conclude this presentation with a brief allusion to preferred formal systems. In our opinion, classical architecture should be inspired by mathematical formalizations based on *graph theory*: Trees growing in parallel and ending in "closed cups," but with infinite "vines" (or connections) among them. They would be immense forests with independent roots, but infinitely connected at the top (they would be Edelman's *re-entry*). On the other hand, quantum architecture should be inspired by *topology* english, or the study of continuous environments in pluridimensional spaces. Unitary topological spaces with boundary, separated at a distance, should "cover" other second-order imaginary spaces. For this, current topology should exert efforts to create new and more specific formal instruments that could serve as model formalizations for quantum-holistic spheres produced in physical ontologies and architectures of living beings. What purpose would this serve? I propose that we continue focusing more on Turing's machine. It will probably continue to be more useful in the short and medium run.

References

1. Damasio, Antonio (1994), *Descartes' Error: Emotion, Reason, and the Human Brain*, New York, Quill/HarperCollins Publishers Inc.
2. Damasio, Antonio (1999), *The Feeling of What Happens: Body and Emotion in the Making of Consciousness*, Florida, Harcourt, Inc.
3. Damasio, Antonio (2003), *Looking for Spinoza: Joy, Sorrow, and the Feeling Brain*, Florida, Harcourt, Inc.
4. Edelman, G.M., Mountcastle, V.B. (1978), *The Mindful Brain: Cortical Organization and the Group-Selective Theory of Higher Brain Function*, MIT Press, Cambridge, Mass.
5. Edelman, G.M. (1987), *Neural Darwinism: The Theory of Neuronal Group Selection*, Basic Books, Nueva York.
6. Edelman, G.M. (1988), *Topobiology: An Introduction to Molecular Embriology*, Basic Books, Nueva York.
7. Edelman, G.M. (1989), *The Remembered Present: A biological Theory of Consciousness*, Basic Books, Nueva York.

8. Edelman, G.M. (1992), *Bright Air; Brilliant Fire: On the Matter of the Mind*, Basic Books, Nueva York.
9. Edelman, G.M., Tononi, G. (2000), *A Universe of Consciousness: How Matter Becames Imagination*, Basic Books, Nueva York.
10. Edelman, G.M. (2001), *Building a Picture of the Brain*, In: Edelman, G.M., Jean-Pierre Changeux (Ed.), *The Brain*, Transaction Publishers, New Brunswick - London.
11. Edelman, G.M. (2004), *Wider than the Sky: the Phenomenal Gift of Consciousness*, Yale University Press, New Haven - London.
12. Edelman, G.M. (2006), *Second Nature: Brain Science and Human Knowledge*, New Haven - London, Yale University Press.
13. Hameroff, Stuart, Penrose, Roger (1996), "Orchestrated Reduction of Quantum Coherence In Brain Microtubules: A Model for Consciousness", In: *Toward a Sciences of Consciousness, The First Tucson Discussions and Debates*, eds. Hameroff, S.R., Kaszniak, A.W. and Scott, A.C., Cambridge, MA: MIT Press, pp. 507-540.
14. Hameroff, Stuart (1996), "Conscious Events As Orchestrated Space-Time Selections", In: *Journal of Consciousness Studies*, 3 (1): 36-53.
15. Hameroff, Stuart (1998), "Quantum Computation In Brain Microtubules? The Penrose-Hameroff 'Orch OR' Model of Consciousness", In: *Philosophical Transactions Royal Society London* (A) 356:1869-1896.
16. Hameroff, Stuart (1998), "Did Consciousness Cause the Cambrian Evolutionary Explosion?", In: Hameroff, S.R., Kaszniak, S.R., Scott, A.C. (eds.), *Toward a Science of Consciousness II: The Second Tucson Discussions and Debates*, Cambridge, MA: MIT Press, pp. 421-437.
17. Hameroff, Stuart (2003), "Time, Consciousness and Quantum Events in Fundamental Spacetime Geometry", In: Bucceriu, R., Saniga, M., *Proceedings of a NATO Advanced Research Workshop*.
18. Hameroff, Stuart (2004), "Quantum States in Proteins and Protein Assemblies: The Essence of Life?", In: *Proceedings of SPIE Conference on Fluctuation and Noise,* Canary Islands, June 2004.
19. Hameroff, Stuart (2006), "Consciousness, Neurobiology and Quantum Mechanics: The Case for a Connection", In: Tuszynski, Jack (ed.), *The Emerging Physics of Consciousness*, Berlin-Heidelberg, Springer-Verlag.
20. Monserrat. J. (1987), *Epistemología Evolutiva y Teoría de la Ciencia*, Publicaciones de la Universidad Comillas, Madrid 1987, 480 págs.
21. Monserrat, J. (1991), "Problema psicofísico y realidad cuántica en la física heterodoxa de David Bohm", In: *Pensamiento*, vol. 47 (1991) 297-312.
22. Monserrat, J. (1995), "Lectura epistemológica de la teoría unificada de la cognición en Allen Newell", In: *Pensamiento*, vol. 51 (1995) 3-42.
23. Monserrat, J. (1995), "¿Está realmente el mundo en mi cabeza? A propósito de J.J. Gibson y D. Marr", In: *Pensamiento*, vol. 51 (1995) 177-213.
24. Monserrat, J. (1996), "Francis Crick y la emergencia de la conciencia visual", In: *Pensamiento*, vol. 52 (1996) 241-252.
25. Monserrat, J. (1998), *La Percepción Visual. La arquitectura del psiquismo desde el enfoque de la percepción visual*, Biblioteca Nueva, Madrid 1998, 640 págs.
26. Monserrat, J. (1999), "Penrose y la mente computacional", In: *Pensamiento*, vol. 55 (1999) 177-216.
27. Monserrat, J. (2000), "Penrose y el enigma cuántico de la conciencia", In: *Pensamiento*, vol. 56 (2000) 177-208.
28. Monserrat, J. (2001), "Engramas neuronales y teoría de la mente", In: , vol. 57 (2001) 176-211.

29. Monserrat, J. (2002), "John Searle en la discusión sobre la conciencia", In: *Pensamiento*, vol. 58 (2002) 143-159.
30. Monserrat, J. (2003), "Teoría de la mente en Antonio R. Damasio", In: *Pensamiento*, vol. 59 (2003) 177-213.
31. Monserrat, J. (2006), "Gerald M. Edelman y su antropología neurológica: Presentación y discusión de su teoría de la mente", In: *Pensamiento*, vol. 62 (2006) 441-170.
32. Penrose, Roger (1989), *The Emperor's New Mind*, Oxford, Oxford University Press.
33. Penrose, Roger (1994), *Shadows of the Mind: A Search for the Missing Science of Consciousness*, Oxford, Oxford University Press.
34. Penrose, Roger (1997), *The Large, the Small and the Human Mind*, Cambridge, Cambridge University Press.
35. Penrose, Roger (2005), *The road to reality: a complete guide to the laws of the universe*, New York, Knopf.
36. Tuszynski, Jack A., Wolff, Nancy (2006), "The Path Ahead", In: Tuszynski, J.A. (ed.), *The Emerging Physics of Consciousness*, Berlin-Heidelberg, Springer-Verlag, pp. 1-25.
37. Tuszynski, Jack A. (ed.) (2006), *The emerging Physics of Consciousness*, Berlin-Heidelberg, Springer-Verlag.
38. Wolff, Nancy, Hameroff, Stuart, "A Quantum Approach to Visual Consciousness", In: *Trends in Cognitive Science*, vol 5, n 11, November 2001, pp. 472-478.

Physical Basis of Quantum Computation and Cryptography

Manuel Calixto

Departamento de Matemática Aplicada y Estadística
Universidad Politécnica de Cartagena
Paseo Alfonso XIII 56
30203 Cartagena, Spain
Manuel.Calixto@upct.es

Abstract. The new Quantum Information Theory augurs powerful machines that obey the "entangled" logic of the subatomic world. Parallelism, entanglement, teleportation, no-cloning and quantum cryptography are typical peculiarities of this novel way of understanding computation. In this article, we highlight and explain these fundamental ingredients that make Quantum Computing potentially powerful.

1 Introduction

Quantum Computing combines two of the main scientific achievements of the 20th century: Information Theory and Quantum Mechanics. Its interdisciplinary character is one of the most stimulating and appealing attributes.

The big success of Computer Science and Artificial Intelligence is linked to the vertiginous technological progress of the last decades. Essentially, computer's power doubles every two years since 1970, according to Moore's law. Extrapolating naively to a near future, this steady exponential growth on the miniaturization of the elementary component (the transistor) would reach the atomic scale by the year 2017. By then, one bit of information could be stored just in one atom. However, we should also start worrying about new and surprising quantum-mechanical effects the arise at atomic scales. Some of them could have a disruptive effect, like the tunnel effect, that put paid to standard computation. Although, instead of fighting against quantum effects, we would be better off allying ourselves with them and thinking of proper alternative architectures adapted to the nanometric scales: the would-be "quantum computer".

Quantum Physics entails a way of processing information which is different from the traditional, classical, methods. The processing of the information carried by the wave function of a quantum physical system is the task of the new *Quantum Information Theory* [1], a perfect marriage between Information Theory and Quantum Mechanics, comparable to the symbiosis between Physics and Geometry that leads to General Relativity. In practical effects, the quantum manipulation of information offers real applications, specially the reliable transmission of information (*Quantum Cryptography*) and potential applications, like

J. Mira and J.R. Álvarez (Eds.): IWINAC 2007, Part I, LNCS 4527, pp. 21–30, 2007.

the design of exponentially fast quantum algorithms (see e.g. [2]) that could threaten the privacy of most of actual business transactions. In fact, the most spectacular discovery in Quantum Computing to date is that quantum computers could efficiently perform some tasks (by efficient we mean that its running time is polynomial in the input size) which are not feasible (i.e. "super-polynomial") on a classical computer. For example, there are cryptographic systems, such as the "public key RSA" [3], whose reliability is based on the assumption that there are no polynomial time (classical) algorithms for integer factoring. However, Peter Shor [4] created an algorithm, to be run on a *quantum computer*, that factorizes integers in polynomial time. Also, there are efficient quantum algorithms for searching [5].

The advantages of Quantum Computing over the classical one rely on two quantum-mechanical properties par excellence, viz: *superposition* (interference or parallelism) end *entanglement*. Quantum superposition allows the possibility of performing simultaneous mathematical operations (equivalent to many classical computers working in parallel); whereas entanglement provides greater quantum correlations in answers than any classical correlation we can imagine.

Unfortunately, the more power we gain, the less stability we get. A quantum computer turns out to be extremely vulnerable, fragile, sensitive to any kind of background noise. Keeping the coherence of several atoms is extremely difficult with actual technology. Other drawback is that we can not amplify a quantum signal, due to the so-called "no-cloning theorem" (there are no perfect quantum copy machines...), thus limiting long-range quantum communications to tens of kilometers; nevertheless, we have quantum teleportation instead (see later). Actually, the impossibility of cloning quantum states has a positive side: the detection of eavesdroppers and the establishment of secure (quantum) communications. Moreover, a certain fault tolerance and quantum error correcting algorithms, together with a big effort in (nano)technology improvement, could make feasible quantum computing in the near future.

By the time being, it is worth analysing the meaning of Quantum Information and the abstract processing of it, disregarding the possible physical support or hardware (i.e., ion trap, nuclear magnetic resonance, laser, etc) that could efficiently accomplish our hypothetic computer in future.

2 Classic Versus Quantum: Bit Versus Qubit

The digital processing of information out of the brain goes through the conversion of messages and signals into sequences of binary digits. A two stable positions *classical* device (like a wire carrying or not electric current) can store one bit of information. Loosely speaking, the manipulation and processing of information comes down to swapping 0's and 1's around though logic gates (viz, NOT, AND, OR). Note that, except for NOT, classical logic gates are *irreversible*; that is, knowing the result $c = a + b$, we can not guess a and b. This loss of information leads to the well known heat dissipation of classical computers. Actually, we could make classical computation reversible, by replacing traditional logic gates by the new ones: NOT, CNOT and CCNOT, in Figure 1, the price being perhaps a waste of memory. However, Quantum Computation must be intrinsically reversible, since it is based on a *unitary* time evolution of the wave function (probability must be conserved), dictated by the Schrödinger equation.

In order to introduce the concept of *qubit*, let me use the following classical analogy. Suppose we drop the electric current in a wire to the limit of not being able to distinguish between cero (0) and positive (1) voltage. We could say then that the "state of electric current flow" of the wire is a statistical mixture $(\psi) = p_0(0) + p_1(1)$ of both possibilities, with probabilities p_0 and p_1. However, we would not gain anything new but just to introduce errors and uncertainty. Quantum Computing would not have any appeal if it wasn't that the quantum state described by the wave function (in Dirac's bracket notation)

$$|\psi\rangle = c_0|0\rangle + c_1|1\rangle \tag{1}$$

(c_0 and c_1 are complex numbers fulfilling $|c_0|^2 + |c_1|^2 = 1$) is not only a statistical mixture with probabilities $p_0 = |c_0|^2$ and $p_1 = |c_1|^2$ but, in addition, it incorporates two important new ingredients: *interference*, or "parallelism", and *entanglement*, or "quantum correlations" (for the last one we actually need two or more qubits, like in the state $|\psi\rangle = |0\rangle|0\rangle + |1\rangle|1\rangle$). The above classical analogy has sense in that the description of physical phenomena starts needing the Quantum Theory as the energy (or action) gap between states (levels, possibilities, etc) becomes smaller and smaller. This happens with more probability in the subatomic world than in the macroscopic world. All the quantum alternatives $\psi_j \sim e^{\frac{i}{\hbar}S_j}$, whose action gap $\Delta S = S_j - S_k$ is of the order of the Planck constant $\hbar$, coexist in some sort of quantum superposition with complex weights, like (1). These quantum alternatives are indistinguishable for a (classical) observer, who does not have access to that particular quantum superposition. In order to observe/measure the actual state, he has to "amplify" the action/energy differences ΔS up to the classical level, that is, up to the limit of being distinguishable by him. In this "amplification" or "measurement" process, the quantum superposition (1) is "destroyed" and only one of the alternatives (e.g., $|0\rangle$ or $|1\rangle$) survives the experience. This is the (standard but not free of controversy) so-called *wave-function collapse* (or measurement process), which raised and keeps raising so many philosophical and interpretation problems in Quantum Mechanics. The coexistence of quantum alternatives gives rise to interference effects that defy the common sense, like the well-known two-splits Young's experiment (see any book on Quantum Mechanics), which highlights the particle-wave duality of the electron.

Thus, the wave function (1) carries an information different from the classic one, which we agree to call *qubit* (quantum bit). Physical devices that store one qubit of information are two-level quantum systems like: spin 1/2 particles and atoms (electrons, silver atoms, etc), polarized light (photons), energy levels of some ions, etc. For example, it is possible to prepare a quantum state like (1) striking a laser beam of proper frequency and duration on some ions.

Note that, with two qubits, we can prepare a register in a quantum superposition of $2^2 = 4$ numbers from 0 up to 3 (we ignore normalization, for simplicity):

$$|\psi\rangle = |a\rangle\,|b\rangle = (|0\rangle + |1\rangle) \otimes (|0\rangle + |1\rangle) = |00\rangle + |01\rangle + |10\rangle + |11\rangle = \sum_{x=0}^{3} |x\rangle. \quad (2)$$

This can be the spin state ($|\downarrow\rangle \equiv |0\rangle$, $|\uparrow\rangle \equiv |1\rangle$) of carbon ($a$) and hydrogen ($b$) nucleus in a chloroform molecule $CHCl_3$. This "toy quantum computer" can implement the CNOT (controlled not) gate in Figure 1, by placing the molecule in an external magnetic field and acting on it with radiowave pulses that flip the spin of the nucleus. Actually, only when the spin of the carbon points in the direction of the external magnetic field (i.e., $|\uparrow\rangle = |1\rangle$), it is possible to flip the spin of the hydrogen. That is, the carbon is the "control" and the hydrogen acts as a XOR gate (see Figure 1). It is proved that, assembling (two-qubit) CNOT and arbitrary one-qubit unitary (quantum) gates is enough to design any classical algorithm like: addition, multiplication, etc (classically, they are the CNOT and CCNOT gates that constitute a universal set).

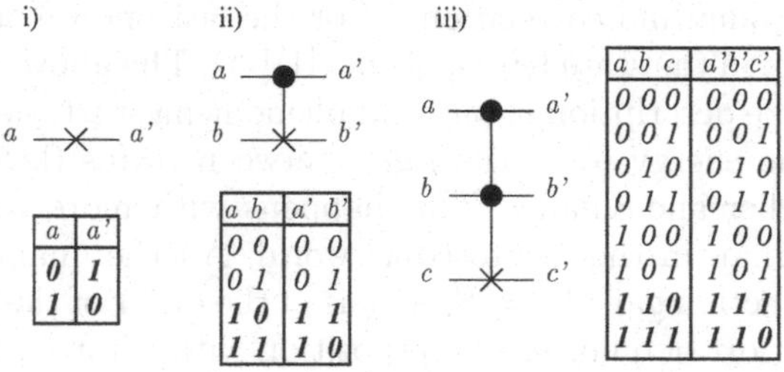

a	a'
0	1
1	0

a b	a' b'
0 0	0 0
0 1	0 1
1 0	1 1
1 1	1 0

a b c	a'b'c'
0 0 0	0 0 0
0 0 1	0 0 1
0 1 0	0 1 0
0 1 1	0 1 1
1 0 0	1 0 0
1 0 1	1 0 1
1 1 0	1 1 1
1 1 1	1 1 0

Fig. 1. Truth tables of the basic reversible gates: NOT, CNOT and CCNOT or Toffoli gates

In order to process more complex quantum information, it is promising to use lineal ion traps (see e.g. [6,7]), where the coupling between electron and vibrational degrees of freedom allows (in principle) the implementation of operations in a multi-qubit register by absorbtion and emission of photons and phonons.

In a four-qubits quantum computer, the application of the unitary operation $U_\oplus$ that implements the adding algorithm modulo 4 between the state (2) and a second one like $|\psi'\rangle = |x'\rangle$, with $x' = 0, \ldots, 3$, gives an output of the form:

$$|\psi\rangle\,|\psi'\rangle = \sum_{x=0}^{3} |x\rangle|x'\rangle \xrightarrow{U_\oplus} \sum_{x=0}^{3} |x\rangle|x \oplus x'\rangle. \quad (3)$$

That is, we have *simultaneously* computed the addition $x \oplus x'$ for four different values of of x, equivalent to four four-bits classical computers working in parallel. This feature is called *quantum parallelism.* However, we can only measure or "amplify" one of the four answers of the output $\sum_{x=0}^{3} |x \oplus x'\rangle \overset{\text{measure}}{\longrightarrow} |x_0 \oplus x'\rangle$. Let us see that it is not exactly superposition or parallelism what makes powerful quantum computation, but it is *entanglement.*

3 Entanglement: EPR Paradox

There are physical situations in which (quantum) particle pairs (or higher groupings) are created as if the state of one member would "instantaneously" determine or influence the state of the other, though they were hundreds of kilometres apart. It is not exactly like having couples of loaded dice that always offer the same face, but much more "intriguing", as we are going to see. For example, spin positron-electron entangled pairs $|EP\rangle = |\uparrow\rangle_e|\downarrow\rangle_p - |\downarrow\rangle_e|\uparrow\rangle_p$ are created in the decay of spin cero neutral particles; also pairs of photons with orthogonal polarizations (V means vertical and H horizontal) $|VH\rangle = |\updownarrow\rangle_1|\leftrightarrow\rangle_2 - |\leftrightarrow\rangle_1|\updownarrow\rangle_2$ are created by striking laser pulses on certain non-linear crystals. These are just particular examples (the so called "singlet states"), but more general situations are also possible.

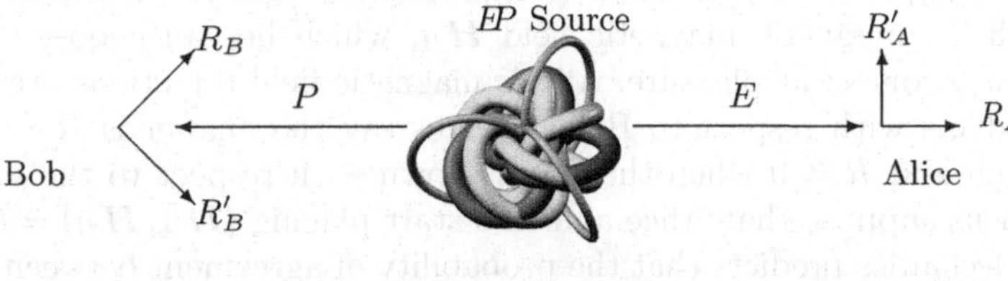

Fig. 2. Measuring entangled pairs $|EP\rangle$

In the case of entangled spins like EP, we propose the following "gedankenexperiment" (imaginary experiment like [8]) depicted in Figure 2. Alice A and Bob B are equipped each of them with magnetic fields $\boldsymbol{H}_A$ and $\boldsymbol{H}_B$, which can be oriented in the directions: $\uparrow, \rightarrow$ and $\nearrow, \searrow$, respectively, like in the Stern-Gerlach experiment for silver atoms. From the result R_A of the electron's (E) spin in the Alice's measurement (which can result in: either parallel $|\uparrow\rangle_e$ or antiparallel $|\downarrow\rangle_e$ to the external magnetic field $\boldsymbol{H}_A$), one can predict with certainty the result R_B of the positron's (P) spin in Bob's measurement, when measuring in the same direction $\boldsymbol{H}_B || \boldsymbol{H}_A$ as Alice (R_B ought to be antiparallel to R_A in this case). This would happen even if Alice and Bob were far away, so that no information exchange between them could take place before each measurement, according to Einstein's causality principle.

In order to motivate the original Einstein-Podolsky-Rosen "paradox" [9], we propose the following classic analogy: let us think of "entangled" pairs of green and red balls, made of metal or wood and whose weight is 0.5 or 1 Kg. The measurement devices (the analog of the magnetic field $\boldsymbol{H}$ directions) can be a flashlight (to measure color), fire (to distinguish metal from wood) and some weighing apparatus. Pairs of balls are "entangled" as: green-red, metal-wood and 0.5-1, and sent each one to Alice and Bob, respectively. Thus, the measurement of a given quality carried out by Alice on one member of the pair, automatically determines the quality of the other member of the pair, even before Bob carries out the corresponding measure. What is then the paradox?. If Alice and Bob are quite far away, so that no message can fly between them while the measures take place, then Bob would never think that the choice of measurement apparatus (flashlight, fire or scales) by Alice on one member of the pair would determine his results (color, fabric and weight) on the other member of the pair. If it were so, then we should start thinking about "telepathy" or "action at a distance", something forbidden by Einstein's Relativity Theory. This situation never would happen in the classic (macroscopic) world, but it is perfectly posible in the quantum (subatomic) arena. Here we have the "esoteric" face of quantum mechanics that upset Einstein. However, let us see that there is nothing mysterious in quantum mechanics when one accepts that, contrary to the classical systems, *subatomic entities have not well defined values of their properties before they are measured*; instead, all posible values must coexist in a quantum superposition like in (1). Indeed, (the following argument is a particular example of Bell's inequalities [10]) let us say that Bob, loyal to the classical mentality, really believes that the positron coming to him (see Figure 2) has a definite spin: either up (parallel) $\uparrow$ or down (antiparallel) $\downarrow$ (but never a mixture...) aligned with his magnetic magnetic field $\boldsymbol{H}_B$, which he can choose either in the direction $\nearrow$ or $\searrow$, at pleasure. Alice's magnetic field directions are rotated $\theta = \pi/4$ radians with respect to Bob's. Let us say the answer is $R = 1$ when the spin is up and $R = 0$ when the spin is down with respect to the magnetic field $\boldsymbol{H}$. Let us suppose that Alice and Bob start placing $(\boldsymbol{H}_A, \boldsymbol{H}_B) = (\rightarrow, \nearrow)$. Quantum Mechanics predicts that the probability of agreement between the answers (R_A, R_B) is $\sin^2(\theta/2) = 0.15$, where θ is the angle between $\boldsymbol{H}_A$ and $\boldsymbol{H}_B$ (that is $\theta = \pi/4$). Bob, who stays quite far away from Alice's place, also thinks that his results R_B are not affected by Alice's choice of measurement direction (either $\uparrow$ or $\rightarrow$). Even more, since he thinks the spin is well defined even before any measurement takes place, he also thinks that the global result would have been (R'_A, R_B), instead of (R_A, R_B), if the choice of measurement had been $(\boldsymbol{H}_A, \boldsymbol{H}_B)' = (\uparrow, \nearrow)$ instead of $(\boldsymbol{H}_A, \boldsymbol{H}_B) = (\rightarrow, \nearrow)$. The agreement between answers (R'_A, R_B) would continue to be the same (15%) since the new angle θ' is the same as before. In the same way, according to "classic" Bob's mentality, if the arrangement were $(\boldsymbol{H}_A, \boldsymbol{H}_B)'' = (\rightarrow, \searrow)$, then the agreement between (R_A, R'_B) would have been again 15%. Taking into account the previous results, and just by simple deduction (transitive property), Bob would then conclude that the agreement between (R'_A, R'_B), in the arrangement $(\boldsymbol{H}_A, \boldsymbol{H}_B)''' = (\uparrow, \searrow)$ would never exceed 15%+15%+15% = 45%. But, on the contrary, the experiment gives 85%, in accordance with Quantum Mechanics, which predicts $\sin^2(\theta'''/2) \simeq 0.85$ for $\theta''' = 3\pi/4$ (to be precise, experiments are not really done with electrons and positrons, but with other spin 1/2 particles or photons, although the same argument applies). The mistake is then to think that, "like balls", electrons have a definite spin (up or down) before the measurement. Otherwise we should start believing in telepathy....

It is clear that these kind of experiences at subatomic level, utterly uncommon in the macroscopic world, could be efficiently used in a future to create really surprising situations. Let us imagine a World-Wide-Web of entangled quantum computers that cooperate performing tasks which are imposible even via satellite. Nowadays, this is just speculation, although there are actual and future applications of entanglement in the field of telecommunications. Le us see some of these implementations of entanglement.

4 Entanglement and Teleportation

One of the most spectacular applications of entanglement is the possibility of transporting a quantum system from one place to another without carrying matter, but just information. Teleporting the polarization state of one photon, like $|\Psi\rangle = c_0 |\updownarrow\rangle + c_1 |\leftrightarrow\rangle$, is nowadays physically realizable thanks to the original idea of Bennet et al [11] and the Innsbruck experiment [12]. However, there is a long way to cover before we can teleport a macroscopic (even a mesoscopic) system. Before we must fight "quantum decoherence" (qubits a fragile and sensitive to any kind of external noise).

Teleportation of one photon goes as follows (see Figure 3). A ultraviolet laser pulse strikes a Barium β-Borate crystal, creating an entangled pair of photons $(F1, F1')$ and other pair $(F2, F2')$ after reflection in a mirror $M1$. The polarizer P prepares $F2$ in the state Ψ, which joins $F1$ through a beam splitter (BS). Then Alice makes a two-qubit measure (also, "coincidence" or Bell's measure) with the photon detectors $D1, D2$. The measurement can have four different answers: $(R_{D_1}, R_{D_2}) = (1,1), (1,0), (0,1), (0,0)$. If both detectors are struck (i.e. the answer is $(1,1)$), Alice tells Bob (through a classic message) that the photon $F1'$ has "transmuted" to the state Ψ, which Bob can verify by using a beam splitter polarizer (BSP), consisting in a calcite crystal. In the other three cases, Alice can always indicate Bob the operation to rotate $F1'$ to Ψ. Thus, we need a two-bits classic message to teleport one qubit (this is some sort of *dense information coding*).

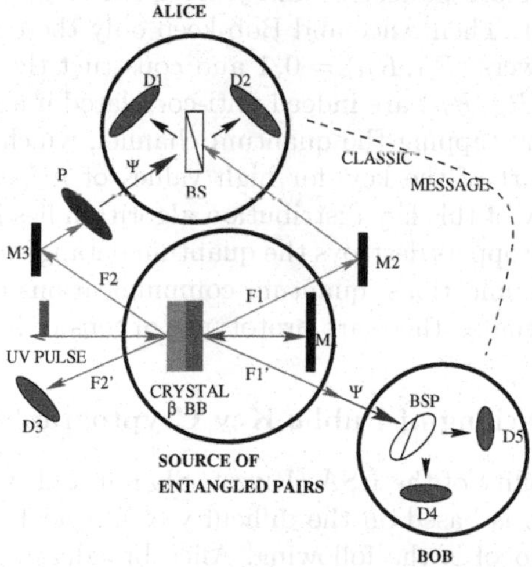

Fig. 3. Quantum teleportation of the polarization state of one photon

Quantum information can not be cloned (*no-cloning quantum theorem*), which can limit long-range quantum communications due to decoherence of quantum signals. However, intermediary teleporting stations can save this obstruction.

However, the impossibility of (perfectly) cloning a quantum signal has a positive side: the detection of eavesdroppers and the establishment of reliable quantum communications.

5 Quantum Cryptography

The basic ingredients to encrypt a secret message M are: a key K (known only by the sender, Alice, and the receiver, Bob) and a cryptographic algorithm E that assigns a cryptogram $C = E_K(M)$ to M though K. The decryption process consists in applying the inverse algorithm $M = E_K^{-1}(C)$. For example, the "one-time pad" algorithm assigns a q-digits $C = \{c_1, \ldots, c_q\}$ (with $c_j = 0, \ldots 2^5 - 1$ the alphabet symbols) to $M = \{m_1, \ldots, m_q\}$ though $K = \{k_1, \ldots, k_q\}$ by using the addition $c_j = m_j \oplus k_j \bmod 32$. The reliability of this simple cryptographic system is guaranteed as long as the key K is randomly generated and not used more than once. The problem is then when Alice and Bob, who are far apart, run out of keys. How to generate new keys overcoming the presence of eavesdroppers?.

5.1 Secure Quantum Private Key Distribution

One possibility is to use entangled pairs [13]). Both can choose the direction of magnetic fields $\boldsymbol{H}$: $\uparrow$ or $\rightarrow$, at pleasure. After measuring n pairs, they broadcast the direction choice of $\boldsymbol{H}$ each time, but not the answer, which can be: $1 = \uparrow$ or $0 = \downarrow$. In average, they should coincide $n/2$ times in the direction choice, for which the answers are perfectly (anti-)correlated $(R_A, R_B) = (1, 0) \equiv 0$ or $(R_A, R_B) = (0, 1) \equiv 1$. Then Alice and Bob keep only these approximately $n/2$ (anti-)correlated answers $(R_A, R_B) = 0, 1$ and construct the key $K = 00101\ldots$ One can prove that (R_A, R_B) are indeed anti-correlated if and only if there has been no eavesdroppers tapping the quantum channel, which can be verified by sacrificing a small part of the key, for high values of n (see [14] for a simple proof). The reliability of this key distribution algorithm lies in the fact that the observation of eavesdroppers destroys the quantum entanglement. Summarizing: unlike classical communications, quantum communications detect the presence of eavesdroppers. Actually, there are prototypes of tens of kilometers long.

5.2 Quantum Cracking of Public Key Cryptographic Systems

Nowadays, the reliability of the RSA (Rivest, Shamir and Adleman) public key cryptographic system is based on the difficulty of integer factoring on classical computers. The protocol is the following. Alice broadcasts her key, consisting of two big integers (s, c), with $c = pq$ the product of two big prime numbers only known by her. Anyone wanting to send her an encrypted message can do it by computing $C = M^s \pmod{c}$. In order to decrypt the message, Alice uses the formula $M = C^t \,(\mathrm{mod}\, c)$, where $t = t(s,p,q)$ can be calculated from the simple equations: $st \equiv 1 \,(\mathrm{mod}\, p-1)$, $st \equiv 1 \,(\mathrm{mod}\, q-1)$. Any other eavesdropper who wants to decrypt the message, firstly has to factorize $c = pq$. To make oneself an idea of the difficulty of this operation, for $c \sim 10^{50}$, and with a rough algorithm, we should make the order of $\sqrt{c} \simeq 10^{25}$ divisions. A quite good classical computer capable to perform 10^{10} divisions per second would last 10^{15} seconds in finding p and q. Knowing that the universe is about $3,8 \cdot 10^{17}$ seconds, this discourages any eavesdropper. Actually, there are more efficient algorithms that reduce the computational time, although it keeps exponentially growing with the input size anyway.

P.W. Shor [4] designed an algorithm, to be run on a quantum computer, that factors in polinomial time $t \sim (\log c)^n$, making factoring a tractable problem in the quantum arena and threatening the security of most of business transactions. The efficiency of the algorithm lies in the quantum mechanical resources: entanglement and parallelism. Essentially, the factoring problem of c reduces to finding the period r of the function $F_c(x) = a^x \,(\mathrm{mod}\, c)$, where a is an arbitrary number between 0 and c (see e.g. [14] for more details). Applying the unitary transformation U_F that implements F [remember the case (3)] to a superposition of $\omega >> r$ numbers x in the first register

$$\sum_{x=0}^{\omega-1} |x\rangle\,|0\rangle \xrightarrow{U_F} \sum_{x=0}^{\omega-1} |x\rangle\,|F_c(x)\rangle \xrightarrow{F_c(x)=u} \sum_{j=0}^{j\simeq\omega/r-1} |x_u + jr\rangle\,|u\rangle \qquad (4)$$

and storing the values $F_c(x)$ in the second register, we entangle both registers. Then measuring the second register, $F_c(x) = u$, we leave the first register in a superposition of $z \simeq \omega/r$ numbers that differ from each other in multiples jr of the period r, which can be obtained by a *quantum Fourier transform*. It is the entanglement between $|x\rangle$ and $|F_c(x)\rangle$ which makes possible the "massive scanning" of the function F_c.

Reliability of RSA and the U.S. Digital Signature Algorithm lie in the fact that factoring and discrete logarithm are intractable problems in classical computers. They are just particular instances of the so-called Hidden Subgroup Problem (see e.g. [2]). This problem encompasses all known "exponentially fast" applications of the quantum Fourier transform.

6 Grover's Quantum Searching Algorithm

Whereas classical searching algorithms need of the order of $P/2$ trials to find an item x_0 in a unstructured list of P items, Grover [5] designed a quantum algorithm that brings the number of trials down to about $\sqrt{P}$ iterations (with success probability of $\sim (P-1)/P$) on the quantum superposition $|\Psi\rangle = \frac{1}{\sqrt{P}} \sum_{x=0}^{P-1} |x\rangle$ of all items (*parallel searching*). Without entering into detail, the searching process consists of enhancing the probability amplitude of $|x_0\rangle$ and dimming the rest in the superposition $|\Psi\rangle$ through consecutive unitary operations. The result is a subtle interference effect that determines x_0 in about $t \simeq (\pi/4)\sqrt{P}$ iterations.

For $P = 4$ (two qubits) the situation is even more surprising: we just need a single trial to turn $|\Psi\rangle$ to $|x_0\rangle$!.

Acknowledgements

Work partially supported by the MCYT and Fundación Séneca under projects FIS2005-05736-C03-01 and 03100/PI/05

References

1. B. Schumacher, Phys. Rev. **A51** 2738 (1995); B. Schumacher & M.A. Nielsen, Phys. Rev. **A54**, 2629 (1996).
2. M. Calixto: On the hidden subgroup problem and efficient quantum algorithms, Fundamental Physics Workshop in honor to A. Galindo, Aula Documental de Investigación, Madrid 2004, Eds. R. F. Alvarez-Estrada, A. Dobado, L. A. Fernández, M. A. Martín-Delgado, A. Munoz Sudupe.
3. A. Menezes, P. van Oorschot, S. Vanstone: Handbook of Applied Cryptography, C.R.C. Press (1997).
4. P.W. Shor, Proceedings of the 35th Annual Symposium on the Theory of Computer Science, Ed. S. Goldwasser (IEEE Computer Society Press, Los Alamitos, CA), pag. 124 (1994).
5. L.K. Grover, Phys. Rev. Lett. **79**, 325 (1997).
6. J.I. Cirac & P. Zoller, Phys. Rev. Lett. **74**, 4091 (1995).
7. J.M. Doyle & B. Friedrich, Rev. Esp. Fis. **13**, 15 (1999).
8. A. Aspect, P. Grangier and G. Roger, Phys. Rev. Lett. **47**, 460 (1981).
9. A. Einstein, B. Podolsky & N. Rosen, Phys. Rev. **47**, 777 (1935).
10. J.S. Bell, Physics **1**, 195 (1964); Rev. Mod. Phys. **38**, 447 (1966).
11. C.H. Bennet, G. Brassard, C. Crepeau, R. Jozsa, A. Peres, & W.K. Wootters, Phys. Rev. Lett. **70**, 1895 (1993)
12. D. Bouwmeester, J.W. Pan, K. Mattle, M. Eibl, H. Weinfurter & A. Zeilinger, Nature **390**, 575 (1997).
13. A. Ekert, Phys. Rev. Lett. **67**, 661 (1991).
14. J. Preskill, Lecture Notes for Physics **229** (1998).

Brain Organization and Computation

Andreas Schierwagen

Institute for Computer Science, Intelligent Systems Department, University of Leipzig, Leipzig, Germany
schierwa@informatik.uni-leipzig.de
http://www.informatik.uni-leipzig.de/~schierwa

Abstract. Theories of how the brain computes can be differentiated in three general conceptions: the algorithmic approach, the neural information processing (neurocomputational) approach and the dynamical systems approach. The discussion of key features of brain organization (i.e. structure with function) demonstrates the self-organizing character of brain processes at the various spatio-temporal scales. It is argued that the features associated with the brain are in support of its description in terms of dynamical systems theory, and of a concept of computation to be developed further within this framework.

1 Introduction

The brain as the basis of cognitive functions such as a thinking, perception and acting has been fascinating scientists for a long time, and to understand its operational principles is one of the largest challenges to modern science.

Only recently, the functional architecture of the brain has gained attention from scientific camps which are traditionally rather distant from neuroscience, i.e. from computer and organization sciences. The reason is that information technology sees an explosion of complexity, forming the basis for both great expectations and worries while the latter come up since software technology is facing a complexity bottleneck [1]. Thus various initiatives started to propagate novel paradigms of Unconventional Computing such as IBM's 'Autonomic Computing' [1], the 'Grand Challenges in Computing Research' in the UK [2], and DFG's 'Organic Computing' (DFG = German Science Foundation) [3].

According to current views, the brain is both a computing and organic entity. The research initiatives mentioned before see therefore the neurosciences as sources of concepts relevant for the new, unconventional computing paradigms envisioned. Hence, the formal concepts which were developed within the Theoretical Neuroscience to describe and understand the brain as an information processing system are of special relevance.

This paper is organized as follows. Section 2 reviews some key features of brain organization (i.e. structure with function). It is followed by Section 3

[1] http://www.research.ibm.com/autonomic
[2] http://www.ukcrc.org.uk/grand_challenges
[3] http://www.organic-computing.org

J. Mira and J.R. Álvarez (Eds.): IWINAC 2007, Part I, LNCS 4527, pp. 31–40, 2007.

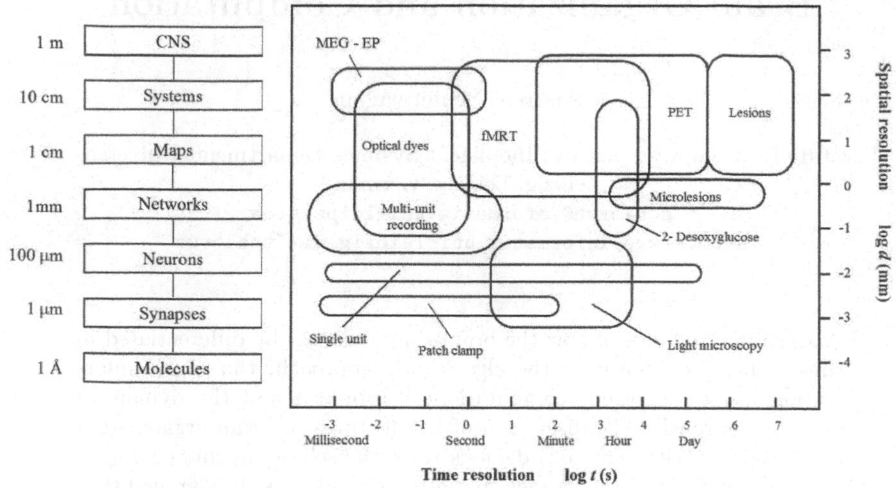

Fig. 1. Levels of brain organization and methods for its investigation. This figure relates the resolution in space and time of various methods for the study of brain function (right) to the scale at which neuronal structures can be identified (left). Adapted from [4]. MEG=magnetoencephalography; EP=evoked potentials; fMRT=functional magnetic resonance tomography; PET=positron emission tomography.

which discusses the different computational approaches developed in Theoretical (Computational) Neuroscience. Questions raised there include the search for the computational unit, the concept of modularity and the development of dynamical systems approaches. We end in Section 4 with some conclusions concerning the needs of a theory of analog, *emergent* computation.

2 Brain Organization and Methods for Investigation

Neuroscientific research is practiced at very different levels extending from molecular biology of the cell up to the behavior of the organism. In the first line, naturally, the neuroscientific disciplines (Neuroanatomy, -physiology, -chemistry and -genetics) are involved, but also Psychology and Cognitive Science. Theoretical Neurobiology (with its subdivisions Computational Neuroscience and Neurocomputing), Physics and Mathematics provide theoretical contributions (e.g. [2,3]). The integration of the results gathered by the disciplines is expected to provide insights in the mechanisms on which the functions of neurons and neural networks are based, and in the long run in those of cognition. The well-grounded and efficient realization of this integration represents one of the greatest challenges of actual neurosciences. New techniques like patch clamp, multi-electrode recording, electroencephalogram (EEG) and imaging methods such as magnetoenzephalography (MEG), positron emission tomography (PET) and functional magnetic resonance tomography (fMRT, nuclear spin tomography) enable investigations on different system levels (Fig. 1), raising again the question of how to integrate conceptionally the results.

The human brain has on the average a mass of 1.4 kg. According to different estimations it contains $10^{11} - 10^{12}$ neurons which differ from other cells of the organism by the pronounced variability of their shapes and sizes (Fig. 2). The individual morphologic characteristics of the neurons are important determinants of neuronal function [5,6,7,8,9], and thus they affect the dynamic characteristics of the neural network, to which they belong, either directly, or by specifying the entire connectivity between the neurons. In neural systems the influences are mutual, so that in general also the global network dynamics affect the connectivity and the form of the individual constituent neurons [5,10].

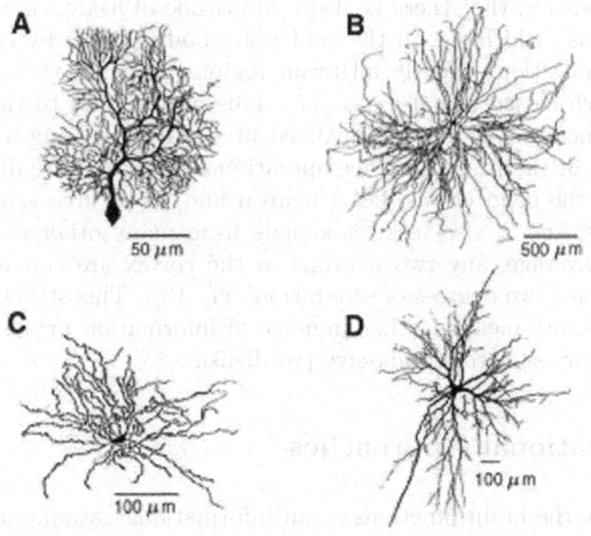

Fig. 2. Examples of dendritic neurons. Dendrites exhibit typical shapes which are used for classification of neurons. A. Purkinje cell from guinea pig cerebellum, B. a-motoneuron from cat spinal chord, C. spiny neuron from rat neostriatum, D. Output neuron from cat superior colliculus. Figures A.-C. from [11], D. from [12].

The specific functions of the brain are essentially based on the interactions each of a large number of neurons by means of their synaptic connections. A mammalian neuron supports between 10^4 and 10^5 synapses whose majority is located on the dendrites. Estimations of the total number of synaptic connections in the human brain amount to 10^{15}. Depending on the effect upon the successor neurons connections are classified as excitatory and inhibitory. The neurons of the cortex are usually assigned to two main categories: the pyramidal cells with a portion of ca. 85%, and the stellate cells with ca. 15% [13]. Pyramidal neurons often have long-range axons with excitatory synapses, and stellate cells with an only locally branched axon often act in an inhibitory manner. The activation status of the pyramidal neurons possibly encodes the relevant information, while the stellate cells raise the difference between center and surround by their inhibitory influence on the local environment, i.e. by lateral inhibition.

On the basis of distribution, density and size of the neuron somata the cortex can be divided in six layers (e.g. [14]). The cell bodies of the pyramidal cells particularly are in the layers III–V, and their apical dendrites extend into the upper layer I. The somata of the stellate cells are mainly in the middle layers III–IV (see (Fig. 3). Efferent connections from the cortex to subcortical and other structures are formed by the axons of the pyramidal cells in layer V; afferences to the cortex mainly come from the thalamus.

Hubel and Wiesel's landmark studies [15,16] of the visual system have led to the assumption that information processing in the brain generally follows a hierarchical principle. Important for the conceptional view on the function of the brain is, however, that there is also a multitude of feedback connections or 'back projections', which e.g. in the geniculate body (CGL) by far outnumber the forward connections. Nearly all brain regions influence themselves by the existence of such closed signal loops [17]. This also applies to the function of the individual neurons, which are involved in signal processing within an area or a subsystem of the brain. Further operational principles are divergence and convergence of the connections, i.e. a neuron and/or an area sends its signals to many others, and it also receives signals from many other neurons and/or areas. On the average, any two neurons in the cortex are connected by only one other neuron ('two degrees of separation', cf. [18]). This structurally caused functional proximity means in the language of information processing that the brain is characterized through massive parallelism.

3 Computational Approaches

Theories of how the brain functions as an informational system are in different ways related to the levels of brain organization. We can differentiate three general conceptions : the algorithmic approach, the neural information processing (neurocomputational) approach and the dynamical approach [19].

The *algorithmic computation* approach attempts to use the formal definition of computation, originally proposed by Turing [20] in order to understand neural computation. Although brains can be understood in some formal sense as Turing machines, it is now generally accepted that this reveals nothing at all of how the brain actually works [19]. Thus, Turing's definition of computation cannot be straightly applied (e.g. [21]).

The *neurocomputational* approach was launched in 1988 by Sejnowski, Koch and Churchland [4]. By stressing the architecture of the brain itself *Computational Neuroscience* was defined by the explicit research aim of "explaining how electrical and chemical signals are used in the brain to represent and process information". In this approach, computation is understood as any form of process in neuronal systems where information is transformed [22]. The 'acid test' for this approach (not passed as yet) is to find a definition for transformation of

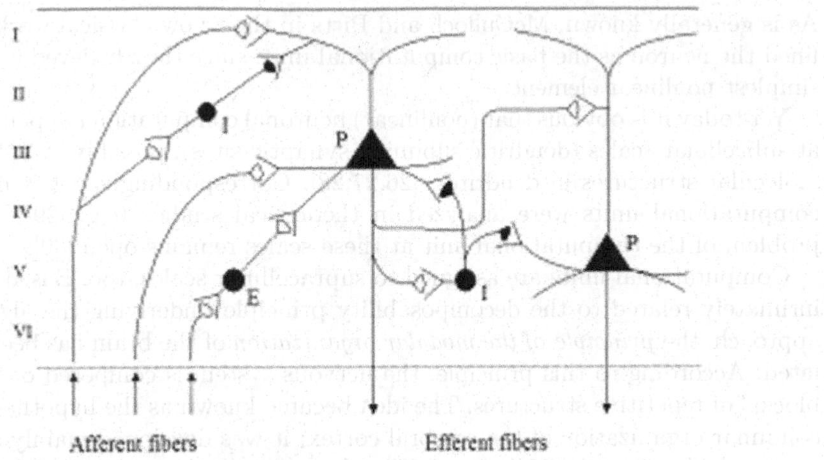

Fig. 3. Scheme of a neuronal circuit in the cerebral cortex. Pyramidal neurons (P — black triangles) receive inputs (either directly via afferent fibers, or from local neurons), generate outputs, and interact with one another. Local neurons (black circles — various types of stellate cells) may be excitatory (E — empty synapse symbols) or inhibitory (I — black synapse symbols). Cortex layers are indicated on the left. Significant variations in cell density, dendritic architecture, and synaptic arrangement enable a vast number of computational possibilities.

information, such that not almost all natural systems count as computational information processors [23,24].

The *dynamical* approach rests on concepts and theories from the sciences (Mathematics, Physics, Chemistry and Biology), and particularly from (Nonlinear) Dynamical Systems Theory. It seeks to understand the brain in terms of analog, rate-dependent processes and physics style models. The brain is considered as a large and complex continuous-time (often also continuous-space) physical system that is described in terms of the dynamics of neural excitation and inhibition.

3.1 Neurocomputational Concepts

While current neurocomputational concepts are of great diversity, most of them are tightly linked to the algorithmic view. The algorithmic as well as the neurocomputational approach attempt to explain properties of the nervous system (e.g., object recognition) in terms of parts of the system (cardinal cells, or 'grandmother neurons'), in accordance with the *decomposition principle* of (linear) Systems Theory. Models of this kind seek to understand on a detailed level how synapses, single neurons, neural circuits and large populations process information. If the information processing capacity of the brain is compared in this way with that of an algorithmic computer, one is confronted with several problems. In the first line, the units of computation are to be determined. The identification of the computational elements, however, is highly controversial.

As is generally known, McCulloch and Pitts in their now classical work [25] defined the neuron as the basic computational unit, since they believed it were the simplest nonlinear element.

Yet today it is obvious that (nonlinear) neuronal computation happens already at subcellular scales (dendritic subunits, synaptic sites), possibly even in supramolecular structures in dendrites. [26,27,28]. Correspondingly, e.g. synapses as computational units were analyzed in theoretical studies (e.g. [29]). But the problem of the computational unit at these scales remains open [30].

Computational units are assigned to supracellular scales, too. Based on ideas intimately related to the decomposability principle underlying the algorithmic approach, the *principle of the modular organization* of the brain has been formulated. According to this principle, the nervous system is composed of 'building blocks' of repetitive structures. The idea became known as the hypothesis of the columnar organization of the cerebral cortex; it was developed mainly after the works of Mountcastle, Hubel and Wiesel, and Szenthágothai (for reviews, see e.g. [31,32,33]).

Referring to and based on these works, the spectacular *Blue Brain Project* was started very recently. According to self-advertisement, the "Blue Brain project is the first comprehensive attempt to reverse-engineer the mammalian brain, in order to understand brain function and dysfunction through detailed simulations" [34]. The central role in this project play 'cortical microcircuits' which have been suggested as modules computing *basic functions*. Indeed, impressive progress has been made in developing computational models for defined 'canonical' microcircuits, especially in the case of online computing on time-varying input streams (see [35] and references therein).

It should be noted, however, that the concept of columnar organization has been questioned by neurobiological experts. Reviewing new findings in different species and cortical areas, it was concluded that the notion of a basic uniformity in the neocortex, with respect to the density and types of neurons per column is not valid for all species [36]. Other experts even more clearly state that it has been impossible to find a canonical microcircuit corresponding to the cortical column [37]. These authors reason that although the column is an attractive idea both from neurobiological and computational point of view, it has failed as an unifying principle for understanding cortical function.

3.2 Concepts from Dynamical Systems Theory

Inconsistencies between neurobiological facts and theoretical concepts are not new in the history of Theoretical Neurobiology. In the case of the column concept they demonstrate that the decomposition principle is possibly not suitable to serve as exclusive guidance principle for the study of information processing in the brain. While the principle of decomposability has a great number of advantages, for example just modularity, many problems in neuroscience seem not decomposable this way. The reason is that brains (like all biological systems) are inherently complex.

An appropriate framework for the description of the behavior of complex systems is represented by the attractor concept of nonlinear dynamical systems theory. Attractors may be informally defined as states of activity toward which a system settles (relaxes) over time. The activity in a neural system is described by a trajectory in the high-dimensional state space, say R^N where N is the number of neurons. Since this state (or phase) space is continuous, the neural system performs an *analog computation* [38]. In this framework, a certain parameter setting (the initial condition) is interpreted as input, the attractor to which the system's state flows as the output, and the flow itself as the process of computation. The criteria of computational complexity developed for digital algorithms are not directly applicable to 'analog algorithms'. Appropriate criteria of 'dynamic complexity' have been suggested: the time of convergence to an attractor within defined error bounds, the degree of stability of the attractor, the pattern of convergence (asymptotic, or oscillatory), type of the attractor (static, periodic, chaotic, stochastic), etc. Important building blocks for a non-standard theory of computation in continuous space and time have been developed by Siegelmann [39] by relating the dynamical complexity of neural networks with usual computational complexity.

While standard artificial neural networks have only point attractors, dynamical systems theory easily handles also cases where the output is a limit cycle or a chaotic attractor. The respective systems, however, have not been considered in computational terms as yet. This holds also for the so-called active, excitable or reaction-diffusion media, of which continuous neural fields are instances (see [40]). These media — spatially extended continua — exhibit a variety of spatio-temporal phenomena. Circular waves, spiral waves, and localized mobile excitations ('bumps') are the most familiar examples. The challenge is to find out how these phenomena can be used to perform useful computations. Generally, data and results are given by spatial defects and information processing is implemented via spreading and interaction of phase or diffusive waves. In several studies it was shown that these media have real capabilities to solve problems of Computational and Cognitive Neuroscience (formation of working memory, preparation and control of saccadic eye movements, emergence of hallucinations under the influence of drugs or the like, 'near-to-death' experiences, for overview see e.g. [41,42] and the references therein) and Artificial Intelligence (navigation of autonomous agents, image processing and recognition, e.g. [43,44]).

4 Conclusions

During the last decade, useful insights on structural, functional and computational aspects of brain networks have been obtained employing network theory [45,46]. From the many investigations in this area (see e.g. [47] for review) we know that the complexity of neural networks is due not only to their size (number of neurons) but also to the interaction of its connection topology and dynamics (the activity of the individual neurons), which gives rise to global states and *emergent behaviors*.

Several attempts were made to substantiate the general idea of computational systems which acquire emergent capabilities during a process of self-organization. The holistic properties of self-organizing systems represent a central intricacy in this respect. There is no 'natural' way to decompose such a system. If a decomposition is made anyhow (e.g. based on anatomical information only), subsystems should at first have a certain behavioral potential (i.e. multi-functionality). Ideas of the unfolding of multi-functionality were subsumed by Shimizu [48] under the term *relational system*. Relational systems obtain their functional properties only during mutual interaction with the other elements of the system while on its part the interactions of the elements depend on the evolving properties of the elements. Thus, an iterative process takes place which is based on principles of self-reference and self-organization. The properties of the system as a whole emerge in such a way that it is able to cope with perturbations from the environment.

An attempt to formalize this concept was undertaken recently [49] using 'chaotic neuromodules'. The results obtained from applications to evolutionary robotics demonstrate the multi-functional properties of coupled chaotic neuromodules but also the limitations of the linear couplings used [50].

A general conclusion to be drawn is that a great deal of progress in Theoretical Neuroscience will depend on tools and concepts made available through the dynamical systems approach to computing. Steps to overcome the existing theoretical restrictions in this area are essential not only for solving the problems in the Neurosciences but also to reach the goals of Unconventional Computing.

References

1. Heylighen, F., Gershenson,C.: The meaning of self-organization in computing. IEEE Intelligent Systems (2003) 72–86
2. Schierwagen, A.: Real neurons and their circuitry: Implications for brain theory. iir-reporte, AdW der DDR, Eberswalde (1989) 17–20
3. Schierwagen, A.: Modelle der Neuroinformatik als Mittler zwischen neurobiologischen Fakten und Kognitionstheorien. In: Maaz, J. (ed.): Das sichtbare Denken Rodopi-Verlag, Amsterdam (1993) 131–152.
4. Senjowski, T.J., Koch C., Churchland, P.S.: Computational Neuroscience. Science 241 (1988) 1299-1306
5. Schierwagen, A.: Growth, structure and dynamics of real neurons: Model studies and experimental results. Biomed. Biochim. Acta 49 (1990) 709–722
6. Schierwagen, A., C. Claus: Dendritic morphology and signal delay in superior colliculus neurons. Neurocomputing 38-40(2001) 343–350.
7. Schierwagen, A., Van Pelt, J.: Synaptic input processing in complex neurons: A model study. In: Moreno-Diaz jr.,R. , Quesada-Arencibia,A., Rodriguez, J.-C. (eds.): CAST and Tools for Complexity in Biological, Physical and Engineering Systems - EUROCAST 2003. IUCTC, Las Palmas (2003) 221–225
8. Van Pelt, J., Schierwagen,A.: Morphological analysis and modeling of neuronal dendrites. Math. Biosciences 188 (2004) 147–155
9. Schierwagen, A., Alpár,A., Gärtner, U.: Scaling properties of pyramidal neurons in mice neocortex, Mathematical Biosciences (2006 doi:10.1016/j.mbs.2006.08.019

10. Van Ooyen, A. (ed.): Modeling Neural Development. MIT Press, Cambridge MA (2003)
11. Segev, I.: Cable and Compartmental Models of Dendritic Trees. In: Bower, J.M., Beeman, D. (eds.): The Book of GENESIS: Exploring Realistic Neural Models with the GEneral NEural SImulation System. Telos/Springer-Verlag, Santa Clara, CA (1998)53–81
12. Schierwagen, A., Grantyn, R.: Quantitative morphological analysis of deep superior colliculus neurons stained intracellularly with HRP in the cat. J. Hirnforsch. 27 (1986) 611-623
13. Braitenberg, V., Schüz, A.: Anatomy of the Cortex: Statistics and Geometry. Springer, Berlin (1991)
14. Creutzfeld, O.: Cortex cerebri. Leistung, strukturelle und funktionelle Organisation der Hirnrinde. Springer-Verlag, Berlin etc. (1983)
15. Hubel D.H., Wiesel, T.N.: Receptive fields, binocular interaction and functional architecture in the cat's visual cortex. J. Physiol. 160 (1962) 106–154
16. Hubel D.H., Wiesel T.N.: Receptive fields and functional architecture of monkey striate cortex. J. Physiol. 195 (1968) 215–243
17. Sporns, O., Tononi, G., Edelman, G.M.: Theoretical neuroanatomy and the connectivity of the cerebral cortex. Behav. Brain Res. 135 (2002) 69–74
18. Watts, D.J., Strogatz, S.H.: Collective dynamics of "small-world networks. Nature 393 (1998) 440-442
19. Churchland, Patricia, and Grush, Rick (1999). Computation and the brain. In: Keil, F., Wilson, R.A. (eds.): The MIT Encyclopedia of Cognitive Sciences, MIT Press, Cambridge, MA (1999)155–158
20. Turing, A.M.: On computable numbers, with an application to the Entscheidungsproblem. Proc. Lond. Math. Soc. 42 (1936) 230–265
21. Mira, J., Delgado, A.E.: On how the computational paradigm can help us to model and interpret the neural function. Natural Computing (2006) DOI 10.1007/s11047-006-9008-6
22. deCharms R.C., Zador, A.M.: Neural representation and the cortical code. Ann.l Rev. Neurosci. 23 (2000) 613–647
23. Searle, J. R.: Is the brain a digital computer? Proc. Amer. Philos. Assoc. 64 (1990) 21-37
24. Grush, R.: The semantic challenge to computational neuroscience. In: Machamer, P.K., Grush, R., McLaughlin, P. (eds.): Theory and method in the neurosciences. University of Pittsburgh Press, Pittsburg (2001)155–172
25. McCulloch WS, Pitts W: A logical calculus of the ideas immanent in nervous activity. Bull. Math. Biol. 52 (1943) 99-115
26. Euler, T., Denk, W.: Dendritic processing. Curr. Opin. Neurobiol. 11 (2001) 415-422
27. Polsky, A., Mel, B.W., Schiller, J.: Computational subunits in thin dendrites of pyramidal cells. Nature Neurosci. 7 (2004) 621-627
28. London, M., Hausser, M.: Dendritic computation. Annu. Rev. Neurosci. 28 (2005) 503-532
29. Maass, W., Zador, A.M.: Synapses as Computational Units. Neural Computation 11 (1999) 903-917
30. Zador, A.M.: The basic unit of computation. Nature Neurosci. 3 (Suppl.) (2000) 1167
31. Hubel, D. H., Wiesel, T. N.: Functional architecture of macaque monkey cortex. Proc. R. Soc. London Ser. B 198 (1977) 1-59

32. Mountcastle, V. B.: The columnar organization of the neocortex. Brain 120 (1997) 701-722
33. Szentágothai J. The modular architectonic principle of neural centers. Rev. Physiol. Bioche. Pharmacol. 98 (1983) 11–61
34. Markram, H.: The Blue Brain Project. Nature Rev. Neurosci. 7 (2006) 153-160
35. Maass, W., Markram, H.: Theory of the computational function of microcircuit dynamics. In S. Grillner and A. M. Graybiel, editors, The Interface between Neurons and Global Brain Function, Dahlem Workshop Report 93. MIT Press, Cambridge, MA (2006) 371–390
36. DeFelipe, J., Alonso-Nanclares, L., Arellano, J.I.: Microstructure of the neocortex: Comparative aspects. J. Neurocytol. 31 (2002) 299-316
37. Horton, J. C., Adams, D. L.: The cortical column: a structure without a function. Phil. Trans. R. Soc. B 360 (2005) 386-62
38. Siegelmann, H. T., Fishman, S.: Analog computation with dynamical systems. Physica D 120 (1998) 214–235
39. Siegelmann, H. T.: Neural Networks and Analog Computation: Beyond the Turing Limit. Birkhauser, Boston (1999)
40. Schierwagen, A., Werner, H.: Analog computations with mapped neural fields. In: Trappl, R.(ed.): Cybernetics and Systems '96, Austrian Society for Cybernetic Studies, Vienna (1996)1084–1089
41. Schierwagen, A., Werner, H.: Fast orienting movements to visual targets: Neural field model of dynamic gaze control. In: 6th European Symposium on Artificial Neural Networks - ESANN 98, D-facto publications Brussels (1998) 91–98
42. Wellner, J., Schierwagen, A.: Cellular-Automata-like Simulations of Dynamic Neural Fields. In: Holcombe, M., Paton,R.C. (eds.): Information Processing in Cells and Tissues, Plenum New York (1998) 295–304
43. Adamatzky, A.: Computing in Nonlinear Media: Make Waves, Study Collisions. In: Kelemen, J., Sosk, P. (eds.): ECAL 2001, LNAI 2159 (2001) 1–10
44. Sienko, T., Adamatzky, A., Rambidi, N.G., Conrad, M. (eds.): Molecular Computing. MIT Press, Cambridge, MA, London, England (2003)
45. Sporns, O., Chialvo, D.R., Kaiser, M., Hilgetag, C.C.: Organization, development and function of complex brain networks. Trends Cogn. Sci. 8 (2004) 418-25.
46. Buzsaki, G., Geisler, C., Henze, D.A., Wang, X.J.: Interneuron diversity series: circuit complexity and axon wiring economy of cortical interneurons. Trends Neurosci. 27 (2004) 186-193
47. Bassett, D.S., Bullmore, E.: Small-world brain networks, Neuroscientist 12 (2006) 512-523
48. Shimizu, H.: Biological autonomy: the self-creation of constraints, Applied Mathematics and Computation, 56 (1993) 177-201
49. Pasemann, F.: Neuromodules: A dynamical systems approach to brain modelling. In: Herrmann, H. J., Wolf, D. E., Poppel, E. (eds.): Supercomputing in Brain Research: From Tomography to Neural Networks. World Scientific, Singapore (1995) 331-348
50. Hülse, M., Wischmann, S., Pasemann, F.: The role of non-linearity for evolved multifunctional robot behavior. In: Moreno, J.M., Madrenas, J., Cosp, J. (eds.): ICES 2005, LNCS 3637, (2005) 108-118

Concepts and Models for the Future Generation of Emotional and Intelligent Systems*

José Antonio Martín H.[1] and Javier de Lope[2]

[1] Dep. Sistemas Informáticos y Computación
Universidad Complutense de Madrid
jamartinh@fdi.ucm.es
[2] Dept. of Applied Intelligent Systems
Universidad Politécnica de Madrid
javier.delope@upm.es

Abstract. In this work, we first present a view of the philosophic study of Intelligent Behavior in a wide sense. We expose some key ideas to understand Intelligence and Rationality in an operational way based on the notions of Prediction and Randomness. In particular, we hypothesize that unpredictability is the key concept of Intelligence while not randomness is the key concept of Rationality. Next we undertake the study of Emotional Behavior discussing the basic principle of emotional attachment which is modeled by means of an operational definition of the Self. We hypothesize that the most basic principles of the Emotional Behavior emerges from a sort of ego-centric mechanism.

1 Introduction

One of the mayor challenges in present and future intelligent systems research is to describe the mechanisms that can lead to behave, act, and feel like Humans or living beings. This field of research has inspired the creation of many concepts and research lines such as Bio-Inspired systems, Bio-Mimetic approaches, Bio-Robotics, the Animat notion and so on.

Anticipatory Behavior could be defined as every kind of behavior which is influenced by any kind of: *knowledge, expectation, prediction, believes or intuition* about the future. But future can be expressed in many terms, for instance, future rewards, future states, future perceptions, future actions, etc. In fact, it is not clear at which level of the animal evolution the concept or at least any vague notion of future emerged. Humans handle a very sophisticated concept of time including past, present and future, but the interplay between cultural evolution and the human concept of time is not well known. Then, claiming that animals like rats, birds, dogs, etc. have a sophisticated notion of future is a hard to maintain hypothesis, also the most common of human behavior does not depend strictly on temporal analytical planning. Taking that into account,

* This work has been partially funded by the Spanish Ministry of Science and Technology, project DPI2006-15346-C03-02.

J. Mira and J.R. Álvarez (Eds.): IWINAC 2007, Part I, LNCS 4527, pp. 41–50, 2007.

any model predictive behavior with the aim to be of high applicability as a general model can not be defined in terms of the very refined cultural human notion of past, present and future. At this time there is a continuously growing body of works explicitly about Anticipatory Behavior with works ranging from initial models [1,2] to most current focused studies [3]. Anticipation as a predictive process plays a mayor role in any Intelligent Behavior, i.e. for taking good decisions we need to predict in some sense the consequences of such decisions.

In the first part of this work Anticipatory Behavior is addressed indirectly trough the study of Intelligence and Rationality. We will see in particular that there is no possibility for an agent to be truly rational or intelligent without a basic anticipatory behavior or in a more general sense a predictive process. In particular we hypothesize that unpredictability is the key concept of Intelligence while not randomness is the key concept of Rationality. But, what does Rationality means? What Intelligence is? Which is the interrelation between Adaptiveness, Rationality and Intelligence?

In the second part of this work, we undertake the study of Emotional Behavior discussing the basic principle of emotional attachment. The main challenge addressed by our research lines is how to express a mechanics of the Emotional Behavior. We hypothesize that such a mechanism of emotions is based on the notion of the Self and that this notion is the basis of the Emotional Attachment Behavior, that is, the root of any Emotional Behavior emerges from a sort of ego-centric mechanism.

2 Rationality

One of the first characteristics of the human behavior that researchers in Artificial Intelligence have tried to emulate was the Rationality. However, what does rationality means? Without entering philosophical depths we can say simply that rationality is about choices. In economics the notion of rationality refers to the fact that an economic agent must behave strictly and permanently maximizing the total income in an economical process.

More close to the Animat approach, we can say that rationality is about action selection and decision making. In [4] there is a relevant list of characteristics of rational behavior in wide sense:

1. It is goal oriented.
2. It exploits opportunities.
3. Looks ahead.
4. It is highly adaptive to unpredictable and changing situations.
5. It is able to realize interacting and conflicting goals.
6. There is a graceful degradation of performance when certain components fail, all of this must be achieved with limited resources and information (e.g. limited experience and knowledge).

Following this line, we encourage ourselves to propose a descriptive and operational definition of rationality.

Definition 1. Rationality is a constrainment of thought and consequently eliminates some freedom. This constrainment or loss of freedom is the result of an evolutionary process by means of natural selection which is oriented to select an order the thought (information processing) in such a way that the principles of cause and effect both innate and learned be satisfied in such a manner that an Animat behaves with at least a conscious indication (justification) that its judgments and its derived actions are `not random events` or has no causal relation with the consequences of its actions or judgments.

It can be seen from this definition that there are some factors that contribute to the degree of rationality, for instance, the vigilance of the consciousness of living organism is gradual and fluctuates between diverse species indeed in the same individual under different situations.

Another gradual factor of rationality derives from the statistical nature of the definition. Following the theory of the "dynamic core" and the theory of "Neuronal Group Selection" [5,6], the fact of the clear differentiation of the randomness depends directly on the entropy of the local micro-system of states associated to the precise contextual moment where the thought occurs, that is, in a specialized neuronal group.

Then, the learning and the knowledge repertoire are also important factors since they allow storing causal relations that allows to decide on the degree of randomness of one operation (behavior and action selection), thus, the greater amount of cause and effect relations the greater potential rationality could have the Animat, of this form, the degree of consciousness, the knowledge repertoire and the entropy of the context determines the degree of rationality of an individual.

This definition is absolutely more general than the definition in Economics science in the sense that an agent can behave rationally indeed when it is continually losing money.

3 Is Water Intelligent?

The question although speaks exclusively about intelligence, as we will se next, is directly related to adaptive behavior, predictive behavior, learning and in general any non inert behavior (Active Behavior). In other words, can water exhibit Intelligence? Is to behave like water the ultimate philosophy to construct the next generation of intelligent systems?

Of course, we can say trivially yes or not, but why? If we get a landscape and we add some water at the top of a mountain, the water goes from the most highest zone to the most lower zone with complete effectiveness and do this "feat" again and again without any mistake in the selection of the more natural (perhaps optimal?) path. This kind of behavior is precisely the one for which we have programmed our 'Intelligent Systems' since the last 50 years with some degree of success, note that this can easily be reformulated as an optimization process, that is, to search for a minimum over an error surface. Water is a source of inspiration in part due to some special features. First, all living organisms in the Earth are based in the Water. The Water has been present in the history of science as a source of discovering, we can remember the famous "eureka". Also the firsts notions of waves propagation where discovered observing water waves and it is very hard to rebate that the study of waves has been one of the most fruitful scientific studies with a lot of theoretical results and practical applications.

It is doubtless that water is intrinsically adaptive, it is, -to quote an example in the picture- as Bruce Lee claims: "formless and shapeless, if you put water into a tea pot it becomes the tea pot". Maybe, this adaptive intrinsic capability of water and the fact that it is a universal solvent, is the reason that makes water the base of all living organisms we know indeed that we expect to discover in other planets. Our intention is to determine if these intrinsic features of water are enough for constructing really intelligent systems.

Traditionally the dominant approach in Artificial Intelligence was symbolic reasoning and logic. Even in Psychology Intelligence has been highly correlated to the capability of abstraction and logical reasoning, not without sufficient critics to this approach which has lead to a measure that has been called even Intelligence Quotient (IQ).

One of the most relevant critics of past approaches to Artificial Intelligence are the works of Brooks and the reactive paradigm of intelligent robotics (in detriment of the deliberative one) with sounded papers like: "Elephants Don't Play Chess!", "Intelligence Without Reason" and "Intelligence Without Representation" [7,8,9]. Only reading the titles we can see the clear difference between past approaches an his claims. The Reactive paradigm has changed the way of doing Robotics. It is a model where the robots directly react to some stimulus from its environment producing a sort of behavior, that is, without reason and representation. Before the age of Reactive Robotics the robot programming tasks where a sort of hard programming techniques mixed with a lot of planning and deliberative rules. After the introduction of the concept of Reactive Behavior, the field of autonomous robotics has gained a great amount of interest and powerful results.

3.1 Information Quantity

We know by the information theory of Shannon [10,11] that the measurement of the information that a message or a system contains can be interpreted as the measurement of the *unexpected*, *unpredictable* and in some sense the *originality* that the message contains. He defined a measure of information content called "self-information" or *surprisal* of a message (m):

$$I(m) = -\log p(m), \tag{1}$$

where $p(m) = Pr(M = m)$ is the probability that message m is chosen from all possible choices in the message space M.

Eq. 1 weights messages with lower probabilities higher in contributing to the overall value of $I(m)$. In other words, infrequently occurring messages are in some sense more valuable.

In this way, if our *message* (m) is interpreted as the *action* (a) (behavior) taken by an agent, we can measure the amount of information in a, that is, how unexpected, unpredictable and original is the action of the agent after the stimulus, this takes us to a measurement of creativity, intelligence and finally active behavior.

3.2 Goal Oriented Active Behavior -Intelligence-

When studying the concept of information quantity and from a behaviorist point of view, we can raise a measurement of the goal oriented active behavior of an agent in a more natural way, measuring the adaptive-creative capabilities of its behavior.

Therefore, we can define the Amount of Active Behavior of an agent in a certain interval (t) of time to fulfill a previously determined goal (J) in a given situation x, such as:

Definition 2 (Amount of Active Behavior). *The amount of information in its actions weighted by the effectiveness of its goal oriented actions, that is: the unexpected, unpredictable and originality weighted by the degree of attainment of its goals.*

Eq. 2 shows a proposed measure of Active Behavior Quantity $\chi(a)$ of an action a_t.

$$\chi_t(a_t) = I(a_t) \times \frac{\partial J_t(x)}{\partial a}, \tag{2}$$

where $J_t(x)$ is the performance index of the goal attainment at time t at situation (x) and $I(a_t)$ is the entropy or information quantity of the action a_t.
That is, given two agents, will show evidence of more Active Behavior the one that whose product between effectiveness and the information quantity of its actions be greater. An immediate consequence of this definition is that:

Proposition 1. A completely predictable action does not demonstrate much Active Behavior still being very effective.

Corollary 1. Then an effective action not necessarily demonstrates Goal Oriented Active Behavior nor Intelligence since it can be simply the most probable action in the system.

The form to measure Goal Oriented Active Behavior is indeed to measure its demonstration, that is, the *originality* weighted by its *effectiveness*.
Thus, we ca reformulate the original question in this way:

Is water really intelligent since it always knows which path must follow with an effectiveness of 100%?

The answer is NO, water is not intelligent since it is not *original*, that is, its actions does not show intelligence because its actions are 100% predictable although are 100% effective.

This formulation is completely related to the ethological and psychological definition of Protean Behavior and Machiavellian Intelligence. Protean Behavior is a concept that defines a class of behavior mainly characterized by adaptively unpredictable actions. Chance became one of the first biologists to recognize the adaptive significance of unpredictable behavior in animals next Humphries and Driver termed this sort of adaptively unpredictable behavior "protean behavior" after the mythical Greek river-god Proteus who eluded capture by continually unpredictably changing form (again water was a source of inspiration).

4 The Mechanics of Emotions

The challenge addressed is how can we express a mechanics of emotions?

We hypothesize that such a mechanism of emotion is based on the notion of *Self* and that this notion is the basis of the Emotional Attachment behavior, of course the universe of the emotions is huge enough to deal with it, so we propose a starting point which is the study of the Emotional Attachment.

In 1905, Albert Einstein introduced its celebrated Theory of Relativity showing that for different reference frames the reality as stated in classical physics could be different if the particularities of the different observers are not taken into account, thus, Einstein's Relativity was the first contribution to the study of the private and subjective perception of the world by different reference frames in a rigorous sense. In this work the concept of reference frame (RF) will be the basis of the mechanical formulation of the notion of Self. In order to address such a mechanical formulation of Emotional Attachment we need to describe a mechanical formulation of the agent's world plus a sort of mechanism that operates on the agent to produce the desired behavior in a 'natural' way, that is, mechanistically.

Let us suppose that an organism has certain perceptual system and this is receiving information continuously from the environment. We claim that the organism is in equilibrium with its environment, that is, for instance it is hidden or it is not in danger, when the change in the input of its perceptual system is constant or very smooth.

The relation between self-safety, survivor and constancy of the input sensations comes from an intuitive observation on adaptation and the theory of evolution. We will call to this principle the "the Principle of Justified Persistence". This principle asserts that:

Proposition 2 (Principle of Justified Persistence). If a living organism is alive in some determined state of its environment, the maximum prior probability to stay alive is obtained when the state of the environment is constant or the change in such state is highly smooth.

In simple words, if you are alive then don't do anything! Indeed if you have a perceptual system and detect some change in the environment try to revert that change as soon as possible. Thus, it is very reasonable that following this basic principle, behaviors like homeostasis has emerged by means of natural selection.

In Fig. 1 we can see how this simple principle can be directly applied to the task of controlling a simple robot.

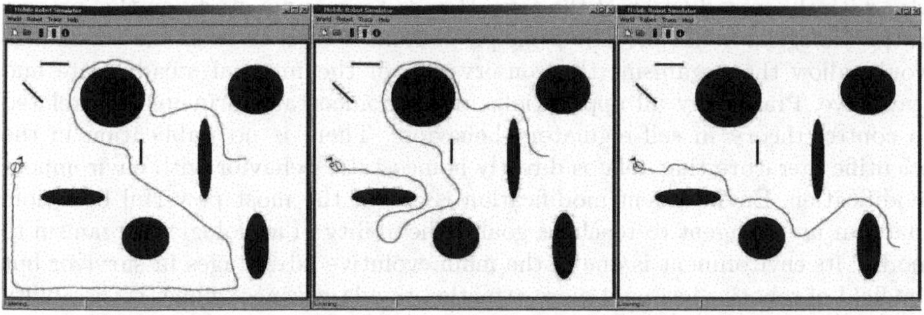

Fig. 1. The Principle of Justified Persistence applied to a mobile robot navigation problem

In this experiment there is nothing in the control program that tells the robot to keep away from obstacles as is the case in most control schemes like the artificial potential field nor any explicit wall following subprogram, indeed we can see that the robot uses the obstacles and walls as its source of constancy. In this model the robot seeks naturally constant signals and tries to maintain this signal as much constant as possible. Thus wall following and tangential obstacle voidance can be explained as an emergent behaviors that are consequence of the Principle of Justified Persistence.

Hence for a single certain system be able to adapt it needs a constant, otherwise, the organism will not be able to adapt because cannot converge towards the nothing. The organism that interacts with its environment by means of its perceptual system needs a "ground" on which to walk, in strict and metaphoric sense. With these basic notions we can promulgate a first law of ego-centric systems:

Law 1. *An organism that is in a steady-state and without internal or external forces that cause disturbances in its environment will remain in a steady-state, that is, without changing its behavior.*

This idea, is exactly equivalent in form to the inertial law of Newton which defines the natural behavior of an inert particle, indeed it is a proclamation of what is natural and what not. Up to date, science is enclosed in this definition of Nature. Nature must be reduced and explained by means of inert particles. But why are we talking about inert particles? And what is the direct relation of inert particles to Robotics? This relation is direct, pure reactive robots are inert particles that react to the stimulus of its environments like any Newtonian particle. In this sense, we can define an autonomous robot as an inertial observer and its systems for perception, reasoning and action oriented to maximize the vital function of the observer.

The vital function is a concept to represent the underlying principles in the biological behavior. The speculations in the search of the principle of the life have been long. Walter Cannon in the year of 1932 [12] tried to come near to the idea with the introduction of the term Homeostasis and a study on this process in the biological behavior. The principle of Homeostasis or resistance to change would allow the organisms the conservation of the internal steady-state and stay alive. Practically all applications of the homeostatic principle are related to control theory in self-regulatory behaviors. There is no publication in the scientific literature that relates directly homeostatic behavior with environment modification. Environment modification is one of the most powerful behaviors that can use an agent to reach its goals. The ability of a biological organism to modify its environment is one of the main evolutive advantages in survivor but the field of robotics has paid more attention to adaptiveness which is the ability to modify the "internal" organization of the agent in detriment of the "external" environment modification behavior.

In recent years the homeostasis as proposed by Cannon has been criticized in main lines by the tendency of the systems to paralyze and some divergences in the field of medicine. In our personal opinion, this criticism comes from the fact of a misunderstanding of the proposed principle, for instance, usually the homeostasis is interpreted like the resistance to change, but to what change?, The internal change? Where the internal and external boundary of the agent is?

If we define an agent like a closed system isolated from the environment, that is, the environment is not a part of the agent then the agent becomes immediately an inert particle that resists changing as any other inert particle from the point of view of the Newtonian mechanics and it is not able to adapt in any way. On the other hand, if the agent is so plastic that doesn't offer any resistance to change the system ends disintegrating itself since it will not have internal cohesion and that would be equivalent to affirm that agent and environment are completely undifferentiated. Then must exists an equilibrium point between the mere resistance to change or inertia and the plasticity to adapt to the environment and the changes occurring on it, thus, we claim for a model of agent where the agent itself is formed by an environment, that is, a kind of field around a reference frame. This field will also be affected by certain "intensity of field" which is modified according to the events experienced by the agent based on relevant sensory information.

Thus this process is a sort of mechanism of environmental formation that, on one hand, acts like a modeling function where the agent differentiates itself from the outer world and on the other hand, it acquires "conscience" about itself (self-recognition). Thus the agent is differentiated gradually from the environment and which determines the intensity of field is the degree of constancy of a sensation.

We can view this process with the following introspection.

-If there is something in the life of each of us, a universal invariant, universal constant or universal reference frame, it is the Self. The Self is the invariant set of perceptions which remains constant throughout all our life, and in the measure in which an "object" appeared more times, that object is more near to the Self, thus, my body has accompanied me during my years and I am able to

perceive that it has changed because I can compare it with other things that have not changed, such as, the Self or something that is more near to the Self, and, I can recognize it as my body because its degree of constancy or change in time has been smooth, thus, the Self is the reference frame of the perceptible universe.-

In the same way, will be observations that ones by frequency, others by intensity and others by relevance, will be bound to the environment with certain distance or difference with respect to the agent's reference frame. Of this form, the agent perceives what it is different from its reference frame adjusted gradually to the intensity field. It is important to note that the inertial law imposes a physical restriction to the perception of an inertial observer.

When we enunciate the inertial law we must indicate with respect to who or what is the movement of the related particle (a reference frame). Such an observer is called an inertial observer and the reference frame used by itself is called inertial reference system. An immediate consequence of the inertial law is that an inertial observer recognizes that a particle is not free, that is, that interacts with another particles when it observes that the speed or the momentum of the particle broke its constancy.

Of which we concluded that the only possible observations are those objects that have a difference with respect to the inertial reference frame of the observer. In case of an adaptive observer as it is the case that we are interested, the effect on the perception is that the observer would be affected by the gradual loss of discrimination on those events that have been part of the reference frame, thus, if the agent has adapted to certain signal or event, this event will tend to undifferentiate of the reference frame and start to gradually loss influence in its behavior, whereas the abrupt new events will be easily detected. These objects that gradually becomes part of the reference frame (Self) are precisely the objects for which the agent "feel" Emotional Attachment. Then, the Agent which has been defined as an environment half inertial half adaptive will tend to keep in a steady state such environment following the Principle of Justified Persistence, and the disturbances that occurs in such environment must be repaired either by means of internal accommodation or adaptation to the new state or by the possibility of the agent to modify its environment. Of this form, we can explain emergent behaviors like the protection of the prole, the emotional attachment and the environment-repairing behavior.

$$\varepsilon(x) = \frac{1}{\frac{\partial P(x)}{\partial t}}, \tag{3}$$

where x is an object in the environment, $\varepsilon(x)$ is the emotional attachment for x which is the result of the inverse partial derivative $\partial P(x)$ of the perceptual function P for x with respect to time dimension ∂t.

5 Conclusions and Further Work

We have exposed some key ideas to understand Intelligence and Rationality in an operational way. Thus, new operational definitions of Rationality and Intelligence has been presented concluding with a formulation of what we call Active Goal Oriented Behavior.

Also we have presented a study about Emotional Behavior discussing the basic principle of Emotional Attachment which was modeled by means of an operational definition of the Self using an introspection. We have presented our hypothesis that asserts that the most basic principles of the Emotions emerge from a sort of ego-centric mechanism. That has lead us into a novel interpretation of the interrelation agent/environment as an adaptive reference frame. In Addition, an experimental result with a simulated robot has been presented in order to appreciate the proposed Principle of Justified Persistence.

Finally the integration of all of this work points to a formulation of a very complex kind of behavior which must involve an *Emotional-Unpredictable-Not Random-Effective Goal Oriented* action selection mechanism.

References

1. Davidsson P. Linearly anticipatory autonomous agents. In *Agents*, pages 490–491, 1997.
2. R. Rosen. *Anticipatory Systems.* Pergamon Press, Oxford, 1985.
3. Butz M., Sigaud O., and Gérard P., editors. *Anticipatory Behavior in Adaptive Learning Systems, Foundations, Theories, and Systems*, volume 2684 of *LNCS.* Springer, 2003.
4. Maes P. The dynamics of action selection. In *Proceedings of the Eleventh International Joint Conference on Artificial Intelligence (IJCAI-89)*, pages 991–997, Detroit, MI, 1989.
5. G. M. Edelman. *Neural Darwinism – The Theory of Neuronal Group Selection.* Basic Books, 1987.
6. Giulio Tononi and Gerald M. Edelman. Consciousness and Complexity. *Science*, 282(5395):1846–1851, 1998.
7. Rodney A. Brooks. Elephants don't play chess. *Robotics and Autonomous Systems*, 6(1&2):3–15, jun 1990.
8. Rodney A. Brooks. Intelligence without reason. In *Proc. 12th Int. Joint Conf. on Artificial Intelligence (IJCAI-91)*, pages 569–595, Sydney, Australia, 1991.
9. Rodney A Brooks. Intelligence without representation. *Artificial Intelligence Journal*, 47:139–159, 1991.
10. C. E. Shannon. A mathematical theory of communication. *The Bell System Technical Journal*, 27(3):379–423, 1948.
11. C. E. Shannon and W. Weaver. *The Mathematical Theory of Communication.* University of Illinois Press, 1949.
12. Cannon Walter. *The wisdom of the body.* W.W. Norton & Company, Inc, 1932.

Modeling Consciousness for Autonomous Robot Exploration

Raúl Arrabales Moreno, Agapito Ledezma Espino,
and Araceli Sanchis de Miguel

Departamento de Informática,
Universidad Carlos III de Madrid,
Avda. de la Universidad 30, 28911 Leganés, Spain
{rarrabal,ledezma,masm}@inf.uc3m.es

Abstract. This work aims to describe the application of a novel machine consciousness model to a particular problem of unknown environment exploration. This relatively simple problem is analyzed from the point of view of the possible benefits that cognitive capabilities like attention, environment awareness and emotional learning can offer. The model we have developed integrates these concepts into a situated agent control framework, whose first version is being tested in an advanced robotics simulator. The implementation of the relationships and synergies between the different cognitive functionalities of consciousness in the domain of autonomous robotics is also discussed.

Keywords: Cognitive Modeling, Consciousness, Attention, Emotions, Exploration.

1 Introduction

Machine Consciousness could be considered as the field of Artificial Intelligence specifically related to the production of conscious processes in engineered devices (hardware and software). Undoubtedly, a multidisciplinary approach is necessary in order to approach such an intricate paradigm. Latest advances and contributions from psychology and philosophy in the scientific study of consciousness have lead computer scientist community to reconsider the possibility of engineering machine consciousness [1].

Although the phenomenal aspects of consciousness are still especially controversial [2][3], we argue that a purely functional approach can be successfully applied in the domain of autonomous robot control. In this work we present a machine consciousness model designed to command an autonomous robot, and the functionality of this model as a solution of the exploration problem. The phenomenal dimension, represented by the question '*Is the robot conscious of the exploration task he is doing?*' is deliberately neglected at this stage of our research.

J. Mira and J.R. Álvarez (Eds.): IWINAC 2007, Part I, LNCS 4527, pp. 51–60, 2007.

In section two we introduce our model and the theories of consciousness in which it is based upon. Section three covers the software architecture where we have integrated the machine consciousness model. In section four we discuss the detailed design and interaction between model components. Finally, we conclude describing salient preliminary results.

2 Evading the Cartesian Theater

Materialist theories of consciousness are not supposed to rely on any link to the soul like the one located by Descartes in the pineal gland [4]. However, the so-called Cartesian materialism associates conscious experience with a concrete place in the brain. The Cartesian theater[1] refers to this materialistic *homunculus*, which would play the role of the director of the brain. In contrast to the Cartesian theater metaphor, there exist other accounts for consciousness based on the idea of interim coalitions of specialized processors running concurrently in our brains. These processors or agents are continuously collaborating and competing for the light of consciousness.

Our model is mainly based on two theories of consciousness: the Global Workspace Theory (GWT) [6] and the Multiple Draft Model (MDM) [2]. GWT depicts a theater where the processors compete for appearing in the scene spotlight, which is the attention focus. Aggregation of processors is produced by the application of contexts. Behind the scenes, context criteria are defined and coordinated (unconsciously) by the director. Context formation mechanisms select the event in the stage that will be illuminated by the spotlight. The MDM adopts the editorial review process metaphor, where coalitions of processors suffer reiterative edition and review until they are presented as the official published conscious content of the mind.

Taking the main ideas from the described metaphors of the mind, we have built a cognitive model of consciousness called CERA (Conscious and Emotional Reasoning Architecture) [7]. Key functionalities of the model can be directly mapped to functional aspects of both GWT and MDM. A layered and modular scheme has been defined, where layers represent levels of control and modules represent cognitive specialized functions. Modules are situated within layers, CERA core layer encloses the key functional modules identified in the mentioned theories of consciousness. This set of functional modules is designed to support the workflows described by consciousness metaphors. Initial version of CERA core layer comprises eight modules: attention, status assessment, global search, preconscious management, contextualization, sensory prediction, memory management, and self-coordination. In this framework, there is no central module representing consciousness. Consciousness is supposed to emerge from the interaction between modules and their management of specialized processors.

[1] The term Cartesian theater was coined by Dennett to define (and reject) the idea of a central point of the brain where all sensory data is projected and conscious experience is produced [5].

3 Software Architecture

CERA has been originally designed to be applied to the domain of autonomous robotics. Therefore, its three layers correspond to different levels of autonomous control. The external layer manages physical robot machinery, and has to be adapted to the particular robot and onboard sensors and actuators being used. Middle layer is called instantiation layer as it encloses the problem-specific components. In the case of unknown environment exploration, instantiation layer contains the map production primitives and robot basic 'innate' behaviors for exploring. Finally, the inner layer contains the mentioned general purpose cognitive functions of consciousness (Fig. 1).

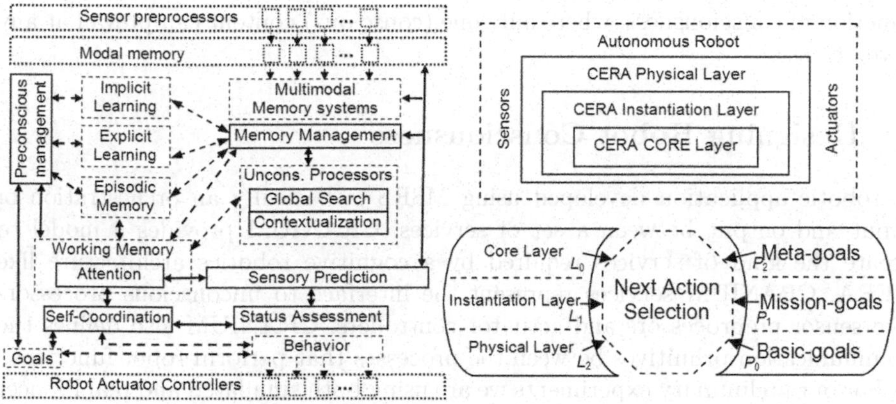

Fig. 1. In the left diagram solid lines represent CERA Core modules. Dashed lines represent CERA instantiation layer (domain-specific modules). Dotted lines represent CERA physical layer. Right diagrams illustrates CERA layered design and next action selection contributions.

Robot behavior is determined by a combination of the three level goals. At the physical level, the integrity of robot hardware is the highest priority (e.g. avoid collisions). Mission goals, unknown environment exploration in our case, are managed at the middle level. The *meta-goals* applied at the core level are related to the emotional dimension of the model of consciousness as explained in the next section.

In order to develop a flexible framework for experimentation with both simulated and real robots, we have integrated CERA into the Microsoft Robotics Studio (MSRS) platform [8]. A key component of MSRS is the Concurrency and Coordination Runtime (CCR) [9], which we use for asynchronous programming and unconscious processors concurrency management. A managed high-performance thread pool dispatches specialized processors tasks. Thread dispatching and asynchronous I/O operations follow diverse coordination patterns as required by CERA core modules.

MSRS is based on a light-weight distributed services-oriented architecture. A MSRS node run a set of services, and nodes can be installed in different machines. Communication and coordination between services is performed using the Decentralized Software Services Protocol (DSSP) [10]. The adaptation of CERA to this environment is the role of CRANIUM (Cognitive Robotics Architecture Neurologically Inspired Underlying Manager). CRANIUM is a wrapper for CERA that provides DSSP services creation and CCR parallel coordination patterns. Basically, CRANIUM is the interface for the creation of unconscious specialized processors and the management of their interactions. Like in a human brain, specialized regions of the brain perform concrete tasks concurrently, and emerging coordination is given by the neural connections between these areas (global access hypothesis) [11]. While CRANIUM provides the underlying neural-like mechanisms, CERA uses these services to produce the integrative function of consciousness, where only one (conscious) content can prevail at any given time.

4 Designing Robot Consciousness

A robotic application developed using MSRS is basically an orchestration of input and output between a set of services. CRANIUM provides a model to create the kind of services required by a cognitive robotics architecture like CERA. CRANIUM services represent the interface to unconscious processors like sensor preprocessors and actuator controllers. CRANIUM also defines the communication primitives between the processes that perform robot functions.

For our preliminary experiments we are using both simulated and real Pioneer P3 DX robots equipped with front and rear bumper arrays and a ring of eight forward ultrasonic transducer sensors (range-finding sonar) (Fig. 2). CRANIUM defines services for acquiring data from bumpers and sonar as well as commanding the differential drive motor system. Equivalent services are available for both real and simulated sensors and actuators.

Fig. 2. Simulated and real Pioneer P3 DX robots

4.1 Physical Layer

CERA Physical Layer subscribes to sensors notifications using CRANIUM. Every time a sensor changes its state, the asynchronous operation is managed by a CERA handler. The process of acquiring a sensor state change corresponds to a minimal perceivable event for the robot. Following Aleksander and Dunmall notation for axioms of neuroconsciousness [12], where A is the agent (the P3 DX robot in our case) and S the sensory-accessible world, these minimal percepts δS_j are the atomic information acquired by sensor handlers. Therefore, these CERA handlers build an internal representation of the percept, called $N(\delta S_j)$. This perception process is twofold, as two differentiable pieces of information are obtained: sensed object or event and its relative position in the world. In a two dimensional world, j has two spatial dimensions, and $(x, y) = (0, 0)$ represents the robot reference system (his subjective point of view).

Measurement of j is provided by each sensor differently. For instance, P3 DX bumper arrays consist of five points of sensing. Bump panels are at angles around the robot (Fig. 3). In this case, j is calculated depending on the bumper panel being pressed.

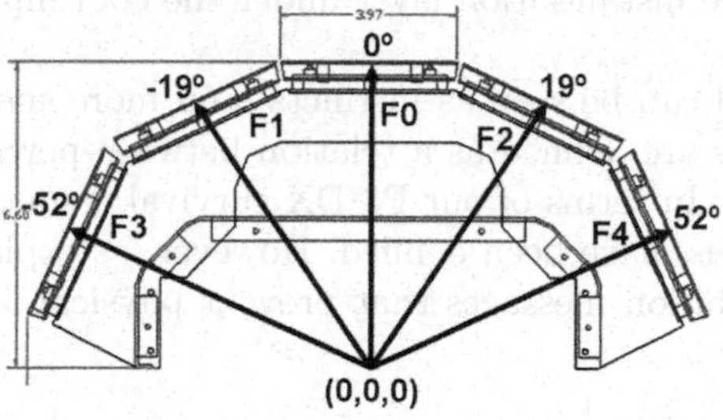

Fig. 3. P3 DX front bumper array consists of five bump panels at angles -52°, -19°, 0°, 19° and 52° to front of the robot. CERA bumper handler detects which bump panels are pressed and assigns values for every j accordingly. The resulting $N(\delta S_j)$ represent a physical obstacle at the relative location j.

As bump panels are a fixed part of the robot body and their activation is on contact, the j value is always the same for each bump panel. However, other sort of sensors would have to calculate relative position of the percept based on its own position or orientation. Like in natural nervous systems, all CERA handlers have to provide the ability to locate the source of the object or event being perceived.

The outputs from sensor handlers are combined into more complex percepts by sensor preprocessors. These preprocessors play the role of specialized group of neurons in charge of unconsciously detecting concrete features or patterns in perceived data. For instance, mammals visual system has specialized neural circuitry for recognizing vertical symmetry, motion, depth, color or shape [13,14]. Analogously, CERA sensor preprocessors provide the robot with feature extraction and recognition mechanisms appropriate for its environment. Some of the

CERA preprocessors that have been already implemented include wall detection and sonar invisible object detection (objects detected by bumper collisions but not detected by sonar or laser range finder).

In addition to sensor handlers and sensor preprocessors, CERA physical layer also contains unconscious processors related to behavior. Robot actuator controllers are defined as per CRANIUM interface to physical P3 DX robot. P3 DX is equipped with two motors (each wheel is connected to its own motor) that contain 500-tick encoders, forming a differential drive platform, where a third unpowered castor wheel provides balance. Initially, three basic actions have been implemented in CERA: stop, move forward, and turn. Move forward operation takes a motor power level for both wheels, and turn operation uses two power levels to apply to each motor in different directions. Thanks to CRANIUM, all the operations triggered by actuator controllers are executed in the context of the CCR dispatcher.

Basic actions are defined as δB_i, where i is the referent indicating the direction of the movement. Following the same notation as used for percepts, the robot representation for basic actions is $N(\delta B_i)$, and $N(B)$ corresponds to robot behavior. The composition of higher level behaviors in terms of physical basic actions is done at the instantiation layer under the coordination of CERA core layer.

Goals at this level can be seen as instincts, and more specifically as survival instincts. Basic goals are defined as a relation between perceptions and actions in the physical layer. In terms of our P3 DX survival a small set of basic goals, like avoiding collisions, have been defined. However, as explained below, higher layers can send inhibition messages that prevent physical layer goals to be accomplished.

4.2 Instantiation Layer

CERA instantiation layer makes use of sensor preprocessors in order to build a mission-specific representation of the world. This layer contains unconscious mission preprocessors, which are designed to recognize mission related objects and events using the perception information obtained in the physical layer. Wall segments and obstacles perceived by sensor preprocessors are internally combined in order to detect corridors or rooms. As percepts coming from the physical layer are j indexed, mission related percepts are built as *M(S)*, a partial description of the sensory accessible world S, where:

$$N(S) = \cup_j N(\delta S_j)$$
$$M(S) \subset N(S)$$

N(S) is the entire representation of the world built by the robot, while *M(S)* could be any subset of *N(S)*. In this case, the $N(\delta S_j)$ components of a concrete set *M(S)* are related due to their source location *j*.

In order to achieve the primary mission goal defined for the present work, unknown environment exploration, a simple two dimensional map representation

Table 1. Goal definition for a single mission (exploration). Layer 0, 1, and 2 refer to physical, instantiation, and core layers respectively. Execution time is discretized in *steps*, *updates* refer to $N(S)$ representation updates, and *mismatches* refer to failures to confirm a past percept, i.e. finding an obstacle where nothing was detected the last time the area was explored. E represents Emotion and n is the number of emotions being considered. Function *Energy* calculates the strength of a given emotion.

Goal	Layer	Description	Evaluation
G_{00}	0	Wander safely	$Eval(G_{00}) = (steps - collisions)/steps$
G_{10}	1	Map the environment	$Eval(G_{10}) = updates/steps$
G_{11}	1	Confirm created map	$Eval(G_{11}) = (updates - mismatches)/steps$
G_{20}	2	Positive emotional state	$Eval(G_{20}) = \sum_n Energy(E_n)$

has been chosen initially. As, for the time being, this is the only aspect of the world that we want the robot to be aware of, this map is actually *N(S)*. The robot keeps this map updated as he explores the world, mapping current perception of walls and obstacles into its two dimensional *N(S)*.

Similarly to percept aggregation, instantiation layer behaviors (called mission behaviors) are composed of the $N(\delta B_i)$ defined in the physical layer. Mission behaviors are the $M(B_i)$ (being $M(B) \subset N(B)$) that better fit mission goals needs. In terms of exploration, different wandering behaviors have been defined in the form of unconscious processors. These $M(B_i)$ compete for selection according to CERA Core layer cognitive rules.

4.3 Core Layer

CERA Core layer can be seen as a control center orchestrating the unconscious processor resources available in lower layers. The cognitive model implemented in this layer is intended to be domain independent, as all problem-specific representations are allocated in the instantiation layer. The general purpose functionality modules available in the core layer operate based on the basic percepts and actions from lower layers. The functionality of CERA Core modules is illustrated below applying the exploration problem.

Attention module is in charge of directing both perception and action. In order to be successful, the robot has to direct its attention to the fulfillment of mission goals, which can be recognized as full or partial solution of the specific problem being tackled. However, CERA design does not follow this strategy directly. Instead of taking mission goals as the drivers for the attentional focus, the meta-goals are considered. Meta-goals are related to the emotional state of the robot, and provide the means to have a general attention mechanism able to deal with multiple missions or different goals of the same mission. The definition of meta-goals characterizes the robot 'personality'. Initially, we have just considered one broad meta-goal: keeping a positive emotional state (Table 1).

Attention module calculates i referents for possible next $M(B_i)$ behaviors. In order to determine which $M(B_i)$ are applicable, contextualization mechanisms are used. The contextualization module provides possible associations between

$N(\delta B_i)$ based on available contextualization criteria. The primary criterion for building wander behaviors is based on the relation between j referent of perceived objects and action i referent. Basically, contextualization criteria for exploring will result in a set of promising directions to continue exploration, e.g. not to pay attention toward directions where an obstacle has been previously detected. Using this technique, attention focus is kept on the $M(S_j)$ perceived in the surrounding of the robot, and a set of possible actions $M(B_i)$ is calculated in that context with the aim of directing sensing.

The initial set of $M(B_i)$ behaviors calculated by CERA Core are considered gaze shifts, and are inspired in eye foveating saccades [15]. The robot is intended to direct its sensors to where relevant perception is predicted to take place. Even though our robot is not equipped with a motorized camera, he can rotate in place operating the differential drive, thus orienting the sonar coverage. Sensory prediction module is always active and listening sensor preprocessors output. Percepts $N(\delta S_j)$ are arranged into sequences, where $N(\delta S_j(t+1))$ is predicted based on past experience. As a first simplistic approach, sensory prediction is based on invariability. Therefore, the sensory prediction module will tell the attention module to direct the i referent of sensing to j locations where $N(\delta S_j(t))$ is different from predicted (or remembered).

As attention is serial, the Attention module has to select a concrete $M(B_i)$ at any given time (which could be composed of one or more $N(\delta B_i)$ and could take several time units to complete). The selection of winning attention focus and its associated behavior is not only based upon the factors explained above. The initial search on $N(S)$ in terms of contextualization criteria and sensory prediction, is extended further on $N(I)$ by the Self-Coordination module. $N(I)$ as defined in [12], is the representation of a imagined world. As both $N(S)$ and $N(I)$ are j-indexed, contextualization mechanisms can apply between perceived and imagined world. Self-Coordination module provides planning capability by searching trajectories in $N(I)$. Search on $N(I)$ is limited in depth and the sensory prediction function is also used to generate imagined perceptions $N(\delta I_j)$. The initial direction of the i referent of the most promising imagined behavior is used to finally select the next behavior to apply.

Evaluation of imagined behaviors is performed taking into account Status Assessment module output. This module implements a model of emotions, where basic emotions are defined and assigned an energy value. Emotions influence cognition, activating or inhibiting perception and action [16]. Additionally, in the context of CERA, emotions are the means to summarize the performance of the robot in terms of goal accomplishment. Consequently, CERA goals are assigned one or more emotional operators, which evaluate the progress being made in the goal achievement (see [7] for a detailed description of CERA emotional operators and associated emotional learning mechanism). Table 1 shows the evaluation functions used for some goals. Making good progress in goal achievement increases the energy of positive emotions like curiosity or joy. On the contrary, failure leads to increases in the energy of negative emotions like fear or anger. Emotional operators establish the relations between goals and specific emotions.

As described by Baars [11], global access is the capacity of accessing any piece of knowledge. The Global Search module is required to index and retrieve any unconscious processor, being a performance aid for the contextualizing function. Analogously, Preconscious management module is designed to be the interface between conscious and unconscious processes. It provides the required environment where different coalitions of unconscious processors can be built in the form of $M(B_i)$ and $M(S_j)$. Also, any 'editorial' review of these draft coalitions is managed in this domain, in order to have a consistent ('conscious') final version. Finally, the Memory Management module serves as an associative database manager, offering an interface to retrieve subsets of $N(S)$ and $N(I)$ related by any contextualization criteria.

5 Conclusion and Future Work

The described CERA architecture presents a novel approach to cognitive robotics where attention can be directed even without information from the real world. $N(I)$ provides a representation that permits the robot to plan possible behaviors. These imagined behaviors are emotionally evaluated the same way that actual performed behaviors. The emotional learning loop is closed when imagined behavior is physically performed, and the real and imagined outcomes are compared. An additional degree of flexibility beneficial to deal with real world is provided by CERA layered design, where Core layer can send inhibition messages that prevent physical layer goals to be accomplished when the threshold of energy of a particular emotion is reached.

There is still countless work to do in order to explore and compare the pros and cons of this kind of cognitive architectures. As of this writing we are working in the improvement of several components of CERA and CRANIUM. The application of forward models is being considered to improve sensory prediction functionality [17]. Additionally, real world experiments require much more effort: three dimensional representation and dealing with imperfect robot odometry.

Multiple mission accomplishment is other area where we believe that CERA can provide a good solution. The approach for this problem would be the creation of multiple instantiation layers. This design permits that the same architecture can be used for other domains, and facilitates the integration of different AI techniques into the unconscious processors.

Other challenges are robot vision and multi-robot collaboration. A pan-tilt onboard camera with foveating capability would increase the perception richness. Coordinated multi-robot exploration is also a challenging problem and very related to the field of autonomous exploration [18]. We believe that the application of inter-subjectivity models might be beneficial in this area.

Acknowledgments. We would like to thank Trevor Taylor from Queensland University of Technology (Brisbane, Australia) and Ben Axelrod from Georgia Institute of Technology (Atlanta, GA) for sharing their Microsoft Robotics Studio *Maze Simulator* code and supporting its application into this research work.

References

1. Holland, O.: Machine Consciousness. Imprint Academic (2003)
2. Dennett, D.: Quining Qualia. In: Consciousness in Modern Science. Oxford Universit Press (1988)
3. Nikolinakos, A.: Dennett on qualia: the case of pain, smell and taste. Philosophical Psychology **13** (2000) 505–522
4. Finger, S.: Descartes and the pineal gland in animals: A frequent misinterpretation. Journal of the History of the Neurosciences **4** (1995) 166–182
5. Dennett, D.C.: Consciousness Explained. Little, Brown and Co., Boston (1991)
6. Baars, B.: A Cognitive Theory of Consciousness. Cambridge University Press, New York (1988)
7. Arrabales, R., Sanchis, A.: A machine consciousness approach to autonomous mobile robotics. In: 5th International Cognitive Robotics Workshop. AAAI-06. (2006)
8. Microsoft, Corp.: Microsoft robotics studio. http://msdn.microsoft.com/robotics/ (2006)
9. Richter, J.: Concurrent affairs: Concurrency and coordination runtime. MSDN Magazine (2006)
10. Nielsen, H., Chrysanthakopoulos, G.: Decentralized software services protocol - dssp. Technical report, Microsoft Corporation (2006)
11. Baars, B.J.: The conscious access hypothesis: Origins and recent evidence. Trends in Cognitive Science (2002) 47–52
12. Aleksander, I., Dunmall, B.: Necessary first-person axioms of neuroconsciousness. In: IWANN. Lecture Notes in Computer Science (2003) 630–637
13. Braitenberg, V.: Vehicles: Experiments in Synthetic Psychology. MIT Press, Cambridge, Mass (1984)
14. Crick, F., Koch, C.: Towards a neurobiological theory of consciousness. Seminars in The Neurosciences **2** (1990)
15. Terzopolous, D.: Vision and action in artificial animals. In: Vision and Action. Cambridge University Press (1998)
16. Ciompi, L.: Reflections on the role of emotions in consciousness and subjectivity, from the perspective of affect-logic. Consciousness & Emotion **2** (2003) 181–196
17. Webb, B.: Neural mechanisms for prediction: do insects have forward models? Trends in Neurosciences **27** (2004) 278–282
18. Burgard, W., Moors, M., Stachniss, C., Schneider, F.E.: Coordinated multi-robot exploration. In: IEEE Transactions on Robotics. Volume 21., IEEE (2005) 376–385

An Insect-Inspired Active Vision Approach for Orientation Estimation with Panoramic Images

Wolfgang Stürzl[1] and Ralf Möller[2]

[1] Robotics and Embedded Systems, Technical University of Munich, Germany
stuerzl@in.tum.de
[2] Computer Engineering, Faculty of Technology, Bielefeld University, Germany
moeller@ti.uni-bielefeld.de

Abstract. We present an insect-inspired approach to orientation estimation for panoramic images. It has been shown by Zeil et al. (2003) that relative rotation can be estimated from global image differences, which could be used by insects and robots as a visual compass [1]. However the performance decreases gradually with the distance of the recording positions of the images. We show that an active vision approach based on local translational movements can significantly improve the orientation estimation. Tests were performed with a mobile robot equipped with a panoramic imaging system in a large entrance hall. Our approach is minimalistic insofar as it is solely based on image differences.

1 Introduction

When leaving the nest and also when returning to it, honey bees and wasps perform peculiar flight maneuvers [2,3]. Cartwright and Collett [4] suggested that honey-bees do this in order to acquire a 'distance-filtered snapshot', i.e. an image that omits nearby landmarks, which could guide them 'to the neighbourhood of the goal from a longer distance than can an unfiltered one'. Their hypothesis has been supported recently by high-speed and high-resolution recordings that show that during these maneuvers wasps keep their head in constant orientation for short time spans [5]. Producing pure translational image motion on their retina, this active vision strategy helps simplifying the estimation of range information. In this paper we describe another benefit of distance-filtered images: emphasising distant objects these images can provide improved orientation estimation.

Having an accurate orientation estimation is important for view-based homing, i.e. the return to a goal position by iteratively comparing the currently perceived images with the goal image, see [6,7,8]. Wrong estimates of current orientation with respect to a reference position are also a main source of errors in path integration.

For a mobile robot it is not straightforward to obtain, after wandering around, images that have the same orientation as a reference image recorded some time ago. Inertial sensors and path integration are known for accumulating errors over time. In indoor environments, compass sensors are often not applicable, and even outdoors slight compass errors will occur.

J. Mira and J.R. Álvarez (Eds.): IWINAC 2007, Part I, LNCS 4527, pp. 61–70, 2007.

1.1 Orientation Estimation for Panoramic Images

Since most insects eyes have very large field of view but usually low resolution (typically ranging from 1° to 5°, see [9]), panoramic images showing the full 360° azimuth are often used as a model of insect vision (see Fig. 1 for an example), and special cameras have been developed to capture such images [10,11].

It has been shown by Zeil et al. [1] that the relative orientation to a reference image $\boldsymbol{I}^a = (I_1^a, I_2^a, \ldots, I_N^a)^\top$ can be estimated by simply calculating the image rotation that minimises the difference to the current image $\boldsymbol{I}^b$, i.e.

$$\hat{\phi}^{ab} = \arg\min_{\phi} \mathrm{SSD}^{ab}(\boldsymbol{\phi}) \ , \tag{1}$$

$$\mathrm{SSD}^{ab}(\boldsymbol{\phi}) = N^{-1} \sum_{i=1}^{N} (I_i^a(\boldsymbol{\phi}) - I_i^b)^2 \ . \tag{2}$$

$\boldsymbol{I}^a(\boldsymbol{\phi})$ denotes image $\boldsymbol{I}^a$ rotated by angle $\phi = |\boldsymbol{\phi}|$ around an axis given by the unit vector $\boldsymbol{\phi}/\phi$. In this paper, the axis is always aligned with the z-axis of the coordinate system, directed opposite to the gravity vector.

It is unclear whether insects are able to rotate images mentally but they certainly can turn on the spot and thereby testing different orientations for the best match. Also most insects are equipped with a compass sense, e.g. the polarised light compass [12], that can give them a first orientation estimate which could then be further improved by image matching.

In (2) all pixels contribute equally. However, it is clear that in case of larger position differences, image parts displaying close objects will not provide meaningful information. Thus the performance usually decreases gradually with the distance of the recording positions, see Fig. 4 a for an example. In the following we describe a method that can improve the orientation estimation.

2 Improving the Orientation Estimation

2.1 Weighted SSD for Orientation Estimation

We define a weighted version of the SSD of two images $\boldsymbol{I}^a$ and $\boldsymbol{I}^b$ as

$$\mathrm{wSSD}^{ab}(\boldsymbol{\phi}) := \Big(\sum_i w_i^{ab}(\boldsymbol{\phi})\Big)^{-1} \sum_i w_i^{ab}(\boldsymbol{\phi})(I_i^a(\boldsymbol{\phi}) - I_i^b)^2 \ . \tag{3}$$

The image shift is estimated as before, see (1),

$$\hat{\phi}^{ab} = \arg\min_{\phi} \mathrm{wSSD}^{ab}(\boldsymbol{\phi}) \ . \tag{4}$$

The weights introduced in (3) are calculated using

$$w_i^{ab}(\boldsymbol{\phi}) = (\mathrm{var}_i^a(\boldsymbol{\phi}) + \mathrm{var}_i^b(\mathbf{0}))^{-1} \ . \tag{5}$$

var_i^{-1} is supposed to give information about the quality of the pixel for rotation estimation, e.g. it should be low if the pixel belongs to an image part showing close objects. For equally weighted pixels, i.e. $\mathrm{var}_i^{a/b} = 1$ for all i, (3) is identical to (2).

2.2 Obtaining Weights by "Active Vision"

In our minimalistic approach we compute var_i-values from pixel differences of images recorded at near-by position with same orientation, i.e.

$$\mathrm{var}_i(\boldsymbol{\phi}) = 1 + |I_i(\boldsymbol{\phi}, \boldsymbol{dx}) - I_i(\boldsymbol{\phi})| + |I_i(\boldsymbol{\phi}, \boldsymbol{dy}) - I_i(\boldsymbol{\phi})| \ . \tag{6}$$

$I_i(\boldsymbol{\phi}, \boldsymbol{dx})$, $I_i(\boldsymbol{\phi}, \boldsymbol{dy})$ are pixels of images recorded at positions shifted by $\boldsymbol{dx}$ or $\boldsymbol{dy}$ relative to the position of image $\boldsymbol{I}(\boldsymbol{\phi})$, without any (in practice with as little as possible) orientation difference. This can be achieved for example using inertial or odometry sensors that work well for short distances. The idea behind (6) is that image parts with large image differences most probably show close objects whereas image parts that correspond to distant objects have small local image differences. Of course, instead of pure pixel differences, less minimalistic methods, e.g. standard optic flow algorithms, could be used to obtain var_i-values.

3 Robot Experiments

We tested our method for orientation estimation using images recorded by a mobile robot equipped with a panoramic imaging system in a large entrance hall, image set 'hall1' [13]. Images were recorded at positions of a 20×10 grid with spacing of 0.5 m. In order to always have additional images at positions $(x + 1, y)$ and $(x, y + 1)$ needed for the computation of var-values using (6), images for orientation estimation were taken only from a 19×9 grid, see Fig. 1. After un-warping and conversion to gray scale, image size is 561×81 pixels. To test the effect of low-pass filtering, Butterworth filters of different cut-off frequencies were applied before un-warping.

3.1 Rotation Error

We define the rotation error as the deviation of the estimated orientation difference $\hat{\phi}^{ab}$ from the true orientation difference ϕ^{ab},

$$|\hat{\phi}^{ab} - \phi^{ab}| \in [0°, 180°] \ . \tag{7}$$

Since the orientation is estimated by shifting the un-warped images horizontally, we measure the orientation error in pixels, i.e.

$$\mathrm{rotErr}^{ab} = |\hat{s}^{ab} - s^{ab}| \in [0\,\text{pixels}, 280.5\,\text{pixels}] \ . \tag{8}$$

The rotation angle can be calculated from the image shift according to $\hat{\phi}^{ab} = \hat{s}^{ab}/(\text{image width}) \times 360°$. Since all images of 'hall1' have approx. the same orientation, the correct image shift is close to zero, i.e. $s^{ab} \approx 0$ and thus $\mathrm{rotErr}^{ab} \approx |\hat{s}^{ab}|$. For sub-pixel estimation of the rotation error, quadratic interpolation was used.

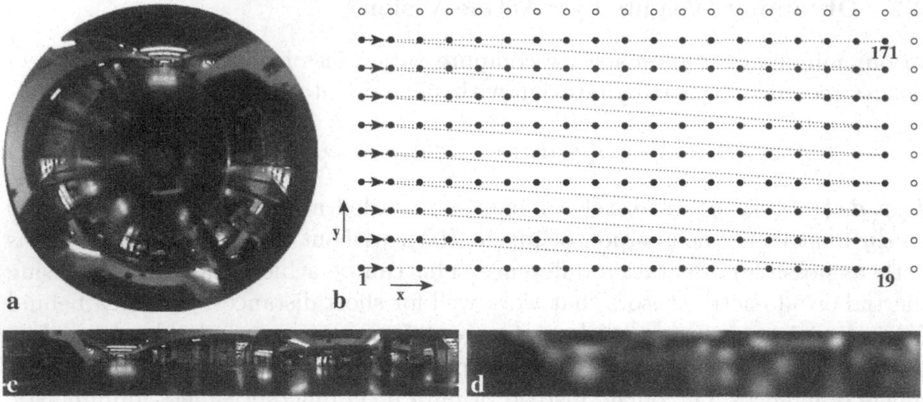

Fig. 1. Panoramic images of 'hall1'. a: Camera image (size 545 × 545 pixels). b: Images were recorded on a 20 × 10 grid (grid spacing is 0.5 m), positions of images used for orientation estimation are marked by black dots (positions 1, 19, and 171 are labelled by corresponding numbers). The additional images recorded at positions marked by circles are only used for calculating var-values, see text. c: Un-warped image at position 1 (size 561 × 81 pixels), elevation ranges from approx. −25° to +25°. Butterworth filter with relative cut-off frequency of 0.4 was applied before un-warping. d: Strong low-pass filtered, un-warped image at position 1 (Butterworth cut-off frequency 0.04).

3.2 Rotation Estimation Using SSD, (1), (2)

An example showing the SSD changing with orientation is given in Fig. 2 a,b for image at position 3. The low-pass filtered image has a broader minimum at $\phi = 0°$. It is also clearly visible that the existence of additional minima requires a starting position not to far away from the global minimum if gradient based methods for rotation estimation are used. Low-Pass filtering can increase the range of suitable starting positions.

Figure 2 c,d show the minimum SSD-value and the corresponding rotation error $|\hat{s}^{ab}|$ using all possible combinations of images, i.e. 171 × 171 rotation estimations. The conspicuous 19 × 19 blocks stem from the sequence in which the images were recorded, see Fig. 1 b. Values on the diagonal are $\min_\phi \mathrm{SSD}^{aa}(\phi) = 0$ and $\mathrm{rotErr}^{aa} = 0$, $a = 1, 2, \ldots, 171$. One can also see that images at positions close to each other usually give small errors. Examples for rotation errors calculated for single reference images are shown in Fig. 4 a,c. They correspond to row (or column) 86 and row (or column) 171 in Fig. 2 d.

3.3 Rotation Estimation Using wSSD, (4) - (6)

For implementing the proposed rotation estimation with weighted SSD, we use two additional images shifted in x, respectively y, by one position on the grid of recording positions, i.e. for the calculation of the var-values (6) we set $\boldsymbol{dx} = (0.5\,\mathrm{m}, 0)$ and $\boldsymbol{dy} = (0, 0.5\,\mathrm{m})$. For a mobile robot this means that while two

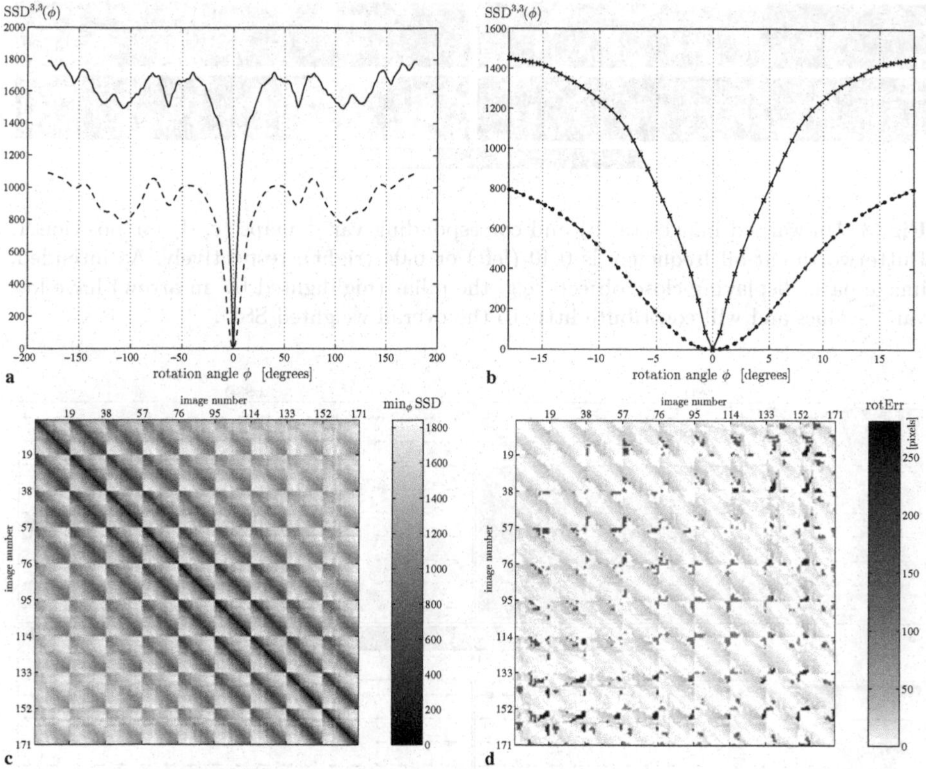

Fig. 2. a: Image difference (SSD) in dependence of rotation angle ϕ for image at position 3 filtered with cut-off frequency 0.40 (continuous curve) and 0.04 (dashed curve). b: Close-up of a) with details of the minimum. Markers (crosses and dots) show SSD-values for integer image shifts. c: Minimum SSD-values with respect to ϕ for all possible combinations of images of 'hall1' (cut-off frequency 0.40). d: Corresponding rotation error coded as gray scale map.

images can be recorded moving straight ahead, the third images requires a 90°-turn. Of course, instead of moving the whole robot it is sufficient to move the camera on top of the robot, see [14] for such a setup.

Examples of var^{-1} maps for image 1 are shown in Fig. 3 c,d. The calculation of var-values in (6) is solely based on image differences. Therefore, only close objects with sufficient contrast will cause high image differences, and the corresponding image parts will receive low weights.

Figure 4 compares rotation errors estimated with SSD (left) and with the proposed active vision strategy (wSSD, right) for reference images at the centre (position 86) and at the top right corner (position 171). The errors for the wSSD-method are clearly smaller.

Results for all pair-wise combinations of unwarped images with cut-off frequency 0.4 (left) and 0.04 (right) using SSD and wSSD are shown in Fig. 5. Since the error matrices are symmetrical $\text{rotErr}^{ab} = \text{rotErr}^{ba}$, and $\text{rotErr}^{aa} = 0$,

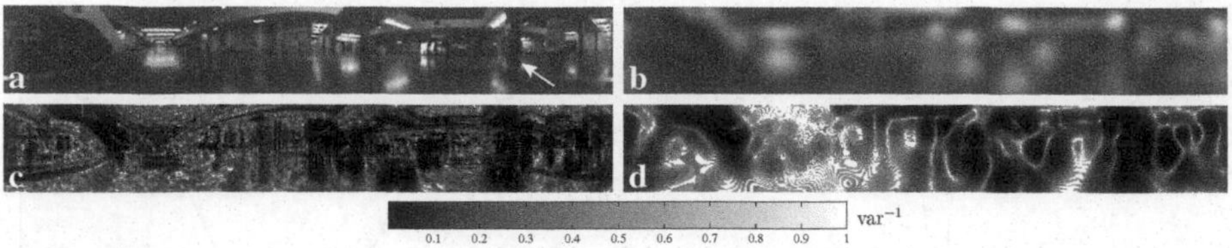

Fig. 3. Un-warped images (a, b) and corresponding var^{-1}-maps (c, d) for position 1. Butterworth cut-off frequency is 0.40 (left) or 0.04 (right), respectively. As intended, image parts displaying close objects, e.g. the pillar (highlighted by an arrow) have low var^{-1}-values and will contribute little to the overall weighted SSD.

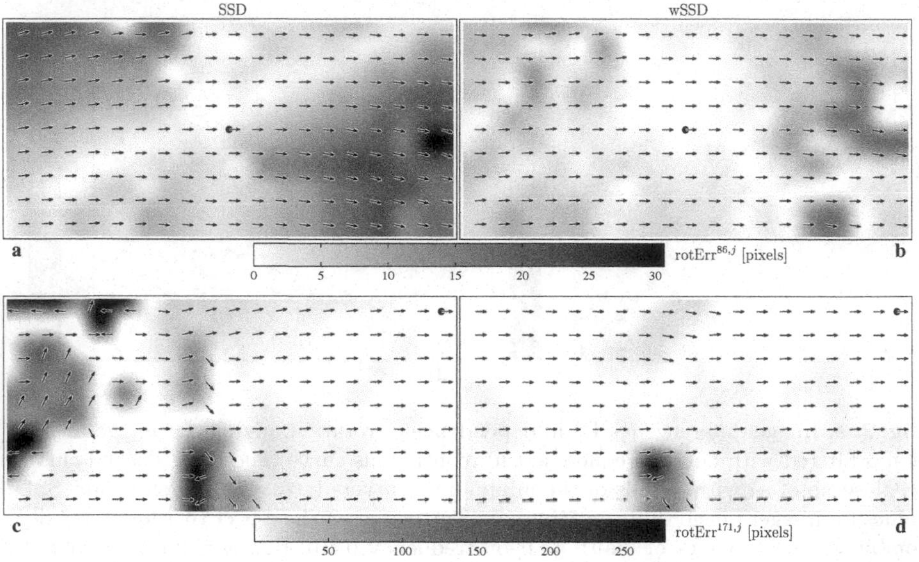

Fig. 4. Rotation errors using standard method (SSD, left) and the weighted SSD (right) for reference images at the centre (position 86, top row) and at the top right corner (position 171, bottom row). The rotation errors $|\hat{s}^{ab}|$ (in pixels) are gray coded (note different scale-bars for a,b and c,d). The vectors depict $(\cos\hat{\phi}^{ab}, \sin\hat{\phi}^{ab})^\top$, $i = 86, 171$. Images with cut-off frequency 0.40 were used.

only the strictly upper triangular matrices were used for calculation of the histograms, giving $(171^2 - 171)/2 = 14535$ values. It is clearly visible that the wSSD-method is superior having significantly smaller errors. Also, large rotation errors > 50 pixels ($\approx 32°$) are highly reduced (see insets). Interestingly (see also Table 1), while strong low-pass filtering reduces the error for the wSSD-method this is not the case for the SSD-method, probably because the number of large rotation errors increases. This finding will be further investigated.

A direct comparison between SSD and wSSD of image 3 and image 22 is shown in Fig. 6. While the global minimum of $\mathrm{wSSD}^{3,22}(\phi)$ is at the correct angle

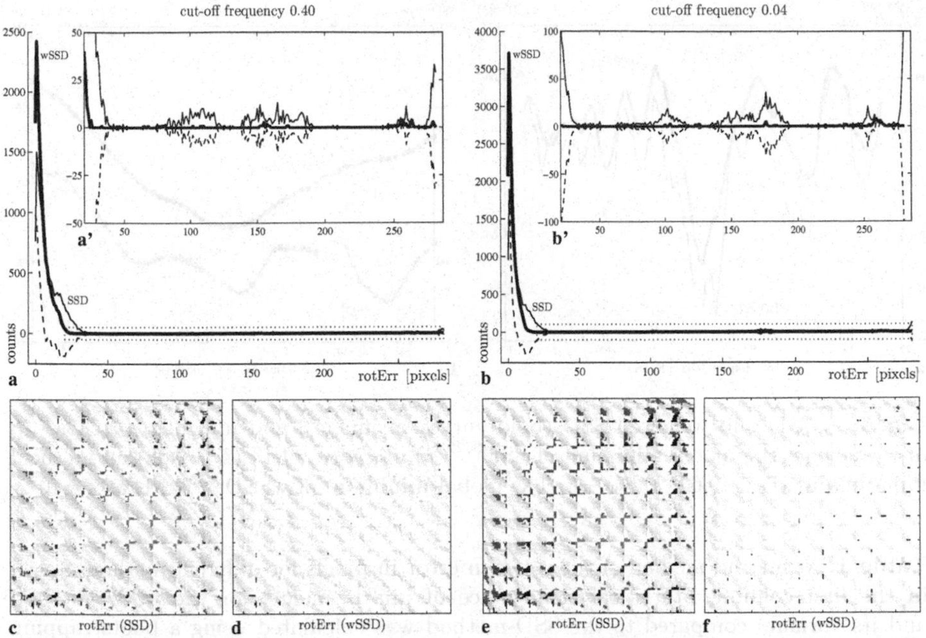

Fig. 5. Rotation errors of the two different methods described in this paper applied to panoramic images that were either filtered with a very weak low-pass (left, cut-off frequency 0.40) or a strong low-pass (right, cut-off frequency 0.04) before un-warping. a,b: Histogram plots of the rotation error (bin width of 1 pixel) for SSD (thin curve) and wSSD (bold), the dashed curves show the difference. The insets (a', b') show details of the rectangular regions marked by dotted lines. Histograms were calculated from a total of 14535 values (note different scaling of the y-axes). c,e: Gray-scale maps of rotation error matrices for SSD method. Mapping of rotation errors to gray-scale as in Fig. 2 d. d,f: Gray-scale maps for wSSD using additional images at positions shifted in x and y.

$\hat{\phi}^{3,22} = 0°$, $\text{SSD}^{3,22}(\psi)$ exhibits much more local minima and a global minimum at $\hat{\phi}^{3,22} \approx -14.1°$.

3.4 Testing Variants of the wSSD-Method

Having shown that the wSSD-method leads to significantly better rotation estimations, see also Table 1 and compare row one and row two, we ask next whether it is possible to simplify the method without loosing much of its benefits.

Movements at Reference Position Only. First we test what we gain if we calculate var-values only for one of the two images that we use for orientation estimation. This image e.g. recorded at the starting or a reference position we will call 'reference image'. For the second image we use equally weighted pixels, i.e. $\text{var}^b = 1$ in (6). Thus the robot would not have to do additional turns to calculate

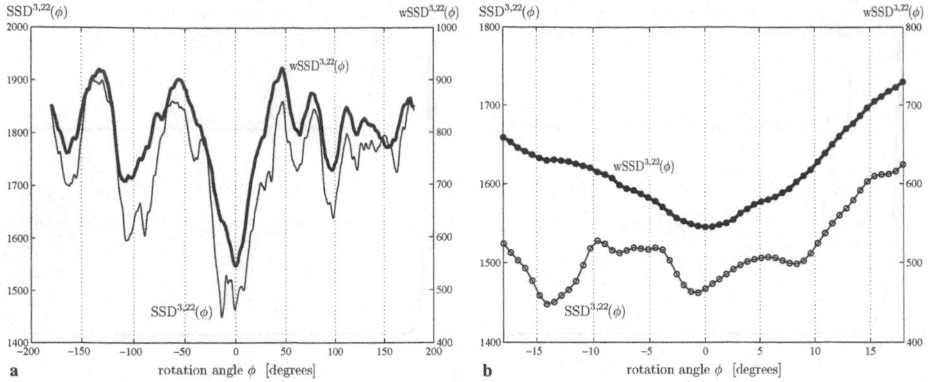

Fig. 6. a: SSD and weighted SSD (bold curve) of image 3 and image 22. b: Close-up of a, showing the global minimum of $\mathrm{SSD}^{3,22}(\phi)$ at $\hat{\phi} \approx -14.1°$ ($\hat{s} = -22$) and a local minimum at $\phi \approx -0.64°$ ($s = -1$). The global minimum of $\mathrm{wSSD}^{3,22}(\phi)$ is at $\hat{\phi} = 0°$.

Table 1. Mean and median of the rotation error in pixels for different methods listed in the first column. Statistical significance of improvements for the wSSD-method and its variants compared to the SSD-method was calculated using a bootstrapping method [15,16]: (**): significant on the $\alpha = 1\,\%$ level, (n.s.): not significant.

	cut-off freq. 0.40		cut-off freq. 0.04	
	mean (signif.)	median (signif.)	mean (signif.)	median (signif.)
SSD	15.4962	6.2537	23.1826	5.0426
wSSD	5.3940 (**)	3.1477 (**)	4.0339 (**)	2.0086 (**)
wSSDRefOnly	16.2611 (n.s.)	5.1471 (**)	9.3525 (**)	2.7219 (**)
wSSDx	8.9818 (**)	4.8174 (**)	8.1686 (**)	3.0061 (**)
wSSDy	12.1413 (**)	4.0660 (**)	5.7159 (**)	2.3350 (**)

var^b during navigation. The results for 'hall1'-images for this simplification are shown in Fig. 7 and Table 1, third row ('wSSDRefOnly'). Compared to the original wSSD-method the performance is clearly worse. As can be seen in the insets, larger rotation errors occur, especially for cut-off frequency 0.4. While the mean rotation error for images with cut-off frequency 0.4 is even slightly higher than for the SSD-method, there is still a significant improvement compared to the SSD-method for images with cut-off frequency 0.04. We conclude that for optimal rotation estimation close objects in both images have to be detected.

Movements in a Single Direction Only. Another way of simplifying the wSSD-method is to use only movements in a single direction and thereby avoiding 90° turns. We tested this wSSD-variant for movements either in x or in y direction, see Table 1, 4th row ('wSSDx') and 5th row ('wSSDy'). Although both variants perform significantly better than the SSD-method, their mean and median rotation errors are higher than for the original wSSD-method. This is

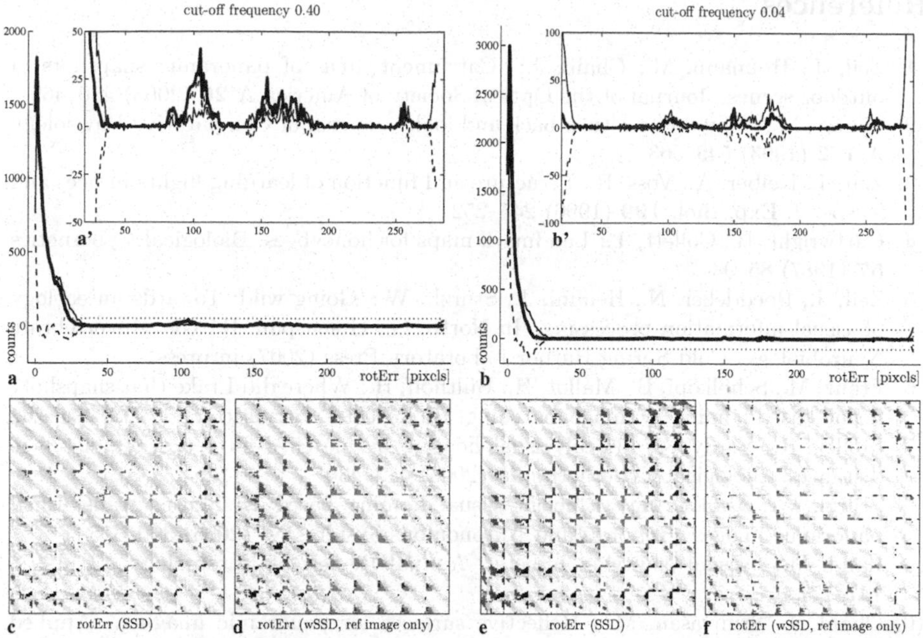

Fig. 7. Results for variant of the wSSD-method were var-values are calculated just for one of the two images used for rotation estimation. Conventions as in Fig. 5.

probably due to the fact that images recorded at positions shifted by a translation movement show very small image shifts in and opposite to the movement direction, i.e. close to the point of expansion and the point of contraction. To reliably detect close objects in all directions, the camera has thus to be moved in non-collinear directions.

4 Discussion and Outlook

We have presented a minimalistic active vision approach for rotation estimation that is solely based on pixel differences and thus, unlike other approaches, e.g. [17], does not involve feature extraction. Although no explicit calculation of the depth structure is done, it is implicitly contained in the var-values calculated from neighbouring images with same orientation.

Despite its simplicity, the proposed method has been shown to significantly improve orientation estimation for panoramic images in an indoor environment. In future work it will be tested in different environments, especially in natural outdoor scenes.

Acknowledgements. We thank Wolfram Schenck for his advice on the bootstrapping method and acknowledge financial support from the 'Deutsche Forschungsgemeinschaft' (grant no. STU 413/1-1 and MO 1037/2-1).

References

1. Zeil, J., Hofmann, M., Chahl, J.: Catchment areas of panoramic snapshots in outdoor scenes. Journal of the Optical Society of America A **20** (2003) 450–469
2. Lehrer, M.: Why do bees turn back and look? Journal of Comparative Physiology A **172** (1993) 549–563
3. Zeil, J., Kelber, A., Voss, R.: Structure and function of learning flights in bees and wasps. J. Exp. Biol. **199** (1996) 245–252
4. Cartwright, B., Collett, T.: Landmark maps for honeybees. Biological Cybernetics **57** (1987) 85–93
5. Zeil, J., Boeddeker, N., Hemmi, J., Stürzl., W.: Going wild: Towards an ecology of visual information processing. In North, G., Greenspan, R., eds.: Invertebrate Neurobiology. Cold Spring Harbor Laboratory Press (2007) in press.
6. Franz, M., Schölkopf, B., Mallot, H., Bülthoff, H.: Where did I take that snapshot? Scene-based homing by image matching. Biological Cybernetics **79** (1998) 191–202
7. Möller, R., Vardy, A.: Local visual homing by matched-filter descent in image distances. Biological Cybernetics **95** (2006) 413–430
8. Stürzl, W., Mallot, H.: Efficient visual homing based on fourier transformed panoramic images. Robotics and Autonomous Systems **54** (2006) 300–313
9. Land, M.: Visual acuity in insects. Annual Review of Entomology **42** (1997) 147–177
10. Chahl, J., Srinivasan, M.: Reflective surfaces for panoramic imaging. Applied Optics **36** (1997) 8275–8285
11. Nayar, S.: Catadioptric omnidirectional camera. In: Computer Vision and Pattern Recognition. (1997) 482–488
12. Horvath, G., Varju, D.: Polarized Light in Animal Vision. Springer (2004)
13. Vardy, A., Möller, R.: Biologically plausible visual homing methods based on optical flow techniques. Connection Science, Special Issue: Navigation **17** (2005) 47–89
14. Chahl, J., Srinivasan, M.: Range estimation with a panoramic visual sensor. Journal of the Optical Society of America A **14** (1997) 2144–2151
15. Efron, B., Tibshirani, R.: An Introduction to the Bootstrap. Chapman & Hall / CRC. (1998)
16. Schatz, A.: Visuelle Navigation mit "Scale Invariant Feature Transform". Master's thesis, Computer Engineering, Faculty of Technology, Bielefeld University (2006)
17. Makadia, A., Geyer, C., Daniilidis, K.: Radon-based structure from motion without correspondences. In: Computer Vision and Pattern Recognition. (2005) 796–803

Natural Interaction with a Robotic Head

O. Déniz, M. Castrillón, J. Lorenzo, and L. Antón-Canalís

Instituto Universitario de Sistemas Inteligentes
y Aplicaciones Numéricas en Ingeniería
Universidad de Las Palmas de Gran Canaria
Edificio Central del Parque Científico-Tecnológico
35017 Las Palmas - Spain
odeniz@dis.ulpgc.es

Abstract. Social robots are receiving much interest in the robotics community. The most important goal for such robots lies in their interaction capabilities. This work describes the robotic head CASIMIRO, designed and built with the aim of achieving interactions as natural as possible. CASIMIRO is a robot face with 11 degrees of freedom. Currently, the robot has audio-visual attention (based on omnidirectional vision and sound localization abilities), face detection, head gesture recognition, owner detection, etc. The results of interviews with people that interacted with the robot support the idea that the robot has relatively natural communication abilities, although certain aspects should be further developed.

1 Introduction

A relatively new area for robotics research is the design of robots that can engage with humans in socially interactive situations. These robots have expressive power (i.e. they all have an expressive face, voice, etc.) as well as abilities to locate, pay attention to, and address people. In humans, these abilities fall within the ambit of what has been called "social intelligence".

Being an emergent field, the number of social robots built seem to increase on a monthly basis, see [7] for a survey. Kismet [3] has undoubtedly been the most influential social robot appeared. It is an animal-like robotic head with facial expressions. Developed in the context of the Social Machines Project at MIT, it can engage people in natural and expressive face-to-face interaction.

It is important to note that inspiration and theories from human sciences has always been involved in the design of these robots, mainly from psychology, ethology and infant social development studies. Kismet, for example, was conceived as a baby robot, its abilities were designed to produce caregiver-infant exchanges that would eventually make it more dexterous. Other authors have taken advantage of autism as an inspiration for building social robots, i.e. by analyzing the significant lack of social abilities that autistic people have, see for example [10].

Careful analysis of the available work leads to the question of whether these and other robots that try to accomplish social tasks have a robust behaviour. Particularly, face recognition (the social ability par excellence) is extremely sensitive to illumination, hair, eyeglasses, expression, pose, image resolution, aging, etc., see [8]. There is the

J. Mira and J.R. Álvarez (Eds.): IWINAC 2007, Part I, LNCS 4527, pp. 71–80, 2007.

impression (especially among the robot builders themselves) that performance would degrade up to unacceptable levels when conditions are different from those used to train or test the implementations. In test scenarios, performance is acceptable. However, it would seem that there is little guarantee that it remains at the same levels for future, unseen conditions and samples. Note that this impression does not appear for other types of robots, say industrial manipulators, where the robot performance is "under control". This leads us to the important question: is building a social robot in any sense different than building other kinds of robots? An answer will be proposed later on.

This document describes CASIMIRO (The name is an Spanish acronym of "expressive face and basic visual processing for an interactive robot), a robot with basic social abilities. CASIMIRO is still under development and its capabilities will be expanded in the future. The paper is organized as follows. Section 2 outlines the conceptual approach taken. Then we briefly describe the implemented perception and action abilities, Sections 4 and 5, and behavior control in Section 6. Experiments are described in Section 7. Finally, we summarize the conclusions and outline future work.

2 Motivation

There are reasons to think that the design of social robots should be qualitatively different than that of other types of robots. The activities and processes that social robots try to replicate are generally of unconscious nature in humans, face recognition being the best example. Nowadays, the existence of unconscious processes in our brain seems to be beyond doubt. Some authors contend that the reason why some mental processes fade into the unconscious is repetition and practice [1]. If this is the case, our social abilities should be certainly more unconscious, as they appear earlier in life. On the other hand, the reason of their well performing may well be the fact that they are unconscious, although we do not delve further on that aspect.

The reproduction of social intelligence, as opposed to other types of human abilities, may lead to fragile performance, in the sense of having very different performances between training/testing and future (unseen) conditions. This limitation stems from the fact that the abilities of the social spectrum are mainly unconscious to us. This is in contrast with other human tasks that we carry out using conscious effort, and for which we can easily conceive algorithms. Thus, a coherent explanation is also given for the truism that says that anything that is easy for us is hard for robots and vice versa.

For some types of robots like manipulators one can extract a set of equations (or algorithms, representations,...) that are known to be valid for solving the task. Once that these equations are stored in the control computer the manipulator will always move to desired points. Sociable robots, however, will require a much more inductive development effort. That is, the designer tests implementations in a set of cases and hopes that the performance will be equally good for unseen (future) cases. Inductive processes crucially depend on *a priori* knowledge: if there is little available one can have good performance in test cases but poor performance in unseen cases (overfitting).

In the field of Inductive Machine Learning, complexity penalization is often used as a principled means to avoid overfitting. Thus, we propose to develop sociable robots starting from simple algorithms and representations. Implementations should evolve mainly through extensive testing in the robot niche (the particular environment and restrictions imposed on the robot tasks, physical body, etc.). Inspiration from human sciences is an asset, though our approach places more emphasis in the engineering decisions taken throughout the robot development process, which depend very much on the niche. The robot CASIMIRO, described in the following sections, has been built following this approach.

3 Robot Overview

This section describes the hardware that constitutes CASIMIRO. Details will be in general left out as the information is mainly technical data. It is important to introduce the hardware at this point because that helps focus the work described in the following sections. CASIMIRO is a robotic face: a set of (9) motors move a number of facial features placed on an aluminium skeleton. It also has a neck that moves the head. The neck has the pan and tilt movements, although they are not independent. The global aspect of the robot is shown in Figure 1.

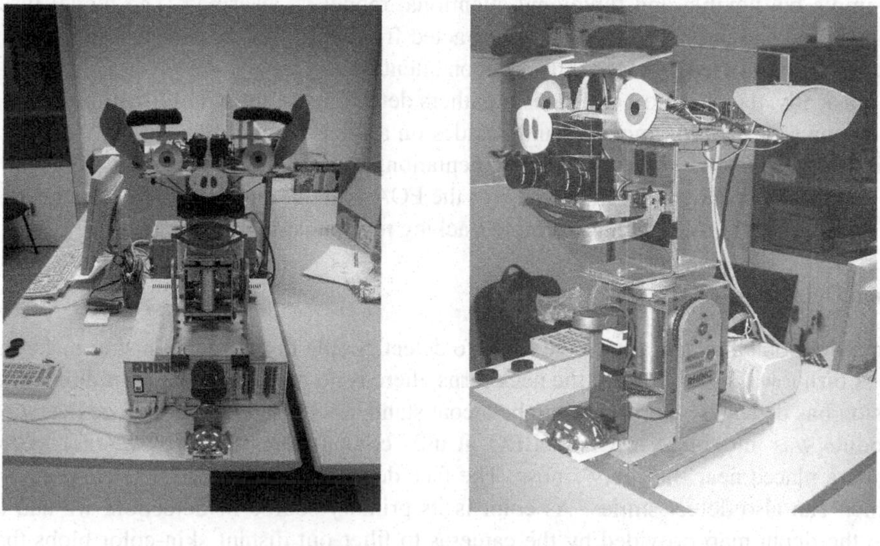

Fig. 1. Global aspect of CASIMIRO

4 Perception Abilities

This section gives an overview of the perceptual abilities implemented in CASIMIRO yet. Due to space constraints details will be omitted.

Omnidirectional Vision

As can be seen in Figure 1 in front of the robot there is an omnidirectional camera. The camera was built using a webcam plus a curved reflective surface. It allows the robot to have a 180° field of view, similar to that of humans. Through adaptive background subtraction, the robot is able to localize people in the surroundings, and pan the neck toward them. The curvature of the mirror allows to extract a rough measure of the distance to the robot.

Sound Localization

The robot has two omnidirectional microphones placed on both sides of the head. The signals gathered by them are amplified and filtered. The direction of the sound source is then estimated by calculating the ITD (*Interaural Phase Delay*) through cross-correlation. The sound localization module only works when the robot's facial motors are not working.

Audio-Visual Attention

The most important goal for social robots lies in their interaction capabilities. An attention system is crucial, both as a filter to center the robot's perceptual resources and as a mean of letting the observer know that the robot has intentionality. In CASIMIRO, a simple but flexible and functional attentional model is described. The model fuses both visual and auditive information extracted from the robot's environment, and can incorporate knowledge-based influences on attention.

Basically, the attention mechanism gathers detections of the omnidirectional vision and sound localization modules and decides on a focus of attention (FOA). Although this can be changed, the current implementation sets the FOA to the visual detection nearest to the sound angle. In other cases the FOA is set to the visual detection nearest to the previous FOA, which is a simple tracking mechanism.

Face Detection

Omnidirectional vision allows the robot to detect people in the scene, just to make the neck turn toward them. When the neck turns, there is no guarantee that omnidirectional vision has detected a person, it can be a coat stand, a wheelchair, etc. A face detection module was integrated in CASIMIRO, it uses color images taken by a color stereo camera placed near the robot's nose. The face detection application is ENCARA [4], which can also detect smiles. As color is its primary source of detection, we had to use the depth map provided by the cameras to filter out distant skin-color blobs that corresponded to furniture, doors, etc. (see Figure 2).

Head Nod and Shake Detection

Voice recognition was not implemented in CASIMIRO. It is estimated that voice recognition errors, dubbed by Oviatt as the Achilles' hell of speech technology, increase a 20%-50% when speech is delivered during natural spontaneous interaction, by diverse speakers or in a natural field environment [9]. The option of making the speaker wear a microphone was discarded from the beginning because it is too unnatural. Due to the fact that (hands-free) speech feedback is very difficult to obtain for a robot, we decided to turn our attention to simpler input techniques such as head gestures. Head nods and shakes are very simple in the sense that they only provide yes/no, understanding/disbelief, approval/disapproval meanings. However, their importance must not be underestimated because of the following reasons: the meaning of head nods and shakes is almost universal, they can be detected in a relatively simple and robust way and they can be used as the minimum feedback for learning new capabilities.

The major problem of observing the evolution of simple characteristics like intereye position or the rectangle that fits the skin-color blob is noise. Due to the unavoidable noise, a horizontal motion (the NO) does not produce a pure horizontal displacement of the observed characteristic, because it is not being tracked. Even if it was tracked, it could drift due to lighting changes or other reasons. The implemented algorithm uses the pyramidal Lucas-Kanade tracking algorithm described in [2]. In this case, there is tracking, and not of just one, but multiple characteristics, which increases the robustness of the system. The tracker looks first for a number of good points to track, automatically. Those points are accentuated corners. From those points chosen by the tracker we can attend to those falling inside the rectangle that fits the skin-color blob, observing their evolution and deciding based on what dimension (horizontal or vertical) shows a larger displacement. Figure 2 shows an example of the system.

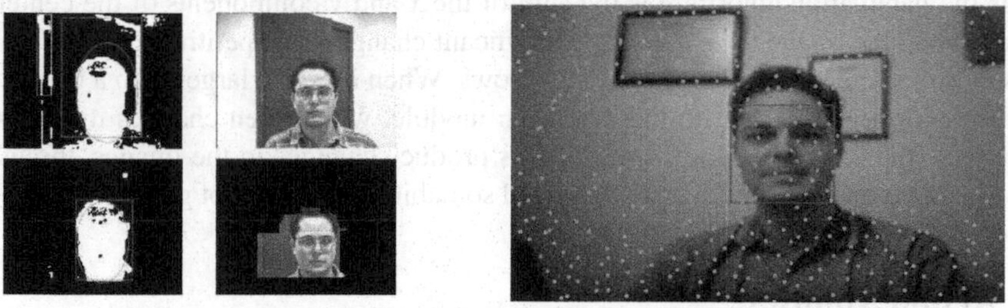

Fig. 2. Left: Top row: skin color detection. Bottom row: skin color detection using depth information. **Right:** Head nod/shake detector.

Memory and Forgetting

In [11] three characteristics are suggested as critical to the success of robots that must exhibit spontaneous interaction in public settings. One of them is the fact that the robot should have the capability to adapt its human interaction parameters based on the outcome of past interactions so that it can continue to demonstrate open-ended behaviour.

CASIMIRO has a memory of the individuals that it sees. Color histograms of (part of) the person's body are used as a recognition technique. Color histograms are simple to calculate and manage and they are relatively robust. The price to pay is the limitation that data in memory will make sense for only one day (at the most), though that was considered sufficient. The region of the person's body from which histograms are calculated depends on the box that contains the face detected by ENCARA. Intersection was used to compare a stored pair of histograms with the histograms of the current image. Memory will be represented in a list of histogram pairs, with data associated to each entry. Each entry in the list is associated to an individual. Currently, the data associated to the individuals are Boolean predicates like "Content", "Greeted", etc.

Memory is of utmost importance for avoiding predictable behaviors. However, memorizing facts indefinitely leads to predictable behaviors too. Behavioral changes occur when we memorize but also when we forget. Thus, a forgetting mechanism can also be helpful in our effort, especially if we take into account the fact that actions chosen by the action-selection module do not always produce the same visible outcome. In our system the power law of forgetting (see [5]) is modelled in the following way. Let $f(t)$ be a forget function, which we use as a measure of the probability of forgetting something: $f(t) = max(0, 1 - t \cdot \exp(-k))$, where k is a constant. We apply the f function to the set of Boolean predicates that the robot retains in memory. When a predicate is to be forgotten, it takes the value it had at the beginning, when the system was switched on.

Habituation

An habituation mechanism developed by the authors was implemented in CASIMIRO, for signals in the visual domain only, i.e. images taken by the stereo camera. The difference between the current and previous frame is calculated. Then it is thresholded and filtered with Open and Close operators. Also, blobs smaller than a threshold are removed. Then the center of mass of the resultant image is calculated. The signal that feeds the habituation algorithm is the sum of the x and y components of the center of mass. When the image does not show significant changes or repetitive movements are present for a while the habituation signal grows. When it grows larger than a threshold, an inhibition signal is sent to the Attention module, which then changes its focus of attention. The neck pan and tilt movements produce changes in the images, though it was observed that they are not periodic, and so habituation does not grow.

5 Action Abilities

Facial Expression

A three-level hierarchy was used to model facial expressions in CASIMIRO. Groups of motors that control a concrete facial feature are defined. For example, two motors are grouped to control an eyebrow. For each of the defined motor groups, the poses that the facial feature can adopt are also defined, like 'right eyebrow raised', 'right eyebrow neutral', etc. The default transitions between the different poses uses the straight line in the space of motor control values.

The designer is given the opportunity to modify these transitions, as some of them could appear unnatural. A number of intermediate points can be put in all along the transition trajectory. Additionally, velocity can be set between any two consecutive points in the trajectory. The possibility of using non-linear interpolation (splines) was considered, although eventually it was not necessary to obtain an acceptable behaviour. The first pose that the modeller must define is the neutral pose. All the defined poses refer to a maximum degree for that pose, 100. Each pose can appear in a certain degree between 0 and 100. The degree is specified when the system is running, along with the pose itself. It is used to linearly interpolate the points in the trajectory with respect to the neutral pose.

As for the third level in the mentioned hierarchy, facial expressions refer to poses of the different groups, each with a certain degree. Currently, CASIMIRO has the following expressions: Neutral, Surprise, Anger, Happiness, Sadness, Fear and Sleep.

Voice Generation

CASIMIRO uses canned text for language generation. A text file contains a list of labels. Under each label, a list of phrases appear. Those are the phrases that will be pronounced by the robot. They can include annotations for the text-to-speech module (a commercially available TTS was used). Labels are what the robot wants to say, for example "greet", "something humorous", "something sad", etc. Examples of phrases for the label "greet" could be: "hi!", "good morning!", "greetings earthling".

The Talk module, which manages TTS, reads the text file when it starts. It keeps a register of the phrases that haven been pronounced for each label, so that they will not be repeated. Given a label, it selects a phrase not pronounced before, randomly. If all the phrases for that label have been pronounced, there is the option of not saying anything or starting again. The Talk module pronounces phrases with an intonation that depends on the current facial expression (see below). This is done by changing the intonation parameters of the TTS.

6 Behavior

Action Selection

CASIMIRO's action selection module is based on ZagaZ [6]. ZagaZ is an implementation of Maes' Behaviour Networks. It has a graphical interface that allows to execute and debug specifications of PHISH-Nets. Specifications have to be compiled before they can be executed. There are two compilation modes: release and debug. The action selection loop in ZagaZ has a period of a few milliseconds for relatively simple networks. It was necessary to introduce a delay of 500 ms on each cycle for the whole system to work well. Behaviors implemented has Boolean inputs like "Frontal Face Detected" which may also correspond to memorized values. The repertory of actions is currently limited to changes in the emotional state (which in turn modifies the displayed facial expression) and commands for talking about something.

Emotions

The Emotions module maintains a position in a 2D valence and arousal space. The module receives messages to shift the current position in one or the two dimensions. The 2D space is divided into zones that correspond to a facial expression. Examples are *happiness*: positive arousal,positive valence; *anger*: negative valence, positive arousal, etc. In order to simplify the module, it is assumed that the expression is given by the angle in the 2D space (with respect to the valence axis), and the degree is given by the distance to the origin. The circular central zone corresponds to the neutral facial expression. When the current position enters a different zone a message is sent to the pose editor so that it can move the face, and to the Talk module so that intonation can be adjusted. A very simple decay is implemented: every once in a while arousal and valence are divided by a factor. This does not change the angle in the 2D space, and thus the facial expression does not change, only the degree.

The emotions that the robot has experienced while interacting with an individual are stored in the memory associated to that individual. Actually, memory is updated periodically with the mean values of arousal and valence experienced with that individual (a running average is used). As for sleep, when the position in the 2D space has been for a certain time in the neutral state arousal is lowered by a given amount (valence will be zero). Besides, sleep has associated a decay factor below 1, so that it tends to get farther the center instead of closer. This way, the emotional state will eventually tend to neutral, and in time to sleep. When the robot is asleep the neck stops working.

7 Evaluation

Figure 3 shows an example interaction session with CASIMIRO. Initially, individual A enters the interaction area. The valence values show that he adopts an uncooperative attitude. The robot tries to ask him if he wants to hear poems, but the individual keeps moving around and the robot has to abort the questions. At time 55 individual A begins to leave the interaction area. The robot is then alone. At time 67 another individual enters the interaction area. This individual B is more cooperative and answers affirmatively to the two questions made by the robot. Individual B leaves the area at around time 126. Then, individual A comes back at time 145 and is recognized by the robot, which avoids greeting him again. Note that upon seeing individual A the robot emotional state turns very negative, for its previous interaction with him ended unsatisfactorily.

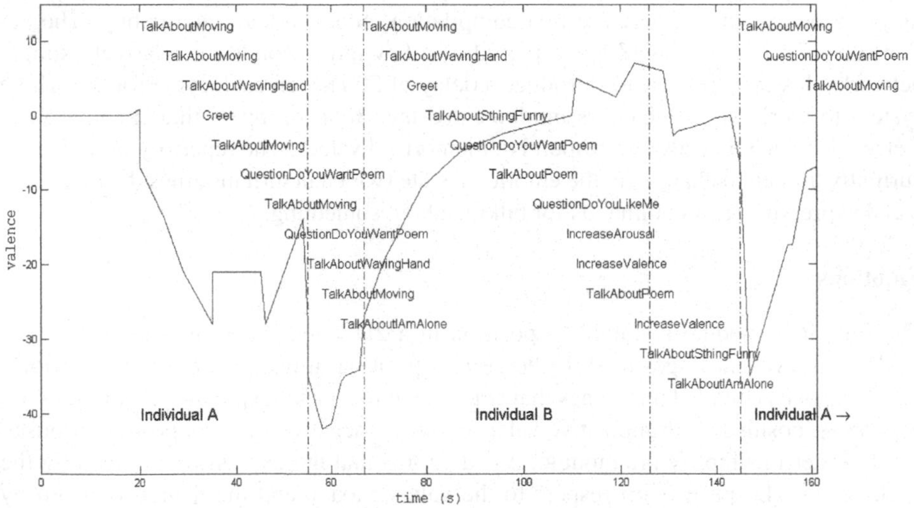

Fig. 3. Example interaction that shows how the robot recognizes people. The figure shows the evolution of the valence values of the robot emotional state and the executed actions.

We arranged for some people to interact with the robot and fill in a questionnaire after the interaction. For each topic, see Table 1, people had to assign a score between 1 (definitely no) to 5 (definitely yes). Out of 19 interviewees, 7 had a PhD in computer science or an engineering, the rest were at least graduate students. Computing knowledgeable individuals were used on purpose to make the test harder for the robot, for these individuals usually know how robots work internally and what programming can do. The point is: it is not what sociability level the robot has, but what sociability level may be perceived as such.

The detailed results of the interviews are omitted here for space reasons. The average score received was *3.54*. Particularly good scores were received by questions 4 and 5 (*4.21* and *4.89*, respectively). For some abilities people were very impressed, like for example with owner recognition. Some asked questions like *'how can the robot do that?'* or *'hey! does it recognize people?'*. We find these results encouraging, given the simplicity of the implemented technique. However, other aspects received less scores, particularly question number 9 (*2.4*, the minimum score received), which suggests aspects that require future development.

Table 1. Questionnaire used to evaluate the robot. The last question allowed a free answer.

Section 1 "I understand the robot":
1. I have understood everything the robot has told me
2. The robot has conveyed its emotions through facial expressions
3. The robot has conveyed its emotions through voice tone
4. The robot pays attention to you
5. The robot is conscious of my presence
6. The robot is conscious of my movements
7. The robot recognizes people
8. This robot is a good starting point for keeping you informed
Section 2 "The robot understands me":
9. The robot understands what people say
10. The robot understands facial expressions
11. The robot knows where I direct my attention
12. This robot may be used to make it learn new things from people
Section 3 "Overall impression":
13. I have not had to make an effort to adapt myself to the robot
14. The robot has not had failures (things that it obviously had to do but it didn't)
15. What do you think the robot should have to be used more frequently?

8 Conclusions and Future Work

This paper has described the current development status of a robot head with basic interaction abilities. The implementation of social abilities in robots necessarily leads to unrobust behaviour, for those abilities are mainly unconscious to us, as opposed to other mental abilities. The approach taken has been to use a complexity penalization approach, as this minimizes error in unseen conditions. CASIMIRO's perceptual abilities include sound localization, omnidirectional vision, face detection, an attention module, memory, habituation, etc. The robot has facial features that can display basic emotional expressions, and it can speak canned text through a TTS. The robot's behaviour is controlled by an action selection module, reflexes and a basic emotional module.

Future work will include research into the possibility of integrating hands-free speech recognition. This is probably one of the most interesting research topics in human-computer interaction. Audio-visual speech recognition, i.e. using both audio and lip motion, is a promising approach that shall be put to practice.

References

1. Bernard J. Baars. *A cognitive theory of consciousness.* Cambridge University Press, NY, 1988.
2. J. Bouguet. Pyramidal implementation of the Lucas Kanade feature tracker. Technical report, Intel Corporation, Microprocessor Research Labs, OpenCV documents, 1999.
3. Cynthia L. Breazeal. *Designing social robots.* MIT Press, Cambridge, MA, 2002.
4. M. Castrillón. *On Real-Time Face Detection in Video Streams. An Opportunistic Approach.* PhD thesis, Universidad de Las Palmas de Gran Canaria, March 2003.
5. H. Ebbinghaus. *Memory. A Contribution to Experimental Psychology.* Teachers College, Columbia University, New York, 1913.
6. D.J. Hernández-Cerpa. Zagaz: Entorno experimental para el tratamiento de conductas en caracteres sintéticos. Master's thesis, Universidad de Las Palmas de Gran Canaria, 2001.
7. M. Lungarella, G. Metta, R. Pfeifer, and G. Sandini. Developmental robotics: a survey. *Connection Science*, 0(0):1–40, 2004.
8. A. Martin, P.J. Phillips, M. Przybocki, and C.I. Wilson. An introduction to evaluating biometric systems. *Computer*, 56:56–63, February 2000.
9. S. Oviatt. Taming recognition errors with a multimodal interface. *Communications of the ACM*, 43(9):45–51, 2000.
10. B. Scassellati. *Foundations for a Theory of Mind for a Humanoid Robot.* PhD thesis, MIT Department of Computer Science and Electrical Engineering, May 2001.
11. J. Schulte, C. Rosenberg, and S. Thrun. Spontaneous short-term interaction with mobile robots in public places. In *Procs. of the IEEE Int. Conference on Robotics and Automation*, 1999.

A Network of Interneurons Coupled by Electrical Synapses Behaves as a Coincidence Detector

Santi Chillemi, Michele Barbi, and Angelo Di Garbo

Istituto di Biofisica CNR, Sezione di Pisa,
Via G. Moruzzi 1, 56124 Pisa, Italy
{santi.chillemi,michele.barbi,angelo.digarbo}@pi.ibf.cnr.it
http://www.pi.ibf.cnr.it

Abstract. Recent experiments show that inhibitory interneurons are coupled by electrical synapses. In this paper the information transmission properties of a network of three interneurons, coupled by electrical synapses alone, are studied by means of numerical simulations. It is shown that the network is capable to transfer the information contained in its presynapstic inputs when they are near synchronous: i.e. the network behaves as a coincidence detector. Thus, it is hypothesized that this property hold in general for networks of larger size. Lastly it is shown that these findings agree with recent experimental data.

1 Introduction

Interneurons innervating the somatic and perisomatic region of pyramidal cells are able to modulate their firing activities [1,2]. Moreover it was found that oscillations, in the gamma frequency band (30 - 100 Hz), are associated to cognitive functions [3]. Paired recording of interneurons have shown that they are interconnected with electrical and inhibitory synapses [4,5]. The relevance of the electrical synapses for the generation of synchronous discharges was shown experimentally: the impairing of the electrical synapses between cortical interneurons disrupts synchronous oscillations in the gamma frequency band [6]. Recently it was also shown that the presence of electrical coupling in a pair of inhibitory interneurons promotes synchronization at all spiking frequencies and this property is enhanced when the strength of the electrical coupling increases [7]. Additional experimental investigations suggest that interneurons play a relevant role in the detection of synchronous activity [8,9]. Moreover, interneurons are involved in the feed-forward inhibition of pyramidal cells, and that is a direct consequence of their fast and reliable response to excitatory inputs [10]. In a previous paper [11] we have investigated the property of a pair of fast spiking interneurons to detect synchronous inputs, here we will try to extend this study to the case of three coupled cells. For semplicity we will consider the case in which the interneurons are connected by electrical synapses alone. A clear motivation to study a small population of interneurons, coupled by electrical coupling alone,

J. Mira and J.R. Álvarez (Eds.): IWINAC 2007, Part I, LNCS 4527, pp. 81–89, 2007.

comes from recent experimental findings [9,12,13,14]. In fact many interneurons of the thalamic reticular nucleus are interconnected by electrical synapses and form clusters that are quite small compared with those in the neocortex; moreover it was shown that the electrical coupling coordinate the rhythmic activity of these neural netwoks [12]. An additional contribution to the synchronization properties of thalamic reticular neurons probably comes from the excitatory inputs that they receive from neocortex and thalamic relay nuclei [12]. Thus, an interesting question is to understand how the firing properties of the coupled interneurons is affected by the time delays of the excitatory inputs they receive. An other example is that of the Inferior Olive region: the corresponding experimental results indicate that gap junctions connecting pair of interneurons play a key role in promoting synchronization[13,14]. In this paper,by using a computational approach, we investigate the capability of a network of three interneurons coupled by electrical synapses alone of transmitting excitatory synaptic inputs.

2 Methods

FS interneurons are not capable of generating repetitive firing of arbitrary low frequency when injected with constant currents [15,16], thereby they have type II excitability[17].

Recent experiments carried out on in vitro fast spiking interneurons reveal that they have high firing rates (up to $\approx 200\ Hz$), average resting membrane potential of -72 mV and input resistance $\approx 89\ M\Omega$; their action potential has a mean half-width $\approx 0.35\ ms$, average amplitude $\approx 61\ mV$ and after-hyperpolarization amplitude $\approx 25\ mV$ [4,5,15].

2.1 Model Description

Here we use the following single compartmental biophysical model of a fast spiking interneuron proposed in [11], well accounting for the features above:

$$C\frac{dV}{dt} = I_E - g_{Na}m^3h(V-V_{Na}) - g_K n(V-V_K) - g_L(V-V_L) + g_{Exc}P(t-t^*) \quad (1)$$

$$\frac{dx}{dt} = \frac{x_\infty - x}{\tau_x}, \quad x_\infty = \frac{\alpha_\infty}{\alpha_\infty + \beta_\infty}, \quad \tau_x = \frac{1}{\alpha_\infty + \beta_\infty}, \quad (x = m, h, n) \quad (2)$$

where $C = 1\ \mu F/cm^2$, I_E is the external stimulation current. The maximal specific conductances and the reversal potentials are respectively: $g_{Na} = 85\ mS/cm^2$, $g_K = 60\ mS/cm^2$, $g_L = 0.15\ mS/cm^2$ and $V_{Na} = 65\ mV$, $V_K =$ -95 mV, $V_L =$ -72 mV. The term $P(t-t^*)$ represents an excitatory pulses starting at time t^* and it is defined by: $P(t-t^*) = H(t-t^*)\{N[e^{-(t-t^*)/\tau_D} - e^{-(t-t^*)/\tau_R}]\}$ where, N is a normalization constant ($| P | \leq 1$), $\tau_D = 2ms$ and $\tau_R = 0.4ms$ are, respectively, realistic values of the decay and rise time constants of the excitatory pulse. The kinetic of the Na^+ current is described by the following activation and deactivation rate variables: $\alpha_m(V) = 3.exp[(V+25)/20]$, $\beta_m(V) = 3.exp[-(V+25)/27]$,

$\alpha_h(V) = 0.026exp[-(V+58)/15]$, $\beta_h(V) = 0.026exp[(V+58)/12]$. The kinetics of the potassium current K^+ is defined by: $\alpha_n(V) = [-0.019(V-4.2)]/\exp[-(V-4.2)/6.4]-1$, $\beta_n(V) = 0.016\exp(-V/vAHP)$,where vAHP $= 13\ mV$. In this model the onset of periodic firing occurs through a subcritical Hopf bifurcation for $I_E \approx 1.47\ \mu A/cm^2$ with a well defined frequency (≈16 Hz)(data not shown).

2.2 Synaptic Coupling and Pulse Timing

The electrical synapses between a pair of interneuron are modeled as follows:

$$I_{El}(i) = -g_{ij}(V_i - V_j), \tag{3}$$

where $g_{ij} = g_{ji}$ is the maximal conductance of the gap junction (in mS/cm^2 unit). In the case of three coupled interneurons the coupling currents are the following:

$$I_{El}(1) = -g_{12}(V_1 - V_2) - g_{31}(V_1 - V_3), \tag{4}$$
$$I_{El}(2) = -g_{12}(V_2 - V_1) - g_{23}(V_2 - V_3), \tag{5}$$
$$I_{El}(3) = -g_{31}(V_3 - V_1) - g_{23}(V_3 - V_2), \tag{6}$$

where $I_{El}(1)$ is the coupling current between interneuron 1 and the last two (2 and 3), etc.. Then, the synaptic coupling currents are introduced by adding each term in the corresponding right hand side of equation 1. For all simulations the adopted value of the parameters g_{ij} were all within the physiological range [4,5,15]. The j-th interneuron receives the excitatory pulse at time t_j with: $t_1 \leq t_2 \leq t_3$ and amplitude (g_{Exc}). Moreover, the time delays between two consecutive pulses will be adopted to be equal: i.e. $t_3 = t_2 + \Delta t$ and $t_2 = t_1 + \Delta t$. In the following the value of parameter t_1 will be set to $200ms$.

2.3 Synaptic Background Activity

Here, in keeping with the experiments, the simulations are carried out to reproduce the membrane potential fluctuations occurring in *in vitro* conditions [15]. Thus, the $j-th$ cell model is injected with a noisy current: $\sigma\xi_j(t)$, $\xi_j(t)$ being an uncorrelated Gaussian random variable of zero mean and unit standard deviation ($<\xi_i(t), \xi_j(t)> = \delta_{ij}, i \neq j = 1,2,3$). The values of the stimulation current I_E and of σ (equal for each cell) are chosen in such a way that no firing occurs in absence of the excitatory pulse. To get an approximation of the firing statistics of each interneuron, the stimulation protocol is repeated ($N_{Trials} = 200$) by using independent realizations of the applied noisy current. We investigate the network of coupled interneuron models in realistic conditions: i.e. when, in the absence of coupling, the firing probability of each cell receiving the excitatory pulse is lower than 1. With this in mind the parameters I_E, σ and g_{Exc} were so chosen that the firing probability of each cell is ≈ 0.75 (see [8]).

2.4 Data Analysis

The data obtained during the simulations will be presented here either as histogram or cumulative spikes count. The spike histogram is obtained as follows: the times of occurrence of spikes t_j (j=1,2,..,N), falling in a given time window are recorded for all trials (N_{Trials}), then the histogram of the t_j values was built by using a bin size of 0.2 ms. Then, the corresponding cumulative spike count up to time T, was built by adding all spikes falling in the bins located before T. Lastly, the firing probability of each cell receiving the excitatory pulses is defined as n_s/N_{Trials}, where n_s represents the total number of spikes generated during all trials.

3 Results

To characterize the firing properties of each cell let us consider, first of all, the case in which the coupling between the cells is set off. The data reported in figure 1 show that the spike histogram exhibits a peak located just to the right of $t_1 = 200ms$ (i.e. the time at which the excitatory pulse is applied). The firing of the interneuron occurs with an estimated latency of about 6 ms.

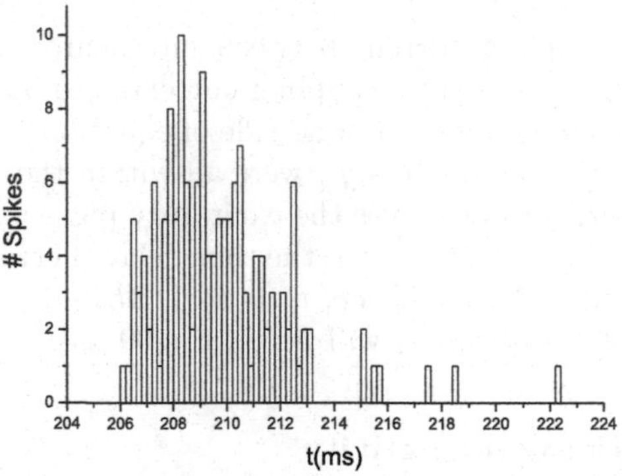

Fig. 1. Spike histogram of a generic interneuron of the network in absence of electrical coupling. The parameter values are: $t_1 = 200ms$, $\Delta t = 0ms$, $g_{Exc} = 5.5\mu A/cm^2$,$\sigma = 0.3\mu A/cm^2$,$I_E = 0.5\mu A/cm^2$,$N_{Trials} = 200$.

Let us introduce the coupling between the interneurons to investigate how the three excitatory pulses are transmitted by the network. To get a more clear understanding these inputs can be thought to be the postsynaptic currents generated by the presynaptic activities of excitatory neural networks. The results, in the case in which the electrical conductances between interneurons are all equal, are reported in figure 2.

When the time delay between the pulse is $\Delta t = 1ms$, the behavior of the cumulative spikes against the time indicate that all cells are responding. The firing

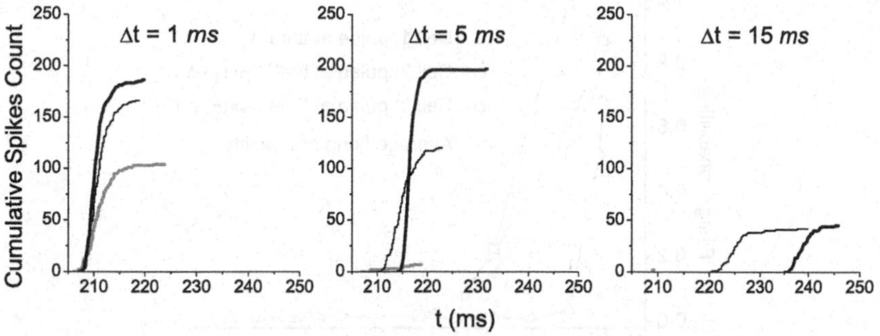

Fig. 2. Information transmission properties of the network for several time delay values. For all panels the parameter values are: $t_1 = 200ms$, $\Delta t = 0ms$, $g_{Exc} = 5.5\mu A/cm^2$, $\sigma = 0.3\mu A/cm^2$, $I_E = 0.5\mu A/cm^2$, $N_{Trials} = 200$, $g_{12} = g_{13} = g_{23} = 0.02mS/cm^2$. The gray line refers to cell 1, the thin black line to cell 2 and the thick black line to cell 3.

probability increases starting from cell 1 to cell 3 (let us remember that cell 1 receive the pulse at time t_1, cell 2 at $t_2 = t_1 + \Delta t$ and cell 3 at $t_3 = t_2 + \Delta t$). In this case the temporal information contained in the excitatory pulses is transmitted by the network of coupled cells by preserving the timing of their inputs. When the time delay between the pulses increases to $\Delta t = 5ms$ there is a reduction of the firing probabilities of cell 1 and 2, while that of cell 3 is practically unchanged. Thus, in this case the overall temporal structure contained in the excitatory pulses is not completely transmitted by the network. The meaning of this last sentence is that the information (i.e. the excitatory pulse) that is transmitted by the cell 1 (by means of generated spikes) is very low compared to that transmitted by cell 2 and 3. In fact the firing in this case are: $p_1 = 0.035, p_2 = 0.6, p_3 = 0.985$ respectively, for cell 1, cell 2 and 3. When the time delay increases to $\Delta t = 15ms$ firing in cell 1 decreases more ($p_1 = 0.01$), while that of cell 2 and 3 is strongly depressed with respect to the cases $\Delta t = 1ms, \Delta t = 5ms$. Increasing the time delay to $\Delta t = 25ms$ slows also the firing of cell 2 and 3 ($p_2 = p_3 = 0.02$).

To get a better understanding of the transmission properties of the network of coupled cells, in figure 3 are reported the corresponding firing probabilities against the time delay between the pulses.

The inspection of these data implies that the transmission of the excitatory inputs occurs when the time delay between them is lower or equal to $1ms$: i.e. the network behaves as a coincidence detector. In other words, the maximal network response occurs when the excitatory inputs are near synchronous. This result is a generalization of that found in the case of two coupled cells in [11]. Moreover the results obtained in the case of two coupled cells can be used to explain why the firing rate of cell 1 is lower of that of cell 2 and that of cell 2 is lower of that of cell 3 (see figure 2). The data reported in figure 3 shows that the firing activity of the network stops when the time delay between the excitatory pulses gets a sufficiently high value. This phenomenon was observed also in the case of two

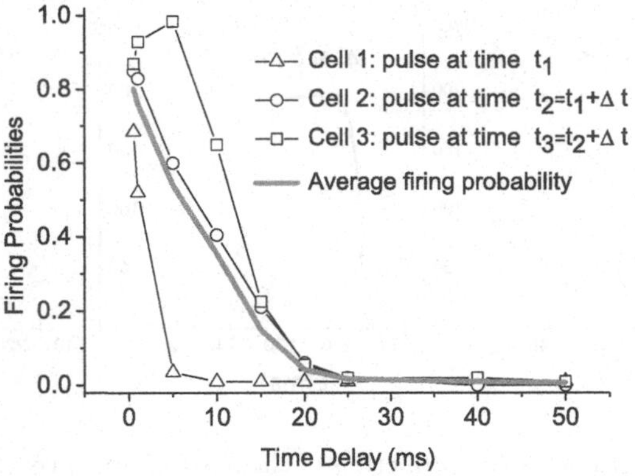

Fig. 3. Firing probabilities of the network of coupled interneurons against the time delay separating the excitatory pulses. The parameter values are:$t_1 = 200ms$, $g_{Exc} = 5.5\mu A/cm^2$, $\sigma = 0.3\mu A/cm^2$, $I_E = 0.5\mu A/cm^2$, $N_{Trials} = 200$, $g_{12} = g_{13} = g_{23} = 0.02mS/cm^2$. The gray line represents the average firing probability of the network.

coupled cells [11] and it was shown, either analytically and numerically, that it occurs because the effective input resistance of each cell decreases when the time delay between the two excitatory pulses is large. This explanation works also in the case of the three coupled cells. A qualitative explanation is the following: let be R_i and $\bar{R}_i$ the effective input resistance of a cell of the network in the cases $\Delta t \simeq 0$ and $\Delta t \gg 1$,respectively. Moreover, let us assume that the dynamical regime of each cell is subthreshold. When it is $\Delta t \simeq 0$ the difference between the membrane potentials of a given cell with the remaining two is smaller than that computed for $\Delta t \gg 1$ (see [11]). Then, in the first case ($\Delta t \simeq 0$)the current fluxes evoked by the excitatory pulse are mediated (mainly) by the capacitive and leakage conductances, while in the second case ($\Delta t \gg 1$) there are additional current fluxes through the electrical synapses. Therefore, it follows that it is $R_i > \bar{R}_i$. Thus, the amplitude of the depolarization evoked by the pulse will be greater in the case $\Delta t \simeq 0$ than that for $\Delta t \gg 1$. An analytical proof will be presented elsewhere. The results presented up to now were obtained in the case $g_{12} = g_{13} = g_{23}$: how change the results when this statement does not hold? The presence of heterogeneity in the electrical coupling conductance between cells is a more realistic representation of a real network of coupled interneurons; thus it is interesting to investigate whether in this case the transfer of information occurs. The corresponding results obtained by using several types of heterogeneity are reported in figure 4.

The inspection of these data clearly shows that the network is able to transfer the information contained in its inputs when they are near synchronous. Thus the network behaves, also in this case, as a coincidence detector as in the case of homogeneous coupling. This finding indicates that this property of the network is robust against the introduction of heterogeneity in the coupling and leads us to hypothesize that it holds also for networks of larger size. The

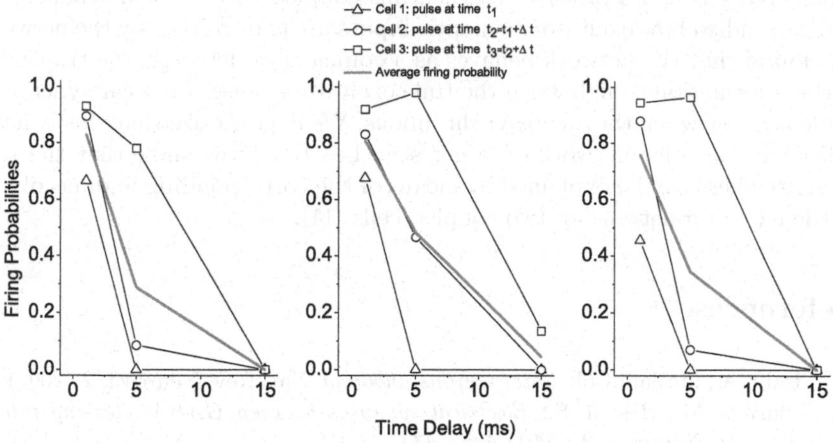

Fig. 4. Firing probabilities of the interneurons against the time delay separating the excitatory pulses for inhomogeneous coupling. The parameter values are: $t_1 = 200ms$, $g_{Exc} = 5.5\mu A/cm^2$, $\sigma = 0.3\mu A/cm^2$, $I_E = 0.5\mu A/cm^2$, $N_{Trials} = 200$. For the left panels it is $g_{12} = 0.02mS/cm^2, g_{13} = g_{23} = 0.08mS/cm^2$; for the middle panel it is $g_{12} = 0.08mS/cm^2, g_{13} = g_{23} = 0.02mS/cm^2$ and for the right panel it is $g_{12} = 0.02mS/cm^2$, $g_{13} = 0.04mS/cm^2, g_{23} = 0.08mS/cm^2$.The gray line represents the average firing probability of the network.

experimental studies on coupled interneurons showed that the presence of the electrical synapses between cells promotes their firing synchrony [4,5,7,8]. In particular it was shown that the presence of electrical coupling in a pair of coupled interneurons confers to the network the capability to detect synchronous inputs. Our results with three coupled interneurons agree with this experimental finding. In the mammalian retina AII Amacrine cells are coupled by electrical synapses and receive excitatory inputs from Rod Bipolar cells [9,18]. In a recent experimental paper were studied the information transmission properties of a pair of coupled AII Amacrine cells [9]: i.e. the firing probability of each cell when receiving excitatory inputs. It was found that when the two excitatory pulses were applied asynchronous the firing probabilities of the cells was low, while it was significantly higher when the two pulses were synchronous. Thus, the results reported here in figures 3 and 4 agree qualitatively with these experimental findings.

4 Conclusions

The excitatory synaptic communication among neurons is the basis for the transmission and coding of the sensory information [10]. This neural activity is modulated by the discharge of inhibitory interneurons [1]. Moreover it is now established that interneurons are coupled also by gap junctions and play a key role for the processing of the neural information [2,4,5]. In this paper we considered a network of three interneurons coupled by electrical synapses alone and we studied how excitatory synaptic inputs are transmitted by the network. It was found that the network behaves as a coincidence detector: the transmission of the information is high when the time excitatory pulses are near synchronous, while it is low when they are asynchrounous. We hypothesize that this behaviour will occurs also for network of larger size. Lastly, it was show that the results presented here can be explained by means of the corresponding findings obtained in the case of a network of two coupled cells [11].

References

1. Fisahn A., McBain C.J.: *Interneurons unbound.* Nat. Rev. Neurosci. **2** (2001) 11-23
2. Galarreta M., Hestrin S.: *Electrical synapses between GABA-releasing interneurons.* Nat. Neurosci. **2** (2001) 425 - 433
3. Csicsvari J., Jamieson B., Wise K.D., Buzsaki G.: *Mechanisms of Gamma Oscillations in the Hippocampus of the Behaving Rat.* Neuron **37** (2003) 311-322
4. Galarreta M., Hestrin S.: *A network of fast-spiking cells in the cortex connected by electrical synapses.* Nature **402** (1999) 72-75
5. Gibson J.R., Beierlein M., Connors B.W.: *Two networks of electrically coupled inhibitory neurons in neocortex.* Nature **402** (1999) 75-79
6. Deans M.R., Gibson J.R., Sellitto C., Connors B.W., Paul D.L.: *Synchronous activity of inhibitory networks in neocortex requires electrical synapses containing connexin36.* Neuron. **31** (2001) 477 - 485
7. Gibson, J.R., Beierlein, M., Connors, B. W.: *Functional Properties of Electrical Synapses between Inhibitory Interneurons of Neocortical Layer 4.* J. Neurophysiol. **93** (2005) 467 - 480
8. Galarreta M., Hestrin S.: *Spike transmission and synchrony detection in networks of GABAergic interneurons* Science **292** (2001) 2295-2299
9. Veruki L.M., Hartveit E.: *All (Rod) amacrine cells form a network of electrically coupled interneurons in the mammalian retina* Neuron **33** (2002) 935-946
10. Povysheva N.V., Gonzalez-Burgos G., Zaitsev A.V., Kroner S., Barrionuevo G., Lewis D.A., Krimer L.S.: *Properties of excitatory synaptic responses in fast-spiking interneurons and pyramidal cells from monkey and rat prefrontal cortex.* Cerebral Cortex **16** (2006) 541 - 552
11. Di Garbo A., Barbi M., Chillemi S.: *Signal processing properties of fast spiking interneurons.* BioSystems **86** (2002) 27 - 37
12. Long M.A., Landisman C.E., Connors B.W.: *Small clusters of electrically coupled neurons generate synchronous rhythms in the thalamic reticular nucleus.* J. Neurosci. **24** (2004) 341 - 349
13. Leznik E., Llinas R.: *Role of gap junctions in the synchronized neuronal oscillations in the inferior olive.* J. Neurophysiol. **94** (2005) 2447 - 2456
14. Placantonakis D.G., Bukovsky A.A., Aicher S.A., Kiem H., Welsh J.P.: *Continuous electrical oscillations emerge from a coupled network: a study of the inferior olive using lentiviral knockdown of connexin36.* J. Neurosci. **26** (2006) 5008 - 5016
15. Galarreta M., Hestrin S.: *Electrical and chemical Synapses among parvalbumin fast-spiking GABAergic interneurons in adult mouse neocortex.* PNAS USA **99** (2002) 12438-12443

16. Erisir A., Lau D., Rudy B., Leonard C. S.: *Function of specific K^+ channels in sustained high-frequency firing of fast-spiking neocortical interneurons.* J. Neurophysiology **82** (1999) 2476-2489
17. Rinzel J., Ermentrout B.: *Analysis of neural excitability and oscillations.* Eds. Koch and Segev, Methods in neural modelling (1989), The MIT Press, Cambridge
18. Strettoi E., Raviola E., Dacheux R.F.: *Synaptic connections of the narrow-field, bistratified rod amacrine cell (All) in the rabbit retina.* J. Comp. Neurol. **325** (1992) 152-168

A Computational Structure for Generalized Visual Space-Time Chromatic Processing

D. Freire-Obregón[1], R. Moreno-Díaz jr.[1], R. Moreno-Díaz[1], G. De Blasio[1], and A. Moreno-Díaz[2]

[1] Institute of Cybernetics, Universidad de Las Palmas de Gran Canaria, Spain
dfreire.ciber@gmail.com, rmorenoj@dis.ulpgc.es, rmoreno@ciber.ulpgc.es, gdeblasio@dis.ulpgc.es

[2] School of Computer Sciences, Universidad Politécnica de Madrid, Spain
amoreno@fi.upm.es

1 Formal Frame

Traditional interpretation of early visual image processing in a retinal level has focused exclusively on spatial aspects of receptive fields (RFs). We have learned recently that RFs are spatiotemporal entities and this characterization is crucial in understanding and modelling circuits in early visual processing ([1]). We present a generalization of the layered computation concept to describe visual information processing in the joint space-time domain.

The starting point is the generalization of the concept of layered computation ([2], [4]). In its original sense, a modular system is structured in layers of similar computing elements. Modules in each layer operate on some input space to provide an output space. Both spaces have a pure physical nature such that input and output lines represent denditric trees, axons or, simply, wires. Outputs can interact with inputs through any type of feedback.

In the generalization, we preserve the existence of layers of computing elements which perform, in general, any type of decision rule or algorithm over both input and output spaces. But the nature of these spaces is changed to be spaces of representation. They are multidimensional spaces where each axis represents possible values of an independent variable which measures an independent possible property of inputs and outputs.

In its formal structure, the input space of the generalized computational frame consists of a spatio-temporal cube or hypercube in which one axis is time and the others correspond to cartesian coordinates (two or three dimensions) or to symbolic space representations. Notice that what it is needed in the input space representation is the possibility to define *distances* as proximities in that space: events close in the space correspond to events close in the property time space.

Operations on the input space act as a generalized convolution over a volume to produce, at each instant and each output point, an output layer to contribute to the output space. In the general case, volumes in the output space are also taken into account by every computational module to decide what the local value of the output layer should be. That is, there is feedback and also, cooperative processing since the outputs of nearby computing units in previous instants are

J. Mira and J.R. Álvarez (Eds.): IWINAC 2007, Part I, LNCS 4527, pp. 90–95, 2007.

used to compute the present output of the unit, due to the space-time character of the feedback sample.

The other main aspect in this paper is the chromatic perception. The retina has basically two types of photoreceptors cells, the cones and the rodes. The rodes are cells that allow us to see in darkness, they still work at low intensities but they are not able to distinguish the color perception. The cones allow us to distinguish colors but its required a high illumination level by them to work. They are three types of cones, those who appreciate the red color, those who appreciate the green color and those who appreciate the blue color. The stimulation of this three basic color allow the generation of all colors perceptible by the human eye.We are going to work with this basic knowledge about chromatic perception.

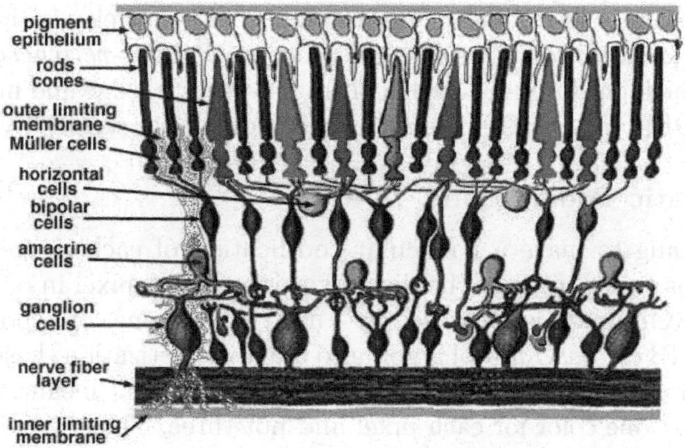

Fig. 1. The layer structure of retina in the classical description

In a computational way, the mixture of light colors, usually red, green and blue (RGB), is made using an additive color system, also referred as the RGB model. The mixture of these three basic colors in a properly way can generate most of the perceptible color by the human eye. That is, for example, a pure red and a light green produce the yellow color. When these colors are mixtured in their highest intensity for each one the result is the white color, the opposite is the black color, that is, the mixture of those colors in their lowest value.

2 The Feedforward Visual Case

In the visual (image processing) case with no feedback the input space comprises a stack of slides from $t = 0$ to $t = -\tau$, where τ is the memory of the input space. The computing units perform convolution-like computation on space-time volumes, which are the spatio-temporal overlapping receptive fields.

In general, operations performed on the space-time receptive field of a computational element can be arbitrary (algorithmic, algebraic, probabilistic, etc.).

What is important in this representation is that no previous outputs can be taken into account to generate a new output slode. That is, no recursive process can be realized by the computing structure. But it is still somehow cooperative as it happens to the visual system of vertebrates, in the sense that the same input data to one unit is shared by all units in the proximities. Cooperativity, as it is well known, is a key factor in reliable layered computation ([3]).

3 Computation and Tool Descriptions

As it corresponds to the visual level with no feedback, we have developed a visual information processing system following the general principle above. It is capable of performing spatio-temporal operations of an arbitrary nature on a stack of images or a video stream. Examples are presented which illustrate various classical filtering space-time effects, linear and non-linear. The novelty is the implementation of the classical neurophysiological concept of *center-periphery* but now in space-time, which produces naturally contrast detection, edge and movement detection, ON, OFF and ON-OFF behaviour, directionality sensitivity and others.

3.1 Chromatic Codification

Our tool is going to make a particular codification of each image before it is processed. This task consist in the discretization of each pixel in only one color. Usually each RGB image is composed by 3 matrix, each one corresponding to one of the basic RGB colors. Our tool is going to make discretization those colors from R,G and B for each pixel to R,G, or B for each pixel. That means, that there is going to be just one color for each pixel and not three. Thus, we will be able to have an image codified into two 2-D matrix, one of those is going to keep the information about de colors, and the other one we will keep the information about the intensities of those colors. This last matrix is the one that is going to be processed. In the next example, we can see that the red color is codified as R–R–R–... leaving in black the spaces for green's and blue's intensities. In the other example we can se the yellow color codified as RG-RG-RG-... where is filled the intensity of the colors that compose the yellow color and leave in black the blue color.

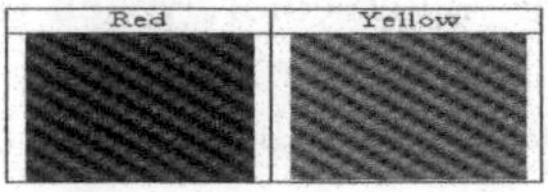

Fig. 2. Chromatic Codification Example

Basically, there are going to be two types of chromatic test: Independent chromatic test and Dependent chromatic test. Independent chromatic test makes a distinction between each color, where a single color only can have an interaction only with this single color and with no other color, that is, the red pixels only have interaction with red, greens with greens and blues with blues. On the other hand the dependent chromatic test allows the relation between every color.

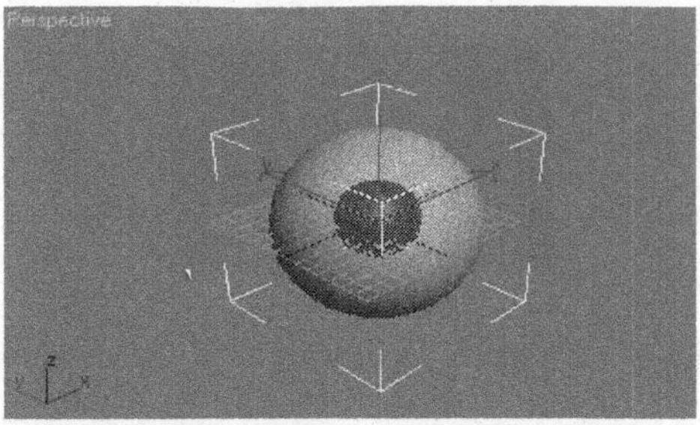

Fig. 3. A Generalized Convolution Center-Periphery forming geometric figures

3.2 Generalized Convolution

The tool allows us the process a set of images using a spatio-temporal structure with the purpose of being able to make a filtered to all images by a generalized convolution forming diverse geometric figures in the nucleus of the set of images.

4 Results

The main test for this paper is a Spatio-Temporal test with chromatic independence. As we can see, we have the following set of images for the input. Each image corresponds to a different period of time, that is, the first image is for time t=0, the second t=1,and so on.

As we can see, the first four images and the last four are in color red. Nevertheless, the four images of the center are yellow with a pink square at the centre.

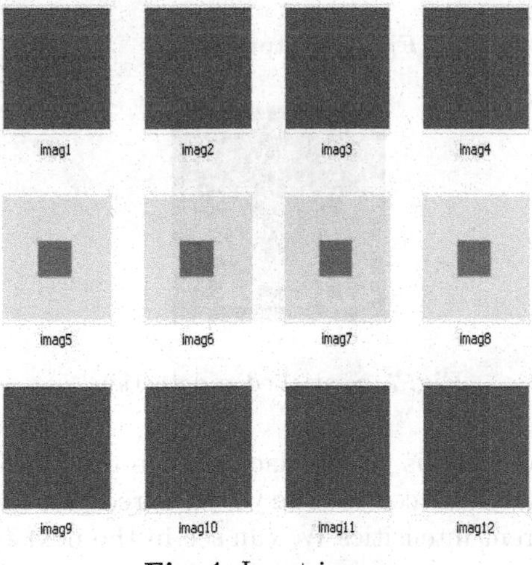

Fig. 4. Input images

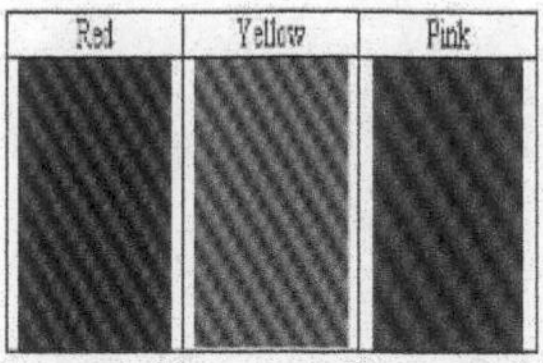

Fig. 5. Chromatic Codification of the example

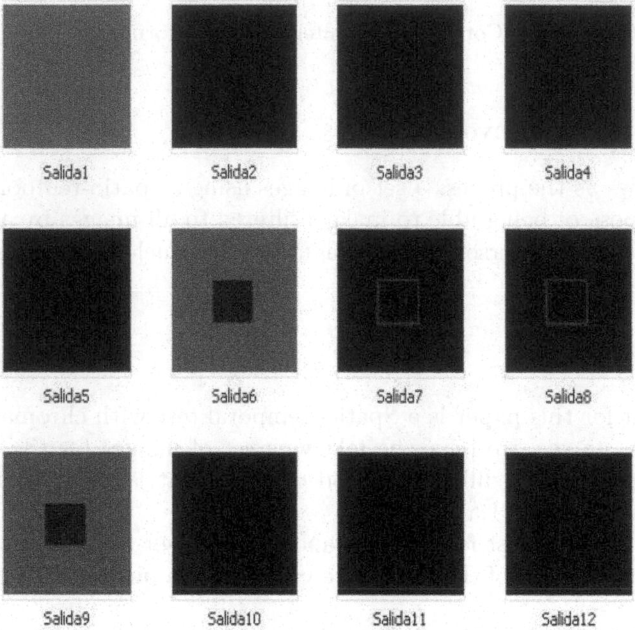

Fig. 6. Output images

Fig. 7. Spatial edge detection

We must realize what are this colors made of: the red is a basic color, but the yellow and pink are not basic colors. The yellow is red plus green, and the pink is red plus blue at different intensities.We can see in the next figure our chromatic codification.

After making a generalized convolution of a simple 3-Dimensional Hexaedrix figure, in this case, a cube of dimension 3x3x3, where the centre is located at the centre pixel of the figure, and everything else is the surround, we have got the following result:

As we can see, we have got an edge detector in a spatial, way as well as in a temporal way. We can appreciate several things:

- For the temporal filter, the red color is totally absorbed because there is no red-transition between images (each color represented in our input set of images have a red component). Otherwise, we have got the chromatic components blue and green of outputs 6 and 9 that show us the temporal transition.
- For the spatial filter we can observe at the output 7 and 8 the edge detector and each color is respecting the edge with other color (green (from yellow) and blue (from pink)) because the have not coincidents chromatic components:

To finish, we would like to comment that the first image is turquese because we are going to start processing the structure in a time equal 0, and to make this we have inserted white images at the beginning of the set of images that we are going to process. Thus, the red component of the white color goes with the red component of the first input image leaving for the first output the blue and green chromatic components of the white color.

5 Future Work

Although the presented results can be considered very interesting from a theoretical point of view, the main objective of this project still to be done is the implementation of different biological retinal models such as rabbit's or frog's retinal models. In the future we will improve this tool making possible the next objectives already mentioned.

References

1. DeAngelis, G.C.; Ohzawa, I.; Freeman, R.D.: Receptive-field Dynamics in the Central Visual Pathways. *Trends Neurosci.* 18, (1995) 18, 451–458
2. McCulloch, W.S.: Embodiments of Mind. The M.I.T. Press: Cambridge, Massachusetts (1965)
3. Mira, J., Delgado, A., Moreno-Díaz, R.: Cooperative Processes in Cerebral Dynamic. In: D.G. Lainiotis, Tzannes (Eds.) Applications of Infromation and Control Systems, D. Reidel Pub. Comp. (1980) 273–280
4. Moreno-Díaz, R., Rubio, E.: A Theoretical Model for Layered Visual Processing. *Int. J. Bio-Med. Comp.* (1979) 10, 134–143

Physiological Laws of Sensory Visual System in Relation to Scaling Power Laws in Biological Neural Networks

Isabel Gonzalo-Fonrodona[1] and Miguel A. Porras[2]

[1] Departamento de Óptica. Facultad de Ciencias Físicas. Universidad Complutense de Madrid. Ciudad Universitaria s/n. 28040-Madrid. Spain
igonzalo@fis.ucm.es

[2] Departamento de Física Aplicada. ETSIM. Universidad Politécnica de Madrid. Rios Rosas 21. 28003-Madrid. Spain

Abstract. Measurements of some visual functions (visual fields, acuity and visual inversion) versus intensity of stimulus, including facilitation, carried out by Justo Gonzalo in patients with central syndrome, are seen to follow Stevens' power law of perception. The characteristics of this syndrome, which reveals aspects of the cerebral dynamics, allow us to conjecture that Stevens' law is in these cases a manifestation of the universal allometric scaling power law associated with biological neural networks. An extension of this result is pointed out.

1 Introduction

Half a century ago, Stevens [1] formulated his well-known relation between sensation or perception P and the physical intensity of a stimulus S, expressed mathematically as a power law of the type

$$P = pS^m \,, \tag{1}$$

where p is a constant and m depends on the nature of the stimulus. This law is regarded as more accurate than the logarithmic Fechner's law, but is not exempt from criticism.

In a different, and somewhat more general context, it was argued that in biological organisms, mass is the determinant factor for the scaling of the physiological behavior. If M is the mass of the organism, many observable biological quantities, for instance Y, are statistically seen to scale with M according to a power law of the form

$$Y = kM^n \,, \tag{2}$$

where k is a constant and, in formal similitude with Stevens' law, the exponent n changes from one observable to another, leading to different (allometric) behavior of observables with respect to mass M. Most of exponents in this law are surprisingly found to be multiples of the power $1/4$. Biological variables that follow these

J. Mira and J.R. Álvarez (Eds.): IWINAC 2007, Part I, LNCS 4527, pp. 96–102, 2007.

quarter-power allometric laws are, for instance, the metabolic rate ($n \simeq 3/4$), lifespan (1/4), growth rate (−1/4), height of trees (1/4), cerebral gray matter (5/4), among many others (see [2,3] and references therein, and also [4,5]).

The allometric scaling laws are supposed to arise from universal mechanisms in all biological systems, as the optimization to regulate the activity of its subunits, as cells. According to West and Brown [6], optimization would be achieved through natural selection by evolving hierarchical fractal-like branching networks, which exchange energy and matter between the macroscopic reservoir and the microscopic subunits. Some examples of these networks are the animal circulatory, respiratory, renal and neural systems, the plant vascular system, etc. The quarter-power allometric laws can be theoretically derived from simple dimensionality reasonings that derive from the geometrical constraints inherent to these networks. As the same authors remark, powers proportional to a multiple of 1/4 would be strictly verified only in ideal biological organisms, while in real organisms the power may slightly depart from these values, since they are affected by stocastic factors, environmental conditions and evolutionary histories. For other authors [7], however, the scaling power laws are valid independently of the network type, and hence also for those without hierarchical or fractal structure.

Though Stevens' law for perception-stimulus relation and biological scaling law with mass relate to different phenomena, their formal similitude indicates a possible connection between them. A more fundamental connection is pointed out here on the basis of the measurements of some visual functions versus stimulation intensity, carried out by Gonzalo [8,9,10] in patients with the central syndrome that he described. After recalling the characteristics of this syndrome, we first verify that the measured data of the visual functions versus stimulus intensity fit well to Stevens' power laws. Second, under reasonable assumptions on the relation between physical stimulus and activated neural mass, we conclude that Stevens' law of perception is, in the cases studies at least, a manifestation of the universal scaling power laws.

2 Characteristics of the Central Syndrome

The central syndrome (or symmetric poly-sensory syndrome) seems to be particularly suitable for the observation of the unfolding of the sensory functions. This syndrome originates from a unilateral parieto-occipital lesion equidistant from the visual, tactile and auditory projection areas, and is featured by [8,10] (a) poly-sensory affection with symmetric bilaterality, (b) functional disgregation of perception, in the sense that sensory qualities are gradually lost in a well-defined order as the stimulus intensity diminishes, and (c) capability to improve the perception by iterative temporal summation and by facilitation through other stimuli, as for instance, strong muscular stress. The syndrome was interpreted [10] as a deficit of cerebral integration due to a deficit of cerebral nervous excitation caused by the loss of a rather unspecific (multisensory) neural mass. This interpretation arises from a model in which functional sensory densities for each sensory system are distributed in gradation through the whole cortex [10,11]. There are other works dealing with this research [12,13,14,15,16,17], or related to it (e.g., [18,19,20]). A close connection can also be established with models based on a distributed character of cerebral processing, its adaptive and long-distance integrative aspects (e.g., [21]).

The remarkable point here is that the central syndrome was explained in terms of a scale reduction of the normal cerebral system caused by the lesion [10,12]. From the concept of dynamic similitude, scaling laws were applied to the sensory decrease —or functional depression— observed in the patients. From the comparison between twelve cases with central syndrome in different degrees, their visual luminosity, acuity and other qualities were found to obey approximate power laws versus their respective visual field amplitudes, with different exponent for each quality, i.e., allometrically [10,12]. This is the formal description of the functional disgregation, or decomposition of the normal perception into its different qualities by their gradual loss, from the most to the less complex ones, as the nervous excitation diminishes, or equivalently, as the magnitude of the lesion grows. The organization of the sensorium can be then visualized, up to a certain extent, as displayed in patients with central syndrome.

For a given individual with central syndrome, the sensory level grows by intensifying the stimulus, or by adding other different stimuli, which are able to compensate for the neural mass lost. This dynamic capability is greater as the neural mass lost is greater, and is null, or extremely low for some functions, in a normal man. [8,10,13].

3 Stevens' Law in Central Syndrome

All experimental data presented are taken from Ref. [8], correspond to two different cases with central syndrome, called M and T (less intense), and refer to the change of visual functions or qualities with intensity of stimulation in a stationary regime. Stevens' law [Eq. (1)] is used to fit the data.

Figure 1(a) shows the experimental data for the visual field amplitude of right eye in cases M, M facilitated by strong muscular stress (40 kg held in his hands), and case T, as functions of the illumination of the test object. The reduction of the visual field in the central syndrome is concentric. In the log-log graphic, Stevens' law [Eq. (1)] yields a straight line of slope equal to the exponent m. As seen, the data fit rather well to Stevens' straight lines not very close to saturation. The slope m of the fitting straight lines is remarkably close to $1/4$ for M and M facilitated, and $1/8$ for T. In Fig. 1(b), similar representation is shown for the visual acuity in central vision, including a normal man. Straight lines with slope $1/4$ fit well to the central part of the data for the two states of case M and for case T, and with slope $1/8$ for normal man.

In another series of experiments, the intensity of light on the test object was kept constant and low, whereas the variable stimulus was the facilitation supplied by muscular stress [Fig. 2(a) and (b)], or by light on the other eye (Fig. 3). Fig. 2(a) shows the measured visual field amplitude of right eye in case M

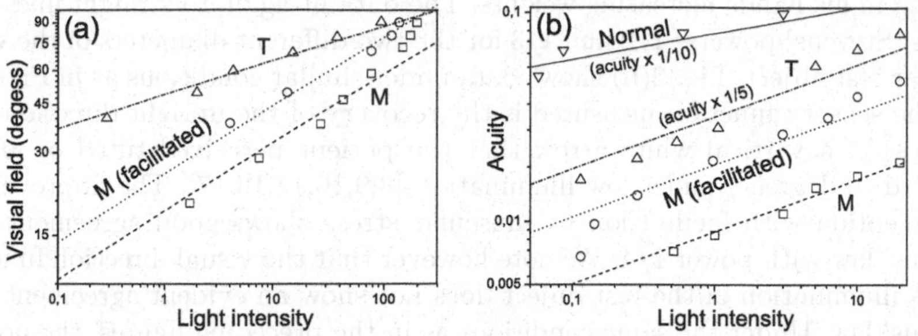

Fig. 1. (a) Visual field of right eye versus relative illumination (test object: 1 cm-diameter, white disk). Squares: M (fitting straight line with slope 1/4). Circles: M facilitated (straight line with slope 1/4). Triangles: T (straight line with slope 1/8). (b) Acuity of right eye versus illumination. Squares: M (fitting with slope 1/4). Circles: M facilitated (fitting with 1/4). Triangles: T (fitting with 1/4). Inverted triangles: Normal man (fitting with 1/8).

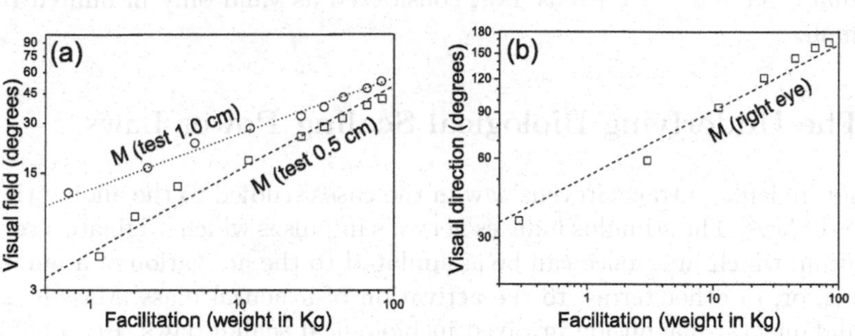

Fig. 2. (a) Visual field amplitude in right eye versus facilitation by muscular stress. Squares: M, 0.5 cm test size (fitting straight line with slope 1/2). Circles: M, 1.0 cm test size (fitting slope 1/3). (b) Visual direction (reinversion) in right eye versus facilitation by muscular stress. Squares: M (fitting slope 1/4).

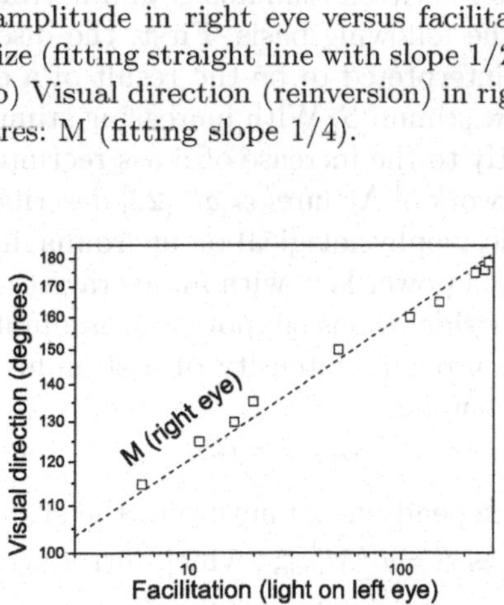

Fig. 3. Visual direction (reinversion) in right eye versus facilitation by illumination on left eye. Squares: M (fitting straight line with slope 1/8).

holding in his hands increasing weights. The data fit again to straight lines with slopes (Stevens' powers) 1/2 and 1/3 for the two different diameters of the white circular test object. Fig. 2(b) shows data under similar conditions as in Fig. 2(a) but the sensory function measured is the recovery of the upright direction (180 degrees) of a vertical white arrow that the patient perceived tilted or almost inverted (0 degrees) under low illumination [8,9,10,15,16,17]. The improvement of perception with facilitation by muscular stress shows good agreement with Stevens' law with power 1/4. We note however that the visual direction function versus illumination of the test object does not show an evident agreement with Stevens' law. Under the same conditions as in the preceding figures, the novelty in Fig. 3 is that facilitation is supplied by illuminating the left eye, which does not see the object. Again, the data can be well fitted with Stevens' law, but now with power 1/8.

The improvement of perception with increasing stimuli in patients with central syndrome is then seen to be describable from Stevens' law of perception, with a noticeable number of exponents around 1/4, as in scaling biological laws. The observed loss of accuracy for very low and very high stimuli reflects the approximate character of Stevens' law, considered as valid only in limited ranges of stimuli.

4 The Underlying Biological Scaling Power Laws

It is not difficult to trace Stevens' law in the cases studied to the allometric scaling power laws. The stimulus induces nervous impulses which originate a cerebral excitation, which in essence can be assimilated to the activation of a number of neurons, or, in other terms, to the activation of a neural mass. Mass is in fact the fundamental magnitude involved in biological scaling laws. Reasonable assumptions on the relation between stimulus S and activated neural mass $M_{\rm neur}$ can be established on the following basis. First, the disorders in patients with central syndrome were interpreted to be the result of a deficit of integration of neural mass $M_{\rm neur}$ at low stimuli S. With increasing stimuli, perception tends to be normal due apparently to the increase of mass recruited. In this sense but in normal man, the recent work of Arthurs *et al.* [22] describes the relation between electrical stimuli and electrophysiological or neuroimaging measures of the human cortical response as a power law, with an average exponent around 1.3 (see also Ref. [23]). These considerations support our assumption that the activated neural mass as a function on the intensity of a stimulus can be approximately described by the power law

$$M_{\rm neur} = \alpha S^{\beta} \,, \tag{3}$$

where β is expected to depend on the nature of the stimulus. Equation (3) can be equivalently written as $S = \alpha' M_{neur}^{1/\beta}$, which introduced into Stevens' law (1) yields

$$P = pS^m = (\alpha' M_{\rm neur}^{1/\beta})^m = kM_{\rm neur}^n \,, \tag{4}$$

with $n = m/\beta$.

The last equality in Eq. (4) is a power law with respect to neural mass, and hence a scaling power law associated to the neural network, the exponent being then expected to fit in many cases to multiples of 1/4, as stated in the introduction.

5 Discussion and Conclusion

From expression (4), and taking the network scaling power laws as more fundamental, it can be concluded that the physiological Stevens' law in cases with central syndrome is a manifestation of the scaling power laws. Increasing intensity of stimulation would result in increasing activated neural mass according to a power law, which in turn leads to an improvement of perception. This conclusion is also supported by the fact that many powers $m = n\beta$ in the experimental fittings with Stevens' law are found to be close to 1/4, since $n = 1/4$ is very often the basic power in scaling laws and β is reported to be close to unity [22].

In normal man, some of the sensory functions analyzed here (amplitude of visual field and direction function) attain permanently its maximum value. Even for stimuli with very low intensity, the cerebral excitation in normal man suffices for the normal value of the sensory function to be reached (e.g., the whole visual field and the upright perception of the visual scene), which can be thought as a "saturated" situation. Other functions as visual acuity, vary in normal man with the intensity of the stimulus, but with a very low exponent as seen in Fig. 1(b), which would reflect a situation close to the "saturation" limit. In this respect, let us mention the scaling power law found between the number of neurons in the visual cortex V1 and the number of neurons in the visual thalamus (roughly the same as the number of retinal ganglion cells) in primates [24], which suggests that visual acuity depends not only on the type of cells activated in retina, but also on an integrative process in the cortex. This was also suggested by the behavior of acuity in central syndrome [8].

As an extension of these results, we finally suggest that Stevens' law of perception also in normal man could originate from the scaling power laws that govern the dynamic behavior of biological neural networks.

References

1. Stevens, S.S.: On the psychophysical law. Psychol. Rev. **64** (1957) 153-181
2. West, G.B., Brown, J.H.: Life's Universal Scaling Laws. Phys. Today. Sept. (2004) 36-42
3. West, G.B., Brown, J.H.: The origin of allometric scalin laws in biology from genomes to ecosystems: towards a quantitative unifying theory of biologica structure and organization. J. Exper. Biol. **208** (2005) 1575-1592
4. Anderson, R.B.: The power law as an emergent property. Mem. Cogn. **29** (2001) 1061-1068
5. Gisiger, T.: Scale invariance in biology: coincidence or footprint of a universal mechanisms? Biol. Rev. **76** (2001) 161-209

6. West, G.B., Brown, J.H.: A general model for the origin of allometric scalling laws in biology. Science **276** (1997) 122-126
7. Banavar, J.R., Maritan, A., Rinaldo, A.: Size and form in efficient transportation networks. Nature **399** (1999) 130-132
8. Gonzalo, J: Investigaciones sobre la nueva dinámica cerebral. La actividad cerebral en función de las condiciones dinámicas de la excitabilidad nerviosa. Publicaciones del Consejo Superior de Investigaciones Científicas, Inst. S. Ramón y Cajal, Madrid, Vol. I (1945), Vol. II (1950). (Avalable in: Instituto Cajal, CSIC, Madrid)
9. Gonzalo, J.: La cerebración sensorial y el desarrollo en espiral. Cruzamientos, magnificación, morfogénesis. Trab. Inst. Cajal Invest. Biol. **43** (1951) 209-260
10. Gonzalo, J.: Las funciones cerebrales humanas según nuevos datos y bases fisiológicas: Una introducción a los estudios de Dinámica Cerebral. Trab. Inst. Cajal Invest. Biol. **44** (1952) 95-157
11. Gonzalo, I., Gonzalo, A.: Functional gradients in cerebral dynamics: The J. Gonzalo theories of the sensorial cortex. In Moreno-Díaz, R., Mira, J. (eds.): Brain Processes, Theories and Models. An Int. Conf. in honor of W.S. McCulloch 25 years after his death. The MIT Press, Massachusetts (1996) 78-87
12. Gonzalo, I.: Allometry in the J. Gonzalo's model of the sensorial cortex. Lect. Not. Comp. Sci (LNCS) **1249** (1997) 169-177
13. Gonzalo, I.: Spatial Inversion and Facilitation in the J. Gonzalo's Research of the Sensorial Cortex. Integrative Aspects. Lect. Not. Comp. Sci. (LNCS) **1606** (1999) 94-103
14. Gonzalo, I., Porras, M.A.: Time-dispersive effects in the J. Gonzalo's research on cerebral dynamics. Lect. Not. Comp. Sci (LNCS) **2084** (2001) 150-157
15. Gonzalo, I., Porras, M.A.: Intersensorial summation as a nonlinear contribution to cerebral excitation. Lect. Not. Comp. Sci. (LNCS) **2686** (2003) 94-101
16. Arias, M., Gonzalo, I.: La obra neurocientífica de Justo Gonzalo (1910-1986): El síndrome central y la metamorfopsia invertida. Neurología **19** (2004) 429-433
17. Gonzalo-Fonrodona, I.: Inverted or tilted perception disorder. Rev. Neurol. **44** (2007) 157-165
18. Delgado, A.E.: Modelos Neurocibernéticos de Dinámica Cerebral. Ph.D.Thesis. E.T.S. de Ingenieros de Telecomunicación, Univ. Politécnica, Madrid (1978)
19. Mira, J., Delgado, A.E., Moreno-Díaz, R.: The fuzzy paradigm for knowledge representation in cerebral dynamics. Fuzzy Sets and Systems **23** (1987) 315-330
20. Mira, J., Manjarrés, A., Ros, S., Delgado, A.E., Álvarez, J.R.: Cooperative Organization of Connectivity Patterns and Receptive Fields in the Visual Pathway: Application to Adaptive Thresholdig. Lect. Not. Comp. Sci. (LNCS) **930** (1995) 15-23
21. Rodríguez, E., George, N., Lachaux, J.P., Martinerie, J., Renault, B., Varela, F.J.: Perceptions shadow: long-distance synchronization of human brain activity. Nature **397** (1999) 430-433
22. Arthurs, O.J., Stephenson, C.M.E., Rice, K., Lupson, V.C., Spiegelhalter, D.J., Boniface, S.J., Bullmore, E.T.: Dopaminergic effects on electrophysiological and functional MRI measures of human cortical stimulus-response power laws. NeuroImage **21** (2004) 540-546
23. Nieder, A., Miller, E.K.: Coding of cognitive magnitude. Compressed scaling of numerical information in the primate prefrontal cortex. Neuron **37** (2003) 149-157
24. Stevens, C.F.: An evolutionary scaling law for the primate visual system and its basis in cortical function. Nature **411** (2001) 193-195

ANF Stochastic Low Rate Stimulation

Ernesto A. Martínez–Rams[1] and Vicente Garcerán–Hernández[2,*]

[1] Universidad de Oriente, Avenida de la América s/n, Santiago de Cuba, Cuba
eamr@fie.uo.edu.cu
[2] Universidad Politécnica de Cartagena, Antiguo Cuartel de Antiguones
(Campus de la Muralla), Cartagena 30202, Murcia, España
vicente.garceran@upct.es

Abstract. Science has been researching on the physiology of the human hearing, and in the last decades, on the mechanism of the neural stimulus generation towards the nervous system. The objective of this research is to develop an algorithm that generalizes the stochastic spike pattern of the auditory nerve fibers (ANF) formulated by Meddis, which fulfils the Volley principle (principle that better describes the operation of the auditory system). The operating principle of the peripheral auditory system together with the models chosen to stimulate the auditory system and the characteristics of the implemented computational model are herein described. The implementation and analysis of the stochastic spike of a simple ANF and the spatial and spatial–temporal stochastic stimulation models demonstrate the superiority of the latter.

1 Introduction

There exist three theories that explain the principle of the cochlear functioning at present: theory of the spatial or tonotopic code (Helmholtz's), theory of the temporal code and the theory of the spatial-temporal code (Volley), [1,2,3,4,5]. The theory of spatial code describes how each zone of the basilar membrane enters resonance at a particular frequency of the sound stimulus. The theory of the temporal code explains how a tone is codified by the stimulation rate of the nervous cells. This last theory does not justify the behaviour of an auditory nerve fiber at high frequency, since the maximum velocity response of only one cell in the auditory system is approximately 1,000 spikes per second. The principle of Volley suggests that, while only one neuron cannot bear the temporal coding of a 20 kHz pure tone, 20 neurons can do so with staggered firing rates, where each neuron would be able to respond in every 20 cycles of the 20kHz pure tone on average. The objective of this research is to develop an algorithm that represents the spike of the auditory nerve fibers (ANF) according to Volley

* We thank the collaboration of Dr. Enrique A. Lopez Poveda, belonging to the Regional Center of Biomedical Research of the Medicine Faculty of the University of Castilla la Mancha, to the Institute of Neurosciences of Castilla León and professor of Salamanca University, Spain. We are also grateful to Dr. Jorge Guilarte Tllez, specialist of Neurophysiologic at Hospital Infantil of Santiago de Cuba, Cuba.

J. Mira and J.R. Álvarez (Eds.): IWINAC 2007, Part I, LNCS 4527, pp. 103–112, 2007.

theory, generalizing the Meddis model. Some authors have presented the model representing the spike produced in the ANF, by making each inner hair cell (IHC) correspond with an ANF [6,7,8,9,10]. Besides this, the present work develops a model which makes each IHC correspond with 20 ANFs.

2 Models Description

The Fig. 1 shows the computer model of the peripheral auditory system, where the sound picked up by the microphone is presented to the outer and middle ear block. The external and middle ear's function [11,12] is to capture the incident sound waves and provide an initial filtering of the signal to increase the sound pressure in the region from 2 kHz to 7 kHz, helping to its location. The ear's performance is modelled by a low-pass filter transfer function where, selecting a 4 kHz resonance frequency, the filter is designed to model a 10 dB frequency peak around 4 kHz, something typical of the external ear's transmission characteristics.

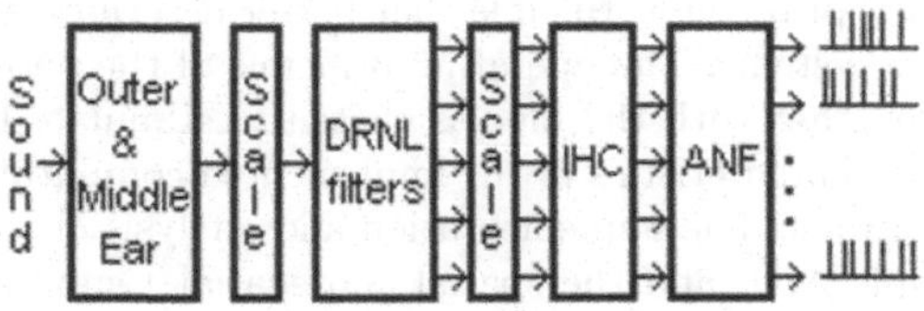

Fig. 1. Computer model of the peripheral auditory system

Then, the signal is scaled and presented to the double resonance non-linear (DRNL) filter [13,15] which represents the function of the basilar membrane (BM), filtering the signal with non-linear characteristic. Once the response from each zone of the BM is obtained, the signal is scaled again and presented to the IHC [8,16]. Lastly, the respective outputs of the IHCs are presented to the ANFs that represent the synaptic union IHC/neuron and are in charge of coding the information going to CNS (Central Nervous System). For the scaling, the input signal is normalized to a RMS value equal to the unit. Then, the signal is scaled multiplying this normalized signal by $X_{0\,dBSPL}10^{L/20}$, being L the level of input signal sound pressure in dB, and $X_{0\,dBSPL}$ for the first scale block equal to $1.4059 \cdot 10^{-8} ms^{-1}$ and in the second scale block a value equal to the unit.

The most significant characteristics in the diagram of Fig. 1 are the following: Increase of the sound pressure in the region from 2kHz to 7kHz; Linear behaviour of the DRNL filter at low levels of input signal (<30 dB SPL), non-linear behaviour of the DRNL filter at intermediate levels of input signal (30 dB - 80 dB SPL), and linear behaviour of the filter DRNL at high levels of input signal (>85 dB SPL) [13]; Recovery of the spontaneous activity in 16.2 ms on average; Synchrony of the auditory nerve response against low frequency stimuli (<1kHz); Adaptation mechanism of short and long duration of the IHCs (see figure 13 in [8]).

3 Stochastic Model for an ANF

Some authors [6,7,8,9,12] have modelled a simple ANF corresponding a unique ANF with each spatial frequency, but the processes occurring to codify the frequencies higher than 1 kHz are not explained.

In fact, the existence of a solid base of neural stimulation patterns generated by the cochlea is a starting point. Although the CNS cannot be regarded as a true model of the processing executed in the auditory center of the human brain, because it is not known how the brain processes information, it is possible to extract from these neural patterns the corresponding information of the applied stimulus.

In our case to check and compare the spatial-temporal model with the spatial model, both were subjected to three sequences of pip tones with the following characteristics: Frequency of the tones: 0.4 kHz, 1 kHz and 3.25 kHz; Duration of the tones: 300 ms; Interval among tones: 300 ms; Intensity of the tones: 5 dB to 80 dB SPL, with steps of 5 dB; Sampling frequency: 8 kHz; Spatial frequency of the BM: 0.33 kHz, 1.02 kHz and 3.27 kHz respectively.

The test to apply consists in comparing the envelope or amplitude level form of the IHC output signal with the firing rate generated by the ANF, whose level of resemblance will show the effectiveness of the implemented pattern. The problem is that the spike signal is a modulated signal whose information (coded sound) is given in the time among pulses, what complicates the recovery of the signal amplitude level. For that reason, the statistical functions will be used to show the most important characteristics of this signal (Post Stimulus Time (PST) and InterSpike Interval (ISI) analysis).

The input of the stochastic ANF model implemented is the neurotransmitter level present in the synapses IHC/neuron. This value scaled by an empirical value h and multiplied by the sample rate of the signal Ts, gives as a result the spike probability Pe of the ANF [8]. Finally, the block or decision logic determines and decides if the spike is produced or not. A discharge occurs if Pe is higher than a specific stochastic value generated with the functions of the $MATLAB^{TM}$ and if the ANF is ready. In this case, a time Tn equal or higher than the refractory period of the ANF should have passed after the last discharge of the neuron (Tr, equal to 1ms).

4 Spatial Stimulation Model and Computer Simulation

In the stochastic pattern of stimulation applying the spatial code theory, each IHC stimulates its corresponding ANF which generates the nervous impulses travelling to the CNS. The Fig. 2a shows the neurotransmitters level present in the synapses IHC/neuron for a pip tone sequence of 1 kHz at the spatial frequency of 1,020 Hz, in which the processes of short and long time adaptation occurring in the IHC can be noticed. In the same figure the ANF spike signal travelling to the CNS is shown.

The first algorithm used to verify the spatial model was the Post Stimulus Time (PST) [7,17], which determines the ANF firing rate, showing the number of

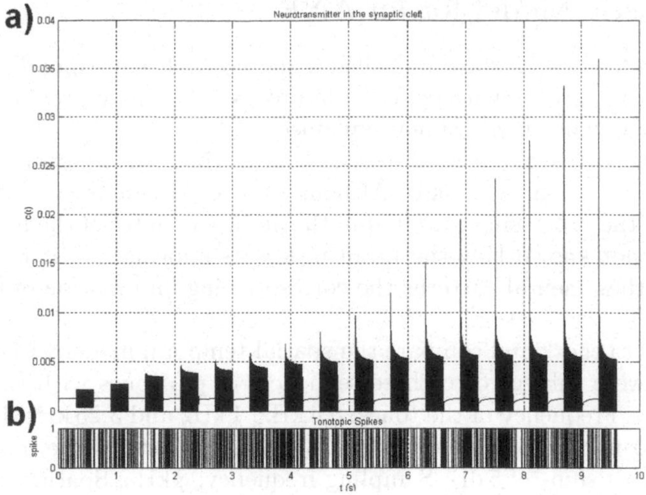

Fig. 2. a) Neurotransmitters level present in the ANF synapses; b) ANF spike

spikes like a function of time whose signal corresponds with the spectral envelope of the stimulus signal applied to the pattern. Figure 3a shows: neurotransmitters level in the synapses (input signal to the model), the ANF firing rate and a comparison between the firing rate and a specific threshold to detect the presence of pip tones. When comparing the firing rate with the neurotransmitters envelope amplitude present in the synapses (Fig. 3a) for levels of stimuli fundamentally over the 25dB SPL, it is possible to observe, subjectively, the similarity between both waves. The signal noise relation (SNR) between the firing rate (Fig. 3b) and the envelope or amplitude of the stimulus signal (Fig. 3a) gave, for 100 simulations, a mean value of 28.0090 dB with 0.22 dB of variance and 0.4724 dB of standard typical deviation, which indicates the similarity between both signals. To carry out these measurements in signals showing different amplitude levels but equal forms or behaviour in the time, both signals were normalized to a value of 1 rms.

Another test on the pattern consisted in recovering the amplitude information from the signal by means of the extraction of the ISI envelope (InterSpike Interval) to estimate the ANF firing rate [10]. The ISI shows the difference in time between two spikes. The greater ISI the smaller signal amplitude, and the smaller ISI the greater signal amplitude. Figure 4 illustrates the input signal to the pattern, the ISI, the firing rate plus a dashed line representing the mean value of the firing rate (threshold) and the comparison between firing rate and threshold to detect the presence or absence of pip tones.

Figure 5a-d depicts the ISI histogram for several cases of the stimuli signals: silence periods and pip tones periods of 1 kHz with intensities of 5 dB, 45 dB and 85 dB SPL. As detected, any discharge is produced under 1 ms corresponding to the refractory period, proving the fulfilment of the ANF recovery after being discharged. Even when there is no sound stimulus (silence) spikes take place.

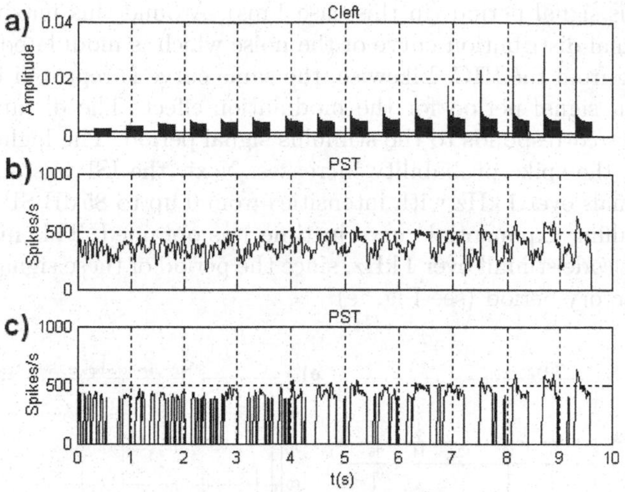

Fig. 3. a) Neurotransmitters level in the synapses; b) ANF firing rate; c) Firing rate compared to a threshold of detection

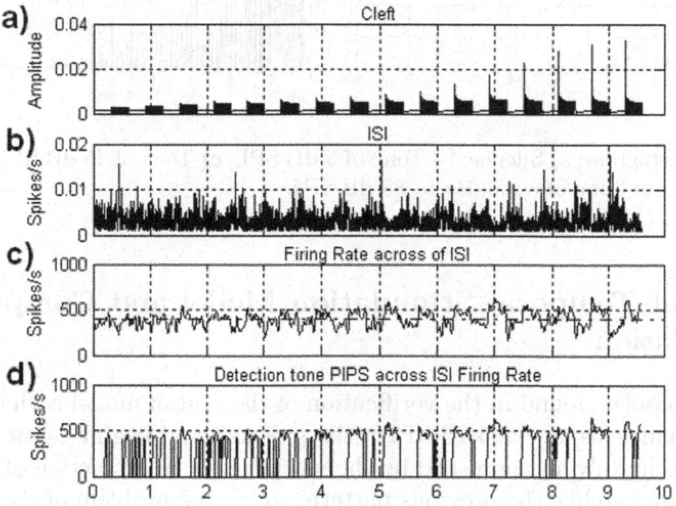

Fig. 4. a) Input signal to the model; b) ISI signal; c) firing rate and threshold level; d) Firing rate compared to threshold level of detection

A falling exponential distribution of the number of spikes in time is also observed. This indicates that the highest spike probability of the neuron is reached 1 millisecond after the last neural spike and an exponential decrease of this probability in time. The exponential form is due to the signal level coming from the IHC which is, in general, smaller than the stochastic value in the modulation process. Besides, the highest spike probability is obtained in the time equivalent to the stimulus signal period (in this case 1 ms). Around this maximum point, there is a normal distribution curve of the noise which is modulated by the signal originating from the IHC. Likewise, the same curve is repeated in multiples of the stimulus signal period for the modulation effect. The distance between successive picks corresponds to the stimulus signal period. The higher the stimulus intensity, the spike probability increases. Next, the ISI was also analysed for stimuli signals over 1 kHz with intensities from 0 up to 85 dB SPL and in all the cases, a falling exponential curve in time was obtained. This means that a neuron cannot code stimuli over 1 kHz, since the period of these signals is shorter than the refractory period (see Fig. 5e).

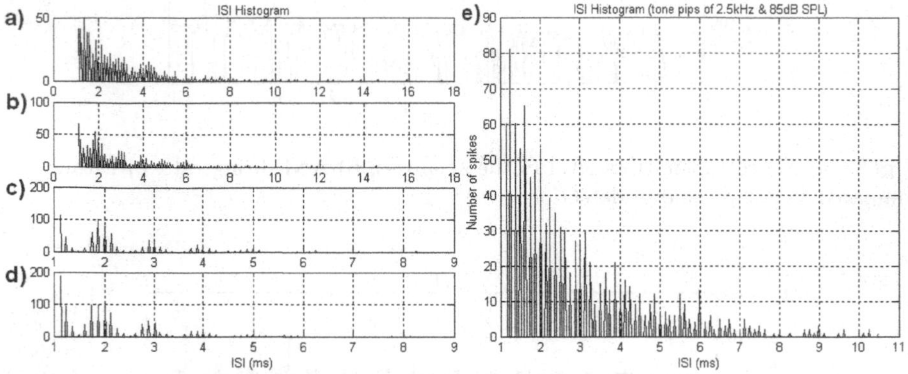

Fig. 5. ISI Histogram: a) Silence; b) Tone of 5 dB SPL; c) Tone of 45 dB SPL; d) Tone of 85 dB SPL; e) Tone of a 2.5kHz to 85 dB SPL

5 Spatial–Temporal Stimulation Model and Computer Simulation

One of the problem found in the verification of the spatial model is that, for low levels of stimulus signal (<35 dB SPL), the presence and beginning of pip tones cannot be accurately distinguished by the methods of the extraction of PST and ISI. This model, unlike the previous pattern, solves the problem of the auditory nervous fiber's behaviour at high frequency. As only an ANF cannot respond to frequencies higher than 1 kHz but a group of them do so, and taking into account that the human peripheral auditory system can perceive a maximum spectral component of 20 kHz, some 20 ANF for each IHC would be necessary to code the sounds at such high frequencies. The graphs exposed over Fig. 6 show the neurotransmitters level in the synapses IHC/neurone for a pip tone sequence of 1 kHz at the spatial frequency of 1,020 Hz, corresponding to silence input signals 5 dB, 45 dB and 80 dB SPL respectively. There, the adaptation processes of short and long duration occurring in the IHC can be noticed. In the graphs placed at the bottom of the same figure, the respective neuron spike signals corresponding to the signal levels shown at the top of same figure, representing the ANF's spikes that will travel to the CNS. In this case, there are 20 neurons associated to the IHC corresponding to the spatial frequency of 1,020 Hz. As the sound intensity increases, the spike probability increases, something noticeable just by looking at the spike density in each graphics.

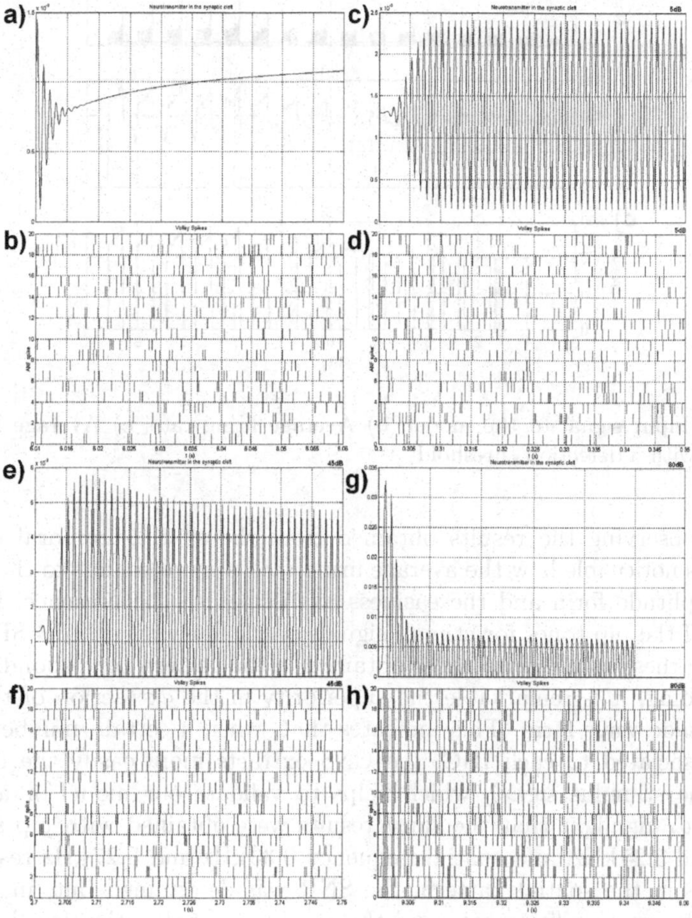

Fig. 6. Neurotransmitter level present in the ANF synapses and Discharge of an assembly of 20 neuron, for: a) b) silence input; c) d) pip tone sequence of 1 kHz, 5 dB; e) f) pip tone sequence of 1 kHz, 45 dB; g) h) pip tone sequence of 1 kHz, 80 dB

To test the spatial–temporal pattern, 20 nervous auditory fibers were associated to each IHC and a firing rate was determined for each ANF (the PST) like in the previous point. Then, averaging the respective values of the PST for each ANF, the average firing rate was determined, which should be similar to the envelope form of the stimulus wave. Figure 7 shows: the neurotransmitter level in the synapses (input signal to the pattern), the average firing rate of ANF spike and a comparison of the average firing rate with a certain threshold in order to detect the presence, beginning or end of a stimulus pip tone.

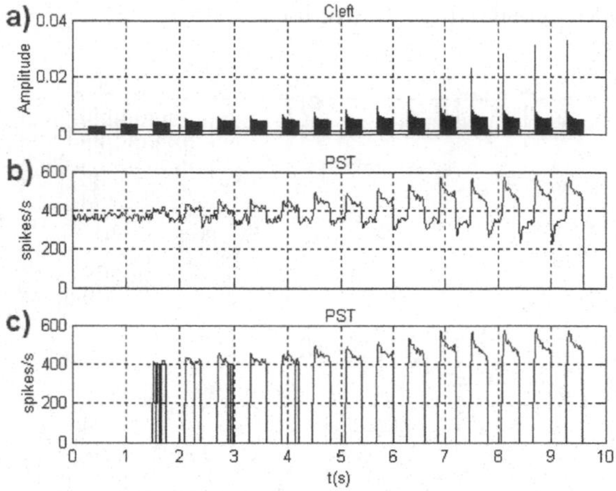

Fig. 7. a) Input signal to the model; b) Average firing rate; c) Average firing rate compared with a detection threshold

When observing the results obtained from the spatial-temporal coding in Fig. 7, it is noticeable how the average firing rate is more similar to the stimulus signal amplitude form and the easiness to distinguish the presence, beginning and end of the pip tones for stimuli signals particularly over 20 dB SPL. When comparing these results with those obtained in the pattern corresponding to the spatial coding, it is remarkable the superiority in the extraction of the stimulus amplitude wave form. This indicates that, when a certain number of ANF (20) is associated to each IHC, the CNS extracts more easily the characteristic of the stimulus signal, which helps to validate our model or algorithm. It is also to highlight that the same results were obtained with pip stimuli at 0.4kHz and 3.25 kHz, and spatial frequency 0.33 kHz and 3.27 kHz respectively. As well as in the spatial pattern, the SNR was determined but, in this case, between the average firing rate and the amplitude of the stimulus signal envelope. After 100 simulations the following values were obtained: a mean value of 31.1684 dB, variance of 0.0092 dB and standard typical deviation of 0.0959 dB, showing the similarity between both signals. Besides, there is an increase of the SNR in 3.1593 dB with regard to the spatial pattern. Both signals were normalized to a value of 1 rms, as well as for the spatial coding. Another test to verify the spatial-temporal model consisted in obtaining the ISI of the signal resulting from a logical operation OR, carried out to the respective outputs of the 20 ANFs associated to an IHC. Figure 8 depicts the histogram for a pip tone of 2.5 kHz (period of 0.4 ms) of intensity 85 dB SPL, where it is observed how the CNS receives spikes corresponding to the 2.5 kHz tone.

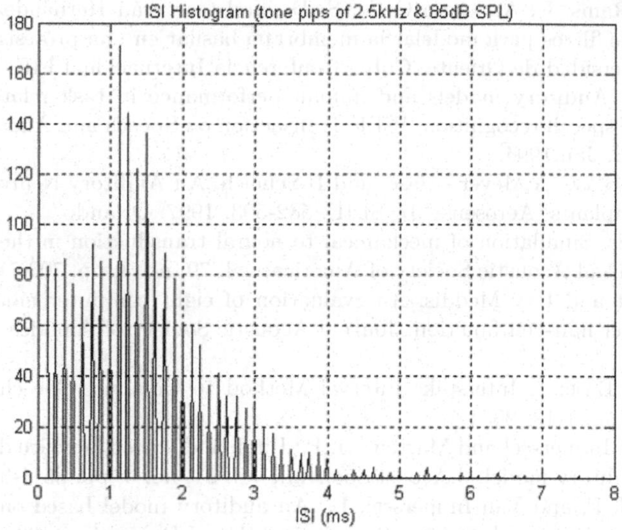

Fig. 8. ISI Histogram of a 2.5kHz tone according to the Volley theory

6 Conclusions

In the course of our research we have tested two stimulation methods. It has been concluded that the stochastic pattern of stimulation, applying the theory of the spatial code, contains information of the coded signal, which is quite similar to the original one, and that a level of SNR 28.0090 dB is obtained. However, with the implementation of the stochastic pattern of spatial–temporal stimulation, a higher similarity was obtained between the recovered signal and the original one, reflected in a stronger signal to noise relationship (31.17 dB SNR). Besides, the stochastic pattern of spatial–temporal stimulation determines more accurately the beginning and end of pip tones, for stimuli fundamentally higher than 20 dB SPL. Moreover, in this research, the implementation of stochastic model of a low firing rate stimulation was achieved fulfilling the Volley theory or spatial–temporal code. Each point of the basilar membrane was associated to not only an ANF, but a group of them (until 20), in order to conform the stochastic model of spatial-temporal stimulation.

References

1. Michael S.Landy: Course of Perception. WEB . 2004.
2. Hearing and Balance. WEB . 22-3-1996.
3. Fundamentals of Hearing & Speech Science. WEB . 2004.
4. Martínez Rams, E. A., Cano Ortiz, S. D., and Garcerán Hernández, V.: Implantes Cocleares: Desarrollo y Perspectivas. Revista Mexicana de Ingeniería Biomédica, vol. XXVII, no. 1, pp. 45-54, June2006.

5. Martínez Rams, E. A., Cano Ortiz, S. D., and Garcerán Hernández, V.: Diseño de banco de filtros para modelar la membrana basilar en una prótesis coclear. 1-6. 2006. Universidad de Oriente, Cuba, Conferencia Internacional FIE 2006.
6. Ghitza O.: Auditory models and human performance in task related to speech coding and speech recognition. IEEE Transaction on Speech and Audio Processing, pp. 115-132, Jan.1994.
7. U.Meyer - B¨ase, A.Meyer - Bäse, and H.Scheich: An Auditory Neuron Models for Cochlea Implants. Aerosense 97, SPIE. 582-593. 1997. Orlando.
8. Ray Meddis: Simulation of mechanical to neural transduction in the auditory recepter. Journal Acoustic Society of America, vol. 79, no. 3, pp. 702-711, Mar.1986.
9. M.J.Hewitt and Ray Meddis: An evaluation of eight computer models of mammalian inner hair-cell function. Journal Acoustic Society of America, pp. 904-917, 1991.
10. U.Meyer - Bäse: A Interspike Interval Method to Compute Speech Signal from Neural Firing. 1-12. 2004.
11. Luc M.Van Immerseel and Martens, J. P.: Pitch and voiced/unvoice determination with an auditory model. J.Acoust.Soc.Am., vol. 91, no. 6, pp. 3511-3526, 1992.
12. Martens, J. P. and Van Immerseel, L.: An auditory model based on the analysis of envelope patterns. Acoustics, Speech, and Signal Processing, 1990.ICASSP-90., 1990 International Conference on, vol. 1 pp. 401-404, 1990.
13. Enrique A.Lopez-Poveda and Ray Meddis: A human nonlinear cochlear filterbank. J.Acoust.Soc.Am., vol. 110, no. 6, pp. 3107-3118, Dec.2001.
14. Reinhold Schatzer, Blake Wilson, Robert Wolford, and Dewey Lawson: Speech Processors for Auditory Prostheses. Sixth Quarterly Progress Report. WEB , 1-30. 2003.
15. Ray Meddis, Lowel P.O'Mard, and Enrique A.Lopez Poveda: A computational algorithm for computing nonlinear auditory frequency selectivity. Journal Acoustic Society of America, vol. 109, no. 6, pp. 2852-2861, June 2001.
16. Alistair McEwan and Andr Van Schaik: A Silicon Representation of the Meddis Inner Hair Cell Model. Proceedings of the ICSC Symposia on Intelligent Systems & Application. 2000.
17. Don H. Johnson: The relationship of post-stimulus time and interval histograms to the timing characteristics of spike trains. Biophysical Journal, vol. 22, pp. 413-430, 1978.

Functional Identification of Retinal Ganglion Cells Based on Neural Population Responses

M.P. Bonomini[1], J.M. Ferrández[1,2], and E. Fernández[1]

[1] Instituto de Bioingeniería, Universidad Miguel Hernández, Alicante
[2] Dpto. Electrónica, Tecnología de Computadoras, Univ. Politécnica de Cartagena, Cartagena, Spain
jm.ferrandez@upct.es

Abstract. The issue of classification has long been a central topic in the analysis of multielectrode data, either for spike sorting or for getting insight into interactions among ensembles of neurons. Related to coding, many multivariate statistical techniques such as linear discriminant analysis (LDA) or artificial neural networks (ANN) have been used for dealing with the classification problem providing very similar performances. This is, there is no method that stands out from others and the right decision about which one to use is mainly depending on the particular cases demands. In this paper, we found groups of rabbit ganglion cells with distinguishable coding performances by means of a simple based on behaviour method. The method consisted of creating population subsets based on the autocorrelograms of the cells and grouping them according to a minimal Euclidian distance. These subpopulations shared functional properties and may be used for functional identification of the subgroups. Information theory (IT) has been used to quantify the coding capability of every subpopulation. It has been described that all cells that belonged to a certain subpopulation showed very small variances in the information they conveyed while these values were significantly different across subpopulations, suggesting that the functional separation worked around the capacity of each cell to code different stimuli. In addition, the overall informational ability of each of the generated subpopulations kept similar. This trend was present for an increasing number of classes until a critical value was reached, proposing a natural value for functional classes.

1 Introduction

The problem of neural cell classification spreads many different aspects, including genetic, morphological or functional characteristics. As Migliore and Shepherd [1] stated, it is very important not only to know the role of a given morphological type of neuron in neural circuits, but also an understanding of the functional phenotype is needed. Since it is not clear how to best classify neurons, many approaches are nowadays combining different techniques to classify morphological, functional, and even genomic features in order to group these cells [2] [3] [4].

J. Mira and J.R. Álvarez (Eds.): IWINAC 2007, Part I, LNCS 4527, pp. 113–123, 2007.

Concerned about this objective, this paper focused on the functional clustering of retinal ganglion cells.

A considerable number of coding studies have focused on single ganglion cell responses [5] [6]. Traditionally, the spiking rate of aisle cells has been used as an information carrier due to the close correlation with the stimulus intensity in all sensory systems. There are, however some drawbacks when analysing single cell firings. Firstly, the response of a single cell cannot unequivocally describe the stimulus since the response from a single cell to the same stimulus has a considerable variability for different presentations. Moreover, the timing sequence differs not only in the time events but also in the spike rates, producing uncertainty in the decoding process. Secondly, the same sequence of neuronal events in an aisle cell may be obtained by providing different stimuli, introducing ambiguity in the neuronal response.

New recording techniques arisen from emerging technologies, allow simultaneous recordings from large populations of retinal ganglion cells. At this time, recordings in the order of a hundred simultaneous spike trains may be obtained. New tools for analysing this huge volume of data must be used and turn out to be critical for proper conclusions. FitzHugh [7] used a statistical analyser which, applied to neural data was able to estimate stimulus features. Different approaches have been proposed on the construction of such a functional population-oriented analizer, including information theory [8] [9], linear filters [10], discriminant analysis [11] and neural networks [12].

Analyzing the neural code, in the context of getting useful information for the clustering algorithm, needs to quantify the amount of information each cell conveys. The goal of this study was to quantify their tendency to group themselves in sets of relatives according to their coding performance, using functional clustering of the autocorrelograms and Information Theory as a tool for providing an empirical value for the goodness of a coding capability. Therefore, a functional separation, or classification based on behaviour, was accomplished and the coding abilities of the subsets cells and the whole cluster determined.

2 Methods

2.1 Experimental Procedures

Extracellular recordings were obtained from ganglion cell populations in isolated superfused albino rabbit (Oryctolagus cuniculus) retina using a rectangular array of 100, 1.5 mm long electrodes, as reported previously [11] [13] [14]. Briefly, after enucleation of the eye, the eyeball was hemisected with a razor blade, and the cornea and lens were separated from the posterior half. The retinas were then carefully removed from the remaining eyecup with the pigment epithelium, mounted on a glass slide ganglion cell side up and covered with a Milipore filter. This preparation was then mounted on a recording chamber and superfused with bicarbonate-buffered Ames medium at 35C. For visual stimulation we used a 17" NEC high-resolution RGB monitor. Pictures were focused with the help of lens onto the photoreceptor layer. The retinas were flashed periodically with full field white light whereas the electrode array was lowered into the retina until a significant number of electrodes detected light evoked single- and multi-unit responses. This allowed us to record with 60-70 electrodes on average during each experiment. The electrode array was connected to a 100 channel amplifier (low and high corner frequencies of 250 and 7500 Hz) and a digital signal processor based data acquisition system. Neural spike events were detected by comparing the instantaneous electrode signal to level thresholds set for each data channel using standard procedures described elsewhere [11] [13] [15]. When a supra-threshold event occurs, the signal window surrounding the event is time-stamped and stored for later, offline analysis. All the selected channels of data as well as the state of the visual stimulus were digitized with a commercial multiplexed A/D board data acquisition system (Bionic Technologies, Inc) and stored digitally. For spike sorting we used a free program, NEV2kit, which has been recently developed by our group [16] and runs under Windows, MacOSX and Linux (source code and documentation is freely available at: http://nev2lkit.sourceforge.net/). NEV2kit loads multielectrode data files in various formats (ASCII based formats, LabView formats, Neural Event Files (.NEV), etc) and is able to sort extracted spikes from large sets of data. The sorting is done using principal component analysis (PCA) and can be performed simultaneously on many records from the same experiment.

2.2 Visual Stimulation

Different experiments were carried out with albino rabbits. The retinas were stimulated with full field flashes at 16 different light intensities within the gray scale. In order to ensure both the number of trials for each intensity was constant and the probabilities of appearance of each intensity was equal, the following procedure was carried out. Firstly, a lookup table with 16 light intensities equally distributed ranging from black (RGB values: 0, 0, 0) to white (RGB values: 255, 255, 255) was constructed. Afterwards, the elements of a list containing 20 repetitions for each of the intensities from the lookup table were relocated by changing their indexes according to a random entry chosen from an uniform distribution. The list was then loaded by a Python script embedded in VisionEgg for presentation of the flashes. Flashes were 300 ms long so that each trial lasted 96 seconds. Figure 1 shows 9 seconds of the light intensity trace for one trial and the intensity histogram for the entire trial.

2.3 Separation into Subpopulations

The separation algorithm consists in calculating autocorrelograms [17] on each of the cells in the dataset, a bin size of 10 ms was used and different time shifts such that the complete flash transitions were included in the analysis. The autocorrelograms then fed a non supervised, partitional clustering method for the creation of a varying number of autocorrelograms groups. We will refer to the number of groups with italic k. The same analysis were carried out for an increasing k at every entire population. We will use the terms class, cluster,

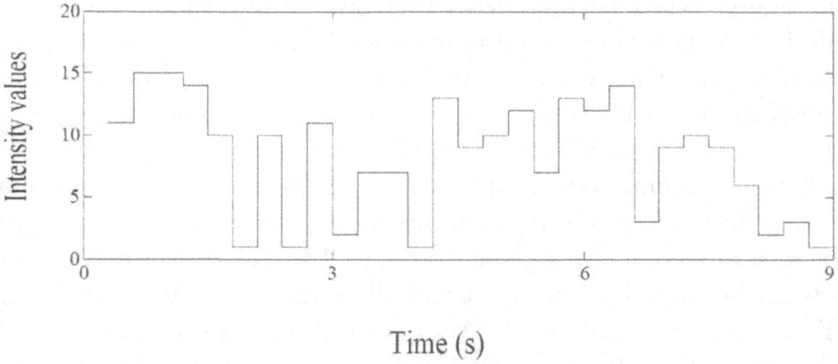

Fig. 1. Visual stimulation. Light intensity trace for 9 seconds of a sample trial containing 30 flashes, each lasting 0.3 ms. Ordinate axis represent the different intensity values (see methods).

subset or group interchangeably. The nearest-neighbour or k-means approach was chosen for the clustering method. This approach decomposes the dataset into a set of disjoint clusters and then minimizes the average squared distance from a cluster centroid among the elements within a cluster, while maximizes this distance when regarding the centroids of the different clusters. This defines a set of implicit decision boundaries that separate the clusters or classes of units according to their periodicity. In this way, we end up with groups of relatives that are a subset of the entire array.

2.4 Information Theory

In 1929, Shannon published "The Mathematical Theory of Communication" [18] where thermodynamic entropy was used for computing different aspects about information transmission, it was known as Information Theory. This computation was also applied for computing the capacity of channels for encoding, transmitting and decoding different messages, regardless of the associated meaning. Information theory may be used as a tool for quantifying the reliability of the neural code just by analysing the relationship between stimuli and responses [19] [20]. This approach allows one to answer questions about the relevant parameters that transmit information as well as addressing related issues such as the redundancy, the minimum number of neurons needed for coding certain group of stimuli, the efficiency of the code, the maximum information that a given code is able to transmit, and the redundancy degree that exists in the population firing pattern [21] [22].

In the present work, the population responses of the retina under several repetitions of flashes were discretized into bins where the firing rates from the cells of the population implement a vector n of spikes counts, with an observed probability P(n). The probability of the occurrence of different stimuli has a known probability P(s). Finally the joint probability distribution is the probability of a global response n and a stimulus s, P(s,n).

The information provided by the population of neurons about the stimulus is given by:

$$I(t) = \sum_{s \in S} \sum_{n} P(s,n) log_2 \frac{P(s,n)}{P(s)P(n)} \quad (1)$$

This information is a function of the length of the bin, t, used for digitising the neuronal ensembles and the number of the stimuli in the dataset.

With the purpose of assessing the quality of the subpopulations obtained, the following procedure was carried out. On every subpopulation generated, the information that single cells conveyed about the stimulus as well as the progression of the mutual information values when increasing the number of cells for each subpopulation was calculated. From these, two informational indicators were constructed: the mean cell information, calculated as the sum of the mutual information of each aisle cell divided by the total number of cells in a particular subpopulation, and the subpopulation information, consisting of the overall mutual information for a subpopulation in which all their cells were taken into account.

3 Results

3.1 Entire Population and Subpopulations Obtained

The generation of subpopulations is illustrated in Figure 2 with an example where the number of clusters was fixed to three. Top panel displays the raster plot of an entire population of ganglion cells while bottom panels (s1, s2 and s3) show the raster plots of the subpopulations obtained by applying the separation method formerly explained using a bin size of 10 ms and a maximum shift of 900 lags. In order to avoid repetitions, the following shortened forms will be defined: the first subpopulation will be referred to as s1, while the second and third one as s2 and s3 respectively. In addition, subpopulations were reordered so that s1 will always account for the subpopulation with fewer cells and subsequent subpopulations will contain an increasing number of cells. Clear differences among the different subpopulations can be perceived. These differences were related mainly with the firing time patterns and the number of cells in each subpopulation. For instance s1, contained very few cells that fired almost constantly during the presentation of the stimuli, s2 contained a considerable number of cells with apparent temporal patterns and s3 was integrated by a higher number of cells which showed a more randomised activity. This behaviour was present for different number of clusters.

3.2 Quality of the Subpopulations: Information Theory Approach

The overall information that each subpopulation accounted for, this is, the subpopulation information, kept similar across classes (one-way ANOVA; p=0.82) while the informative value of the individual cells, summarised by the mean cell

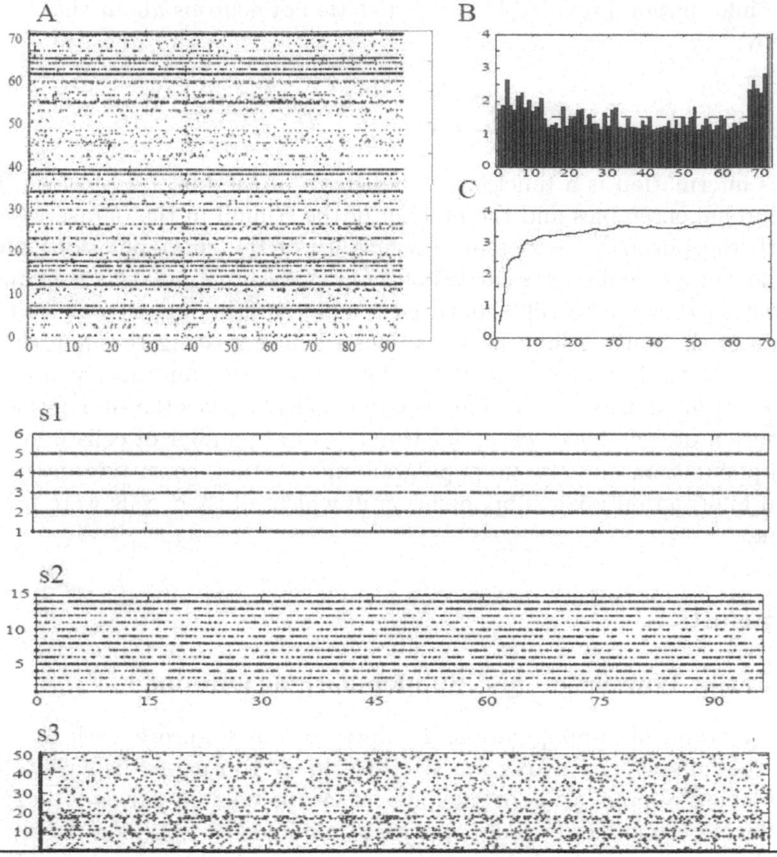

Fig. 2. Example of simultaneously recorded extracellular responses from a population of rabbit ganglion cells to a trial of random full field flashes with 16 different intensities (see methods). (A) Original population raster plot. Each dot represents a single spike. (B) Mutual information values for each cell in the recording (bars) and for the whole population (last gray bar). The overall mean cell information is showed as a dashed line. (C) Accumulative mutual information for an increased number of cells. In this example the number of cluster was fixed to three and the lower panel shows the raster plots for each subpopulation (named s1, s2 and s3).

information significantly varied (one-way ANOVA; $p \leq 0.0005$). Figure 3 shows the information that each cell conveyed about the stimulus. Notice the difference in the MCI (dashed line) and SI (last bar) values.

Notice the difference in the mean cell information value (dashed line) from the subpopulation information (last bar), which is equalized across subpopulations around a similar value to that of the original population. Surprisingly, the subpopulation formed by the very few continuously firing cells gave the higher mean cell information (2.420.10 bits; MSE), above the overall value from the whole population (1.480.01 bits; MSE) (Figure 2, right top panel, dashed line).

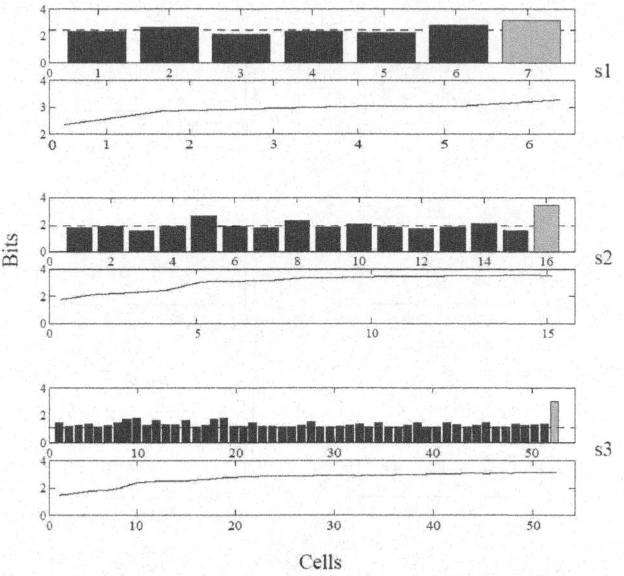

Fig. 3. Information about the stimulus for the subpopulations showed in Fig 2. Upper panels show the mutual information values for each cell in the recording (bars) and for the whole population (last gray bar). The overall mean cell information in each case is showed as a dashed line. Lower panel present the accumulative mutual information for an increased number of cells. Note the relationship between number of cells and mean cell information.

Moreover, s2, kept the moderately informative cells (1.880.07 bits; M SE) and s3 grouped the many worst cells on a mean information basis (1.330.02 bits; MSE). Interestingly, the fewer cells a subpopulation contained, the higher the mean cell information resulted at a certain subpopulation, while the number of cells did not affect the subpopulation information. Figure 4 shows how this behaviour holds for an increasing number of subpopulations generated up to five classes. In the left column the Mean cell information is shown. It can be observed that it decreases as the number of cells in a certain subpopulation increases. However, in the right column the Subpopulation information is plotted. This information keeps invariant for any cell number. Different trials are represented with different markers.

It is evident the differences in the mean cell information through out the different subpopulations arisen at any number of clusters, while the subpopulation information turned up nearly constant for all the cases. For the case in which 4 subpopulations are originated, the latter trend starts to suffer whilst for five subsets such a behaviour is completely vanished. Thus, there is a maximum number of subpopulations which might optimise a clustering strategy that is able to split the entire population on a smart informational basis. Such critical value for the number of groups (k) can be appreciated on Figure 5, which shows that the individual components of every subpopulation have similar information values.

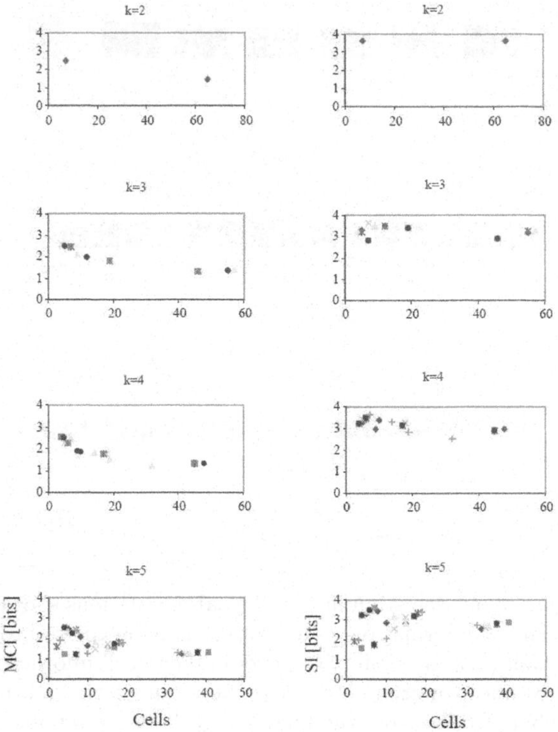

Fig. 4. Cell number versus coding quality. Left column: Mean cell information (MCI), Right column: Subpopulation information (SI). Note that MCI decreases as the number of cells in a certain subpopulation increases whereas SI keeps invariant for any cell number. Different trials are represented with different markers on each panel.

Thus for k=4 although the two central clusters start to overlap, the goodness in the group separation it is still evident. However, for five sets, the clustering behaviour becomes totally blurred.

4 Discussion

A new method for defining subsets in a population of neuronal responses has been defined. It permits pruning and classifyng the relevant elements of the visual system, getting insight the neural code more accurately. It has been shown that the generated subsets share their own coding behaviour, quantified by information theory, identifying different subset encoders with different temporal responses. This has been observed for different number of clusters. Also, the close relation between the coding goodness of a subset and the mean information of its cells has been observed using the proposed clustering algorithms.

A clear trend in the clustering strategy was present in all the subpopulations generated for a number of classes (k) less than five. This is due to the fact that

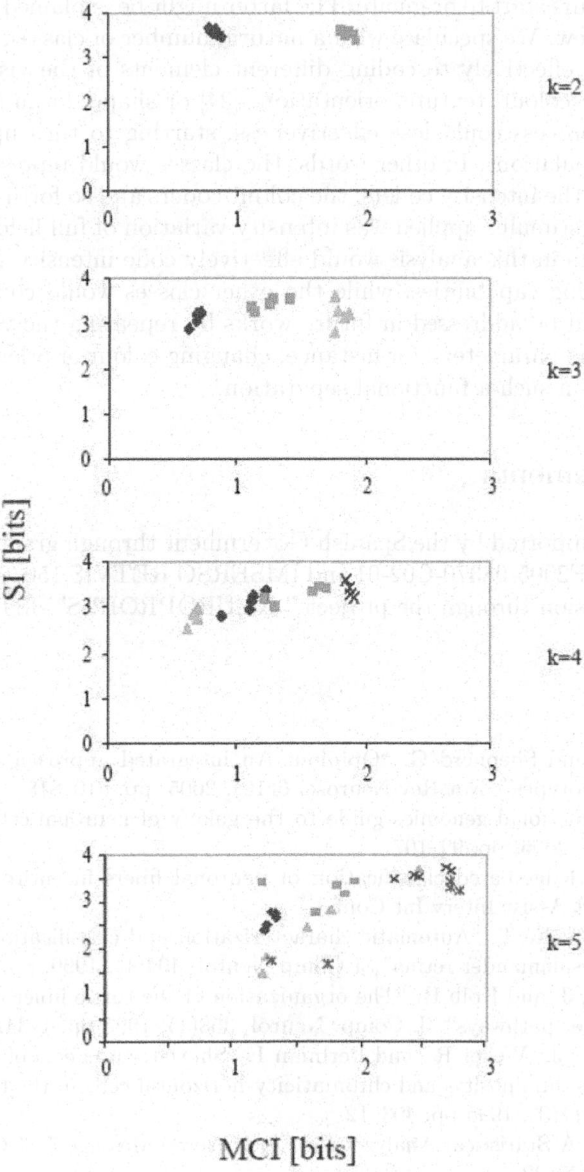

Fig. 5. Relationships between number of clusters and mean cell information values. Different trials are represented with different markers. Note that the individual components of each subpopulation have similar subpopulation information values and that there is a maximum number of clusters which optimise the classification strategy. Thus for k=4 although the two central clusters start to overlap, the group separation it is still apparent. For k=5, the clustering behaviour becomes totally blurred.

there exists a critic k value for which subpopulations with similar sizes but very different behaviours start to originate. The latter might be explained from a functional point of view. We speculate with a natural number of classes, where every class contributes effectively to coding different elements of the visual scenario such as intensity, colour, texture, orientation [24] or shape. From that number on, the coding process could lose effectiveness, starting to turn up redundant classes or subpopulations. In other words, the classes would represent different kind of cells, like the intensity coders, the colour coders and so forth. Taking into account that the stimulus applied was intensity variation of full field flashes, the best coder subsets in this analysis would effectively code intensity. In fact, they get the best coding capabilities, while the other classes would code other features. This should be addressed in future works by repeating the visual stimuli with other varying parameters, for instance, changing colour or orientation lines, in order to confirm such a functional separation.

Acknowledgements

This work was supported by the Spanish Government through grants NAN2004-09306-C05-4, SAF2005-08370-C02-01 and IMSERSO RETVIS 150/06 and by the European Comission through the project "'NEUROPROBES"' IST-027017.

References

1. Migliore, M. and Shepherd G. "Opinion: An integrated approach to classifying neuronal phenotypes", Nat Rev Neurosci 6(10), 2005, pp. 810–818.
2. Diaz E. "A functional genomics guide to the galaxy of neuronal cell types", Nat Neurosci, 9(1), 2006, pp. 99–107.
3. Xia Y. "Knowledge-based classification of neuronal fibers in entire brain", Med Image Comput Assist Interv Int Conf.
4. Costa, L. and Velte T. "Automatic characterization and classification of ganglion cells from the salamander retina", J Comp Neurol, 404(1), 1999, pp. 33–51.
5. Ammermuller, J. and Kolb H. "The organization of the turtle inner retina. I. ON- and OFF-center pathways", J. Comp. Neurol, 358(1), 1995 pp. 1–34.
6. Ammermuller , J., Weiler R., and Perlman I. "Short-term effects of dopamine on photoreceptors, luminosity- and chromaticity-horizontal cells in the turtle retina", Vis Neurosci, 12(3), 1995 pp. 403–12.
7. Fitzhugh, R. "A Statistical Analyzer for Optic Nerve Messages", J Gen Phyosiol, 41, 1958 pp. 675–92.
8. Rieke, F., et al. "Spikes: Exploring the Neural Code", ed. M. Press, 1997, Cambridge.
9. JGolledge, H.D., et al. "Correlations, feature-binding and population coding in primary visual cortex", Neuroreport, 14(7), 2003 pp. 1045–1050.
10. Warland, D., Reinagel P., and Meister M. "Decoding Visual Information from a Population of Retinal Ganglion Cells", J Neurophysiol, 78, 1997, pp. 2336–2350
11. Fernández, E., et al. "Population Coding in spike trains of sinultaneosly recorded retinal ganglion cells Information", Brain Res, 887, 2000, pp. 222-229.

12. Ferrández, J., et al. "A Neural Network Approach for the Analysis of Multineural Recordings in Retinal Ganglion Cells: Towards Population Encoding", Lecture Notes on Computer Sciences, 1999, pp.289–298.
13. Normann, R., et al. "High-resolution spatio-temporal mapping of visual pathways using multi-electrode arrays", Vision Res, 41, 2001, pp.1261–75.
14. Ortega, G., et al. "Conditioned spikes: a simple and fast method to represent rates and temporal patterns in multielectrode recordings", J Neurosci Meth, 133, 2004, pp. 135–141.
15. Shoham S., Fellows M., and Normann R. "Robust, automatic spike sorting using mixtures of multivariate t-distributions", J Neurosci Meth, 127, 2003, pp. 111–122.
16. Bongard, M., Micol D., and Fernández E. "Nev2lkit: a tool for handling neuronal event files", http://nev2lkit,sourceforge,net/.
17. Bonomini, M. P., Ferrández J.M., Bolea J.A., and Fernández E. "RDATA-MEANS: An open source tool for the classification and management of neural ensemble recordings", J Neurosci Meth, 148, 2005 pp. 137–146.
18. Shannon, C. "A Mathematical Theory of Communication", Bell sys Tech, 27, 1948, pp. 379–423.
19. Borst, A. and Theunissen F. " Information Theory and Neural Coding", Nature Neurosci, 2(11), 1999, pp. 947–957.
20. Amigo, J.M., et al. "On the number of states of the neuronal sources", Biosystems, 68(1), 2003, pp. 57–66.
21. Panzeri, S., Pola G., and Petersen R.S. "Coding of sensory signals by neuronal populations: the role of correlated activity", Neuroscientist, 9(3), 2003, pp. 175–180.
22. Pola, G., et al. "An exact method to quantify the information transmitted by different mechanisms of correlational coding", Network, 14(1), 2003, pp. 35–60.
23. McClelland, J. and Rumelhart D. "Explorations in Parallel Distributed Processing", ed. M. Press. 1986, Cambridge.
24. Kang, K., Shapley R.M., and Sompolinsky H. "Information tuning of populations of neurons in primary visual cortex", J Neurosci, 24(15), 2004, pp. 3726-3735.

Towards a Neural-Networks Based Therapy for Limbs Spasticity

Alexandre Moreira Nascimento[1], D. Andina[2], and Francisco Javier Ropero Peláez[1]

[1] Politechnique School of the University of São Paulo
alexandrenascimento@ig.com.br
[2] Group for Automation and Soft Computing, Technical Univ. of Madrid (GASC/UPM)
d.andina@upm.es, fjavier@usp.br

Abstract. This article presents a neural network model for the simulation of the neurological mechanism that produces limbs hiper-rigidity (spasticity). In this model, we take into account intrinsic plasticity, which is the property of biological neurons that consists in the shifting of the action potential threshold according to experience. In accordance to the computational model, a therapeutic technique for diminishing limbs spasticity is proposed and discussed.

Keywords: Neural networks, hypertonia, spasticity intrinsic plasticity, muscle, treatment.

1 Introduction

Limbs spasticity is a severe problem that results from brain or spinal cord injury. Initially, the patient loses the movement of the limbs, presenting a very low muscular tone (hypotonia). Gradually (in a few weeks) muscular tone increases, so that a extreme rigidity of the limbs (hypertonia) follows the episode of hypotonia. The spastic or hyper-rigid limb produces pain, limbs deformation and, in some occasions, bones detachment from their joints. According to literature [1,2], spasticity might be produced by two hypothetical causes: a) Gradual increase in muscle spindle sensitivity b) Increased excitability of central neurons involved in the reflex arc. Microneurographic studies [1, 2] in humans and neurophysiological experiments in animals have found no abnormality in the sensitivity of muscle spindles so that the other alternative, the increased excitability of central neurons involved in spinal stretch reflex, appears to be the main responsible of spasticity. Pierrot-Dseilligny and Mazieres [2] proposed that the disruption of the descending motor inputs produces spasticity. Several theories were proposed for explaining how this occurs. Among them, the sprouting of intraspinal synapses that substitutes the degenerating ones from motor inputs may contribute to the increment of the excitability in spinal neurons. Alternatively we propose that the recently discovered property called intrinsic plasticity [3, 4, 5] might also explain

J. Mira and J.R. Álvarez (Eds.): IWINAC 2007, Part I, LNCS 4527, pp. 124–131, 2007.

spasticity. According to this property, neurons that are intensely stimulated increase their action potential threshold so that their firing probability is lowered in the future. Conversely, if a neuron is under-stimulated, its action potential threshold diminishes so that its probability of firing increases in the future. According to this, after a brain or spinal injury, the lack of functional inputs to spinal neurons makes these neurons to be initially silent giving rise to hypotonia. Gradually, these initially silent neurons become more and more active due to the diminution of their action potential threshold (according to intrinsic plasticity) thereby producing hypertonia and spasticity. Although intrinsic plasticity was initially detected in cortical neurons [4], now there is compelling evidence that this same process exist in spinal cord neurons [6, 7]. In the following pages we present a neural network model of the neurons involved in spinal cord reflexes in which intrinsic plasticity is introduced in alfa and gamma motor neurons. This model confirms our theoretical supposition that spasticity is related to intrinsic plasticity. Several treatments for spasticity has been proposed, from pharmacological treatments like botulinum toxin intending to relax the muscle, to chirurgical solutions by cutting tendons. Unfortunately there is no available therapy aiming to the more probable cause of spasticity, the abnormally high excitability of motor neurons. Our computational model simulations suggest a solution for quieting these abnormally excited neurons. Taking into account that alfa motor neurons have lost its main input from the brain, we suggest to highly stimulate their ancillary input, from Ia neurons. This can be done by quickly tapping the muscle. We simulated this tapping process in the computer model, arriving to the results that will be discussed in the paper.

2 Methods and Materials

2.1 Biological Mechanism

The skeleton's muscles have two types of sensorial receptors, the muscular spindles and the Golgi tendon organ [8]. In this work we do not deal with the Golgi tendon organ, which is important for protecting the muscle when carrying very heavy charges. We will focus in the muscular spindle sensory receptors that have an important role in the control system loop that is used by the muscle to follow brain commands for executing a muscular action. The spindle is a cylindric structure, thicker in its central portion. Inside, it has two or more muscular fibers functionally specialized as stretch mechanical receptors. These fibers are called intrafusal. The remaining muscular fibers are called extrafusal and are the responsible of the muscle strength. Inside the spindle, each Ia axon is coiled around the central portion of the intrafusal fiber. The fiber's central part is a non-contractional segment. When it is strained the spiral is distorted generating action potentials that are sent to the spinal cord. In each extremity of the central region, the intrafusal fiber possesses a contractile element that receives the axons coming from gamma motor-neurons in the spinal cord. The action potentials arriving at these axons provoke the stretch of the intrafusal fiber central part. When the stretch of the muscle opposes to this contraction, terminals Ia generates a burst of action potentials. Through Ia neurons the nervous system constantly receives information about the degree of stretch of the different muscles. The sensitivity of this stretch sensor is regulated through gamma neurons.

2.2 Model of the Circuit Involved in Motor Control

The motor control of the muscle can be modeled as it is shown in figure 1. In this model a single muscle is being represented by two parallel springs, with spring constants K_1 and K_2 respectively. The first spring represents the extrafusal muscle fiber and the latter the intrafusal muscle fiber. Alpha neurons and gamma neurons relays brain commands to the muscle. Alpha neurons relays strength commands and gamma neurons displacement commands.

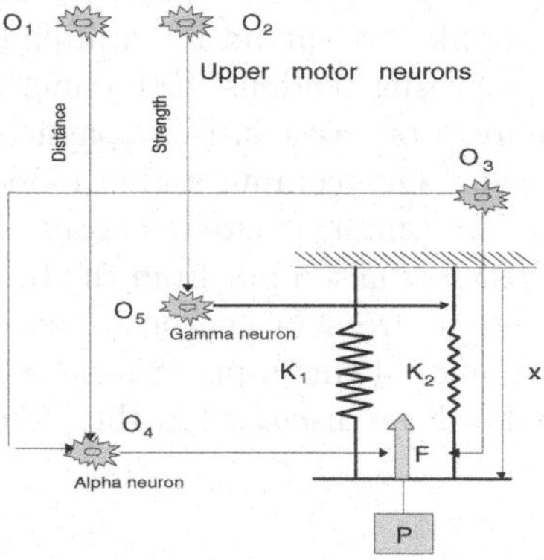

Fig. 1. Simplified scheme of the neural circuit involved in motor control. The muscle has been substituted by two springs representing extrafusal and intrafusal fibers. O_3 represents the output of the, so called, Ia neurons whose axons are coiled in intrafusal fibers.

Considering F, the force applied by the extrafusals fibers, and K_1, the spring constant of extrafusal fibers, the displacement X can be calculated by the following expression:

$$X = F/K_1$$

The force f of the spring, representing the intrafusal fibers, is calculated according to the following equation, where X_o is the length of the spring without applying any weight.

$$f = k_2(X - X_o)$$

In the model, the output O_3 of the alfa motor neuron regulates the extrafusal fibers strength by altering their spring constant, K_1. For simplicity, and since there is a direct proportion between O_3 and K_1, we made $K_1 = O_3$ in the model. Intrafusal fibers length is inversely proportional to the gamma motor command. Accordingly and for simplicity we made $X_o = 1/O_4$.

$$f = k_2(F/O_3 - 1/O_4)$$

2.3 Mathematical Model of the Neuron

Rate-code neurons are used in our model. Their inputs I_j are modulated by the synaptic weights w_{ij}. The activation, a_i, of the neuron is calculated by:

$$a = \sum_{j=1}^{j=n} W_{ij} I_j$$

Where each of the weights are calculated like the conditional probability of the neuron´s output given the input [9]:

$$W_{ij} = P(O_i/I_j)$$

which is asymptotically equivalent to the following rule:

$$\Delta W_{ij} = \varepsilon I_j (O_i - W_{ij})$$

where ε is the learning constant which is usually a small decimal number. A non-linear function (sigmoid) yields the output of the neuron in terms of a_i $O_i = f(a_i)$:

$$O = \frac{1}{1+e^{-25(a-shift)}}$$

Where shift is a sigmoid displacement parameter that is included for modeling intrinsic plasticity (see Figure 2).

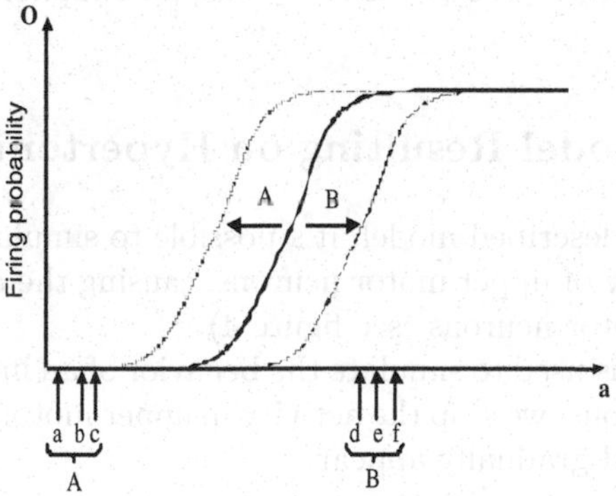

Fig. 2. Intrinsic plasticity consists in the gradual leftward or rightward shift of the neuron's activation function (here a sigmoid) for a sustained trend of low (cases a, b and c) or high (cases d, e, f) activations of the neuron respectively

The *shift* parameter is 1 when the sigmoid function is maximally dislocated to the right and zero when it is maximally dislocated to the left.

$$shift_t = \frac{\xi O_{t-1} + shift_{t-1}}{1+\xi}$$

2.4 Simplified Neural Model of Muscle Response

The diagram in figure 3 represents the neural network used in our computational model.

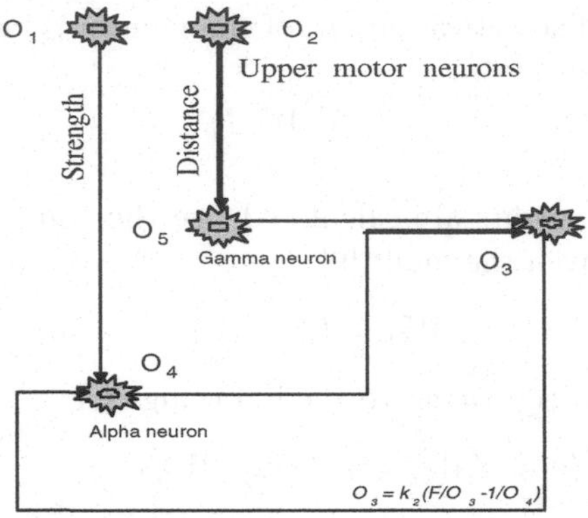

Fig. 3. The more complex system in Figure 1 is here simplified so that the muscle's (Ia neurons and extrafusal and intrafusal fibers) response is embedded inside a virtual neuron whose ouput is O3

The output O_3 of the virtual neuron represents the overall muscle computation, or in other words the firing probability of Ia neurons in terms of O_4 and O_5.$O_3 = k_2 f = k_2(F/O_3 - 1/O_4)$ where k_2 was, for simplicity, arbitrarily made equal to one.

3 Spasticity Model Resulting on Hypertonia

Using the previously described model, it's possible to simulate spasticity, which results from the injury of upper motor neurons, causing the absence of inputs to alpha and gamma motor neurons (see figure 4).

Initially the model is used to simulate the behavior of an intact nervous system and, after 1250 iterations, we stop the activity in upper motor neurons. Spasticity and hypertonia should gradually appear.

3.1 Model of a Therapy for Attenuation of Spasticity

After spasticity is installed in the model, it is possible to simulate the application of a therapy based on a physical stimulation of the hypertonic muscles. The aim is to stimulate the alpha neuron through the feedback O_3 from Ia neurons. The application of the stimulus (tapping the muscle) drives alpha neurons to a

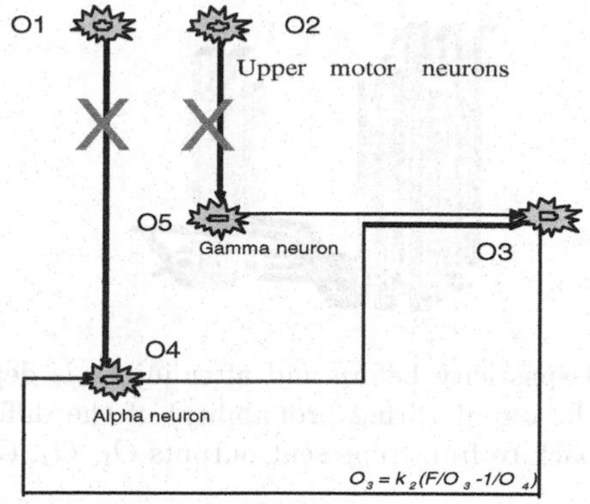

Fig. 4. Simulation of an injury in spinal or brain areas. Alfa and gamma neurons are detached from upper motor neurons.

new state in which the sigmoid is shifted rightwards so that the probability of firing the neuron (that causes spasticity) is reduced. In this way, although the brain stimuli to alfa neurons was eliminated due to the injury, it is possible to provide an alternative means of stimulation through a feedback mechanism from Ia neurons (O_3) thereby reducing muscle's rigidity.

3.2 Software for the Simulation of the Proposed Model

The software that simulates the proposed model was developed in Matlab. The first 1250 iterations (horizontal axis) represents the normal situation in which upper motor neurons are normally sending commands to alpha and gamma neurons in the spinal cord. In the following iterations, activity from upper motor neurons is canceled as occurs after a cerebral or spinal injury. Finally, a therapeutic procedure is applied from iteration 3000, by repeatedly stimulating Ia neuron. The therapeutic procedure ends at iteration 3000 and, from iteration 3000 ahead, it is possible to see the consequences of treatment.

4 Results

With this simulation, between iteration 1 and 1250, upper neurons stimulate alpha and gamma neurons by using a square shaped wave. After the 1250th iteration, upper neurons activity ceases and the basal activity of alpha (O_4) and gamma (O_5) neurons gradually increases so that the muscle gradually becomes hypertonic as shown by the gradual elevation of the two frontward ribbons in figure 5.

This increment of the basal activity of alpha and gamma neurons is the consequence of the leftward shifting of their sigmoidal activation functions as is shown in figure 6 where the vertical axis represents the sigmoid shift. After the 3000th iteration, a square shaped stimulation is applied in Ia fibers, as depicted by the

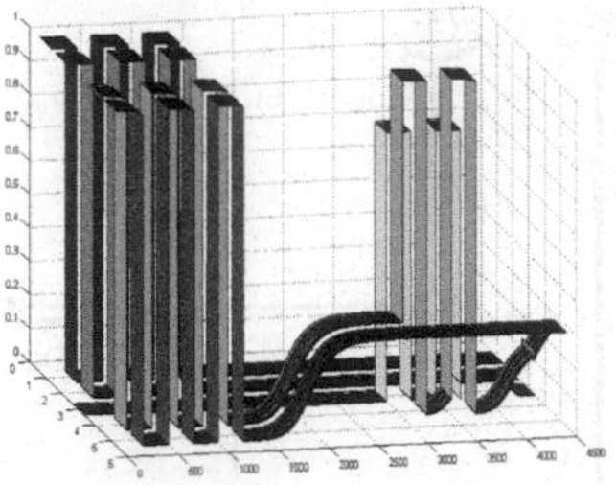

Fig. 5. The evolution of spasticity before and after injury is depicted in this figure. Vertical axis represent the ouput (firing probability) of the different neurons in the network. Ribbons from back to front represent outputs O_1, O_2, O_3, O_4 and O_5.

third ribbon in the figure, which finally, causes the alpha neuron to diminish its basal activation when the stimulation is no more produced (see ribbon corresponding to O_4 in figure 5. The diminution of O_4 basal activity leads to the disappearance of the muscle's rigidity.

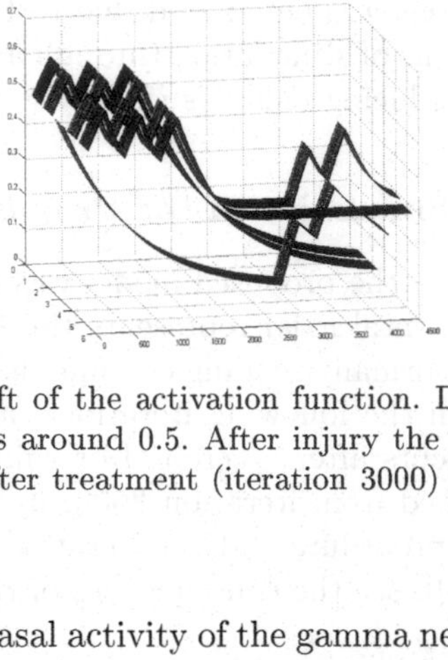

Fig. 6. Evolution of the shift of the activation function. During the first 1250 alpha and gamma neuron's shift is around 0.5. After injury the shift decreases so that the sigmoid moves leftwards. After treatment (iteration 3000) only the shift of the alpha and Ia neuron increase.

While this occurs, the basal activity of the gamma neuron continues the same, meaning that the stimulation does not influence gamma neurons. After ceasing the therapeutical procedure in which Ia neurons are no longer stimulated, hypertonia comes back (See evolution of the curves in figure 5 after iteration 3500) This means that the improvement is transitory: it lasts while the therapeutical stimulation is performed. For a more sustained recovery, reconnection of upper motor neurons is necessary, by means of other type of therapies.

5 Conclusion

Analyzing the simulation results, it can be concluded that it is possible to simulate the neuron control mechanism of the limbs. The model makes possible to test our hypothesis for explaining spasticity as a result of intrinsic plasticity that, in the long run, makes non-stimulated neurons become extremely sensitive so that they fire even without upper motor neuron's inputs. The proposed model allows to test possible therapies evaluating their performance and anticipating the efficacy of these therapies along time. We proposed a therapy in which Ia neurons are mechanically stimulated. It was shown that alpha neurons becomes quieter, although the therapy does not produce any effect upon gamma neurons. Despite the therapy might provide some relieve and comfort to the patient by reducing the contraction of extrafusal fibers, a synergetic approach that restore the connectivity with upper motor neurons is encouraged for obtaining more sustained results.

References

1. D. Burke. Critical examination of the case for and against fusimotor involvement in disorders of muscle tone. *Adv Neurol*, 39:133 – 155, 1983.
2. Pierrot-Deseilligny and L. E., Mazieres. Spinal mechanism underlying spasticity. *Clinical neurophysiology in spasticity*, In: Delwaide, P.J. and Young, R.R., eds. Amsterdam: Elsevier, 1985.
3. Desai N. Homeostatic plasticity in the cns: synaptic and intrinsic forms. *Journal of Physiology Paris*, (97):391 – 402, 2003.
4. Rutherford L. C. Desai, N. S. and G. Turrigiano. Plasticity in the intrinsic excitability of cortical pyramidal neurons. *Nature Neuroscience*, 2(6):515 – 520, 1999.
5. G.G. Turrigiano and S.B. Nelson. Homeostatic plasticity in the developing nervous system. *Nature Reviews in Neuroscience*, 5:97 – 106, 2004.
6. N. Chub and M.J. O'Donovan. Blockade and recovery of spontaneous rhythmic activity after application of neurotransmitter antagonists to spinal networks of the chick embryo. *Journal of Neuroscience*, 18:294 – 306, 1998.
7. Obrietan K. Van Den Pol, A.N. and A. Belousov. Glutamate hyperexcitability and seizure-like activity throughout the brain and spinal cord upon relief from chronic glutamate receptor blockade in culture. *Neuroscience*, 74:653 – 674, 1996.
8. Connors B.W. Bear, M.F. Paradise, m.a. *Neuroscience. Exploring the Brain. Lippincott, Williams and Wilkins. USA*, 2001.
9. J. R. Pelaéz and J. R. C. Piqueira. Biological clues for up-to-date artificial neurons". In: Andina,D. and Phan., D.T. (Eds.). Computational Inteligence: for Engineering and Manufacturing. 1 ed. Berlin: Springer-Verlag:131 – 146, 2006.

A Bio-inspired Architecture for Cognitive Audio

Pedro Gómez-Vilda[1], José Manuel Ferrández-Vicente[2], Victoria Rodellar-Biarge[1], Agustín Álvarez-Marquina[1], and Luis Miguel Mazaira-Fernández[1]

[1] Grupo de Informática Aplicada al Tratamiento de Seal e Imagen, Facultad de Informática, Universidad Politécnica de Madrid, Campus de Montegancedo, s/n, 28660 Madrid

[2] Dpto. Electrónica, Tecnología de Computadoras, Univ. Politécnica de Cartagena, 30202, Cartagena

pedro@pino.datsi.fi.upm.es

Abstract. A comprehensive view of speech and voice technologies is now demanding better and more complex tools amenable of extracting as much knowledge about sound and speech as possible. Many knowledge-extraction tasks from speech and voice share well-known procedures at the algorithmic level under the point of view of bio-inspiration. The same resources employed to decode speech phones may be used in the characterization of the speaker (gender, age, speaking group, etc.). Based on these facts the present paper examines a hierarchy of sound processing levels at the auditory and perceptual levels on the brain neural paths which can be translated into a bio-inspired audio-processing architecture. Through this paper its fundamental characteristics are analyzed in relation with current tendencies in cognitive audio processing. Examples extracted from speech processing applications in the domain of acoustic-phonetics are presented. These may find applicability in speaker's characterization, forensics, and biometry, among others.

1 Introduction

Bio-inspired Speech Processing is the treatment of speech following paradigms used by the human sound perception system, which has specific structures for this purpose. An open question is if bio-inspiration is convenient for specific tasks as Speech Recognition [4], [6]. The answer is that bio-inspiration may offer alternative ways to implement specific functions in speech processing, helping to improve the performance of conventional methods. Cognitive Audio as a whole and Speech Understanding in particular are specific application paradigms involving capabilities such as location and movement detection, source enhancement and separation, source identification, speaker's identification, recognition of discourse, detection of emotions in voice, etc. These are part of what is known as Cognitive Audio, in the sense that the understanding of the surrounding world by human beings is strongly dependent on them. The purpose of the present paper is to provide a hierarchical description of speech processing by bio-inspired methods discussing the fundamentals of speech understanding, devising a general bio-inspired architecture for Cognitive Audio.

J. Mira and J.R. Álvarez (Eds.): IWINAC 2007, Part I, LNCS 4527, pp. 132–142, 2007.

2 Fundamentals of Speech Production and Perception

Speech production is based in the flow of air through the larynx, inducing a vibration of the vocal folds known as glottal source, which is modulated by the pharyngeal, oral and nasal cavities and radiated through the mouth. This is the basis for voiced speech, although some sounds are produced without the vibration of the vocal cords, as in the case of unvoiced or whispered speech. The tongue position relative to lips, teeth, palate or velum, and the opening of the constrictions produce resonances which change the perception of the sound produced. These resonances called formants, and give a good description of the message issued (acoustic-phonetic decoding) as well as the speaker's personality. Formants are labeled in order of increasing frequency, F_1 being the lowest, in the range of 250-900 Hz. F_2 sweeps a wider range, from 600 to 3600 Hz. Formants F_3, and higher may be also present in voiced speech, however, the lowest two formants give a good description of vowel-like phonemes (see Fig.1).

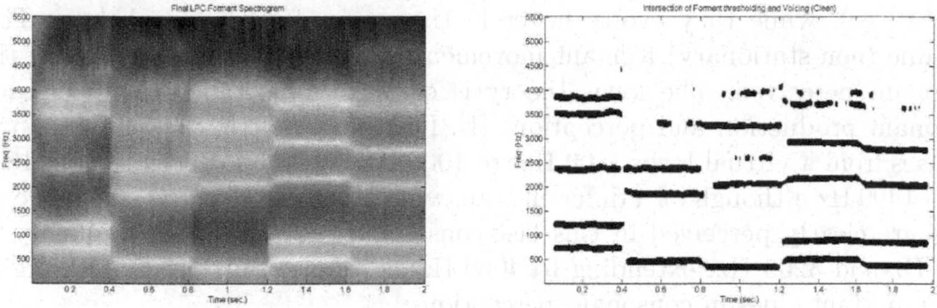

Fig. 1. Left: Spectrogram of the five vowels in Spanish (/a/, /e/, /i/, /o/, /u/) from a male speaker, obtained by Adaptive Linear Prediction. Right: Formant plot for the same recording.

The first two formants for /a/ appear at 750 and 1250 Hz, F1 decreasing to 500 Hz and 400 Hz for /e/ and /i/, whereas F_2 climbs to 1800 and 2100 Hz for the same vowels. F_1 shows the same values for /o/ and /u/ as for /e/ and /i/, whereas F_2 moves to 900 and 850 Hz. This is seen in the vowel triangle represented in Fig.2, plotting F_2 vs F_1.

Vowels appear as stable narrow-band patterns, known as Characteristic Frequencies (CF). Consonant behavior is rather different, as these sounds are produced by constrictions of the articulation organs, resulting in vowel formant transitions, this phenomenon known as co-articulation (in voicing), or in the production of noise bursts (in unvoiced sounds). Formant transitions from stable CF positions to new CF positions (virtual loci) are known as FM (frequency modulation) components. The presence of wide bands, generally above 2000 Hz are known as noise bursts (NB) or also as blips. These patterns (CF, FM and NB) bear important communication clues [20], and in the human brain certain types of neurons are specifically tuned to detect each one of them. An example for consonant sounds is given in Fig.3.

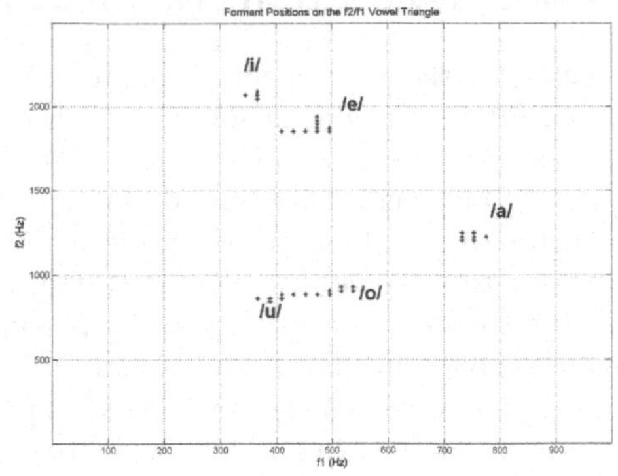

Fig. 2. Vowel triangle showing the five vowels in Spanish given in Fig.1. Vowel positions are designated by the corresponding labels.

The first two formants move to the CF's of /a/ for /ba/ (0.5 sec) and /da/ (0.8-1.0 sec), while they evolve faster in the cases of /dzha/ and /ga/. This dynamic (non-stationary) formant movement is related to the character of the consonant perceived. The locus theory is of most importance to understand consonant production and perception [1]. It is observed that F_1 climbs up in all cases from a virtual locus (800 Hz) to 1000 Hz, while F_2 descends from 1800 Hz to 1400 Hz although at a different rate, which for /dzha/ is the steepest one. Blips are clearly perceived in this last consonant (1.22 sec) at a frequency of 2500 Hz and 3200 Hz, extending to 4000 Hz as a noise burst (NB). NB's give also important clues in consonant perception.

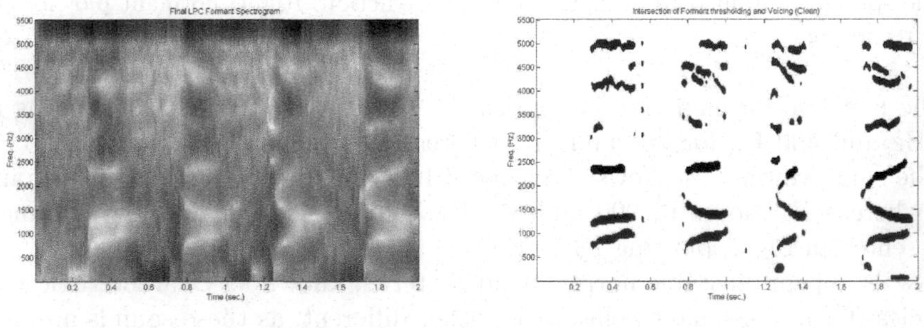

Fig. 3. Left: LPC spectrogram of the syllables /ba/, /da/, /dzha/, /ga/ from the same speaker. Right: Formant plots.

Similar patterns are observed for the unvoiced consonants /pa/, /ta/, /tsha/ and /ka/, as shown in Fig.4 (left). In this case the formant onsets are sharper, but the general tendencies of the formats are similar to the cases of /ba/, /da/, /dzha/ and /ga/. Another important clue for the perception of the consonant in the case of /ca/ is the presence of a column of blips just before t=1 sec (at 1000, 1800, 2500, 3000, 3500, 4000, and 4500 Hz). Two blips are also present

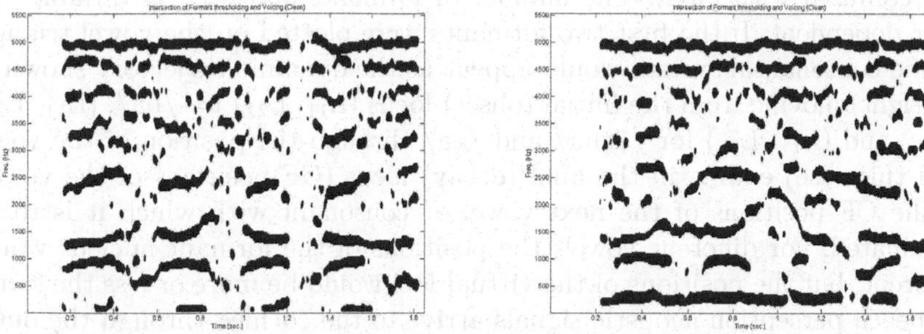

Fig. 4. Left: Formant plots of the syllables /pa/, /ta/, /tsha/, /ka/ from the same speaker. Right: Formant plots of the syllables /ma/, /na/, /nja/, /nga/.

before the onset of the two first formants in the case of /ka/. Finally, in Fig.4 (right) four nasals are presented, corresponding to the four positions studied before. In this case, besides observing more similar dynamic behaviour for the first two formants, a nasalization bar appears at a frequency around 300 Hz. A Generalized Phoneme Description may be derived, as shown in Fig.5 (left) where the temporal patterns of a typical phoneme are shown based on a nuclear vowel system and the pre-onset and post-decay positions.

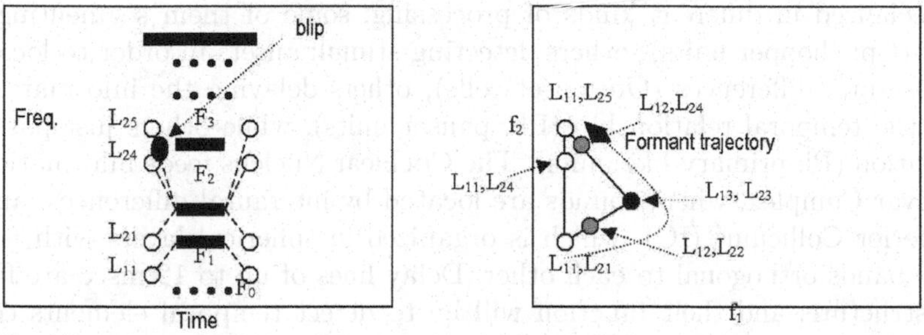

Fig. 5. Left: Generalized Phoneme Description. Right: Loci of the GPD on the vowel triangle. White circles indicate the positions of the loci. The dark dot gives the position of the specific vowel modeled (/a/ in the present case).

The description is based on a vowel nucleus defined by formants F_1 and F_2. These are characterized by well defined relatively stable CF positions. The onset is marked by formant F1 moving from a specific locus (L_{11}) to the final CF position (positive FM). The formant F_2 may move from a low frequency locus (L_{21}) (positive FM) or from high frequency ones (L_{24}, L_{25}) (negative FM) depending on the specific articulation place of the front consonant. Blips appear mainly in palatal articulations, and possibly extend to wide-band patterns (with frequencies above 3000 Hz). Loci in the decay side evolve to next vowel or consonant articulation places. Nasalization appears as a low formant F_0 which must not be confused with pitch. The number of formants above F_3 is variable and speaker dependent. If the first two formants were plotted on the vowel triangle a specific consonantal system would appear as the dynamic trajectory shown in Fig.5 (right) moving from the initial (onset) locus (L_{11}, L_{21}) for /ba/, (L_{11}, L_{24}) for /da/ and (L_{11}, L_{25}) for /dzha/ and /ga/ through the position of the vowel (/a/ in this case) ending in the final (decay) locus (CF positions of the vowel) or in the CF positions of the next vowel or consonant with which it is to be co-articulated. For different vowels the positions of the formant nucleus would be different, but the positions of the virtual loci would be more or less the same.

In speech perception acoustic signals arrive to the cochlea through the outer and middle ear. Important processing is produced in the basilar membrane, along the cochlea or inner ear (see Fig.6), operating as a filter bank. Low frequencies produce maximum excitation in the apical end of the membrane, while high frequencies produce maximum excitation towards the basal area. These peak locations code different frequency stimuli present in speed inducing the excitation of transducer cells (hair-cells) at different positions along the cochlea which will be responsible for the mechanical to neural transduction process that propagates electrical impulses to higher neural centers through auditory nerve fibers, each one being specialized in the transmission of a different characteristic frequency (CF). CF fibers tend to respond to each of the spectral components (F_0, F_1, F_2...) of speech in distinct groups, within each group the interpeak intervals represent the period of the corresponding spectral components [16]. The next processing centre is the cochlear nucleus (CN) where different types of neurons are specialized in different kinds of processing, some of them segmenting the signals (Cp: chopper units), others detecting stimuli onsets in order to locate it by inter-aural differences (On: onset cells), others delaying the information to detect the temporal relationship (Pb: pauser units), while others just pass the information (Pl: primary-like units). The Cochlear Nucleus feeds information to the Olivar Complex, where sounds are located by interaural differences, and to the Inferior Colliculus (IC), which is organized in spherical layers with isofrequency bands orthogonal to each other. Delay lines of up to 12 msec are found in its structure, and their function will be to detect temporal elements coded in acoustic signals (CF and FM components). This centre sends information to the thalamus (Medial Geniculated Body) which acts as a relay station for prior representations (some neurons exhibit delays of a hundred milliseconds), and as a tonotopic mapper of information arriving to cortex, where high level processing takes place. It seems that the neural tissue in the brain is organized as ordered feature maps [16] according to this sensory specialization. The specific location of the neural structures in the cortex responsible for speech processing and understanding is not well defined as the subjects of experimentation have been mainly animals. Although recent reports on speech brain center research using Nuclear Magnetic Resonance have been published, these studies do not reach single neuron resolutions yet. In cats neurons have been found that fire when FM-like frequency transitions are present (FM elements) [10], while in macaque some neurons respond to specific noise bursts (NB components) [14].

Other neurons are specialized to detect the combinations among these elements [19]. In humans, evidence exists of a frequency representation map in the Heschl circumvolution [15] and of a secondary map with word-addressing capabilities [12]. A good description of the structures involved and their functionality is given in [5].

3 A Bio-inspired Architecture for Speech Processing

A plausible structure to implement some of the described speech processing capabilities of the human auditory system can be proposed as in Fig.6. A bio-inspired speech processing architecture should start with a structure for transforming a time-domain signal into its most important frequency components. This is usually done by the Cochlear System and the Cochlear Nucleus. To implement this functionality some choices are available, as filterbanks, gammatones, FFT or LPC, among others. Gammatone filters [8] are the closest to the operation of the biological system, as they have been designed to mimic the space-time behavior of the real cochlea. Nevertheless, as the processing is based in formant-like pattern detection, LPC has been selected as it is specialized in formant detection by all-pole inverse filtering.

The characteristics of the Linear Prediction inverse filtering proposed in the present work are the following:

- It is based on lattice filters, which are highly efficient in the modeling of the vocal tract by high-order all-pole transfer functions. The time evolution of the vocal tract transfer function due to articulation is tracked using an adaptive implementation of the lattice filters.
- A combined iterative method derived originally from the study of the glottal pathology [3] is used to de-couple the glottal source spectral behavior from that of the vocal tract by means of paired lattices. This is especially important to obtain an accurate formant description of speech in the frequency domain.

The left hand-side plots in Fig.1 and Fig.3 have been obtained using such methodology. Once the power spectral density of the vocal tract transfer function has been decoupled from the glottal source formant features are to be estimated from the resulting spectrogram, defined as:

$$X(m,n) = 20 \cdot log_{10} \left| \sum_{k \in arg\{V\}} V(k)x(n+k)e^{-jmk\Omega\tau} \right| \tag{1}$$

where *x(n)* is the speech signal, *V(k)* is a specific framing window, and τ and Ω are the resolutions in time and frequency. The representation *X(m,n)* can be seen as a two-dimensional image, indexed by time *(n)* and frequency *(m)*. This means that many tools devised for image processing can be used for the detection of time-frequency features, as CF or FM patterns. The first basic operation on the LPC spectrogram will be to enhance formant trajectories. This is carried out

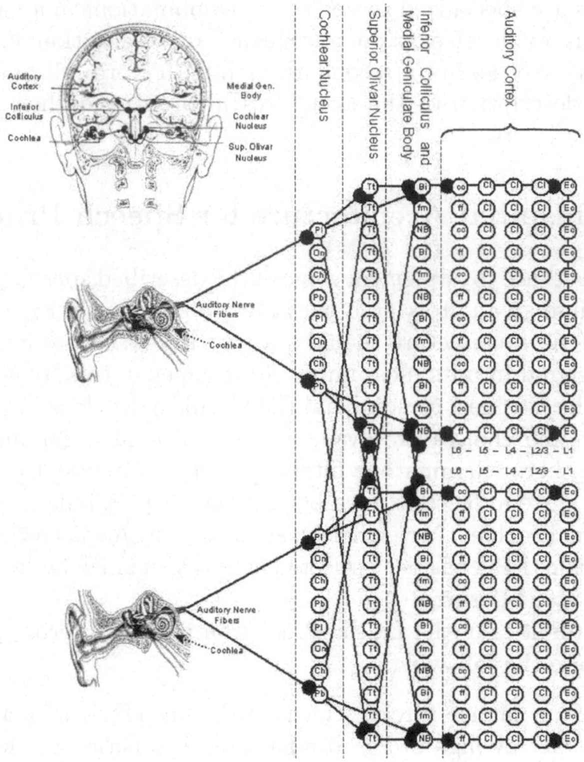

Fig. 6. Bioinspired architecture of the auditory system to model speech recognition functionalities. Two earlike-like frequency coders will separate sounds into auditory-like stimuli to represent both ears. These stimuli are timely encoded by the Primary-Like (Pl), Onset (On), Chopper (Cp) and Pauser-Build Up (Pb) neurons in the Cochlear Nucleus. These connect to Tono-Topic neurons (Tt or CF cells) in the Superior Olivar Nucleus. The outputs create contacts with the Inferior Colliculus and the Medial Geniculate Body. It seems that binaural information is coded in Bi neurons, while certain neurons can be found sensitive to frequency transitions (fm). Noise Bursts may be also detected at this level by NB neurons. Finally the stimuli reach the 6-layer columnar structure of the Auditory Cortex (Cl units). Certain neurons code associations of characteristic frequency stimuli (cc) while others code associations of frequency transitions (ff).

using a simple bio-inspired algorithm to mimic lateral inhibition, this mechanism being active in certain neuron associations in the Inferior Colliculus [16]. The proposed algorithm is expressed as:

$$\hat{X}(m,n) = \sum_{i=-1}^{1} w(i)X(m-i,n) \tag{2}$$

where the respective weights are w_{-1}=w_1=-1/2 and w_0=1. This filter has to be applied to each column of the LPC spectrogram using buffering, to avoid data corruption.

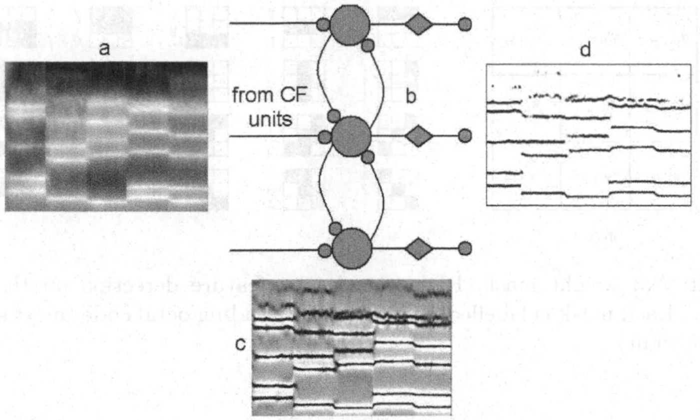

Fig. 7. Formant Detection for the utterance /aeiou/: The speech spectral density (a) is processed by columns of neurons implementing lateral inhibition (b), producing differentially expressed formant lines (c), which are transformed into narrow formant trajectories (d) after non-linear saturation

A columnar organization of the filter for three cells is shown in Fig.7. It may be seen that the lateral inhibition filter produces sharp estimations of the spectral peaks (see Fig.7.b). The whitish bands surrounding the formants are due to the characteristic "mexican hat" response of the filter. The final formant distribution is given in Fig.7.c after adaptive thresholding. The problem of feature detection in formant spectrograms is related to a well known one in Digital Image Processing (DIP) [9]. A classical methodology is based on the use of reticule masks on the image matrix *X(m,n)*:

$$\tilde{X}(m,n) = \sum_{i=-1}^{1} \sum_{j=0}^{2} w_{i,j} X(m-i, n-j) \tag{3}$$

where $w_{i,j}$ is a 3x3 mask with a specific pattern and a specific set of weights. The lateral-inhibition filtering given in (2) is a special case of (3) where the weights of columns *j=1,2* have been filled with zeros. It may be shown that the generic filtering in (3) is equivalent to a liftering process [2].

There are two important concepts linked to mask design: the specific pattern in time-frequency, and the specific weight adjustment. The basic cells for formant trajectory processing shown in Fig.8 have been derived from the neural structures of the auditory centers in brain, as presented in section 2. To produce unbiassed results, the weight associated to each black square is fixed to $+1/s_b$ and the weight associated to white squares is fixed to $-1/s_w$, s_b and s_w being the number of squares in black or white found in a 3x3 mask, respectively. Weight adjustment may be pre-assigned or adaptive. In this last case a database of spectrograms and 9:18:1 MLP structures for the training of each cell [7] can be used. The results presented here are based on pre-assigned weights, for the sake of brevity. The indexing in the time domain is intended to preserve the causality principle,

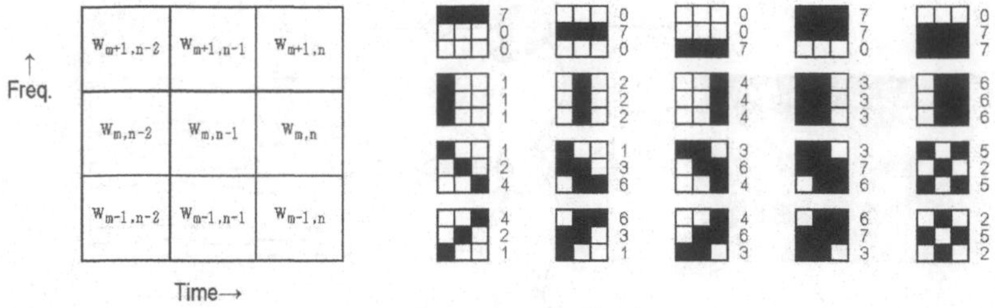

Fig. 8. Left: 3x3 weight mask. Right: Masks for feature detection on the formant spectrogram. Each mask is labelled with the corresponding octal code (most significant bits: bottom-right).

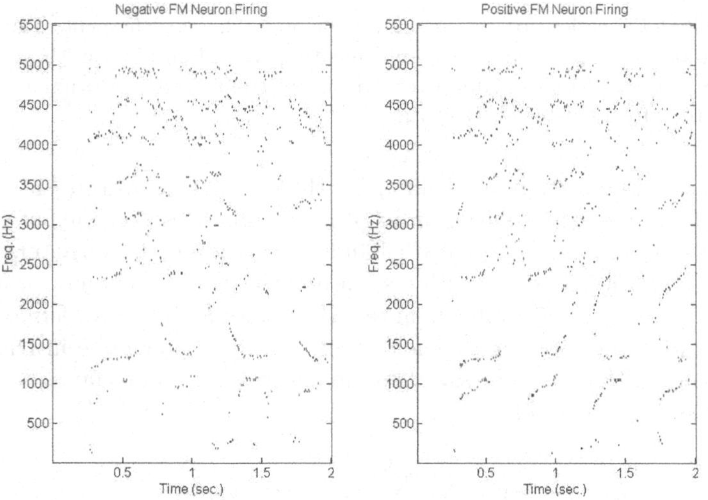

Fig. 9. Dynamic formant detection for the utterance /ba/, /da/, /dzha/, /ga/. Left: Firing of negative FM units. Right: Firing of positive FM units.

whereas the indexing in the frequency domain is established to ease operations as the detection of dynamic changes (found in FM neurons). The upper row of masks 700-077 is intended for formant detection, mimicking CF neurons. The next descending row 111-666 is intended for the detection of wide-band features, as noise bursts or *blips*, specific in fricative consonant detection. Masks 124-376 and 421-673 are designed to detect specific ascending or descending trajectories, as in positive-negative FM neurons. The results of scanning the CF units output for the uterance /ba/, /da/, /dzha/, /ga/ given in Fig.3 can be seen in Fig.9. The characteristic ascending and descending patterns for F_1-F_2 can be clearly appreciated.

4 Conclusions

Through the present work several aspects have to be concluded. First of all, it has been shown that formant-based speech processing may be carried out combining classical techniques as inverse filtering and bio-inspired techniques, as lateral inhibition. Good estimations of formant spectrograms may be of most interest for phonetic studies or forensic sciences, for instance. The use of adaptive filtering and the separation of the glottal influence render clear and neat formant spectrograms, showing the specific features considered as most relevant for the assignment of phonetic features and phonetic decoding in speech perception. This could open the possibility to designing new paradigms in formant-based speech recognition, as well as in speaker's identification and verification tasks. The experiments presented in dynamic formant detection using well-known bio-inspired mask processors (CF and FM units) show the power of these techniques, which offer also interesting properties under the computational point of view. The structures studied correspond roughly to the processing centres in the Olivar Nucleus and the Inferior Colliculus. The study of short-time memory-like structures found in the upper levels of the brain, and especially the columnar structures of the Auditory Cortex [11] using low order regressors, fundamental for phonemic parsing and the adaptive building-up of grammar structures [21] [18] deserve an extensive attention out of the scope of the present work and is left for future research.

Acknowledgements

This work is being funded by grants TIC2003-08756, TEC2006-12887-C02-00 from Plan Nacional de I+D+i, Ministry of Education and Science, CCG06-UPM/TIC-0028 from CAM/UPM, and by project HESPERIA (http.// www.proyecto-hesperia.org) from the Programme CENIT, Centro para el Desarrollo Tecnológico Industrial, Ministry of Industry, Spain.

References

1. Delattre, P., Liberman, A., Cooper, F. "Acoustic loci and transitional cues for consonants", J. Acoust. Soc. Am., Vol. 27, 1955, pp. 769–773.
2. Deller, J. R., Proakis, J. G., and Hansen, J. H. "Discrete-Time Processing of Speech Signals", Macmillan, NY, 1993.
3. Gómez, P., Godino, J. I., Alvarez, A., Martínez, R., Nieto, V., Rodellar, V. "Evidence of Glottal Source Spectral Features found in Vocal Fold Dynamics", Proc of the ICASSP'05, 2005, pp. 441–444.
4. Hermansky, H. "Should Recognizers Have Ears?", ESCA-NATO Tutorial and Research Workshop on Robust Speech Recognition for Unknown Communication Channels, Pont--Mousson, France, 17-18 April, 1997, pp. 1–10.
5. Ferrández, J. M. "Study and Realization of a Bio-inspired Hierarchical Architecture for Speech Recognition", Ph.D. Thesis (in Spanish), Universidad Politécnica de Madrid, 1998.

6. Gómez, P., Martínez, R., Rodellar, V., Ferrández J. M. "Bio-inspired Systems in Speech Perception: An overview and a study case", IEEE/NML Life Sciences Systems and Applications Workshop (by invitation), National Institute of Health, Bethesda, Maryland, July 13-14, 2006.
7. Haykin, S. "Neural Networks - A comprehensive Foundation", Prentice-Hall, Upper Saddle River, NJ, 1999.
8. Irino, T., and Patterson, R. D. "A time-domain, level-dependent auditory filter: the gammachirp", J. Acoust. Soc. Am., vol. 101, no. 1, January 1997, pp. 412–419.
9. Jahne, B. "Digital Image Processing", Springer, Berlin, 2005.
10. Mendelson J. R., Cynader, M. S. "Sensitivity of Cat Primary Auditory Cortex (AI) Neurons to the Direction and Rate of Frequency Modulation", Brain Research, 327, 1985, pp. 331–335.
11. Mountcastle, V. B. "The columnar organization of the neocortex", Brain 120, 1997, pp. 701–722.
12. Ojemann, G. A. "Organization of language cortex derived from investigation during neurosurgery", Sem Neuros, 2, 1990, pp. 297–305.
13. O'Shaughnessy, D. "Speech Communication", IEEE Press, Park Avenue, NY, 2000.
14. Rauschecker, J. P., Tian, B., Hauser, M. "Processing of Complex Sounds in the Macaque Nonprimary Auditory Cortex", Science, vol. 268, 7 April 1995, pp. 111–114.
15. .Sams, M., Salmening, R. "Evidence of sharp frequency tuning in human auditory cortex", Hearing Research, 75, 1994, pp. 67–74.
16. Schreiner, C.E. "Time Domain Analysis of Auditory-Nerve Fibers Firing Rates", Curr. Op. Neurobiol., 5, 1995, pp. 489–496.
17. Secker, H., and Searle, C. "Study and Realization of a Bio-inspired Hierarchical Architecture for Speech Recognition", J. Acoust. Soc. Am. 88 (3), 1990, pp. 1427–1436.
18. Sejnowski, T. J. and Rosenberg, C. R. "Parallel networks that learn to pronounce English text", Complex Systems 1, 1987, pp. 145–168.
19. Suga, N. "Cortical Computational Maps for Auditory Imaging", Neural Networks, 3, 1990, pp. 3–21.
20. Suga, N. "Basic Acoustic Patterns and Neural Mechanism Shared By Humans and Animals for Auditory Perception: A Neuroethologists view", Proceedings of Workshop on the Auditory bases of Speech Perception, ESCA, July 1996, pp. 31–38.
21. Waibel A. "Neural Network Approaches for Speech Recognition", Advances in Speech Signal Processing, S. Furui and M. M. Sondhi, Eds. Dekker, NY, 1992, pp. 555–597.

An Adaptable Multichannel Architecture for Cortical Stimulation

J.M. Ferrández[1,2], E. Liaño[1], M.P. Bonomini[1], and E. Fernández[1]

[1] Instituto de Bioingeniería, Universidad Miguel Hernández, Alicante
[2] Dpto. Electrónica, Tecnología de Computadoras, Univ. Politécnica de Cartagena
Biomedical Technologies S.L.
jm.ferrandez@upct.es

Abstract. An architecture for a cortical stimulator with visual neuroprosthetic purposes is presented. This device uses a 3D penetrating multielectrode array, which will be implanted in V1, offering different signal amplitude sets with the programable current source module. This electrode array has been proved for injecting current (charge) in a safety, secure and precise way during animal acute experimentation. The dynamic characteristic of the stimulator provide the possibility to adapt the current level to the different electrodes and tissue impedances. The architectureis based on a microprocessor circuit with programmable waveforms with a transistor based current injection stage. With the proposed system, a wide stimuli set can be used for obtaining the optimal parameters to use in a visual neuroprosthesis using as input a retinomorphic system. The histological results validate the stimulation and implantation procedures.

1 Introduction

Neural stimulation provides a new emerging approach to motor and sensory prostheses. Implantable stimulators such as pacemakers [1], cochlear implants [2], deep brain [3], and spinal cord devices [4]improve the quality of life of cardiac patients, deaf people, Parkinson disease affected, and decrease the pain in some diseases, so nowadays they are implanted regularly in most hospitals. However visual prosthesis are in a promising initial stage. Electrical stimulation of the visual cortex produces localized visual perceptions called phosphenes [5] [6]. The retinotopic organization of primary visual cortex would produce an ordered arrangement of phosphenes by stimulating this area through the stimulation of spatially distributed electrodes.

There exists a vast number of scientific areas which need to establish a synergic cooperation in order to give sight inside some of the tasks to perform: they include neural coding analysis, system modeling, hardware implementation, biocompability of materials, rehabiliation, etc. The neural stimulator must drive an array of electrodes implanted in neural tissue in a safe way, so balanced pulses are required, using a minimal size and low power consumption prosthesis.

In a stimulator circuit, the objective is to transfer energy from the device to the tissue in a precisely controlled process: the stimulator produces energy, which

J. Mira and J.R. Álvarez (Eds.): IWINAC 2007, Part I, LNCS 4527, pp. 143–152, 2007.

origins a voltage across the electrodes, passing current through the tissue, and the rest of the energy dissipates as heat in the system and should be minimize in order to improve the stimulation efficiency. So it is important to study the relation stimulator/electrodes defining the appropriate stimulation and configuration patterns.

The duration, frequency and either the voltage or the current amplitude must be controlled very accurately by the stimulator, however the tissue impedances are highly variable. Current injection provides more control over the injected charge and behaves similar to the natural process which elicits an action potential in excitable tissues. So, this approach has been widely used in cochlear or retinal implants.

In this paper, a multiplexed current source architecture based on transistors is presented. It drives a penetrating 3D multielectrode array (the Utah Multielectrode Array). It will be implanted in V1 , visual cortex (Fig 1) offering different signal amplitude sets with the programmable current source device. This specification provides the possibility to adapt the current level to different electrodes impedances and to a particular person characteristics. This is a crucial process due to the different phosphenes threshold (variability) required by the potential users, and because of an adaptation process to the current levels provided by the brain.

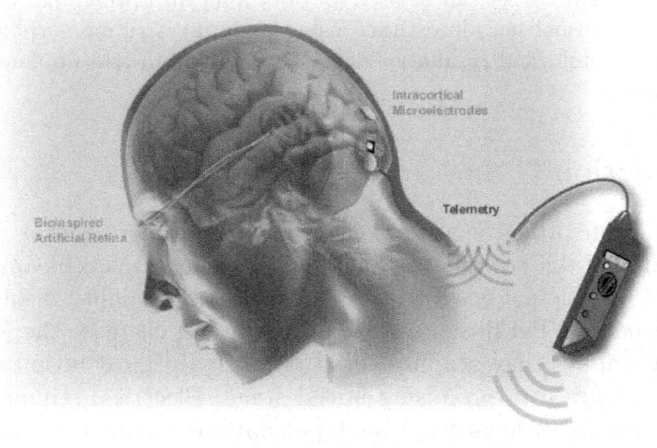

Fig. 1. The proposed visual prosthesis

2 The Electrode Array

Present design and manufacturing of intracortical electrodes may be considered as a real choice for neuroprosthetics devices intended to the recovery of motor and sensory abilities. Electrical stimulation of single or small populations of neurons in the central nervous system using penetrating microelectrodes arrays requires current pulsing for eliciting action potentials.

However, this approach must fulfil several conditions:

1. Alive tissue must accept the device *(biocompatibility)*
2. Electrical stimulation of nerve tissue must be induced by means of a charge displacement *(polarization)*, which is performed efficiently through current stimulation.
3. Current injection in neural tissue must be effective and safe inside a working zone delimited by electrolysis reactions nature.

The intracortical multielectrode array uses a three-dimensional architecture based on a silicon Pt/Ir ended needle array (Utah Electrode Array-UEA (Fig 2 left)) [7]. Its tip has been designed to be located in cortex layer IV, where afferents from he lateral geniculate nucleus (LGN) arrive (Fig 2 right). This electrode array has been used extensively in acute and chronic recording experiments. It has been implanted for years and it has the FDA autorization for its use in humans.

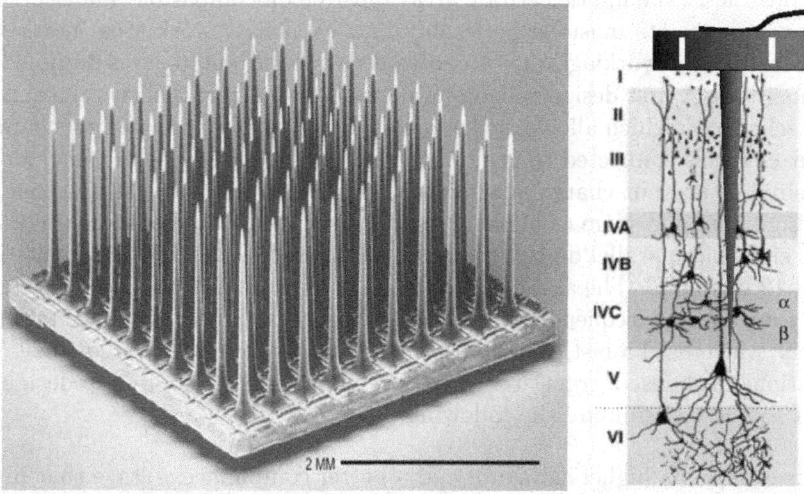

Fig. 2. Left: Scanning electron micrograph of the Utah Electrode Array. Right: Tip location in a cortical implanted electrode.

The multielectrode array is nearly the most relevant part in a neuroprosthesis because it must must provide enough charge to evoke action potentials, the tissue must accept the device with few reaction, so biocompatible coverings are required, it must communicate with the stimulator, and it has to be durable in order to be implanted for many years. The Utah multielectrode array verify this requirements.

3 The Transistor Based Current Source Stage

The analog current source circuit delivers stimulation to the neurons via the electrode array. It has been used current source because a precise charge delivery is a crucial parameter for eliciting an electrical stimulation of neuron populations. A voltage source cannot maintain constant current or constant charge delivery while tissue or electrode impedance is changing over time, and hence it will produce changes in the current, charge, delivered. Charge injection involve electron transferring across the electrode-tissue interface and therefore need that some chemical species be either oxidized or reduced. Metal electrodes must inject charge predominantly by faradic processes because the charge required to elicit a physiological response far exceeds that available from a capacitive mechanism. Faradic processes may be reversible or not. In reversible faradic processes the chemical reactions produce no new chemical species in the electrode boundaries, then the system remains equal with a charge-balanced current stimulus. However, irreversible faradic processes involve production of chemical species that do not remain bounded to the electrode surface, these reactions lead to electrode corrosion by electrolysis reactions. A precise current source control prevent an irreversible electrode dissolution. As a conclusion, for a durable implantation of such a prosthesis, the multielectrode array must be biocompatible and electrical-charge displacements must be preferably in a capacitive work area, never in a irreversible-faradic working zone. A constant current stimulator was designed using transistors. A first design was op-amp based [8]. Figure 3 shows the current source schematic, which allocates the analog circuitry. Two exhausting channels (output channels connected to ground through a low resistance) are included if it's desired to prevent charge accumulation on electrode before stimulating. In this design a unique stage to allocate digital and analog electronics is used, and former eight bits parallel digital to analog converter, DAC08, is substituted by a serial 12 bits DAC. The aim of this change is to compare the power consumption of the new high frequency device controlling a serial DAC, with the previous topology, which used a fast parallel TTL DAC with a lower clock frequency. The motivations to choose a transistor based design instead of an op-amp design for current source and sink are the following:

1. It is obtained a higher slew rate and a better compliance voltage that in the operational choices, and discarding expensive power operational amplifiers.
2. As the topology uses a totem-pole configuration without polarization, no power consumption is drawn when no stimulation is delivered (in this case both BJT are open).
3. This last point gives an extra security aspect to the topology, because when the input voltage is inside the band-gap BJT-base polarization voltages, no current is drained. So, unexpected voltage fluctuations on the inputs must overrun 0.8V to able current injection or sinking (a high security margin).
4. Better compliance towards power ratio, than in operational amplifier topologies.

It has been appreciated the quality of current pulses delivered in accuracy and rise and fall transition times. The current source converts an input analog voltage to a current output over the load connected to StChannel closed to ground. A power drain test was carried out wit this new topology and resulted that for

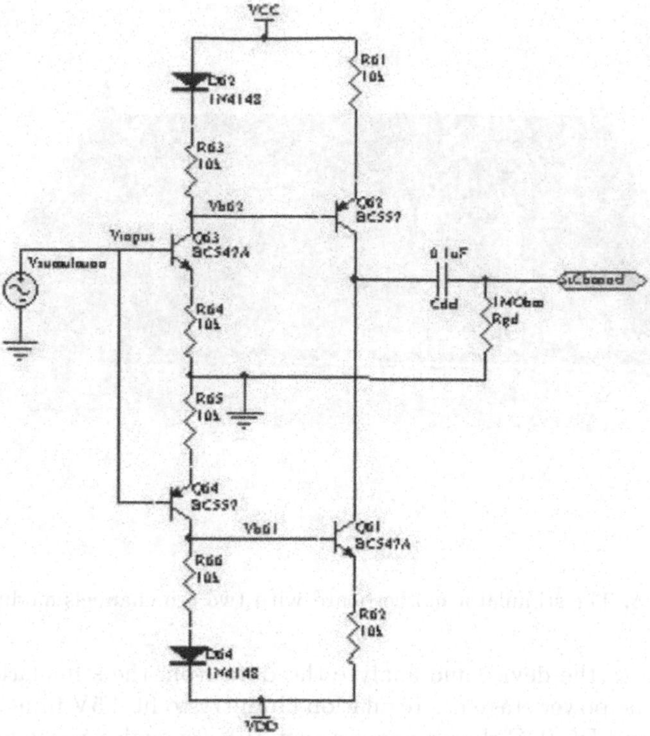

Fig. 3. The transistor-based current source circuit

a clock, fCLK=16MHz, the complete device draws +5V, 4.4mA (the op-amp based design consumed 20 mA using fCLCK=4MHz) . The power consumption saved by the commutation BJT-based current source and a serial CMOS DAC is high enough to able the device to drive such a high frequency 12 bits serial DAC instead a parallel one.

4 The Control Stage Submodule

The digital stage comprises the 100 channels motherboard microprocessor-based backplane, with the 10 channels digital submodules. Each submodule includes an Arizona Microchip PIC16F877, the power stage and an I2C read only memory for logging purposes. Figure 4 shows the layout of this motherboard with two submodules.

The microprocessors provides 8 kBytes Flash Program Memory, 368 Byte RAM Data Memory, 256 Byte EEPROM Data Memory, In-Circuit Serial Programming. (ICSP) via two pins (it is important to reprogram the chip on the board when using SMT technology), and a wide operating voltage range: 2.0V to 5.5V. It is oriented to a stand-alone operation, but includes communication circuitry to link with a host PC on USB port, in order to download the

Fig. 4. The stimulator motherboard with two ten channels modules

configuration to the device and analyze the data from the stimulation and the electrodes. The power stage has regulation circuitry to fit +5V from an external 9V battery, and DC/DC charge pump to get 10V (enough voltage for delimitation of maximum compliances). Also has an inverter to draw -5V to the analog stage if desired in future applications. The eeprom used is an I2C 64K CMOS memory for storing the relevant parameters of the stimulation, impedances, time, battery level. For each channel, the stimulator is capable of delivering 200 μA, 150 μA, 100 μA, 75 μA, 50 μA and 25 μA biphasic pulses (anodic-first or cathodic-first) at programmable duration and latencies. Each channel was capacitively-coupled to ensure the delivery of charge-balanced pulses. This control submodule is customizable from a Windows-based graphical user interface (GUI) (Fig. 5) where the amplitude, frequency and phase width for each electrode can be defined. Pulse width values from the system are 50, 100, 150, 200 ms, the frequencies are 25 Hz, 50 Hz, 75 Hz and 100 Hz, and as mentioned, the amplitudes cover 200 μA, 150 μA, 100 μA, 75 μA, 50 μA and 25 μA. Once defined the pulse waveform for each electrode and if it is active or not and sent to the device, the program can start or stop the stimulation. It also includes built-in safety features that consist in the ability to sense the failure of any of the output drivers or other modes of operation that could result in charge imbalance and tissue damage. The motherboard receive the configuration using a USB connection, and it downloads the stimulation parameters to each individual 10 channel submodule. Each submodule provides current to the electrodes using a round-robin scheme, so only one of the electrodes is stimulating at a time for each submodule. If 10 submodules are connected, the stimulation will consist in the sequential stimulation of 10 simultaneous electrodes.

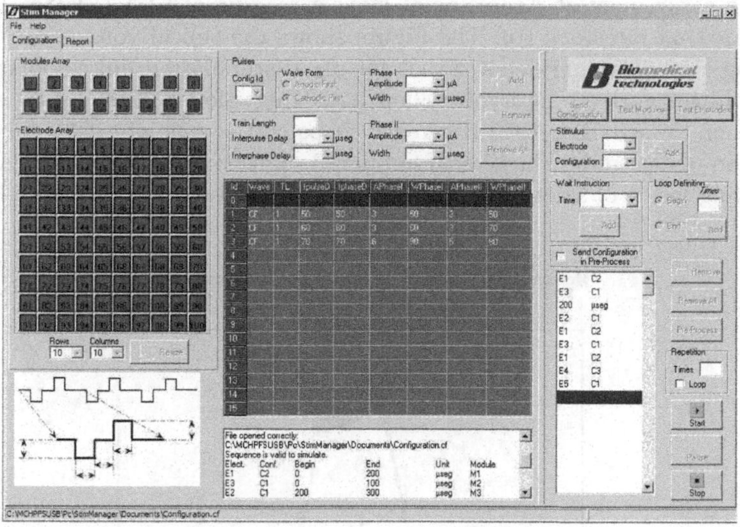

Fig. 5. The graphical user interface software

5 Results

In order to characterize the stimulation parameters, the multielectrode array was inmersed in saline solution, the array was dipped into a ringer bath, with an Ag/AgCl reference electrode closing the circuit. A shunt resistance (Rshunt) of 150 Ω was added to give information about the current trespassing the interface.

The current waveform used for stimulation is an Anodic-First (AF) bipolar current pulse delivered at 25 Hz. This signal allows the re-polarization of the neural media in the second phase, therefore, it is easy to monitor the evoked spikes, minimizing stimulus artifacts, and as it was described earlier, it is more adequate to stimulate by means of constant current stimulus because it is more reliable to control the charge injection through any electrode-tissue interface. Because charge injection becomes independent of tissue impedances, an increase in safety is achieved.

In the following Figure an AF current pulse is shown with its associated delivered voltage waveform observed in the medium. A predominant capacitive behavior is observed since voltage linearly increases with the onset of the current waveform and a consequent negative voltage slope appears while the cathodic pulse. For larger amounts of charge, diffusion components become significant, therefore, a more resistive behaviour is observed which compromises the electrode integrity by means of no-reversible REDOX processes.

Three stimulation signal points are selected to get voltage information about Access potential and Capacitive cycles at interface. They are referred in the Figure as V Access, V anodic and V cathodic. The voltage measurement at the end of current injection gives idea about the charge storage at interface, and the measurement at Access point gives a reading of ESA (Electrochemical surface area) at electrode tip. The Figure shows the tipical voltage response of a working electrode, and denotes that for usual stimulation pulses polarization access voltages work on the 1 volt region, achieving safety values. Total electrode polarization towards Ag/AgCl Reference keeps on the zone of +/-3V for a large population of electrodes (large charge pulses). This voltage compliance ensures the working inside of safety margins delimited by water electrolysis levels.

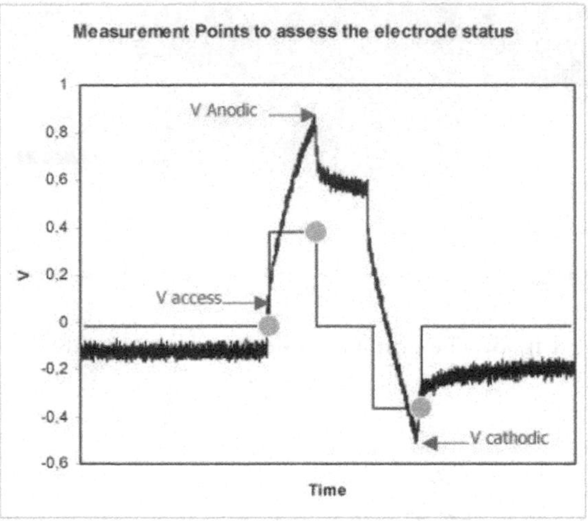

Fig. 6. Anodic First current source pulse and corresponding voltage waveform observed in the medium

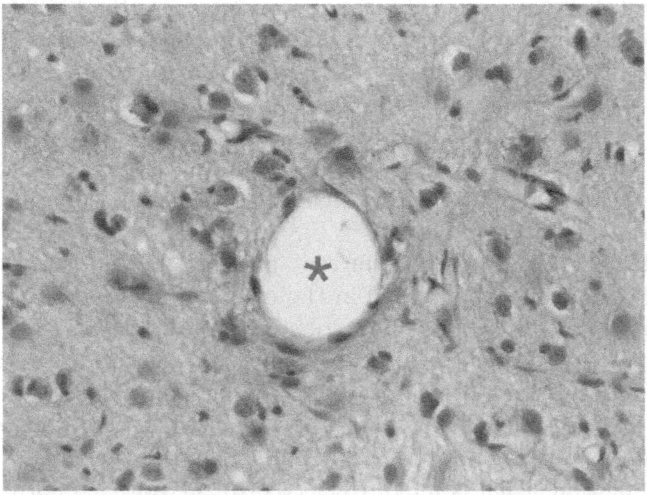

Fig. 7. Histological analysis of the acute response of rabbit cortical tissue to array implantation and stimulation

This device has also been implanted in rabbits and it is currently being used in animal experimentation using acute and chronic stimulation for obtaining histological data. In Figure 7 the histological analysis of the acute response of rabbit cortical tissue is shown. The multielectrode array was implanted and the stimulator provided an anodic first 30 microAmps biphasic pulse during 64 hours. The hole marked shows the tip of the electrode that was implanted using a pneumatic insertion device. No evidence neither microhemorraghes nor inflammatory reactions has been observed.

6 Conclusions

A cortical stimulation device has been developed for neuroprosthetic purposes. In order to achieve the requirements of constant current through variable impedances, a transistor-based current source circuit has been used which feeds a penetrating 3D multielectrode array. This electrode array has been proved for injecting current, charge, in a safety, secure and precise way. The control architecture consists in a microprocessor based motherboard customizable with programmable waveforms which can be configured using a PC connection (USB). It drives 10 channel submodules, including the current source circuitry and the measurement functionality operating in parallel, providing stimulation to 10 simultaneous electrodes. With the proposed system, a wide stimuli set can be used for testing and obtaining the optimal parameters to used in a visual neuroprosthesis using as input a retinomorphic response of camera images. The results concluded that the main advantage using a transistor based current source stimulator is a high slew-rate limited only by the DAC slew-rate, and a very low power consumption. It has been stimulating continuously during 240 h using a 9V battery. A noise voltage reduction has also been observed, this is a very important issue in little current delivering. The main disadvantage is less linearity for little currents, and higher clock frequency needed due to serial digital to analog coversion. Nowadays, a new module with RF link and new electrodes is being designed. The histological data obtained in animal acute experimentation provided the working parameters for a precise and safety stimulation.

Acknowledgements

This work was supported by the Spanish Government through grants NAN2004-09306-C05-4, SAF2005-08370-C02-01 and IMSERSO RETVIS 150/06 and by the European Comission through the project "'NEUROPROBES"' IST-027017.

References

1. Woollons, D.J. "To beat or not to beat: the history and development of heart pacemakers", IEEE Engineering Science and Education Journal, 4(6), 1995, pp. 259–269.
2. G. Loeb "Cochlear prosthetics", Annual Review in Neuroscience, vol. 13, 1990, pp. 357–371.

3. Medtronic “Deep Brain Stimulation”, http://www.medtronic.com/
4. North RB, Kidd DH, Olin JC, Sieracki JM. “Spinal cord stimulation electrode design: prospective, randomized, controlled trial comparing percutaneous and laminectomy electrodes-part I: technical outcomes”, Neurosurgery. 2002 Aug, 51(2), pp. 381–9.
5. Schmidt E.M., Bak M.J. et al “Feasibility of a Visual Prosthesis for the Blind Based on Intracortical Stimulation of the Visual Cortex”, Brain 119, 1996, pp. 507–522.
6. Normann R., Maynard E., Guillory K., Warren D “Cortical Implants for the Blind”, IEEE Spectrum, May, 1996, pp. 54–59.
7. Campbell P., Jones K., Huber R., Horch K. Normann R “A Silicon-Based, Three-Dimensional Neural Interface:Manufacturing Processes for an Intracortical Electrode Array”, IEEE Transactions on Biomedical Engineering, Vol.38 N 8 , pp. 758–767.
8. Ferrández J.M., Bonomini P., Fernández E. “A Multiplexed Current Source Portable Stimulator Architecture for a Visual Cortical Neuroprosthesis”, Lect Not Comp Sci 3561, Springer Verlag, 2005, pp. 310–318.

Spiking Neural P Systems. Power and Efficiency

Gheorghe Păun

Institute of Mathematics of the Romanian Academy
PO Box 1-764, 014700 Bucureşti, Romania
george.paun@imar.ro, gpaun@us.es

Abstract. This is a brief survey of spiking neural P systems, a branch of membrane computing recently introduced with motivation from neural computing. Basic ideas, examples, some results, and several research topics are presented.

1 The General Framework

Membrane computing is a branch of natural computing which abstracts computing models from the architecture and the functioning of the living cells and from cells cooperation in tissues, organs or other structures. The obtained models, called P systems, were shown to be Turing equivalent and able to efficiently solve computationally hard problems, and also very useful in devising models for biological phenomena and for applications in economics, linguistics, computer science, optimization. As basic sources of information, the reader is referred to the monograph [23], the volume [11], as well as to the web page of membrane computing, from `http://psystems.disco.unimib.it`.

A constant interest of membrane computing was to also capture features of the way the neurons "compute", alone or organized in neural nets, and a major contribution in this respect was the introduction of so-called *spiking neural P systems*, [18], much investigated in the last years.

The most intuitive way to introduce spiking neural P systems (in short, SN P systems) is by watching the movie available at `http://www.igi.tugraz.at/tnatschl/spike_trains_eng.html`, in the web page of Wofgang Maass, Graz, Austria: neurons are sending to each others *spikes*, electrical impulses of identical shape (duration, voltage, etc.), with the information "encoded" in the frequency of these impulses, hence in the time passed between consecutive spikes. For neurologists, this is nothing new, related drawings already appear in papers by Ramón y Cajal, a pioneer of neuroscience at the beginning of the last century, but in the recent years "computing by spiking" became a vivid research area, with the hope to lead to a neural computing "of the third generation" – see [12], [21], etc.

For membrane computing it is rather natural to incorporate the idea of spiking neurons (already neural-like P systems exist, based on different ingredients – see [23], efforts to compute with a small number of objects were recently made in

J. Mira and J.R. Álvarez (Eds.): IWINAC 2007, Part I, LNCS 4527, pp. 153–169, 2007.

several papers – see, e.g., [2], using the time as a support of information, for instance, taking the time between two events as the result of a computation, was also considered – see [3]), but still important differences exist between the general way of working with multisets of objects in the compartments of a cell-like membrane structure – as in membrane computing – and the way the neurons communicate by spikes. A way to answer this challenge was proposed in [18]: neurons as single membranes, placed in the nodes of a graph corresponding to synapses, only one type of objects present in neurons, the spikes, with specific rules for handling them, and with the distance in time between consecutive spikes playing an important role (e.g., the result of a computation being defined either as the whole spike train of a distinguished output neuron, or as the distance between consecutive spikes). Details will be given immediately.

What is obtained is a computing device whose behavior resembles the process from the neuron nets and which is meant to generate strings or infinite sequences (like in formal language theory), to recognize or translate strings or infinite sequences (like in automata theory), to generate or accept natural numbers, or to compute number functions (like in membrane computing). Results of all these types will be mentioned below. Nothing is said here, because nothing was done so far, about using such devices in "standard" neural computing applications, such as pattern recognition. Several open problems and research topics will be mentioned below (a long list of such topics, prepared for the Fifth Brainstorming Week on Membrane Computing, Sevilla, January 29-February 2, 2007, can be found in [24]).

It is worth mentioning here that "general" membrane computing is now an area of intense research related to applications, mainly in biology/medicine, but also in economics, distributed evolutionary computing, computer graphics, etc. (see [11], [27], [28]), but this happens after a couple of years of research of a classic language-automata-complexity type; maybe this will be the case also for the spiking neural P systems, which need further theoretical investigation before passing to applications.

2 An Informal Overview – With an Example

Very shortly, an SN P system consists of a set of *neurons* (cells, consisting of only one membrane) placed in the nodes of a directed graph and sending signals (*spikes*, denoted in what follows by the symbol a) along *synapses* (arcs of the graph). Thus, the architecture is that of a tissue-like P system, with only one kind of objects present in the cells. The objects evolve by means of *spiking rules*, which are of the form $E/a^c \rightarrow a; d$, where E is a regular expression over $\{a\}$ and c, d are natural numbers, $c \geq 1, d \geq 0$. The meaning is that a neuron containing k spikes such that $a^k \in L(E), k \geq c$, can consume c spikes and produce one spike, after a delay of d steps. This spike is sent to all neurons to which a synapse exists outgoing from the neuron where the rule was applied. There also are *forgetting rules*, of the form $a^s \rightarrow \lambda$, with the meaning that $s \geq 1$ spikes are removed, provided that the neuron contains exactly s spikes. We say that the rules "cover" the neuron, all spikes are taken into consideration when using a rule.

The system works in a synchronized manner, i.e., in each time unit, each neuron which can use a rule should do it, but the work of the system is sequential in each neuron: only (at most) one rule is used in each neuron. One of the neurons is considered to be the *output neuron*, and its spikes are also sent to the environment. The moments of time when a spike is emitted by the output neuron are marked with 1, the other moments are marked with 0. The binary sequence obtained in this way is called the *spike train* of the system – it might be infinite if the computation does not stop.

Figure 1 recalls an **example** from [18], and this also introduces the standard way to represent an SN P system (note that the output neuron, σ_7 in this case, is indicated by an arrow pointing to the environment), and a simplification in writing the spiking rules: if we have a rule $E/a^c \to a; d$ with $L(E) = \{a^c\}$, then we write simply $a^c \to a; d$. If all rules are of this form, then the system is called *bounded* (or *finite*), because it can handle only finite numbers of spikes in the neurons (this is the case with the system given in Figure 1).

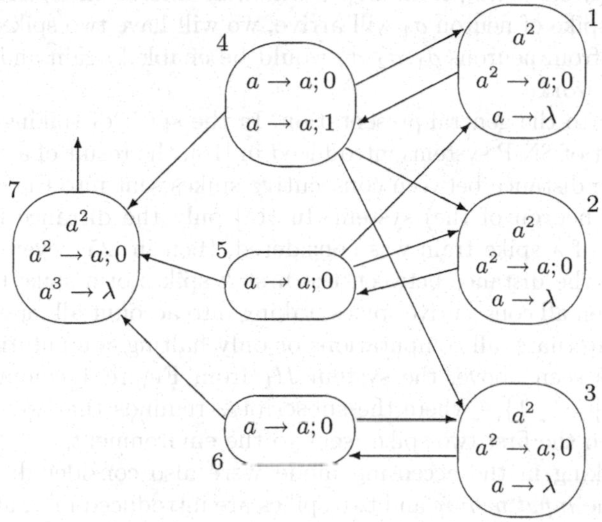

Fig. 1. An SN P system generating all even natural numbers

In the beginning, only neurons $\sigma_1, \sigma_2, \sigma_3$, and σ_7 contain spikes, hence they fire in the first step – and spike immediately. In particular, the output neuron spikes, hence a spike is also sent to the environment. Note that in the first step we cannot use the forgetting rule $a \to \lambda$ in $\sigma_1, \sigma_2, \sigma_3$, because we have more than one spike present in each neuron.

The spikes of neurons $\sigma_1, \sigma_2, \sigma_3$ will pass to neurons $\sigma_4, \sigma_5, \sigma_6$. In step 2, $\sigma_1, \sigma_2, \sigma_3$ contain no spike inside, hence will not fire, but $\sigma_4, \sigma_5, \sigma_6$ fire. Neurons σ_5, σ_6 have only one rule, but neuron σ_4 behaves non-deterministically, choosing between the rules $a \to a; 0$ and $a \to a; 1$. Assume that for $m \geq 0$ steps we use here the first rule. This means that three spikes are sent to neuron σ_7, while each of neurons $\sigma_1, \sigma_2, \sigma_3$ receives two spikes. In step 3, neurons $\sigma_4, \sigma_5, \sigma_6$ cannot fire, but all $\sigma_1, \sigma_2, \sigma_3$ fire again. After receiving the three spikes, neuron σ_7 uses its forgetting rule and gets empty again. These steps can be repeated arbitrarily many times.

In order to have neuron σ_7 firing again, we have to use sometimes the rule $a \to a; 1$ of neuron σ_4. Assume that this happens in step t (it is easy to see that $t = 2m + 2$, for some $m \geq 2$ as above). This means that at step t only neurons σ_5, σ_6 emit their spikes. Each of neurons $\sigma_1, \sigma_2, \sigma_3$ receives only one spike – and forgets it in the next step, $t + 1$. Neuron σ_7 receives two spikes, and fires again, thus sending the second spike to the environment. This happens in moment $t + 1 = 2m + 2 + 1$, hence between the first and the second spike sent outside have elapsed $2m + 2$ steps, for some $m \geq 0$. The spike of neuron σ_4 (the one "prepared-but-not-yet-emitted" by using the rule $a \to a; 1$ in step t) will reach neurons $\sigma_1, \sigma_2, \sigma_3$, and σ_7 in step $t + 1$, hence it can be used only in step $t + 2$; in step $t + 2$ neurons $\sigma_1, \sigma_2, \sigma_3$ forget their spikes and the computation halts. The spike from neuron σ_7 remains unused, there is no rule for it. Note the effect of the forgetting rules $a \to \lambda$ from neurons $\sigma_1, \sigma_2, \sigma_3$: without such rules, the spikes of neurons σ_5, σ_6 from step t will wait unused in neurons $\sigma_1, \sigma_2, \sigma_3$ and, when the spike of neuron σ_4 will arrive, we will have two spikes, hence the rules $a^2 \to a; 0$ from neurons $\sigma_1, \sigma_2, \sigma_3$ would be enabled again and the system will continue to work.

Let us return to the general presentation. In the spirit of spiking neurons, in the basic variant of SN P systems introduced in [18], the result of a computation is defined as the distance between consecutive spikes sent into the environment by the (output neuron of the) system. In [18] only the distance between the first two spikes of a spike train was considered, then in [25] several extensions were examined: the distance between the first k spikes of a spike train, or the distances between all consecutive spikes, taking into account all intervals or only intervals that alternate, all computations or only halting computations, etc.

Therefore, as seen above, the system Π_1 from Figure 1 computes the set $N_2(\Pi_1) = \{2n \mid n \geq 1\}$ – where the subscript 2 reminds that we consider the distance between the first two spikes sent to the environment.

Systems working in the accepting mode were also considered: a neuron is designated as the *input neuron* and two spikes are introduced in it, at an interval of n steps; the number n is accepted if the computation halts.

Two main types of results were obtained: computational completeness in the case when no bound was imposed on the number of spikes present in the system, and a characterization of semilinear sets of numbers in the case when a bound was imposed (hence for finite SN P systems).

Another attractive possibility is to consider the spike trains themselves as the result of a computation, and then we obtain a (binary) language generating device. We can also consider input neurons and then an SN P system can work as a transducer. Such possibilities were investigated in [26]. Languages – even on arbitrary alphabets – can be obtained also in other ways: following the path of a designated spike across neurons, or using *extended* rules, i.e., rules of the form $E/a^c \to a^p; d$, where all components are as above and $p \geq 1$; the meaning is that p spikes are produced when applying this rule. In this case, with a step when the system sends out i spikes, we associate a symbol b_i, and thus we get a language over an alphabet with as many symbols as the number of spikes simultaneously produced. This case was investigated in [9].

The proofs of all computational completeness results known up to now in this area are based on simulating register machines. Starting the proofs from small universal register machines, as those produced in [20], one can find small universal SN P systems. This idea was explored in [22] – the results are recalled in Theorem 4.

In the initial definition of SN P systems several ingredients are used (delay, forgetting rules), some of them of an unrestricted form (general synapse graph, general regular expressions). As shown in [15], rather restrictive normal forms can be found, in the sense that some ingredients can be removed or simplified without losing the computational completeness. For instance, the forgetting rules or the delay can be removed, both the indegree and the outdegree of the synapse graph can be bounded by 2, while the regular expressions from firing rules can be of very simple forms.

There were investigated several other types of SN P systems: with several output neurons ([16], [17]), with a non-synchronous use of rules ([4]), with an exhaustive use of rules (whenever enabled, a rule is used as much as possible for the number of spikes present in the neuron, [19]), with packages of spikes sent along specified synapse links ([1]), etc. We refer the reader to the bibliography of this note, with many papers being available at [27].

3 A Formal Definition

We introduce the SN P systems in a general form, namely, in the extended (i.e., with the rules able to produce more than one spike) computing (i.e., able to take an input and provide an output) version.

A computing extended *spiking neural P system*, of degree $m \geq 1$, is a construct of the form

$$\Pi = (O, \sigma_1, \ldots, \sigma_m, syn, in, out), \text{ where:}$$

1. $O = \{a\}$ is the singleton alphabet (a is called *spike*);
2. $\sigma_1, \ldots, \sigma_m$ are *neurons*, of the form $\sigma_i = (n_i, R_i), 1 \leq i \leq m$, where:
   a) $n_i \geq 0$ is the *initial number of spikes* contained in σ_i;
   b) R_i is a finite set of *rules* of the following two forms:
      (1) $E/a^c \to a^p; d$, where E is a regular expression over a and $c \geq p \geq 1$, $d \geq 0$;
      (2) $a^s \to \lambda$, for $s \geq 1$, with the restriction that for each rule $E/a^c \to a^p; d$ of type (1) from R_i, we have $a^s \notin L(E)$;
3. $syn \subseteq \{1, 2, \ldots, m\} \times \{1, 2, \ldots, m\}$ with $i \neq j$ for all $(i, j) \in syn$, $1 \leq i, j \leq m$ (*synapses* between neurons);
4. $in, out \in \{1, 2, \ldots, m\}$ indicate the *input* and the *output* neurons, respectively.

The rules of type (1) are *firing* (we also say *spiking*) *rules*, those of type (2) are called *forgetting* rules. An SN P system whose firing rules have $p = 1$ (they produce only one spike) is said to be of the *standard* type (non-extended).

The firing rules are applied as follows. If the neuron σ_i contains k spikes, and $a^k \in L(E), k \geq c$, then the rule $E/a^c \rightarrow a^p; d \in R_i$ can be applied. This means consuming (removing) c spikes (thus only $k - c$ spikes remain in σ_i; this corresponds to the right derivative operation $L(E)/a^c$), the neuron is fired, and it produces p spikes after d time units (a global clock is assumed, marking the time for the whole system, hence the functioning of the system is synchronized). If $d = 0$, then the spikes are emitted immediately, if $d = 1$, then the spikes are emitted in the next step, etc. If the rule is used in step t and $d \geq 1$, then in steps $t, t+1, t+2, \ldots, t+d-1$ the neuron is *closed* (this corresponds to the refractory period from neurobiology), so that it cannot receive new spikes (if a neuron has a synapse to a closed neuron and tries to send a spike along it, then that particular spike is lost). In the step $t + d$, the neuron spikes and becomes again open, so that it can receive spikes (which can be used starting with the step $t+d+1$, when the neuron can again apply rules). Once emitted from neuron σ_i, the spikes reach immediately all neurons σ_j such that $(i, j) \in syn$ and which are open, that is, the p spikes are replicated and each target neuron receives p spikes; spikes sent to a closed neuron are "lost".

The forgetting rules are applied as follows: if the neuron σ_i contains exactly s spikes, then the rule $a^s \rightarrow \lambda$ from R_i can be used, meaning that all s spikes are removed from σ_i.

If a rule $E/a^c \rightarrow a^p; d$ of type (1) has $E = a^c$, then we write it in the simplified form $a^c \rightarrow a^p; d$.

In each time unit, if a neuron σ_i can use one of its rules, then a rule from R_i *must* be used. Since two firing rules, $E_1/a^{c_1} \rightarrow a^{p_1}; d_1$ and $E_2/a^{c_2} \rightarrow a^{p_2}; d_2$, can have $L(E_1) \cap L(E_2) \neq \emptyset$, it is possible that two or more rules can be applied in a neuron, and in that case, only one of them is chosen non-deterministically. Note however that, by definition, if a firing rule is applicable, then no forgetting rule is applicable, and vice versa.

Thus, the rules are used in the sequential manner in each neuron, at most one in each step, but neurons function in parallel with each other. It is important to notice that the applicability of a rule is established based on the *total* number of spikes contained in the neuron.

The initial configuration of the system is described by the numbers $n_1, n_2, \ldots, n_m$, of spikes present in each neuron, with all neurons being open. During the computation, a configuration is described by both the number of spikes present in each neuron and by the state of the neuron, more precisely, by the number of steps to count down until it becomes open again (this number is zero if the neuron is already open). Thus, $\langle r_1/t_1, \ldots, r_m/t_m \rangle$ is the configuration where neuron σ_i contains $r_i \geq 0$ spikes and it will be open after $t_i \geq 0$ steps, $i = 1, 2, \ldots, m$; with this notation, the initial configuration is $C_0 = \langle n_1/0, \ldots, n_m/0 \rangle$.

A computation in a system as above starts in the initial configuration. In order to compute a function $f : \mathbf{N}^k \longrightarrow \mathbf{N}$, we introduce k natural numbers $n_1, \ldots, n_k$ in the system by "reading" from the environment a binary sequence $z = 10^{n_1-1}10^{n_2-1}1 \ldots 10^{n_k-1}1$. This means that the input neuron of Π receives a spike in each step corresponding to a digit 1 from the string z and no spike otherwise. Note that we input exactly $k+1$ spikes, i.e., after the last spike we assume that no further spike is coming to the input neuron. (Another possibility is to consider k input neurons and to introduce each $n_i, 1 \leq i \leq k$, as the distance between two spikes which enter the ith input neuron.) The result of the computation is also encoded in the distance between two spikes: we impose the restriction that the system outputs exactly two spikes and halts (sometimes after the second spike), hence it produces a train spike of the form $0^{b_1}10^{r-1}1b^{b_2}$, for some $b_1, b_2 \geq 0$ and with $r = f(n_1, \ldots, n_k)$ (the system outputs no spike a non-specified number of steps from the beginning of the computation until the first spike).

The previous definition covers many types of systems/behaviors. If the neuron σ_{in} is not specified, then we have a generative system: we start from the initial configuration and we collect all results of computations, which can be the distance between the first two spikes (as in [18]), the distance between all consecutive spikes, between alternate spikes, etc. (as in [25]), or it can be the spike train itself, either taking only finite computations, hence generating finite strings (as in [5], [9], etc.), or also hon-halting computations (as in [26]). Similarly, we can ignore the output neuron and use an SN P system in the accepting mode: a number introduced in the system as the distance between two spikes entering the input neuron is accepted if and only if the computation halts. In the same way we can accept input binary strings or strings over arbitrary alphabets. In the second case, a symbol b_i is taken from the environment by introducing i spikes in the input neuron.

4 Three Examples

Not all types of SN P systems mentioned above will be discussed below, and only some of them are illustrated in this section.

The first example, borrowed from [5], is presented in Figure 2, and it is meant to generate binary strings.

Its evolution can be analyzed on a transition diagram as that from Figure 3, which is a very useful tool for studying systems with a bounded number of spikes present in their neurons: because the number of configurations reachable from the initial configuration is finite, we can place them in the nodes of a graph, and between two nodes/configurations we draw an arrow if and only if a direct transition is possible between them. In Figure 3, also the rules used in each neuron are indicated, with the following conventions: for each r_{ij} we have written only the subscript ij, with **31** being written in boldface, in order to indicate that a spike is sent out of the system at that step; when a neuron $\sigma_i, i = 1, 2, 3$, uses no rule, we have written $i0$, and when it spikes (after being closed for one step), we write is.

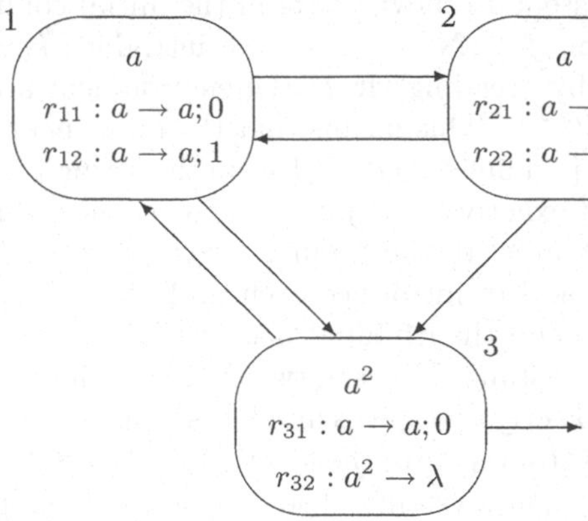

Fig. 2. The initial configuration of system Π_2

We do not enter into details concerning the paths in this diagram. Anyway, the transition diagram of a finite SN P system can be interpreted as the representation of a non-deterministic finite automaton, with C_0 being the initial state, the halting configurations being final states, and each arrow being marked with 0 if in that transition the output neuron does not send a spike out, and with 1 if in the respective transition the output neuron spikes; in this way, we can identify the language generated by the system. In the case of the finite SN P system Π_2, the generated language is $L(\Pi_2) = (0^*0(11 \cup 111)^*110)^*0^*(011 \cup 0(11 \cup 111)^+(0 \cup 00)1)$.

The next example, given in Figure 4, is actually of a more general interest, as it is a part of a larger SN P system which simulates a register machine. The figure presents the module which simulates a SUB instruction; moreover, it does it without using forgetting rules (the construction is part of the proof that forgetting rules can be avoided – see [15]).

The idea of simulating a register machine $M = (n, H, l_0, l_h, R)$ (number of registers, set of labels, initial label, halt label, set of instructions) by an SN P system Π is to associate a neuron σ_r with each register r and a neuron σ_l with each label l from H (there also are other neurons – see the figure), and to represent the fact that register r contains the number k by having $2k$ spikes in neuron σ_r. Initially, all neurons are empty, except neuron σ_{l_0}, which contains one spike. During the computation, the simulation of an instruction $l_i : (\mathtt{OPP}(r), l_j, l_k)$ starts by introducing one spike in the corresponding neuron σ_{l_i}, and this triggers the module associated with this instruction.

For instance, in the case of a subtraction instruction $l_i : (\mathtt{SUB}(r), l_j, l_k)$, the module is initiated when a spike enters the neuron σ_{l_i}. This spike causes neuron σ_{l_i} to immediately send a spike to the neurons $\sigma_{l_{i1}}, \sigma_{l_{i2}}$, and σ_r. If register r is not empty, then the rule $a(aaa)^+/a^3 \to a; 0$ will be applied and the spike emitted will cause neurons $\sigma_{l_{i3}}, \sigma_{l_{i5}}$, and finally neuron σ_{l_j} to spike. (In this process, neuron $\sigma_{l_{i4}}$ has two spikes added during one step and it cannot spike.) If register r is empty, hence neuron σ_r contains only the spike received from σ_{l_i},

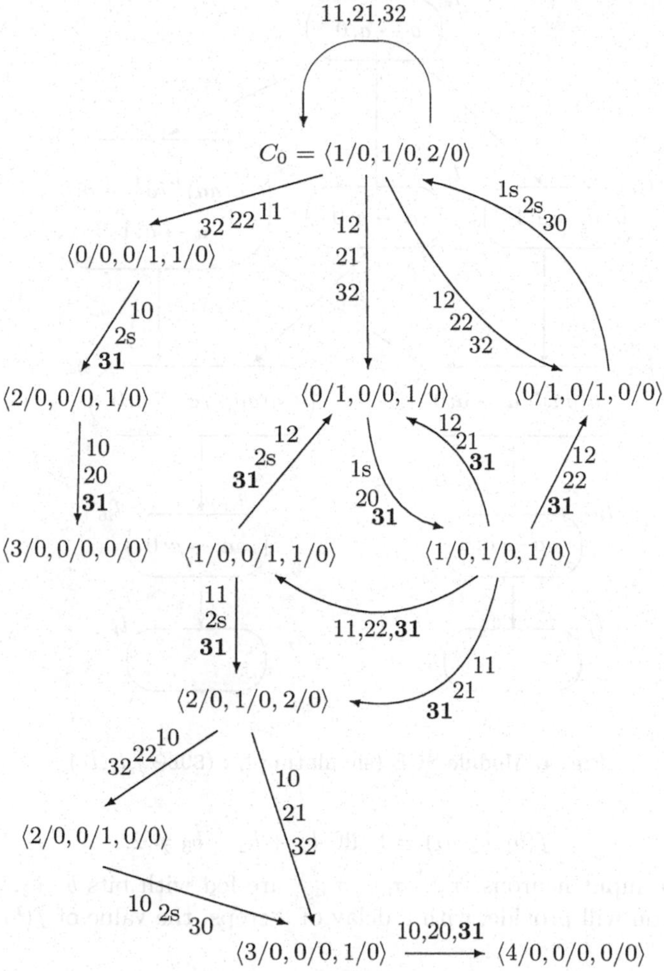

Fig. 3. The transition diagram of system Π_2

then the rule $a \rightarrow a; 1$ is applied and the subsequent spikes will cause neurons $\sigma_{l_{i4}}, \sigma_{l_{i6}}$, and finally neuron σ_{l_k} to spike. (In this process, neuron $\sigma_{l_{i3}}$ has two spikes added during one step and does not spike.) After the computation of the entire module is complete, each neuron is left with either zero spikes or an even number of spikes, allowing the module to be run again in a correct way.

The third example deals with an SN P system used as a transducer, and it illustrates the following result from [26]: *Any function* $f : \{0,1\}^k \longrightarrow \{0,1\}$ *can be computed by an SN P system with k input neurons (also using further* $2^k + 4$ *neurons, one being the output one).*

The idea of the proof of this result is suggested in Figure 5, where a system is presented which computes the function $f : \{0,1\}^3 \longrightarrow \{0,1\}$ defined by

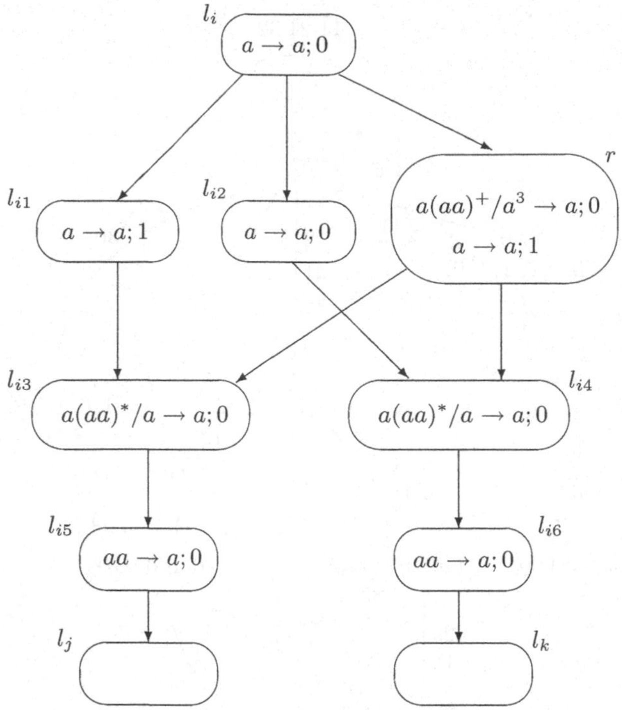

Fig. 4. Module SUB (simulating l_i : (SUB$(r), l_j, l_k$)

$$f(b_1, b_2, b_3) = 1 \text{ iff } b_1 + b_2 + b_3 \neq 2.$$

The three input neurons, $\sigma_{in_1}, \sigma_{in_2}, \sigma_{in_3}$, are fed with bits b_1, b_2, b_3, and the output neuron will provide, with a delay of 3 steps, the value of $f(b_1, b_2, b_3)$.

5 Some Results

There are several parameters describing the complexity of an SN P system: number of neurons, number of rules, number of spikes consumed or forgotten by a rule, etc. Here we consider only some of them and we denote by $N_2SNP_m(rule_k, cons_p, forg_q)$ the family of all sets $N_2(\Pi)$ computed as specified in Section 3 by SN P systems with at most $m \geq 1$ neurons, using at most $k \geq 1$ rules in each neuron, with all spiking rules $E/a^c \to a; t$ having $c \leq p$, and all forgetting rules $a^s \to \lambda$ having $s \leq q$. When any of the parameters m, k, p, q is not bounded, it is replaced with $*$. When we work only with SN P systems whose neurons contain at most s spikes at any step of a computation (*finite* systems), then we add the parameter $bound_s$ after $forg_q$. (Corresponding families are defined for other definitions of the result of a computation, as well as for the accepting case, but the results are quite similar, hence we do not give details here.)

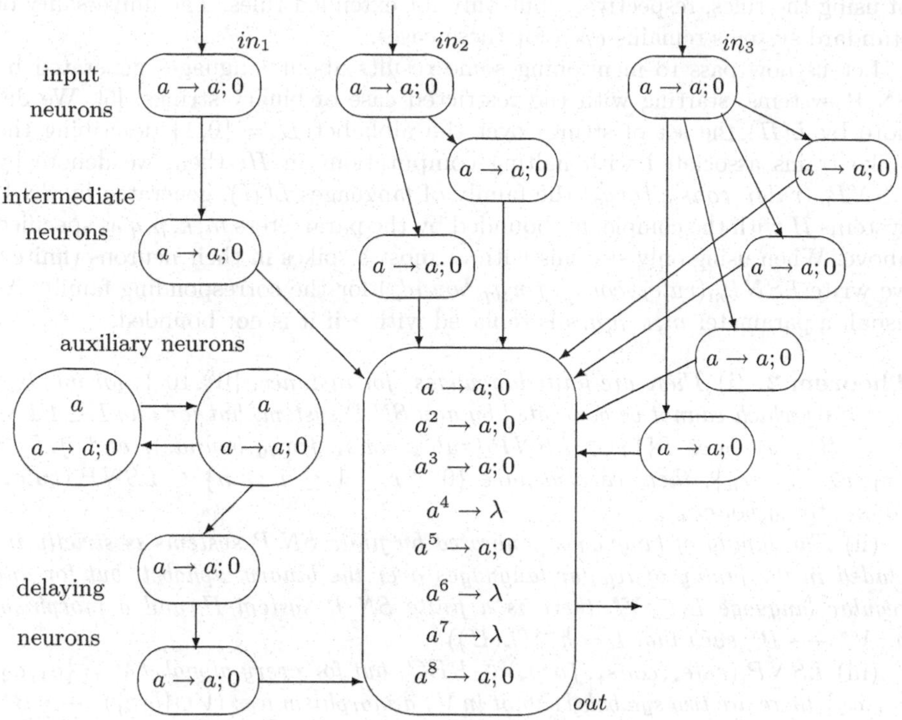

Fig. 5. Computing a Boolean function of three variables

By $NFIN, NREG, NRE$ we denote the families of finite, semilinear, and Turing computable sets of (positive) natural numbers (number 0 is ignored); they correspond to the length sets of finite, regular, and recursively enumerable languages, whose families are denoted by FIN, REG, RE. We also invoke below the family of recursive languages, REC (those languages with a decidable membership).

The following results were proved in [18] and extended in [25] to other ways of defining the result of a computation.

Theorem 1. (i) $NFIN = N_2SNP_1(rule_*, cons_1, forg_0) = N_2SNP_2(rule_*, cons_*, forg_*)$.

(ii) $NRE = N_2SNP_*(rule_k, cons_p, forg_q)$ *for all* $k \geq 2, p \geq 3, q \geq 3$.

(iii) $NSLIN = N_2SNP_*(rule_k, cons_p, forg_q, bound_s)$, *for all* $k \geq 3$, $q \geq 3$, $p \geq 3$, *and* $s \geq 3$.

Point (ii) was proved in [18] also for the accepting case, and then the systems used can be required to be deterministic (at most one rule can be applied in each neuron in each step of the computation). In turn, universality results were proved in [19] and [4] also for the exhaustive and for the non-synchronized modes of using the rules, respectively, but only for extended rules. The universality of standard systems remains *open* for these cases.

Let us now pass to mentioning some results about languages generated by SN P systems, starting with the restricted case of binary strings, [5]. We denote by $L(\Pi)$ the set of strings over the alphabet $B = \{0, 1\}$ describing the spike trains associated with halting computations in Π; then, we denote by $LSNP_m(rule_k, cons_p, forg_q)$ the family of languages $L(\Pi)$, generated by SN P systems Π with the complexity bounded by the parameters m, k, p, q as specified above. When using only systems with at most s spikes in their neurons (finite), we write $LSNP_m(rule_k, cons_p, forg_q, bound_s)$ for the corresponding family. As usual, a parameter m, k, p, q, s is replaced with $*$ if it is not bounded.

Theorem 2. (i) *There are finite languages (for instance, $\{0^k, 10^j\}$, for any $k \geq 1$, $j \geq 0$) which cannot be generated by any SN P system, but for any $L \in FIN$, $L \subseteq B^+$, we have $L\{1\} \in LSNP_1(rule_*, cons_*, forg_0, bound_*)$, and if $L = \{x_1, x_2, \ldots, x_n\}$, then we also have $\{0^{i+3}x_i \mid 1 \leq i \leq n\} \in LSNP_*(rule_*, cons_1, forg_0, bound_*)$.*

(ii) *The family of languages generated by finite SN P systems is strictly included in the family of regular languages over the binary alphabet, but for any regular language $L \subseteq V^*$ there is a finite SN P system Π and a morphism $h : V^* \longrightarrow B^*$ such that $L = h^{-1}(L(\Pi))$.*

(iii) *$LSNP_*(rule_*, cons_*, forg_*) \subset REC$, but for every alphabet $V = \{a_1, a_2, \ldots, a_k\}$ there are two symbols b, c not in V, a morphism $h_1 : (V \cup \{b, c\})^* \longrightarrow B^*$, and a projection $h_2 : (V \cup \{b, c\})^* \longrightarrow V^*$ such that for each language $L \subseteq V^*$, $L \in RE$, there is an SN P system Π such that $L = h_2(h_1^{-1}(L(\Pi)))$.*

These results show that the language generating power of SN P systems is rather eccentric; on the one hand, finite languages (like $\{0, 1\}$) cannot be generated, on the other hand, we can represent any RE language as the direct morphic image of an inverse morphic image of a language generated in this way. This eccentricity is due mainly to the restricted way of generating strings, with one symbol added in each computation step. This restriction does not appear in the case of extended spiking rules. In this case, a language can be generated by associating the symbol b_i with a step when the output neuron sends out i spikes, with an important decision to take in the case $i = 0$: we can either consider b_0 as a separate symbol, or we can assume that emitting 0 spikes means inserting λ in the generated string. Thus, we both obtain strings over arbitrary alphabets, not only over the binary one, and, in the case where we ignore the steps when no spike is emitted, a considerable freedom is obtained in the way the computation proceeds. This latter variant (with λ associated with steps when no spike exits the system) is considered below.

We denote by $LSN^eP_m(rule_k, cons_p, prod_q)$ the family of languages $L(\Pi)$, generated by SN P systems Π using extended rules, with the parameters m, k, p, q as above.

The next counterparts of the results from Theorem 2 were proved in [9].

Theorem 3. (i) $FIN = LSN^eP_1(rule_*, cons_*, prod_*)$ *and this result is sharp, as* $LSN^eP_2(rule_2, cons_2, prod_2)$ *contains infinite languages.*

(ii) $LSN^eP_2(rule_*, cons_*, prod_*) \subseteq REG \subset LSN^eP_3(rule_*, cons_*, prod_*)$; *the second inclusion is proper, because* $LSN^eP_3(rule_3, cons_4, prod_2) - REG \neq \emptyset$; *actually,* $LSN^eP_3(rule_3, cons_6, prod_4)$ *contains non-semilinear languages.*

(iii) $RE = LSN^eP_*(rule_*, cons_*, prod_*)$.

It is an *open problem* to find characterizations or representations in this setup for families of languages in the Chomsky hierarchy different from FIN, REG, RE. We close this section by mentioning the results from [22]:

Theorem 4. *There are universal computing SN P systems with (i) standard rules and 84 neurons and with (ii) extended rules and 49 neurons, and there are universal SN P systems used as a generators of sets of numbers with (iii) standard rules and 76 neurons and with (iv) extended rules and 50 neurons.*

These values can probably be improved (but the feeling is that this improvement cannot be too large).

Tool-kits for handling strings or infinite sequences, on the binary or on the arbitrary alphabet, are provided in [26] and [10]. For instance, in this latter paper one gives constructions of SN P systems for computing the union and concatenation of two languages generated by SN P systems, the intersection with a regular language, while the former paper shows how length preserving morphisms (codings) can be computed; the problem remains *open* for arbitrary morphisms, Kleene $*$, inverse morphisms.

An interesting result is reported in [6]: SAT can be decided in constant time by using an arbitrarily large pre-computed SN P system, of a very regular shape (in what concerns the synapse graph) and with empty neurons, after plugging the instance of size (n, m) (n variables and m clauses) of the problem into the system, by introducing a polynomial number of spikes in (polynomially many) specified neurons. This way of solving a problem, by making use of a pre-computed resource given for free, on the one hand, resembles the supposed fact that only part of the brain neurons are active (involved in "computations") at each time, on the other hand, is not very common in computability and requests further research efforts (what kind of pre-computed resource is allowed, so that no "cheating" is possible? how the given resource should be activated? define and study complexity classes for this framework).

6 Using the Rules in an Exhaustive Way

An essential difference between general P systems and SN P systems is the fact that the latter systems use the rules in a sequential way in each neuron. Introducing parallelism in SN P systems is not only natural from a theoretical point of view, but also presumably useful in order to add computational efficiency to these systems. A proposal in this respect was recently made in [19], and we recall here some details from this paper.

One considers SN P systems of the form introduced above, but with only one type of rules, namely, of the form $E/a^c \rightarrow a^p; d$, with the components as usual and p also allowed to be zero – hence also the forgetting rules are controlled by regular expressions. For all forgetting rules we impose to have $d = 0$ (no delay is possible in this case – we do not consider the possibility to close the neuron when using a forgetting rule).

However, essential now is not the form of rules, but the way they are used: in the exhaustive way, in the following sense. If a rule $E/a^c \rightarrow a^p; d$ is associated with a neuron σ_i which contains k spikes, then the rule is enabled (we also say *fired*) if and only if $a^k \in L(E)$. Using the rule means the following. Assume that $k = sc + r$, for some $s \geq 1$ (this means that we must have $k \geq c$) and $0 \leq r < c$ (the remainder of dividing k by c). Then sc spikes are consumed, r spikes remain in the neuron σ_i, and sp spikes are produced and sent to the neurons σ_j such that $(i, j) \in syn$ (as usual, this means that the sp spikes are replicated and exactly sp spikes are sent to each of the neurons σ_j). In the case of the output neuron, sp spikes are also sent to the environment. Of course, if neuron σ_i has no synapse leaving from it, then the produced spikes are lost.

The computations proceed as in the SN P systems with usual rules, and a spike train is associated with each computation by writing 0 for a step when no spike exits the system and 1 with a step when one or more spikes exit the system. Then, a number is associated – and said to be generated/computed by the respective computation – with a spike train containing at least two occurrences of the digit 1, in the form of the steps elapsed between the first two occurrences of 1 in the spike train. Number 0 is ignored.

For an SN P system Π, we denote by $N_2^{ex}(\Pi)$ the set of numbers computed by Π in this way, and by $N_2SNP_m^{ex}(rule_k, cons_q, forg_r)$ we denote the family of all sets $N_2^{ex}(\Pi)$ generated by SN P systems with at most $m \geq 1$ neurons, using at most $k \geq 1$ rules in each neuron, with all spiking rules $E/a^c \rightarrow a^p; t$ having $c \leq q$, and all forgetting rules $E/a^c \rightarrow \lambda$ having $c \leq r$. When any of the parameters m, k, q, r is not bounded, then it is replaced with $*$.

The following simple observations are rather useful:

1. If an SN P system Π has only rules of the form $a^c \rightarrow a^p; d$ and forgetting rules $a^s \rightarrow \lambda$, then in each neuron each rule can be used exactly once, hence in this case the exhaustive mode coincides with the sequential mode.
2. There are constructions in [18] where one uses neurons with two spikes and two rules of the form

$$a^2/a \rightarrow a; 0, \qquad a \rightarrow a; n,$$

such that this neuron spikes twice, at interval of n steps. In the exhaustive mode, when enabled, the first rule will consume both spikes, but the functioning of the neuron in the exhaustive mode can be the same as in the constructions from [18] if we start with three spikes and use the rules

$$a^3/a^2 \rightarrow a; 0, \qquad a \rightarrow a; n$$

(the first rule consumes two spikes and the second rule consumes the third spike; each rule is used only once).

Using these observations, several examples and results from [18] can be carried to the exhausting case. In particular, this is true for the characterizations of finite sets of numbers (they equal the sets generated by SN P systems with one or two neurons) and for semilinear sets of numbers (their family is equal to the family of sets of numbers generated by SN P systems with bounded neurons).

We do not enter into details, but we just mention that the examples from Figures 3 (generating all even numbers) and 5 (generating all natural numbers) from [18], as well as all Lemmas 9.1–9.6 from [18] are valid also for the exhaustive way of using the rules, via the previous two observations, and this directly leads to the two characterization results mentioned above.

Somewhat expected, also the equivalence with Turing machines is obtained:

Theorem 5. $NRE = N_2SNP^{ex}_*(rule_k, cons_q, forg_r)$ *for all* $k \geq 5, q \geq 5$, $r \geq 1$.

The SN P systems with the exhaustive use of rules are computationally complete also in the accepting case, even when using only deterministic systems.

Many issues remain to be investigated for this new class of SN P systems. Practically, all questions considered for sequential SN P systems are relevant also for the exhaustive case. We just list some of them: associating strings to computations; finding universal SN P systems, if possible, with a small number of neurons; handling strings or infinite sequences over binary or arbitrary alphabets (both input and output neurons are considered); restricted classes of systems (e.g., with a bounded number of spikes present at a time in any neuron) or versions of output (taking k neurons as output neurons and thus producing vectors of dimension k of natural numbers). In the bibliography below we indicate papers dealing with each of these issues for the case of usual SN P systems. Proof techniques from these papers might be useful also in the exhaustive case, while the results proved in these papers should be checked to see whether their counterparts hold true also for exhaustive SN P systems.

Another interesting issue is that of using the parallelism present in our systems in order to solve computationally hard problems in a polynomial time. Usual SN P systems are probably not able to do this (unless an arbitrarily large workspace is freely available/precomputed, initiated in polynomial time, and self-activated during the computation, as proposed in [6]). Is the parallelism of SN P systems with the exhaustive use of rules useful in this respect?

7 Plenty of Research Topics

Many problems were already mentioned above, many others can be found in the papers listed below, and further problems are given in [24]. We recall only some general ideas: bring more ingredients from neural computing, especially related to learning/training/efficiency; incorporate other facts from neurobiology, such as the role played by astrocytes, the way the axon not only transmits impulses, but also amplifies them; consider not only "positive" spikes, but also inhibitory impulses; define a notion of *memory* in this framework, which can be read without being destroyed; provide ways for generating an exponential working space (by splitting neurons? by enlarging the number of synapses?), in such a way to trade space for time and provide polynomial solutions to computationally hard problems; define systems with a dynamical synaptic structure; compare the SN P systems used as generators/acceptors/transducers of infinite sequences with other devices handling such sequences; investigate further systems with exhaustive and other parallel ways of using the rules, as well as systems working in a non-synchronized way; find classes of (accepting) SN P systems for which there is a difference between deterministic and non-deterministic systems; find classes which characterize levels of computability different from those corresponding to finite automata (semilinear sets of numbers or regular languages) or to Turing machines (recursively enumerable sets of numbers or languages).

We close with a more technical idea: use more general types of rules, for instance, of the form $E/a^n \rightarrow a^{f(n)}; d$, where f is a partial function from natural numbers to natural numbers (maybe with the property $f(n) \leq n$ for all n for which f is defined), and used as follows: if the neuron contains k spikes such that $a^k \in L(E)$, then c of them are consumed and $f(c)$ are created, for $c = \max\{n \in \mathbf{N} \mid n \leq k$, and $f(n)$ is defined$\}$; if f is defined for no n smaller than or equal to k, then the rule cannot be applied. This kind of rules looks both adequate from a neurobiological point of view (the sigmoid excitation function can be captured) and mathematically powerful.

References

1. A. Alhazov, R. Freund, M. Oswald, M. Slavkovik: Extended variants of spiking neural P systems generating strings and vectors of non-negative integers. In [14], 123–134.
2. A. Alhazov, R. Freund, A. Riscos-Nunez: One and two polarizations, membrane creation and objects complexity in P systems. *Proc. SYNASC 05*, Timişoara, IEEE Press, 2005, 385–394
3. M. Cavaliere, R. Freund, A. Leitsch, Gh. Păun: Event-related outputs of computations in P systems. *Proc. Third Brainstorming Week on Membrane Computing*, Sevilla, 2005, RGNC Report 01/2005, 107–122.
4. M. Cavaliere, O. Egecioglu, O.H. Ibarra, M. Ionescu, Gh. Păun, S. Woodworth: Unsynchronized spiking neural P systems. Submitted, 2006.
5. H. Chen, R. Freund, M. Ionescu, Gh. Păun, M.J. Pérez-Jiménez: On string languages generated by spiking neural P systems. In [13], Vol. I, 169–194.
6. H. Chen, M. Ionescu, T.-O. Ishdorj: On the efficiency of spiking neural P systems. In [13], Vol. I, 195–206, and *Proc. 8th Intern. Conf. on Electronics, Information, and Communication*, Ulanbator, Mongolia, June 2006, 49–52.
7. H. Chen, M. Ionescu, A. Păun, Gh. Păun, B. Popa: On trace languages generated by spiking neural P systems. In [13], Vol. I, 207–224, and *Proc. DCFS2006*, Las Cruces, NM, June 2006.
8. H. Chen, T.-O. Ishdorj, Gh. Păun: Computing along the axon. In [13], Vol. I, 225–240, and *Progress in Natural Computing*, in press.
9. H. Chen, T.-O. Ishdorj, Gh. Păun, M.J. Pérez-Jiménez: Spiking neural P systems with extended rules. In [13], Vol. I, 241–265.

10. H. Chen, T.-O. Ishdorj, Gh. Păun, M.J. Pérez-Jiménez: Handling languages with spiking neural P systems with extended rules. *Romanian J. Information Sci. and Technology*, 9, 3 (2006), 151–162.
11. G. Ciobanu, Gh. Păun, M.J. Pérez-Jiménez, eds.: *Applications of Membrane Computing*, Springer, Berlin, 2006.
12. W. Gerstner, W Kistler: *Spiking Neuron Models. Single Neurons, Populations, Plasticity.* Cambridge Univ. Press, 2002.
13. M.A. Gutiérrez-Naranjo et al., eds.: *Proceedings of Fourth Brainstorming Week on Membrane Computing*, Febr. 2006, Fenix Editora, Sevilla, 2006.
14. H.J. Hoogeboom, Gh. Păun, G. Rozenberg, A. Salomaa, eds.: *Membrane Computing, International Workshop, WMC7, Leiden, The Netherlands, 2006, Selected and Invited Papers*, LNCS 4361, Springer, Berlin, 2007.
15. O.H. Ibarra, A. Păun, Gh. Păun, A. Rodríguez-Patón, P. Sosik, S. Woodworth: Normal forms for spiking neural P systems. In [13], Vol. II, 105–136, and *Theoretical Computer Sci.*, to appear.
16. O.H. Ibarra, S. Woodworth: Characterizations of some restricted spiking neural P systems. In [14], 424–442.
17. O.H. Ibarra, S. Woodworth, F. Yu, A. Păun: On spiking neural P systems and partially blind counter machines. In *Proceedings of Fifth Unconventional Computation Conference, UC2006*, York, UK, September 2006.
18. M. Ionescu, Gh. Păun, T. Yokomori: Spiking neural P systems. *Fundamenta Informaticae*, 71, 2-3 (2006), 279–308.
19. M. Ionescu, Gh. Păun, T. Yokomori: Spiking neural P systems with exhaustive use of rules. *Intern. J. Unconventional Computing*, 2007.
20. I. Korec: Small universal register machines. *Theoretical Computer Science*, 168 (1996), 267–301.
21. W. Maass, C. Bishop, eds.: *Pulsed Neural Networks*, MIT Press, 1999.
22. A. Păun, Gh. Păun: Small universal spiking neural P systems. In [13], Vol. II, 213–234, and *BioSystems*, in press.
23. Gh. Păun: *Membrane Computing. An Introduction.* Springer, Berlin, 2002.
24. Gh. Păun: Twenty six research topics about spiking neural P systems. Available at [27], 2006.
25. Gh. Păun, M.J. Pérez-Jiménez, G. Rozenberg: Spike trains in spiking neural P systems. *Intern. J. Found. Computer Sci.*, 17, 4 (2006), 975–1002.
26. Gh. Păun, M.J. Pérez-Jiménez, G. Rozenberg: Infinite spike trains in spiking neural P systems. Submitted 2005.
27. The P Systems Web Page: `http://psystems.disco.unimib.it`.
28. The Sheffield P Systems Applications Web Page: `http://www.dcs.shef.ac.uk/~marian/PSimulatorWeb/P_Systems_applications.htm`

Solving Subset Sum in Linear Time by Using Tissue P Systems with Cell Division

Daniel Díaz-Pernil, Miguel A. Gutiérrez-Naranjo,
Mario J. Pérez-Jiménez, and Agustín Riscos-Núñez

Research Group on Natural Computing
ETS Ingeniería Informática - University of Sevilla
{sbdani,magutier,marper,ariscosn}@us.es

Abstract. Tissue P systems with cell division is a computing model in the framework of Membrane Computing based on intercellular communication and cooperation between neurons. The ability of cell division allows us to obtain an exponential amount of cells in linear time and to design cellular solutions to **NP**-complete problems in polynomial time. In this paper we present a solution to the Subset Sum problem via a family of such devices. This is the first solution to a numerical **NP**-complete problem by using tissue P systems with cell division.

1 Introduction

In the cell-like model of P systems [6], membranes are hierarchically arranged in a tree-like structure. Its biological inspiration comes from the morphology of cells, where small vesicles are surrounded by larger ones. This biological structure can be abstracted into a tree-like graph, where the root represents the skin of the cell (i.e. the outermost membrane) and the leaves represent membranes that do not contain any other membrane (elementary membranes). Besides, two nodes in the graph are connected if they represent two membranes such that one of them contains the other one.

Recently, new models of P systems have been explored. One of them is the model of *tissue P systems* where the tree-like membrane structure is not considered anymore, being replaced by a general graph.

This model has two biological inspirations (see [4]): intercellular communication and cooperation between neurons. The common mathematical model of these two mechanisms is a net of processors dealing with symbols and communicating these symbols along channels specified in advance. The communication among cells is based on symport/antiport rules, which were introduced as communication rules for P systems in [5]. In symport rules objects cooperate to traverse a membrane together in the same direction, whereas in the case of antiport rules, objects residing at both sides of the membrane cross it simultaneously but in opposite directions.

This paper is devoted to the study of the computational efficiency of tissue P systems with cell division. In literature different models of cell-like P systems

J. Mira and J.R. Álvarez (Eds.): IWINAC 2007, Part I, LNCS 4527, pp. 170–179, 2007.

have been successfully used in order to design efficient solutions to **NP**-complete problems (see, for example, [2] and the references therein). These solutions are obtained by generating an exponential amount of workspace in polynomial time and using parallelism to check simultaneously all the candidate solutions.

From the seminal definition of tissue P systems [3,4], several research lines have been developed and other variants have arisen (see [1] and references therein). One of the most interesting variants of tissue P systems was presented in [8], where the definition of tissue P systems is combined with the one of P systems with active membranes, yielding *tissue P systems with cell division*. The biological inspiration is clear: alive tissues are not *static* networks of cells, since cells are duplicated via mitosis in a natural way. One of the main features of such tissue P systems with cell division is related to their computational efficiency. In [8], a polynomial-time solution to the **NP**-complete problem SAT is shown, and in [1] a linear-time solution for the 3-COL problem was presented. In this paper we go on with the research in this model and present a linear-time solution to another well-known numerical **NP**-complete problem: the Subset Sum problem.

The paper is organised as follows: first we recall some preliminary concepts and the definition of tissue P systems with cell division. Next, recognising tissue P systems are briefly described. A linear–time solution to the Subset Sum problem is presented in the following section, including a short overview of the computation and of the necessary resources. Finally, some conclusions and lines for future research are presented.

2 Preliminaries

In this section we briefly recall some of the concepts used later on in the paper.

An *alphabet*, Σ, is a non empty set, whose elements are called *symbols*. An ordered sequence of symbols is a *string*. The number of symbols in a string u is the *length* of the string, and it is denoted by $|u|$. As usual, the empty string (with length 0) will be denoted by λ. The set of strings of length n built with symbols from the alphabet Σ is denoted by Σ^n and $\Sigma^* = \cup_{n\geq 0}\Sigma^n$. A *language* over Σ is a subset from Σ^*. A *multiset* m over a set A is a pair (A, f) where $f : A \rightarrow \mathbb{N}$ is a mapping. If $m = (A, f)$ is a multiset then its *support* is defined as $supp(m) = \{x \in A \,|\, f(x) > 0\}$ and its *size* is defined as $\sum_{x\in A} f(x)$. A multiset is empty (resp. finite) if its support is the empty set (resp. finite). If $m = (A, f)$ is a finite multiset over A, then it will be denoted as $m = \{\{a_1, \ldots, a_k\}\}$, where each element a_i occurs $f(a_i)$ times. Multisets can also be represented as strings in a natural way.

In what follows we assume the reader is already familiar with the basic notions and the terminology underlying P systems. For details, see [7].

3 Tissue P Systems with Cell Division

In the first definition of the model of tissue P systems [3,4] the membrane structure did not change along the computation. The main features of tissue P systems with cell division, from the computational point of view, are that cells obtained by division have the same labels as the original cell, and if a cell is divided, then its interaction with other cells or with the environment is blocked during the mitosis process. In some sense, this means that while a cell is dividing it closes the communication channels with other cells and with the environment. This features imply that the underlying graph is dynamic, as nodes can be added during the computation by division and the edges can be deleted/re-established for dividing cells.

Formally, a *tissue P system with cell division* of initial degree $q \geq 1$ is a tuple of the form $\Pi = (\Gamma, w_1, \ldots, w_q, \mathcal{E}, \mathcal{R}, i_0)$, where:

1. Γ is a finite *alphabet*, whose symbols will be called *objects*.
2. $w_1, \ldots, w_q$ are strings over Γ.
3. $\mathcal{E} \subseteq \Gamma$.
4. $\mathcal{R}$ is a finite set of rules of the following form:
   (a) *Communication rules*: $(i, u/v, j)$, for $i, j \in \{0, 1, \ldots, q\}, i \neq j, u, v \in \Gamma^*$.
   (b) *Division rules*: $[a]_i \rightarrow [b]_i[c]_i$, where $i \in \{1, 2, \ldots, q\}$ and $a, b, c \in \Gamma$.
5. $i_0 \in \{0, 1, 2, \ldots, q\}$.

A tissue P system with cell division of degree $q \geq 1$ can be seen as a set of q cells labelled by $1, 2, \ldots, q$. We shall use 0 as the label of the environment, and i_0 for the output region (which can be the region inside a cell or the environment). Despite the fact that cell-like models include an explicit description of the initial membrane structure, this is not the case here. Instead, the underlying graph expressing connections between cells is implicit, being determined by the communication rules (the nodes are the cells and the edges indicate if it is possible for pairs of cells to communicate directly).

The strings $w_1, \ldots, w_q$ describe the multisets of objects placed initially in the q cells of the system. We interpret that $\mathcal{E} \subseteq \Gamma$ is the set of objects placed in the environment, each one of them in an arbitrarily large amount of copies.

The communication rule $(i, u/v, j)$ can be applied over two cells i and j such that u is contained in cell i and v is contained in cell j. The application of this rule means that the objects of the multisets represented by u and v are interchanged between the two cells.

The division rule $[a]_i \rightarrow [b]_i[c]_i$ can be applied over a cell i containing object a. The application of this rule divides this cell into two new cells with the same label. All the objects in the original cell are replicated and copied in each of the new cells, with the exception of the object a, which is replaced by the object b in the first new cell and by c in the second one.

Rules are used as usual in the framework of membrane computing, that is, in a maximally parallel way (a universal clock is considered). In one step, each object in a membrane can only be used for one rule (non-deterministically chosen when there are several possibilities), but any object which can participate in a rule of any form must do it, i.e, in each step we apply a maximal set of rules. This way of applying rules has only one restriction: when a cell is divided, the division rule is the only one which is applied for that cell in that step; the objects inside that cell do not move in that step.

4 Recognising Tissue P Systems with Cell Division

NP-completeness has been usually studied in the framework of *decision problems*. Let us recall that a decision problem is a pair (I_X, θ_X) where I_X is a language over a finite alphabet (whose elements are called *instances*) and θ_X is a total boolean function over I_X.

In order to study the computational efficiency, a special class of tissue P systems is introduced in [8]: *recognising*[1] *tissue P systems*. The key idea is the same one as from cell-like recognising P systems, that were introduced in [9] as the natural framework to study and solve decision problems within Membrane Computing. Note that deciding whether an instance of a problem has an affirmative or negative answer is equivalent to deciding if a string belongs or not to the language associated with the problem.

In literature, recognising cell-like P systems are associated in a natural way with P systems with *input*. The data related to an instance of the decision problem need to be provided to the P system in order to compute the appropriate answer. This is done by codifying in unary form each instance as a multiset placed in an *input membrane*. The output of the computation (yes or no) is sent to the environment. In this way, cell-like P systems with input and external output are devices which can be seen as black boxes, in the sense that the user provides the data before the computation starts, and then waits *outside* the P system until it sends to the environment the output in the last step of the computation.

A recognising tissue P system with cell division of degree $q \geq 1$ is a tuple $\Pi = (\Gamma, \Sigma, w_1, \dots, w_q, \mathcal{E}, \mathcal{R}, i_{in}, i_0)$, where

- $(\Gamma, w_1, \dots, w_q, \mathcal{E}, \mathcal{R}, i_0)$ is a tissue P system with cell division of degree $q \geq 1$ (as defined in the previous section).
- The working alphabet Γ has two distinguished objects yes and no, present in at least one copy in an initial multiset $w_1, \dots, w_q$, but not present in $\mathcal{E}$.
- Σ is an (input) alphabet strictly contained in Γ.
- $i_{in} \in \{1, \dots, q\}$ is the input cell.
- The output region i_0 is the environment.
- All computations halt.
- If $\mathcal{C}$ is a computation of Π, then either the object yes or the object no (but not both) must have been released into the environment, and only in the last step of the computation.

The computations of the system Π with input $w \in \Gamma^*$ start from a configuration of the form $(w_1, w_2, \dots, w_{i_{in}} w, \dots, w_q; \mathcal{E})$, that is, after adding the multiset w to the contents of the input cell i_{in}. We say that the multiset w is *recognised* by Π if and only if the object yes is sent to the environment, in the last step of all its associated computations. We say that $\mathcal{C}$ is an accepting (resp. rejecting) computation if the object yes (resp. no) appears in the environment associated with the corresponding halting configuration of $\mathcal{C}$.

[1] In [8] they were called *recognizer* tissue P systems.

Definition 1. *We say that a decision problem* $X = (I_X, \theta_X)$ *is solvable in polynomial time by a family* $\mathbf{\Pi} = \{\Pi(n) : n \in \mathbb{N}\}$ *of recognising tissue P systems with cell division if the following holds:*

- *The family* $\mathbf{\Pi}$ *is* polynomially uniform *by Turing machines, that is, there exists a deterministic Turing machine working in polynomial time which constructs the system* $\Pi(n)$ *from* $n \in \mathbb{N}$.
- *There exists a pair* (cod, s) *of polynomial-time computable functions over* I_X *such that:*
  - *for each instance* $u \in I_X$, $s(u)$ *is a natural number and* $cod(u)$ *is an input multiset of the system* $\Pi(s(u))$;
  - *the family* $\mathbf{\Pi}$ *is* polynomially bounded *with regard to* (X, cod, s), *that is, there exists a polynomial function* p, *such that for each* $u \in I_X$ *every computation of* $\Pi(s(u))$ *with input* $cod(u)$ *is halting and, moreover, it performs at most* $p(|u|)$ *steps;*
  - *the family* $\mathbf{\Pi}$ *is* sound *with regard to* (X, cod, s), *that is, for each* $u \in I_X$, *if there exists an accepting computation of* $\Pi(s(u))$ *with input* $cod(u)$, *then* $\theta_X(u) = 1$;
  - *the family* $\mathbf{\Pi}$ *is* complete *with regard to* (X, cod, s), *that is, for each* $u \in I_X$, *if* $\theta_X(u) = 1$, *then every computation of* $\Pi(s(u))$ *with input* $cod(u)$ *is an accepting one.*

In the above definition we have imposed to every tissue P system $\Pi(n)$ a *confluent* condition, in the following sense: every computation of a system with the *same* input multiset must always give the *same* answer. The pair of functions (cod, s) are called a *polynomial encoding* of the problem in the family of P systems.

We denote by $\mathbf{PMC}_{TD}$ the set of all decision problems which can be solved by means of recognising tissue P systems with cell division in polynomial time.

5 The Subset Sum Problem

The Subset Sum problem is the following one: *Given a finite set* A, *a weight function,* $w : A \to \mathbb{N}$, *and a constant* $k \in \mathbb{N}$, *determine whether or not there exists a subset* $B \subseteq A$ *such that* $w(B) = k$.

Next, we shall prove that the Subset Sum problem can be solved in a linear time by a family of recognising tissue P systems with cell division. We shall address the resolution via a brute force algorithm.

We will use a tuple $(n, (w_1, \ldots, w_n), k)$ to represent an instance of the problem, where n stands for the size of $A = \{a_1, \ldots, a_n\}$, $w_i = w(a_i)$, and k is the constant given as input for the problem.

Theorem 1. `SUBSET SUM` $\in \mathbf{PMC}_{TD}$.

Proof. Let $A = \{a_1, \ldots, a_n\}$ be a finite set, $w : A \longrightarrow \mathbb{N}$ a weight function with $n = |A|$ and $k \in \mathbb{N}$. Let $g : \mathbb{N} \times \mathbb{N} \to \mathbb{N}$ be a *function* defined by $g(n, k) =$

$((n+k)(n+k+1)/2)+n$. This function is primitive recursive and bijective between $\mathbb{N}^2$ and $\mathbb{N}$ and computable in polynomial time. Let us denote by $u = (n,(w_1,\ldots,w_n),k)$, where $w_i = w(a_i)$, $1 \le i \le n$, the given instance of the problem. We define the polynomially computable function $s(u) = g(n,k)$.

We will provide a family of tissue P systems where each P system solves all the instances of the `SUBSET SUM` problem with the same size. The weight function w of the concrete instance will be provided via an input multiset determined via the function $cod(u) = \{\{v_i^j : w(a_i) = j \wedge 1 \le i \le n\}\} \cup \{\{q^k\}\}$, where v_i^j (i.e., j copies of object v_i) represents that j is the weight of the element a_i.

Next, we will provide a family of recognising tissue P systems with cell division which solve the `SUBSET SUM` problem in linear time. For each $(n,k) \in \mathbb{N}^2$ we will consider the system $\Pi(n,k) = (\Gamma, \Sigma, \omega_1, \omega_2, \mathcal{R}, \mathcal{E}, i_{in}, i_0)$, where

- $\Gamma = \Sigma \cup \{A_i, B_i, : 1 \le i \le n\}$
  $\cup \{a_i : 1 \le i \le n + \lceil \log n \rceil + \lceil \log(k+1) \rceil + 11\}$
  $\cup \{c_i : 1 \le i \le n+1\}$
  $\cup \{d_i : 1 \le i \le \lceil \log n \rceil + \lceil \log(k+1) \rceil + 4\}$
  $\cup \{e_i : 1 \le i \le \lceil \log n \rceil + 1\}$
  $\cup \{B_{ij} : 1 \le i \le n \wedge 1 \le j \le \lceil log(k+1) \rceil + 1\}$
  $\cup \{b, D, p, g_1, g_2, f_1, T, S, N, \mathbf{yes}, \mathbf{no}\}$
- $\Sigma = \{q\} \cup \{v_i : 1 \le i \le n\}$
- $\omega_1 = a_1\, b\, c_1\, \mathbf{yes}\, \mathbf{no}$
- $\omega_2 = D A_1 \cdots A_n$
- $\mathcal{R}$ is the following set of rules:
  1. *Division rules:*
     $r_{1,i} \equiv [A_i]_2 \rightarrow [B_i]_2[\lambda]_2$ for $i = 1,\ldots,n$
  2. *Communication rules:*
     $r_{2,i} \equiv (1, a_i/a_{i+1}, 0)$ for $i = 1,\ldots,n + \lceil \log n \rceil + \lceil \log(k+1) \rceil + 10$
     $r_{3,i} \equiv (1, c_i/c_{i+1}^2, 0)$ for $i = 1,\ldots,n$
     $r_4 \equiv (1, c_{n+1}/D, 2)$
     $r_5 \equiv (2, c_{n+1}/d_1 e_1, 0)$
     $r_{6,i} \equiv (2, e_i/e_{i+1}^2, 0)$ for $i = 1,\ldots,\lceil \log n \rceil$
     $r_{7,i} \equiv (2, d_i/d_{i+1}, 0)$ for $i = 1,\ldots,\lceil \log n \rceil + \lceil \log(k+1) \rceil + 3$
     $r_{8,i} \equiv (2, e_{\lceil \log n \rceil + 1} B_i/B_{i1}, 0)$ for $i = 1,\ldots,n$
     $r_{9,i,j} \equiv (2, B_{ij}/B_{ij+1}^2, 0)$ for $i = 1,\ldots,n$, $j = 1,\ldots,\lceil \log(k+1) \rceil$
     $r_{10,i} \equiv (2, B_{i\lceil \log(k+1) \rceil + 1} v_i/p, 0)$ for $i = 1,\ldots,n$
     $r_{11} \equiv (2, pq/\lambda, 0)$
     $r_{12} \equiv (2, d_{\lceil \log n \rceil + \lceil \log(k+1) \rceil + 4}/g_1 f_1, 0)$
     $r_{13} \equiv (2, f_1 p/\lambda, 0)$
     $r_{14} \equiv (2, f_1 q/\lambda, 0)$
     $r_{15} \equiv (2, g_1/g_2, 0)$
     $r_{16} \equiv (2, g_2 f_1/T, 0)$
     $r_{17} \equiv (2, T/\lambda, 1)$
     $r_{18} \equiv (1, bT/S, 0)$
     $r_{19} \equiv (1, S\mathbf{yes}/\lambda, 0)$

$r_{20} \equiv (1, a_{n+\lceil \log n \rceil + \lceil \log(k+1) \rceil + 11} b / N, 0)$
$r_{21} \equiv (1, N\mathtt{no}/\lambda, 0)$

- $\mathcal{E} = \Gamma - \{\mathtt{yes}, \mathtt{no}\}$
- $i_{in} = 2$, is the input cell
- $i_0 = env$, is the output cell

The design is structured in the following stages:

- *Generation Stage*: The initial cell labelled by 2 is divided into two new cells; and the divisions are iterated n times until a cell has been produced for each possible candidate solution. Simultaneously to this process, two counters (c_i and a_i) evolve in the cell labelled by 1: the first one controls the step in which the communication between cells 2 starts and the second one will be useful in the output stage.
- *Pre–checking Stage*: When this stage starts, we have 2^n cells labelled by 2, each of them encoding a subset of the set A. In each such a cell, as many objects p as the weight of the corresponding subset will be generated. Recall that there are k copies of the object q in each cell labelled by 2 (since they were introduced as part of the input multiset).
- *Checking Stage*: In each cell labelled by 2, the number of copies of objects p and q are compared. The way to do that is removing from the cell in one step all possible pairs (p, q). After doing so, if some objects p or q remain in the cell, then the cell was not encoding a solution of the problem; otherwise, the weight of the subset of A encoded on the cell equals to k and hence it encodes a solution to the problem.
- *Output Stage*: The system sends to the environment the right answer according to the results of the previous stage:
  - *Answer yes:* After the checking stage, there is a cell labelled by 2 without objects p nor q. In this case, such a cell sends an object T to the cell 1. This object T causes the cell 1 to expel an object yes to the environment (see rules r_{18} and r_{20}).
  - *Answer no:* In each cell labelled by 2 there exists an object p or q. In this case, no object T arrives to the cell labelled by 1 and an object no is sent to the environment.

The non-determinism of this family of recognising tissue P systems with cell division lies in the division rules. These division rules are not competitive: the non-determinism is due to the order in which the rules are applied. When the generation stage ends, the same configuration is reached regardless the order of application of the division rules: 2^n cells labelled by 2, each of them with the codification of a different subset of A.

6 An Overview of the Computation

First of all, we recall the polynomial encoding of the Subset Sum problem in the family $\mathbf{\Pi}$ constructed in the previous section. Let $u = (n, (w_1, \dots, w_n), k)$ be an instance of the problem, $s(u) = g(n,k)$ and $cod(u) = \{\{v_i^j : w(a_i) = j \wedge 1 \leq i \leq n\}\}$.

Next, we describe informally how the recognising tissue P system with cell division $\Pi(s(u))$ with input $cod(u)$ works. Let us start with the *generation stage*. Recall that if a division rule is triggered, the communication rules cannot be simultaneously applied. In this stage we have two parallel processes:

- On the one hand, in the cell labelled by 1 we have two counters: a_i, which will be used in the answer stage and c_i, which will be multiplied until getting 2^n copies in exactly n steps.
- On the other hand, in the cell labelled by 2, the division rules are applied. For each object A_i (which codifies a member of the set A) we obtain two cells labelled by 2: One of them has an element B_i and the other does not.

When all divisions have been done, after n steps, we will have 2^n cells with label 2 and each of them will contain the encoding of a subset of A. At this moment, the generation stage ends and the pre-checking stage begins.

For each cell 2, an object D is changed by a copy of the counter c. In this way, in the cell 1 2^n copies of D will appear and, in each cell labelled by 2 there will be an object c_{n+1}. The occurrence of such object c_{n+1} in the cells 2 will produce the apparition of two counters:

(a) The counter d_i lets the checking stage start, since it produces the apparition of the objects g_1 and f_1 after $\lceil \log n \rceil + \lceil \log(k+1) \rceil + 4$ steps.
(b) The counter e_i will be multiplied for obtaining n copies of $e_{\lceil \log n \rceil + 1}$ in the step $n + \lceil \log n \rceil + 2$. Then, we trade objects $e_{\lceil \log n \rceil + 1}$ and B_i against B_{i1} for each element A_i in the subset associated with the membrane. After that, for each $1 \leq i \leq n$ we get $k+1$ copies of $B_{i\lceil \log(k+1) \rceil + 1}$. Then for each element A_i in the subset associated with the membrane we get $\min\{k+1, w(a_i)\}$ copies of object p, in the step $n + \lceil \log n \rceil + \lceil \log(k+1) \rceil + 5$.

The checking takes place in the step $n + \lceil \log n \rceil + \lceil \log(k+1) \rceil + 6$, when all pairs of objects p and q present in any cell labelled by 2 are sent to the environment. In this way, if the weight of the subset associated with a cell is equal to k, then no object p or q remains in this cell in the next step. Otherwise, if the encoding is not exactly of weight k, then at least one object p or q will remain in the cell. In the next step the answer stage starts. Two cases must be considered for each cell:

- If no object p or q remain in the cell, the object f_1 does not evolve, g_1 evolves to g_2, and in the step $n + \lceil \log n \rceil + \lceil \log(k+1) \rceil + 8$ the objects f_1 and g_2 are traded by T with the environment. In the next step T is sent to the cell 1, and in the step $n + \lceil \log n \rceil + \lceil \log(k+1) \rceil + 10$, the objects T and b are sent to the environment traded by S. Finally in the step $n + \lceil \log n \rceil + \lceil \log(k+1) \rceil + 11$ the objects S and $\mathtt{yes}$ are sent to the environment.
- If any object p or q remains in the cell, such object is sent to the environment together with the object f_1. This causes that the object b still remains in

the cell 2 after the step $n + \lceil \log n \rceil + \lceil \log(k+1) \rceil + 10$. In this way, the objects b and $a_{n+\lceil \log n \rceil + \lceil \log(k+1) \rceil + 11}$ are traded by the object N with the environment, and in the step $n + \lceil \log n \rceil + \lceil \log(k+1) \rceil + 12$ the objects N and **no** are sent to the environment.

6.1 Necessary Resources

Next, we show that the family $\mathbf{\Pi} = \{\Pi(g(n,k)) \ : \ n, k \in \mathbb{N}\}$ defined in Theorem 1 is polynomially uniform by Turing machines. To this aim we are going to show that it is possible to build $\Pi(g(n,k))$ in polynomial time with respect to the size of u.

It is easy to check that the rules of a system $\{\Pi(g(n,k)) \ : \ n, k \in \mathbb{N}\}$ of the family are defined recursively from the values n and k. Besides, the necessary resources to build an element of the family are of polynomial order with respect to the same:

- Size of the alphabet: $n \cdot \lceil \log(k+1) \rceil + 6n + 2\lceil \log(k+1) \rceil + 3\lceil \log n \rceil + 28 \in O(n \cdot \log k)$
- Initial number of cells: $2 \in \theta(1)$.
- Initial number of objects: $n + 6 \in \theta(n)$.
- Number of rules: $n \cdot \lceil \log(k+1) \rceil + 5n + 2\lceil \log(k+1) \rceil + 3\lceil \log n \rceil + 26 \in O(n \cdot \log k)$
- Maximal length of a rule: 3.

7 Conclusions and Future Work

Natural Computing studies new computational paradigms inspired from various well-known natural phenomena in physics, chemistry and biology. This paper is devoted to a new field in Natural Computing: the study of the structure and functioning of cells as living organisms able to process and generate information.

Membrane Computing is a new cross-disciplinary field of Natural Computing which has reached an important success in its short life. In these years many results have been presented related to the computational power of membrane devices, but up to now no implementation *in vivo* or *in vitro* has been carried out. This paper deals with the study of *algorithms* to solve well-known problems and in this sense it is a theoretical result, mainly related to computational efficiency. Moreover, this paper represents a new step in the study of algorithms in the framework of P systems because it exploits tissue P Systems with Cell Division (a variant poorly studied) to solve an **NP**-complete problem.

The basic idea is to consider a distributed and parallel computing device, structured as the cells of a tissue, by means of arrangement of cells where various chemicals (we call them *objects*, to be free of any interpretation) evolve according to local reaction rules. Because the chemicals from the compartments of a cell are swimming in an aqueous solution, the data structure we consider is that of a *multiset* – a set with multiplicities associated with its elements. Also, in analogy with what happens in a cell, the rules are applied in a parallel and a non–deterministic manner.

P systems are computational devices whose power has to be studied in a deeper extent. In the last years, several papers have explored this power, both in the framework of cell-like P systems and tissue-like P systems with membrane creation. These papers have shown that **NP**-complete problems are solvable (in polynomial time) by families of recognising P systems in such models. In this paper we have shown that numerical **NP**-complete problems can also be solved (in polynomial time) by families of recognising tissue P systems with Cell Division, in a uniform way. The specific techniques for designing solutions to concrete problems (generation, evaluation, checking, and output stages) are quite different from a P system model to another, so the simulation of one model in the other one is not a trivial question.

Other lines to follow in the future are the extension of the techniques presented in this paper for the study of other numerical **NP**-complete problems and to develop a software for simulating these computational processes.

Acknowledgement

The authors wish acknowledge the support of the project TIN2006-13425 of the Ministerio de Educación y Ciencia of Spain, cofinanced by FEDER funds, and the support of the project of excellence TIC-581 of the Junta de Andalucía.

References

1. Díaz-Pernil, D., Gutiérrez-Naranjo, M.A., Pérez-Jiménez, M.J., Riscos-Núñez,A. A linear–time tissue P system based solution for the 3–coloring problem. *Theoretical Computer Science*, to appear.
2. Gutiérrez-Naranjo, M.A., Pérez-Jiménez, M.J. and Romero-Campero, F.J. A linear solution for QSAT with Membrane Creation. *Lecture Notes in Computer Science* **3850**, (2006), 241–252.
3. Martín Vide, C., Pazos, J., Păun, Gh. and Rodríguez Patón, A. A New Class of Symbolic Abstract Neural Nets: Tissue P Systems. *Lecture Notes in Computer Science* **2387**, (2002), 290–299.
4. Martín Vide, C., Pazos, J., Păun, Gh. and Rodríguez Patón, A. Tissue P systems. *Theoretical Computer Science*, **296**, (2003), 295–326.
5. Păun, A. and Păun, Gh. The power of communication: P systems with symport/antiport. *New Generation Computing*, **20**, 3, (2002), 295–305.
6. Păun, Gh. Computing with membranes. *Journal of Computer and System Sciences*, **61**, 1, (2000), 108–143.
7. Păun, Gh. *Membrane Computing. An Introduction.* Springer–Verlag, Berlin, (2002).
8. Păun, Gh., Pérez-Jiménez, M.J. and Riscos-Núñez, A. Tissue P System with cell division. In Gh. Păun, A. Riscos-Núñez, A. Romero-Jiménez and F. Sancho-Caparrini (eds.), *Second Brainstorming Week on Membrane Computing*, Sevilla, Report RGNC 01/2004, (2004), 380–386.
9. Pérez-Jiménez, M.J., Romero-Jiménez, A. and Sancho-Caparrini, F. A polynomial complexity class in P systems using membrane division. In E. Csuhaj-Varjú, C. Kintala, D. Wotschke and Gy. Vaszyl (eds.), *Proceedings of the 5th Workshop on Descriptional Complexity of Formal Systems, DCFS 2003*, (2003), 284–294.

On a Păun's Conjecture in Membrane Systems

Giancarlo Mauri[2], Mario J. Pérez-Jiménez[1], and Claudio Zandron[2]

[1] Research Group on Natural Computing
Department of Computer Science and Artificial Intelligence. University of Sevilla
Avda. Reina Mercedes s/n, 41012 Sevilla, Spain
marper@us.es
[2] Dipartimento di Informatica, Sistemistica e Comunicazione. Universitá di Milano–Bicocca
via Bicocca degli Arcimboldi 8, I–20126, Milano, Italy
mauri/zandron@disco.unimib.it

Abstract. We study a Păun's conjecture concerning the unsolvability of **NP**–complete problems by polarizationless P systems with active membranes in the usual framework, without cooperation, without priorities, without changing labels, using evolution, communication, dissolution and division rules, and working in maximal parallel manner. We also analyse a version of this conjecture where we consider polarizationless P systems working in the minimally parallel manner.

1 Introduction

Every deterministic Turing machine working in polynomial time can be simulated in polynomial time by a family of recognizing P systems using only basic rules, that is, evolution, communication, and rules involving dissolution [14]. If a decision problem is solvable in polynomial time by a family of recognizing P systems (using only basic rules), then there exists a deterministic Turing machine solving it in polynomial time [20]. As a consequence of these results, the class of all decision problems solvable in polynomial time by this kind of P systems is equal to the standard complexity class **P** [5]. For that reason, recognizing P systems constructing in polynomial time an exponential workspace, expressed in the number of objects, cannot solve **NP**–complete problems in polynomial time (unless **P** = **NP**).

Hence, in order to efficiently solve **NP**–complete problem by P systems it seems necessary to be able to construct an exponential workspace (expressed by the number of membranes) in polynomial time. These models abstract the way of obtaining new membranes through the processes of *mitosis* (membrane division) and *autopoiesis* (membrane creation).

P systems with active membranes (using division rules) have been successfully used to efficiently solve **NP**–complete problems. The first solutions were given constructing a P system associated with each instance of the problem due to the systems lack of an input membrane. Actually, we say that this kind of solutions are *semi–uniform* if the following is true: (a) there exists a deterministic Turing

J. Mira and J.R. Álvarez (Eds.): IWINAC 2007, Part I, LNCS 4527, pp. 180–192, 2007.

machine working in polynomial time which constructs the P system processing an instance of the problem (we say the family of P systems associated with all the instances is *polynomially uniform by Turing machines*); and (b) the instance of the problem has an affirmative answer if and only if every computation of the P system associated with it is an accepting computation (we say the P system is *confluent*).

The first semi–uniform polynomial–time solutions of computationally hard decision problems were given by Gh. Păun [11,12], C. Zandron et al. [20], S.N. Krishna et al. [7], and A. Obtulowicz [8]. In 2003, P. Sosik [19] gave a semi–uniform polynomial–time solution to **QSAT**, a well known **PSPACE**–complete problem.

There is another way to solve decision problems by P systems when we consider the possibility to have an input membrane in the systems in which we can introduce objects before the system starts to work. In this case, all instances of a decision problem having the same size (according to a prefixed polynomial time criterion) are processed by the same system.

P systems with active membranes have also been successfully used to design uniform polynomial–time solutions to some well-known **NP**–complete problems, such as **SAT** [17], *Subset Sum* [15], *Knapsack* [16], *Partition* [6], and the *Common Algorithmic Problem* [18].

All papers mentioned above deal with P systems with three polarizations using only division of elementary membranes (in [19] also division for non–elementary membranes are permitted), and working in the *maximal parallelism* in using the rules, that is, in each step, the assignment of objects to the rules to be applied is maximal, no further rule can be applied in any region. The number of polarizations can be decreased to two [1] without loss of efficiency.

It seems clear that the usual framework of P systems with active membranes to solve decision problems is too powerful from the complexity point of view. Then, it would be interesting to analyse which features allows to P systems with active membranes, but without polarizations, to still get polynomial–time solutions to computationally hard problems, and what features, once removed, only allows to obtain polynomial–time solutions to tractable problems, in the classical sense.

The present paper is a contribution to the problem of describing borderlines between tractability and intractability in terms of descriptional resources required in (recognizing) membrane systems using division rules.

The paper is organized as follows. In the next section we present the Păun's conjecture concerning polarizationless P systems with active membranes with three electrical charges and working in the maximally parallel mode. Also we provide some partial solutions to this conjecture by using the notion of dependency graph associated with a P system. Section 3 is devoted to formulate a new version of the Păun's conjecture, addressing P systems working in the minimally parallel mode. We give some partial solutions to this new version. Some conclusions and open problems are given in the last Section.

2 A Păun's Conjecture

Usual P systems with active membranes use three *electrical charges* for membranes, controlling the application of the rules which, basically, can be of the following types: *evolution rules*, by which single objects evolve to a multiset of objects, *communication rules*, by which an object is introduced in or expelled from a membrane, maybe modified during this operation into another object, *dissolution rules*, by which a membrane is dissolved, under the influence of an object, which may be modified into another object by this operation, and *membrane division rules* (both for elementary and non-elementary membranes, or only for elementary membranes).

Definition 1. *A P system with polarizationless active membranes of the initial degree $n \geq 1$ is a tuple of the form $\Pi = (\Gamma, H, \mu, \mathcal{M}_1, \ldots, \mathcal{M}_n, R, h_o)$, where:*

1. *Γ is the alphabet of objects;*
2. *H is a finite set of labels for membranes;*
3. *μ is a membrane structure, consisting of n membranes having initially neutral polarizations, injectively labeled with elements of H;*
4. *$\mathcal{M}_1, \ldots, \mathcal{M}_n$ are strings over Γ, describing the multisets of objects placed in the n initial regions of μ;*
5. *R is a finite set of developmental rules, of the following forms:*
   (a) *$[a \to v]_h$, for $h \in H, a \in \Gamma, v \in \Gamma^*$ (object evolution rules).*
   (b) *$a[\ ]_h \to [b]_h$, for $h \in H, a, b \in \Gamma$ (in communication rules).*
   (c) *$[a\ ]_h \to b[\ ]_h$, for $h \in H, a, b \in \Gamma$ (out communication rules).*
   (d) *$[a\ ]_h \to b$, for $h \in H, a, b \in \Gamma$ (dissolution rules)*
   (e) *$[a\ ]_h \to [b\ ]_h[c\ ]_h$, for $h \in H, a, b, c \in \Gamma$ (weak division rules for elementary or non-elementary membranes).*
   (a) *$h_o \in H$ or $h_o = env$ indicates the output region (in this case, usually h_o do not appear in the description of the system).*

Also, we can consider rules of the form $[[\ ]_{h_1}[\ ]_{h_2}]_{h_3} \to [[\ ]_{h_1}]_{h_3}\ [[\ ]_{h_2}]_{h_3}$, where h_1, h_2, h_3 are labels: if the membrane with label h_3 contains other membranes than those with labels h_1, h_2, these membranes and their contents are duplicated and placed in both new copies of the membrane h_3; all membranes and objects placed inside membranes h_1, h_2, as well as the objects from membrane h_3 placed outside membranes h_1 and h_2, are reproduced in the new copies of membrane h_3. These rules are called *strong division rules for non–elementary membranes.*

Using the *maximally parallel manner*, at each computation step (a global clock is assumed) in each region of the system we apply the rules in such a way that no further rule can be applied to the remaining objects or membranes. In each step, each object and each membrane can be involved in only one rule.

A halting computation provides a result given by the number of objects present in region h_o at the end of the computation; this is a region of the system if $h_o \in H$ (and in this case, for a computation to be successful, exactly one membrane with label h_o should be present in the halting configuration), or it is the environment if $h_o = env$.

We denote by $\mathcal{AM}^0$ the class of recognizing polarizationless P systems with active membranes, and we denote by $\mathcal{AM}^0(\alpha, \beta)$, where $\alpha \in \{-d, +d\}$ and $\beta \in \{-ne, +new, +nes\}$, the class of all recognizing P systems with polarizationless active membranes such that: (a) if $\alpha = +d$ (resp. $\alpha = -d$) then dissolution rules are permitted (resp. forbidden); and (b) if $\beta = +new$ or $+nes$ (resp. $\beta = -ne$) then division rules for elementary and non–elementary membranes, weak or strong (respectively only division rules for elementary) are permitted.

The class of all decision problems solvable in uniform (resp. semi–uniform) way, and in polynomial time by a family of recognizing membrane systems is denoted by $\mathbf{PMC}_{\mathcal{R}}$ (resp. $\mathbf{PMC}^*_{\mathcal{R}}$)

Proposition 1. *For each* $\alpha \in \{-d, +d\}$, $\beta \in \{-ne, +new, +nes\}$, *and* $\epsilon = *, \lambda$, *we have:*

1. $\mathbf{PMC}_{\mathcal{AM}^0(\alpha,\beta)} \subseteq \mathbf{PMC}^*_{\mathcal{AM}^0(\alpha,\beta)}$
2. $\mathbf{PMC}^{\epsilon}_{\mathcal{AM}^0(\alpha,-ne)} \subseteq \mathbf{PMC}^{\epsilon}_{\mathcal{AM}^0(\alpha,+new)}$
3. $\mathbf{PMC}^{\epsilon}_{\mathcal{AM}^0(\alpha,-ne)} \subseteq \mathbf{PMC}^{\epsilon}_{\mathcal{AM}^0(\alpha,+nes)}$
4. $\mathbf{PMC}^{\epsilon}_{\mathcal{AM}^0(-d,\beta)} \subseteq \mathbf{PMC}^{\epsilon}_{\mathcal{AM}^0(+d,\beta)}$

where $\epsilon = *$ *(respectively* ϵ = *empty string) means that the complexity classes are associated with semi–uniform (respectively, uniform) solutions.*

At the beginning of 2005, Gh. Păun (problem **F** from [13]) wrote: *My favorite question (related to complexity aspects in P systems with active membranes and with electrical charges) is that about the number of polarizations. Can the polarizations be completely avoided? The feeling is that this is not possible – and such a result would be rather sound: passing from no polarization to two polarizations amounts to passing from non–efficiency to efficiency.*
That is, formally we can formulate the called conjecture of Păun as follows:

> *The class of all decision problems solvable in polynomial time by polarizationless P systems with active membranes using evolution, communication, dissolution and division rules for elementary membranes (working in the maximally parallel mode) is equal to the class* **P**

This conjecture can be expressed in terms of complexity classes in P systems as follows: $\mathbf{P} = \mathbf{PMC}_{\mathcal{AM}^0(+d,-ne)} = \mathbf{PMC}^*_{\mathcal{AM}^0(+d,-ne)}$
Next, we study possible answers to the conjecture of Păun.

2.1 A Partial Affirmative Answer

Let us recall that using the concept of *dependency graph* associated with a P system, a partial affirmative answer to the Păun's conjecture can be given.

Let Π be a P system whose working alphabet is Γ and the set of labels is H, and we denote by *env* the label of the environment. The *dependency graph associated with* the system Π is the directed graph G_{Π} whose nodes are the pairs $(a, h) \in \Gamma \times (H \cup \{env\})$ such that the object a in membrane (maybe the environment) labelled by h either triggers a rule or it is produced by a rule, and $((a, h), (a', h'))$ is an arc in the graph if there exists a rule r of Π such that the object a in membrane labelled by h produces the object a' in membrane (maybe the environment) labelled by h' by the application of rule r.

It can be proved that there exists a deterministic Turing machine that constructs the dependency graph, G_Π, associated with Π, in polynomial time, that is, in a time bounded by a polynomial function depending on the total number of rules and the maximum length of the rules (see [4]).

Let Δ_Π be the set whose elements are the pairs $(a,h) \in \Gamma \times (H \cup \{env\})$ such that there exists a path (within the dependency graph) from (a,h) to $(\mathtt{yes}, env)$. Having in mind that the *reachability problem* (see chapter 1 from [10]) can be solved by a search algorithm running in polynomial (quadratic) time, there exists a deterministic Turing machine that constructs the set Δ_Π in polynomial time, that is, in a time bounded by a polynomial function depending on the total number of rules and the maximum length of the rules (see [4]).

Let $\mathbf{\Pi} = \{\Pi(n) : n \in \mathbf{N}\}$ be a family of recognizing P systems with input membranes (not using dissolution rules) solving a decision problem X, in a uniform way. Let (cod, s) be a polynomial encoding associated with that solution. An instance u of the problem will be accepted by the system $\Pi(s(u))$ with input $cod(u)$ if and only if there is an object a in a membrane h of the initial configuration of the system such that there exists a path in the associated dependency graph from (a,h) to $(\mathtt{yes}, env)$. As a consequence of the previous results we have the following:

Theorem 1. *For each $\beta \in \{-ne, +new, +nes\}$, we have $\mathbf{P} = \mathbf{PMC}_{\mathcal{AM}^0(-d,\beta)}$.*

Similar characterizations of **P** can be obtained when we deal with semi–uniform solutions in the framework of recognizing polarizationless P systems with active membranes, and where dissolution rules are forbidden. The proofs are similar, it is enough to consider the system $\Pi(u)$, for each instance u of the decision problem, instead of the system $\Pi(s(u))$ with input the multiset $cod(u)$.

Theorem 2. *For each $\beta \in \{-ne, +new, +nes\}$, we have $\mathbf{P} = \mathbf{PMC}^*_{\mathcal{AM}^0(-d,\beta)}$.*

2.2 A Partial Negative Answer

It has been shown ([4]) that the class of decision problems solvable in polynomial time in a semi–uniform way by families of recognizing polarizationless P systems with active membranes where dissolution rules are permitted, and using division rules for elementary and non–elementary membranes, contains the standard complexity class **NP**.

Theorem 3. *We have the following:*

1. $\mathbf{SAT} \in \mathbf{PMC}^*_{\mathcal{AM}^0(+d,+nes)}$
2. $\mathbf{NP} \cup \mathbf{co\text{–}NP} \subseteq \mathbf{PMC}^*_{\mathcal{AM}^0(+d,+nes)}$

Moreover, it has been obtained an efficient *uniform* solution for the **QSAT**–problem ([2]) in this framework.

Theorem 4. *We have the following:*

1. $\mathbf{QSAT} \in \mathbf{PMC}_{\mathcal{AM}^0(+d,+nes)}$.
2. $\mathbf{PSPACE} \subseteq \mathbf{PMC}_{\mathcal{AM}^0(+d,+nes)}$.

Hence, in the framework of polarizationless P systems with active membranes and working in the maximally parallel mode, dissolution rules play a crucial role from the computational efficiency point of view. Specifically, if dissolution rules are forbidden then it is not possible to solve **NP**–complete problems in polynomial time (unless $\mathbf{P} = \mathbf{NP}$). Nevertheless, if dissolution rules are permitted then it is possible to efficiently solve computationally hard problems.

That is, the Păun's conjecture has a (partial) negative answer (assuming that $\mathbf{P} \neq \mathbf{NP}$). Nevertheless, the answer will be (partially) affirmative if dissolution rules are forbidden.

3 A New Version of a Păun's Conjecture

Recently, in [3] a more relaxed strategy of using the rules was introduced, the so-called *minimal parallelism*: in each region where at least a rule can be applied, at least one rule must be applied (if there is no conflict with the objects), without any other restriction. This introduces an additional degree of non-determinism in the system evolution.

P systems with active membranes working in the *minimally parallel mode* means the following:

- All the rules of any type involving a membrane h form the set R_h, this means all the rules of the form $[a \to v]_h$, all the rules of the form $a[\ ]_h \to [b]_h$, and all the rules of the form $[a]_h \to z$ and $[a]_h \to [b]_h[c]_h$, with the same h, constitute the set R_h.
- If a membrane h appears several times in a given configuration of the system, then for each occurrence of the membrane we consider a different set R_h.
- Then, in each step, from each set R_h associated with each membrane labelled by $h \in H$, from which at least a rule *can* be used, at least one rule *must* be used (if there is no conflict with the objects; for example, if we have only an object a in membrane h and we have an evolution rule $[a \to b]_h$ and a send–in rule $a[\ ]_{h'} \to [c]_{h'}$, being h' the label of a membrane immediately inside membrane h, then we can apply at least a rule from R_h and from $R_{h'}$, but we will apply only one between these two rules).

Of course, as usual for P systems with active membranes, each membrane and each object can be involved in only one rule, and the choice of rules to use and of objects and membranes to evolve is done in a non-deterministic way. In each step, the use of rules is done in the bottom-up manner (first the inner objects and membranes evolve, and the result is duplicated if any surrounding membrane is divided).

In this kind of P systems still universality and semi–uniform polynomial–time solutions to **SAT** were obtained in the new framework by using P systems with active membranes, with three polarizations [3].

Theorem 5. *The* **SAT** *problem can be solved in a semi–uniform way and in a linear time by polarization P systems with active membranes, without dissolution rules and using (weak) division for non–elementary membranes, and working in the minimally parallel mode.*

Next, we define new classes of P systems related to $\mathcal{AM}^0$. Let $\alpha \in \{-d, +d\}$, $\beta \in \{-ne, +new, +nes\}$ and $\gamma \in \{m, M, md, Md\}$. Then we denote by $\mathcal{AM}^0(\alpha, \beta, \gamma)$ the class of recognizing P systems with polarizationless active membranes such that:

- $\alpha = +d$: dissolution rules are permitted.
- $\alpha = -d$: dissolution rules are forbidden.
- $\beta = +new$: division rules for elementary and (weak) non–elementary membranes are permitted.
- $\beta = +nes$: division rules for elementary and (strong) non–elementary membranes are permitted.
- $\beta = -ne$: only division rules for elementary membranes are permitted.
- $\gamma = m$: working in the minimally parallel mode.
- $\gamma = md$: working in the deterministic minimally parallel mode.
- $\gamma = M$: working in the maximally parallel mode.
- $\gamma = Md$: working in the deterministic maximally parallel mode.

Proposition 2. *We have the following:*

1. $\mathbf{PMC}^{\epsilon}_{\mathcal{AM}^0(\alpha,\beta,md)} \subseteq \mathbf{PMC}^{\epsilon}_{\mathcal{AM}^0(\alpha,\beta,m)}$
2. $\mathbf{PMC}^{\epsilon}_{\mathcal{AM}^0(\alpha,\beta,m)} \subseteq \mathbf{PMC}^{\epsilon}_{\mathcal{AM}^0(\alpha,\beta,M)}$
3. $\mathbf{PMC}^{\epsilon}_{\mathcal{AM}^0(\alpha,\beta,Md)} \subseteq \mathbf{PMC}^{\epsilon}_{\mathcal{AM}^0(\alpha,\beta,M)}$

where $\epsilon = *$ *or* $\epsilon =$ *empty string.*

We can formulate the Păun's conjecture in P systems working in the minimally parallel mode.

> *The class of all decision problems solvable in polynomial time by polarizationless P systems with active membranes using evolution, communication, dissolution and division rules for elementary membranes (working in the minimally parallel mode) is equal to the class* **P**

This conjecture can be expresses in terms of complexity classes in P systems as follows: $\mathbf{P} = \mathbf{PMC}_{\mathcal{AM}^0(+d,-ne,m)} = \mathbf{PMC}^{*}_{\mathcal{AM}^0(+d,-ne,m)}$.
Next, we study possible answers to the new version of Păun's conjecture.

3.1 A Partial Affirmative Answer

Let us recall that through the concept of *dependency graph* associated with a P system, we have given a partial affirmative answer to the Păun's conjecture related with P systems working in the maximally parallel mode, and without using dissolution rules.

Let us also recall that, in order to give that answer, the main property that the dependency graph associated with a P system must satisfy is the following:

every computation of the system is an accepting computation if and only if there exists an object a in an initial membrane (labelled by h) of the system such that there exists a path (within the dependency graph) from (a, h) to $(\mathtt{yes}, env)$.

When a polarizationless P system with active membranes works in the minimally parallel mode, in each transition step we can think that objects are assigned to rules, non–deterministically choosing the rules and the objects assigned to each rule, according to the semantic of the minimally parallel mode. The objects which remain unassigned are left where they are, and they are passed unchanged to the next configuration (and belonging to the same membrane because dissolution rules are not permitted). So, the above property is satisfied by the computations of this kind of P systems working in the minimally parallel mode (without using dissolution rules) because we can pass from a node (a, h) to another node (a', h') in the dependency graph if and only if there exists a transition step producing (a', h') from (a, h).

Hence, we have a negative answer to the to the Păun's conjecture related with P systems working in the minimally parallel mode (deterministically or not), and without using dissolution rules.

Theorem 6. $\mathbf{P} = \mathbf{PMC}^{\epsilon}_{\mathcal{AM}^0(-d,\beta,md)} = \mathbf{PMC}^{\epsilon}_{\mathcal{AM}^0(-d,\beta,m)}$, *where* $\epsilon = *$ *or* $\epsilon =$ *empty string.*

3.2 A Partial Negative Answer

Next, we give a semi–uniform linear–time solution to **SAT** by using polarizationless P systems with active membranes working in the minimally parallel mode, and now using dissolution rules.

Theorem 7. **SAT** *can be solved in a semi–uniform way and in a linear time by polarizationless P systems with active membranes, using evolution, communication, dissolution and (weak) division for non–elementary membrane rules, and working in the deterministic minimally parallel mode.*

Proof. Let $\varphi = C_1 \wedge \cdots \wedge C_m$ be a propositional formula in conjunctive normal form, such that each clause C_j, $1 \leq j \leq m$, is of the form $C_j = y_{j,1} \vee \cdots \vee y_{j,k_j}$, $k_j \geq 1$, for $y_{j,r} \in \{x_i, \neg x_i \mid 1 \leq i \leq n\}$, and being $\{x_1, \ldots, x_n\}$ the set of variables of φ. For each $i = 1, 2, \ldots, n$, let us denote

$$t(x_i) = \{C_j \mid \text{there is } r, 1 \leq r \leq k_j, \text{ such that } y_{j,r} = x_i\},$$
$$f(x_i) = \{C_j \mid \text{there is } r, 1 \leq r \leq k_j, \text{ such that } y_{j,r} = \neg x_i\}.$$

That is, $t(x_i)$ (respectively, $f(x_i)$) is the set of clauses which assume the value *true* when x_i is *true* (resp. when x_i is false). Obviously, these sets have at most m elements.

We construct a recognizing P system

$$\Pi(\varphi) = (\Gamma, H, \mu, \mathcal{M}_0, \mathcal{M}_1, \ldots, \mathcal{M}_m, \mathcal{M}_p, \mathcal{M}_q, \mathcal{M}_r, \mathcal{M}_s, \mathcal{M}_{s'}, \mathcal{R})$$

associated with the formula φ as follows:

$$\Gamma = \{a_i, f_i, t_i \mid 1 \le i \le n\} \cup \{c_j, d_j \mid 1 \le j \le m\} \cup \{p_i \mid 1 \le i \le 2n+7\}$$
$$\cup \{q_i \mid 1 \le i \le 2n+1\} \cup \{r_i \mid 1 \le i \le 2n+5\} \cup \{b_1, b_2, y, \mathbf{yes}, \mathbf{no}\},$$
$$H = \{0, 1, 2, \ldots, m, p, q, r, s, s'\},$$
$$\mu = [_s[_{s'}[_p\ ]_p[_0[_q\ ]_q[_r]_r[_1\ ]_1[_2\ ]_2 \cdots [_m\ ]_m]_0]_{s'}]_s,$$
$$\mathcal{M}_p = p_1,\ \mathcal{M}_q = q_1,\ \mathcal{M}_r = r_1,\ \mathcal{M}_0 = a_1, \mathcal{M}_s = \mathcal{M}_{s'} = \mathcal{M}_j = \lambda,\ (1 \le j \le m)$$

The set of evolution rules, $\mathcal{R}$, consists of the following rules:

(a) $[p_i \to p_{i+1}]_p$, for all $1 \le i \le 2n+6$
$[q_i \to q_{i+1}]_q$, for all $1 \le i \le 2n$
$[r_i \to r_{i+1}]_r$, for all $1 \le i \le 2n+4$

(b) $[a_i]_0 \to [f_i]_0[t_i]_0$, for all $1 \le i \le n$
$[f_i \to f(x_i)a_{i+1}]_0$ and $[t_i \to t(x_i)a_{i+1}]_0$, for all $1 \le i \le n-1$,
$[f_n \to f(x_n)]_0$; $[t_n \to t(x_n)]_0$

(c) $c_j[\ ]_j \to [c_j]_j$ and $[c_j]_j \to d_j$, for all $1 \le j \le m$.

(d) $[q_{2n+1}\ ]_q \to q_{2n+1}[\ ]_q$; $[q_{2n+1} \to b_1^m]_0$

(e) $b_1[\ ]_j \to [b_1]_j$; and $[b_1]_j \to b_2$, for all $1 \le j \le m$,
$[b_2\ ]_0 \to b_2$; $[p_{2n+7}\ ]_p \to p_{2n+7}[\ ]_p$; $[p_{2n+7}\ ]_{s'} \to \mathbf{no}[\ ]_{s'}$; $[\mathbf{no}\ ]_s \to \mathbf{no}[\ ]_s$
$[r_{2n+5}\ ]_r \to r_{2n+5}$; $[r_{2n+5}\ ]_0 \to y[\ ]_0$; $[y\ ]_{s'} \to \mathbf{yes}$; $[\mathbf{yes}\ ]_s \to \mathbf{yes}[\ ]_s$

An overview of the computation of $\Pi(\varphi)$

The rules of type (a) are used for evolving general counters p_i, q_i and r_i in membranes labelled by p, q and r, respectively, making possible the correct synchronization.

In parallel with these rules, the non–elementary membrane 0 evolves by means of the rules of the type (b). In step $2i-1$ $(1 \le i \le 2n)$, object a_i produces the division of the membrane 0 (with f_i, t_i corresponding to the truth values *false*, *true*, respectively, for variable x_i). In step $2i$ we introduce inside membrane 0 the clauses satisfied by x_i or $\neg x_i$, respectively. Let us recall that when we divide membrane 0, all inner objects and membranes are replicated. At the end of this phase, all 2^n truth assignments for the n variables are generated and they are encoded in membranes labeled by 0.

In parallel with the process of membrane division, in the odd steps (until step $2n+1$), if a clause C_j is satisfied by the previously expanded variable, then the corresponding object c_j enters membrane j, by means of the first rule of the type (c), producing their dissolution in the next step by means of the second rule of that type and sending objects d_j to membrane 0.

In step $2n+2$, in each membrane 0, the counters q_i and r_i follow evolving and the second rule of the type (d) produces m copies of the object b_1.

Thus, the configuration $\mathcal{C}_{2n+2}$ of the system obtained after $2n+2$ steps, consists of 2^n copies of membrane 0, each of them encoding a truth assignment of the variables associated with φ, and containing the membrane q empty, the membrane r with the object r_{2n+3}, possible objects c_j and d_j, $1 \leq j \leq m$, as well as copies of only membranes with labels $1, 2, \ldots, m$ corresponding to clauses which were not satisfied by the truth assignment generated in that copy of membrane 0. Also, in that configuration the membrane p contains the object p_{2n+3} and membranes s' and s are empties.

Hence, formula φ is satisfied if and only if there is a membrane 0 where all membranes $1, 2, \ldots, m$ have been dissolved. In order to check this last condition, we proceed as follows.

In step $2n+3$ we use the first rule of the type (e) which introduces objects b_1 in each membrane j which has not been dissolved. In parallel, the counters p and r follow evolving. In step $2n+4$ objects b_1 in membrane j (in each such membrane appearing in the configuration $\mathcal{C}_{2n+2}$) dissolve these membranes producing object b_2 in membrane 0. Therefore, the presence of objects b_2 in membrane 0 of the configuration $\mathcal{C}_{2n+4}$ means that the truth assignment encoded by that membrane makes the formula false.

In step $2n+5$ the counter r_{2n+5} exits from membrane r and, simultaneously, each membrane 0 containing an object b_2 is dissolved by the third rule of the type (e). Hence, formula φ is satisfied if and only if in the configuration $\mathcal{C}_{2n+5}$ there exists a membrane 0 that has not been dissolved.

In step $2n+6$, the counter p_i evolves to p_{2n+7} in membrane p, and if there is a membrane 0 that has not been dissolved, the object r_{2n+5} sends to membrane s' an object y. On the contrary, only the counter p_i evolves.

In step $2n+7$ the counter p_{2n+7} exits from membrane p to membrane s', by applying the first rule of the type (f). Moreover, in that step, if the formula φ is satisfiable then an object y dissolves the membrane s' by applying the sixth rule of the type (f) producing an object **yes** in the skin, that in step $2n+8$ is sent to the environment; and the system halts. On the contrary, if membrane s' has not been dissolved, in step $2n+8$ by applying the second rule of the type (f) the object p_{2n+7} exits from membrane s' producing an object no in the skin, and in step $2n+9$ sends to the environment an object no; then, the system halts.

The system $\Pi(\varphi)$ uses $9n+2m+18$ objects, $m+6$ initial membranes, containing in total 4 objects, and $8n+4m+21$ rules. The length of any rule is bounded by $m+3$. Clearly, all computations stop (after at most $2n+9$ steps) and all give the same answer, **yes** or no, to the question whether formula φ is satisfiable.

Corollary 1. $\mathbf{NP} \cup \mathbf{co\text{-}NP} \subseteq \mathbf{PMC}^*_{\mathcal{AM}^0(+d,+new,md)}$.

As a consequence of this result, we have polarizationless P systems with active membranes, using dissolution rules and (weak) division for non–elementary membranes, and working in the deterministic minimally parallel mode are able to give semi–uniform solutions of **NP**–complete problems. That is, we have a (partial) negative answer to the new version of the Păun's conjecture.

The following picture describes the results obtained until now related to Păun's conjecture in both modes, where $-u$ (resp. $+u$) means semi–uniform (resp. uniform) solutions, $-ne$ (resp. $+ne$) means using division only for elementary membranes (resp. division for elementary and non–elementary membrane, *strong* in the maximal parallelism and *weak* in the minimal parallelism). Through this graph, we try to specify whether or not it is possible to solve computationally hard problems by recognizing P systems of the class associated with each node. The direction of each arrow shows a relation of inclusion, and each blue node provides an open question.

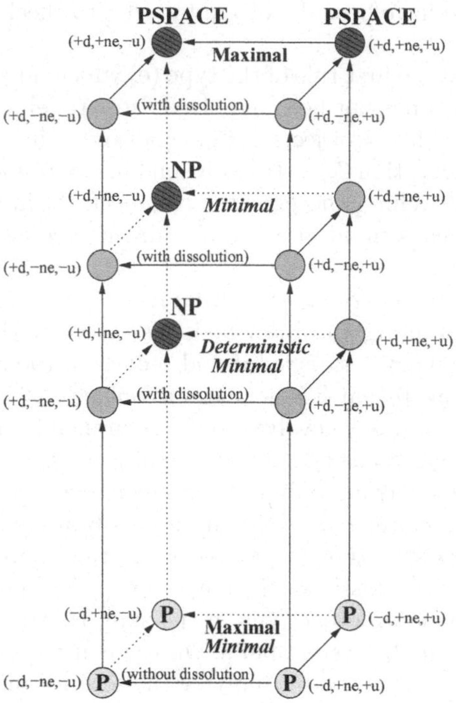

4 Conclusions and Open Problems

A conjecture of Păun, related to the impossibility to solve NP–complete problems in polynomial–time by means of polarizationless P systems with active membranes, is studied in this paper. Partial solutions are given within the usual framework of P systems working in the maximally parallel mode. As a consequence of the results obtained, the crucial role played by dissolution rules when we try to solve computationally hard problems, is highlighted.

Besides, a new version of that conjecture is formulated, this time associated to polarizationless P systems with active membranes working in the minimally parallel mode. Other partial solutions also arise from this new version and once again, dissolution rules are shown to be a singular ingredient which gives a borderline between efficiency (the possibility of solving computationally hard problems using feasible membrane computing resources) and non–efficiency.

Finally, we propose some open problems.

1. $\mathbf{NP} \cup \mathbf{co\text{-}NP} \subseteq \mathbf{PMC}^{\epsilon}_{\mathcal{AM}^0(+d,-ne,m)}$?
2. $\mathbf{NP} \cup \mathbf{co\text{-}NP} \subseteq \mathbf{PMC}_{\mathcal{AM}^0(+d,+new,md)}$?
3. $\mathbf{PMC}_{\mathcal{AM}^0(\alpha,\beta,m)} \subseteq \mathbf{PMC}_{\mathcal{AM}^0(\alpha,\beta,Md)}$?
4. Consider other ingredients in the framework $\mathcal{AM}^0(-d,\beta,\gamma)$ that permit to solve **NP**–complete problems.
5. Study the computational efficiency of the class $\mathcal{AM}^0$ **with** evolution rules with length 2 (or **with** communication rules without evolution of objects).
6. Study the computational efficiency of the class $\mathcal{AM}^0$ **without** evolution rules (or **without** communication rules).

Acknowledgments

The work of the first and the third authors was partially supported by MiUR under the project "Azioni Integrate Italia-Spagna - Theory and Practice of Membrane Computing". The work of the second author was supported by the project TIN2006–13425 of the Ministerio de Educación y Ciencia of Spain, co–financed by FEDER funds, the project of Excellence TIC–581 of the Junta de Andalucía, and the Acción Integrada Hispano–Italiana HI2005–0194.

References

1. A. Alhazov, R. Freund. On efficiency of P systems with active membranes and two polarizations. In [9], 81–94.
2. A. Alhazov, M.J. Pérez–Jiménez. Uniform solution to QSAT using polarizationless active membranes. In M.A. Gutiérrez, Gh. Păun, A. Riscos, F.J. Romero (eds.) *Proceedings of the Fourth Brainstorming Week on Membrane Computing, Volume I*, Fénix Editora, Sevilla, 2006, 29-40.
3. G. Ciobanu, L. Pan, Gh. Păun, M.J. Pérez-Jiménez. P systems with minimal parallelism. *Theoretical Computer Sci.*, to appear.
4. M.A. Gutiérrez–Naranjo, M.J. Pérez–Jiménez, A. Riscos–Núñez, F.J. Romero–Campero. On the power of dissolution in P systems with active membranes. *Lecture Notes in Computer Science*, **3850** (2006), 224–240.
5. M.A. Gutiérrez–Naranjo, M.J. Pérez–Jiménez, A. Riscos–Núñez, F.J. Romero–Campero, A. Romero–Jiménez. Characterizing tractability by cell-like membrane systems. In K.G. Subramanian, K. Rangarajan, M. Mukund (eds.) *Formal models, languages and applications*, World Scientific, Series in Machine Perception and Artificial Intelligence - Vol. 66, 2006, pp. 137–154.
6. M.A. Gutiérrez-Naranjo, M.J. Pérez-Jiménez, A. Riscos-Núñez. A fast P system for finding a balanced 2-partition. *Soft Computing*, **9**, 9 (2005), 673–678.
7. S.N. Krishna, R. Rama. A variant of P systems with active membranes: Solving NP–complete problems. *Romanian Journal of Information Science and Technology*, **2**, 4 (1999), 357–367.
8. A. Obtulowicz. Deterministic P systems for solving SAT problem. *Romanian Journal of Information Science and Technology*, **4**, 1–2 (2001), 551–558.
9. G. Mauri, Gh. Păun, M.J. Pérez-Jiménez, G. Rozenberg, A. Salomaa (eds.). *Membrane Computing, International Workshop, WMC5, Milano, Italy, 2004, Selected Papers*, Lecture Notes in Computer Science, **3365**, Springer-Verlag, Berlin, 2005.

10. C.H. Papadimitriou. *Computational Complexity*, Addison–Wesley, Massachussetts, 1995.
11. Gh. Păun. P systems with active membranes: Attacking **NP**–complete problems. *Journal of Automata, Languages and Combinatorics*, **6**, 1 (2001), 75–90.
12. Gh. Păun. Computing with membranes: Attacking **NP**–complete problems. In I. Antoniou, C. Calude, M.J. Dinneen (eds.) *Unconventional Models of Computation*, Springer–Verlag, London, 2000, 94–115.
13. Gh. Păun. Further twenty six open problems in membrane computing. In M.A. Gutiérrez, A. Riscos, F.J. Romero, D. Sburlan (eds.) *Proceedings of the Third Brainstorming Week on Membrane Computing*, Fénix Editora, Sevilla, 2005, 249-262
14. M.J. Pérez–Jiménez, A. Romero–Jiménez, F. Sancho–Caparrini. The P versus NP problem through cellular computing with membranes. In N. Jonoska, Gh. Paun, Gr. Rozenberg (eds.) *Aspects of Molecular Computing*, Lecture Notes in Computer Science, Springer-Verlag, Berlin, **2950** (2004), 338-352.
15. M.J. Pérez-Jiménez, A. Riscos-Núñez. Solving the Subset-Sum problem by active membranes. *New Generation Computing*, **23**, 4 (2005), 367–384.
16. M.J. Pérez-Jiménez, A. Riscos-Núñez. A linear–time solution to the Knapsack problem using P systems with active membranes. In C. Martín-Vide, Gh. Păun, G. Rozenberg, A. Salomaa (eds.) *Membrane Computing*, Lecture Notes in Computer Science, Springer–Verlag, Berlin, **2933** (2004), 250–268.
17. M.J. Pérez-Jiménez, A. Romero-Jiménez, F. Sancho-Caparrini. A polynomial complexity class in P systems using membrane division. In E. Csuhaj-Varjú, C. Kintala, D. Wotschke, G. Vaszil (eds.) *Proceedings of the 5th Workshop on Descriptional Complexity of Formal Systems, DCFS 2003*, Computer and Automation Research Institute of the Hungarian Academy of Sciences, Budapest, 2003, pp. 284-294.
18. M.J. Pérez-Jiménez, F.J. Romero–Campero. Attacking the Common Algorithmic Problem by recognizer P systems. In M. Margenstern (ed.) *Machines, Computations and Universality*, Lecture Notes in Computer Science, Springer–Verlag, Berlin, **3354** (2005), 304-315.
19. P. Sosik. The computational power of cell division. *Natural Computing*, **2**, 3 (2003), 287–298.
20. C. Zandron, C. Ferretti, G. Mauri. Solving NP–complete problems using P systems with active membranes. In I. Antoniou, C. Calude, M.J. Dinneen (eds.) *Unconventional Models of Computation*, Springer–Verlag, Berlin, 2000, 289–301.

A Parallel DNA Algorithm Using a Microfluidic Device to Build Scheduling Grids

Marc García-Arnau[1], Daniel Manrique[2], and Alfonso Rodríguez-Patón[2]

[1] Artificial Intelligence Department, Universidad Politécnica de Madrid, Boadilla del Monte s/n, 28660 Madrid, Spain
mgarnau@dia.fi.upm.es

[2] Artificial Intelligence Department, Universidad Politécnica de Madrid, Boadilla del Monte s/n, 28660 Madrid, Spain
{dmanrique,arpaton}@fi.upm.es

Abstract. Microfluidic systems, which constitute a miniaturization of a conventional laboratory to the dimensions of a chip, are expected to become the key support for a revolution in the world of biology and chemistry. This article proposes a parallel algorithm that uses DNA and such a distributed microfluidic device to generate scheduling grids in polynomial time. Rather than taking a brute force approach, the algorithm presented here uses concatenation and separation operations to gradually build the DNA strings that represent a Multiprocessor Task scheduling problem grids. The microfluidic device used makes for an autonomous system, also enabling it to solve the problem without the need of external control.

1 Introduction

In 1994 Leonard Adleman proved empirically what Richard Feynman had postulated several decades earlier: the chemical and electrical properties of matter give molecules the natural ability to make massively parallel calculations. So, for the first time, Adleman used DNA strands to solve an instance of the Hamiltonian path problem on a 7-node graph [1]. This fired the starting gun for a new branch of research known as biomolecular computing. A year later, Richard Lipton put forward a DNA computational model that generalized the techniques employed by Adleman, which he used to solve an instance of the SAT [2]. Since then, many researchers have exploited DNA's potential for solving computationally difficult problems (class NP-complete). Two good examples can be found in [3] and [4]. NP-complete problems have two prominent features: 1) there are as yet no polynomial algorithms to solve them, and 2) all their "yes" instances can be verified efficiently [5].

Microfluidic systems, also called microflow reactors or "lab-on-a-chip" (LOC), are passive fluidic devices built on a chip layer which is used as a substrate. They are basically composed of cavities or microchambers between which liquid can move along the microchannels that link them. Therefore, controlled chemical reactions can be carried out in each cavity independently, that is, in parallel.

J. Mira and J.R. Álvarez (Eds.): IWINAC 2007, Part I, LNCS 4527, pp. 193–202, 2007.

These systems are tantamount to the miniaturization of a conventional laboratory to the dimensions of a chip and are expected to become the key support for a revolution in the world of biology and chemistry. New emerging disciplines, as synthetic biology and systems biology, demand computer scientists to contribute in the resolution of new challenging problems such as drug discovery, the understanding of computational processes in cells or the analysis of genetic pathways controlling biological processes in living organisms. In all those problems, microfluidic systems implementing parallel algorithms may play an important role [6,7]. These needs have lead to a fast development of microfluidic-based technology in the last years. For instance, a microfluidic chip for automated nucleid acid purification from bacterial or mammalian cells is constructed in [8]. The use of microfluidic systems to implement automated DNA sequencing devices is also been considered in [9]. Furthermore, the role played by valves and pumps in these kind of devices is studied in [10,11]. Finally in [12], the possibility of implementing microfluidic memory and control devices is presented. Besides that, some work also exists using microfluidic systems to attack computationally hard problems. Thus, a microflow reactor is used in [13] to solve the Minimum Clique Problem using a brute force strategy and codifying each possible subgraph as a DNA strand. Furthermore, two microfludic systems each implementing DNA algorithms are proposed in [14] and [15] for the Hamiltonian path problem and the shortest common superstring problem, respectively. Moreover, a microfluidic DNA computer is used in [16] to solve the satisfability problem.

The present work deals with a classic scheduling problem: optimal scheduling of tasks with precedence constraints in a multiprocessor scenario. This paper starts from the previous resolution of the problem of getting all independents sets of a dependency graph, and proposes a parallel DNA algorithm using a microfluidic device to build optimal scheduling grids. To do so, it uses a constructive approach that has nothing to do with traditional brute force strategies, gradually putting together correct solutions and removing invalid results along the way. The remainder of this article is organized as follows. Section 2 presents the problem. Section 3 describes the parallel algorithm and the microfluidic support system. An example of how the proposed system runs is given in section 4. Finally, section 5 sets out the final remarks.

2 Multiprocessor Task Scheduling Problem

Let T be a set of tasks with single execution times. Let $G(T, E)$ be a direct acyclical graph (dag), which we will term dependency graph that establishes a partial order $\prec$ on T. Each graph node represents a task, each arc a relation of precedence between two tasks. If there is a path between two tasks t_i and t_j, t_j is said to be a descendant of t_i and, therefore, its execution will not be able to start until t_i has been run ($t_i \prec t_j$). Finally, let the positive integers M and d be the number of the processors and the scheduling deadline, respectively. A schedule for the tasks of T on M processors that respects the partial order established by the graph G is a function $f : T \rightarrow \mathbb{Z}^+$ that satisfies [17]:

$$If\ (t_i \prec t_j),\ then\ f(t_i) < f(t_j) \quad (1)$$
$$For\ all\ u \leq d,\ |t_i : f(t_i) = u| \leq M \quad (2)$$

This schedule's end time will be $\max_{t_i \in T} f(t_i)$. A schedule of $|T|$ tasks on M processors that respects the dependency graph can actually be represented as a scheduling grid A of size $N \times M$, (with $N \leq |T|$) where each element $A(k, m_r) = t_i$, $(1 \leq k \leq N)$, $(1 \leq r \leq M)$ indicates that task t_i has been scheduled to run on processor m_r at time k ($t_i \in T \cup \emptyset$, where the symbol $\emptyset$ represents the null process). A scheduling grid A will be valid if it meets the following conditions: a) it does not contain repeated elements (with exception of the null process $\emptyset$), b) it contains all the tasks in the set T (it is complete) and c) it satisfies both the constraints expressed in (1) and (2). A valid grid will be optimal if it also maximises the schedule's parallelism, that is, it has the least possible number of rows Figure 1.

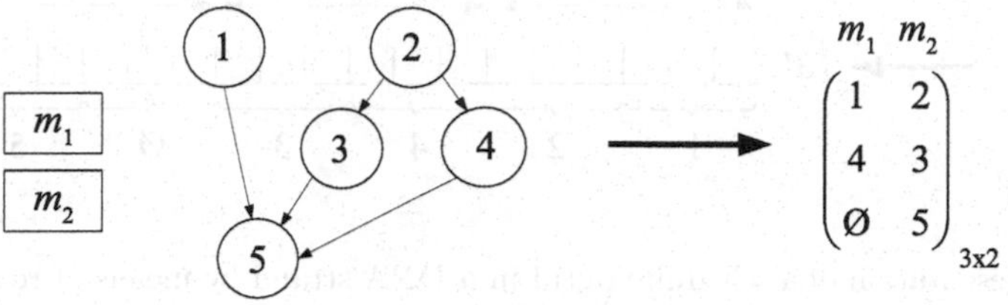

Fig. 1. Dependency graph associated with the set of tasks $T = \{1, 2, 3, 4, 5\}$ together with an optimal scheduling grid for a two-processor scenario m_1 and m_2

To achieve such maximum parallelism in a schedule, it is essential to observe the partial order $\prec$ between the tasks established by the graph G. We know that if a task t_j is the descendant of another task t_i, then t_j will not be able to be executed in parallel with (or, of course, before) t_i. If we call two tasks without a relation of precedence in the partial order $\prec$ independent, then the maximum set of tasks that can be run in parallel with a given task t_i is made up of the maximum independent set of tasks containing t_i. Therefore, the greatest possible parallelism between all the problem tasks is given by all the maximal independent sets obtained from the dependency graph. An independent set s_i is maximal if there is no other greater set v such that $s_i \subset v$.

Getting all these s_i is an NP-complete problem. There are multiple classical algorithms that have addressed its solution as it has also been found to be a problem besetting some genomics efforts, such as mapping genome data [18]. However, the proposals for solving it from the molecular or membrane computing paradigms are very few. For instance, [19] proposes a P system with active membranes to solve it. Recently, our group has developed a DNA algorithm to solve the minimum clique cover problem for graphs, which is an equivalent problem, albeit on the complementary graph [20].

The algorithm proposed below is based on this result, that is, on getting the maximal independent sets s_i from the dependency graph to design a system capable of building optimal scheduling grids for our problem. For the example graph of Figure 1, those sets are: $s_1 = \{1, 2\}$, $s_2 = \{1, 3, 4\}$ and $s_3 = \{5\}$.

3 A Parallel DNA Algorithm and a Microfluidic Device

3.1 System Design

We use a single DNA sequence to codify each task of the set T. A scheduling grid A is represented as a task sequence, composed of N subsequences of M elements each, where N is the schedule size (number of rows in A) and M is the number of processors. An example is shown in Figure 2.

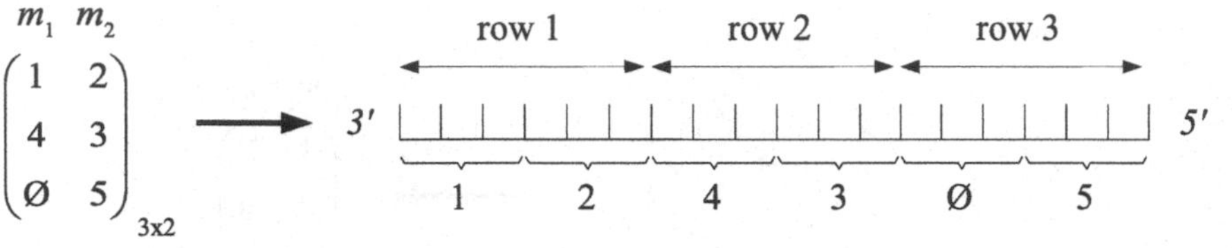

Fig. 2. Representation of a scheduling grid in a DNA strand by means of row concatenation. ($N = 3$ and $M = 2$).

The present algorithm builds the possible scheduling grids in parallel (concatenation operation), removing any that are invalid as they are detected (separation operation). Both operations are based on DNA's natural property of hybridization through complementarity. Additionally, these operations are supported in this case not by test tubes but by the microfluidic device's microchambers and microchannels around which the scheduling grids under construction circulate.

We propose a two-layer system architecture S. It is composed of three different subsystems S_i, each of which is associated with one of the maximum independent sets s_i of G. In turn, each subsystem S_i contain several interconnected nodes v_j, which correspond to the tasks of their associated set s_i. If $|s_i| < M$, a new element $v_\emptyset$, corresponding to the null process, is added to the subsystem S_i. This way, symbols $\emptyset$ can be used to fill the grid positions that cannot house any other task in the set T. Figure 3 a) illustrates this point.

The system S works as follows. All its subsystems S_i operate in parallel and synchronously to add whole rows to the scheduling grids they receive. Once the respective row has been added, each S_i sends the resulting grids to their neighbouring subsystems which continue the operation. This process is done N times until the grid is complete (with $N \leq |T|$). Within a subsystem S_i, a row is generated in M steps that are taken by its nodes v_j working in parallel to add, one by one, each of the M tasks of which that row is composed. The total number of steps that it takes to build the scheduling grids in the system S is,

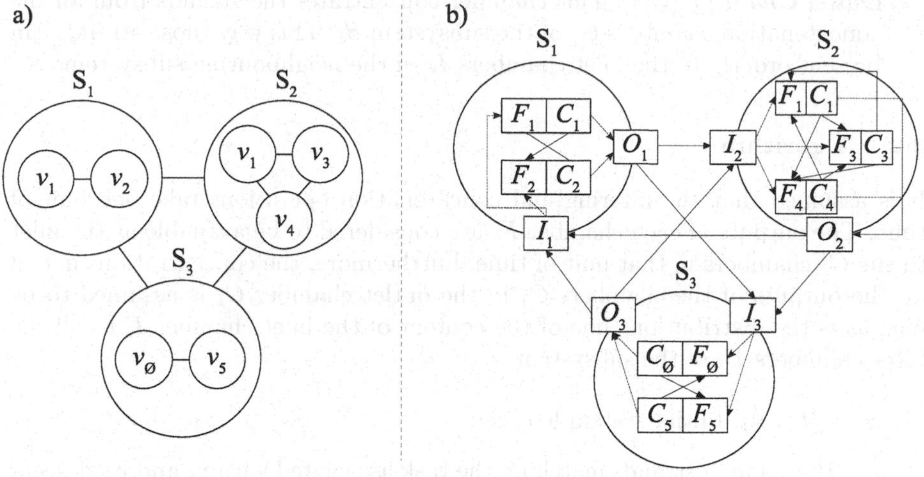

Fig. 3. a) Diagram of the two-layer system S associated with the example in Fig. 1. It is composed of three subsystems each with several nodes: $S_1 = \{v_1, v_2\}$, $S_2 = \{v_1, v_3, v_4\}$, $S_3 = \{v_5, v_\emptyset\}$. **b)** Design of the microfluidic device for the system S shown in Fig. 3 a). Each element v_j is composed of a filter chamber F_j and a concatenation chamber C_j. Additionally, all the subsystems S_i have an inlet chamber I_j and an outlet chamber O_i.

therefore, $N \times M$. The system stops when it detects that the scheduling grids already contain all the tasks of the set T. For this to be possible, those grids (DNA strands) need to be sequenced after each iteration of system S.

In the following, we detail how the microfluidic device has been designed to implement the functionality described. Each of the elements v_j making up a subsystem S_i is composed of a filter chamber F_j and a concatenation chamber C_j. Additionally, each subsystem S_i has an inlet chamber I_i and an outlet chamber O_i. All the chambers linked by microchannels have pumps to circulate the fluid in the direction of the arrows (Fig. 3 b)). The function and connectivity of each chamber type of the proposed device is:

- *Filter Chamber* (F_j): This chamber receives the strings from the chambers C_p of all the neighbouring v_p or from the inlet chamber I_i of the subsystem S_i. It retains the strings that already contain the task associated with v_j (repeats) or any of its descendent tasks. It sends the other strands to its associated concatenation chamber C_j.
- *Concatenation Chamber* (C_j): This chamber receives the strings from its associated F_j chamber. It concatenates the task associated with v_j with the strings. It sends the resulting strands to the chambers F_p of all its neighbouring v_p or to the outlet chamber O_i of the subsystem S_i.
- *Inlet Chamber* (I_i): This chamber receives and concentrates the flow of strands from the chambers O_r of the neighbouring subsystems S_r. This way all filtering chambers F_j of the the subsystem S_i are loaded orderly.

– *Outlet Chamber* (O_i) : This chamber concentrates the strands from all the concatenation chambers C_j of the subsystem S_i. This way those strands can be sent orderly to the inlet chambers I_r of the neighbouring subsystems S_r.

3.2 Algorithm

It is assumed that the filtering and concatenation operations take one unit of time. The outputs of each chamber F_j are considered to be available at the inlet to the C_j chambers in that unit of time. Furthermore, the concentration time of all the outputs of the chambers C_j in the outlet chamber O_i is assumed to be one, as is the distribution time of the content of the inlet chamber I_i to all the filter chambers F_j of the subsystem S_i:

1. *Step* ($t = 0$). Initial system loading:
   - Put enough strands matching the task associated with v_j and its descendent tasks into the filter chambers F_j of each S_i, (except in $F_\emptyset$).
   - Put enough copies of the strand of the task associated with v_j, and enough copies of the auxiliary strands and enzymes to allow concatenation into the concatenation chambers C_j of each subsystem S_i.
   - Put enough copies of the strands of all the tasks associated with v_j of that subsystem into the inlet chambers I_i of each subsystem S_i.
2. *Steps* ($t = 1$) *to* ($t = N$). While the strings are incomplete:
   - *For* ($n = 1$) *to* ($n = M$) *do*
     - *Computation*: For all elements v_j of all subsystems S_i in parallel, do a filter operation in $F_{j(n)}$ and a concatenate operation in $C_{j(n)}$.
     - *Internal communication*: For all v_j of all subsystems S_i in parallel, pump the results of chambers $C_{j(n)}$ to the chambers $F_{p(n+1)}$ of their neighbouring v_p:

$$Input(F_{p\,(n+1)}) = \bigcup_{\forall\, p \neq j} Output(C_{j\,(n)})$$

   - Load chambers O_i with the content of all chambers C_j of all subsystems S_i in parallel:

$$O_{i(t)} = \bigcup_{j} C_{j(t)}$$

   - Communication between subsystems: Load the chambers I_i of all the subsystems S_i with the flow from the chambers O_r of the neighbouring subsystems S_r in parallel. Divide the content of I_i among the chambers F_j of all their v_j:

$$I_{i(t+1)} = \bigcup_{\forall\, r \neq i} O_{r(t)}$$

3. *Step* ($t = N + 1$). System output:
   - The chambers O_i already contain the strings of length $N \times M$, with all the tasks of T. The system stops and returns the resulting schedules output.

$$Output(S) = \bigcup_i O_{i(t)}$$

This is, therefore, a polynomial algorithm in terms of number of tasks and processors. It takes $N \times M$ steps (with $N \leq |T|$) to build the problem-solving grids.

4 Example

The system completes a total of $3 \times 2 = 6$ iterations before stopping and returning the problem-solving scheduling grid:

- $\boxed{t_0}$ (Initial loading of I_i and F_i)

  $I_1(t_0) = \{1, 2\}$ $\quad I_2(t_0) = \{1, 3, 4\}$ $\quad I_3(t_0) = \{5, \emptyset\}$

- $\boxed{t_1}$ (Execution of the subsystems S_i and final loading of O_i)

S_1	S_2	S_3
$F_1 = I_1(t_0)$ $C_1 = \{21\}$ $F_2 = I_1(t_0)$ $C_2 = \{12\}$	$F_1 = I_2(t_0)$ $C_1 = \{31, 41\}$ $F_3 = I_2(t_0)$ $C_3 = \{13, 43\}$ $F_4 = I_2(t_0)$ $C_4 = \{14, 34\}$	$F_5 = I_3(t_0)$ $C_5 = \{\emptyset 5\}$ $F_\emptyset = I_3(t_0)$ $C_\emptyset = \{5\emptyset\}$

$O_1(t_1) = \{12, 21\}$ $\quad O_2(t_1) = \{31, 41, 13, 43, 14, 34\}$ $\quad O_3(t_1) = \{5\emptyset, \emptyset 5\}$

- $\boxed{t_2}$ (Communication between subsystems S_i (loading of I_i and of F_j). Execution of subsystems S_i)

  $I_1(t_2) = \{31, 41, 13, 43, 14, 34, 5\emptyset, \emptyset 5\}$
  $I_2(t_2) = \{12, 21, 5\emptyset, \emptyset 5\}$
  $I_3(t_2) = \{12, 21, 31, 41, 13, 43, 14, 34\}$

S_1	S_2	S_3
$F_1 = I_1(t_2)$ $C_1 = \{431, 341\}$ $F_2 = I_1(t_2)$ $C_2 = \{5\emptyset 2, \emptyset 52\}$	$F_1 = I_2(t_2)$ $C_1 = \{-\}$ $F_3 = I_2(t_2)$ $C_3 = \{123, 213\}$ $F_4 = I_2(t_2)$ $C_4 = \{124, 214\}$	$F_5 = I_3(t_2)$ $C_5 = \{125, 215, 315, 415,$ $135, 435, 145, 345\}$ $F_\emptyset = I_3(t_2)$ $C_\emptyset = \{12\emptyset, 21\emptyset, 31\emptyset, 41\emptyset,$ $13\emptyset, 43\emptyset, 14\emptyset, 34\emptyset\}$

– $\boxed{t_3}$ (Execution of the subsystems S_i and final loading of O_i)

S_1	S_2	S_3
$F_1 = C_2(t_2)$ $C_1 = \{-\}$ $F_2 = C_1(t_2)$ $C_2 = \{-\}$	$F_1 = C_3(t_2) + C_4(t_2)$ $C_1 = \{-\}$ $F_3 = C_1(t_2) + C_4(t_2)$ $C_3 = \{1243, 2143\}$ $F_4 = C_1(t_2) + C_4(t_2)$ $C_4 = \{1234, 2134\}$	$F_5 = C_\emptyset(t_2)$ $C_5 = \{12\emptyset5, 21\emptyset5, 31\emptyset5, 41\emptyset5, 13\emptyset5, 43\emptyset5, 14\emptyset5, 34\emptyset5\}$ $F_\emptyset = C_5(t_2)$ $C_\emptyset = \{125\emptyset, 215\emptyset, 315\emptyset, 415\emptyset, 135\emptyset, 435\emptyset, 145\emptyset, 345\emptyset\}$

$O_1(t_3) = \{-\}$
$O_2(t_3) = \{1243, 2143, 1234, 2134\}$
$O_3(t_3) = \{12\emptyset5, 21\emptyset5, 31\emptyset5, 41\emptyset5, 13\emptyset5, 43\emptyset5, 14\emptyset5, 34\emptyset5, 125\emptyset, 215\emptyset, 315\emptyset, 415\emptyset, 135\emptyset, 435\emptyset, 145\emptyset, 345\emptyset\}$

– $\boxed{t_4}$ (Communication between subsystems S_i (loading of I_i and of F_j). Execution of subsystems S_i)

$I_1(t_4) = \{1243, 2143, 1234, 2134, 12\emptyset5, 21\emptyset5, 31\emptyset5, 41\emptyset5, 13\emptyset5, 43\emptyset5, 14\emptyset5, 34\emptyset5, 125\emptyset, 215\emptyset, 315\emptyset, 415\emptyset, 135\emptyset, 435\emptyset, 145\emptyset, 345\emptyset\}$
$I_2(t_4) = \{12\emptyset5, 21\emptyset5, 31\emptyset5, 41\emptyset5, 13\emptyset5, 43\emptyset5, 14\emptyset5, 34\emptyset5, 125\emptyset, 215\emptyset, 315\emptyset, 415\emptyset, 135\emptyset, 435\emptyset, 145\emptyset, 345\emptyset\}$
$I_3(t_4) = \{1243, 2143, 1234, 2134\}$

S_1	S_2	S_3
$F_1 = I_1(t_4)$ $C_1 = \{-\}$ $F_2 = I_1(t_4)$ $C_2 = \{-\}$	$F_1 = I_2(t_4)$ $C_1 = \{-\}$ $F_3 = I_2(t_4)$ $C_3 = \{-\}$ $F_4 = I_2(t_4)$ $C_4 = \{-\}$	$F_5 = I_3(t_4)$ $C_5 = \{12435, 21435, 12345, 21345\}$ $F_\emptyset = I_3(t_4)$ $C_\emptyset = \{1243\emptyset, 2143\emptyset1234\emptyset, 2134\emptyset\}$

– $\boxed{t_5}$ (Execution of subsystems S_i and final loading of O_i)

S_1	S_2	S_3
$F_1 = C_2(t_4)$ $C_1 = \{-\}$ $F_2 = C_1(t_4)$ $C_2 = \{-\}$	$F_1 = C_3(t_4) + C_4(t_4)$ $C_1 = \{-\}$ $F_3 = C_1(t_4) + C_4(t_4)$ $C_3 = \{-\}$ $F_4 = C_1(t_4) + C_3(t_4)$ $C_4 = \{-\}$	$F_5 = C_\emptyset(t_4)$ $C_5 = \{1243\emptyset5, 2143\emptyset5, 1234\emptyset5, 2134\emptyset5\}$ $F_\emptyset = C_5(t_4)$ $C_\emptyset = \{12435\emptyset, 21435\emptyset, 12345\emptyset, 21345\emptyset\}$

$O_1(t_5) = \{-\}$
$O_2(t_5) = \{-\}$
$O_3(t_5) = \{1243\emptyset5, 2143\emptyset5, 1234\emptyset5, 2134\emptyset5, 12435\emptyset, 21435\emptyset, 12345\emptyset, 21345\emptyset\}$

– $\boxed{t_6}$ (End of execution. Results output)
When the algorithm finished, we got eight strings codifying eight combinations of the problem-solving scheduling grid:

$$\begin{array}{c} m_1 \;\; m_2 \\ \begin{pmatrix} 1 & 2 \\ 4 & 3 \\ \emptyset & 5 \end{pmatrix} \end{array} \begin{array}{c} m_1 \;\; m_2 \\ \begin{pmatrix} 2 & 1 \\ 4 & 3 \\ \emptyset & 5 \end{pmatrix} \end{array} \begin{array}{c} m_1 \;\; m_2 \\ \begin{pmatrix} 1 & 2 \\ 3 & 4 \\ \emptyset & 5 \end{pmatrix} \end{array} \begin{array}{c} m_1 \;\; m_2 \\ \begin{pmatrix} 2 & 1 \\ 3 & 4 \\ \emptyset & 5 \end{pmatrix} \end{array} \begin{array}{c} m_1 \;\; m_2 \\ \begin{pmatrix} 1 & 2 \\ 4 & 3 \\ 5 & \emptyset \end{pmatrix} \end{array} \begin{array}{c} m_1 \;\; m_2 \\ \begin{pmatrix} 2 & 1 \\ 4 & 3 \\ 5 & \emptyset \end{pmatrix} \end{array} \begin{array}{c} m_1 \;\; m_2 \\ \begin{pmatrix} 1 & 2 \\ 3 & 4 \\ 5 & \emptyset \end{pmatrix} \end{array} \begin{array}{c} m_1 \;\; m_2 \\ \begin{pmatrix} 2 & 1 \\ 3 & 4 \\ 5 & \emptyset \end{pmatrix} \end{array}$$

5 Concluding Remarks

This article proposes a parallel algorithm that uses DNA and a distributed microfluidic device to generate scheduling grids in polynomial time. Microfluidic systems are passive fluidic devices built on a chip layer which is used as a substrate. They contain cavities, microchannels, pumps and valves that allow controlled chemical reactions to be carried out independently, that is, in parallel.

The algorithm described in this paper takes a constructive approach based on concatenation and filter operations to get optimal scheduling grids. This uses fewer strings than would be necessary if we tried to generate those grids by brute force. Although, from the computational point of view, a tough combinatorial problem has to be solved beforehand, this algorithm constitutes an interesting approach to the possibilities brought by microfluidic systems inherent parallelism. From now on, with the advent of synthetic and systems biology, computer scientists biologists and chemists will deal toguether with the resolution of new and challenging problems, in which microfluidic systems implementing parallel algorithms may play an important role.

The evolution of these systems since Adleman's and Lipton's early experiments prove that they constitute a promising and highly interesting technology for implementing distributed DNA algorithms. The structure of microfluidic systems can be exploited to design topologies, and these topologies can then be used to implement automated bioalgorithms on a miniaturized scale, as shown in this paper.

Acknowledgements

This research has been partially funded by the Spanish Ministry of Science and Education under project TIN2006-15595.

References

1. Adleman, L. M.: Molecular computation of solutions to combinatorial problems. Science. **266** (1994) 1021–1024
2. Lipton, R. J.: DNA solution of hard computational problems. Science. **268** (1995) 542–545
3. Ouyang, Q., Kaplan, Peter D., Liu, S., Libchaber, A.: DNA Solution of the Maximal Clique Problem. Science. **278** (1997) 446–449
4. Sakamoto, K., Gouzu, H., Komiya, K., Kiga, D., Yokohama, S., Yokomori, T., Hagiya, M.: Molecular Computation by DNA Hairpin Formation. Science. **288** (2000) 1223–1226

5. Garey, M. R., and Johnson, D. S.: Computers and Intractability, A Guide to the theory of NP-completeness. W. H. Freeman, San Francisco (1979).
6. Dittrich, P. S., and Manz, A.: Lab-on-a-chip: microfluidics in drug discovery. Nature Reviews Drug Discovery. **5** (2006) 210–218
7. David, N. Breslauer, Philip, J., Lee and Luke P. Lee: Microfluidics-based systems biology. Molecular Biosystems. **2** (2006) 97–112
8. Hong, J. W., Studer, V., Hang, G., Anderson, W. F., Quake, S. R.: A nanoliter-scale nucleic acid processor with parallel architecture. Nature Biotechnology. **22** (2004) 435–439
9. Kartalov, E. P., Quake, S. R.: Microfluidic device reads up to four consecutive base pairs in DNA sequencing-by-synthesis. Nucleic acids research. **32** (2004) 2873–2879
10. Grover, W. H., Mathies, R. A.: An integrated microfluidic processor for single nucleotide polymorphism-based DNA computing. Lab Chip. **5** (2005) 1033–1040
11. Van Noort, D., Landweber, L. F.: Towards a re-programmable DNA Computer. Natural Computing. **4** (2005) 163–175
12. Groisman, A., Enzelberger, M., Quake, S. R.: Microfluidic memory and control devices. Science. **300** (2003) 955–958
13. McCaskill, J. S.: Optically programming DNA computing in microflow reactors. Biosystems. **59** (2001) 125–138
14. Ledesma, L., Pazos, J., Rodríguez-Patón, A.: A DNA Algorithm for the Hamiltonian Path Problem Using Microfluidic Systems. In N. Jonoska, Gh. Paun and G. Rozenberg, Eds., "Aspects of Molecular Computing - Essays dedicated to Tom Head on the occasion of his 70th birthday", LNCS 2950, Springer-Verlag. (2004) 289–296
15. Ledesma, L., Manrique, D., Rodríguez-Patón, A.: A Tissue P System and a DNA Microfluidic Device for Solving the Shortest Common Superstring Problem. Soft Computing. **9** (2005) 679–685
16. Livstone, M., Weiss, R., Landweber, L.: Automated Design and Programming of a Microfluidic DNA Computer. Natural Computing. **5** (2006) 1–13
17. Papadimitriou, C. H. and Steiglitz, K.: Combinatorial optimization: algorithms and complexity. Prentice-Hall (1982) with Ken Steiglitz; second edition by Dover (1998).
18. Butenko, S. and Wilhelm, W. E.: Clique-detection Models in Computational Biochemistry and Genomics. European Journal of Operational Research (2006). To appear.
19. Head, T.: Aqueous simulations of membrane computations. Romanian J. of Information Science and Technology. **5** (2002) 355–364
20. García-Arnau, M., Manrique, D., Rodríguez-Patón A.: A DNA algorithm for solving the Maximum Clique Cover Problem. Submitted.

P System Models of Bistable, Enzyme Driven Chemical Reaction Networks

Stanley Dunn[1] and Peter Stivers[2]

[1] Department of Biomedical Engineering
smd@occlusal.rutgers.edu
[2] Department of Mathematics
Rutgers University, Piscataway, NJ
pstivers@eden.rutgers.edu

Abstract. In certain classes of chemical reaction networks (CRN), there may be two stable states. The challenge is to find a model of the CRN such that the stability properties can be predicted. In this paper we consider the problem of building a P-system designed to simulate the CRN in an attempt to determine if the CRN is stable or bistable. We found that for the networks in [2] none of the bistable CRN would have a bistable P-system by stoichiometry alone. The reaction kinetics must be included in the P-system model; the implementation of which has been considered an open problem. In this paper we conclude that a P-system for a CRN in m reactants and n products has at most $2(m^2+mn)$ membranes and $6(m^2+mn)$ rules. This suggests that P-system models of a chemical reaction network, including both stoichiometry and reaction kinetics can be built.

1 Introduction

Craciun et al [2] considered the problem of understanding and characterizing enzyme driven chemical reaction networks (CRNs) that exhibit bistability - the capability that a CRN has to switch between two or more stable states. There is evidence that metabolic pathways can exhibit bistability, i.e., chemical signaling that causes the network to switch between two different stable steady states.

The common understanding of bistability is that there is a pathway in the network where the product of one reaction inhibits or promotes another reaction. Craciun et al argue that that bistability can result from very simple chemistry: Common enzymatic mechanisms and mass action kinetics have the capacity to demonstrate bistability in certain situations. By capacity, we mean that there exists combinations of rate constants and substrate supply rates that the CRN has at least two stable states.

Craciun et al develop a model called the *species-reaction*, or *SR* graph that models relationships between each species and the reactions in which it is a reactant or product. The main result of the paper [2] is a theorem that relates the number and type of cycles in the graph to the number of steady states for the CRN.

J. Mira and J.R. Álvarez (Eds.): IWINAC 2007, Part I, LNCS 4527, pp. 203–213, 2007.

If it is the case that there is a relationship between a given CRN and a description of the network architecture (such as the species-reaction graph in [2]), then it is reasonable to assume that there is a P system that simulates the chemical reaction network and that this simulation can be used to *experimentally* determine if there are multiple steady states for the CRN. Determining whether or not a CRN has the capacity for bistability using the *SR* graph amounts to finding cycles in the graph. Clearly, such a computation must be computable by a P-system. The only question that remains is (1) the particular type of P-system and (2) the representation of the CRN.

There are two parts to the description of a CRN: The stoichiometry (the mol balance) and the mass balance. Both have a precise relationship to the network structure, but the dynamic behavior of CRNs are governed by mass-action kinetics. Although the obvious plan of attack would be to model the CRN by the mass-action kinetics, it remains an unsolved problem as to how to implement this model [5].

In [1], Cazzaniga and colleagues study the behavior of P systems based on the Gillespie algorithm [4] for solving mass-action dynamics of chemical reaction networks. They compare a multi-compartmental approach with a dynamic probabilistic approach. In the former, the authors assume that the Gillespie algorithm runs in each compartment and that the assumptions have all been satisfied. In the latter algorithm, multiple, synchronized processes each running Gillespie's algorithm simulate the chemical reaction network. Cazzaniga et al use the quorum sensing problem as an example and show the results for implementations of both algorithms. While the results were substantively the same, the authors point out that the multi-compartmental approach has the distinct advantage of a single time stream and therefore system behavior. While the parallel design is computationally appealing, it isn't necessarily the best approach for reaching a descriptive model.

In [7] Jimenez and colleagues study P systems based on mass-action dynamics, which should mirror the (multi-compartmental) Gillespie approach in [1]. The rules in this model allow for reactions to take place with a rate constant k. Jimenez et al associate probabilities with each rule that are based on mass-action kinetics. Presumably, the authors are referring to the Gillespie algorithm, but it is not clear from the paper. They, too, use the Vibrio Fisheri quorum sensing problem as an example of their model. Jimenez et al also demonstrate the quorum sensing behavior, albeit with smaller populations. It is not clear if the semantics of their rules follow that of the Gillespie algorithm [4].

While much work has been done in the area of P systems with mass action rules, the number of examples for which they have been applied has been limited. Jimenez and co-workers applied their algorithm to the problem of the EGFR signaling cascade [8], but there are not too many other instances of how these models can be applied to real biochemical problems.

Our goal in this paper is to apply the model of P system with rule semantics based on mass action kinetics to the problem of analyzing the stability of enzyme driven chemical reaction networks. The data is from Craciun et al [2] and the simulator is based on the algorithm of Romero-Campero and Gheorghe [3]; we have used both Scilab and MATLAB implementations. If the rule semantics are truly mass-action dynamics, then with a sufficiently large number of trials, the networks should exhibit uni-stable or bistable behavior, as appropriate.

2 P-System Models of CRN Stoichiometry

None of the implementations reviewed permit fractional transport; this is in part the open problem outlined by Paun [5] on the implementation of ordinary differential equations. Therefore, we shall refer to these systems as P systems with rule semantics based on mass-action kinetics are really rules with semantics based on stoichiometry.

Given that the *SR* graphs of Craciun et al are not based on mass-action kinetics, but are based on the relationships of chemical entities to the reaction in which they participate and that these models show some promise for showing the capacity for bistability, it is reasonable to consider P-systems as models of these CRNs since they do not require mass-action kinetics. Also, the *SR* graph bears a close relationship to the stoichiometric relationships of the CRN and it is these relationships that we will implement in a P-system. Thus, we consider instead a P-system model of a CRN based on the stoichiometry of the network.

The PNAS article by Craciun hints that stoichiometric relationships can be used to determine if a CRN is stable or bistable. Their method is based on network models of the CRN, which suggests that simulation or analysis of the dynamic behavior is not necessary; another way of putting this would be that the results suggest that a static analysis of the CRN is both necessary and sufficient.

Our first question was to determine whether or not a P system model of the CRN stoichiometry can be used to determine if the CRN is stable, or bistable as in some of the cases in [2]. We used a stochastic model of P systems, based on the work of [1,7] to implement the stoichiometric simulation. This implementation is a multi-compartment Gillespie algorithm that was originally used to solve the quorum sensing problem. In the current problem, a single compartment model was used with rules that are the stoichiometry from [2] as shown in Table 1 below.

Figure 1 shows the results of a six minute simulation of the fourth CRN, the first of nine that exhibits bistability. The two stable states for CRN 4 are in Table 2.We repeated this simulation a number of times and there was no simulation in which the output tended toward the second state; in each case, the state of the P system tended toward the first steady state.

The networks in Table 1 are 1 compartment each. The difficulty with simulating these systems in overcoming the problem of modeling the replenishment of the substrate and inhibitor during the simulation. Our solution to this problem was to run the simulation for a short period of time, replenish the substrate and then continue the simulation for the unit of time during which there is no replenishment. The reaction and replenishment rates are given in the supplementary material to [2].

Table 1. Chemical Reaction Networks and their bistability from [2]

	Network	Bistability
1	$E+S \underset{k_2}{\overset{k_1}{\rightleftharpoons}} ES \overset{k_3}{\rightarrow} E+P$	No
2	$E+S \underset{k_2}{\overset{k_1}{\rightleftharpoons}} ES \overset{k_3}{\rightarrow} E+P \qquad E+I \overset{k_4}{\rightarrow} EI$	No
3	$E+S \underset{k_2}{\overset{k_1}{\rightleftharpoons}} ES \overset{k_3}{\rightarrow} E+P \qquad ES+I \overset{k_4}{\rightarrow} ESI$	No
4	$E+S \underset{k_2}{\overset{k_1}{\rightleftharpoons}} ES \overset{k_3}{\rightarrow} E+P \qquad E+I \underset{k_5}{\overset{k_4}{\rightleftharpoons}} EI$ $ES+I \underset{k_7}{\overset{k_6}{\rightleftharpoons}} ESI \underset{k_9}{\overset{k_8}{\rightleftharpoons}} EI+S$	Yes
5	$E+S_1 \underset{k_2}{\overset{k_1}{\rightleftharpoons}} ES_1 \qquad S_2+ES_1 \underset{k_4}{\overset{k_3}{\rightleftharpoons}} ES_1S_2 \overset{k_5}{\rightarrow} E+P$	No
6	$E+S_1 \underset{k_2}{\overset{k_1}{\rightleftharpoons}} ES_1 \qquad E+S_2 \underset{k_4}{\overset{k_3}{\rightleftharpoons}} ES_2$ $S_2+ES_1 \underset{k_6}{\overset{k_5}{\rightleftharpoons}} ES_1S_2 \underset{k_8}{\overset{k_7}{\rightleftharpoons}} S_1+ES_2 \qquad ES_1S_2 \overset{k_9}{\rightarrow} E+P$	Yes

Table 2. The steady state concentrations for the bistable CRN number 4 from [2]

Species	Steady State 1	Steady State 2
E	0.1875	2.3135 E-15
S	1	1096.6331
I	1	1
P	1733.2661	637.6330

Table 3. Mass-action kinetics for the fourth chemical reaction network in [2]

$$
\begin{aligned}
[\dot{S}] &= -k_1[E][S] + k_2[ES] + k_8[ESI] - k_9[EI][S] \\
[\dot{E}] &= -k_1[E][S] + k_2[ES] + k_3[ES] - k_4[E][I] + k_5[EI] \\
[\dot{I}] &= -k_4[E][I] + k_5[EI] - k_6[ES][I] + k_7[ESI] \\
[\dot{ES}] &= k_1[E][S] - k_2[ES] - k_3[ES] - k_4[ES][I] + k_5[ESI] \\
[\dot{EI}] &= k_4[E][I] - k_5[EI] + k_8[ESI] - k_9[EI][S] \\
[\dot{ESI}] &= k_6[ES][I] - k_7[ESI] - k_8[ESI] + k_9[EI][S] \\
[\dot{P}] &= k_3[ES]
\end{aligned}
$$

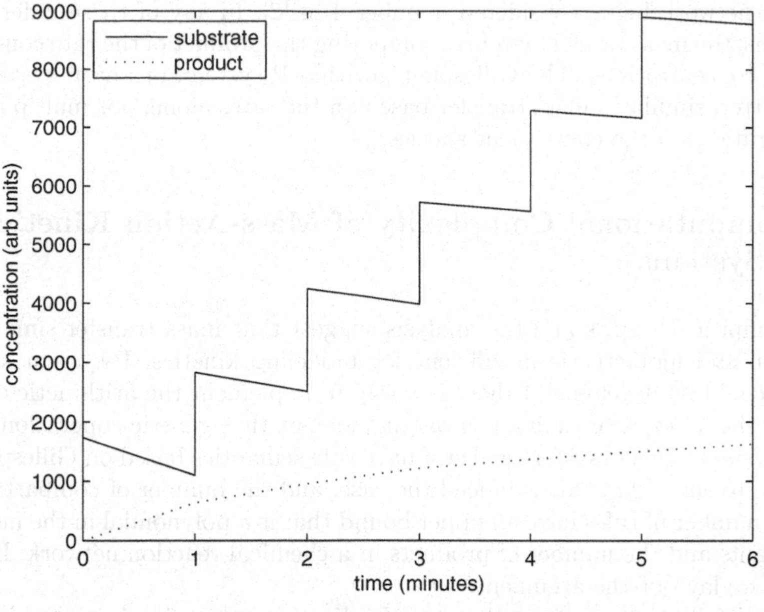

Fig. 1. Substrate and Product concentration as a function of time for CRN 4. The product concentration reaches the steady state concentration of 1733.2661. The staircase substrate concentration is due to the replenishment.

3 P-System Models of Mass-Action Kinetics

Because of the restricted model of mass-action kinetics, fractional mass transport cannot occur and chemical reaction networks that should exhibit bistable behavior do not. The problem with implementing P-system models is that the P-system (such as [1,7]) has mass transfer of whole, or integer units.

This returns us to Paun's open problem X in [5]. In order to model true mass action kinetics, there must exist a model in which fixed or floating point arithmetic us used to represent the real numbers in the mass-action kinetics. We considered a model of fixed point arithmetic in our Gillespie mass-action P system, but regardless of the resolution, i.e., regardless of the number of fixed point digits, the simulation of the fourth CRN in Table 1 showed convergence to the first of the two steady states.

Although we have no proof, we conjecture that a P-system simulation based on stoichiometry, with or without fixed point representation of the rate constants, cannot simulate bistable behavior exhibited by the underlying CRN. Although we can argue that the P-system simulates mass action, it does not simulate mass-action kinetics because the P-system does not implement multiplication[1].

[1] One might argue that the additions are simulated by the transfer of a molecule between compartments.

The table below shows the mass-action kinetic rate laws for the first bistable reaction network being considered, number 4 in [2]. In any of these differential equations, the mass transfer requires computing the product of the rate constants and the concentrations. The Gillespie algorithm P-system, or any P-system for that matter, simulates mass transfer based on the rates alone, not multiplied by concentrations of the constituent species.

4 Computational Complexity of Mass-Action Kinetic P-Systems

These empirical results and the analysis suggest that mass transfer simulation based on stoichiometry is insufficient for modeling kinetics. P-system kinetic models could be developed if there is a way to implement the arithmetic operations in the kinetics, meaning can one implement the (generic) operation $k_i[S]$ in a P-system. This P-system will still have rule semantics based on Gillespie [4]. We hope to show that this is indeed the case, and the number of compartments and the number of rules have an upper bound that is a polynomial in the number of reactants and the number of products in a chemical reaction network. In this section, we lay out the argument.

In what follows, we assume that a multiplication can be implemented in a P-system with 2 membranes and 6 rules and that an addition can be implemented in a P-system that has 1 membrane and 1 rule [9]. Furthermore, we shall assume that a chemical reaction network

$$A + B \rightarrow C \tag{1}$$

$$A \rightarrow B + C \tag{2}$$

in which the species B is both a reactant and a product shall not occur unless one of the reactions is part of a reversible reaction. That is, we shall assume that no chemical species can occur as both a reactant and a product in a given chemical reaction. First we shall describe our notation

- Let C be a system of k chemical reactions, $|C| = k$
- Let S be the set of chemical species in these reactions;
- Let $R \subseteq S$ be the set of reactants $= \{r_1, r_2, r_3, \ldots, r_R\}$
- Let $P \subseteq S$ be the set of products $= \{p_1, p_2, p_3, \ldots, p_P\}$
- Let $r_i \in R$ and $p_i \in P$
- Let e be the non-empty set $\{r_i, \mathrm{p}_i\}$ of reactants and products for a chemical reaction in the set $C = \{\{r_1, p_1\}, \{r_2, p_2\}, \{r_3, p_3\} \ldots \{r_k, p_k\}\}$
- Let m_k to be the number of reactants for the first k reactions: $m_k = |r_1 \cup r_2 \cup \ldots \cup r_k|$ and let n_k be the number of products in these first k reactions: $n_k = |p_1 \cup p_2 \cup \ldots \cup p_k|$

Then $m_{\mathrm{k}+1} = |r_1 \cup r_2 \cup \ldots \cup r_{(\mathrm{k}+1)}|$ and $n_{\mathrm{k}+1} = |p_1 \cup p_2 \cup \ldots \cup p_{(\mathrm{k}+1)}|$

There are three variables that describe the size of the CRN: The number of reactants m, the number of products n and the number of reactions k. First we show that as the size of the CRN increases, the number of reactants and/or products must increase. The consequence is that the bound on the size of the P-system as a function of the size of the CRN need only be expressed as a function of m and n, with k being subsumed in the analysis.

Lemma 1. *If a chemical reaction e_{k+1} is added to the system such that $|C \cup e_{k+1}| = k+1$, then both m_k and n_k are monotonically increasing with limits $\lim_{k\to\infty} m_k = \lim_{k\to\infty} n_k = |S|$. That is, any increase in the number of reactions in a system C* must *lead to an increase in the total number of reactants and/or the total number of products in the system until $m_k = n_k = |S|$.*

Proof.

Part A: The base case of two reactions

Suppose that the system C has two reactions e_1 and e_2. If both reactions have the same set of species, $r_1 \cup p_1 = r_2 \cup p_2$, then either $r_1 \neq r_2$, $p_1 \neq p_2$ or both since otherwise, they would be the same reaction e_1. There are three cases to consider for each of m and n:

- If $m_1 < m_2$ then there is an additional unique reactant in e_2 and therefore the total number of reactants has increased.
- If $m_1 = m_2$ and since $e_1 \neq e_2$ then $n_1 < n_2$ and therefore the total number of products has increased.
- The case where $m_1 > m_2$ cannot occur since it implies that $|r_1 \cup r_2| < |r_1|$, which says at least $r_2 = \emptyset$ which cannot occur.
- If $n_1 < n_2$ then there is an additional unique product in e_2 and therefore the total number of products has increased.
- If $n_1 = n_2$ and since $e_1 \neq e_2$ then $m_1 < m_2$ and therefore the total number of reactants has increased.
- The case where $n_1 > n_2$ cannot occur since it implies that $|p_1 \cup p_2| < |p_1|$, which says at least $p_2 = \emptyset$, which cannot occur.

Thus, we have shown that any increase in the number of reactions in a system C must lead to an increase in the total number of reactants and/or the total number of products in the system.

Part B: The induction step

Consider now the case of adding a reaction to a system C where $|C| = k > 2$. Neither the set of reactants r_{k+1} nor the set of products p_{k+1} can be the empty set.

Suppose we add a new reaction $e_{k+1} = \{r_{k+1}, p_{k+1}\}$ such that the products in the set p_{k+1} have already appeared in C , then $n_k = n_{k+1}$. Then, there must be an increase in the number of reactants, $m_{k+1} > m_k$ until $m_{k+1} = |S|$. Therefore, either the number of reactants has increased, or the number of reactants has reached the limit $|S|$.

Suppose we add a new reaction $e_{k+1} = \{r_{k+1}, p_{k+1}\}$ such that the reactants in the set r_{k+1} have already appeared in C , then $m_k = m_{k+1}$. Therefore, there must be an increase in the number of products, $n_{k+1} > n_k$ until $n_{k+1} = |S|$. Therefore, either the number of products has increased, or the number of products has reached the limit $|S|$.

Lastly, suppose we add a new reaction $e_{k+1} = \{r_{k+1}, p_{k+1}\}$ such that $m_{k+1} > m_k$ and $n_{k+1} > n_k$. Thus, the number of reactants and products has increased, or the number of reactants and products has reached the limit $|S|$.

Therefore, for each increase in k, there is an increase in either m or n or both. An increase in the number of reactions in a system C leads to an increase in the total number of reactants and/or the total number of products in the system until $m_k = n_k = |S|$.

Theorem 1. *A P system with k reactions of m reactants and n products will require at most* $2(m^2 + mn)$ *membranes and* $6(m^2 + mn)$ *rules to compute the multiplications in the reaction rate equations.*

Proof. The consequence of Lemma 1 is that one need not consider the number of reactions in a system C, but only the total number of unique reactants and products in the system. Therefore, we can prove the statement above by induction on m and n only.

Part A: The base case $m = n = 1$

Suppose there is one reaction $A \to B$ with the reaction rate equations $\frac{d[B]}{dt} = k_1[A]$ and $\frac{d[A]}{dt} = -k_1[B]$. Without regard for the sign of the rate constant, there is one multiplication in each equation. A P-system for a single multiplication has 2 membranes and 6 rules; this CRN requires 4 membranes and 12 rules as predicted by the theorem.

Part B: The induction step

Suppose we have a system C for which the theorem holds. We now add 1 reactant, 1 product or both.

Adding 1 reactant: If a new reactant is added to the system C , then there is a new reaction rate equation (RRE) that is added to the system. Also, there is now an additional multiplication in the original RREs as the concentration of the new species is an additional term in the product of the reactants. This additional multiplication will introduce a change to the P system model of the system C:

The system C has $2(m^2 + mn)$ membranes, and with the additional reactant:

$$2(m_0^2 + m_0 * n) + 2(m_0 + n + m_0 + 1) = 2((m + 1)^2 + (m + 1)n)$$
$$2(m_0^2 + m_0 * n + m_0 + n + m_0 + 1) = 2((m + 1)^2 + (m + 1)n)$$
$$2(m_0^2 + 2 * m_0 + 1 + m_0 * n + n) = 2((m + 1)^2 + (m + 1)n)$$
$$2((m_0 + 1)^2 + (m_0 + 1)n) = 2((m + 1)^2 + (m + 1)n)$$

The system C has $6(m^2 + mn)$ rules and with the additional reactant:

$$\begin{aligned} 6(m_0^2 + m_0 * n) + 6(m_0 + n + m_0 + 1) &= 6((m+1)^2 + (m+1)n) \\ 6(m_0^2 + m_0 * n + m_0 + n + m_0 + 1) &= 6((m+1)^2 + (m+1)n) \\ 6(m_0^2 + 2 * m_0 + 1 + m_0 * n + n) &= 6((m+1)^2 + (m+1)n) \\ 6((m_0+1)^2 + (m_0+1)n) &= 6((m+1)^2 + (m+1)n) \end{aligned}$$

Adding 1 product: If a new product is added to the system C, then there is a new reaction rate equation that is added to the system. Also, there is now an additional multiplication in the original RREs as the concentration of the new species is an additional term in the product of the (reaction) products. This additional multiplication will introduce a change to the P system model of the system C:

The system C has $2(m^2 + mn)$ membranes and with the additional product:

$$\begin{aligned} 2(m_0^2 + m_0 * n) + 2 * m_0 &= 2(m^2 + m(n+1)) \\ 2(m_0^2 + m_0 * n + m_0) &= 2(m^2 + m(n+1)) \\ 2(m_0^2 + m_0(n+1)) &= 2(m^2 + m(n+1)) \end{aligned}$$

The system C has $6(m^2 + mn)$ rules and with the additional product:

$$\begin{aligned} 6(m_0^2 + m_0 * n) + 6(m_0) &= 6((m+1)^2 + (m+1)n) \\ 6(m_0^2 + m_0 * n + m_0) &= 6((m+1)^2 + (m+1)n) \\ 6(m_0^2 + m_0(n+1)) &= 6((m+1)^2 + (m+1)n) \\ 6((m_0+1)^2 + (m_0+1)n) &= 6((m+1)^2 + (m+1)n) \end{aligned}$$

Adding 1 reactant and 1 product: If a new reactant and a new product are both added to the system C , then there is a new reaction rate equation for each. Also, there are now additional multiplications in the original RREs as the concentration of the new species are additional terms in the product of the (reaction) reactants or products, respectively. This additional multiplication will introduce a change to the P system model of the system C:

The system C has $2(m^2 + mn)$ membranes, and with the additional reactant and product:

$$\begin{aligned} &2(m_0^2 + m_0 * n_0) + 2(m_0 + m_0 + 1 + n_0 + m_0 + 1) = \\ &\qquad 2((m+1)^2 + (m+1)(n+1)) \\ &2(m_0^2 + m_0 * n_0 + m_0 + m_0 + 1 + n_0 + m_0 + 1) = \\ &\qquad 2((m+1)^2 + (m+1)(n+1)) \\ &2(m_0^2 + 2 * m_0 + 1 + m_0 * n_0 + n_0 + m_0 + 1) = \\ &\qquad 2((m+1)^2 + (m+1)(n+1)) \\ &2((m_0+1)^2 + (m_0+1)(n_0+1)) = 2((m+1)^2 + (m+1)(n+1)) \end{aligned}$$

The system C has $6(m^2 + mn)$ rules, and with the additional reactant and product:

$$6(m_0^2 + m_0 * n_0) + 6(m_0 + m_0 + 1 + n_0 + m_0 + 1) = 6((m+1)^2 + (m+1)(n+1))$$
$$6(m_0^2 + m_0 * n_0 + m_0 + m_0 + 1 + n_0 + m_0 + 1) = 6((m+1)^2 + (m+1)(n+1))$$
$$6(m_0^2 + 2 * m_0 + 1 + m_0 * n_0 + n_0 + m_0 + 1) = 6((m+1)^2 + (m+1)(n+1))$$
$$6((m_0+1)^2 + (m_0+1)(n_0+1)) = 6((m+1)^2 + (m+1)(n+1))$$

Thus the theorem holds. □

There is not as straightforward a rule for the additional addition operations as there is for multiplications. This is because the number of additions depends on the number of reactions in which a particular specie appears. It is clear, however, that the number of rules and membranes required for an addition is less than those required for multiplications. Based on the assumption that each specie appears only once in each reaction, then there is an upper bound on the number of additions in each ODE in the mass-action kinetics. Therefore, the number of membranes and rules required to implement the additions for the mass-action kinetics is also polynomial in the number of reactants, products and rules. It is possible to develop a P system that models both stoichiometry and true mass-action kinetics.

5 Conclusions

The use of P-systems to model chemical and biochemical systems has been studied for some time. Although the limitations have been understood [5], there have been several attempts to develop models that engender mass-action kinetics [1,7]. Although the papers suggest that the semantics are mass-action, the rules in these systems do not permit for fractional mass transport.

This is born out by the present application, using mass-action P-systems to simulate enzyme driven chemical reaction networks that exhibit bistability [2]. We found that there is at least one case where the network is bistable, but the mass-action P-system could only reach one of the two steady states. There is clearly a limitation to these stoichiometric mass-action models.

To fully implement mass-action kinetics, the P-system would have to implement all of the operations to solve the ODEs that model the kinetics. In this paper we considered the representation of the ODEs, which would require multiplications and additions that are a function of the number of reactants, products and reactions. The main result of this paper is that the number of multiplications and additions have polynomial bounds and therefore, can be computed in a P-system.

The plan for future work includes implementing the complete stoichiometric and mass-action P-system and to demonstrate that these P-systems can demonstrate the bistability of the networks such as those in [2]. Clearly, elucidating other non-linear properties of chemical reaction networks is a topic worthy of further consideration.

References

1. Cazzaniga P, Pescini D, Romero-Campero F.J., Besozzi D and Mauri G. Stochastic Approaches in P Systems for Simulating Biological Systems. Fourth Brainstorming Week on Membrane Computing, Sevilla (Spain), Jan 30 - Feb 3, 2006, 145-165.
2. Craciun G, Tang Y and Feinberg M. Understanding bistability in complex enzyme-driven reaction networks. PNAS 2006; 103; 8697-8702.
3. http://www.dcs.shef.ac.uk/~marian/PSimulatorWeb/PSystemMF.htm, last accessed 18 February 2007.
4. Gillespie D.T. Exact stochastic simulation of coupled chemical reactions. Journal of Physical Chemistry, 81; 2340-2361.
5. Paun Gh. Further Twenty Six Open Problems in Membrane Computing. Third Brainstorming Meeting on Membrane Computing, Sevilla (Spain), Feb. 2005
6. Paun Gh and Perez-Jimenez M.J. Membrane computing: Brief introduction, recent results and applications. BioSystems Vol 85, No 1, pp 11-22
7. Perez-Jimenez M.J. and Romero-Campero F.R. Modeling Vibrio fisheri's behaviour Using P Systems. Systems Biology Workshop, ECAL 2005, Sept. 2005.
8. Perez-Jimenez M.J. and Romero-Campero F.R. Modeling EGFR signaling cascade using continuous membrane systems. In G. Plotkin (Ed.) Proceedings of the Third International Workshop on Computational Methods in Systems Biology 2005 (CMSB 2005), 118-129.ling cascade using continuous membrane systems. In G. Plotkin (Ed.) Proceedings of the Third International Workshop on Computational Methods in Systems Biology 2005 (CMSB 2005), 118-129.
9. http://psystems.ieat.ro/, last accessed 18 February 2007.

A Novel Improvement of Neural Network Classification Using Further Division of Partition Space

Lin Wang[1], Bo Yang[1,⋆], Zhenxiang Chen[1], Ajith Abraham[2], and Lizhi Peng[1]

[1] School of Information Science and Engineering, University of Jinan, Jinan, China
[2] Centre for Quantifiable Quality of Service in Communication Systems, Norwegian University of Science and Technology, Norway
yangbo@ujn.edu.cn

Abstract. Further Division of Partition Space (FDPS) is a novel technique for neural network classification. Partition space is a space that is used to categorize data sample after sample, which are mapped by neural network learning. The data partition space, which are divided manually into few parts to categorize samples, can be considered as a line segment in the traditional neural network classification. It is proposed that the performance of neural network classification could be improved by using FDPS. In addition, the data partition space are to be divided into many partitions, which will attach to different classes automatically. Experiment results have shown that this method has favorable performance especially with respect to the optimization speed and the accuracy of classified samples.

Keywords: Classification, neural network, partition space, further division.

1 Introduction

Classification is an important research area in data mining. In supervised classification tasks, a classification model is usually constructed according to a given training set. Once the model has been built, it can map a test data to a certain class in the given class set. Many classification techniques including decision tree [1, 2], neural network (NN) [3], support vector machine (SVM) [4, 5], rule based classifiers systems etc. have been proposed. Among these techniques, decision tree is simple and easy to be comprehended by human beings. SVM is a new machine learning method developed on the Statistical Learning Theory. SVM is gaining popularity due to many attractive features, and promising empirical performance. SVM is based on the hypothesis that the training samples obey a certain distribution which restricts its application scope. Neural network classification, which is supervised, has been proved to be a practical approach with lots

⋆ Corresponding author.

J. Mira and J.R. Álvarez (Eds.): IWINAC 2007, Part I, LNCS 4527, pp. 214–223, 2007.

of success stories in several classification tasks. However, its training efficiency is usually a problem, which is the current focus of our research in this paper.

In conventional neural network classification, the partition space is a line segment between 0 and 1. A sample from the original data set is mapped to this line segment by the neural network. The sample will be deemed as class 0 if it is more close to 0 than to 1. Otherwise, the sample will be deemed to belong to class 1. If the partition space is divided into partitions, and the mapped sample gets close to these partitions freely, the mapping relationship (using a neural network) could be formed easily. Therefore, the performance of neural network classification including training speed and accuracy could be improved. This is the basic inspiration of our research.

Samples are mapped to the partition space by neural networks, and then partition space is assigned by the distribution of mapped samples. If neural networks are optimized, its corresponding signed partition space will be formed. Particle Swarm Optimization(PSO) is used to optimize neural network in FDPS for its predominant features. The problem, to assign the divided partitions in partition space by a certain category, once perplexed us. Very soon it was found that the category of majority in one partition should control the partition that it belongs to. Process that assigns partitions by category of majority is called color partitions. The traditional neural network classification could be regarded as a specific example of FDPS. Its partition space is divided into 2 partitions, and its dimension of partition vector is 1.

This paper is arranged as follows. Particle swarm optimization is outlined in Section 2. Section 3 illustrates the detailed method of FDPS. Section 4 outlines and discusses our experimental results followed by conclusions in Section 5.

2 Particle Swarm Optimization (PSO)

Particle swarm optimization is a population based stochastic optimization technique [6], inspired by social behavior of bird flocking or fish schooling. There are two reasons that PSO is attractive. There are few parameters to be adjusted and usually PSO achieves better results in a faster, cheaper way compared with other methods.

PSO is initialized with a population of random solutions and searches for optima by updating generations. The potential solutions, called particles, fly through the problem space by following the current optimum particles. Each particle keeps track of its coordinates in the problem space that are associated with the best solution (fitness) it has achieved so far. (The fitness value is also stored.) This value is called *pbest*. Another "best" value that is tracked by the particle swarm optimizer is the best value, obtained so far by any particle in the neighbors of the particle. This location is called *lbest*. When a particle takes all the population as its topological neighbors, the best value is a global best and is called *gbest*.

The particle swarm optimization concept consists of, at each time step, changing the velocity of (accelerating) each particle toward its *pbest* and *lbest* locations (local version of PSO). Acceleration is weighted by a random term, with separate random numbers being generated for acceleration towards *pbest* and *lbest* locations.

3 Further Division of Partition Space (FDPS)

In this section, FDPS is explained in detail. At first, we describe how the data set is mapped by using a neural network in FDPS. We then illustrate how to divide partition space into partitions and how to distribute mapped samples over these partitions.In order to color these partitions fairly, the reason why weights of classes are needed in FDPS is explained in the Subsection 3.3. The formula for getting the weight is also illustrated. Then we narrate the details of coloring partitions which is the kernel part of the training in FDPS. After partitions are colored, neural network and its corresponding colored partition space will be used in the evaluation of training and in the classification of new samples. The whole training algorithm is described in Subsection 3.7.

3.1 Mapping Data Set

Mapping is the process to transform the training data set from original space to partition space. The mapping relation we used is the back propagation learning method for artificial neural networks. The dimension of input data set vector is defined as n and dimension of partition vector is defined as m. Every element of partition vector is mapped by an isolated neural network from the same input vector $\boldsymbol{I}$. The *i th* back propagation neural network structure which is used in FDPS shown in Figure 1.

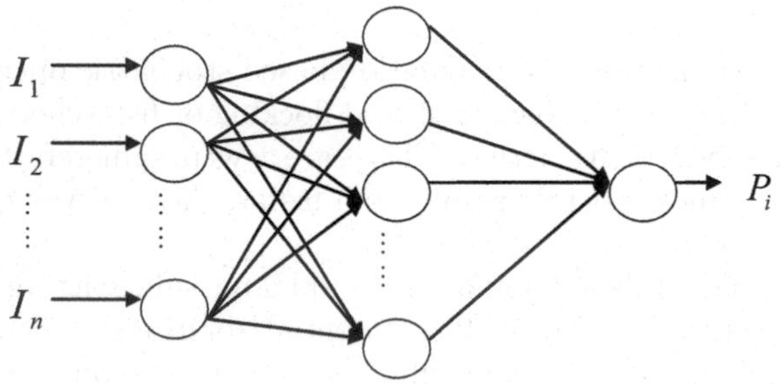

Fig. 1. The structure of *i th* neural network

Where the target partition space vector is $\boldsymbol{P}$. $\boldsymbol{P}_i$ is the *i th* element of $\boldsymbol{P}$ and $\boldsymbol{I}$ is an input data set vector. The mapping formula is:

$$P_i = NN_i\left(\boldsymbol{I}\right), i = 0...m-1 \quad (1)$$

Whole data set is mapped to partition space using (1).

3.2 Division of Partition Space

The partition space is divided into many partitions. If m is equal to 1, the partition space is one-dimensional and every partition is a line segment. If m is equal to 2, the partition space is two-dimensional and every partition is a rectangle. If m is equal to 3, the partition space is three-dimensional and every piece is cube, and so on.

$$TotalPartitionNumber = \prod_{d=1}^{m} partitionsnumber_d \tag{2}$$

$partitionsnumber_d$ is the number of partitions in d axes, $TotalPartitionNumber$ is the number of partitions in the whole partition space. Figure 2 illustrates this concept.

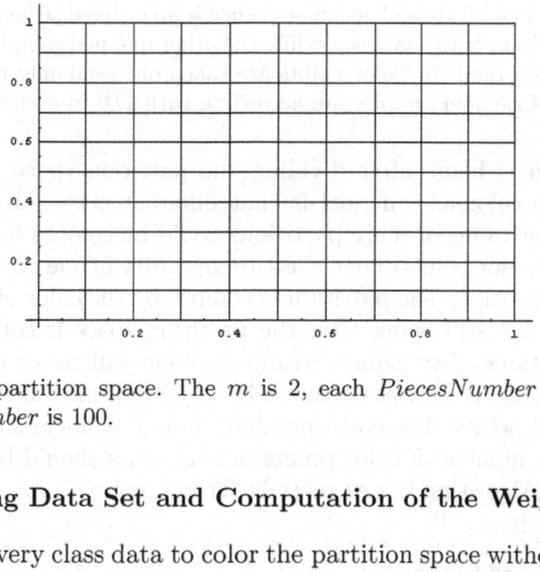

Fig. 2. Divided partition space. The m is 2, each $PiecesNumber$ is 10, and so the $TotalPiecesNumber$ is 100.

3.3 Analyzing Data Set and Computation of the Weight of Classes

It is unfair for every class data to color the partition space without using weight, because the number of each class data in the data set is different. A class will control most area in the partition space, if its number is much bigger than any other in the same data set. In order to solve the problem, a weight is needed. The weight of the first class should be smaller than second one, if the number of the first one is bigger than the second. The proportion of class c in the whole data set is defined as R_c:

$$R_c = \frac{Num_c}{Num_{total}} \tag{3}$$

Where number of samples in class c is Num_c, and Num_{total} is the total number of samples in the whole data set. The weight of class c is calculated by:

$$W_c = \frac{1}{R_c} \tag{4}$$

3.4 Coloring Partitions

When every sample in the data set is mapped to the partition space by neural networks, parts of which are used to color the partitions,is related to proportion DP. These selected mapped samples are called color points. Color points are taken out from each class in the mapped data set by the proportion DP. There are two methods to generate the sequence of color points. Order sequence and random sequence could be used for generation in one class.

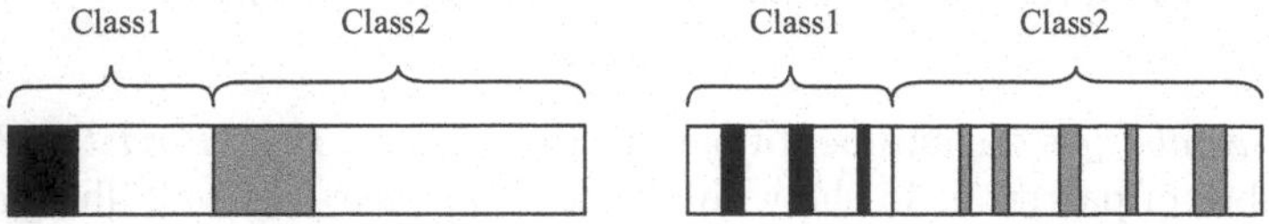

Fig. 3. This is a mapped training data set. The shadowed parts are color points with the proportion DP in each class. The left sequence is an ordered sequence. Color points are in the front position of the data set, while the other mapped samples followed after. The right sequence is random. Color points are taken out randomly in each class, but the total number of random samples are according with DP in each class.

Every partition is blank after dividing the partition space into partitions. Each of them are colored. Our goal is that different classes are to be colored with different colors. One or more partitions could be colored by one class. The principle is that if color points of one class are majority in one partition, this class will control the partition. The partition is colored by the color of this class. The partitions, which are still blank after the partition space is colored, are called unclassified partitions. Test points (training point will never drop into these pieces, if they do, the piece will be colored), which fall into these partitions are called unclassified points. The corresponding sample is unclassified. If the data set is uneven, the number of color points of each class should be multiplied by its weight. Color Algorithm in one piece is shown as;

```
Majority: =0;
For C=0 to Number of Classes-1
  Begin
  If Color points' number of class c in this partition*Weight of class C>
  Number of color points of class Majority in this partition *Weight of
  class Majority
  Then Majority: =C;
  End;
partition Color:=Color(Majority);
```

All the partitions should run this algorithm to color themselves. For example: in the training data set, the number of points of class 1 is twice as much as class 0. We calculate W_0=3 and W_1=1.5 and set each $PartitionNumber$ to 10 and m=2. So $TotalPartitionNumber$ is 100. Figure 4 illustrates the distribution of color points and the coloring of the partition space.

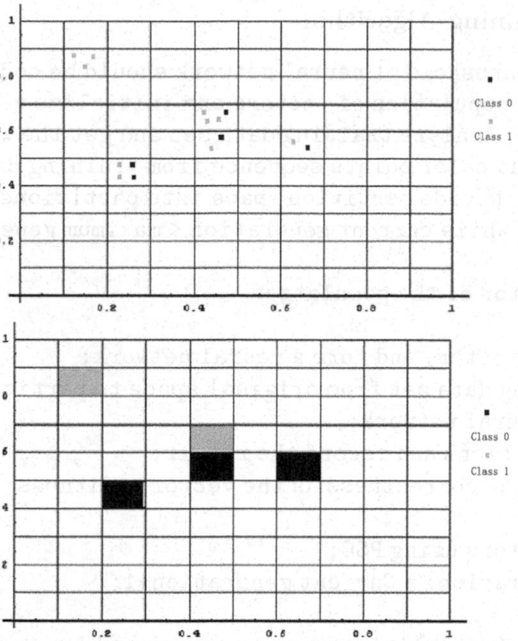

Fig. 4. The figure on top illustrates the distribution of color points. The partition space should be colored as per bottom figure.

3.5 Calculation of Correctness

The performance of neural network is to evaluated, because the network should be optimized to achieve the best performance. FDPS is supervised, and so the evaluation of neural network should be calculated at each generation. The correctness is calculated by the following formula, where L is the total number of samples of training data set:

$$Right = \begin{cases} 1, Color(point) = Color(partition) \\ 0, \qquad\qquad\qquad Else. \end{cases} \tag{5}$$

$$Correctness = \frac{\sum_{i=0}^{L-1} Right_i}{L} \tag{6}$$

3.6 Categorizing New Sample by Using FDPS

After training, an optimized neural network and its corresponding colored result (the colored partition space) is produced. A new sample should be mapped by this neural network and then, its category should be judged according to the color of the partition, which the corresponding mapped point of this sample belongs to. If the partition color is not blank, this sample is categorized by the corresponding class of the color. Otherwise, partition color is blank and the sample is unclassified in FDPS.

3.7 FDPS Training Algorithm

```
The weight and threshold of neural network should be coded, to form
a vector; Form a population of vectors and initialize vectors by
random constant; Analyze training data set and get the weight of
classes; Take out color points sequence from training data set
according to DP; Divide partition space into partitions; Current
generation: =0; While current generation < maximum generation do
  Begin
  For every vector of the population
    Begin
    Decode the vector, and form a neural network;
    Map training data set from original space to partition
    space by neural network;
    Color partitions and record the result;
    Calculate the correctness as the vector's fitness;
    End;
   Optimize vectors using PSO;
   Current generation: = Current generation+1;
   End.
The goal is to get the elitist neural network and its corresponding
colored results.
```

4 Experiments and Results

In order to evaluate the performance of this algorithm, four criterions are defined.

Training Accuracy (TA), a method to adapt training data to achieve better TA.

$$TA = \frac{NumberOfCorrectlyClassifiedSamplesInTrainingDataSet}{NumberOfTotalSamplesInTrainingDataSet} \tag{7}$$

Generalization Accuracy (GA), a method to obtain better generalization capability.

$$GA = \frac{NumberOfCorrectlyClassifiedSamplesInTestDataSet}{NumberOfTotalSamplesInTestingDataSet} \tag{8}$$

Accuracy of Classified Samples (ACS), displays the classification ability of the method on the data which is to be categorized. On the condition that a sample could be classified by a method, higher values of ACS ensures higher classification accuracy.

$$ACS = \frac{NumberOfCorrectlyClassifiedSamplesInTestDataSet}{NumberOfClassedSamplesInTestingDataSet} \tag{9}$$

Proportion of Unclassified Samples (PUS) illustrate unclassified proportion of data sets. Lower PUS represents more samples to be be categorized. The PUS in traditional neural network classification is 0, because all of the samples are categorized by this method. In FDPS, PUS is not 0 for the reason that some samples are mapped to unclassified partitions.

$$PUS = \frac{NumberOfUnclassedSamplesInTestingDataSet}{NumberOfTotalSamplesInTestingDataSet} \tag{10}$$

Experiments need many trails for one data set. So Mean (M) and Standard Deviation (SD) are also needed for performance evaluation. Many data sets have been used for testing and evaluation of FDPS procedure. This paper uses the breast cancer data as the representative data set. Breast cancer is the most common cancer in women in many countries. Most breast cancers are detected as a lump/mass on the breast, or through self examination or mammography [7,8,9]. The Wisconsin breast cancer data set has 30 attributes and 569 instances of which 357 are of benign and 212 are of malignant type. The data set is randomly divided into a training data set and a test data set. The first 285 data is used for training and the remaining 284 data is used for testing. Binary classification is adopted in this research.

Experiments were conducted to evaluate TA, GA and ACS and the training speed. Ten trails were conducted and the mean and standard deviation is reported. We used a three-layered back propagation neural network with 30 hidden neurons. Parameters used for this data set are: DP=0.5, m=2, $PartitionNumber_0$ =10, $PartitionNumber_1$=10, Max Generation=1000, Population Size=50,ϕ_1= 0.05,ϕ_2= 0.05 and VMAX=1.2. Table 1 and Figure 5 illustrate the overall performance of the FDPS.

Some accuracy results are shown in Table 1. The average performance of both TA and ACS of FDPS is obviously higher than the traditional method. But, average GA is lower in FDPS, at the same time SD of GA is higher. The maximum GA in FDPS is higher than maximum GA in the traditional method, while the minimum GA in FDPS is lower than the traditional method.

Table 1. Performance results for breast cancer data

	Traditional Method			FDPS			
	TA	GA	ACS	TA	GA	ACS	PUS
1	93.21%	93.21%	93.21%	94.28%	94.64%	94.64%	0%
2	94.28%	93.93%	93.93%	93.57%	87.50%	87.5%	0%
3	92.85%	90.35%	90.35%	94.64%	91.43%	92.42%	1.07%
4	93.57%	91.79%	91.79%	95.00%	93.21%	93.21%	0%
5	92.85%	92.14%	92.14%	95.35%	90.72%	91.70%	1.07%
6	93.21%	91.79%	91.79%	94.64%	90.00%	90.97%	1.07%
7	93.21%	88.21%	88.21%	95.00%	93.21%	93.21%	0%
8	93.21%	93.21%	93.21%	94.64%	89.64%	89.64%	0%
9	94.28%	89.64%	89.64%	94.64%	89.64%	89.96%	0.36%
10	91.42%	92.5%	92.5%	96.07%	93.57%	94.24%	0.71%
M	93.21%	**91.68%**	91.68%	**94.78%**	91.36%	**91.75%**	0.43%
SD	0.81%	1.78%	1.78%	0.66%	2.25%	2.25%	0.50%

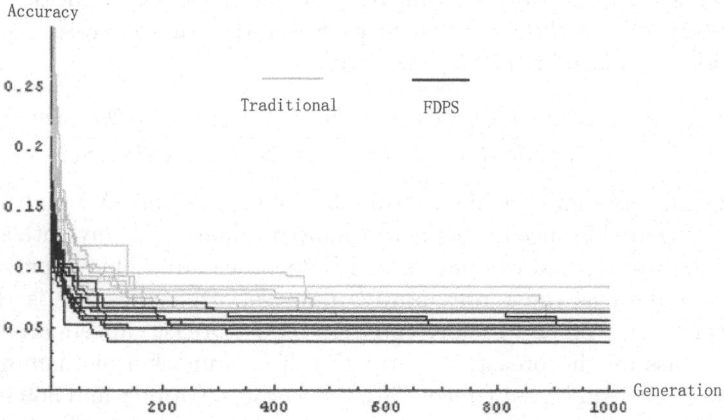

Fig. 5. Speed of convergence with Y axis representing (1.0 - TA)

Figure 5 illustrates that FDPS performed very well and has converged within the first 400 generations. During the training process, accuracy of FDPS is optimized faster than the traditional method. The final accuracy of FDPS is better than traditional method, which clearly means that the FDPS method aids the neural network learning process. It has faster optimization speed than the direct approach because mapping of neural network in FDPS is flexible and unrestricted.

5 Conclusions

This paper proposes novel technique to improve traditional neural network classification. This technique which maps data sample freely and easily is based on further division of partition space. From the experiment results, it is evident that performance measures and optimization speed of FDPS is better and faster than the traditional method. There are still some limitations in FDPS. The mean of GA is lower in FDPS, notwithstanding that the standard deviation is larger than the traditional method (leading to the best GA in FDPS is higher). More experiments on different data sets are required to better analyze FDPS's performance.

Acknowledgments. This research was supported by the National Natural Science Foundation of China under Grant No.60573065; the Natural Science Foundation of Shandong Province of China under Grant No.Z2006G03. Ajith Abraham is supported by the Centre for Quantifiable Quality of Service in Communication Systems, Centre of Excellence, appointed by The Research Council of Norway, and funded by the Research Council, NTNU and UNINETT.

References

1. J. R. Qinlan: Introduction of decision trees. Machine Learning.**1**(1986)86-106.
2. Freund Y:Boosting a weak learning algorithm by majority.Information Computation.**121**(1995)256-285.
3. Lu Hongjun, Setiono Rudy, Liu Huan: Effect data mining using neural networks. IEEE Transaction on knowledge and data engineering. **8**(1996)957-961.
4. B. E. Boser, I. M. Guyon, V. N. Vapnik:A training algorithm for optimal margin classifiers.Proceedings of the 5th Annual ACM Workshop on Computational Learning Theory, ACM Press.(1992)144-152.
5. V. N. Vapnik:The Nature of Statistical Learning Theory.Springer Verlag.(1995).
6. J. Kennedy and R. C. Eberhart:A new optimizer using paritcle swarm theory in Proceeding of the Sixth Int. Symposium on Micromachine and Human Science, Nagoya, Japan.(1995)39-43.
7. DeSilva, C.J.S. et al.: Artificial Neural networks and Breast Cancer Prognosis.The Australian Computer Journal. **26**(1994)78-81.
8. Shieu-Ming Chou, Tian-Shyug Lee, et al.:Mining the breast cancer pattern using artificial neural networks and multivariate adaptive regression splines.Expert Systems with Applications.**27**(2004)133-142.
9. Ravi Jain and Ajith Abraham, A Comparative Study of Fuzzy Classifiers on Breast Cancer Data, Australiasian Physical And Engineering Sciences in Medicine, Australia, **27** (4), 2004, pp. 147-152.

Morphisms of ANN and the Computation of Least Fixed Points of Semantic Operators

Anthony Karel Seda

Department of Mathematics and Boole Centre for Research in Informatics,
University College Cork, Cork, Ireland
a.seda@ucc.ie

Abstract. We consider a notion of morphism of neural networks and develop its properties. We show how, given any definite logic program P, the least fixed point of the immediate consequence operator T_P can be computed by the colimit of a family of neural networks determined by P.

1 Introduction

It is an important, challenging and interesting problem to integrate symbolic models of computation and natural models of computation. A particular instance of this problem is the integration of logic programming, viewed as logic-based symbolic computation, and artificial neural networks (ANN), viewed as models inspired by the brain; we refer to the introduction and references of [2] for a discussion of the issues involved and their ramifications.

One much-studied aspect of this question is the computation by ANN of the operators Ψ_P, such as the immediate consequence operator T_P, determined by the standard semantics of logic programs P. This particular problem is usually considered from two points of view: exact computation when P is propositional, and approximation when P is first-order.

In [2] and elsewhere, approximation of Ψ_P by ANN is the main focus, whereas in [6] we considered, in the same context, convergence and approximation of interpretations and of programs relative to certain metrics and pseudometrics. In this paper, we consider the question of "convergence" and "limits", in some sense, of the approximating ANN. In fact, we consider their colimits in the categorical sense of [4], and the thinking here is motivated by the following two observations. First, morphisms of ANN should reflect the structure of ANN, and one should expect isomorphic ANN to compute the same function. (Alternatively, one might formulate a definition of morphism and of isomorphism along the lines of bisimulation, that is, based on "states".) Second, the (co)limit ANN should exactly compute the least fixed points of Ψ_P when P is first order.

In summary, we adopt here a rather strong definition of morphism reflecting the structure of ANN and closely related to the computational capabilities of the networks we employ, see Definition 2, and we consider its basic properties. The results are satisfactory, see Propositions 1, 2 and 3. Furthermore, the results on colimits give appropriate versions of convergence and "limit" for neural networks,

J. Mira and J.R. Álvarez (Eds.): IWINAC 2007, Part I, LNCS 4527, pp. 224–233, 2007.

see Propositions 7, Corollary 1, Proposition 8 and Corollary 2. In particular, Proposition 7 and Corollary 1 show how, given any definite first-order logic program P, one may construct an infinite 3-layer feedforward ANN, using only binary threshold units and weights of value 0 or 1, which computes the least fixed point of T_P.

2 Neural Networks and Their Morphisms

Definition 1. A *neural network* $\mathcal{N} = (\mathcal{N}, \mathcal{N}^{\text{in}}, \mathcal{N}^{\text{out}}, \mathcal{C}, \mathcal{W}, \mathcal{H}, \Theta)$ is a septuple consisting of a weighted digraph together with the following extra structure. We let $\mathcal{N}$ denote both the neural network and its set of *nodes, neurons or units* N_i, where i belongs to some countable (possibly infinite) index set $\mathcal{I}$; the terms node, neuron and unit are used interchangeably. The set $\mathcal{N}^{\text{in}} \subseteq \mathcal{N}$ is a distinguished set of *input neurons*, and $\mathcal{N}^{\text{out}} \subseteq \mathcal{N}$ is a distinguished set of *output neurons.* We denote by c_{ij} the (digraph) *connection* between neurons N_i and N_j, if such exists, and by $\mathcal{C}$ the set of all the given connections c_{ij}. We denote by $w_{ji} \in \mathbb{R}$ the *weight* of a connection c_{ij}, and by $\mathcal{W}$ the set of all the weights. With each neuron N_i we associate an *output function* or *activation function* denoted by $h_i : \mathbb{R} \to \mathbb{R}$, and a *threshold* $\theta_i \in \mathbb{R}$; we denote by $\mathcal{H}$ the set of these functions and by Θ the set of the thresholds. Given a connection c_{ij} between N_i and N_j, we call N_i the *source* $\text{src}(c_{ij})$ and N_j the *target* $\text{trg}(c_{ij})$ of the connection. We sometimes denote typical neurons simply by N or M, rather than by N_i or M_i, with corresponding notation c_{NM}, w_{MN}, h_N, and θ_N then being employed.

We work with discrete equidistant moments of time $t = 0, 1, 2, \ldots$. Then the *(activation) potential* p_i for a neuron N_i at time t is given by

$$p_i(0) := 0, \qquad p_i(t) := \left(\sum_{k \in \mathcal{I}_i} w_{ik}\, \sigma_k(t) \right) - \theta_i + \iota_i(t) \quad \text{if } t > 0, \tag{1}$$

where $\mathcal{I}_i$ is the set of all those indices k for which there is a connection c_{ki} from N_k to N_i; $\sigma_k(t)$ is the *output value* emitted by the neuron N_k at time t; and $\iota_i(t)$ is an external input received by N_i at time t (we set $\iota_i(t) = 0$ if $N_i \notin \mathcal{N}^{\text{in}}$). Indeed, the output value $\sigma_i(t)$ of a neuron N_i at time t is computed by

$$\sigma_i(0) := 0 \ , \qquad \sigma_i(t) := h_i(p_i(t-1)) \quad \text{if } t > 0. \tag{2}$$

As usual, units are mainly distinguished by the nature of their output function h, but the only one we need to distinguish here is the *binary threshold* unit, wherein h_N is a threshold function H; thus, $H(x) = 1$ if $x \geq 0$ and is 0 otherwise.

Definition 2. Suppose that $\mathcal{N}$ and $\mathcal{M}$ are neural networks. A *homomorphism* or just *morphism* $\Phi : \mathcal{N} \to \mathcal{M}$ consists of a pair of mappings ϕ (the *node mapping of* Φ) and $\hat{\phi}$ (the *connection mapping of* Φ) satisfying the following properties.
(1) ϕ maps nodes N of $\mathcal{N}$ to nodes $\phi(N)$ of $\mathcal{M}$, and $\hat{\phi}$ maps each weighted connection $N_1 \xrightarrow{w} N_2$ in $\mathcal{N}$ to a weighted connection $\phi(N_1) \xrightarrow{\hat{\phi}(w)} \phi(N_2)$ in $\mathcal{M}$, where $\hat{\phi}(w) = w$. Thus, $\hat{\phi}$ preserves orientation of connections and their weights.

(2) Activation functions and thresholds satisfy $h_{\phi(N)} = h_N$ and $\theta_{\phi(N)} = \theta_N$ for all nodes N in $\mathcal{N}$.

Notice that a morphism of neural networks induces a morphism of the underlying (weighted) digraphs.

Definition 3. Suppose that $\mathcal{N}$ is a neural network. The morphism $Id_{\mathcal{N}} : \mathcal{N} \to \mathcal{N}$, where the node mapping is the identity mapping on $\mathcal{N}$ and the connection mapping is the identity on $\mathcal{C}$, is called the *identity* morphism on $\mathcal{N}$.

Definition 4. Suppose that $\Phi_1 : \mathcal{N}_1 \to \mathcal{N}_2$ and $\Phi_2 : \mathcal{N}_2 \to \mathcal{N}_3$ are morphisms of neural networks. Then the composition $\Phi_2 \circ \Phi_1 : \mathcal{N}_1 \to \mathcal{N}_3$ is the morphism of neural networks in which (i) the node mapping is the composite $\phi_2 \circ \phi_1$ of the node mappings of Φ_1 and Φ_2, and (ii) the connection mapping is the composite $\hat{\phi}_2 \circ \hat{\phi}_1$ of the connection mappings of Φ_1 and Φ_2.

Definition 5. Suppose that $\mathcal{N}_1$ and $\mathcal{N}_2$ are neural networks. An *isomorphism* $\Phi : \mathcal{N}_1 \to \mathcal{N}_2$ is a morphism $\Phi : \mathcal{N}_1 \to \mathcal{N}_2$ such that there is a morphism $\Phi' : \mathcal{N}_2 \to \mathcal{N}_1$ with the property that $\Phi' \circ \Phi = Id_{\mathcal{N}_1}$ and $\Phi \circ \Phi' = Id_{\mathcal{N}_2}$.

Remark 1. (1) Both $Id_{\mathcal{N}}$ and $\Phi_2 \circ \Phi_1$ satisfy Definition 2 and therefore both are indeed morphisms.
(2) A morphism $\Phi : \mathcal{N}_1 \to \mathcal{N}_2$ of neural networks is an isomorphism if and only if its node mapping and its connection mapping are both bijections.
(3) An isomorphism $\Phi : \mathcal{N}_1 \to \mathcal{N}_2$ of neural networks induces an isomorphism of the underlying (weighted) digraphs.
(4) An isomorphism $(\phi, \hat{\phi})$ of digraphs induces an isomorphism Φ of neural networks provided (i) there is a connection $N_1 \xrightarrow{w} N_2$ in $\mathcal{N}_1$ iff $\phi(N_1) \xrightarrow{\hat{\phi}(w)} \phi(N_2)$ is a connection in $\mathcal{N}_2$ and $\hat{\phi}(w) = w$, and (ii) activation functions and thresholds satisfy $h_{\phi(N)} = h_N$ respectively $\theta_{\phi(N)} = \theta_N$ for all nodes N in $\mathcal{N}$.

3 Morphisms of Multilayer FNNs

Let $\mathcal{N}$ be a neural network such that there is a partition $\mathcal{L}_1, \mathcal{L}_2, \dots, \mathcal{L}_n$ of its set $\mathcal{N}$ of neurons satisfying $\mathcal{L}_1 = \mathcal{N}^{\text{in}}$, $\mathcal{L}_n = \mathcal{N}^{\text{out}}$ and $n \geq 3$. In this case, we call $\mathcal{N}$ an *n-layer feedforward neural network (FNN)* if each connection $c_{ij} \in \mathcal{C}$ is such that $\text{src}(c_{ij}) \in \mathcal{L}_m$ and $\text{trg}(c_{ij}) \in \mathcal{L}_{m+1}$, for some m with $1 \leq m \leq n-1$. Such networks are referred to as *multilayer feedforward neural networks* in general, and $\mathcal{L}_m$ is referred to as a *layer*, the *m-th layer* or *layer m*. In particular, layer 1 constitutes the layer of input neurons, layer n the layer of output neurons, and layers 2 to $n-1$ are referred to as layers of *hidden neurons*. We think of the units in $\mathcal{L}_m$ as indexed by a set $|\mathcal{L}_m|$, and if this set is finite of size r_m, say, we take $|\mathcal{L}_m|$ to be $\{1, 2, \dots, r_m\}$; otherwise we take $|\mathcal{L}_m|$ to be $\mathbb{N}$. Finally, we take all external inputs $\iota_i(t)$ to be zero in the case of an FNN (see Equation (1)).

We denote a typical unit in layer m by N_k^m, for $k \in |\mathcal{L}_m|$ and $m = 1, \dots, n$. However, we often denote a typical input unit N_k^1 by I_k, for $k \in |\mathcal{L}_1|$, and similarly we often denote a typical output unit N_i^n by O_i, for $i \in |\mathcal{L}_n|$.

The following restriction is sometimes imposed in the literature.

Condition 1. Each neuron in layer m is connected (possibly with weight zero) to every neuron in layer $m+1$, for $m = 1, \ldots, n-1$. There are no other connections unless the network is *made recurrent*. By this latter term, we mean that the input and output layers are in one-to-one correspondence, and each output unit is connected with weight one to the corresponding input unit.

We denote by w_{ji}^m, for $m = 1, \ldots, n-1$, the weight of the connection c_{ij}^m from unit N_i^m in layer m to unit N_j^{m+1} in layer $m+1$. Note that the networks we are currently considering compute functions $f : \mathbb{R}^{c(|\mathcal{L}_1|)} \to \mathbb{R}^{c(|\mathcal{L}_n|)}$ when applied to real vector inputs, where $c(A)$ denotes the cardinality of a set A.

Proposition 1. Suppose that $\mathcal{N}$ and $\mathcal{M}$ are multilayer FNNs satisfying Condition 1 (not necessarily having the same number of layers) and that $\Phi : \mathcal{N} \to \mathcal{M}$ is a morphism of neural networks. Then the following statements hold.
(1) All units in a given layer in $\mathcal{N}$ are mapped by Φ into a common layer in $\mathcal{M}$.
(2) A path of connections of length l in $\mathcal{N}$ is mapped by Φ to a path of connections of length l in $\mathcal{M}$.

Thus, under Condition 1, a morphism cannot make "vertical" identifications of neurons, that is, cannot identify neurons in different layers, but can only make identifications of neurons in the same layer (horizontal identifications).

Proof. (1) Suppose that N_i^m and N_j^m are units in layer m in $\mathcal{N}$, and that N_i^m is mapped to $\phi(N_i^m) = M_1$ in layer m_1 in $\mathcal{M}$ and that N_j^m is mapped to $\phi(N_j^m) = M_2$ in layer m_2 in $\mathcal{M}$, where $m_1 \neq m_2$; suppose further that $m_1 > m_2$, with a similar argument if $m_2 > m_1$.

Case 1. Layer m is not the output layer. Then there is a layer $m+1$ in $\mathcal{N}$. Consider connections c_{i1}^m from N_i^m to N_1^{m+1}, and c_{j1}^m from N_j^m to N_1^{m+1}. Unit $\phi(N_1^{m+1})$ cannot belong to layer m_1 otherwise the connection $\hat{\phi}(c_{i1}^m)$ is a connection between two units in the same layer m_1. Similarly, $\phi(N_1^{m+1})$ cannot belong to layer $m_1 + k$ for any $k \geq 2$, if such layers exist, otherwise we would have a connection $\hat{\phi}(c_{i1}^m)$ between units in layers m_1 and $m_1 + k$ with $k \geq 2$, which is impossible. Hence, $\phi(N_1^{m+1})$ must belong to layer $m_1 + 1$. But then $\hat{\phi}(c_{j1}^m)$ is a connection between units in layers m_2 and $m_1 + 1$, and this also is impossible.

Case 2. Layer m is the output layer. Then there is a layer $m-1$ in $\mathcal{N}$. Consider the connections c_{1i}^{m-1} from N_1^{m-1} to N_i^m, and c_{1j}^{m-1} from N_1^{m-1} to N_j^m. The unit $\phi(N_1^{m-1})$ cannot belong to layer m_1 otherwise the connection $\hat{\phi}(c_{1i}^{m-1})$ is a connection between two units in the same layer m_1. Similarly, $\phi(N_1^{m-1})$ cannot belong to layer $m_1 + k$ for any $k \geq 1$, if such layers exist, because then we would have a connection $\hat{\phi}(c_{1i}^{m-1})$ directed from a unit in layer $m_1 + k$ with $k \geq 1$ towards a unit in layer m_1, which is impossible. In fact, because of the connection c_{1j}^{m-1} it is clear that $\phi(N_1^{m-1})$ must belong to layer $m_2 - 1$. But this is impossible otherwise we have a connection $\hat{\phi}(c_{1i}^{m-1})$ from a unit in layer $m_2 - 1$ to a unit in layer m_1 where $m_1 > m_2$. These contradictions establish (1).

(2) Suppose that $N_1 \xrightarrow{c_{12}} N_2 \xrightarrow{c_{23}} N_3 \xrightarrow{c_{34}} \cdots \xrightarrow{c_{(l-1)(l)}} N_l \xrightarrow{c_{l(l+1)}} N_{l+1}$ is a path of length l in $\mathcal{N}$. Consecutive nodes N_j and N_{j+1} must belong to consecutive layers in $\mathcal{N}$ because of the connection $c_{j(j+1)}$ between them. But then $\hat{\phi}(c_{j(j+1)})$ must be the connection in $\mathcal{M}$ between $\phi(N_j)$ and $\phi(N_{j+1})$. Hence, $\phi(N_j)$ and $\phi(N_{j+1})$ belong to consecutive layers in $\mathcal{M}$, and the result follows.

Remark 2. (1) The proof of this result does not use the weights and associated apparatus of a neural network, only the structure pertaining to the units (arranged in layers) and properties of digraphs.
(2) The homomorphic image of a multilayer FNN is itself a multilayer FNN.

Proposition 2. Suppose that $\Phi : \mathcal{N} \to \mathcal{M}$ is an isomorphism of multilayer FNNs $\mathcal{N}$ and $\mathcal{M}$. In the presence of Condition 1 the following statements hold.
(1) $\mathcal{N}$ and $\mathcal{M}$ have the same number n, say, of layers.
(2) ϕ maps layer m in $\mathcal{N}$ bijectively onto layer m in $\mathcal{M}$ for $1 \leq m \leq n$. In particular, ϕ maps the input layer of $\mathcal{N}$ onto the input layer of $\mathcal{M}$, and maps the output layer of $\mathcal{N}$ onto the output layer of $\mathcal{M}$.
(3) Let N be any neuron in $\mathcal{N}$. For the same input given to $\mathcal{N}$ and $\mathcal{M}$ at time $t = 0$, the outputs of N and $\phi(N)$ at any later time t are equal.
(4) $\mathcal{N}$ and $\mathcal{M}$ compute the same function.

Proof. Statement (1) follows immediately from (2) of Proposition 1 on considering paths of maximum length in $\mathcal{N}$ and $\mathcal{M}$, and Statement (2) follows from (1) of Proposition 1 given that ϕ is a bijection on the sets of nodes in $\mathcal{N}$ and $\mathcal{M}$.
(3) Consider a neuron N in $\mathcal{N}$ and its image $\phi(N)$ in $\mathcal{M}$. The claim is clearly true at time $t = 0$. Suppose that the claim is true at times $1, \ldots, t-1$. Then, by the properties of an isomorphism and the current hypothesis, we have
$p_N(t-1) = \left(\sum_{M \in \mathcal{I}_N} w_{NM}\, \sigma_M(t-1)\right) - \theta_N$
$= \left(\sum_{\phi(M) \in \mathcal{I}_{\phi(N)}} w_{\phi(N)\phi(M)}\, \sigma_{\phi(M)}(t-1)\right) - \theta_{\phi(N)}$
$= \left(\sum_{M' \in \mathcal{I}_{\phi(N)}} w_{\phi(N)M'}\, \sigma_{M'}(t-1)\right) - \theta_{\phi(N)} = p_{\phi(N)}(t-1).$
Thus, $\sigma_N(t) = h_N(p_N(t-1)) = h_{\phi(N)}(p_{\phi(N)}(t-1)) = \sigma_{\phi(N)}(t)$, and (3) follows.
(4) That $\mathcal{N}$ and $\mathcal{M}$ compute the same function follows immediately from (3).

We next consider 3-layer FNN. Here, a version of Proposition 2 holds under the following condition which is much weaker than Condition 1.

Condition 2. (1) No input unit and no hidden unit has no output connections.
(2) No hidden unit and no output unit has no input connections.

Proposition 3. Suppose that Condition 2 is satisfied by 3-layer FNN $\mathcal{N}_1$ and $\mathcal{N}_2$. Then an isomorphism $\Phi : \mathcal{N}_1 \to \mathcal{N}_2$ maps input units to input units, output units to output units, and hidden units to hidden units. Furthermore, isomorphic 3-layer FNN $\mathcal{N}_1$ and $\mathcal{N}_2$ satisfying Condition 2 compute the same function.

Proof. **Claim (1)** Φ maps input units in $\mathcal{N}_1$ to input units in $\mathcal{N}_2$.

Proof of claim. (a) Suppose that N_1 is an input neuron in $\mathcal{N}_1$ and is mapped to a hidden unit $M_1 = \phi(N_1)$ in $\mathcal{N}_2$. By the assumption (1), there is a connection $M_1 \to M_2$ in $\mathcal{N}_2$, where M_2 is an output unit. Clearly $\phi^{-1}(M_2)$ cannot be an input nor an output unit in $\mathcal{N}_1$ because of the connection $M_1 \to M_2$. Hence $\phi^{-1}(M_2)$ is a hidden unit and we have a connection $N_1 \to \phi^{-1}(M_2)$. By assumption (1), there is a connection $\phi^{-1}(M_2) \to N_2$, say, in $\mathcal{N}_1$, and N_2 is an output unit. But this leads to a connection $M_2 \to \phi(N_2)$ in $\mathcal{N}_2$, which is impossible since M_2 is an output unit. Thus, $\phi(N_1) = M_1$ cannot be a hidden unit.
(b) Suppose that N_1 is an input unit in $\mathcal{N}_1$ and is mapped to an output unit $M_1 = \phi(N_1)$ in $\mathcal{N}_2$. By our assumption (1), there is a connection $N_1 \to N_2$ in $\mathcal{N}_1$ for some N_2. But then we have a connection $\phi(N_1) \to \phi(N_2)$ in $\mathcal{N}_2$, that is, a connection $M_1 \to \phi(N_2)$ which is clearly impossible since M_1 is an output unit. Thus, Claim (1) is established.

That Φ maps output units in $\mathcal{N}_1$ to output units in $\mathcal{N}_2$ and maps hidden units in $\mathcal{N}_1$ to hidden units in $\mathcal{N}_2$ follows similarly. The remaining statement of the result now follows in the same way as Proposition 2 from this point on.

Remark 3. The activation potential of a neuron in a neural network has the form $p_i(t) = \left(\sum_{k \in \mathcal{I}_i} w_{ik}\, \sigma_k(t)\right) - \theta_i + \iota_i(t)$, by Equation (1). So, adding connections with weight zero does not affect any computations. Thus, Conditions 1 and 2 are not very restrictive in relation to computation by neural networks.

4 Colimits of 3-Layer FNN and Least Fixed Points

Let P be an arbitrary definite logic program, let ground(P) denote the set of all ground instances of clauses in P, let I denote the least fixed point of T_P and let $l : B_P \to \mathbb{N}$ be a *level mapping* where B_P denotes the Herbrand base for P, see [3] for undefined terms and notation relating to logic programming. (We assume l has the property: given $n \in \mathbb{N}$, we can effectively find the set of all $A \in B_P$ such that $l(A) = n$, see [6]; then, given any $n \in \mathbb{N}$, the set of all atoms A such that $l(A) = n$ is finite.) If $l(A) = n$, then we say that the *level* of A is n. Using l, we fix an ordering on $B_P = (A_1, A_2, A_3, \ldots)$ in which those terms of B_P of level 0 are listed first, followed by those of level 1, followed by those of level 2, etc.

We will need certain of the steps made in proving [6, Theorem 3.19], and for convenience we sketch them here. We are working specifically with T_P in this paper for reasons of space limitation, but similar results can be established for quite general operators Ψ_P.

Proposition 4. Suppose that $A \in T_P \uparrow k$, see [3]. Then there is a clause $A \leftarrow$ body in ground(P) such that A does not occur in body and body $\models T_P \uparrow (k-1)$.

Proof. Clearly, $k \geq 1$. Suppose that $A \in T_P \uparrow k_0 = T_P(T_P \uparrow (k_0 - 1))$ and that k_0 is the smallest natural number with this property. Then there is a clause $A \leftarrow$ body in ground(P) such that $T_P \uparrow (k_0 - 1) \models$ body. By definition of k_0, we have $A \notin T_P \uparrow (k_0 - 1)$ and hence A does not occur in body. Finally, by monotonicity, we obtain that $T_P \uparrow (k-1) \models$ body, as required.

Since P is definite, we have

$$T_P \uparrow 0 \subseteq T_P \uparrow 1 \subseteq \cdots \subseteq T_P \uparrow n \subseteq \cdots \subseteq I = \bigcup_{n=1}^{\infty} T_P \uparrow n,$$

where $T_P \uparrow n$ denotes the n-th upward power $T_P^n(\emptyset)$ of T_P.

Given $n \in \mathbb{N}$, there are only finitely many atoms[1] $A_1, A_2, \ldots, A_m \in I$ with $l(A_i) \leq n$ for $i = 1, \ldots, m$, and, by directedness, there is (a smallest) $k = k_n \in \mathbb{N}$ such that $A_1, A_2, \ldots, A_m \in T_P \uparrow k_n$. Consider the atom A_i, where $1 \leq i \leq m$, and the following sequence of steps.

Step 1 We have $A_i \in T_P \uparrow k_n = T_P(T_P \uparrow (k_n - 1))$. Therefore, there is a clause

$$A_i \leftarrow A_i^1(1), \ldots, A_i^{m(i)}(1)$$

in ground(P) such that $A_i^1(1), \ldots, A_i^{m(i)}(1) \in T_P \uparrow (k_n - 1)$. (There may be many such clauses, including unit clauses possibly, and we select one of them.)

Step 2 Because $A_i^1(1), \ldots, A_i^{m(i)}(1) \in T_P \uparrow (k_n - 1) = T_P(T_P \uparrow (k_n - 2))$, there are clauses in ground(P) as follows:

$$\begin{aligned} A_i^1(1) &\leftarrow A_{i,1}^1(2), \ldots, A_{i,1}^{m(i,1)}(2) \\ A_i^2(1) &\leftarrow A_{i,2}^1(2), \ldots, A_{i,2}^{m(i,2)}(2) \\ \vdots \; &\leftarrow \qquad \vdots \\ A_i^{m(i)}(1) &\leftarrow A_{i,m(i)}^1(2), \ldots, A_{i,m(i)}^{m(i,m(i))}(2), \end{aligned}$$

where each of the atoms $A_{i,j}^r(2)$ in each of the bodies belongs to $T_P \uparrow (k_n - 2)$.

Step 3 Because each of the $A_{i,j}^r(2)$ in the previous step belongs to $T_P \uparrow (k_n - 2) = T_P(T_P \uparrow (k_n - 3))$, we have a finite collection of ground clauses (one for each of the $A_{i,j}^r(2)$ in Step 2) the first of which has the form

$$A_{i,1}^1(2) \leftarrow A_{i,1,1}^1(3), \ldots, A_{i,1,1}^{m(i,1,1)}(3),$$

where each new body atom belongs to $T_P \uparrow (k_n - 3)$, and so on.

At each stage in this process we select a program clause in which the head of the clause does not occur in the body by means of Proposition 4.

This process terminates producing unit clauses in its last step. Let $P_{i,n}$ denote the (finite) subset of ground(P) consisting of all the clauses which result; it is clear that $T_{P_{i,n}} \uparrow k_n$ consists of the heads of all the clauses in $P_{i,n}$. We carry out this construction for $i = 1, \ldots, m$ to obtain programs $P_{1,n}, \ldots, P_{m,n}$ such that, for $i = 1, \ldots, m$, $T_{P_{i,n}}(T_{P_{i,n}} \uparrow k_n) = T_{P_{i,n}} \uparrow k_n$ (indeed, $T_{P_{i,n}} \uparrow k_n$ is the least fixed point of $T_{P_{i,n}}$ by Kleene's theorem), $A_i \in T_{P_{i,n}} \uparrow k_n$, and $T_{P_{i,n}} \uparrow r \subseteq T_P \uparrow r \subseteq I$ for all $r \in \mathbb{N}$. Let $\overline{P}_n$ denote the program $P_{1,n} \cup \ldots \cup P_{m,n}$. Then $\overline{P}_n$ is a finite subprogram of ground(P), and $T_{P_{i,n}} \uparrow k_n \subseteq T_{\overline{P}_n} \uparrow k_n \subseteq T_P \uparrow k_n \subseteq I$ for

[1] Notice that, depending on l, there may be no atoms A with $l(A) \leq n$; this case is handled by the abuse of notation obtained by allowing m to be 0.

$i = 1, \ldots, m$. Furthermore, $A_1, \ldots, A_m \in T_{\overline{P}_n} \uparrow k_n$, and $T_{\overline{P}_n} \uparrow k_n$ is the least fixed point of $T_{\overline{P}_n}$, see [6]. Moreover, having determined $\overline{P}_n$, we assume that, for each $n \in \mathbb{N}$, $\overline{P}_{n+1}$ consists of the clauses in $\overline{P}_n$ together with extra clauses which deal with atoms of level $n+1$ as above; thus, $\overline{P}_n \subseteq \overline{P}_{n+1}$ for each $n \in \mathbb{N}$.

Proposition 5. We have $\bigcup_{n=1}^{\infty} T_{\overline{P}_n} \uparrow k_n = I$.

Proof. Clearly, $\bigcup_{n=1}^{\infty} T_{\overline{P}_n} \uparrow k_n \subseteq I$. If $A \in I$, then $l(A) = n$ for some n and hence $A \in T_{\overline{P}_n} \uparrow k_n$. Therefore, $I \subseteq \bigcup_{n=1}^{\infty} T_{\overline{P}_n} \uparrow k_n$.

For a definite program P, let $lfp(T_P)$ denote the least fixed point of T_P, and let $\mathcal{P}$ denote $\bigcup_{n=1}^{\infty} \overline{P}_n$, as defined above. Then we have the following result.

Proposition 6. For any definite program P, we have $lfp(T_{\mathcal{P}}) = lfp(T_P) = I$.

Proof. Since $\mathcal{P} \subseteq P$, we have $lfp(T_{\mathcal{P}}) \subseteq lfp(T_P)$.

Conversely, suppose that $A \in lfp(T_P)$. Then $l(A) = n$ for some n and hence $A \in T_{\overline{P}_n} \uparrow k_n$. Therefore, $A \in T_{\mathcal{P}} \uparrow k_n \subseteq lfp(T_{\mathcal{P}})$, as required.

Having obtained the programs $\overline{P}_n$ for each $n \in \mathbb{N}$, we apply the algorithm established by the proof of [2, Theorem 3.2] to $\overline{P}_n$.

Theorem 1. ([2, Theorem 3.2]) For each propositional program P (not necessarily definite), there exists a 3-layer FNN, having only binary threshold units, which computes T_P.

By means of this result, we obtain for each $n \in \mathbb{N}$ a 3-layer FNN $\mathcal{N}_n$ which computes $T_{\overline{P}_n}$. Furthermore, the construction of $\overline{P}_n$ and the properties of the algorithm given by [2, Theorem 3.2] show that we can construct a well-defined, possibly infinite, 3-layer FNN $\mathcal{N}$ as follows: (i) a unit is in $\mathcal{N}$ if and only if it is in the corresponding layer of $\mathcal{N}_n$ for large enough n; (ii) there is a connection c_{ij} from N_i to N_j in $\mathcal{N}$ if and only if c_{ij} is a connection of the same weight from N_i to N_j in $\mathcal{N}_n$ for large enough n; the threshhold of a unit in $\mathcal{N}$ is equal to its threshhold in any $\mathcal{N}_n$ containing it; all units in $\mathcal{N}$ are binary threshold units.

Proposition 7. The network $\mathcal{N}$ computes $T_{\mathcal{P}}$.

Proof. Let $A \in B_{\mathcal{P}}$ and let J be an arbitrary interpretation for $\mathcal{P}$ (defined on the Herbrand base $B_{\mathcal{P}}$ of $\mathcal{P}$). We input J to $\mathcal{N}$.
Case 1. $A \notin I$. In this case, A is not the head of any clause in $\mathcal{P}$, and so $T_{\mathcal{P}}(J)(A) =$ false. On the other hand, there are no connections of weight 1 to the unit representing A in the output layer of $\mathcal{N}$. Therefore, $\mathcal{N}$ also outputs false as the value of the unit representing A.
Case 2. $A \in I$. Suppose $l(A) = n$. Then A is the head of exactly one clause $A \leftarrow$ body in $\mathcal{P}$, and this clause is in $\overline{P}_n$. Suppose first that $T_{\mathcal{P}}(J)(A) =$ true. Then $J \models$ body and hence $J|_{B_{\overline{P}_n}} \models$ body. Therefore, $T_{\overline{P}_n}(J)(A) =$ true. Since $\mathcal{N}_n$ computes $T_{\overline{P}_n}$, it outputs value true in the unit representing A. Hence, $\mathcal{N}$ outputs value true in the unit representing A, as required.

Suppose finally that $T_{\mathcal{P}}(J)(A) = \text{false}$. Then $J \not\models \text{body}$, and therefore $J|_{B_{\overline{P}_n}} \not\models \text{body}$. Thus, we have $T_{\overline{P}_n}(J)(A) = \text{false}$ since the clause $A \leftarrow \text{body}$ is the only one in $\mathcal{P}$ with head A and hence is the only one in $\overline{P}_n$ with head A. Since, again, $\mathcal{N}_n$ computes $T_{\overline{P}_n}$, it outputs value false in the unit representing A. Hence, $\mathcal{N}$ outputs value false in the unit representing A, as required.

Corollary 1. Suppose the definite logic program P is given. Then, when made recurrent, $\mathcal{N}$ computes $I = lfp(T_P)$.

Proof. Since $\mathcal{N}$ computes $T_{\mathcal{P}}$, when made recurrent it computes each iterate of $T_{\mathcal{P}}$ applied to the empty interpretation. The result now follows.

It follows from the construction of $\mathcal{N}$ that the inclusion $\Phi_{n,m} : \mathcal{N}_n \to \mathcal{N}_m$ is a morphism of neural networks whenever $n \leq m$, and the inclusion $I_n : \mathcal{N}_n \to \mathcal{N}$ is also a morphism for each n, and we have the following result.

Proposition 8. The neural network $\mathcal{N}$ is a colimit of the diagram

$$\mathcal{N}_1 \xrightarrow{\Phi_{1,2}} \mathcal{N}_2 \xrightarrow{\Phi_{2,3}} \mathcal{N}_3 \quad \cdots \quad \mathcal{N}_n \xrightarrow{\Phi_{n,n+1}} \mathcal{N}_{n+1} \cdots$$

Proof. It is clear that the diagram

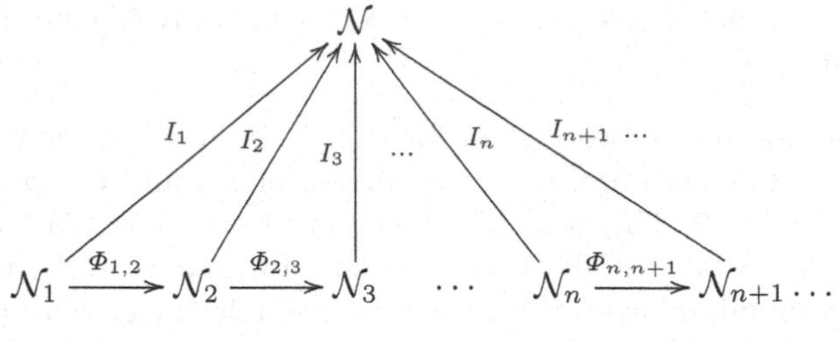

is commutative and is a cocone. Suppose that

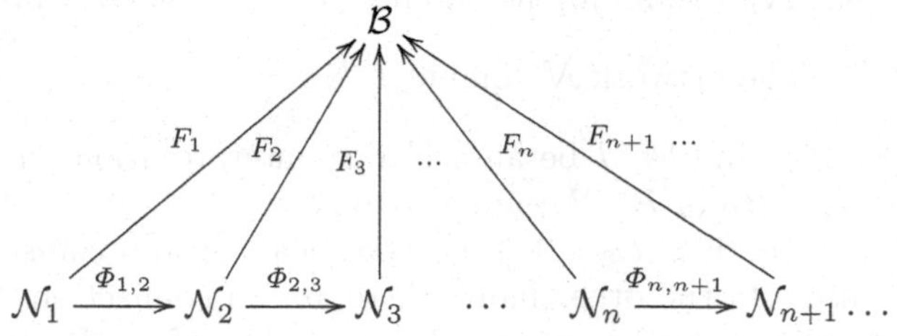

is another cocone. Define $\Phi : \mathcal{N} \to \mathcal{B}$ as follows. Let $N \in \mathcal{N}$ be a unit. Then $N \in \mathcal{N}_n$ for some n. Define ϕ by $\phi(N) = f_n(N)$. Suppose that $N \in \mathcal{N}_m$ with $n < m$. Then $f_m(N) = f_m \circ \phi_{n,m}(N) = f_n(N)$, so ϕ is well-defined. Likewise, if $c \in \mathcal{N}$ is a connection, then $c \in \mathcal{N}_n$ for some n. Again, we define $\hat{\phi}$ by $\hat{\phi}(c) = \hat{f}_n(c)$, and $\hat{\phi}$ is well-defined. Moreover, it is clear that $\Phi : \mathcal{N} \to \mathcal{B}$ is a morphism of neural networks and hence is a cocone morphism.

Suppose $\Psi : \mathcal{N} \to \mathcal{B}$ is another cocone morphism. Let $N \in \mathcal{N}$ be a unit. Then $N \in \mathcal{N}_n$ for some n. But $\psi \circ i_n = f_n$ and so $f_n(N) = \psi \circ i_n(N) = \psi(N)$. Therefore, $\phi(N) = \psi(N)$ and hence $\phi = \psi$. Similarly, $\hat{\phi} = \hat{\psi}$ and so $\Phi = \Psi$.

Thus, there is a unique morphism $\Phi : \mathcal{N} \to \mathcal{B}$, and therefore $\mathcal{N}$ is a colimit of the diagram given in the statement of the proposition, as required.

Corollary 2. For any definite program P, any colimit of the diagram

$$\mathcal{N}_1 \xrightarrow{\Phi_{1,2}} \mathcal{N}_2 \xrightarrow{\Phi_{2,3}} \mathcal{N}_3 \quad \ldots \quad \mathcal{N}_n \xrightarrow{\Phi_{n,n+1}} \mathcal{N}_{n+1} \cdots,$$

computes the least fixed point of T_P.

Proof. This result follows from Proposition 3, the fact that all colimits are isomorphic, the construction of $\mathcal{N}$, and Corollary 1.

5 Conclusions and Further Work

(1) A number of other authors have considered categorical aspects of neural networks, see [1,5]. It is of interest to investigate categorical properties of categories of neural networks and of natural models in general relative to appropriate notions of "morphism".

(2) Finite ANN of the type we are considering cannot compute T_P nor its least fixed point for arbitrary definite programs P. However, it is of interest to avoid using infinite neural networks, if possible, by emulating them by networks of bounded size allowing unbounded iteration coded up in the weights (not restricted to 0 and 1) with the weights changing at each iteration. Well-known work of Sontag and Siegelmann on realizing conventional programming ideas within networks, and proving universality results, suggests that this may be possible.

References

1. Healy, M.J., Caudell, T.P.: Neural networks, knowledge, and cognition: A mathematical semantic model based upon category theory. Technical Report UNM: EECE-TR-04-020, School of Engineering, New Mexico (2004).
2. Hitzler, P., Hölldobler, S., Seda, A.K.: Logic programs and connectionist networks. Journal of Applied Logic **2**(3) (2004) 245–272.
3. Lloyd, J.W.: Foundations of Logic Programming. Springer, Berlin (1987).
4. MacLane, S.: Categories for the Working Mathematician. Volume 5 of Graduate Texts in Mathematics. Springer-Verlag (1971).
5. Pfalzgraf, J.: Modelling connectionist networks: Categorical, geometric aspects (Towards "Homomorphic Learning"). In: Proceedings of Computing Anticipatory Systems: CASYS 2003, Liège, Belgium, August 11–16, 2003. 1–15.
6. Seda, A.K.: On the integration of connectionist and logic-based systems. In Hurley, T., Mac an Airchinnigh, M.M., Schellekens, M., Seda, A.K., Strong, G., eds.: Proceedings of MFCSIT'04, Trinity College Dublin, July, 2004. Volume 161 of Electronic Notes in Theoretical Computer Science (ENTCS)., Elsevier (2006) 109–130.

Predicting Human Immunodeficiency Virus (HIV) Drug Resistance Using Recurrent Neural Networks

Isis Bonet[1], María M. García[1], Yvan Saeys[2], Yves Van de Peer[2], and Ricardo Grau[1]

[1] Center of Studies on Informatics, Central University of Las Villas, Cuba
{isisb,mmgarcia,rgrau}@uclv.edu.cu
[2] Department of Plant Systems Biology, Flanders Interuniversity Institute for Biotechnology (VIB), Ghent University, Belgium
yvan.saeys@ugent.be, yves.vandepeer@psb.ugent.be

Abstract. Predicting HIV resistance to drugs is one of many problems for which bioinformaticians have implemented and trained machine learning methods, such as neural networks. Predicting HIV resistance would be much easier if we could directly use the three-dimensional (3D) structure of the targeted protein sequences, but unfortunately we rarely have enough structural information available to train a neural network. Fur-thermore, prediction of the 3D structure of a protein is not straightforward. However, characteristics related to the 3D structure can be used to train a machine learning algorithm as an alternative to take into account the information of the protein folding in the 3D space. Here, starting from this philosophy, we select the amino acid energies as features to predict HIV drug resistance, using a specific topology of a neural network. In this paper, we demonstrate that the amino acid ener-gies are good features to represent the HIV genotype. In addi-tion, it was shown that Bidirectional Recurrent Neural Networks can be used as an efficient classification method for this prob-lem. The prediction performance that was obtained was greater than or at least comparable to results obtained previously. The accuracies vary between 81.3% and 94.7%.

1 Introduction

The Human Immunodeficiency Virus (HIV) is one of the main causes of death in the world. The HIV is a human pathogen that infects certain types of lymphocytes called T-helper cells, which are important to the immune system. Without a sufficient number of T-helper cells, the immune system is unable to defend the body against infections.

It is a great challenge for scientists to design an effective drug against HIV. Nevertheless, some approved antiretroviral drugs are currently available for the treatment of HIV infection. Most of them focus on two of the most important viral enzymes, namely Protease and Reverse transcriptase.

Several statistical techniques and machine learning algorithms have been used to predict HIV resistance in silico, such as cluster analysis and linear discriminant

J. Mira and J.R. Álvarez (Eds.): IWINAC 2007, Part I, LNCS 4527, pp. 234–243, 2007.

analysis, as described by Sevin [1]. A simple metric to predict the Protease inhibitors resistance has been proposed by Scmidt et al. (2000) [2]. Wang and Larder (2003) used neural networks to predict resistance to the Protease inhibitor Lopinavir [3]. In [4] was used decision trees while in [5] was used decision trees and the k-nearest neighbor technique (KNN) to predict the resistance of protease inhibitors. Recently, convex optimization techniques have been used together with Least Absolute Shrinkage, and Selection Operator (LASSO) and Support Vector Ma-chine (SVM) models for regression or classification of the protease and reverse transcriptase resistance [6][7].

In this paper, we will focus on the study of seven Protease inhibitors. The contact energy of the amino acids will be used to describe the sequences, and bidirectional recurrent neuronal networks are suggested for the analysis of sequences and resistance. The performance will be compared to that of other machine learning methods.

2 Methods

2.1 Datasets

There are several databases with available information about HIV protease and its resistance associated with drugs. We used the Stanford HIV Resistance Database Protease (http://hivdb.stanford.edu/cgi-bin/PIResiNote.cgi) to develop our strategy because it is the one mostly used in the literature. This database contains information about the genotype and phenotype for seven of the mostly used protease inhibitors: amprenavir (APV), atazanavir (ATV), nelfinavir (NFV), ritonavir (RTV), saquinavir (SQV), lopinavir (LPV) and indinavir (IDV). The genotype is documented for the mutated positions and con-sequent changed amino acids. The phenotype is represented by the resistance-fold based on the concentration of the drug to inhibit the viral protease.

Cases with unknown changes were discarded in order to eliminate missing values in learning, and seven databases were constructed, one for each drug. We took as reference sequence (wild type) the HXB2 protease and built the mutants by changing the amino acid in the corresponding reported mutated positions. For the resistance-fold we used the cut-off value of 3.5 as previously reported in the literature for these drugs [8][5]. If the drug resistance is greater than the cut-off, the mutant is classified as resistant and otherwise as susceptible.

2.2 Feature Representation

One of the most important steps to apply a classification method is to find good features to represent the input information. In some approaches the simple representation of each sequence position by a binary vector of 20 elements (i.e. the amino acids) has been used. In that case a value of 1 is given to the analyzed amino acid position and a 0 to all the others. Mutual information profiles have also been used to represent each sequence of the Protease enzyme [4].

We used features close to the 3D-structure to represent the primary sequence. In particular we chose the amino acid contact energy as an adequate representation because it determines the (un)folding of the protein. The contact energy changes the protein structure and that the substitution of a simple amino acid is enough to observe this [9][10]. For this reason the energy is used to represent the amino acids of the Protease sequence.

We analyze two feature representations:

- The energy associated with each amino acid, which we will refer to as *Energy*.
  $Energy : A \rightarrow \mathbb{R}$
  where A is the set of 20 amino acids and $\mathbb{R}$ is the set of real numbers
- The variation of the energy with regard to the wild type, i.e. the energy difference between the positions in the analyzed sequence and the corresponding position in the wild type, or vice versa. This variation is called $\triangle Energy$.
  $\triangle Energy(A_i) = Energy(AW_i) - Energy(A_i)$

where AW_i is the amino acid in the position i of the wild type sequence, and A_i is the amino acid in the position i of the mutated sequence.

2.3 Problem Formulation

The problem was transformed into seven similar classification problems of two classes.

For each problem the target function is defined as:
$F : C \rightarrow O$
O= {resistant, susceptible}
where $C \subseteq \mathbb{R}^{99}$, because the database consists of sequences of the same length, namely 99 amino acids. Each element of C is a protease sequence identified by an amino acid vector. All amino acids are represented by their *Energy* or $\triangle Energy$ which is, in both cases, a real value.

Finally, after having designed the classification task we proceed to choose an appropriate classification method.

2.4 Classification Methods

We used several classification methods such as Support Vector Machines (SVM), MultiLayer Perceptrons (MLP) and Bidirectional Recurrent Neural Networks (BRNN).

SVM. The SVM is a technique developed by Vapnik in 1995 from statistical learning theory[11]. SVMs have become an important machine learning technique for many pattern recognition problems, especially in computational biology. For SVM training and testing we used the LIBSVM software library available at http://www.csie.ntu.edu.tw/cjlin/libsvm [12].

MLP. The Multilayer Perceptron (MLP) [13] is a type of artificial neural network that simulates one of the countless functions of our nervous system: classification. Consequently, it structurally and functionally simulates part of the nervous system. This was one of the reasons for choosing a MLP to solve this problem. We used the standard Backpropagation algorithm with some heuristics [14], in order to achieve a higher efficiency, accelerating its convergence speed.

BRNN. Recurrent neural networks were originally created to analyze time series in which the present moment is influenced by the past and the future [15]. We used them here to analyze one-dimensional spatial sequences but the idea is essentially similar. There are several topologies for recurrent networks that have been used in the literature to solve different problems. A bidirectional dynamic topology is described and used for prediction of secondary structures by Baldi [16]. We decided to use a neural network topology where the sequence is analyzed in three parts with identical length , so that the processing of the middle part is influenced by the first and third part. Simultaneously, these extreme parts are influenced by the middle. In this way the training of the network represents the nature of the problem a little better.

Figure 1 shows an example of this topology for our problem. The network has 33 input neurons and 2 output neurons. It has context layers backward (HB), forward (HF) and the hidden layer (HO). In other words, this topology consists of two context blocks, one of them with recurrence to the left and another one with recurrence to the right. For each sequence we refer to s as the middle part, while we refer to left as the information received from layer HB (subsequence "$s-1$") and right as the information received from layer HF (subsequence "$s+1$"). But it should be noted that s is also considered the "right part" of the sequence $s-1$ and the "left part" of sequence $s+1$.

Table 1. Number of neurons associated to the context layers backward (HB), forward (HF) and to the hidden layer (HO) for each neuronal network

Drug	HB=HF	HO
SQV	11	11
LPV	11	11
RTV	20	20
APV	20	20
IDV	27	20
ATV	32	32
NFV	20	20

To HB and HF we developed several tests always assigning the same weight to the pattern from the previous subsequence and to the pattern from the posterior subsequence, according to the given drug. Table 1 shows the numbers of neurons of the topology with which we obtained the best results. In this case the 33

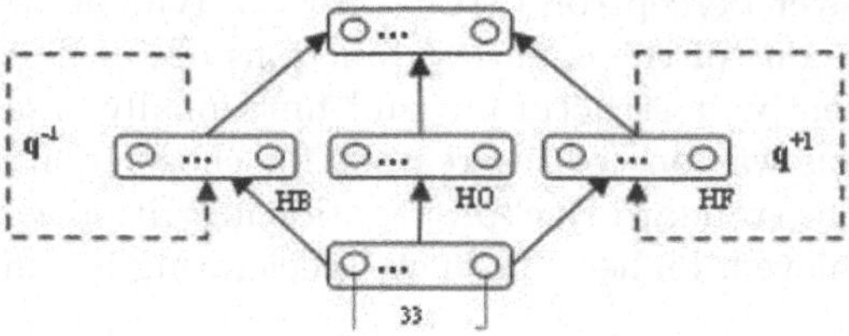

Fig. 1. Bidirectional Recurrent Neuronal Network Topology. Each of the arrows from layer to layer means that there are connections of all the neurons of the origin layer with all the neurons of the destination layer. The discontinuous arrows represent the connections between the parts, the shift operator q^{+1} means that the connection is from a left immediate part, and the shift operator q^{-1}means that the connection comes from the right immediate part.

inputs are real values and the outputs are (0,1) or (1,0), meaning resistant and susceptible, respectively.

A classification problem is not the typical problem to solve in this kind of network. We do not have an output associated with each subsequence. But we considered the three parts of a sequence associated with the same output. Specifically the sequence was divided in three parts, that is, 33 inputs in each part and three outputs- one output for each part.

As training algorithm of this network, the Backpropagation Through Time (BPTT) was used [17]. As target function we used Cross-Entropy and as output activation function we used a Softmax function.

3 Results

As explained above, we used three different methods (SVM, MLP and BRNN) to predict resistance of HIV sequences using seven inhibitors. We compared the results with those published previously [4][5]. All results were averaged using 10-fold crossvalidation.

We used MLP and SVM to compare with the results obtained up to now to demonstrate that the *Energy* as well as $\triangle$*Energy* are adequate feature representations for the resistance prediction.

In table 2 the columns 1, 2 and 3 correspond to the results reported by James (2004) using KNN, the classic decision tree using ID3 and a variant of a decision tree developed respectively. The column 4 represents the results obtained by Beerenwinkel et al. (2002) using a classic decision tree.The columns 5 and 6 show our results using a MLP and the columns 7 and 8 show our results using SVM. After analyzing these results we can see that the feature representation using *Energy* as well as the representation using $\triangle$*Energy* are appropriate to describe the sequence in this task because both are similar to the previous results.

Table 2. Prediction performance of methods used before: KNN, several decision trees and the prediction performance using MLP. Prediction performance is measured in terms of accuracy.

	1	2	3	4	5	6	7	8
					MLP		SVM	
Drug	KNN	Dtree	NewDtree	Dtree[1]	*Energy*	Δ*Energy*	*Energy*	Δ*Energy*
SQV	81.7	80	85.7	87.5	85.47	87.88	87.82	85.23
LPV	81.1		89.5		92.33	87.88	88.57	88.57
RTV	82	89	89.5	89.8	90.92	90.71	91.83	92.15
APV	80.9	75.8	75.8	87.4	82.17	80.65	82.30	83.64
IDV	80.6	85	85.5	89.1	86.96	92.55	91.51	92.57
ATV					80.00	74.16	74.38	72.72
NFV	73.6	91.8	93.7	88.5	86.63	87.13	84.86	84.86

As explained earlier, we used BRNN to solve the problem. The network has three output values during the predicting process, that is, the output is a vector with three components, because an output is obtained for each part. As in the other techniques used, we represent the output with the two values explained before - resistance and susceptible. The difference with regard to the other methods is that, now we will have three outputs in the prediction.

The BPTT algorithm is based on the unfolding and folding process. For each case in the training dataset, in the forward process the network is unfolding as a classical feedforward network and executes the Backpropagation algorithm to obtain the corresponding output as is shown in Figure 2. In the backward process the network is folding again to turn back as the beginning (Fig. 1) in order to update the weights [17].

As is illustrated in figure 2, we split the instances in the database. Figure 2 shows the first step to the BPTT and the processing of the outputs in order to use this network in this classifica-tion problem. The prediction is divided in three tasks. The first task is to split the sequence in three parts, representing the three entries to parts the network. The second task is to unfold the network and to obtain the three outputs for this input, and the third task is to compute the final output - to represent the resistance or not of this protein - as an adequate combination of the three previous outputs.

In our training dataset we represented the class in the three outputs (output 1, output 2, output3) using the same value, that means, 1 for resistance and 0 for susceptible. But in the prediction we can obtain different outputs for each part, i.e. the output is a vector of three coordinates: (O_1, O_2, O_3), where $O_i \in \{0, 1\}$

[1] Note the results of classical decision trees by James are different than those of Beerenwinkel due to a different number of cases; Beerenwinkel et al. used more cases to decision tree training. The columns 5 and 6 show our results using a MLP.

Table 3. Classification performance using bidirectional recurrent neural networks

	BRNN	
Drug	mode	middle output
SQV	91.16	91.16
LPV	94.42	94.39
RTV	93.42	94.73
APV	89.25	88.71
IDV	92.55	92.55
ATV	82.67	81.33
NFV	94.06	93.07

Now the problem is the following: once the network has finished its prediction and we have its vectorial output, we need to select one of its components as the sole final output. In this paper we will deal with two of several variants to obtain one output from the three outputs. A first output variant is the mode of the three outputs and a second variant is the output corresponding to the middle time (output from time t=2). In the case of the first variant we are obtaining the value that is more frequent at the three parts and that gives the same weight to all parts of the sequence. In the second case it is valid to remember that this middle output was influenced by the other two parts. For this reason it presumably has more information about the whole sequence than the other ones.

For this method we took as feature values the $\triangle Energy$ as is shown in Table 3. Similar results were obtained using the selection variant of the mode of the three outputs as well as the variant using the output of the middle time.

We also used statistical methods to analyze the results. A Friedman two-way ANOVA test was used to compare the results of Table 2 in order to validate the accuracy of the MLP. This test showed that there are significant differences between the methods. The Friedman test demonstrated that methods 3 and 5 (as referred in Table 2) are better than the rest. A Wilcoxon test ratified that these two algorithms (3 and 5) are similar. A Friedman two-way ANOVA test was also used to compare the different SVM used in this work and a Wilcoxon test was applied to compare with the results obtained by other authors (reported in Table 2). With these tests we do not obtain significant differences between the results of these algorithms.

We proceeded in the same way with the BRNN. We compared the results of this method with the best result in previous works, and with the best result using MLP. The statistical tests yielded significant differences between the accuracy of these techniques.

We used these statistics also to compare the two different kinds of final output processing (mode and middle output). In this case we demonstrated that the results of both BRNN are an appropriated method to the problem, because the accuracy is greater than or similar to the rest.

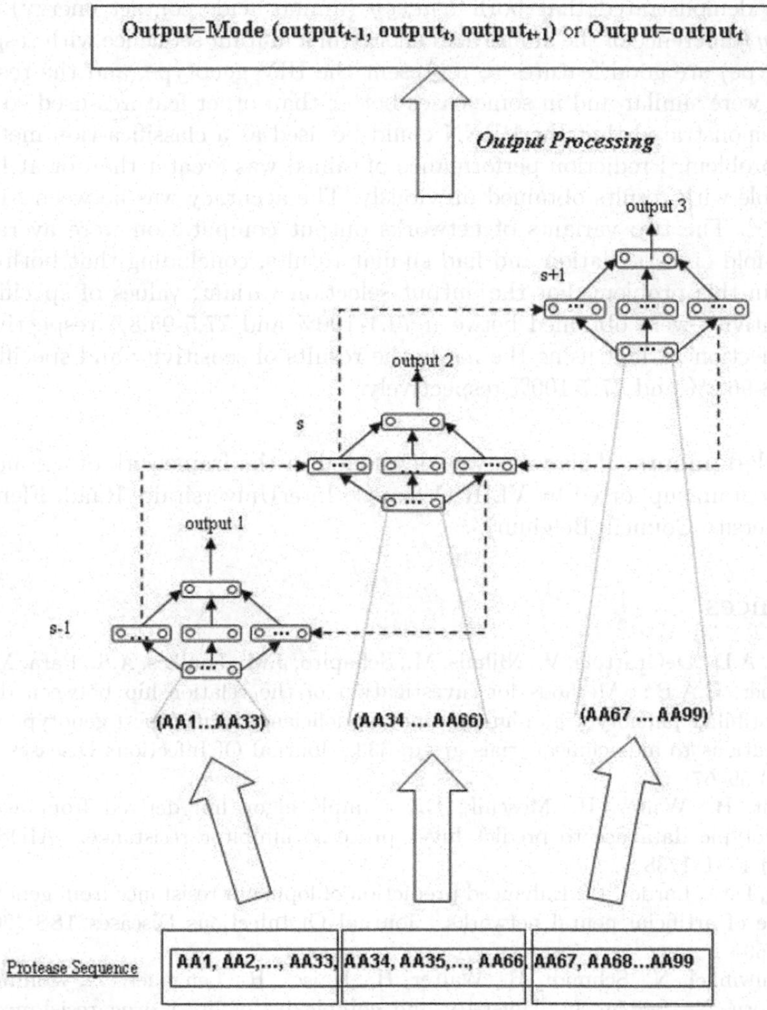

Fig. 2. Bidirectional Recurrent Neuronal Network Unfolding

4 Conclusions

In this paper we analyzed a recurrent neural network with an appropriate topology to analyze sequences in classification problems. In particular, we studied the problem to predict the Human Immunodeficiency Virus Drug Resistance. Amino acid energies of the Protease were used as features to represent the sequence with characteristics related to their 3D structure. A comparative evaluation of a selection of machine learning algorithms was performed, demonstrating the reliability of both the use of energy as features and the use of recurrent neural network as predictors.

It was demonstrated that both *Energy* (amino acids contact energy) and $\triangle Energy$ (difference of the amino acid energy in a mutant sequence with respect to wild type) are good features to represent the HIV genotype, and the results obtained were similar and in some cases better than other features used so far. It was demonstrated that the BRNN could be used as a classifica-tion method for this problem. Prediction performance obtained was greater than or at least comparable with results obtained previously. The accuracy was between 81.4% and 94.7%. The two variants of networks output computation were averaged using 10-fold crossvalidation and had similar results, concluding that both can be used in this problem. For the output selection variant, values of specificity and sensitivity were obtained between 74.1-100% and 77.5-95.8% respectively. In the selection variant using the mode the results of sensitivity and specificity were 84.8-96.2% and 77.7-100% respectively.

Acknowledgments. This work was developed in the framework of a collaboration program supported by VLIR (Vlaamse InterUniversitaire Raad, Flemish Interuniversity Council, Belgium).

References

1. Sevin, A.D., DeGruttola, V., Nijhuis, M., Schapiro, J.M., Foulkes, A.S., Para, M.F., Boucher, C.A.B.: Methods for investigation of the relationship between drug-susceptibility phenotype and human immunodeficiency virus type 1 genotype with applications to aids clinical trials group 333. Journal Of Infectious Diseases **182** (2000) 59–67
2. Scmidt, B., Walter, H., Moschik, B.: Simple algorithm derived from ageno-/phenotypic database to predict hiv-1 protease inhibitor resistance. AIDS **14** (2000) 1731–1738
3. Wang, D.C., Larder, B.: Enhanced prediction of lopinavir resistance from genotype by use of artificial neural networks. Journal Of Infectious Diseases **188** (2003) 653–660
4. Beerenwinkel, N., Schmidt, B., Walter, H., Kaiser, R., Lengauer, T., Hoffmann, D., Korn, K., Selbig, J.: Diversity and complexity of hiv-1 drug resistance: A bioinformatics approach to predicting phenotype from genotype. PNAS **99** (2002) 8271–8276
5. James, R.: Predicting Human Immunodeficiency Virus Type 1 Drug Resistance from Genotype Using Machine Learning. Msc thesis, University of Edinburgh (2004)
6. Rabinowitz, M., Myers, L., Banjevic, M., Chan, A., Sweetkind-Singer, J., Haberer, J., McCann, K., Wolkowicz, R.: Accurate prediction of hiv-1 drug response from the reverse transcriptase and protease amino acid sequences using sparse models created by convex optimization. Bioinformatics **22** (2006) 541–549
7. Cao, Z.W., Han, L.Y., Zheng, C.J., Ji, Z.L., Chen, X., Lin, H.H., Chen, Y.Z.: Computer prediction of drug resistance mutations in proteins. Drug Discovery Today **10** (2005) 521–529
8. Beerenwinkel, N., Daumer, M., Oette, M., Korn, K., Hoffmann, D., Kaiser, R., Lengauer, T., Selbig, J., Walter, H.: Geno2pheno: estimating phenotypic drug resistance from hiv-1 genotypes. Nucl. Acids Res. **31** (2003) 3850–3855

9. Miyazawa, S., Jernigan, R.L.: Protein stability for single substitution mutants and the extent of local compactness in the denatured state. Protein Eng. **7** (1994) 1209–1220
10. Miyazawa, S., Jernigan, R.L.: Residue potentials with a favorable contact pair term and an unfavorable high packing density term, for simulation and threading. J. Mol. Biol. **256** (1996) 623–644
11. Vapnik, V.: The Nature of Statistical Learning Theory. Springer-Verlag, New York (1995)
12. Chang, C., Lin, C.: Libsvm (2001)
13. Rumelhart, D.E., Hinton, A.G.E., Williams, A.R.J.: Learning internal representations by error propagation. Volume Parallel distributed processing: explorations in the microstructure of cognition, vol. 1: foundations. MIT Press (1986) 318–362
14. Bonet, I., Daz, A., Bello, R., Sardia, Y.: Learning optimization in a mlp neural network applied to ocr. MICAI 2002: Advances in Artificial Intelligence. LNAI **2313** (2002) 292–300
15. Tsoi, A., Back, A.: Discrete time recurrent neural network architectures: A unifying review. Neurocomputing **15** (1997) 183–223
16. Baldi, P., Soren, B.: Bioinformatics: The Machine Learning Approach. 2 edn. MIT Press. (2001)
17. Werbos, P.J.: Backpropagation through time: What it does and how to do it. Volume 78., IEEE (1990) 1550–1560

Error Weighting in Artificial Neural Networks Learning Interpreted as a Metaplasticity Model*

Diego Andina[1], Aleksandar Jevtić[1], Alexis Marcano[1], and J.M. Barrón Adame[2]

[1] Universidad Politécnica de Madrid, Spain
d.andina@gc.ssr.upm.es, a.jevtic@gc.ssr.upm.es, a.marcano@gc.ssr.upm.es
[2] Universidad de Guanajuato, Mexico
badamem@salamanca.ugto.mx

Abstract. Many Artificial Neural Networks design algorithms or learning methods imply the minimization of an error objective function. During learning, weight values are updated following a strategy that tends to minimize the final mean error in the Network performance. Weight values are classically seen as a representation of the synaptic weights in biological neurons and their ability to change its value could be interpreted as artificial plasticity inspired by this biological property of neurons. In such a way, metaplasticity is interpreted in this paper as the ability to change the efficiency of artificial plasticity giving more relevance to weight updating of less frequent activations and resting relevance to frequent ones. Modeling this interpretation in the training phase, the hypothesis of an improved training is tested in the Multilayer Perceptron with Backpropagation case. The results show a much more efficient training maintaining the Artificial Neural Network performance.

Keywords: Neural Networks, Backpropagation Training Algorithm, Metaplasticity, Binary Detection.

1 Introduction

The idea proposed is to improve the basic error minimization algorithm used to train an Artificial Neural Network (ANN) [1] manipulating the error objective function in order to give more relevance to less frequent training patterns and to subtract relevance to the frequent ones. So, if the objective is to minimize an expected error E_M defined by the following expression:

$$E_M = \varepsilon \{E(x)\} \tag{1}$$

where X is a random variable of the training input vectors $x = (x_1, x_2, ..., x_n)$, $(x \in R^n)$, where R^n is the n-dimensional space and $E(x)$ is the expression of

* This research has been supported by the National Spanish Research Institution "Comisíon Interministerial de Ciencia y Tecnología - CICYT" as part of the project AGL2006-12689/AGR.

J. Mira and J.R. Álvarez (Eds.): IWINAC 2007, Part I, LNCS 4527, pp. 244–252, 2007.

a given error criterion as a function of the inputs applied in ANN training to update its weights in each training iteration step, then

$$E_M = \int_{R^n} E(x) f(x) dx = \int_{R^n} e(x) dx \tag{2}$$

$$E_M = \int_{R^n} \frac{e(x)}{f_X^*(x)} f_X^*(x) dx = \varepsilon^* \left\{ \frac{e(x)}{f_X^*(x)} \right\} \tag{3}$$

From statistical inference theory applied to eq. (3), an estimator of E is given by:

$$\hat{E}_M = \frac{1}{N} \sum_{k=1}^{N} \frac{e(x_k)}{f_X^*(x_k)} \tag{4}$$

where $x_k^*, k = 1, 2, ..., N$, are independent sample vectors whose *pdf* is $f_X^*(x)$, that we call Weighting Function and $f_X^*(x)$ can be arbitrarily chosen by the designer if $f_X^*(x) \neq 0$ wherever $e(x) \neq 0$, $\forall x \in R^n$. Note that from eq. (3) $f_X^*(x)$ is ideally given by [1]:

$$(f_X^*(x))_{opt} = \frac{1}{E_M} e(x) \tag{5}$$

2 Weighting Operation

What lies in eq. (4) is that an error objective function $E(x_k)$ can be weighted by a proper function $w^*(x_k)$ without affecting the final error objective. In fig. 1 we present a block diagram for the Weighted Training.

On the hypothesis that by giving more relevance in weight update to less frequent activations and resting relevance to frequent ones, Metaplasticity [2][3] is being modelled and therefore training can be improved, we test for the case of a Multilayer Perceptron (MLP) the following weight functions:

$$w_X^*(x) = A\sqrt{(2\pi)^{N_1}} \cdot e^{B\sum_{i=1}^{8} x_i^2} \tag{6}$$

and

$$w_X^*(x) = C\hat{y} \tag{7}$$

In eq. (6) an inverse Gaussian function is proposed as weighting function as a standard assumption for the weighting function. In eq. (7) the network output is used and advantage is taken from the inherent a posteriori probabilities estimation for each input class of MLP outputs, so the statistical distribution of training patterns is used to quantify how frequent a pattern is. A, B, and C are parameters to be adjusted according to the training to converge. N_1 represents the number of MLP inputs.

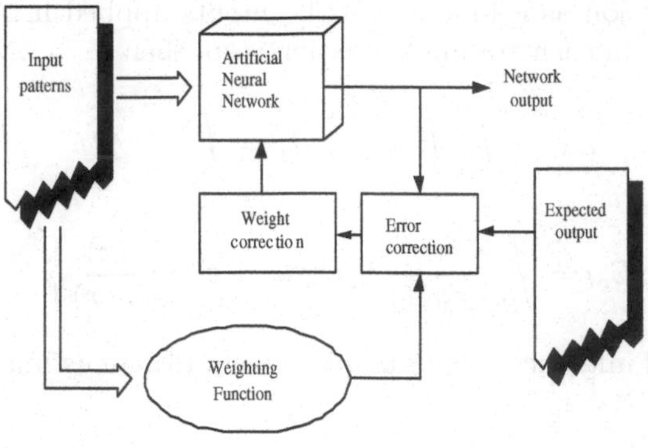

Fig. 1. Weighted training cycle

3 Computer Results

Experiments have been carried out in order to evaluate the Backpropagation with Weighting (BPW) algorithm [4]. The main objective of these experiments is the preliminary evaluation of the weighting function capabilities and limits in particular cases. Note that the range of implementation of the idea proposed is very wide as many ANNs learning methods imply the minimization of an error function. We present the results obtained from training 100 MLPs applying a BPW algorithm consisting in Least Mean Square (LMS) criterion modified by the proposed weighting functions.

3.1 General Characteristics of the Experiments

The ANNs applied are MLPs with structure 16/8/1 (that is 16 inputs, and one hidden layer of 8 units). The choice of the structure and the rest of the parameters of the network was the optimal solution for the given example application [1][5]. The activation function is sigmoidal with scalar output in the range (0,1) and it is the same for all the neurons.

For the training of the network we used balanced patterns of two classes, being class H_0 noise patterns and being class H_1 signal received with additive Gaussian noise. These patterns configure the problem of signal detection noise and the ANN acts as a binary detector. The application of the ANN is an elemental radar detection problem [5][6] when the basic parameter for the patterns is the Signal-to-Noise ratio, *SNR*, and the performance of the detectors is evaluated in terms of the Neyman-Pearson criterion. That is, maximizing probability of detection, P_d, (the probability of classifying correctly the patterns belonging to the class H_1) for a fixed false alarm probability, P_{fa} (the probability of classifying erroneously the patterns belonging to the class H_0). In the radar literature, performance is evaluated through the Detection curves (P_d vs. *SNR*), so we use these detection curves to present the results of our method.

In each experiment 100 networks were trained in order to achieve mean results that does not depend on initial random value of the weights of the ANN. Fig. 2 shows the error evolution comparison of the network trained with BPW and classical BP training. Error is calculated as the rate of misclassified patterns of the test set out of the total number of patterns. We can notice that the BPW training algorithm requires much less iterations to consider an ANN trained than the classical BP does, which shortens the total time of training.

In the following experiments, two different criterions were applied to stop the training: in one case it was stopped when the error reached zero (denoted as is_m) and in the other the training was conducted with a fixed number of 3000 patterns ($3is_m$). When inverse Gaussian function (6) was applied as Weighting function, the training was conducted as usual BP training, that is, maintaining the expression of the error function to be minimized from the beginning to the end of the training. Fig. 3 shows an example of NN training using only weighting function (6). But in the case of (7) the weighting function is not valid until the output of the network is a sufficiently good approximation of the *a posteriori* probabilities of the inputs. In the first iterations $\hat{y}$ can tend to values very close to zero and the MLP does not learn. So, in this second case, function (6) was applied till the error probability achieved a value in the range of 0.1-0.2 and then switched to function (7) till the end of training. Fig. 4 shows the error evolution during the network training phase for the second case. As usual, three set of patterns have been used to design the network. A training set (composed of patterns of $SNR = 13.2dB$ for class H_1), a test set to calculate the error during training and a validation set to obtain the detection curves.

The detection probability for three different false alarm probability (probability of "decide H_0 when input corresponds to H_1") values related to the SNR are shown in fig. 5, 6 and 7, respectively.

The red line represents the theoretical maximum by Marcum theorem [6]. The green line represents average performance for the networks that were trained until the error probability reached zero and the blue line is used for the networks trained with the fixed number of patterns. False alarm probabilities, P_{fa}, of

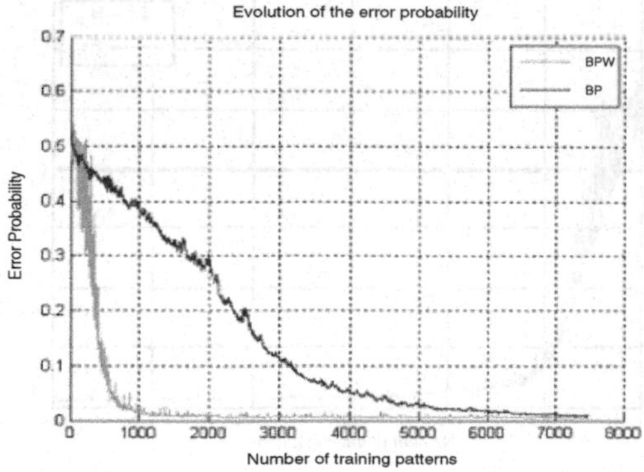

Fig. 2. Classification error in training phase of BP and BPW

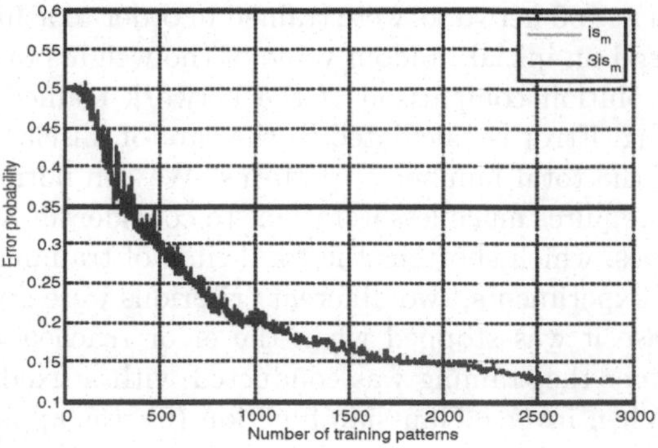

Fig. 3. Classification error in training phase with only one weighting function. $3is_m$ and is_m have the same evolution.

10^{-2}, 10^{-3} and 10^{-4} have been considered. For the detection probability that corresponds to the false alarm probability of 0.01, we find that the results are noticeably better if the NNs were trained with the fixed number of patterns (3000) for all the values in relation to the *SNR* between 0 and 8 *dB*. In the case of false alarm probability of 0.001 and 0.0001 we also get better results for training a network with the fixed number of patterns and the curve (blue) is much closer to the theoretical one (red). For the high *SNR* values the results could be improved, which could make a part of the future lines of investigation for this application.

Fig. 8 shows the results obtained for setting the threshold for changing the weighting functions at 0.15. Again, we considered two criterions for stopping the training of a network, when error reaches zero and with the fixed number of patterns. We can see that the decision to change the weighting function when the threshold 0.15 was reached gave also satisfactory results. Again, the experiments were carried out for false alarm probability values 10^{-2}, 10^{-3} and 10^{-4}. The

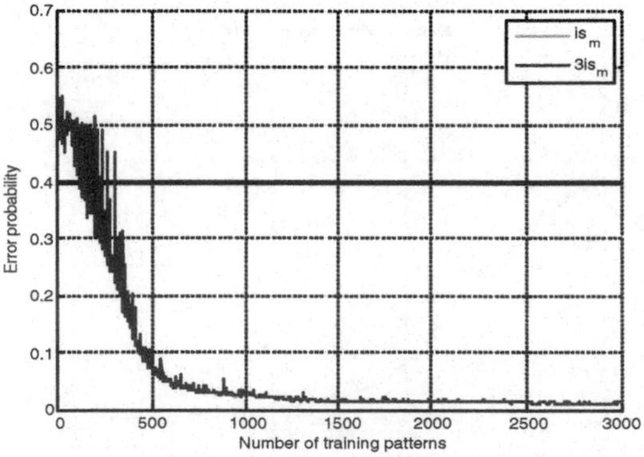

Fig. 4. Classification error in training phase, threshold 0.2. $3is_m$ and is_m have the same evolution.

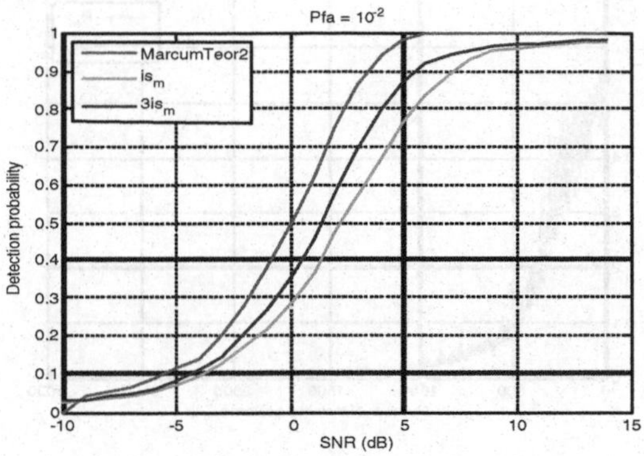

Fig. 5. Detection probability, P_{fa}=10^{-2}, threshold 0.2

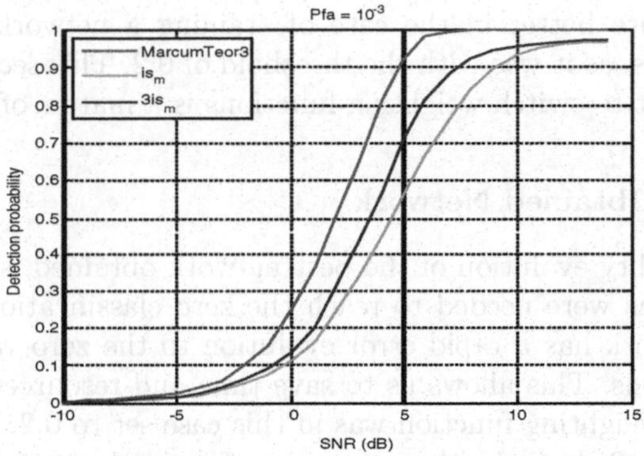

Fig. 6. Detection probability, P_{fa}=10^{-3}, threshold 0.2

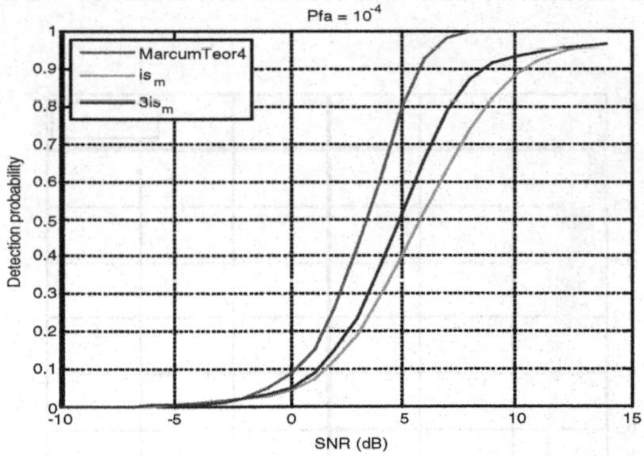

Fig. 7. Detection probability, P_{fa}=10^{-4}, threshold 0.2

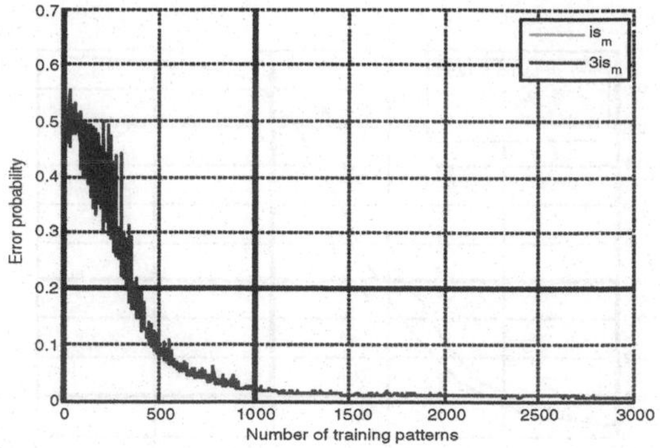

Fig. 8. Classification error in training phase, threshold 0.15. $3is_m$ and is_m have the same evolution.

results obtained are better in the case of training a network with the fixed number of patterns, as it was with the threshold of 0.2. This seems to show that the optimum point to switch weighting functions is a matter of study.

3.2 The Best Obtained Network

The error probability evolution of the best network obtained is shown in fig. 9. Only 355 iterations were needed to reach the zero classification error. We can see that the network has a rapid error evolution to the zero value, with a low number of iterations. This allows us to save time and resources. The threshold for changing the weighting function was in this case set to 0.2.

Fig. 10, 11 and 12 show the characteristics of trained network for false alarm probabilities, P_{fa}, of 10^{-2}, 10^{-3} and 10^{-4}, respectively. We can see that the distance between two curves is less than 1 dB. Even though the number of iterations used was small, we could continue the training with a fixed number

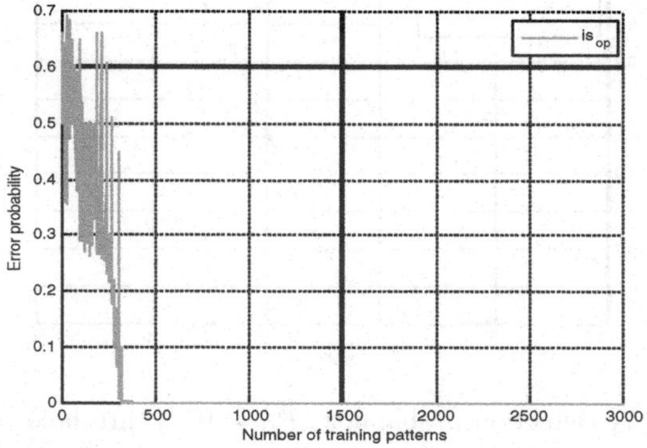

Fig. 9. Classification error in training phase, threshold 0.2, the best case.

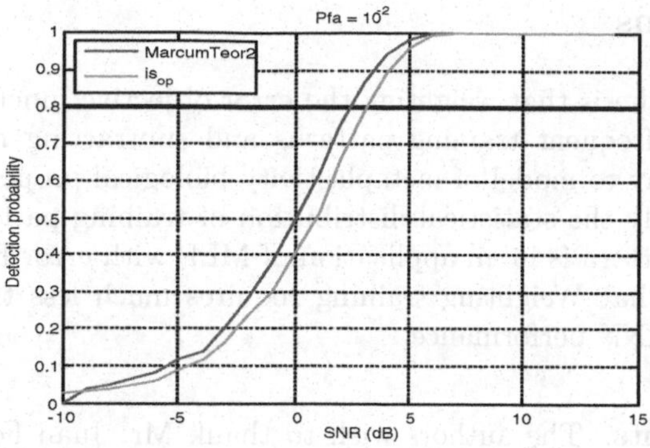

Fig. 10. Detection probability, P_{fa}=10^{-2}, threshold 0.2, the best case

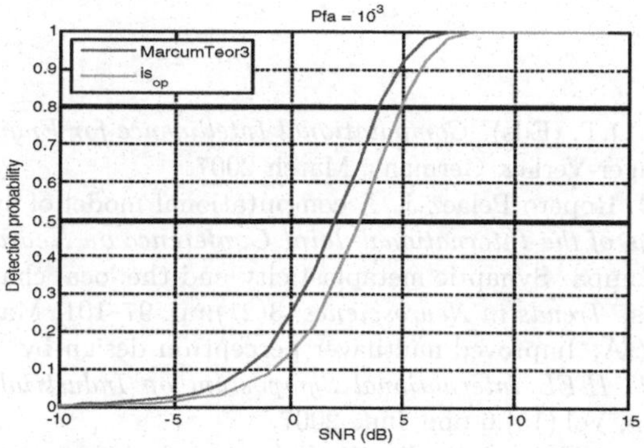

Fig. 11. Detection probability, P_{fa}=10^{-3}, threshold 0.2, the best case

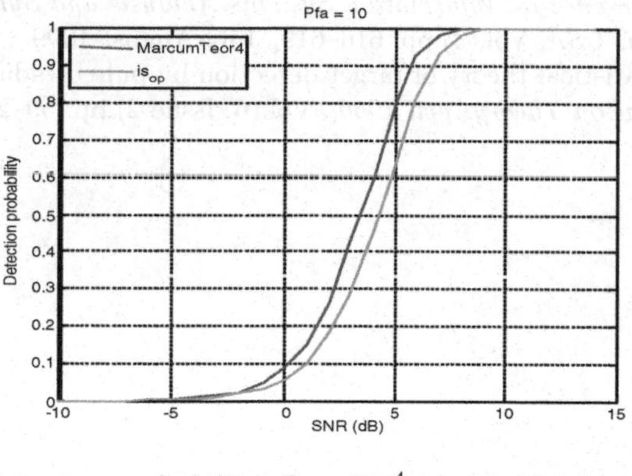

Fig. 12. Detection probability, P_{fa}=10^{-4}, threshold 0.2, the best case

of patterns and obtain values even closer to the theoretical maximum. These results support, one more time, the superiority of the performance of NNs trained applying BPW criterion with two weighting functions.

4 Conclusions

We test the hypothesis that weighting the error objective function giving more relevance to less frequent training patterns and subtracting relevance to frequent ones is a way to model of metaplasticity biological properties in artificial neurones. We apply the statistical distribution of training patterns to quantify how frequent a pattern is in an application of MLP with error Backpropagation training, finding that Weighting training requires much less training patterns maintaining the ANN performance.

Acknowledgments. The authors wish to thank Mr. Juan Fombellida, graduated student from Technical University of Madrid, Spain, for his work in the realization and discussion of the experiments.

References

1. Andina D. Pham D.T. (Eds). *Computational Intelligence for Engineering and Manufacturing.* Springer-Verlag, Germany, March 2007.
2. Godoy Simoes M. Ropero Pelaez J. A computational model of synaptic metaplasticity. *Proceedings of the International Joint Conference on Neural Networks*, 1999.
3. P. Friedrich P. Tompa. Synaptic metaplasticity and the local charge effect in postsynaptic densities. *Trends in Neuroscience*, 3(21):pp. 97–101, May 1998.
4. Andina D. Jevtić A. Improved multilayer perceptron design by weighted learning. *Proc. of the 2007 IEEE International Symposium on Industrial Electronics ISIE 2007, Vigo, Spain*, Vol.(15):6 pp., June 2007.
5. Alarcón M.J. Torres-Alegre S. Andina D. Behavior of a binary detector based on an ann digital implementation in the presence of seu simulations. *Proc. of III World Multiconference on Systemics, Cybernetics and Informatics (SCI'99) and 5th International Conference on Information Systems Analysis and Synthesis (ISAS'99), Orlando, Florida, USA*, Vol.(3):pp. 616–619, July-August 1999.
6. Marcum J. A statistical theory of target detection by pulsed radar. *IEEE Transactions on Information TheoryApril 1960.*, Vol.(6, Issue 2):pp. 59–267, April 1960.

A First Approach to Birth Weight Prediction Using RBFNNs

A. Guillén[1], I. Rojas[2], J. González[2], H. Pomares[2], and L. J. Herrera[2]

[1] Department of Informatics
University of Jaen. Spain
[2] Department of Computer Architecture and Computer Technology
Universidad de Granada. Spain

Abstract. This paper presents a first approach to try to determine the weight of a newborn using a set of variables determined uniquely by the mother. The proposed model to approximate the weight is a Radial Basis Function Neural Network (RBFNN) because it has been successfully applied to many real world problems. The problem of determining the weight of a newborn could be very useful by the time of diagnosing the gestational diabetes mellitus, since it can be a risk factor, and also to determine if the newborn is macrosomic. However, the design of RBFNNs is another issue which still remains as a challenge since there is no perfect methodology to design an RBFNN using a reduced data set, keeping the generalization capabilities of the network. Within the many design techniques existing in the literature, the use of clustering algorithms as a first initialization step for the RBF centers is a quite common solution and many approaches have been proposed. The following work presents a comparative of RBFNNs generated using several algorithms recently developed concluding that, although RBFNNs that can approximate a training data set with an acceptable error, further work must be done in order to adapt RBFNN to large dimensional spaces where the generalization capabilities might be lost.

1 Introduction

The problem of predicting the weight of a newborn using some parameters measured from the mother translates into the problem of approximating a function. Formally, a function approximation problem can be formulated as, given a set of observations $\{(\boldsymbol{x}_k; y_k); k = 1, ..., n\}$ with $y_k = F(\boldsymbol{x}_k) \in \mathbb{R}$ and $\boldsymbol{x}_k \in \mathbb{R}^d$, it is desired to obtain a function $\mathcal{G}$ so $y_k = \mathcal{G}\ (\boldsymbol{x}_k) \in \mathbb{R}$ with $\boldsymbol{x}_k \in \mathbb{R}^d$.

Designing an RBF Neural Network (RBFNN) to approximate a function from a set of input-output data pairs, is a common solution since this kind of networks are able to approximate any function [4,12]. Once this function is learned, it will be possible to generate new outputs from input data that were not specified in the original data set, making possible to predict the weight of a newborn.

J. Mira and J.R. Álvarez (Eds.): IWINAC 2007, Part I, LNCS 4527, pp. 253–260, 2007.

The most important information that could be obtained is the fetal macrosomia, this is, a a birth weight of more than 4,000 g. The macrosomia is difficult to predict and clinical and ultrasonographic estimates tend to have errors [15]. Furthermore, the weight of the fetus is a risk factor for several diseases such us gestational diabetes mellitus [3], therefore, if we are able to approximate the weight of the newborn, we will know in advance one of the many elements that are used to identify diseases.

The rest of the paper is organized as follows, Section 2 describes briefly the RBFNN model, Section 3 introduces the algorithms used to design the RBFNNs to predict the newborn weight and Section 4 shows the results. Finally, in Section 5, conclusions are drawn.

2 RBFNN Description

An RBFNN (Figure 1) $\mathcal{F}$ with fixed structure to approximate an unknown function F with n entries and one output starting from a set of values $\{(\boldsymbol{x}_k; y_k); k = 1, ..., n\}$ with $y_k = F(\boldsymbol{x}_k) \in \mathbb{R}$ and $\boldsymbol{x}_k \in \mathbb{R}^d$, has a set of parameters that have to be optimized:

$$\mathcal{F}(\boldsymbol{x}_k; C, R, \Omega) = \sum_{j=1}^{m} \phi(\boldsymbol{x}_k; \boldsymbol{c}_j, r_j) \cdot \Omega_j \quad (1)$$

where $C = \{\boldsymbol{c}_1, ..., \boldsymbol{c}_m\}$ is the set of RBF centers, $R = \{r_1, ..., r_m\}$ is the set of values for each RBF radius, $\Omega = \{\Omega_1, ..., \Omega_m\}$ is the set of weights and $\phi(\boldsymbol{x}_k; \boldsymbol{c}_j, r_j)$ represents an RBF. The activation function most commonly used for classification and regression problems is the Gaussian function because it is continuous, differentiable, it provides a softer output and improves the interpolation capabilities [2,14].

The procedure to design an RBFNN starts by setting the number of RBFs in the hidden layer, then the RBF centers $\boldsymbol{c}_j$ must be placed and a radius r_j has to be set for each of them. Finally, the weights Ω_j can be calculated optimally by solving a linear equation system [5].

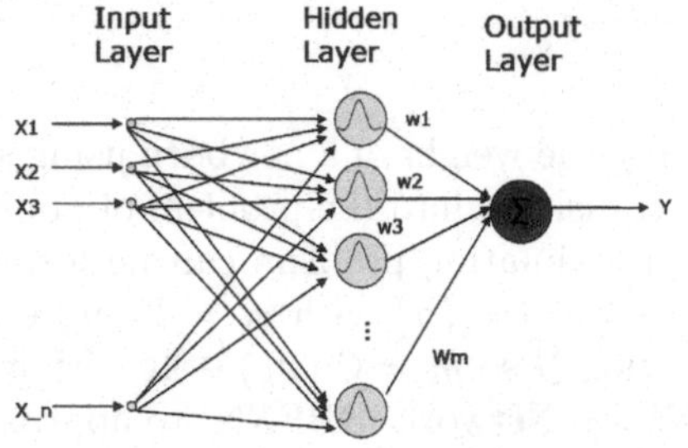

Fig. 1. A Radial Basis Function Neural Network

3 Algorithms for Designing RFBNNs

This section presents the algorithms used to design the RBFNNs that predict the newborn weight. Some of these algorithms have been recently developed showing a better performance than classical algorithms used up to date.

3.1 Fuzzy C-Means (FCM)

This algorithm presented in [1] uses a fuzzy partition of the data where an input vector belongs to several clusters with a membership value. It defines an objective distortion function to be minimized is:

$$J_h(U, C; X) = \sum_{k=1}^{n} \sum_{i=1}^{m} u_{ik}^h \|\boldsymbol{x}_k - \boldsymbol{c}_i\|^2 \tag{2}$$

where $X = \{\boldsymbol{x}_1, \boldsymbol{x}_2, ..., \boldsymbol{x}_n\}$ are the input vectors, $C = \{\boldsymbol{c}_1, \boldsymbol{c}_2, ..., \boldsymbol{c}_m\}$ are the centers of the clusters, $U = [u_{ik}]$ is the matrix where the degree of membership is established by the input vector to the cluster, and h is a parameter to control the degree of the partition fuzziness. After applying the least square method to minimize the function in Equation 2, we get the equations to reach the solution trough an iterative process.

3.2 Improved Clustering for Function Approximation Algorithm: ICFA

This algorithm uses the information provided by the objective function output in such a way that the algorithm will place more centers where the variability of the output is higher instead of where there are more input vectors.

In order to make the centers closer to the areas where the target function is more variable, a change in the similarity criteria used in the clustering process it is needed. To consider these situations, the parameter w is introduced (4) to modify the values of the distance between a center and an input vector. w will measure the difference between the estimated output of a center and the output value of an input vector. The smaller w is, the more the distance between the center and the vector will be reduced. This distance is calculated now by modifying the norm in the euclidean distance:

$$d_{kj} = \|\boldsymbol{x}_k - \boldsymbol{c}_j\|^2 \cdot w_{kj}^2. \tag{3}$$

To fulfill this task, the CFA algorithm defines a set $O = \{o_1, ..., o_m\}$ that represents a hypothetic output for each center.

$$w_{kj} = \frac{|F(\boldsymbol{x}_k) - o_j|}{\max\limits_{i=1}^{n} \{F(\boldsymbol{x}_i)\} - \min\limits_{i=1}^{n} \{F(\boldsymbol{x}_i)\}} \tag{4}$$

where $F(\boldsymbol{x})$ is the function output and o_j is the estimated output of $\boldsymbol{c}_j$.

Thus, the objective function to be minimize is redefined as:

$$J_h(U,C,W) = \sum_{k=1}^{n}\sum_{i=1}^{m} u_{ik}^h \|\boldsymbol{x}_k - \boldsymbol{c}_i\|^2 w_{ik}^2 \tag{5}$$

This function is minimized using an alternating optimization procedure in the same way as in the FCM algorithm, although new equations are needed to calculate the positions of the centers, the membership values and the expected output values:

$$u_{ik} = \left(\sum_{j=1}^{m}\left(\frac{d_{ik}}{d_{jk}}\right)^{\frac{2}{h-1}}\right)^{-1} \qquad \boldsymbol{c}_i = \frac{\sum_{k=1}^{n} u_{ik}^h \boldsymbol{x}_k w_{ik}^2}{\sum_{k=1}^{n} u_{ik}^h w_{ik}^2}$$

$$\boldsymbol{o}_i = \frac{\sum_{k=1}^{n} u_{ik}^h Y_k d_{ik}^2}{\sum_{k=1}^{n} u_{ik}^h d_{ik}^2} \tag{6}$$

where d_{ij} is the weighted euclidean distance between center i and input vector j, and $h > 1$ is a parameter that allow us to control how fuzzy will be the partition and usually is equal to 2.

The ICFA algorithm performs a migration step with the objective of reducing the global distortion of the partition by putting closer two centers. It performs a pre-selection of the centers, to decide what centers will be migrated, it is used a fuzzy rule that selects centers that have a distortion value above the average. By doing this, centers that do not add a significant error to the objective function are excluded because their placement is correct and they do not need help from other center. The center to be migrated will be the one that has assigned the smallest value of distortion and the destination of the migration will be the center that has the biggest value of distortion. If the total distortion of the partition has nor decreased after the migration, the centers remain at their original positions. The idea of a migration step was introduced in [13] as an extension of Hard C-means.

3.3 Fuzzy Possibilistic CFA

The algorithm that is used in the design is an adaptation of the one presented in [9] but modifying the way the input data is partitioned. As classical clustering algorithms, the proposed algorithm defines a distortion function that has to be minimized. The distortion function is based in a fuzzy-possibilistic approach as it was presented in [6], although the migration step remains the same as for ICFA. The function is:

$$J_h(U,C,T,W;X) = \sum_{k=1}^{n}\sum_{i=1}^{m} (u_{ik}^{h_f} + t_{ik}^{h_p}) D_{ikW}^2 \tag{7}$$

restricted to the constraints: $\sum_{i=1}^{m} u_{ik} = 1 \ \forall k = 1...n$ and $\sum_{k=1}^{n} t_{ik} = 1 \ \forall i = 1...m$.

As the previous approaches, the final position of the centers is reached by an alternating optimization approach where all the elements defined in the function to be minimized are updated iteratively using the equations obtained by differentiating $J_h(U,T,C,W;X)$ with u_{ik}, t_{ik}, $\boldsymbol{c}_i$ and o_i.

3.4 Possibilistic Centers Initializer (PCI)

This algorithm [8] adapts the algorithm proposed in [9] using a mixed approach between a possibilistic and a fuzzy partition, combining both approach as it was done in [16]. The objective function to be minimized is defined as:

$$J_h(U^{(p)},U^{(f)},C,W;X) = \sum_{k=1}^{n}\sum_{i=1}^{m}(u_{ik}^{(f)})^{h_f}(u_{ik}^{(p)})^{h_p}D_{ikW}^2 + \sum_{i=1}^{m}\eta_i\sum_{k=1}^{n}(u_{ik}^{(f)})^{h_f}(1-u_{ik}^{(p)})^{h_p} \tag{8}$$

where $u_{ik}^{(p)}$ is the possibilistic membership of x_k in the cluster i, $u_{ik}^{(f)}$ is the fuzzy membership of $\boldsymbol{x}_k$ in the cluster i, D_{ikW} is the weighted euclidean distance, η_i is a scale parameter that is calculated by: $\eta_i = \frac{\sum_{k=1}^{n}(u_{ik}^{(f)})^{h_f}\|\boldsymbol{x}_k - \boldsymbol{c}_i\|^2}{(u_{ik}^{(f)})^{h_f}}$

This function is obtained by replacing de distance measure in the FCM algorithm by the objective function of the PCM algorithm, obtaining a mixed approach. The scale parameter determines the relative degree to which the second term in the objective function is compared with the first. This second term forces to make the possibilistic membership degree as big as possible, thus, choosing this value for η_i will keep a balance between the fuzzy and the possibilistic memberships.

3.5 Output Value-Based Initializer (OVI)

This algorithm [7] changes the perspective of the previous ones, the idea is to think the output space as a flat surface ($Y(\boldsymbol{x}_k) = 0$), where some of the values of this surface have been modified by an n-dimensional element, obtaining y_k. From this, it can be assumed that the most common value of the target function is constant and equal 0. The preprocessing of the output is performed by making the most frequent output value equal 0. This can be easily performed by calculating the fuzzy mode of the output values and subtracting it to each y_k. Once this situation is achieved, all the most common values that have an output around 0 will not affect the distortion significantly so the centers will be mostly influenced by the input vectors with high output values.

This distortion function combines the information provided by a coordinate in the input vector space and its corresponding output in such a way that, if a neuron is near an input vector and the output of the input vector is high, the activation value of that neuron respect that input vector will be high. The distortion function is defined as:

$$\delta = \sum_{k=1}^{n}\sum_{i=1}^{m} D_{ik}^2 a_{ik}^l |Y_k^p| \tag{9}$$

where D_{ik} represents the euclidean distance from a center c_i to an input vector $\boldsymbol{x}_k$, a_{ik} is the activation value that determines how important the input vector $\boldsymbol{x}_k$ is for the center $\boldsymbol{c}_i$, l is a parameter to control the degree of overlapping between the neurons, Y_k is the preprocessed output of the input vector $\boldsymbol{x}_k$, and p allows the influence of the output when initializing the centers to increase or decrease.

The OVI algorithm calculates how much an input vector will activate a neuron in function of its output value. From this, the value of the radius can be set as the distance to the farthest input vector that activates a center. In order to do that, a threshold has to be established to decide when an input vector activates or does not activate a center.

Using the values of the A matrix, an activation threshold ($\vartheta_{overlap}$) that allows us to calculate the distance to the farthest input vector that activates a neuron can be established. The proposed algorithm selects the radius for each center independently of the positions of the other centers, unlike in the KNN heuristic [11], and it allows to maintain an overlap between the RBFs, unlike in the CIV heuristic [10].

Each radius is defined as:

$$r_i = \max\{\ D_{ik}\ /\ a_{ik} > \vartheta_{overlap}\ ,\ 1 \le i \le m\ ,\ 1 \le k \le n\ \} \tag{10}$$

The selection of a threshold makes the algorithm more flexible, because it can increase or decrease the degree of overlap between the RBFs.

4 Experimental Results

The data used for the experiments was provided by the Preventative Medicine Department at the University of Granada and consists in a cohort of 969 pregnant women considering the following parameters: number of cigarettes smoked during the pregnancy, mother's weight at the beginning and at the end of the pregnancy, gestation days and the mother's age. A set of 500 randomly chosen elements from the original set was used for training and the rest for test.

To compare the results provided by the different algorithms, it will be used the normalized root mean squared error (NRMSE) which is defined as:

$$\textbf{NRMSE} = \sqrt{\frac{\sum\limits_{k=1}^{n}\left(y_k - \mathcal{F}(\boldsymbol{x}_k; C, R, \Omega)\right)^2}{\sum\limits_{k=1}^{n}\left(y_k - \bar{Y}\right)^2}} \tag{11}$$

where $\bar{Y}$ is the average of the outputs of the target function, in this case, the final weight of the newborn.

The radii of the RBFs were calculated using the k-neighbors algorithm with k=1, except for the OVI algorithm which uses its own technique as described above. The weights were calculated optimally by solving a linear equation system.

Table 1 shows the approximation errors for the training and test data sets. As the results show, the performance of the algorithms is quite similar although the OVI algorithm seems to perform better than the rest probably as a consequence of its own method to calculate the radii. All the algorithms are able to fit the training set with a reasonable error for any number of centers with no improvement of error when increasing the number of RBFs. Unfortunately, the test errors are unacceptable for all the algorithms showing how the RBFNNs loose their generalization capabilities in high dimensional spaces using a reduced amount of data.

Table 1. Mean of the approximation error (NRMSE) for the training and test sets

Training

Clusters	**FCM**	**ICFA**	**FPCFA**	**PCI**	**OVI**
5	0.639	0.652	0.642	0.638	0.635
6	0.680	0.642	0.641	0.644	0.640
7	0.675	0.633	0.629	0.636	0.625
8	0.674	0.632	0.653	0.669	0.617
9	0.644	0.655	0.631	0.629	0.619
10	0.644	0.623	0.645	0.650	0.631

Test

Clusters	**FCM**	**ICFA**	**FPCFA**	**PCI**	**OVI**
5	4.583	4.461	4.518	4.575	4.545
6	4.569	4.606	4.499	4.625	4.619
7	4.639	4.550	4.634	4.479	4.579
8	4.593	4.573	4.575	4.451	4.621
9	4.580	4.570	4.577	4.614	4.607
10	4.628	4.588	4.597	4.601	4.591

5 Conclusions

This work has presented an application of RBFNNs to a real world problem: the prediction of a newborn's weight. This parameter could be quite useful in the diagnosis of macrosomia which can lead to complications at the childbirth and also to be considered in the diagnosis of other diseases which has the newborn weight as a risk factor. The RBFNNs were designed using a classical methodology where the centers of the RBFs are initialized using clustering techniques and applying local search algorithms. The results showed how, for the training set, the RBFNNs were able to approximate reasonably well the weights although test errors become unacceptable, independently of the algorithm used. This results cheer to keep on researching on this subject although aiming at other aspects such us how the Radial Basis Function behave in high dimensional spaces.

References

1. J. C. Bezdek. *Pattern Recognition with Fuzzy Objective Function Algorithms.* Plenum, New York, 1981.
2. A. G. Bors. Introduction of the Radial Basis Function (RBF) networks. *OnLine Symposium for Electronics Engineers*, 1:1–7, February 2001.
3. R. Dyck, H. Klomp, L.K. Tan, R.W. Turner, and M.A. Boctor. A comparison of rates, risk factors, and outcomes of gestational diabetes between aboriginal and non-aboriginal women in the Saskatoon Health District. *Diabetes Care*, 25:487–493, 2002.
4. A. Gersho. Asymptotically Optimal Block Quantization. *IEEE Transanctions on Information Theory*, 25(4):373–380, July 1979.
5. J. González, I. Rojas, J. Ortega, H. Pomares, F.J. Fernández, and A. Díaz. Multiobjective evolutionary optimization of the size, shape, and position parameters of radial basis function networks for function approximation. *IEEE Transactions on Neural Networks*, 14(6):1478–1495, November 2003.
6. A. Guillén, I. Rojas, J. González, H. Pomares, L. J. Herrera, and A. Prieto. A Fuzzy-Possibilistic Fuzzy Ruled Clustering Algorithm for RBFNNs Design. *Lecture Notes in Computer Science*, 4259:647–656, 2006.
7. A. Guillén, I. Rojas, J. González, H. Pomares, L.J. Herrera, and A. Prieto. Supervised RBFNN Centers and Radii Initialization for Function Approximation Problems. In *International Joint Conference on Neural Networks, 2006. IJCNN '06*, pages 5814–5819, July 2006.
8. A. Guillén, I. Rojas, J. González, H. Pomares, L.J. Herrera, O. Valenzuela, and A. Prieto. A Possibilistic Approach to RBFN Centers Initialization. *Lecture Notes in Computer Science*, 3642:174–183, 2005.
9. A. Guillén, I. Rojas, J. González, H. Pomares, L.J. Herrera, O. Valenzuela, and A. Prieto. Improving Clustering Technique for Functional Approximation Problem Using Fuzzy Logic: ICFA algorithm. *Lecture Notes in Computer Science*, 3512:272–280, June 2005.
10. N. B. Karayiannis and G. W. Mi. Growing radial basis neural networks: Merging supervised and unsupervised learning with network growth techniques. *IEEE Transactions on Neural Networks*, 8:1492–1506, November 1997.
11. J. Moody and C.J. Darken. Fast learning in networks of locally-tunned processing units. *Neural Computation*, 1(2):281–294, 1989.
12. J. Park and J. W. Sandberg. Universal approximation using radial basis functions network. *Neural Computation*, 3:246–257, 1991.
13. G. Patanè and M. Russo. The Enhanced-LBG algorithm. *Neural Networks*, 14(9):1219–1237, 2001.
14. I. Rojas, M. Anguita, A. Prieto, and O. Valenzuela. Analysis of the operators involved in the definition of the implication functions and in the fuzzy inference proccess. *International Journal of Approximate Reasoning*, 19:367–389, 1998.
15. M. A. Zamorski and W.S. Biggs. Management of Suspected Fetal Macrosomia. *American Family Physician*, 63(2), January 2001.
16. J. Zhang and Y. Leung. Improved possibilistic C–means clustering algorithms. *IEEE Transactions on Fuzzy Systems*, 12:209–217, 2004.

Filtering Documents with a Hybrid Neural Network Model

Guido Bologna, Mathieu Boretti, and Paul Albuquerque

University of Applied Science HES-SO, Laboratoire d'Informatique Industrielle
Rue de la Prairie 4, 1202 Geneva, Switzerland
Guido.Bologna@hesge.unige.ch

Abstract. This work presents an application example of text document filtering. We compare the DIMLP neural hybrid model to several machine learning algorithms. The clear advantage of this neural hybrid system is its transparency. In fact, the classification strategy of DIMLPs is almost completely encoded into the extracted rules. During cross-validation trials and in the majority of the situations, DIMLPs demonstrated to be at least as accurate as support vector machines, which is one of the most accurate classifiers of the text categorization domain. In the future, in order to further increase DIMLP accuracy, we believe that common sense knowledge could be easily inserted and refined with the use of symbolic rules.

1 Introduction

Until the end of the eighties, document classification was mainly a knowledge engineering process. In practice, symbolic rules describing how to classify documents under the existing categories were first determined by experts and then used in classification algorithms. In the nineties, categorization of documents by machine learning models was strongly investigated; in many cases the accuracy of document classification systems reached human performance [6]. A clear advantage of the machine learning approach is that the intervention of knowledge experts is avoided, thus saving time and financial ressources.

In this work we present a comparison study between several machine learning models applied to a particular document classification problem. For the first time we generate symbolic rules from the DIMLP neural hybrid system [2] in a classification task related to text categorization. We focus on a filtering application dataset of small size, as we would like to mimic a situation for which one would quickly annotate a small set of text documents using labels such as *relevant* or *non-relevant.*

In the majority of the experiments, DIMLP networks were at least as accurate as SVMs with the advantage of offering a starting point for explaining neural network classifications to text mining experts. In the remaining sections, section two describes the machine learning models used for the comparison, section three gives a description of DIMLP networks and its rule extraction algorithm, section four describes how to index text documents, section five illustrates the experiments and is followed by a conclusion.

J. Mira and J.R. Álvarez (Eds.): IWINAC 2007, Part I, LNCS 4527, pp. 261–271, 2007.

2 Naive Bayes, IB1, C4.5 and SVMs

In this work, we compare DIMLP bagged networks to Support Vector Machines (SVMs), Naive Bayes, IB1 and C4.5 Decision Trees. In the following paragraphs we shortly describe the models.

2.1 Naive Bayes Classifier

Without loss of generality, let us consider a classification problem with class C and class $\bar{C}$, which is the complement of C. Given a document D of class C, the Bayes Theorem states that

$$p(C|D)p(D) = p(D|C)p(C); \tag{1}$$

with $p()$ denoting the probability, which is equivalent to

$$p(C|D) = \frac{p(C)}{p(D)}p(D|C). \tag{2}$$

Similarly, with class $\bar{C}$ we obtain

$$p(\bar{C}|D) = \frac{p(\bar{C})}{p(D)}p(D|\bar{C}). \tag{3}$$

A document is modelled according to a set of words; given class C, the probability of occurrence of word i is defined as $p(w_i|C)$. Further, by the strong hypothesis of independence of words, we obtain that the probability of a given document D, given a class C is

$$p(D|C) = \prod_i p(w_i|C) \tag{4}$$

and

$$p(D|\bar{C}) = \prod_i p(w_i|\bar{C}). \tag{5}$$

Dividing (2) by (3) gives

$$\frac{p(C|D)}{p(\bar{C}|D)} = \frac{p(C)}{p(\bar{C})} \prod_i \frac{p(w_i|C)}{p(w_i|\bar{C})}. \tag{6}$$

If the numerator of the right hand side of (6) is greater than the denominator, the class of document D is C, else the class is the complement of C. Note that probabilities of words given classes C and $\bar{C}$ are estimated with the use of the training set.

2.2 IB1

Instance based learning (IB1) [1] corresponds to the well known nearest neighbour classifier. A new sample is classified according to the class of the nearest training case belonging to the training set. Many variants of this model have been defined; for instance it is possible to take into account k neighbours and to assign the class most represented in the neighbourhood. Note that larger k values help to reduce the effects of noisy points within the training data set.

2.3 Decision Trees

A tree is a recursive structure. Each intermediary node of the tree is divided into several partitions. For binary trees, two partitions are determined at each node. The splitting criterion is based on the "discriminatory power" of one or several combined variables. Many segmentation criteria have been proposed; the most famous are Shannon entropy, Gini coefficients with many variants and the Khi squared criterion. An important question during tree induction is the size of the obtained structure, in order to obtain a good classifier. As a consequence many tree pruning techniques have been established. C4.5 is one of the most popular binary decision trees [5].

2.4 Support Vector Machines

Amongst the best models for text categorization SVMs have been often referred by many authors [8]. This model tries to determine a discriminant frontier between two classes achieving maximum separation also denoted as the *margin*. In practice for a linear classifier the separation border corresponds to a hyperplane.

Specifically, let us consider a linearly separable problem with data points of the form $\{(x_1, l_1), ..., (x_N, l_N)\}$, with x_i representing an n dimensional data vector and l_i being a learning value belonging to $\{-1, 1\}$. We would like the SVM linear classifier to determine the optimal hyperplane of the form

$$wx - b = 0, \tag{7}$$

with vector w being perpendicular to the separating hyperplane. It can be shown that the margin has length $2/|w|$. It is possible to maximize the margin by minimizing $|w|$.

Learning is achieved by solving a quadratic programming optimization problem. With the use of non-linear kernels (analogous to non-linear transfer functions in neural networks) it is possible to create non-linear classifiers with maximum margin discriminatory frontiers.

3 DIMLPs with Bagging

The Discretized Interpretable Multi-Layer Perceptron (DIMLP) is a special neural network model for which symbolic rules are generated to explain the knowledge embedded within the connections and the activations of neurons. With respect to the training set, the degree of matching between network responses and extracted rules is 100%. Moreover, the computational complexity of the rule extraction algorithm scales in polynomial time with the dimensionality of the problem, the number of training examples, and the size of the network. In this work, we give a short description of the DIMLP model; the interested reader will find more details in [2].

3.1 A Single DIMLP Network

The DIMLP architecture differs from the standard multi-layer perceptron architecture in two main ways :

1. Each neuron in the first hidden layer is connected to only one input neuron.
2. The activation function used by the neurons of the first hidden layer is the staircase function instead of the sigmoid function.

Staircase activation functions create linear discriminant frontiers between several classes. Moreover, with the special pattern of connectivity between the input layer and the first hidden layer, linear frontiers are parallel to the axis defined by input variables. The training phase is carried out by varying the weights in order to minimize the usual Sum Squared Error Function by a modified back-propagation algorithm [2].

3.2 Rule Extraction

The key idea behind rule extraction is the precise localization of discriminant frontiers. In a standard multi-layer perceptron discriminant frontiers are not linear [2]; further, their precise localization is not straightforward. The use of staircase activation functions turns discriminant frontiers into well determined hyper-planes.

Generally, the rule extraction task corresponds to the resolution of a covering problem where discriminant hyper-planes represent frontiers between regions of different classes. Discriminant hyper-planes are precisely determined by weight values connecting an input neuron to a hidden neuron. For each neuron of the first hidden layer the number of stairs related to the staircase function, corresponds to the number of possible discriminant hyper- planes. The rule extraction algorithm checks whether a hyper-plane frontier is effective or not in a given region of the input space.

The *relevance* of a discriminant hyper-plane corresponds to the number of points viewing this hyper-plane as the transition to a different class. In the first step of the covering algorithm the relevance of discriminant hyper-planes is estimated from all available examples and DIMLP responses. In practice, for all examples each input variable x_i is varied by the quantity δ_i, such as δ_i depends on the localization of next virtual hyper-plane.

Once the relevance of discriminant hyper-planes has been established a special decision tree is built according to the *highest relevant hyper-plane* criterion.

In other terms, during tree induction in a given region of the input space the hyper-plane having the largest number of points viewing this hyper-plane as the transition to a different class is added to the tree.

Each path between the root and a leaf of the obtained decision tree corresponds to a rule. At this stage rules are disjointed and generally their number is large, as well as their number of antecedents. Therefore, a pruning strategy is applied to all rules according to the most enlarging pruned antecedent criterion. The use of this heuristic involves that at each step the pruning algorithm prunes the rule antecedent which most increases the number of covered examples without altering DIMLP classifications. Note that at the end of this stage, rules are no longer disjointed and unnecessary rules are removed.

When it is no longer possible to prune any antecedent or any rule, again, to increase the number of covered examples by each rule all thresholds of remaining antecedents are modified according to the *most enlarging* criterion. More precisely, for each attribute new threshold values are determined according to the list of discriminant hyper-planes. At each step, the new threshold antecedent which most increases the number of covered examples without altering DIMLP classifications is retained.

The general algorithm is summarized below.

1. Determine relevance of discriminant hyper-planes using available examples and DIMLP classifications.
2. Build a decision tree according to the highest relevant hyper-plane criterion.
3. Prune rule antecedents according to the most enlarging pruned antecedent criterion.
4. Prune unnecessary rules.
5. Modify antecedent thresholds according to the most enlarging criterion.

Generally, the search for the minimal covering is an NP-hard problem. However, as the presented rule extraction algorithm uses several heuristics the overall computational complexity to find a sub-optimal solution is polynomial with respect to the number of inputs, stairs and examples. Therefore, even for reasonably large datasets and large DIMLP networks the rule extraction problem is tractable. Note that extracted rules exactly represent neural network responses.

3.3 Bagging

Bagging is based on resampling techniques. Assuming that q is the size of the original training set, bagging generates for each classifier q examples drawn with replacement from the original training set. As a consequence, for each network many of the generated examples may be repeated while others may be left out.

The rule extraction technique presented in 3.2 can be applied to any DIMLP architecture having as many hidden layers as desired. In fact, the first hidden layer determines the exact location of hyper-planes, whereas all other layers just switch on or off discriminant frontiers in a given region of the input space. Since the overall combination of DIMLP networks is again a DIMLP network we can use the same rule extraction technique. Here, all virtual hyper-planes defined by all DIMLP networks are taken into account. The previous rule extraction technique (cf. 3.2) is only modified in step 1. This step becomes:

1. Determine virtual hyper-planes from all DIMLP networks.

4 Document Indexing

A typical representation of a text document is given by a histogram of words. This is also the well known *bag of words* representation. As an alternative, it is possible to constitute vectors of word frequencies which are more robust with text documents of different lengths. Several authors found that more sophisticated representations than this did not yield significantly better results [3]. For instance, sentences rather than individual words have been used [7], but results were not encouraging.

A more sophisticated indexing function is represented by the *tfidf* function.

$$tfidf(w_k, d_j) = Card(w_k, d_j) \cdot \log \frac{N}{Card(D(w_k))}; \tag{8}$$

with $Card(w_k, d_j)$ representing the number of times w_k appears in document d_j, N being the number of documents in the training set and $Card(D(w_k))$ the number of documents in the training set in which w_k occurs. Two features are embodied into this function. On one hand, the more often a word appears in a document, the more often it is representative of its content, while on the other hand the more documents a word appears in, the less discriminating it is. Several variants of the $tfidf$ function have been established; essentially the differences between these functions reside in the logarithms and normalization factors.

Typical vectors describing texts have thousands of components. Several strategies have been proposed, in order to reduce vector dimensionality. For instance, neutral words such as articles, prepositions conjunctions and stop words are almost always removed. It is also possible to group words that share the morphological root; this is called *stemming.* While it is not completely clear whether stemming is effective or not, the propensity to adopt it, is frequent.

Another simple approach consists in retaining the words that occur in the greatest number of text documents. As a result, it is possible to reach a dimensionality reduction factor of ten with no loss in accuracy [9]. A typical rule that has been used by many authors is to remove all words that occur at most k times with k ranging from one to five [6].

Word selection is able to be performed according to information-theoretic term selection functions. One of the most popular is based on the χ^2 statistic. The key idea is to measure how dependent a word is to a class; more particularly, thus words with no dependence on any classes are removed. Other well known selection functions, such as *Information gain* and *Mutual information* are reported by Sebastiani [6].

5 Experiments

We retrieved from the Reuters internet site (www.reuters.com) a set of 330 text documents belonging to several classes : economy, sport, politics, health, technology, entertainment, science, and weather. From this sample set we created two other datasets. The first denoted as *Dataset_1* contained 110 samples, the second which is called *Dataset_2* had 220 samples and finally the whole sample set is named *Dataset_3*. Note that all text documents were of comparable length.

The goal was to filter the class of the text documents related to economy. Note also that in these three datasets, half of the economic text documents belonged to the economy class. Performance was estimated by average values of accuracy, recall and precision on ten repetitions of ten fold cross-validation. Learners have been set with default learning parameters. SVMs, Naive Bayes, IB1, and C4.5 were run with the Weka package[1], while DIMLPs were retrieved from the "Neural Hybrid Systems" internet site[2]. Note that the default architecture for the DIMLP neural network had a number of neurons in the first hidden layer equal to the number of inputs (without the second hidden layer). DIMLPBT denotes an ensemble of 25 DIMLP networks trained with bagging. Finally, during cross-validation trials, the number of components in the input vectors depended on the words present in the training set.

5.1 Rough Input Vectors

Text documents were indexed according to word occurrence; no other processing was performed, not even for the suppression of stop words. This experimented setup gave the highest dimensional input vectors. Tables 1, 2 and 3 summarize cross-validation results on the three datasets. Naive Bayes classifiers, SVMs and

Table 1. Average predictive accuracy using rough input vectors

	Dataset_1	Dataset_2	Dataset_3
Avg. dim.	6185.5 ± 1.9	9405.3 ± 3.3	11360.7 ± 2.6
Naive Bayes	**92.0 ± 1.3**	**89.2 ± 0.9**	90.0 ± 0.3
IB1	56.7 ± 1.0	67.5 ± 1.2	72.0 ± 1.3
C4.5	76.3 ± 1.6	78.1 ± 2.2	85.9 ± 1.3
SVM	90.3 ± 0.8	87.3 ± 1.2	90.0 ± 0.7
DIMLPBT	88.4 ± 1.9	88.6 ± 0.9	**91.5 ± 0.5**

DIMLPBTs were significantly more accurate than IB1 and C4.5 decision trees. Contrary to the curse of dimensionality phenomenon, these two models clearly improved their predictive accuracy from a training set containing 100 samples to a training set including 300 training cases. Note also that this improvement was much more moderate for DIMLPBT, while SVMs were stable.

[1] http://www.cs.waikato.ac.nz/ml/weka/

[2] http://www.wv.inf.tu-dresden.de/~borstel/sycosy/doku.php?id=software

Table 2. Average predictive recall using rough input vectors

	Dataset_1	Dataset_2	Dataset_3
Naive Bayes	91.4 ± 2.1	88.5 ± 0.9	90.0 ± 0.9
IB1	**100 ± 0.0**	47.8 ± 2.3	54.8 ± 2.5
C4.5	73.6 ± 3.6	75.6 ± 3.3	86.0 ± 2.1
SVM	97.6 ± 1.0	**90.4 ± 1.1**	**92.2 ± 0.8**
DIMLPBT	92.2 ± 2.1	88.8 ± 1.1	90.3 ± 0.9

Table 3. Average predictive precision using rough input vectors

	Dataset_1	Dataset_2	Dataset_3
Naive Bayes	**93.1 ± 2.1**	**90.4 ± 1.4**	90.3 ± 1.0
IB1	53.9 ± 0.7	88.3 ± 0.2	90.7 ± 0.7
C4.5	80.8 ± 1.2	80.5 ± 2.9	86.4 ± 1.7
SVM	86.6 ± 0.8	86.0 ± 1.0	88.7 ± 0.9
DIMLPBT	87.0 ± 2.0	89.2 ± 1.3	**92.8 ± 0.3**

Amongst the three most accurate models, the better predictive average recall was obtained by SVMs, whereas better average predictive precision was obtained by Naive Bayes and DIMLPBT.

5.2 Removing Non-frequent and Rare Words

The second series of experiments concerned dimensionality reduction by removing articles, adjectives, conjunctions, prepositions, stop words and words that appeared at most once in every text document belonging to the training set. In the results shown in table 4 the dimensionality reduction factor was at most 4.5, with DIMLPBT reaching the highest accuracy. Note also that with respect to table 1, DIMLPBT accuracy improved for the first and the second dataset.

Table 4. Average predictive accuracy with the suppression of rare words

	Dataset_1	Dataset_2	Dataset_3
Avg. dim.	1385.1 ± 0.7	2264.4 ± 0.9	2929.6 ± 0.9
Naive Bayes	90.9 ± 1.5	88.2 ± 0.8	88.9 ± 0.5
IB1	58.0 ± 1.4	57.7 ± 0.8	70.4 ± 1.3
C4.5	80.0 ± 2.0	79.1 ± 1.2	83.8 ± 1.7
SVM	90.7 ± 1.8	87.6 ± 0.9	90.0 ± 0.6
DIMLPBT	**91.3 ± 0.8**	**89.9 ± 0.7**	**91.6 ± 0.6**

5.3 Dimensionality Reduction by Abstraction of Words

We tried to further reduce the dimensionality with the use of abstraction. The idea was to replace for instance a cat by an animal, as an animal is a much more

Table 5. Average predictive accuracy with the abstraction of words

	Dataset_1	Dataset_2	Dataset_3
Avg. dim.	401.3 ± 0.3	670.3 ± 0.3	909.6 ± 0.3
Naive Bayes	79.6 ± 1.5	80.3 ± 1.8	80.0 ± 0.6
IB1	61.3 ± 1.4	68.1 ± 1.0	68.3 ± 0.5
C4.5	66.8 ± 3.7	73.3 ± 3.8	74.8 ± 1.8
SVM	$\mathbf{86.4 \pm 1.2}$	82.4 ± 1.7	84.7 ± 0.7
DIMLPBT	84.3 ± 1.9	$\mathbf{86.0 \pm 0.6}$	$\mathbf{86.5 \pm 0.8}$

abstract notion than that of a cat. WordNet [4] allowed us to make this transformation. Obviously, this process reduces the input vector dimensionality, as several words tend to be transformed into the same word. The results illustrated in table 5 show that predictive accuracy was lower.

Thus, the transformation of words into a more abstract representation is not beneficial in this context.

5.4 Rule Extraction from DIMLPs

Table 6 illustrates the results of rule extraction from DIMLPBT networks with input vectors containing non-relevant and non-frequent words (cf. table 4). Fidelity corresponds to the degree of matching between rules and networks on the testing set. The accuracy is given for those testing cases for which rules and networks agree. The number of antecedents per ruleset is related to the complexity of the classifier generated for each cross validation trial. As expected, complexity increased with the number of training cases.

Table 6. Description of the rules extracted from DIMLPBT with input vectors containing non-relevant words

	Dataset_1	Dataset_2	Dataset_3
Fidelity	81.2	88.7	92.4
Accuracy	92.1	90.8	92.4
Nb of rules	24.1	30.2	34.7
Nb. of ant.	68.6	101.8	126.9

Table 7 illustrates the results of rule extraction from DIMLPBT networks with input vectors containing abstracted words. Rule predictive accuracy was lower compared to the previous table.

Below we give a few examples of typical rules for class "Economy"; note that rule antecedents are given in terms of word occurrence.

1. $(forecast \geq 1) \Rightarrow$ Class = *Economy*
2. $(cents \geq 1) \Rightarrow$ Class = *Economy*

Table 7. Description of the rules extracted from DIMLPBT with input vectors containing abstracted words

	Dataset_1	Dataset_2	Dataset_3
Fidelity	83.9	89.7	92.1
Accuracy	86.6	87.5	87.8
Nb of rules	18.9	29.6	37.0
Nb. of ant.	55.5	107.8	145.1

3. $(dow \geq 1) \Rightarrow$ Class = *Economy*
4. $(percent \geq 1)(prices \geq 1) \Rightarrow$ Class = *Economy*
5. $(company \geq 1)\ (fell \geq 1) \Rightarrow$ Class = *Economy*

These rule examples were 100% accurate during a particular cross-validation trial. Word "dow" in the second rule represents the "Dow Jones", which is a very important economic index. Moreover "cents" is related to dollars in the second rule, while "company" and "fell" are also very important words in the context of economy.

6 Conclusion

We presented several machine learning models trained to filter three datasets of small size. Even with very high dimensional vectors DIMLPs, SVM and Naive Bayes classifiers were able to reach satisfactory accuracy. In order to reduce input dimensionality, we removed words which occur at most once. As expected, average predictive accuracy was roughly very close for the best models with a dimensionality reduction factor equal to 4.5 at most. We tried further, word abstraction; however predictive accuracy started to degrade too much. In the future, in order to further increase DIMLPBT predictive accuracy, we believe that in addition to training cases, common sense knowledge could be easily inserted refined and extracted with the use of symbolic rules.

Acknowledgements

We would like to thank Régis Boesch and Nabil Abdennadher for having provided us with strong computing power for the calculation of cross-validation statistics. Finally, we would like to thank the RCSO-TIC of the University of Applied Sciences of Western Switzerland for financing this research.

References

1. Aha, D.W., Kibler, D., Albert, M.K. (1991). Instance-based Learning Algorithms. *Machine Learning, 6 (1)*, 37–66, Springer Netherlands.
2. Bologna, G. (2004). Is it Worth Generating Rules from Neural Network Ensembles ? *Journal of Applied Logic, 2 (3)*, 325–348.

3. Dumais, S.T., Platt, J., Heckerman, D., Sahami, M. (1998). Inductive Learning Algorithms and Representations for Text Categorization. *Proc. of CIKM-98, 7th ACM Int. Conf. on Information and Knowledge Management,* 148–155.
4. Miller, G.A. (1995). WordNet: a Lexical Database for English. *Commun. ACM, 38 (11),* ACM Press, New York, NY, USA.
5. Quinlan, J.R. (1993). *Programs for Machine Learning.* San Mateo, Calif., Morgan Kaufmann Publishers.
6. Sebastiani, F. (2002). Machine Learning in Automated Text Categorization. *ACM Comput. Surv., 34 (1),* 1–47, ACM press, New-York, USA.
7. Tzeras, K., Hartmann, S. (1993). Automatic Indexing Based on Bayesian Inference Networks. *Proc. of SIGIR-93, 16th ACM Int. Conf. on Research and Development in Information Retrieval,* 22–34.
8. Vapnik, V. (1995). *The Nature of Statistical Learning Theory.* Springer, N.Y.
9. Yang, Y., Pedersen, J.O. (1997). A Comparative Study on Feature Selection in Text Categorization. *Proc. of ICML-97, 14th Int. Conf. on Machine Learning,* 412–420.

A Single Layer Perceptron Approach to Selective Multi-task Learning

Jaisiel Madrid-Sánchez, Miguel Lázaro-Gredilla, and Aníbal R. Figueiras-Vidal*

Department of Signal Theory and Communications,
Universidad Carlos III de Madrid
Avda. Universidad 30, 28911, Leganés (Madrid), Spain
{jaisiel,miguel,arfv}@tsc.uc3m.es

Abstract. A formal definition of task relatedness to theoretically justify multi-task learning (MTL) improvements has remained quite elusive. The implementation of MTL using multi-layer perceptron (MLP) neural networks evoked the notion of related tasks sharing an underlying representation. This assumption of relatedness can sometimes hurt the training process if tasks are not truly related in that way. In this paper we present a novel single-layer perceptron (SLP) approach to selectively achieve knowledge transfer in a multi-tasking scenario by using a different notion of task relatedness. The experimental results show that the proposed scheme largely outperforms single-task learning (STL) using single layer perceptrons, working in a robust way even when not closely related tasks are present.

1 Introduction

One of the first and most extended MTL implementations is based on the use of multi-layer neural networks [2,4,14,17]. This approach allow us to exploit task relatedness by sharing the internal representation at the hidden level of the network, so it is implicitly assumed that tasks are going to present a correlated internal structure. However, to improve the generalization accuracy, it will be completely necessary to provide a set of learning tasks that are known to be appropriated. To relax this hard requirement, it would be desirable to propose algorithms that can discover the relation between multiple learning tasks, so that a positive knowledge transfer occurs during the training process. These frameworks are typically named by *selective transfer* approaches.

We remark the works developed by Daniel Silver and Sebastian Thrun in the context of selective multi-task learning neural networks [17,18]. In both frameworks, it is assumed that tasks will constructively interact during the training of a standard MTL scheme by using an a priori measure of task relatedness,

* This work has been partly supported by CM project S-0505/TIC/0223 and MEC grant TEC 2005-00992. The work of J. Madrid-Sánchez was also supported by the Chamber of Madrid Community and European Social Fund by a scholarship.

J. Mira and J.R. Álvarez (Eds.): IWINAC 2007, Part I, LNCS 4527, pp. 272–281, 2007.

i.e., the main role in the selective transfer process is played by a previously defined metric of tasks relatedness. In other words, it is assumed that the measure of task relatedness is good enough to force resulting tasks to "see" the whole information (weighted or not) of each other during the training process.

Instead, in this paper we introduce a new framework for doing selective transfer by using a different notion of tasks relatedness from the typical one in MTL neural networks. This alternative definition is inspired in Hierarchical Bayes for multi-task learning [11], where it is considered that closely related tasks have the same whole representation. In particular, following a viewpoint closer to the human brain behavior [13,19], the learner is provided with a common representation that exploits similarities between tasks (related information), and a private part (not accessible to the rest of tasks) that captures the specific structure of each task (unrelated information).

Since tasks following this new viewpoint of task relatedness are not supposed to share any internal representation, it is possible to extend multi-task learning to single-layer neural networks. We use coupling parameters weighting importance of the common and the private representation as a function of the grade of task relatedness. This allows us to automatically achieve selective transfer, not requiring any a priori measure of similarity. Moreover, the use of linear learners may be of great utility when tasks within the domain are linear/quasi-linear, or they present a high dimensional input space [16].

The rest of the paper is organized as follows. In Section 2, we introduce the notation used throughout the paper and, in order to assert the basis of our method, we review two main approaches that appeared in the machine learning literature to model task relatedness. In Section 3 we set a framework to achieve selective transfer using a single-layer perceptron. Section 4 presents some experiments over two domains showing that the proposed framework leads to a significant improvement in generalization performance when compared to single-task learning. Finally, in Section 5 we present our conclusions and some lines for further work.

2 Notation and Setup

We assume the following setup. We consider that the domain has T learning tasks and that all data from the tasks come from the same space $X \times Y$. For simplicity we assume that $X \subset \mathbb{R}^d$ and $Y \subset \mathbb{R}$. For each task the learner is provided with M input-output independent trials according to some (unknown) probability distribution P_t on $X \times Y$, so the total data available by a supervised multi-task learning is

$$\{\{(\mathbf{x}_{11}, y_{11}), ..., (\mathbf{x}_{M1}, y_{M1})\}, ..., \{(\mathbf{x}_{1T}, y_{1T}), ..., (\mathbf{x}_{MT}, y_{MT})\}\}.$$

In addition, the learner is provided with an *action* space A, an *error function* $\mathcal{E} : Y \times A \to [0, L]$ for some positive real number L, and a *decision* or *hypothesis space* $\mathcal{H}$ containing functions $h : X \to A$.

We assume that the learner is trained within an environment where P_t is different for each task but P_t's are related in any grade. The goal is to learn T

functions $h_1, h_2, ..., h_T$ such that $h_t(\mathbf{x}_{it}) \approx y_{it}$, or equivalently the error function $\mathcal{E}(h_t(\mathbf{x}_{it}), y_{it}) \approx 0$.

All applications considered in Section 4 are included in a simple version of this setup, consisting of using the same input data $\mathbf{x}_{it}$ for all the tasks. That is, for every $i \in \{1, ..., M\}$ vector $\mathbf{x}_{it}$ is the same for all $t \in \{1, ..., T\}$, but the output targets y_{it} differ for each t.

2.1 Models for Task Relatedness

All frameworks for multi-task learning are based on some definition of the notion of task relatedness formalized through the design of a multi-task learning approach. In this sense, we will pay attention to two models of task relatedness widely used in the Machine Learning context:

1. As MTL neural networks considers so far, tasks may be related assuming that they are going to share a common underlying representation. This is the case of [1] and, specially, [2,4,17] where backpropagation multi-task learning schemes are employed.

   To formalize notation for MTL using multi-layer neural networks, as done in [2] we consider that the hypothesis space $\mathcal{H} : X \to A$ is split into two sections: $\mathcal{H} = \mathcal{G} \circ \mathcal{F}$ where $\mathcal{F} : X \to V$ and $\mathcal{G} : V \to A$, where V is an arbitrary set[1]. We simplify the notation writing

   $$X \xrightarrow{\mathcal{F}} V \xrightarrow{\mathcal{G}} A$$

   $\mathcal{F}$ is called the *representation space* and an individual member f of $\mathcal{F}$ is called an *internal representation* or just a *representation*. On the other hand, $\mathcal{G}$ is called the *output function space*.

   Based on the information about the domain knowledge, contained in the samples of all tasks, the learner searches for a good representation $f \in \mathcal{F}$, and then searches $\mathcal{G} \circ f$ for an adequate individual output function $g_t \in \mathcal{G}$ or hypothesis $h_t = g_t \circ f \in \mathcal{H}$ with small error, for each t, $1 \leq t \leq T$.
2. Besides, in some implementations of MTL task relatedness is modeled through assuming that all functions learned are close to some model [7,12,20]. That is, this viewpoint considers that closely related tasks are going to share a same whole representation. This is the case for functions capturing preferences in user's modeling problems [6].

   In this case, we assume that there is not a representation space, $\mathcal{F}$, shared by all the tasks, so the hypothesis space is not divided into two sections, that is, using the same nomenclature as before, the hypothesis space is equal to the output function space, $\mathcal{H} = \mathcal{G} : X \to A$. Consequently, the learner just searches $\mathcal{G}$ (or $\mathcal{H}$) for a good external representation, given by g_t.

   Under this notion, selective transfer becomes an inherent feature since representation is split in a common model, extracted from the domain knowledge, and an individual (private) model for each task, both of them with significance depending on the grade of task relatedness.

[1] That is, $\mathcal{H} = \{g \circ f : g \subset \mathcal{G}, f \in \mathcal{F}\}$.

3 Selective Transfer by Coupling Common/Private Representation

Following the second notion of task relatedness, Hierarchical Bayesian methods [11,12] assume that all task parameters, $\mathbf{w}_t$, come from a particular Gaussian probability distribution. This implies that all $\mathbf{w}_t$ are "close" to some mean parameter, $\mathbf{w}_c$ (the mean of the Gaussian distribution).

In this section we follow the intuition of Hierarchical Bayes, assuming the model proposed in [6] by which all $\mathbf{w}_t$ are written, for every $t \in \{1, ..., T\}$, as

$$\mathbf{w}_t = \mathbf{w}_c + \mathbf{w}_{p,t} \ . \tag{1}$$

When private vectors $\mathbf{w}_{p,t}$ are "small", tasks will be similar to each other. So, as the second notion of task relatedness defines, tasks are related in a way that the true models are all close to some canonical model $\mathbf{w}_c$, which plays the role of the mean of the Gaussian assumed by Hierarchical Bayes.

3.1 Single Layer Perceptron for MTL

The method proposed in this subsection is built upon an approach based on convex optimization for modeling consumer heterogeneity in conjoint estimation [8]. Specifically, this approach starts from an individual-level Ridge-Regression (RR) [10], which minimizes a convex cost function with respect to task parameters $\mathbf{w}_t$. In particular, this cost function is parametrized by a positive weight λ that is typically set using cross validation. For every task the following cost function is independently minimized:

$$\min_{\mathbf{w}_t} \left\{ \frac{1}{\lambda} \sum_{i=1}^{M} (h_t(\mathbf{x}_{it}) - y_{it})^2 + \|\mathbf{w}_t\|^2 \right\} \tag{2}$$

where $h_t(\mathbf{x}_{it}) = \mathbf{x}_{it}\mathbf{w}_t$ results in a convex problem.

As we see, the cost function is composed of a first part that measures the fit, and a regularization term, $\mathbf{w}_t^T \mathbf{w}_t = \|\mathbf{w}_t\|^2$. A positive parameter λ defines the trade-off between fit and complexity.

To pool information across tasks, [8] presents a coupled cost function. Inspired by (1), they propose modeling heterogeneity and pooling information across tasks by *shrinking* the final models, $\mathbf{w}_t$, towards a mean model, $\mathbf{w}_c$. The following convex optimization problem is considered:

$$\min_{\mathbf{w}_t, \mathbf{w}_c} \left\{ \frac{1}{\lambda} \sum_{t=1}^{T} \sum_{i=1}^{M} (h_t(\mathbf{x}_{it}) - y_{it})^2 + \sum_{t=1}^{T} \|\mathbf{w}_t - \mathbf{w}_c\|^2 \right\} \tag{3}$$

using a linear function $h_t(\mathbf{x}_{it}) = \mathbf{x}_{it}\mathbf{w}_t$.

Again, this cost function consists of two parts. The first part refers to fit and the second one makes a regularization but, unlike the individual-level RR cost function (2) where final models are shrinked towards zero, by shrinking final models towards vector $\mathbf{w}_c$. Higher values of λ result in more homogeneous estimates.

Next, we proceed to build a MTL implementation using a single layer perceptron neural network upon the framework above explained. Using the later notation, in this case the goal of the learner is to selectively search only in the space $\mathcal{G}$ (previously called output function space) for an hypothesis (or output function), that is:

$$h_t(\cdot) = g_t(\cdot) = \rho_c g_c(\cdot) + \rho_p g_{p,t}(\cdot) \tag{4}$$

where ρ_c and ρ_p are coupling parameters weighting significance of the common and private representation in the total model ($g_c(\cdot)$ and $g_{p,t}(\cdot)$, respectively).

As equation (4) shows, in this work we assume that common/private coupling factors are the same for all tasks, i.e. $\rho_{c,t} = \rho_c$ and $\rho_{p,t} = \rho_p$, so we just take into account the global positive knowledge transfer caused by the interaction of the set of tasks, instead of each particular knowledge transfer. In this case we easily select parameters ρ_c and ρ_p by cross-validation. Nevertheless, our algorithm can be directly extended to the case of multiple coupling factors, $\rho_{c,t}$ and $\rho_{p,t}$ for every $t \in \{1, ..., T\}$, by using a more robust process to select these parameters.

In particular, we establish a non-convex problem by replacing the lineal function with the widely used in SLP hyperbolic tangent activation function [5] in (3). Following (4), we introduce coupling parameters so the next regularized SLP cost function depending on final models, $\mathbf{w}_t$, and common model, $\mathbf{w}_c$, is minimized:

$$\min_{\mathbf{w}_t, \mathbf{w}_c} \Bigg\{ C\Big(\mathcal{E}\Big(h_t(\mathbf{x}_{it}), y_{it}\Big), \mathbf{w}_t, \mathbf{w}_c\Big) :=$$

$$:= \sum_{t=1}^{T} \sum_{i=1}^{M} (h_t(\mathbf{x}_{it}) - y_{it})^2 + \lambda \left[\rho_c \|\mathbf{w}_c\|^2 + \rho_p \sum_{t=1}^{T} \|\mathbf{w}_t - \mathbf{w}_c\|^2 \right] \Bigg\} \tag{5}$$

for all $i \in \{1, 2, ..., M\}$ and $t \in \{1, 2, ..., T\}$, and using $h_t(\mathbf{x}_{it}) = \tanh(\mathbf{x}_{it}\mathbf{w}_t)$.

Again, λ is a regularization parameter. For simplicity in the selection of the coupling parameters, after preliminary results, we use in this work coupling parameters depending on a common coupling factor γ, specifically $\rho_c = \frac{1}{\gamma}$ and $\rho_p = \frac{1}{1-\gamma}$. Clearly, parameter γ will take care of the grade of task relatedness (under the second viewpoint) by a regularization constraint in the way that when $\gamma \approx 1$, the problem (5) will tend to make the models to be the same (the private model $\mathbf{w}_t - \mathbf{w}_c = \mathbf{w}_{p,t}$ is close to zero, thus $\mathbf{w}_t = \mathbf{w}_c$), solving one only single-task learning (finding $h_t = g_t = g_c$ having $g_{p,t} = 0$ for every $t \in \{1, 2, ..., T\}$); and for a value close to zero, $\gamma \approx 0$, (5) will tend to make all tasks unrelated ($\mathbf{w}_c$ equal to zero), solving the T tasks independently by T independent single-task learnings (finding $h_t = g_t = g_{p,t}$ having $g_c = 0$).

The non-convex problem (5) can be minimized by a gradient-based learning. Using gradient descent minimization, we have the following stochastic updates for $\mathbf{w}_t$ and $\mathbf{w}_c$ (leaving out the dependences on i for clarity):

$$\mathbf{w}_t = \mathbf{w}_t - \eta \left[(h_t - y_t)(1 - h_t^2)\mathbf{x}_{in} + \lambda \frac{1}{1-\gamma}(\mathbf{w}_t - \mathbf{w}_c) \right] \tag{6}$$

$$\mathbf{w}_c = \mathbf{w}_c - \eta\lambda \left[\frac{1}{\gamma}\mathbf{w}_c + \frac{1}{1-\gamma}\sum_{t=1}^{T}(\mathbf{w}_t - \mathbf{w}_c) \right] \tag{7}$$

η being the learning rate.

4 Experiments

We run two experiments, consisting of a synthetic domain and a character recognition problem, respectively. Both scenarios are provided with binary tasks that intuitively seem to keep any similarity (but we do not know the grade of relatedness and the sort of relatedness if there is any). Since there is not any previous idea about significance of tasks within each domain we consider that all tasks are equally significant, having into account the global performance of the domain instead of focusing on the individual performance of a selected main task. Nevertheless, the framework with priority tasks would remain the same, being feasible the introduction of weighting error parameters in order to control interaction of secondary tasks over the primary one.

We compare the performance of the single-layer multi-task approach here proposed (noted by **SLP-MTL** in tables) with respect to its single-task version, this last by using both the traditional single-layer perceptron cost function (**SLP-STL** in tables) and taking into account a regularized approximation (**SLP-STL (reg)** in tables), that is $C = \sum_{i=1}^{M}(h_t(\mathbf{x}_{it}) - y_{it})^2$ and $C = \sum_{i=1}^{M} (h_t(\mathbf{x}_{it}) - y_{it})^2 + \lambda \left\| \mathbf{w}_t \right\|^2$, respectively, where $h_t(\mathbf{x}_{it}) = \tanh(\mathbf{x}_{it}\mathbf{w}_t)$. Performance of solving all tasks with the same representation (noted by **One SLP**), that is, an individual SLP for all tasks (SLP-MTL with $\gamma = 1$), will be also presented as extra information.

Experiments are carried out using different number of training samples and different tasks, so that we can explore in an clearer way the advantages of the MTL approach.

The coupling parameter, γ, has been selected in all schemes by cross-validation. In particular, due to the reduced number of training samples utilized in all domains (when MTL makes specially sense [4]), a leave-one-out procedure has been followed. Values explored have been the same for all domains. Specifically, we test the following values of the tasks-coupling parameter, γ: 0.01, 0.1, 0.3, 0.5, 0.7, 0.9, 0.99 (0 would be the STL case and 1 the one of all tasks using the same representation). We test several values of the regularizer parameter, λ, as well (namely, 1, 1e-2, 1e-4).

All learners are trained choosing a linearly decreased learning rate, its initial value being 0.4. Moreover, a previous convergence analysis has been carried out in order to select an appropriate number of epochs.

Finally, results are averaged over 50 repetitions in all scenarios, including the statistical significance of the difference of error rates in our multi-tasking approach and those corresponding to STL approximations, measured by using the Kruskal-Wallis test [9].

4.1 Description of Datasets

Balloon. Extracted from [3], this domain was proposed by the psychologist Michael Pazzani in order to evaluate the influence of prior knowledge on concept acquisition [15].

There are four data sets representing different conditions of an experiment inflating a balloon. The number of possible instances are 16. We consider several scenarios using 7, 10 and 13 training samples, and the remaining for test. The train and test sets have been randomly varied for each repetition.

Character Recognition. This database, collected in [3], was generated by using a first order theory which describes the structure of ten capital letters of the English alphabet. The capital letters represented are the following: A, C, D, E, F, G, H, L, P, R. Each instance is structured and is described by a set of segments (lines).

Since each segment may belong to several letters, instead of assigning each set of segments to a letter, the goal of the learner will be to identify the set of letters that are described by a certain segment (line). This way we are promoting output tasks to be more related.

Again, we consider a variable number samples (10, 50, 100) for the training set, and a fixed number of test samples (specifically, 100).

4.2 Discussion

In this subsection we present in tables global test error rates (in %) of the methods above explained over the three scenarios studied. For the Balloon problem we also present a table showing correlation coefficients between all tasks, so that we can compare with the value of coupling parameter γ.

Balloon. From results in Table 1 we draw the following commentaries:

- SLP-MTL dramatically improves global performance of any STL approach. Best error rates are obtained when we use a few training samples, tending to STL performance as the learner is provided with more information about tasks (obviously, if all available data are present during the training process generalization capacity becomes useless).
- Sharing the whole representation leads to very bad performance, which indicates that tasks are not closely related as second notion of tasks relatedness defines.
- Selective transfer seems to be efficiently achieved by SLP-MTL practically in all cases. In fact, the coupling factor is larger when correlation between tasks is too.

Table 1. Test error rates (in %) of single-layer perceptron approaches in Balloon. Averaged performance of SLP-MTL at the 5% significance level (with respect to best STL approach) is in bold. The coupling factor is shown in brackets.

Tasks	Data (train/test)	One SLP	SLP-STL	SLP-STL (reg)	SLP-MTL (γ)
1,2	**7/9**	44.667	11.750	21.778	**3.125** (0.9)
1,2	**10/6**	45.500	2.563	6.333	**0.250** (0.9)
1,2	**13/3**	45.846	0	0	0 (0.9)
1,2,3	**7/9**	46.74	11.833	22.889	**8.958** (0.7)
1,2,3	**10/6**	47.556	3.750	9.222	**1.583** (0.7)
1,2,3	**13/3**	47.478	0.167	1.111	0.042 (0.7)
1,2,3,4	**7/9**	50.200	15.469	28.444	**12.656** (0.9)
1,2,3,4	**10/6**	51.300	6.906	18.750	**5.031** (0.99)
1,2,3,4	**13/3**	53.683	3.313	10.167	2.969 (0.5)

Table 2. Correlation coefficients for tasks in Balloon domain. Significant correlations are in bold (probability of getting a correlation as large as the observed value by random chance less than 0.05).

Task 1	Task 2	Task 3	Task 4	
1	**0.34**	0.04	0.25	**Task 1**
	1	0.03	**0.65**	**Task 2**
		1	**0.66**	**Task 3**
			1	**Task 4**

Character Recognition. The following conclusions emerge from the results in Table 5:

- The proposed method significantly overcomes any STL approach in practically all cases.
- In case of using the same representation for all tasks we get quite good performance, specially when 5 and 10 tasks are considered (computing correlation coefficients, most of the letters have a significant value larger than 0.5).
- Correlation between letters A and C (tasks 1 and 2 respectively) is the lowest (namely, less than 0.1), so the test error of One SLP remains high. Nevertheless, STL approaches are significantly improved by the selective transfer method by using low values of the coupling factor.
- When all capital letters are considered there are interactions between two sets of letters not very correlated, but having closely correlated letters each one (specifically, a set with straight lines (E,F,H,L) against a set with curved lines (C,D,P,R,G)), so the coupling factor keeps a medium value.

Table 3. Test error rates (in %) of single-layer perceptron approaches in Character Recognition. Averaged performance of SLP-MTL at the 5% significance level (with respect to best STL approach) is in bold. The coupling factor is shown in brackets.

Tasks	Data (train/test)	One SLP	SLP-STL	SLP-STL (reg)	SLP-MTL (γ)
1,2	**10/100**	24.267	25.217	22.383	**21.450** (0.1)
1,2	**50/100**	26.167	23.833	23.067	**20.650** (0.3)
1,2	**100/100**	28	24.167	22.450	**20.867** (0.3)
1,...,5	**10/100**	28.200	30.300	28.487	28.086 (0.7)
1,...,5	**50/100**	28.200	30.573	28.707	**27.666** (0.7)
1,...,5	**100/100**	31.107	32.967	28.727	**27.213**(0.7)
1,...,10	**10/100**	30.900	32.473	31.620	**30.636** (0.5)
1,...,10	**50/100**	32.170	34.757	32.750	**29.603** (0.3)
1,...,10	**100/100**	32.240	35.610	31.497	**29.889** (0.5)

5 Conclusions

In this work we have proposed a single-layer neural network approach to selectively train several tasks that are related in any grade, by using a notion of task relatedness that does not require to share a common underlying representation, but a coupled common and private representation.

Experimental results show significant improvements of our method with respect to STL approaches. Moreover, in order to promote a positive knowledge transfer, a coupling factor that takes into account the grade of task relatedness is found in all cases.

As interesting topics for further research, we remark the extension of the proposed selective transfer framework to multi-layer neural networks by using both a common and private internal representation. Doing selective transfer as a function of the notion of task relatedness instead of the grade of task relatedness results in a straightforward implementation as well.

References

1. Argyriou, A., Evgeniou, T., Pontil, M.: Multi-Task Feature Learning. In: NIPS 2006, Vancouver, B.C., Canada (4-7 December 2006) (to appear)
2. Baxter, J.: Learning Internal Representations. In: COLT, ACM Press (1995) 311-320
3. Blake, C.L., Merz. C.J.: UCI Repository of Machine Learning Databases (1998)
4. Caruana, R.: Multitask Learning. Machine Learning **28** (1997) 41-75
5. Duda, R., Hart, P.: Pattern Classification and Scene Analysis. Wiley-Interscience, New York, USA (1973)
6. Evgeniou, T., Pontil, M.: Regularized Multitask Learning. In: SIGKDD, Seattle, Washington, USA (22-25 August 2004)

7. Evgeniou, T., Micchelli, C.A., Pontil, M.: Learning Multiple Tasks with Kernel Methods. Journal of Machine Learning Research **6** (2005) 615-637
8. Evgeniou, T., Pontil, M., Toubia, O.: A Convex Optimization Approach to Modeling Heterogeneity in Conjoint Estimation. Marketing Science (to appear)
9. Gibbons, J., Chakraborti, S.: Nonparametric Statistical Inference. 3 edn. Marcel Dekker, New York, USA (1992)
10. Hastie, T., Tibshirani, R., Friedma, J.H.: The Elements of Statistical Learning. Springer, New York, USA (2003)
11. Heskes, T.: Empirical Bayes for learning to learn. In Langley, P. ed.: ICML-2000. Morgan Kaufmann Publishers, Inc. (2000) 367-374
12. Bakker, B., Heskes, T.: Task clustering and gating for Bayesian multitask learning. Journal of Machine Learning Research **4** (2003) 83-99
13. Jovicich, J.: Brain Areas Specific for Attentional Load in a Motion-Tracking Task. Journal of Cognitive Neuroscience **13** (2001) 1048-1058
14. Liao, X., Carin, L.: Radial Basis Function Network for Multi-task Learning. In Weiss, Y., Schölkopf, B. and Platt, J. eds.: Advances in NIPS. Volume 18., MIT Press (2006) 795-802
15. Pazzani, M.: The influence of prior knowledge on concept acquisition: Experimental and computational results. Journal of Experimental Psychology: Learning, Memory and Cognition **17**(3) (1991) 416-432
16. Rao, C.: Linear statistical inference and its applications. John Wiley, New York, USA (1965)
17. Silver, D.: Selective Transfer of Neural Network Task Knowledge. PhD Thesis, University of Western Ontario (2000)
18. Thrun, S., OSullivan, J.: Clustering learning tasks and the selective crosstask transfer of knowledge. In Thrun, S. and Pratt, L.Y. eds.: Learning to Learn. Kluwer Academic Publishers (1998) 235-257
19. Tomasi, D., Ernst, T., Caparelli, E., Chang, L.: Practice-induced changes of brain function during visual attention: A parametric fMRI study at 4 Tesla. NeuroImage, **23**(4) (2004) 1414-1421
20. Yu, K., Tresp, V., Schwaighofer, A.: Learning Gaussian processes from multiple tasks. In: ICML 2005, ACM Press (2005) 1012-1019

Multi-task Neural Networks for Dealing with Missing Inputs

Pedro J. García-Laencina[1], Jesús Serrano[1],
Aníbal R. Figueiras-Vidal[2], and José-Luis Sancho-Gómez[1]

[1] Dpto. Tecnologías de la Información y las Comunicaciones
Universidad Politécnica de Cartagena.
Plaza del Hospital, 1, 30202, Cartagena (Murcia), Spain
pedroj.garcia@upct.es
[2] Dpto. Teoría de la Señal y Comunicaciones
Universidad Carlos III de Madrid.
Avda. de la Universidad, 30, 28911, Leganés (Madrid), Spain

Abstract. Incomplete data is a common drawback in many pattern classification applications. A classical way to deal with unknown values is missing data estimation. Most machine learning techniques work well with missing values, but they do not focus the missing data estimation to solve the classification task. This paper[1] presents effective neural network approaches based on Multi-Task Learning (MTL) for pattern classification with missing inputs. These MTL networks are compared with representative procedures used for handling incomplete data on two well-known data sets. The experimental results show the superiority of our approaches with respect to alternative techniques.

1 Introduction

As humans, we can receive data through our senses, and after processing the obtained information, we are able to identify the data source. For example, many of us can recognize voices over a poor telephone line, distinguish the grapes used to make a wine, or identify several species of flowers. Pattern classification is the discipline of building machines to classify data (patterns) based on either a priori knowledge or on statistical information extracted from the patterns.

A complete pattern classification system consists of a group of sensors that gather the observations to be classified; a feature extraction mechanism that computes numeric or symbolic information from the observations; and a classification scheme for assigning a class to each observation, relying on the extracted features. In the past forty years, a large number of Artificial Neural Networks (ANNs) models have been proposed for performing pattern classification tasks [1]. ANNs can recognize patterns working simultaneously with continuous, binary, ordinal and nominal data.

[1] This work is partially supported by Ministerio de Educación y Ciencia under grant TEC2006-13338/TCM, and by Consejería de Educación y Cultura de Murcia under grant 03122/PI/05.

J. Mira and J.R. Álvarez (Eds.): IWINAC 2007, Part I, LNCS 4527, pp. 282–291, 2007.

Missing or incomplete data is an usual drawback in many real-world applications. Data may contain unknown features due to different reasons, e.g., data collection procedure can be imperfect, sensor failures producing a distorted or unmeasurable value, data occlusion by noise, non-response in surveys [2]. Handling missing data has become in a fundamental requirement for pattern classification, because an inappropriate missing data treatment may cause large errors or false results on classification.

One of the most recommended ways for dealing with unknown features is missing data imputation. Imputation is a generic term for filling in unknown features with plausible values provided by a missing data estimator [2]. In this approach, missing features are estimated from the available data, and after the imputation is done, a classifier is trained using the edited training set (i.e., complete patterns and incomplete vectors with imputed values). Sharpe *et al.* use FFNN (Feed-Forward Neural Network) models to estimate incomplete data [3]. In this approach, known as Reduced Neural Networks (Reduced NN), a set of FFNNs is created, where each one of them is trained to learn each possible combination of features with unknown values using as inputs the remaining complete features. After imputations are done using these networks, the edited set is used to train an ANN classifier. Unfortunately, this method requires a huge number of neurons with an increasing number of incomplete features. Another widely used approach is replacing missing feature values by values from the features of their K nearest neighbors (KNN), in the available complete training data; and after that, a machine learning algorithm classifies the obtained complete set [4]. Other proposed solutions are based on Gaussian Mixture Models (GMM) trained by Expectation-Maximization (EM) algorithm [5]. It works efficiently in many situations, but the main disadvantages of GMM-EM are that it is necessary to assume an underlaying data distribution, and they are not able to estimate missing values in the operation mode, only during the training stage.

However, neither of them can provide a solution that includes a desirable characteristic: a missing data imputation oriented to solve the classification problem. Up to now, most used techniques in pattern classification with missing data divide the problem in two separated and isolated tasks, classification task and imputation tasks, what are solved by different learners. In this work, effective approaches to classify incomplete input vectors using FFNN schemes based on Multi-Task Learning (MTL) are presented, where the estimation of missing values is oriented by the learning of the classification task. In particular, this paper extends our previous research works on MTL and missing data [6,7]. Our method utilizes the incomplete features as extra tasks, and learns them at the same time with the main classification task. The remaining of this paper is organized as follows. Section 2 introduces the basic MTL notions in FFNN. Next, in Section 3, proposed method is presented, and its training and operation mode are explained. In Section 4, different MTL schemes based on private subnetworks are described. Section 5 shows obtained results on two well-known data sets, and discusses the relative advantages and drawbacks of our approaches over other representatives methods. Finally, conclusions and future works end this paper.

2 Multi-Task Neural Networks

Human learning frequently involves learning several tasks simultaneously; in particular, humans compare and contrast related tasks for solving a problem. For example, reading and writing are difficult tasks for children, but it can be easier when both tasks are learned at the same time, i.e., parallel learning of both tasks contribute to the learning of each one of them.

In recent years, several machine learning methods have been proposed to learn from multiple tasks [7,8,9,10]. This approach to learning is named Multi-Task Learning (MTL). The basic idea is that a task will be learned better and/or faster if can leverage the information contained in the training signals of other related tasks during learning. This paper is based on the MTL method used by Caruana [8], where tasks share a hidden layer of neurons in a FFNN. The task which is desired to be learned better is called the primary or main task, and the tasks what are used as hints by the main task are referred to as the secondary or extra tasks. With respect to weights, weights of the first layer are updated depending on the error of all tasks; while output layer weights are only influenced by task associated error.

3 A MTL Neural Network Approach for Classification with Incomplete Data

Assume a set of N labeled incomplete input vectors,

$$\mathcal{D} = \{(\mathbf{x}^{(n)}, \mathbf{t}^{(n)}, \mathbf{m}^{(n)}) : \mathbf{x}^{(n)} \in \mathbb{R}^d, x_i^{(n)} \text{ is missing} \Longleftrightarrow m_i^{(n)} = 1\}_{n=1}^{N} \quad (1)$$

where $\mathbf{x}^{(n)}$ is the n-th input vector composed of d features; labeled as $\mathbf{t}^n$ in a 1-of-c codification, with c possible classes; and $\mathbf{m}^{(n)}$ indicates what input features are unknown in $\mathbf{x}^{(n)}$. Moreover, the vector $\mathbf{a} = [a_1, a_2, ..., a_m]$ is defined, whose components are the m incomplete attributes in the data set.

In a MTL approach, this problem is composed of two kind of different tasks to be learned: a main classification task; and m secondary imputation tasks associated to each feature with missing values. Working in this manner, in a first MTL approach, we can consider an ANN with a hidden layer of common neurons that learn in parallel all tasks. Figure 1(a) shows this approach. Output layer has c outputs units for classification task, and m imputation outputs for each one of the incomplete features. In the hidden layer, the number of neurons depends on the problem to be solved. The inputs units of this network are the d input features, and also, the c components of the classification target vector. This approach uses linear inputs, modified hyperbolic tangent as activation function $g(\cdot)$ in all neurons, and linear outputs. Figure 1(b) shows the implemented MTL neuron for dealing with missing inputs. Hidden neurons are different to the classical neuron, because they compute different outputs for the distinct tasks to be learned, i.e., it computes a different sum-product of its weights and its inputs for each output unit. In particular, they do not include in the sum product the

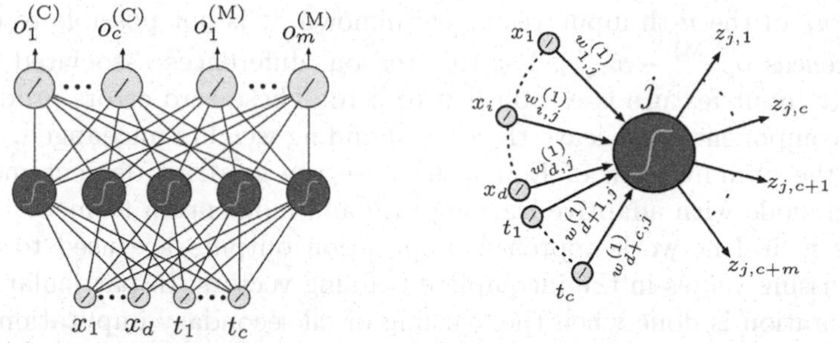

(a) MTL network with a common hidden layer.

(b) MTL neuron for handling missing data.

Fig. 1. MTL network with a common hidden layer that learns the classification and the imputation tasks at the same time. Biases are implicit for simplicity.

input signal that they have to learn in the corresponding output unit. It is done to avoid direct connections to map the input as output [8]. The j-th hidden neuron outputs are computed according with the following expressions,

For $k = 1, ..., c$

$$z_{j,k} = g\left(\sum_{i=1}^{d} w_{i,j}^{(1)} x_i + w_{0,j}^{(1)}\right) \tag{2}$$

For $k = c+1, ..., c+m$

$$z_{j,k} = g\left(\sum_{\substack{i=1 \\ i \neq a_{k-c}}}^{d} w_{i,j}^{(1)} x_i + \sum_{i=1}^{c} w_{d+i,j}^{(1)} t_i + w_{0,j}^{(1)}\right) \tag{3}$$

Finally, the outputs $\mathbf{o} = [\mathbf{o}^{(C)}, \mathbf{o}^{(M)}]$ are obtained by a linear combination of the outputs of the hidden neurons using a second layer of processing units.

3.1 Training Stage

Before training, all weights are initialized randomly, and also, the training set is normalize to zero mean and unit variance.

During the training stage, the network weights are updated iteratively in order to minimize an error function. In this paper, the Sum-of-Squares-Error (SSE) function is used,

$$E = \frac{1}{2}\sum_{n=1}^{N}\left(\|\mathbf{o}^{(n,C)} - \mathbf{t}^{(n)}\|^2 + \sum_{k=1}^{m}\left(o_k^{(n,M)} - x_{a_k}^{(n)}\right)^2\right) \tag{4}$$

where $\mathbf{o}^{(n,C)}$ and $o_k^{(n,M)}$ are, respectively, the classification output and the k-th imputation output obtained by the input vector $\mathbf{x}^{(n)}$. This error function depends on the differences between obtained outputs and desired outputs. If the attribute a_k of the n-th input vector is unknown, it is not possible to compute the differences $o_k^{(n,\mathrm{M})} - x_{a_k}^{(n)}$. For this reason, differences associated to every incomplete input feature is established to zero. These zero errors produce zero gradient components that leave the corresponding weights unchanged. With respect to the minimization of Equation 4, we use gradient descent method in sequential mode with adaptive learning rate and momentum term.

In this neural network approach, imputation outputs are used to estimate and fill missing values in the incomplete training vectors. In particular, missing data imputation is done when the learning of all secondary imputation tasks is stopping. Since classification and imputation tasks are learned by the same MTL network, learning of the main classification task affects to these imputed values, and so, this imputation is oriented to solve the classification task.

3.2 Operation Stage

During the operation stage, the classification performance and generalization capabilities of the MTL approach are evaluated on a new set of incomplete input vectors. The classification process at operation phase is shown in the procedure below.

Algorithm 1. Operation Stage.

Require: $\mathbf{x}^{(n)}$ (input vector), $\mathbf{m}^{(n)}$ (missing data indicator vector), c (number of classes)
if $\mathbf{x}^{(n)}$ is incomplete **then**
  for $k = 1$ to c **do**
    Set up to zero the incomplete values labeled by $\mathbf{m}^{(n)}$
    Check the k-th class labeled in $\mathbf{t}^{(n)}$ as extra input, $[\mathbf{x}^{(n)}, \mathbf{t}^{(n)}]$
    Forward propagation, and filling in the missing values using $\mathbf{o}^{(n,\mathrm{M})} \rightarrow \tilde{\mathbf{x}}^{(n)}$
    Forward propagation with $\tilde{\mathbf{x}}^{(n)} \rightarrow \mathbf{o}^{(n,\mathrm{C})}$
    Measure the similarity between $\mathbf{o}^{(n,\mathrm{C})}$ and $\mathbf{t}^{(n)}$
  end for
  Choose the most consistent class of all possible targets
else
  Forward propagation with $\mathbf{x}^{(n)} \rightarrow$ Classify with $\mathbf{o}^{(n,\mathrm{C})}$
end if

If $\mathbf{x}^{(n)}$ is completely known, the MTL network directly classifies using the classification output $\mathbf{o}^{(n,\mathrm{C})}$. When an incomplete pattern is presented to the network, the classification process of the MTL approach is different, because the missing data estimation is done by means of $\mathbf{o}^{(\mathrm{M})}$. But these outputs depend on $\mathbf{t}^{(n)}$ as part of the input, and this class information is not available during the operation stage. In order to solve it, all possible classes are checked, and the most *consistent* class is selected. The consistency of the class labeled in $\mathbf{t}^{(n)}$ is a measure of the difference between $\mathbf{t}$ and the output $\mathbf{o}^{(\mathrm{C})}$, which is obtained by the MTL network after imputation is realized using the corresponding $\mathbf{o}^{(\mathrm{M})}$.

4 Exploiting Private Subnetworks

Previous works have extended the basic MTL scheme using private or specific subnetworks [6,7]. As we can see in Figure 2(a), a first approach is adding a private subnetwork to learn only the classification task. This private subnetwork is connected to all input features, but not to the extra inputs (classification target). It learns only the main task; and so, the use of a private subnetwork helps to get a better generalization of the classification task. Hidden neurons in this subnetwork work as a classical artificial neuron [1], computing the same sum-product for all outputs. Private subnetwork size has to be optimized to provide a good classification performance. On the other hand, the number of hidden neurons in the common subnetwork depends on all tasks, because it supports the knowledge transfer between tasks.

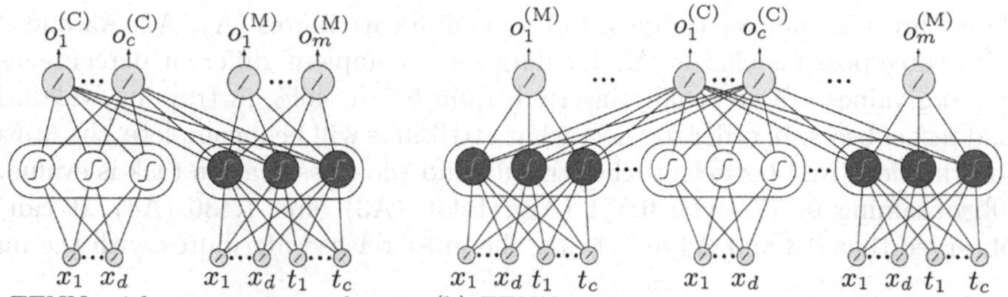

(a) FFNN with a common subnetwork, and a private subnetwork.

(b) FFNN with m common subnetworks and a private subnetwork.

Fig. 2. First, it is showed a MTL network with a hidden layer that learns the classification and the imputation tasks at the same time, and a private subnetwork used only by the main task; secondly, a MTL network with a private subnetwork used only by the main task, and m common subnetworks, which learn the classification and the corresponding imputation task simultaneously

The two solutions proposed in Figures 1 and 2(a) have a common subnetwork what is shared by all tasks. MTL makes the assumption that the learned tasks are related, and these relations are what contribute to the learning [8]. This default assumption allows unrelated tasks to decrease the generalization performance across all learned tasks in the MTL network causing a loss of knowledge for some tasks, what should be avoided [9]. For this reason, it is essential to consider the relation between tasks and how each task is learned during the training of the MTL networks [9]. An alternative to the previous approaches, see Figure 2(b), is using a common subnetwork for each imputation task, which learns the main task and the associated secondary task. Thus, learning between unrelated secondary tasks is avoided, and each one of them is guided by the learning of the classification task, i.e., common weights of each private subnetwork are influenced by the learning of the main task and the corresponding imputation task. In this scheme, the drawback is the high number of neurons that are required.

5 Simulation Results

In order to test the MTL networks introduced in this paper, two well-known datasets are used, *Iris Data* and *Pima Indians Data* [11]. *Iris* data set is randomly divided into three subsets: 1/3 instances of the dataset are used as training set, 1/6 instances are used as validation set and the rest 1/2 are used as test set. Such process is repeated twelve times and twelve groups of training, validation and test subsets are generated, and also ten simulations have been made in each group of subsets. In contrast, the *Pima* data set has been previously divided into training and test sets. The different MTL networks showed in Figure 1, 2(a) and 2(b) are respectively labeled as MTL-A, MTL-B and MTL-C. These networks are compared with three representative methods [3,4,5], described in Section 1.

5.1 Iris Data

There are 150 samples composed of four input attributes (A1, A2, A3 and A4) with three possible classes. As this data set is complete, different percentages of missing values are artificially inserted, from 5% to 40%, in training, validation and test subsets. In order to select what attributes will be incomplete, the Mutual Information (MI) between each attribute and the classification task is evaluated [9], obtaining 0.877 (A1), 0.511 (A2), 1.446 (A3), and 1.436 (A4). It can be observed that A3 and A4 are clearly the most related attributes with the main

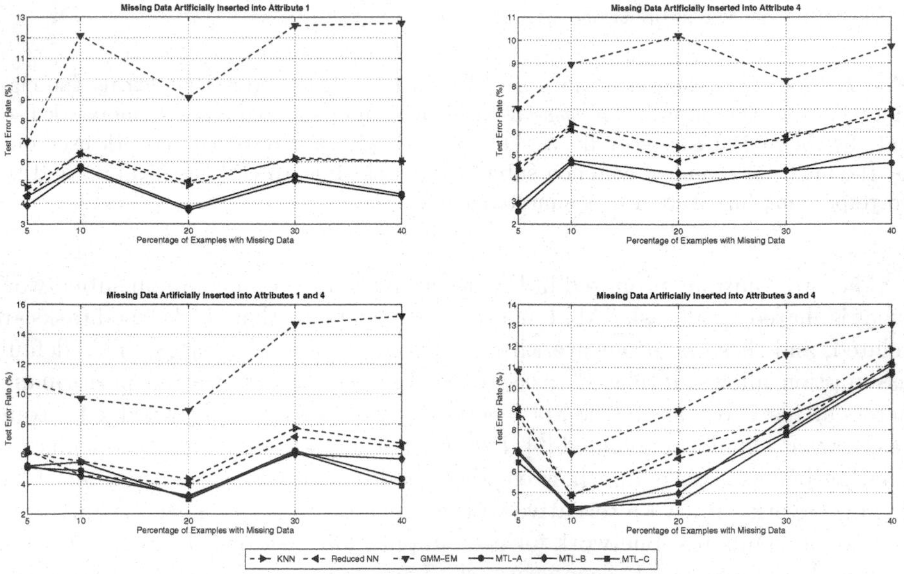

Fig. 3. Obtained results on *Iris* data set using KNN imputation, Reduced NNs, GMM trained with EM, and the different proposed MTL schemes. These graphs show the evolution of the test error rate with respect to the percentage of incomplete samples in the selected attributes.

(classification) task. Considering this fact, missing values have been introduced according with the following combinations of attributes: A1, A4, A1-A4 and A3-A4. Figure 3 shows the average accuracy rates using KNN imputation, Reduced NN, GMM-EM and the different proposed MTL schemes. When there are only a secondary task, i.e., an incomplete attribute, MTL-B and MTL-C schemes are the same, as the two first graphs.

Considering the results shown in Figure 3, proposed MTL schemes outperform the obtained classification accuracy by the other methods in the most simulations. When missing values occur in related attributes, the MTL networks perform better than the other procedures. Only with 30% and 40% in the most related attributes, in training, validation and test subsets, the MTL advantages are not so clear. Figure 4 shows the evolution of the training cost (SSE) for each task when 30% of missing data is artificially inserted in the attributes A1 and A4 using the three proposed networks. As we can see in this figure, the secondary task associated to the attribute A4 is learned better and faster than the other secondary task, because the attribute A4 is more related with the main one than A1. Also, it is clear to see how the training error associated to the main task decreases in a sudden way when the first imputation is done. This imputation is done when the learning of the secondary tasks is stopping. Each one of the following imputations gradually affects less the learning of the classification task because the learning of the secondary task is also stopped gradually. Comparing the proposed networks, the use of a private subnetwork to learn specifically the main task gets a better and faster learning in the classification task.

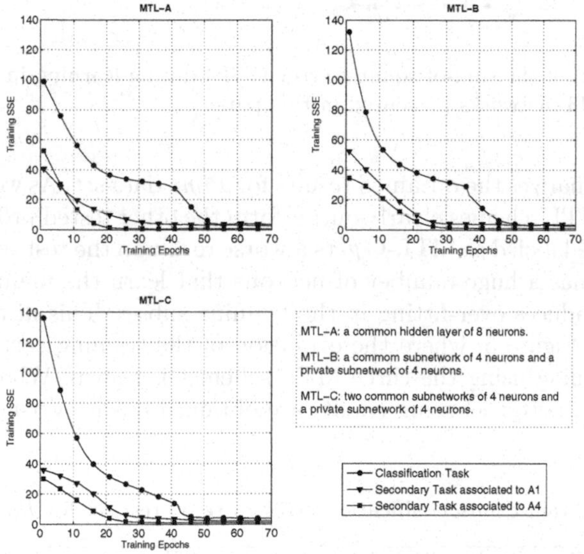

Fig. 4. Evolution of the sum-of-squares error (SSE) during learning in the *Iris* data set with 30% of missing data in the attributes A1 and A4 using the MTL networks presented in this paper

5.2 Pima Indians Data

This real problem consists of a training set with 300 cases, where 100 cases present unknown inputs; and a test set of 332 samples. In particular, three attributes are incomplete: the attributes A3, A4 and A5, with 4.33%, 32.67% and 1.00% as missing data percentages respectively. The MI measured between each one of them and the classification task are: A3, 0.111; A4, 0.232; and A5, 0.534.

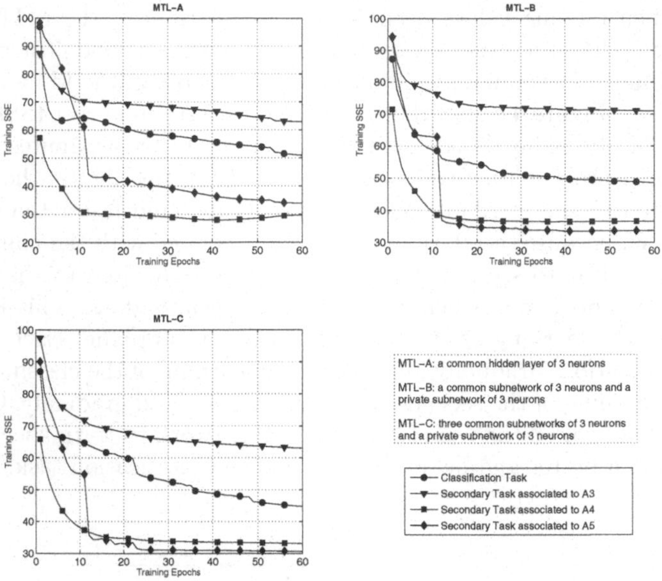

Fig. 5. Evolution of the sum-of-squares error (SSE) during learning in the *Pima* data set using the MTL networks presented in this paper

Table 1 summarizes the obtained results for *Pima* data set. As we can see on it, the proposed MTL schemes clearly outperform the other tested procedures. Only the architecture labeled as MTL-C gets a worse result on the test set because this MTL network has a huge number of neurons that learn the main classification task, and it produces over-fitting in the training subset. This disadvantage can be observed in Figure 5, where the evolution of the training cost for each task during the learning using the three MTL schemes is shown. When the MTL-C network is used, better learning results are obtained but it over-fits the training subset.

Table 1. Obtained misclassification error rates on test set for *Pima Indians*

Method	KNN	Reduced NNs	GMM-EM	MTL-A	MTL-B	MTL-C
% Error	21.02 ± 0.33	19.76 ± 0.58	21.98 ± 0.01	19.68 ± 0.18	19.58 ± 1.38	22.65 ± 1.76

6 Conclusions and Future Works

Different neural network schemes to classify incomplete patterns have been presented in this paper. Unlike other methods, classification and missing data estimation are combined in only one network using common and private subnetworks in MTL schemes. To do this, we have used the classification as main task and each incomplete feature as a secondary task; and also, a novel artificial neuron model has been implemented in order to learn all these tasks. Outputs that learn incomplete features are used to estimate missing values during learning process. Imputed values are those that contribute to improve the classification accuracy, because the learning of imputation tasks is oriented by the learning of the main task. Another great improvement is obtained when classification targets are used as extra inputs to learn the imputation tasks. During the operation phase, class information is not available, and the most consistent class is chosen. The effectiveness of our MTL networks has been justified empirically in two well-known databases.

In the future it would be valuable to set the number of neurons in each subnetwork dynamically using constructive methods, to implement a relation measure between tasks, and extending the presented approach to different machines [10].

References

1. Bishop, C.M.: Neural Networks for Pattern Recognition. Oxford University Press, Oxford, UK (1995)
2. Little, R.J.A., Rubin, D.B.: Statistical Analysis with Missing Data. 2 edn. John Wiley & Sons, New Jersey, USA (2002)
3. Sharpe, P.K., Solly, R.J.: Dealing with missing values in neural network-based diagnostic systems. Neural Computing and Applications **3** (1995) 73–77
4. Batista, G.E.A.P.A., Monard, M.C.: A study of k-nearest neighbour as an imputation method. In Abraham, A., del Solar, J.R., Köppen, M., eds.: HIS. Volume 87 of Frontiers in Artificial Intelligence and Applications., IOS Press (2002) 251–260
5. Ghahramani, Z., Jordan, M.I.: Supervised learning from incomplete data via an EM approach. In Cowan, J.D., Tesauro, G., Alspector, J., eds.: Advances in NIPS. Volume 6., Morgan Kaufmann Publishers, Inc. (1994) 120–127
6. García-Laencina, P.J., Sancho-Gómez, J.L., Figueiras-Vidal, A.R.: Pattern classification with missing values using multitask learning. In: IJCNN, Vancouver, BC, Canada, IEEE Computer Society (2006) 3594–3601
7. García-Laencina, P.J., Figueiras-Vidal, A.R., Serrano-García, J., Sancho-Gómez, J.L.: Exploiting multitask learning schemes using private subnetworks. In Cabestany, J., Prieto, A., Hernández, F.S., eds.: IWANN. Volume 3512 of Lecture Notes in Computer Science., Springer (2005) 233–240
8. Caruana, R.: Multitask Learning. PhD thesis, Carnegie Mellon University (1997)
9. Silver, D.L.: Selective Transfer of Neural Network Task Knowledge. PhD thesis, University of Western Ontario (2000)
10. Evgeniou, T., Micchelli, C., Pontil, M.: Learning multiple tasks with kernel methods. Journal of Machine Learning Research **6** (2005) 615–637
11. D.J. Newman, S. Hettich, C.B., Merz, C.: UCI repository of machine learning databases (1998)

Theoretical Study on the Capacity of Associative Memory with Multiple Reference Points

Enrique Mérida-Casermeiro[1], Domingo López-Rodríguez[1], Gloria Galán-Marín[2], and Juan M. Ortiz-de-Lazcano-Lobato[3]

[1] Department of Applied Mathematics, University of Málaga, Málaga, Spain
{merida,dlopez}@ctima.uma.es
[2] Department of Electronics and Electromechanical Engineering, University of Extremadura, Badajoz, Spain
gloriagm@unex.es
[3] Department of Computer Science and Artificial Intelligence, University of Málaga, Málaga, Spain
jmortiz@lcc.uma.es

Abstract. An extension to Hopfield's model of associative memory is studied in the present work. In particular, this paper is focused in giving solutions to the two main problems present in the model: the apparition of spurious patterns in the learning phase (implying the well-known and undesirable effect of storing the opposite pattern) and the problem of its reduced capacity (the probability of error in the retrieving phase increases as the number of stored patterns grows). In this work, a method to avoid spurious patterns is presented and studied, and an explanation to the previously mentioned effect is given. Another novel technique to increase the capacity of a network is proposed here, based on the idea of using several reference points when storing patterns. It is studied in depth, and an explicit formula for the capacity of the network is provided. This formula shows the linear dependence of the capacity of the new model on the number of reference points, implying the increase of the capacity in this model.

1 Introduction

Associative memory has received much attention for the last two decades. Though numerous models have been developed and investigated, the most influential is Hopfield's associative memory, based on his bipolar model (BH) [1]. This kind of memory arises as a result of his studies on collective computation in neural networks.

Hopfield's model consists in a fully-interconnected series of bi-valued neurons (outputs are either -1 or $+1$). Neural connection strength is expressed in terms of weight matrix $W = (w_{i,j})$, where $w_{i,j}$ represents the synaptic connection between neurons i and j. This matrix is determined in the learning phase by applying Hebb's postulate of learning [2], and no further synaptic modification is considered later.

Two main problems arise in this model: the apparition of spurious patterns and its low capacity.

Spurious patterns are stable states, that is, local minima of the corresponding energy function of the network, not associated to any stored (input) pattern. The simplest, but

J. Mira and J.R. Álvarez (Eds.): IWINAC 2007, Part I, LNCS 4527, pp. 292–302, 2007.

not the least important, case of apparition of spurious patterns is the fact of storing, given a pattern, its opposite, i.e. both X and $-X$ are stable states for the net, but only one of them has been introduced as an input pattern.

The problem of spurious patterns is very fundamental for cognitive modelers as well as practical users of neural networks.

Many solutions have been suggested in the literature. Some of them [3,4] are based on introducing asymmetry in synaptic connections.

However, it has been demonstrated that synaptic asymmetry does not provide by itself a satisfactory solution to the problem of spurious patterns, see [5,6]. Athithan [7] provided a solution based on neural self-interactions with a suitably chosen magnitude, if Hebb's learning rule is used, leading to the near (but not) total suppression of spurious patterns.

Crick [8] suggested the idea of unlearning the spurious patterns as a biologically plausible solution to suppress them. With a physiological explanation, they suggest that spurious patterns are unlearned randomly by human brain during sleep, by means of a process that is the reverse of Hebb's learning rule. This may result in the suppression of many spurious patterns with large basins of attraction. Experiments have shown that their idea leads to an enlargening of the basins for correct patterns along with the elimination of a significant fraction of spurious patterns [9]. However, a great number of spurious patterns with small basins of attraction do survive. Also, in the process of indiscriminate reverse learning, there is a finite probability of unlearning correct patterns, what makes this strategy unacceptable.

On the other hand, the capacity parameter α is usually defined as the quotient between the maximum number of patterns to load into the network, and the number of used neurons that achieve an acceptable error probability in the retrieving phase, usually $p_e = 0.01$ or $p_e = 0.05$. It was empirically shown that this constant is approximately $\alpha = 0.15$ for BH (very close to its actual value, $\alpha = 0.1847$, see [4]). The meaning of this capacity parameter is that, if the net is formed by N neurons, a maximum of $K \leq \alpha N$ patterns can be stored and retrieved with little error probability.

McEliece [10] showed that an upper bound for the asymptotic capacity of the network is $\frac{1}{2 \log N}$, if most of the input (prototype) patterns are to remain as fixed points. This capacity decreases to $\frac{1}{4 \log N}$ if every pattern must be a fixed point of the net.

By using Markov chains to study capacity and the recall error probability, Ho et al. [11] showed results very similar to those obtained by McEliece, since for them it is $\alpha = 0.12$ for small values of N, and the asymptotical capacity is given by $\frac{1}{4 \log N}$.

Kuh [12] manifested roughly similar estimations by making use of normal approximation theory and the theorems about exchangeables random variables.

In this work, a multivalued generalization of Hopfield's model (called MREM) is studied as an associative memory, and a technique to totally avoid the apparition of spurious patterns (in both models, BH and MREM) is explained in terms of the decrease of the energy function associated to patterns.

The main contribution of this paper consists in an extension of these models as associative memories to overcome the problem of the reduced capacity, by using a new technique which ensures the linear increase of the capacity.

2 The MREM Model

Let us consider a recurrent neural network formed by N neurons, where the state of each neuron $i \in \mathcal{I} = \{1, \ldots, N\}$ is defined by its output V_i taking values in any finite set $\mathcal{M} = \{m_1, m_2, \ldots, m_L\}$. This set does not need to be numerical.

The state of the network, at time t, is given by a N-dimensional vector, $\boldsymbol{V}(t) = (V_1(t), V_2(t), \ldots, V_N(t)) \in \mathcal{M}^N$. Associated to every state vector, an energy function, is defined:

$$E(\boldsymbol{V}) = -\frac{1}{2}\sum_{i=1}^{N}\sum_{j=1}^{N} w_{ij} f(V_i, V_j) + \sum_{i=1}^{N} \theta_i(V_i) \tag{1}$$

where $w_{i,j}$ is the weight of the connection from the j-th neuron to the i-th neuron, $f : \mathcal{M} \times \mathcal{M} \to \mathbb{R}$ can be considered as a measure of similarity between the outputs of two neurons, usually verifying the conditions mentioned in [13]:

1. For all $x \in \mathcal{M}$, $f(x, x) = c \in \mathbb{R}$.
2. f is a symmetric function: for every $x, y \in \mathcal{M}$, $f(x, y) = f(y, x)$.
3. If $x \neq y$, then $f(x, y) \leq c$.

and $\theta_i : \mathcal{M} \to \mathbb{R}$ are the threshold functions. Since thresholds will not be used for content addressable memory, henceforth we will consider θ_i be the zero function for all $i = 1, \ldots, N$.

The introduction of this similarity function provides, to the network, of a wide range of possibilities to represent different problems [13,14]. So, it leads to a better and richer (giving more information) representation of problems than other multivalued models, as SOAR and MAREN [15,16], since in those models most of the information enclosed in the multivalued representation is lost by the use of the signum function that only produces values in $\{-1, 0, 1\}$.

The energy function characterizes the dynamics of the net, as happened in BH. In every instant, the net evolves to reach a state of lower energy than the current one.

In this work, we have considered discrete time and semi-parallel dynamics, where only one neuron is updated at time t. The next state of the net will be the one that achieves the greatest descent of the energy function by changing only one neuron output.

Let us consider a total order in $\mathcal{M}$. The potential increment when a-th neuron changes its output from V_a to $l \in \mathcal{M}$ at time t, is

$$U_{a,l}(t) = -\sum_{i=1}^{N} \left[w_{i,a} \cdot \left(f(V_i(t), l) - f(V_i(t), V_a(t))\right)\right] \tag{2}$$

(due to the similarity conditions imposed to f).

We use the following updating rule for the neuron outputs:

$$V_a(t+1) = \begin{cases} l, & \text{if } U_{a,l}(t) \geq U_{b,k}(t) \forall k \in \mathcal{M} \text{ and } \forall b \in \mathcal{I} \\ V_a(t), & \text{otherwise} \end{cases} \tag{3}$$

This means that each neuron computes in parallel the value of a L-dimensional vector of potentials, related to the energy decrement produced if the neuron state is changed.

The only neuron changing its current state is the one producing the maximum decrease of energy.

It has been proved that the MREM model with this dynamics always converges to a minimal state [13]. This result is particularly important when dealing with combinatorial optimization problems, where the application of MREM has been very fruitful [13,17].

If function $f(x, y) = 2\delta_{x,y} - 1$, which equals 1 if and only if its two parameters coincide, and -1 in the rest of cases, is used and $\mathcal{M} = \{-1, 1\}$, MREM reduces to Hopfield's model. So, MREM is a powerful generalization of BH and other multivalued models, because it is capable of representing the information more accurately than those models.

2.1 MREM as Auto-associative Memory

Now, let $\{X^{(k)} : k = 1, \ldots, K\}$ be a set of patterns to be loaded into the neural network. Then, in order to store a pattern, $X = (X_i)_{i \in \mathcal{I}}$, components of the W matrix must be modified in order to make X the state of the network with minimal energy.

As pointed out in [13], since energy function is defined as in Eq. (1), we calculate $\frac{\partial E}{\partial w_{ij}} = -\frac{1}{2} f(V_i, V_j)$ and we modify the components of matrix W in order to reduce the energy of state $\boldsymbol{V} = X$ by the rule $\Delta w_{i,j} = -\alpha \frac{\partial E}{\partial w_{i,j}} = \frac{\alpha}{2} f(X_i, X_j)$ for some $\alpha > 0$. For simplicity, we can consider $\alpha = 2$, resulting:

$$\Delta w_{i,j} = f(X_i, X_j) \tag{4}$$

and considering that, at first, $W = 0$, that is, all the states of the network have the same energy and adding over all the patterns, the next expression is obtained:

$$w_{i,j} = \sum_{k=1}^{K} f(X_i^{(k)}, X_j^{(k)}) \tag{5}$$

Equation (5) is a generalization of Hebb's postulate of learning, because the weight w_{ij} between neurons is increased in correspondence with their similarity.

It must be pointed out that, when bipolar neurons and the product function $f(x, y) = xy$ are used, the well-known learning rule of patterns in the Hopfield's network is obtained. This is also achieved by the use of the function $f(x, y) = 2\delta_{x,y} - 1$. In the rest of this work we will consider the use of this function for our study.

Analogously to BH, the network is initialized with the known part of the pattern to be retrieved (called probe). The network dynamics will converge to a stable state minimizing the energy function, and it will be the answer of the network.

3 How to Avoid Spurious States

When a pattern X is loaded into the network, by modifying weight matrix W, not only the energy corresponding to state $\boldsymbol{V} = X$ is decreased. This fact can be explained in terms of the so-called associated vectors.

Definition 1. *Given a state V, its* associated matrix *is defined as $G_V = (g_{i,j})$ such that $g_{i,j} = f(V_i, V_j)$.*

Its associated vector *is $A_V = (a_k)$, with $a_{j+N(i-1)} = g_{i,j}$, that is, it is built by expanding the associated matrix as a vector of N^2 components.*

Lemma 1. *The increment of energy of a state V when pattern X is loaded into the network, by using Eq. (4), is given by:*

$$\Delta E(\boldsymbol{V}) = -\frac{1}{2} < A_X, A_V >$$

where $< \cdot, \cdot >$ denotes the usual inner product.

Lemma 2. *Given a state vector $\boldsymbol{V}$, we have $A_V = A_{-V}$. So $E(\boldsymbol{V}) = E(-\boldsymbol{V})$.*

These two results explain why spurious patterns are loaded into the network.

It must be noted that, in MREM, the number of spurious patterns appearing after the load of a vector into the net is greater than the corresponding in BH.

An important remark has to be done at this point: With this notation, the expression of the energy function can be rewritten as:

$$E(\boldsymbol{V}) = -\frac{1}{2} \sum_{k=1}^{K} < A_{X^{(k)}}, A_V >$$

Since all associated vectors are vectors of N^2 components taking value in $\{-1, 1\}$, their norms are equal, $||A_V||_E = N$ for all $\boldsymbol{V}$. This result implies that what is actually stored in the network is the orientation of the vectors associated to loaded patterns.

From the above expression for the increment of energy, and using that components of associated vectors are either -1 or 1, the following expression for the decrease of energy when a pattern is loaded is obtained:

$$-\Delta E(\boldsymbol{V}) = \frac{1}{2}(N - 2d_H(\boldsymbol{V}, X))^2 \tag{6}$$

where $d_H(\boldsymbol{V}, X)$ is the Hamming distance between vectors $\boldsymbol{V}$ and X.

After this explanation, we propose a solution for this problem:

Definition 2. *The* augmented pattern *$\hat{X}$, associated to $X \in \mathcal{M}^N$, is defined by appending to X the possible values of its components, that is, if $\mathcal{M} = \{m_1, \ldots, m_L\}$, then $\hat{X} = (X_1, \ldots, X_N, m_1, \ldots, m_L\}$. Particularly:*

- *In case of bipolar outputs, $\mathcal{M} = \{-1, 1\}$, and it is $\hat{X} = (X_1, \ldots, X_N, -1, 1)$.*
- *If $\mathcal{M} = \{1, \ldots, L\}$, then $\hat{X} = (X_1, \ldots, X_N, 1, 2, \ldots, L)$.*

By making use of augmented patterns, the problem of spurious patterns is solved, as stated in the next result, which is easy to prove:

Theorem 1. *The function Ψ that associates an augmented pattern to its corresponding associated vector is injective.*

Then, in order to store a pattern X, it will suffice to load its augmented version, which will be the unique state maximizing the decrease of energy.

It must be noted that it will only be necessary to consider N neurons, their weights, and the weights corresponding to the last L neurons, that remain fixed, and do not need to be implemented.

4 Associative Memory with Multiple Reference Points

In Hopfield's classical model, the unique reference point is the origin in $\mathbb{R}^N$, that is, patterns are not shifted or translated to achieve better results. As the network stores the orientations of the associated vectors, it could be useful to shift patterns by different amounts in order to be capable of distinguishing them more accurately.

In this work, let us consider $\mathcal{M} = \{1, \ldots, M\}$. So, to load the set $\{X^{(k)} : k = 1, \ldots, K\}$, we use as reference points $O^{(1)}, \ldots, O^{(Q)} \in \mathcal{M}^N$. This means that what the net is going to store is the set of augmented patterns related to $X^{(k)} - O^{(q)}$, for each k and q.

As $X_i^{(k)} - O_i^{(q)} \in \{1-M, \ldots, -1, 0, 1, \ldots, M-1\} = \mathcal{M}'$, the augmented pattern associated to $X^{(k)} - O^{(q)}$ will be (using the same notation for simplicity)

$$X^{(k)} - O^{(q)} = (X_1^{(k)} - O_1^{(q)}, \ldots, X_N^{(k)} - O_N^{(q)}, 1 - M, \ldots, M - 1\}$$

In addition, we will refer to the components of the above vector as

$$(X^{(k)} - O^{(q)})_i = \begin{cases} X_i^{(k)} - O_i^{(q)} & i \leq N \\ i - (N + M) & N + 1 \leq i \leq N + 2M - 1 \end{cases}$$

Let us also denote $L = 2M - 1$, the cardinal of the set $\mathcal{M}'$.

By extending what was exposed in Sec. 2, a new energy function is introduced:

$$E(V) = -\frac{1}{2} \sum_{q=1}^{Q} \sum_{i=1}^{N+L} \sum_{j=1}^{N+L} w_{i,j}^{(q)} f((V - O^{(q)})_i, (V - O^{(q)})_j) \tag{7}$$

where

$$w_{i,j}^{(q)} = \sum_{q=1}^{Q} \sum_{k=1}^{K} f((V - O^{(q)})_i, (V - O^{(q)})_j)$$

The above expression can be rewritten in the following terms:

$$E(V) = \sum_{q=1}^{Q} E_q(V)$$

The expression of this new energy function implies that each neuron will be able to perform a more complex process in each step, since it has to take into account Q reference points, available in its own local memory. Thus, we are increasing the complexity of the neuron model, that will lead to a higher performance of the net in terms of an increase of its capacity, as we will prove in the next Section.

We must observe that this new model, with $Q = 1$, reduces to MREM standard associative memory.

5 Capacity of the New Model

The capacity of the network is a measure of the amount of patterns that can be introduced into the network such that at the retrieving phase the probability of error does not exceed a threshold, p_e. The aim of this section is to present a study of the network capacity (similar to [4]) that provides us with an exact or very approximate expression for the capacity parameter α.

Let us suppose that K patterns $X^{(1)}, \ldots, X^{(K)}$, have been loaded into the network, and that state vector $\boldsymbol{V}$ matches a stored pattern, $X^{(k_0)}$. Suppose that state $\boldsymbol{V}'$ coincides to $\boldsymbol{V}$ except in one component. Without loss of generality, this component can be assumed to be the first one, that is, $V_i = V_i'$ if $i > 1$ and $V_1 \neq V_1'$.

By denoting as $D = \Delta E = E(\boldsymbol{V}') - E(\boldsymbol{V})$ the energy increment between these two states $\boldsymbol{V}$ and $\boldsymbol{V}'$, the pattern $X^{(k_0)}$ is correctly retrieved when pattern $\boldsymbol{V}'$ is introduced into the net if $D > 0$, because this condition implies that $\boldsymbol{V}$ is a fixed point for the dynamics that are being used. So, in order to calculate the error probability in the retrieval phase, the probability $P(D < 0)$ must be computed.

But

$$D = \Delta E = \sum_{q=1}^{Q} E_q(\boldsymbol{V}') - \sum_{q=1}^{Q} E_q(\boldsymbol{V}) = \sum_{q=1}^{Q} (E_q(\boldsymbol{V}') - E_q(\boldsymbol{V})) = \sum_{q=1}^{Q} \Delta E_q \quad (8)$$

and thus we have to compute $D_q = \Delta E_q$.

To this end, we present some technical results which will guide us to the main result of this section, the capacity of the network with multiple reference points. Proofs for these results will be omitted due to the limitation in the length of this paper. The reader can refer to [18] for a detailed proof in the case of Hopfield's bipolar model with multiple reference points.

Lemma 3. *We have*

$$D_q = N + 3 - \sum_{i=2}^{N} \phi_i + \sum_{k \neq k_0} \sum_{i=2}^{N+L} \xi_i \quad (9)$$

where

1. *ϕ_i is a random variable (r. v.) with mean $E(\phi_i) = 1 - 4\frac{2M^2+1}{3M^3}$ and variance $V(\phi_i) = 8\frac{2M^2+1}{3M^3}\left(1 - 2\frac{2M^2+1}{3M^3}\right)$, for all $i \leq N$.*
2. *ξ_i is another r. v. with mean $E(\xi_i) = 0$ and variance $V(\xi_i) = 8\frac{2M^2+1}{3M^3}$, for every i.*

The exact formula for D is given in the following lemma, by making use of this last result applied to Eq. (8).

Lemma 4. *For $N \geq 30$ and $Q \geq 1$,*

$$D = Q(N + 3) + \Omega$$

where Ω is a Gaussian r. v. with mean

$$\mu = Q(N - 1)(4\frac{2M^2 + 1}{3M^3} - 1)$$

and variance given by

$$\sigma^2 = 8Q\frac{2M^2 + 1}{3M^3}\left((N - 1)(1 - 2\frac{2M^2 + 1}{3M^3}) + (N + 2M - 2)(K - 1)\right)$$

This result allows us to calculate

$$P(D < 0) = P(Q(N+3) + \Omega < 0) = P(\Omega < -Q(N+3)) =$$

$$= P\left(\frac{\Omega - \mu}{\sigma} < \frac{-Q(N+3) - \mu}{\sigma}\right)$$

Since $Z = \frac{\Omega-\mu}{\sigma}$ is a Gaussian with mean 0, and variance 1, there exists one unique $z_\alpha \in \mathbb{R}$ such that $P(Z < z_\alpha) = p_e$. For example, for $p_e = 0.05$, it is $z_\alpha = -1.645$, and for $p_e = 0.01$, it is $z_\alpha = -2.326$. By using that $P(Z < z_\alpha) = p_e = P(D < 0)$, we arrive at

$$\frac{-Q(N+3) - \mu}{\sigma} = z_\alpha \tag{10}$$

The next step is to use the proper definition of the parameter of capacity α. It is the quotient between the number of patterns and the number of neurons which achieve an error probability lower than p_e. So, $\alpha = \frac{K}{N}$, that is, $K = \alpha N$.

By combining Eq. (10) and the above expression for K, we get the following result:

Theorem 2. *The capacity of the network as associative memory with multiple reference points is given by*

$$\alpha = \frac{1}{N}\left[1 + \frac{1}{V}\left(\frac{T^2}{z_\alpha^2} - U\right)\right] \tag{11}$$

where

$$T = \sqrt{Q}\left((N+3) + (N-1)(4\frac{2M^2+1}{3M^3} - 1)\right)$$

$$U = 8(N-1)\frac{2M^2+1}{3M^3}(1 - 2\frac{2M^2+1}{3M^3})$$

and

$$V = 8(N + 2M - 2)\frac{2M^2+1}{3M^3}$$

From this theorem, some important corollaries can be stated:

Corollary 1. *The capacity of the network is asymptotically increasing with the number of reference points.*

This result can be deduced from the fact that Eq. (11) can be rewritten in the following terms:

$$\alpha = R_0(M, N) + R_1(M, N)Q$$

with $R_1(M, N) > 0$. So, given N and M, α is an increasing function for values of $Q \geq 1$. This implies that, by increasing the number of reference points, capacity greater than 1 may be achieved, as can be verified in Fig. 1. It must be remembered that the maximum capacity in BH was 1.

Corollary 2. *Given M and Q, there exists a positive constant $\alpha_{\min}$ such that $\alpha \geq \alpha_{\min}$ for all N.*

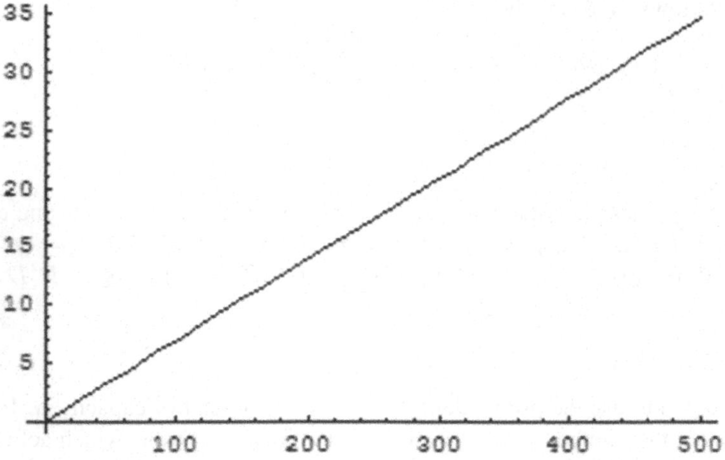

Fig. 1. Capacity of a network with $N = 40$ neurons and $M = 4$ states, as a function of Q, which varies from 1 to 500

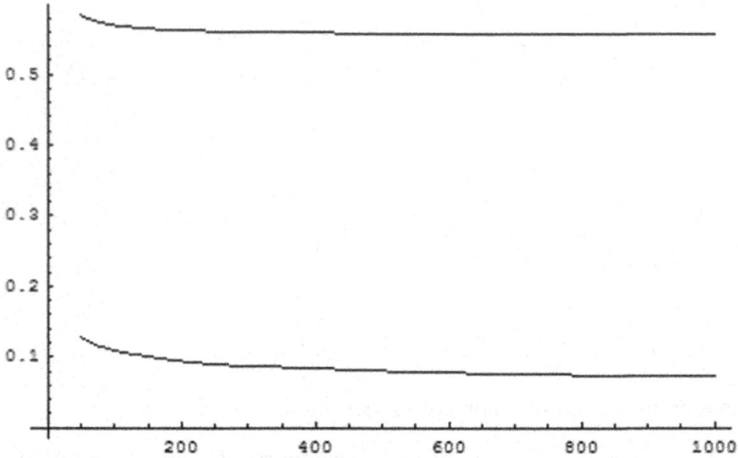

Fig. 2. Capacity of the bipolar network with $N \in \{50, \dots, 1000\}$ neurons and $Q = 4$, compared with the bound given by McEliece (lower graph)

This corollary can be proved just by noting that the capacity is a decreasing function of N, the number of neurons, so we can fix M and Q and calculate the value $\alpha_{\min} = \lim_{N\to\infty} \alpha = \frac{2Q}{z_\alpha^2} \frac{2M^2+1}{3M^3}$. If we consider $M = 2$ and $Q = 2$ and $p_e = 0.01$, a value of $\alpha_{\min} = 0.2772$ is obtained, and for $p_e = 0.05$, a value of $\alpha_{\min} = 0.5543$. In Fig. 2, the value of α for a bipolar network is compared with the upper bound given by [10], and it is shown that $\alpha > \frac{1}{2 \log N}$, meaning a great improvement on the capacity of the net, since this upper bound tends to 0 as N approaches ∞, and our technique ensures a minimum positive amount of capacity for the net.

6 Conclusions and Future Work

In this paper, an extension to Hopfield's associative memory has been studied to overcome some of the most important problems or lacks it possesses: spurious patterns and low capacity.

A method to avoid the apparition of spurious patterns has been presented. This method also explains the well-known (and undesirable) phenomenon of storing the opposite of a pattern.

A new technique to increase the network capacity as a content-addressable memory has also been proposed, based on the use of multiple reference points, which contributes many new possibilities of study and research.

Our future work covers several aspects of these methods:

- Find the optimal configuration of $O^{(q)}$ for a given set of patterns (randomly distributed or with a specific distribution), that is, the distribution of $O^{(q)}$ which discriminates most the patterns and makes the net achieve the maximum possible capacity.
- Consider a mix of fixed and random reference points.

References

1. Hopfield, J.: Neural networks and physical systems with emergent collective computational abilities. Volume 79. (1982) 2254 – 2558
2. Hebb, D.O.: The Organization of Behavior. Wiley (1949)
3. Parisi, G.: Asymmetric neural networks and the process of learning. J. Phys. A: Math. and Gen. **19** (1986) L675 – L680
4. Hertz, J.A., Grinstein, G., Solla, S.A.: Heidelberg Colloquium on Glassy Dynamics. In:. Springer-Verlag (1987)
5. Treves, A., Amit, D.J.: Metastable states in asymmetrically diluted hopfield networks. J. Phys A: Math. and Gen. **21** (1988) 3155 – 3169
6. Singh, M.P., Chengxiang, Z., Dasgupta, C.: Analytic study of the effects of synaptic asymmetry. Phys. Rev. E **52** (1995) 5261 – 5272
7. Athithan, G., Dasgupta, C.: On the problem of spurious patterns in neural associative models. IEEE Transactions on Neural Networks **8** (1997) 1483 – 1491
8. Crick, F., Mitchinson, G.: The function of dream sleep. Nature **304** (1983) 111 – 114
9. Hemmen, J.L.v., Kuhn, R.: Collective Phenomena in Neural Networks. Springer-Verlag (1991)
10. McEliece, R.J., Posner, E.C., Rodemich, E.R., Venkatesh, S.S.: The capacity of the hopfield associative memory. IEEE Transactions on Information Theory **IT-33** (1987) 461 – 482
11. Ho, C.Y., Sasase, I., Mori, S.: On the capacity of the hopfield associative memory. In: Proceedings of IJCNN 1992. (1992) II196 – II201
12. Kuh, A., Dickinson, B.W.: Information capacity of associative memory. IEEE Transactions on Information Theory **IT-35** (1989) 59 – 68
13. Mérida-Casermeiro, E., Muñoz Pérez, J.: Mrem: An associative autonomous recurrent network. Journal of Intelligent and Fuzzy Systems **12** (2002) 163 – 173
14. Mérida-Casermeiro, E., Galán-Marín, G., Muñoz Pérez, J.: An efficient multivalued hopfield network for the travelling salesman problem. Neural Processing Letters **14** (2001) 203 – 216

15. Erdem, M.H., Ozturk, Y.: A new family of multivalued networks. Neural Networks **9** (1996) 979 – 989
16. Ozturk, Y., Abut, H.: System of associative relationships (soar). (1997)
17. Mérida-Casermeiro, E., Muñoz Pérez, J., Domínguez-Merino, E.: An n-parallel multivalued network: Applications to the travelling salesman problem. Computational Methods in Neural Modelling, Lecture Notes in Computer Science **2686** (2003) 406 – 413
18. López-Rodríguez, D., Mérida-Casermeiro, E., Ortiz-de Lazcano-Lobato, J.M.: Hopfield network as associative memory with multiple reference points. In Ardil, C., ed.: International Enformatika Conference. (2005) 62 – 67

Classification and Diagnosis of Heart Sounds and Murmurs Using Artificial Neural Networks

Juan Martínez-Alajarín[1], José López-Candel[2], and Ramón Ruiz-Merino[1]

[1] Universidad Politécnica de Cartagena, Cartagena 30202, Spain
juanc.martinez@upct.es
[2] Hospital General Universitario Reina Sofía, Murcia 30003, Spain

Abstract. Cardiac auscultation still remains today as the basic technique to easily achieve a cardiac valvular diagnosis. Nowadays, auscultation can be powered with automated computer-aided analysis systems to provide objective, accurate, documented and cost-effective diagnosis. This is particulary useful when such systems offer remote diagnosis capabilities. ASEPTIC is a telediagnosis system for cardiac sounds that allows the analysis of remote phonocardiographic signals. The pattern recognition stage of ASEPTIC is presented in this paper. It is based in feature selection from the cardiac events, and classification using a multilayer perceptron artificial neural network trained with Levenberg-Marquardt algorithm for fast convergence. Three categories of records have been considered: normal, with holosystolic murmur, and with midsystolic murmur. Experimental results show high correct classification rates for the three categories: 100%, 92.69%, and 97.57%, respectively.

1 Introduction

Although in the last 30 years cardiac auscultation has been replaced by modern techniques (mainly echocardiography) to diagnose the valvular state of the heart, it is still widely used as a screening technique. Nowadays, efforts are mainly conducted to develop computer systems that can aid the physician to diagnose the state of the heart and provide rapid, accurate, objective, documented and cost-effective diagnosis [1]. This is specially useful in rural areas [2], because of the lack of physicians and high cost modern techniques.

One of such systems is ASEPTIC (Aided System for Event-based Phonocardiographic Telediagnosis with Integrated Compression) [3], which is a complete system for telediagnosis of the cardiovascular condition by analysing phonocardiographic (PCG) signals generated by the heart. The analysis is performed using only the PCG signal, without needing auxiliary signals like ECG or pulse. ASEPTIC includes a processing stage that analyzes the PCG recordings, and a compression/decompression stage [4] for an efficient transmission of the PCG signals to and from the processing unit. After signal conditioning and basic preprocessing, ASEPTIC detects the individual cardiac events, and then pattern recognition is performed using the features extracted from the cardiac events and a

J. Mira and J.R. Álvarez (Eds.): IWINAC 2007, Part I, LNCS 4527, pp. 303–312, 2007.

neural classifier (multilayer perceptron artificial neural network with Levenberg-Marquardt training algorithm [5]). Although a variety of classification methods exists, neural classifiers have been widely used for PCG because of their good performance and learning capabilities [6,7,8].

In these article we describe the feature extraction and neural classification of the information extracted from the PCG in order to obtain the diagnosis of the heart condition. Section 2 provides a brief description of the operations that are performed during the analysis, and how the information that contains the PCG signal is prepared for classification. In Section 3, full details of the feature extraction and classification using an artificial neural network (ANN) are presented. Section 4 reports experimental results, and final conclusions are given in Section 5.

2 System Overview

The processing stage of ASEPTIC is arranged as a modular hierarchical structure with four abstraction levels (Figure 1). The first three levels perform, respectively, signal conditioning, delimitation of the relevant segments of the PCG to be analyzed, and identification of the cardiac events. The fourth level computes several features from the events previously detected, and these features are used as inputs of an ANN to classify them in several predefined categories (pathologies).

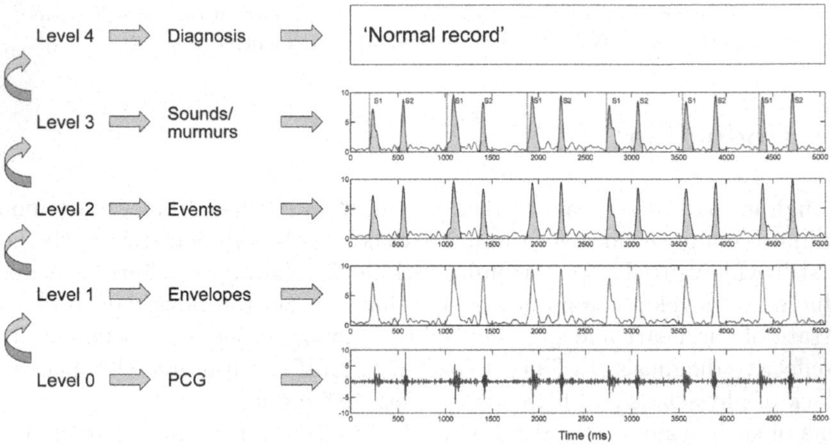

Fig. 1. Processing levels of the hierarchical structure of ASEPTIC. Level 0 (PCG signal) does not form part of the processing levels, but is only the input signal to the hierarchy.

Initially, the PCG is decimated by a factor 2 (from 8000 samples/s to 4000 samples/s), and scaled in the range [+1,-1] by dividing the PCG by its maximum absolute value. The resulting signal is digitally filtered using two IIR (Infinite Impulse Response) Chebyshev type I 3rd. order filters, with cutting frequencies $f_{c1} = 40\,\text{Hz}$ (high-pass) and $f_{c2} = 800\,\text{Hz}$ (low-pass). Finally, three instantaneous magnitudes are derived from the PCG (instantaneous amplitude, IA, energy, IE, and frequency, IF [9]) and their envelopes are computed using a moving average filter [10].

The autocorrelation signal of the product of the three envelopes provides a symmetric signal with its maximum in the central point. The main relative maximum peaks detected in either of the halves of the autocorrelation signal are used to define the average cardiac rhythm as the mean value of the segments defined by these main peaks. An events detection method is then used to detect the basic cardiac events [11]. This method is based in the detection of the relative maxima in the amplitude envelope and the computation of a set of associated points. They define the temporal limits of the events, and a basic identification of them as *sounds* or *murmurs* is provided .

From the computed average cardiac rhythm and the detected events, the PCG is segmented in cardiac cycles, beginning with a first heart sound (S1). Then the detected events are identified using the following information: events duration, amplitude and maximum frequency, relative distance between events, number of events in the cardiac cycle, and situation of the middle point of the event (only for murmurs). Algorithms have been developed to identify the following events: S1, S2, S3, S4, midsystolic clicks (MSC), and murmurs. In this last case, information about the relative situation of the murmur in the cardiac cycle is also provided (early/mid/late/holo and systolic/diastolic). Identification is based in three methods, used sequentially until one of them provides the identification of all the events in the cardiac cycle: energy envelope, spectral-based energy tracking [12], and the application of the IF to the A5 subband of the wavelet decomposition of the PCG.

3 Neural Classification of Heart Sounds and Murmurs

After the processing of levels 1, 2 and 3 of the hierarchy, the diagnostic category of the record is often restricted to a subset of all the possible categories considered. Pattern recognition performed in level 4 allows to define precisely the category, and provides a diagnosis combining these results with physiological and auscultation information.

Pattern recognition consists of two steps: firstly, a feature set is computed from the detected events; those features which provides more information are selected, and also redundant information is removed. Secondly, the selected features are used as inputs of an ANN, which provides the category to which the analyzed PCG record belongs to, and defines the pathology.

3.1 Feature Extraction

For each cardiac event, a set of 13 features has been extracted from temporal measurements and from the envelopes of instantaneous amplitude, energy and frequency:

- duration of the event (d),
- average value of the event in the IA envelope ($avgIA$),
- maximum value of the event in the IA envelope ($maxIA$),
- normalized standard deviation of the event in the IA envelope ($nsdIA$),
- area enclosed by the event in the IA envelope, which equals the average value multiplied by the duration of the event ($areaIA = d \cdot avgIA$),
- average, maximum, normalized standard deviation, and area enclosed of the event in the IE envelope ($avgIE$, $maxIE$, $nsdIE$, and $areaIE$),
- average, maximum, normalized standard deviation, and area enclosed of the event in the IF envelope ($avgIF$, $maxIF$, $nsdIF$, and $areaIF$),

Although the IA and IE signals are obviously related (since energy is the square value of the amplitude), both signals are needed because each of them provides important information during events delimitation and identification (depending on which kind of events must be highlighted, the IA or IE signals are used). It is thus expected that the set of 13 features will provide some redundant information. Principal Component Analysis (PCA) has been used to find which features are the most discriminant and also to remove redundant information. A new set of non-correlated features is then obtained. Each new feature is called a *principal component* (PC), and they are sorted from higher variance (PC1) to lower variance (PC13). Usually, more than 90% of the variance of the original feature set can be achieved with only a few PCs. This allows to select a reduced feature set used to classify the PCG records, thus avoiding a large dimension feature space (*curse of dimensionality*) that would make the design of the classifier more complex, specially when the training data set is not very large.

3.2 Classification

Classification is used to refine the event identification information extracted in level 3 of the processing hierarchy. The classifier will provide the category or pathology to which the PCG record belongs to, which is used as the diagnosis of the analysis system. Additional information like physiological data of the patient, or data coming from the auscultation process also could be used to refine the diagnosis of the classifier.

After a feature set has been extracted from each event and the most discriminant features have been selected using PCA, the selected features for all the events in a specific cardiac cycle form the cardiac cycle feature set vector. This vector will be used to establish the pathology of the PCG record. For a set of several cardiac cycles (training data set), a matrix of data is obtained: each row provides the feature vector of each cardiac cycle (which is called a *pattern*), and each column provides a specific feature for the different cardiac cycles. This data matrix will be used to train an heteroassociative classifier to match each feature vector with its corresponding pathology.

From the many existing methods for pattern recognition, the method chosen for classification has been a neural network because of its learning capabilities and generalization of information. In particular, the multilayer perceptron

(MLP) has been used, with the Levenberg-Marquardt [5] training algorithm, which reduces the learning time considerably with respect to the backpropagation algorithm.

The structure of the neural network consists of three layers with feedforward connections:

- the input layer, with N_{feat} neurons, one for each feature of the cardiac cycle,
- the middle (or hidden) layer, with 40 neurons, and
- the output layer, with N_{cat} neurons, one for each category or class.

The activation functions have been the hyperbolic tangent for the hidden layer, and the lineal function for the output layer. The following parameters have been used to train the network: Levenberg-Marquardt training algorithm, target error of 0.0001, learning factor of 0.01, and maximum number of iterations of 1000. The training of the network consisted in two stages: 1) one half of the patterns (set 1) was used to train the network and the other half (set 2) was used to test the network; 2) the sets were swapped (set 2 for training and set 1 for testing). Finally, results were merged, obtaining a matrix of results whose size equals the total number of patterns (cardiac cycles).

As a previous step to the training of the network, the input data were normalized (mean = 0 and standard deviation = 1) to avoid that differences of the data ranges can have influence in the training. After the training, when the input data were presented at the network, they were also normalized with the same scale factors that the training data.

4 Experimental Results

Diagnosis of heart sounds using the previous methodology has been tested with a database of 94 heart cycles of real PCG records extracted from [13], belonging to three different categories: normal records (with only the first and second heart sounds, S1 and S2), records with holosystolic murmur (HSM), and records with midsystolic murmur (MSM). The number of cardiac cycles for each category has been the following: 31 (S1-S2), 26 (HSM), and 37 (MSM). In the normal records, two events per cardiac cycle were detected (S1 and S2). For the records with murmurs, three events per cycle were detected (S1, S2 and the murmur). Thus, the total number of detected events was 251.

Initially, PCA was applied to the matrix with the 13 features extracted from the 251 events. Table 1 shows the variance of the 13 new variables (principal components), and the percentage that they represent with respect to the total variance, sorted from higher to lower variance.

Figure 2 shows the contribution of each original feature to the principal component with the highest variance (PC1), which owns more than 50% of the variance of the whole feature set. In this figure it can be seen that the contribution of all features is quite similar except for $nsdIA$, $nsdIE$ and $nsdIF$, which have a lower contribution. From now on, these three features will be discarded.

Table 1. Variance and its percentage (from the total variance) for the principal components obtained from PCA

PC	Variance	Variance (%)
1	7.1777	55.2134
2	2.3064	17.7413
3	1.9384	14.9107
4	0.6623	5.0945
5	0.5738	4.4137
6	0.0955	0.7343
7	0.0879	0.6758
8	0.0690	0.5304
9	0.0486	0.3740
10	0.0181	0.1395
11	0.0158	0.1215
12	0.0043	0.0332
13	0.0023	0.0176

Then, PCA was used again to analyze the remaining 10 features. PC1 was represented versus PC2 for the 4 types of detected events (S1, S2, HSM and MSM) for 6 different feature sets, which are shown in Figure 3: a) $\{ d, avgIA, maxIA, areaIA \}$, b) $\{ d, avgIE, maxIE, areaIE \}$, c) $\{ d, avgIF, maxIF, areaIF \}$, d) $\{ d, avgIA, avgIF, areaIA, areaIF \}$, e) $\{ d, avgIE, avgIF, areaIE, areaIF \}$, f) $\{ d, avgIF, areaIA, areaIF \}$. Event duration was included in all the feature sets since it is the only temporal feature (although the area enclosed by the event contains also this information in an implicit way).

A comparison between amplitude features and energy features (Figures 3a-3b and 3d-3e) reveals that, although it is possible to discriminate between normal sounds and murmurs for all the cases, those figures with amplitude features provide better separation between MSM and HSM. Besides, the best separation between events is achieved always for those sets with IF features (Figures 3c, 3d and 3f). Separation between S1 and S2 was not possible at all for the six data sets.

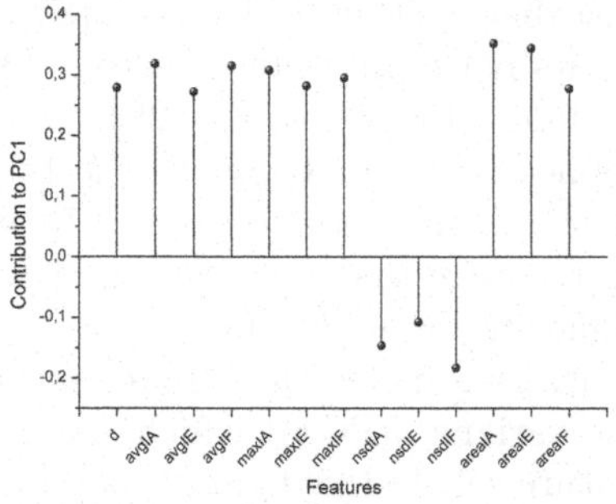

Fig. 2. Contribution of the original features to PC1

These results show the importance of the features extracted from the IF envelope, since they allow to characterize quantitatively the different events and discriminate between the different types.

For classification, 5 features have been selected for each event: d, $avgIA$, $avgIF$, $areaIA$, and $areaIF$, which corresponds to the case with the biggest separation between HSM and MSM (Figure 3d). The feature vector for each cardiac cycle has been obtained using the 5 previous features for the two main sounds (S1 and S2), the systole and the diastole. When an event (usually a murmur) exists in the systole or the diastole, the features for that period (systole or diastole) have been those of the event. In that case, the duration of the event in the systole or diastole has been expressed in percentage with respect to the total duration of the period ($\frac{\text{event duration}}{\text{period duration}} \cdot 100$), instead of using the duration of the event measured in time units. When a period does not contain cardiac events, their five features are 0. Since the records used for classification do not include diastolic murmurs (so there is not any event in the diastole), the 5 diastolic features (all of them 0's) have not been used, since they are the same for all the records and do not provide any discriminant information.

A 15-element feature vector is then formed:

- records without murmurs: [d_{S1}, $avgIA_{\text{S1}}$, $avgIF_{\text{S1}}$, $areaIA_{\text{S1}}$, $areaIF_{\text{S1}}$, 0, 0, 0, 0, 0, d_{S2}, $avgIA_{\text{S2}}$, $avgIF_{\text{S2}}$, $areaIA_{\text{S2}}$, $areaIF_{\text{S2}}$]
- records with murmurs: [d_{S1}, $avgIA_{\text{S1}}$, $avgIF_{\text{S1}}$, $areaIA_{\text{S1}}$, $areaIF_{\text{S1}}$, $\frac{d_{\text{murmur}} \cdot 100}{d_{\text{systole}}}$, $avgIA_{\text{murmur}}$, $avgIF_{\text{murmur}}$, $areaIA_{\text{murmur}}$, $areaIF_{\text{murmur}}$, d_{S2}, $avgIA_{\text{S2}}$, $avgIF_{\text{S2}}$, $areaIA_{\text{S2}}$, $areaIF_{\text{S2}}$]

Thus, a feature matrix with 94 rows and 15 columns is formed, each row being a pattern for the input to the neural network. The input layer to the neural network has $N_{\text{feat}} = 15$ neurons plus the bias (unit-input neuron used to set the offset at the input of each neuron), and the output layer includes $N_{\text{class}} = 3$ neurons, since three categories or classes have been used: normal records (C_1), records with HSM (C_2), and records with MSM (C_3). A winner-take-all approach has been used for the output layer, where only one neuron (corresponding to the class of the murmur) will take the value of 1, and the other two neurons will take 0 value.

The complete pattern set has been divided in set 1 and set 2, where both contain representative patterns in all the feature space. A full training of the neural network consists really of two steps: 1) the ANN is trained with set 1 and validated with set 2, and 2) the training and validation sets are swapped. In this form, all the patterns can be validated although they are never used simultaneously as training patterns and validation patterns. Results from the two steps are merged, so a 94-by-3 matrix is obtained. These results show the category assigned by the ANN for each cardiac cycle.

Confusion matrix can be obtained from the result matrix, and contains all the possibilities of categories assigned by the ANN for each pattern. Since the number of patterns is not large, averaging over 10 trainings have been used to improve the accuracy. Convergence of the ANN with the Levenberg-Marquardt

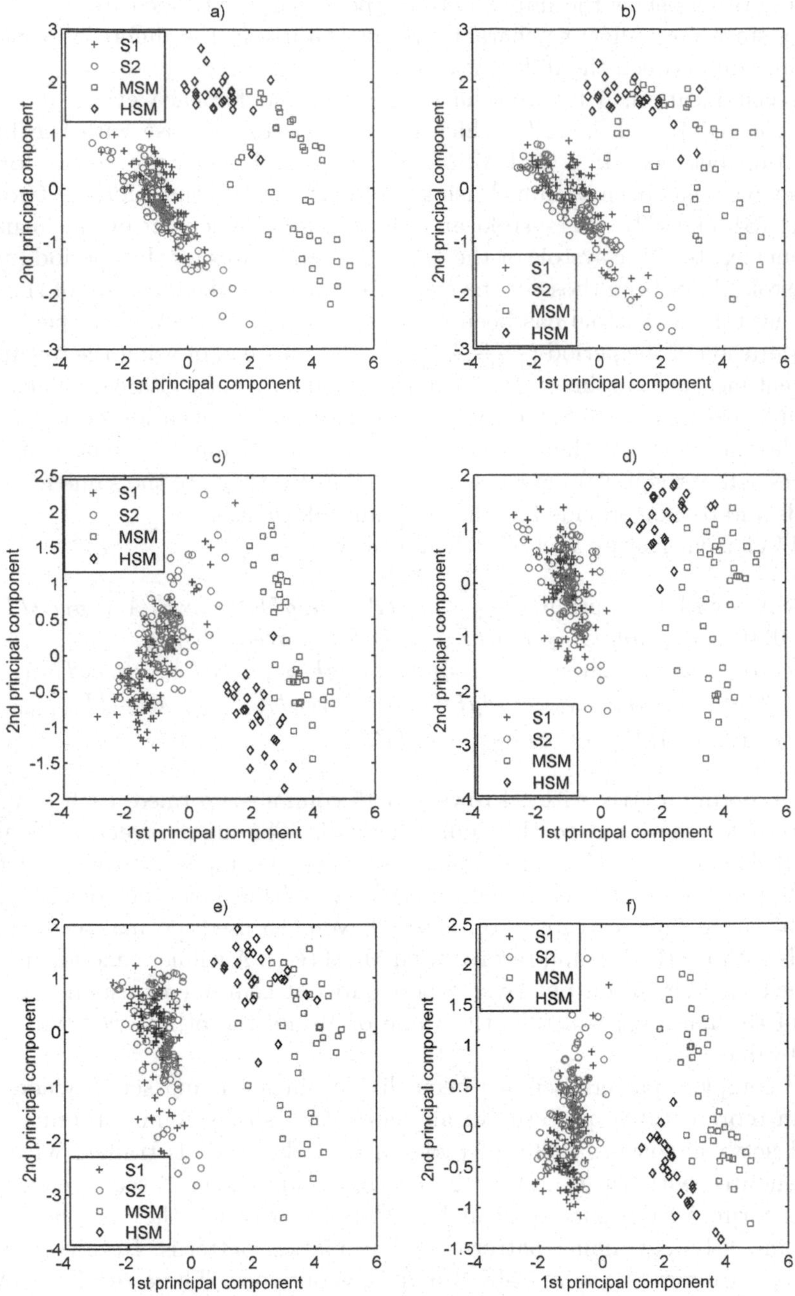

Fig. 3. Representation of PC1 versus PC2 for different feature sets: a) { d, $avgIA$, $maxIA$, $areaIA$ }, b) { d, $avgIE$, $maxIE$, $areaIE$ }, c) { d, $avgIF$, $maxIF$, $areaIF$ }, d) { d, $avgIA$, $avgIF$, $areaIA$, $areaIF$ }, e) { d, $avgIE$, $avgIF$, $areaIE$, $areaIF$ }, f) { d, $avgIF$, $areaIA$, $areaIF$ }

algorithm has been very fast, needing only 3 or 4 iterations in all cases (3.25 iterations in average). The average time for each of the 10 trainings has been 4.66 seconds, using a PC laptop with Intel Pentium IV 2.8 GHz processor.

The resulting confusion matrix is shown in Table 2. This matrix shows in rows the real category to where a cardiac cycle belongs to (C_i), and in columns the category where it has been classified by the ANN (C_i'). Ideally, for a perfect classification, the main diagonal cells should be 100%, and the rest of cells should be 0%. Percentage of results over the number of records for each category has been also indicated in brackets.

Table 2. Confusion matrix where results have been averaged for 10 trainings. Rows (C_i) represent the real category of the pattern, and columns (C_i') represent the category where the pattern has been assigned by the classifier. Percentage is shown in brackets.

Class	C_1'	C_2'	C_3'
C_1	31.0 (100.00%)	0.0 (0.00%)	0.0 (0.00%)
C_2	0.2 (0.77%)	24.1 (92.69%)	1.7 (6.54%)
C_3	0.1 (0.27%)	0.8 (2.16%)	36.1 (97.57%)

Table 2 shows that all the cardiac cycles of class C_1 have been classified correctly. There is also a small percentage of cycles with murmur (0.77% of patterns in C_2, and 0.27% of patterns in C_3) that have been incorrectly assigned to class C_1. The hit rate for the classification has been of 92.69% and 97.57% for classes C_2 and C_3, respectively. 6.54% of the cardiac cycles with HSM were classified as MSM (in class C_3), whereas 2.16% of the cycles with MSM where classified as HSM (in class C_2).

5 Conclusions

The description of pattern recognition for phonocardiography using an artificial neural network as classifier has been presented in this article. Classification is performed for a three category set of records: normal, HSM, and MSM records. After basic preprocessing and individual events detection, a set of 13 features is extracted for each event. PCA has been used to remove redundant information and to select the 5 most discriminant features. The analysis made with PCA revealed that features extracted from the instantaneous frequency envelope were among the most discriminant features, allowing clear separation between non-murmur and murmur records, and between holosystolic and midsystolic murmurs. Finally, a multilayer perceptron was used to classify the feature vector for each cardiac cycle. The Levenberg-Marquardt training algorithm has provided good convergence, and target error was reached after very few iterations. Results achieved a very high hit rate, with 100% correct classification for normal records.

It is worth noting that the diagnosis system uses only the PCG signal, so signal acquisition remains as simple as possible, without needing additional synchronization signals. Although results have been very promising for three categories, it is planned to increase the record database with more pathologies and with more records per pathology, trying to cover the most frequent valvular pathologies.

Acknowledgments

This work has been supported by Ministerio de Ciencia y Tecnología of Spain under grants TIC2003-09400-C04-02 and TIN2006-15460-C04-04.

References

1. Watrous, R.L.: Computer-aided auscultation of the heart: From anatomy and physiology to diagnostic decision support. In: Proceedings of the 28th IEEE EMBS Annual International Conference, New York City (USA) (2006) 140–143
2. de Vos, J.P., Blanckenberg, M.M.: Automated pediatric cardiac auscultation. IEEE Transactions on Biomedical Engineering **54**(2) (February 2007) 244–252
3. Martínez-Alajarín, J., López-Candel, J., Ruiz-Merino, R.: ASEPTIC: Aided system for event-based phonocardiographic telediagnosis with integrated compression. Computers in Cardiology **33** (2006) 537–540
4. Martínez-Alajarín, J., Ruiz-Merino, R.: Wavelet and wavelet packet compression of phonocardiograms. Electronics Letters **40**(17) (2004) 1040–1041
5. Hagan, M.T., Menhaj, M.B.: Training feedforward networks with the Marquardt algorithm. IEEE Transactions on Neural Networks **5**(6) (November 1994) 989–993
6. Ölmez, T., Dokur, Z.: Classification of heart sounds using an artificial neural network. Pattern Recognition Letters **24** (2003) 617–629
7. Turkoglu, I., Arslan, A., Ilkay, E.: An intelligent system for diagnosis of the heart valve diseases with wavelet packet neural networks. Computers in Biology and Medicine **33** (2003) 319–331
8. Gupta, C.N., Palaniappan, R.: Neural network classification of homomorphic segmented heart sounds. Applied Soft Computing **7** (2007) 286–297
9. Sharif, Z., Zainal, M.S., Sha'ameri, A.Z., Salleh, S.H.S.: Analysis and classification of heart sounds and murmurs based on the instantaneous energy and frequency estimations. In: Proceedings of the IEEE TENCON. Volume 2. (2000) 130–134
10. Liang, H., Lukkarinen, S., Hartimo, I.: Heart sound segmentation algorithm based on heart sound envelogram. In: Computers in Cardiology. (1997) 105–108
11. Martínez-Alajarín, J., Ruiz-Merino, R.: Efficient method for events detection in phonocardiographic signals. Proceedings of SPIE **5839** (June 2005) 398–409
12. Haghighi-Mood, A., Torry, J.N.: A sub-band energy tracking algorithm for heart sound segmentation. In: Computers in Cardiology. (1995) 501–504
13. Mason, D.: Listening to the heart. F. A. Davis Co. (2000)

Requirements for Machine Lifelong Learning

Daniel L. Silver and Ryan Poirier

Jodrey School of Computer Science
Acadia University
Wolfville, Nova Scotia, Canada B4P 2R6
danny.silver@acadiau.ca

Abstract. A system that is capable of retaining learned knowledge and selectively transferring portions of that knowledge as a source of inductive bias during new learning would be a significant advance in artificial intelligence and inductive modeling. We define such a system to be a machine lifelong learning, or ML3 system. This paper makes an initial effort at specifying the scope of ML3 systems and their functional requirements.

1 Introduction

Over the last ten years progress has been made in machine learning and statistical modeling that exhibit aspects of knowledge retention and inductive transfer. These represent advances in inductive modeling that move beyond *tabula rasa* learning and toward machines capable of lifelong learning [17]. Henceforth, this article will refer to such as *machine lifelong learning* (ML3) systems. Despite the progress that has been made, there is need for a clear definition of the knowledge retention and inductive transfer problem. Toward that end, this paper makes an initial effort at specifying the scope of ML3 systems and their functional requirements.

2 Scope of ML3 Systems

The constraint on a learning system's hypothesis space, beyond the criterion of consistency with the training examples, is called *inductive bias* [4]. An inductive bias of a learning system can be expressed as the system's preference for one hypothesis over another, for example Occam's Razor suggests a bias for simple over more complex hypotheses. Inductive bias is essential for the development of a hypothesis with good generalization from a practical number of examples [5]. Ideally, a lifelong learning system can select its inductive bias to tailor the preference for hypotheses according to the task being learned [18]. One type of inductive bias is knowledge of the task domain. The retention and use of domain knowledge as a source of inductive bias remains an unsolved problem in machine learning.

In [14,15] *knowledge-based inductive learning* is defined as an ML3 approach that uses knowledge of the task domain as a source of inductive bias. Figure 1

J. Mira and J.R. Álvarez (Eds.): IWINAC 2007, Part I, LNCS 4527, pp. 313–319, 2007.

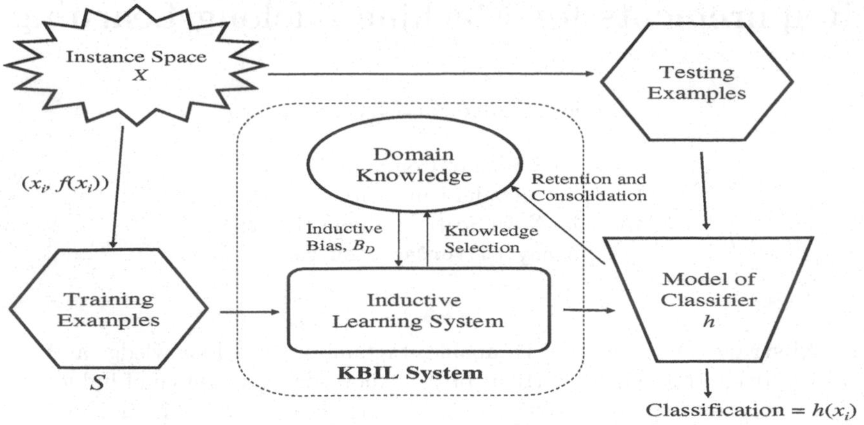

Fig. 1. The framework for knowledge based inductive learning

provides the framework for knowledge based inductive learning. As with a standard inductive learner, training examples are used to develop a hypothesis of a classification task. However, unlike a standard learning system, knowledge from each hypothesis is saved in a long-term memory structure called *domain knowledge*. When learning a new task, aspects of domain knowledge are selected to provide a positive inductive bias to the learning system. The result is a more accurate hypothesis developed in a shorter period of time. The method relies on the transfer of knowledge from one or more prior secondary tasks, stored in domain knowledge, to the hypothesis for a new primary task. The problem of selecting an appropriate bias becomes one of selecting the most related task knowledge for transfer.

An ML3 system is typically composed of short-term and long-term components and/or exhibits short-term and long-term processes. Although two phases of learning may not be necessary, it is frequently required so as to ensure that long-term domain knowledge is not corrupted by inaccurate short-term learning. The following three sections outline general requirements for ML3 systems and specific requirements for long-term retention of learned knowledge and short-term learning with inductive transfer.

3 General Requirements

3.1 Form of knowledge Retention

Learned knowledge can be stored in functional or representational form within a ML3 [14]. The simplest method of retaining task knowledge in functional form is to save the respective training examples. Other methods of retaining functional knowledge involve the storage or modelling of search parameters such as the learning rate in neural networks. An advantage of retaining functional knowledge, particularly the retention of the actual training examples, is the accuracy and purity of the knowledge. Disadvantages of retaining functional knowledge are the large amount of storage space that it requires and difficulties in using such knowledge during future learning.

Alternatively, a description of an accurate hypothesis developed from the training examples can be retrained. We define this to be a representational form of knowledge retention. The description of a decision tree or a neural network are examples of representations. The advantages of retaining representational knowledge is its compact size relative to the space required for the original training examples and its ability to generalize beyond those examples. The disadvantage of retaining representational knowledge is the potential loss of accuracy from the original training examples.

3.2 Form of knowledge Transfer

The form in which task knowledge is retained can be separated from the form in which it is transferred. For example, the retained hypothesis representation for a learned task can be used to generate functional knowledge in the form of training examples [10,15].

Representational transfer involves the direct or indirect assignment of known task representation to the model of a new target (or primary) task [14]. In this way the learning system is initialized in favour of a particular region of hypothesis space of the modeling system [9,11,16]. Representational transfer often results in substantially reduced training time with no loss in the generalization performance of the resulting hypotheses.

In contrast to representational transfer, functional transfer employs the use of implicit pressures from training examples of related tasks [1], the parallel learning of related tasks constrained to use a common internal representation [2,3], or the use of historical training information from related tasks [17,6]. These pressures reduce the effective hypothesis space in which the learning system performs its search. This form of transfer has its greatest value in terms of increased generalization performance from the resulting hypotheses.

3.3 Input and Output Type, Complexity and Cardinality

The output representation of a system capable of retaining and transferring knowledge should not be constrained to a particular data type. A ML3 system should be capable of predicting class categories and real-value outputs including scalar values as well as vectors.

An ML3 should be capable of dealing with its environment over a lifetime with a fixed number of inputs and outputs for the task domain(s) under study. Certain inputs or outputs might go unused for many tasks of a domain early in the learning system's lifetime only to be used quite frequently later in life. The rationale for this requirement is not to constrain an ML3 system to a fixed amount of internal representation (this could change over time) but to ensure a consistent interface with the environment and with other entities such as a software agent, a application program or a human user.

3.4 Scalability

A ML3 system must be capable of scaling up to large numbers of inputs, outputs, training examples and learning tasks. Preferably, both the space and time complexity of the learning system grows polynomially in all of these factors.

3.5 Accumulation of Practice

A ML3 system should facilitate the practice of a task. The system's normal methods should retain and transfer knowledge from one learning episode of a task to another such that the generalization accuracy of the long-term hypothesis for the task increases. But, how can a ML3 system determine from the training examples that it is practicing a task it has previously learned versus learning a new but closely related task [7,12]. We have come to the conclusion that a ML3 system should not have to be explicit in this determination. Rather, the similarity, or relatedness, of a set of training examples to that of prior domain knowledge should be implicit; each training example should be able to draw upon those aspects of domain knowledge that are most related. This suggests that domain knowledge should be seen as continuum as apposed to a set of disjoint tasks.

4 Requirements for Long-Term Retention of Learned Knowledge

4.1 Effective Retention

A ML3 system should resist the introduction and accumulation of domain knowledge error. Only hypotheses with an acceptable level of generalization accuracy should be retained else, once saved in long-term memory, the error from a hypothesis may be transferred to future hypotheses. A ML3 system must be concerned with this systemic growth in error over its lifetime. Similarly, The process of retaining a new hypothesis should not reduced its accuracy or that of prior hypotheses existing in long-term memory. In fact, the integration or consolidation of new task knowledge should increase the accuracy of related prior knowledge.

4.2 Efficient Retention

A ML3 system should be efficient in its use of long-term memory (efficient in space). In particular, the system should make use of memory resources such that the duplication of information is minimized. A representational form of task knowledge will be more space efficient than a functional form because of the reasons cited in Section 3.1. A ML3 system should also be computationally efficient (efficient in time) when storing learned knowledge in long-term memory. Ideally, retention should occur during short-term learning, however, in order to ensure effective retention (integration and reduction of error) this is rarely possible.

4.3 Effective Indexing

A ML3 must be capable of selecting the appropriate prior knowledge for inductive transfer during short-term learning. This requires that a ML3 be capable of indexing into long-term memory for task knowledge that is most related to the primary task. Typically, primary task knowledge will arrive in the form of training examples (functional knowledge) and no representational knowledge will be provided. This requires design choices in the construction of the ML3 system. The system must either use functional examples to select related domain knowledge or generate a hypothesis representation for the primary task to estimate its similarity to existing domain knowledge representation.

4.4 Efficient Indexing

A ML3 system must make the selection of related knowledge as rapid as possible. Preferably, the computational time for indexing into domain knowledge should be no worse than polynomial in the number of tasks having been stored. Experimentation has shown that a representational form of retained knowledge (*e.g.* graph of a decision tree) can be more efficiently indexed than a functional form (*e.g.* examples used to train the decision tree) [8].

4.5 Meta-knowledge of the Task Domain

In most cases, it will be necessary for a ML3 system to determine and retain meta-knowledge of the task domain. For example, it may be necessary to estimate the probability distribution over the input space so as to manufacture appropriate functional examples from retained task representation [15]. Alternatively, it may be necessary to retain characteristics of the learning process (learning curve, error rate) for each task.

5 Requirements for Short-Term Learning with Inductive Transfer

5.1 Effective Learning

The inductive transfer (bias) from long-term memory should never decrease the generalization performance of a hypothesis developed by a ML3 system. A ML3 system should produce a hypothesis for the primary task that meets or exceeds the generalization performance of that developed strictly from the training examples. There is evidence that the functional form of knowledge transfer somewhat surpasses that of representation transfer in its ability to produce more accurate hypotheses [3,13]. Starting from a prior representation can limit the development of novel representation required by the hypothesis for the primary task. In terms of neural networks this representational barrier manifests itself in terms of local minimum.

5.2 Efficient Learning

Inductive transfer from long-term memory should not increase the computational time for developing a hypothesis for the primary task as compared to using only the training examples. In fact, inductive transfer should reduce training time. In practice this reduction is rarely observed because of the computation required to index into prior domain knowledge. In terms of memory (space), there will typically be an increase in complexity as prior domain knowledge must be used during the learning of the new task. Our research has shown that a representational form of knowledge transfer will be more efficient than a functional form (supplemental training examples) [13].

Sections 4.3, 4.4, 5.1 and 5.2 indicate an interesting dichotomy between effective and efficient inductive transfer. Effective learning requires functional transfer whereas efficient learning requires representation transfer.

5.3 Transfer Versus Training Examples

A ML3 must take into consideration the estimated sample complexity and number of available examples for the primary task and the generalization accuracy and relatedness of retained knowledge in long-term memory. During the process of inductive transfer a ML3 must weigh the relevance and accuracy of retained knowledge along side that of the information resident in the training examples.

6 Conclusion

This paper has outlined the scope and functional requirements for a ML3 system. A ML3 system can retain and transfer knowledge in either representational or functional form. A ML3 system should have no bounds on input and output variable type and complexity and it should be scalable in terms of number of inputs, outputs, number of training examples and learning tasks. A ML3 should facilitate the practice of a task and treat domain knowledge as a continuum of tasks rather than a set of disjoint tasks.

Efficient long-term retention of learned knowledge should cause no loss of prior task knowledge, no loss of new task knowledge, and an increase in the accuracy of old tasks if the new task being retained is related. A ML3 must be capable of efficiently selecting the most effective prior knowledge for inductive transfer during short-term learning.

Efficient short-term learning with inductive transfer should produce a hypothesis for a primary task that meets or exceeds the generalization performance of a hypothesis developed from only the training examples. Experimental results indicate that effective learning excels under functional transfer whereas efficient learning requires representation transfer. Lastly, we point out that a ML3 must weigh the relevance and accuracy of retained knowledge along side that of the available training examples for the primary task.

References

1. Yaser S. Abu-Mostafa. Hints. Neural Computation, 7:639671, 1995.
2. Jonathan Baxter. Learning internal representations. Proceedings of the Eighth International Conference on Computational Learning Theory, 1995.
3. Richard A. Caruana. Multitask learning. Machine Learning, 28:4175, 1997.
4. Tom. M. Mitchell. The need for biases in learning generalizations. Readings in Machine Learning, pages 184191, 1980. ed. Jude W. Shavlik and Thomas G. Dietterich.
5. Tom M. Mitchell. Machine Learning. McGraw Hill, New York, NY, 1997.
6. D.K. Naik and Richard J. Mammone. Learning by learning in neural networks. Artificial Neural Networks for Speech and Vision, 1993.
7. Robert OQuinn. Knowledge Transfer in Artificial Neural Networks. Honours Thesis, Jodrey School of Computer Science, Acadia University,Wolfville, NS, 2005.
8. Ryan Poirier and Daniel L. Silver. Effect of curriculum on the consolidation of neural network task knowledge. Proc. of IEEE International Joint Conf. on Neural Networks (IJCNN 05), 2005.
9. Mark Ring. Learning sequential tasks by incrementally adding higher orders. Advances in Neural Information Processing Systems 5, 5:155122, 1993. ed. C. L. Giles and S. J. Hanson and J.D. Cowan.
10. Anthony V. Robins. Catastrophic forgetting, rehearsal, and pseudorehearsal. Connection Science, 7:123146, 1995.
11. Jude W. Shavlik and Thomas G. Dietterich. Readings in Machine Learning. Morgan Kaufmann Publishers, San Mateo, CA, 1990.
12. Daniel L. Silver and Ricchard Alisch. A measure of relatedness for selecting consolidated task knowledge. Proceedings of the 18th Florida Artificial Intelligence Research Society Conference (FLAIRS05), pages 399404, 2005.
13. Daniel L. Silver and Peter McCracken. Selective transfer of task knowledge using stochastic noise. In Yang Xiang and Brahim Chaib-draa, editors, Advances in Artificial Intelligence, 16th Conference of the Canadian Society for Computational Studies of Intelligence (AI2003), pages 190205. Springer-Verlag, 2003.
14. Daniel L. Silver and Robert E. Mercer. The parallel transfer of task knowledge using dynamic learning rates based on a measure of relatedness. Connection Science Special Issue: Transfer in Inductive Systems, 8(2):277294, 1996.
15. Daniel L. Silver and Robert E. Mercer. The task rehearsal method of life-long learning: Overcoming impoverished data. Advances in Artificial Intelligence, 15th Conference of the Canadian Society for Computational Studies of Intelligence (AI2002), pages 90101, 2002.
16. Satinder P. Singh. Transfer of learning by composing solutions for elemental sequential tasks. Machine Learning, 1992.
17. Sebastian Thrun. Lifelong learning algorithms. Learning to Learn, pages 181209.1 Kluwer Academic Publisher, 1997.
18. Paul E. Utgoff. Machine Learning of Inductive Bias. Kluwer Academc Publisher, Boston, MA, 1986.

Multitask Learning with Data Editing*

Andrés Bueno-Crespo[1], Antonio Sánchez-García[2],
Juan Morales-Sánchez[3], and José-Luis Sancho-Gómez[3]

[1] Dpto. Informática de Sistemas,
Universidad Católica San Antonio, Murcia, Spain
abueno@pdi.ucam.edu
[2] Área Técnica de Estudios Avanzados y Tratamiento Digital de Señales,
S.A. de Electrónica Submarina (SAES), Cartagena (Murcia), Spain
[3] Dpto. Tecnologías de la Información y las Comunicaciones,
Universidad Politécnica de Cartagena, Cartagena (Murcia), Spain

Abstract. In real life, the task learning is reinforced by the related tasks that we have learned or that we learn at the same time. This scheme applied to Artificial Neural Networks (ANN) is known with the name of Multitask Learning (MTL). So, the information coming from the related secondary tasks provide a bias to the main task, which improves its performances versus a Single-Task Learning (STL) scheme. However, this implies a bigger complexity. Data Editing procedures are used to reduce the algorithmic complexity, obtaining an outstanding samples set from the original set. This edited set gets the performance very fast. In this paper we combine MTL with Data Editing, so we can approach the small samples set training in an MTL scheme.

1 Introduction

Multitask Learning (MTL) allows to learn a task (main task) using information from secondary tasks related with the main one. The purpose of these secondary tasks is to help to the main task to improve its performances. Caruana et al. [1] use the MTL procedure to train an ANN to order the pneumonia patients list, obtaining better results that with the STL scheme. Caruana also proposes other alternatives of algorithms for MTL [2,3,4]. Ghosn and Bengio [5] work about the bias learning and how these tasks make weight share domain. Silver and Mercer create relations between tasks to know the best task that can be used as secondary task in the MTL training process [6].

Unfortunately, MTL increments the algorithm complexity. In this sense, sample selection procedures may be used to obtain new samples sets to work in a MTL scheme improving the performance with a small samples set. The proposed method combine the fast performances of Data Editing procedures and the information provided from the related tasks. Using related tasks to train an ANN has

* This work is partially supported by Ministerio de Educación y Ciencia under grant TEC2006-13338/TCM, and by Consejería de Educación y Cultura de Murcia under grant 03122/PI/05.

J. Mira and J.R. Álvarez (Eds.): IWINAC 2007, Part I, LNCS 4527, pp. 320–326, 2007.

been proved to be useful to improve the main task performances. This is because information provided by the secondary tasks is used as an inductive bias for the main task [5]. Thus, ANNs are trained using samples from all different tasks, producing a bias in the solution obtained when only samples of the main task are used. In this work, we use the MTL method developed by Caruana [1,2,3,4]. Figure 1 shows this MTL scheme, where all tasks (each one of them is associated to a network output) share a hidden layer of neurons in an ANN.

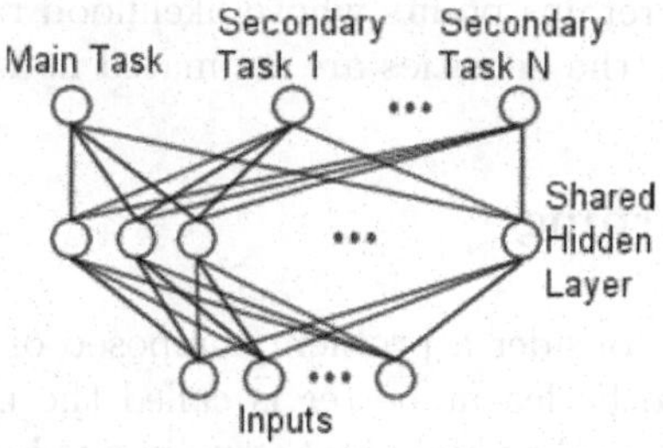

Fig. 1. MTL scheme. All tasks share the hidden layer, and the N secondary tasks help the main one to improve its performance.

The rest of the paper is organized as follows. Section 2 explains the Data Editing procedure. In Section 3, we set a MTL method with Data Editing. Section 4 presents the experiment results over a two-class classification problem showing the performance improvement of our proposed approach compared to other MTL and STL procedures. Finally, the main conclusions end this paper.

2 Data Editing

Data Editing has been developed to avoid a common drawback about both Kernel and K-Nearest Neighbor (K-NN) methods. These procedures take too long to compute and need too much storage for the whole training set. However, in many problems, it is only necessary to retain a small fraction of the training set to approximate very well the decision boundary of the K-NN classifier. There are many editing algorithms. Ripley [7] shows the performance of Data Editing in a STL scheme. Choi et al. [8] work with a condensed set showing some modifications from the original Data Editing procedure. Here, we use the Condensed-NN algorithm proposed by Hart [9], and it is specified as follows:

1. Divide the current patterns into a store and a grabbag. One possible partition is to put the first point in the store, the rest in the bag.
2. Classify each sample in the grabbag by the 1-NN rule using the store as training set. If the sample is incorrectly classified transfer it to the store.
3. Return to Step 2 unless no transfer occurred or the grabbag is empty.
4. Returns the store.

The selected samples by the Condensed-NN algorithm can be used to form a new secondary task. The samples of this reduced set are very close to the decision frontier, and so they will performance very fast, helping the principal task performance. In some cases, a lot of overlapping samples are in the decision frontier. So we need to combine the well-known $\{K, L\}$-NN algorithm and the Condensed-NN algorithm described previously. The $\{K, L\}$-NN is used to remove the overlapping data so the Condensed-NN algorithm obtains a better outstanding group of samples. To our purpose we also use the Hand & Batchelor algorithm [10]. This method only retains points whose likelihood ratio against every class exceeds some threshold (the densities are estimated non-parametrically).

3 Multitask Learning

In this work, we firstly consider a problem composed of two related tasks. The task which is desired to be learnt better is called the main task and the task whose training data are used as hints by the main one is called as the secondary task. In order to make the content of the paper easier, the nomenclature used in the following sections is showed in the Table 1.

Table 1. Nomenclature of mathematical symbol used in this paper

$\mathbf{X} = \{\mathbf{x}^n\}_{n=1\dots N}^N$	Input set
$\mathbf{T} = \{t^n\}_{n=1\dots N}^N$	Target set
$\mathcal{M} = \{\mathbf{X}^m, \mathbf{T}^m\}$	Data set to learning the main task
$\mathcal{S} = \{\mathbf{X}^s, \mathbf{T}^s\}$	Data set to learning the secondary task
$\mathbf{x}^n = [x_1^n, x_2^n, \dots, x_d^n]$	Input vector
t_k^n	Target of the $\mathbf{x}^n$ corresponding to the k-th task
y_k^n	Network output of the $\mathbf{x}^n$ corresponding to the k-th task
z_j^n	Output for the j-th neuron corresponding to the $\mathbf{x}^n$

In a STL framework, an ANN is trained to learn only the main task, i.e., the network is trained using only the information corresponding to the main task, the set $\mathcal{M}$. In contrast, when MTL is used, an ANN learns all tasks at the same time using all training data, i.e., the sets $\mathcal{M}$ and $\mathcal{S}$. In this work, we also consider that the input data are the same for both tasks, i.e., $\mathbf{X}^m = \mathbf{X}^s = \mathbf{X}$.

When we use Data Editing to reduce the training data set corresponding to the main task, i.e., the set $\mathcal{M}$ is reduced to $\mathcal{M}_{ed}$, the STL presents some difficulties to get an optimal performance. To reinforce it, we propose an MTL framework, in which the information provided by a secondary related task is used to improve the performance of the main one (the extension to more tasks is obvious). In this approach, the secondary task is obtained from the edited set $\mathcal{S}_{ed}$. This information works like a bias for the main task, so it is necessary that the secondary task must be related with the main one.

The original MTL scheme is modified to handle all tasks. In our approach, data samples are divided into two sets, the data set for the main task ($\mathcal{M}$ or $\mathcal{M}_{ed}$ if the edited sets are used) and the data set corresponding to the secondary task ($\mathcal{S}$ or $\mathcal{S}_{ed}$). Most input vectors belong to both sets, but there are samples that belong only to the main task and others to the secondary related tasks. So, some modifications to the original Back-Propagation (BP) algorithm are necessaries. If we use linear activation functions in the output units, and considering the sum-of-squares error as the error function to be minimized during the learning process, the error corresponding to the n-th pattern is given by:

$$E^n = \frac{1}{2}\sum_{k=1}^{c}(y_k^n - t_k^n)^2 \quad (1)$$

where $c - 1$ is the number of secondary tasks related with the main one.

As it has been already mentioned, network weights are calculated using BP algorithm with gradient descent optimization. We use this notation for weights: w_{ji} denotes a weight going from the input unit i to hidden unit j; and so, w_{kj} denotes a weight in the second layer going from hidden neuron j to output unit k. The derivatives of (1) with respect to the first-layer and second-layer weights are respectively given by,

$$\frac{\partial E^n}{\partial w_{ji}} = \delta_j^n x_i^n, \qquad \frac{\partial E^n}{\partial w_{kj}} = \delta_k^n z_j^n \quad (2)$$

where δ's are the *errors* used for the weights actualization in the BP algorithm.

Weights that connect each output unit to hidden neurons are only influenced by errors produced by the corresponding task, being updated by

$$\delta_k^n = y_k^n - t_k^n \quad (3)$$

While the first layer weights are updated depending on the error of all tasks:

$$\delta_j^n = z_j^n(1 - z_j^n)\sum_{k=1}^{c}\alpha_k^n w_{kj}\delta_k^n \quad (4)$$

where α_k^n represents the k-th bias for the n-th input pattern. This bias α_k^n is equal to 0 or 1 depending on the n-th pattern is associated to the k-th task. For example, if the n-th vector belongs to all sets (i.e., it belongs to the main task and also to all secondary tasks), α_k^n is equal to 1 for all k.

In this paper, the experiment is composed of two tasks ($c = 2$), and so, the implemented MTL network has two outputs, $k = 1$ for the main task and $k = 2$ for the secondary task. Therefore, the equation (4) can be written as

$$\delta_j^n = z_j^n(1 - z_j^n)(\alpha_1^n w_{j,1}\delta_1^n + \alpha_2^n w_{j,2}\delta_2^n) \quad (5)$$

4 Experiment

In order to test the proposed approach, a two-class two-dimensional problem is used. This problem has been artificially generated, resulting two different sets,

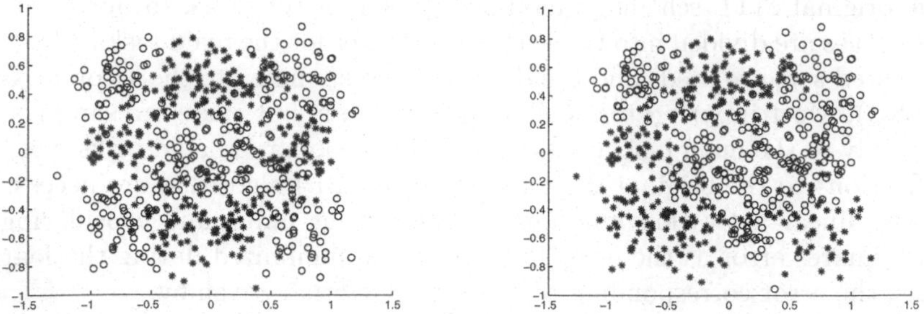

(a) Main task. (b) Secondary related task.

Fig. 2. Training sets of a two-class decision problem

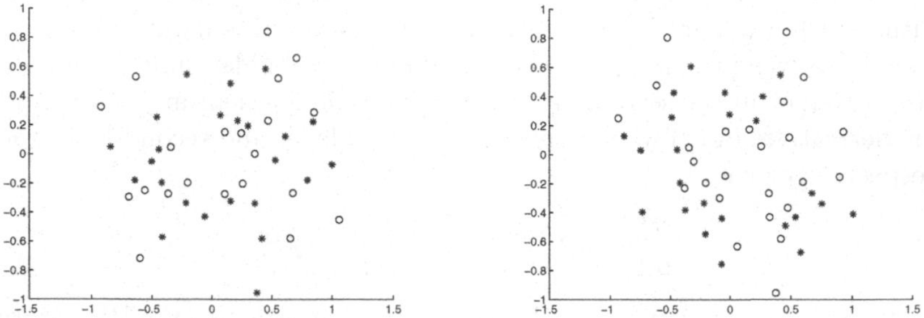

(a) Selected samples from the main task. (b) Selected samples from the related secondary task.

Fig. 3. Selected samples from data editing procedure used for the main task and the secondary one

one of them is used as the main task and the another set as the secondary one. Figure 2(a) shows the set corresponding to the main task, and Figure 2(b) the data for the secondary one. It is clear to see that both tasks are clearly related, and also, most of the input data belong to the two data sets. In each case, we have 600 samples for the training set, 600 samples for the validation set, and 6000 samples for the test set. The two classes are equiprobable in all sets.

In addition, two reduced sets are obtained with Data Editing. Figure 3 shows these edited data sets. From the 600 samples that compose the training sets (Figure 2), the number of samples have decreased to 88, where 45 patterns belong at the main task (Figure 3(a)) and 49 patterns to the secondary one ((Figure 3(b)). How it can be appreciated, the edited set (Figure 3) is a small fraction of the original set (Figure 2) composed by samples close to the boundary. The error propagation has not repercussion if the sample does not belong to a determinate task. This reduction of the data set makes the performance very fast.

We compare the the main task performance of the STL and MTL over all sets (i.e. non-edited and edited data sets). The implemented networks have two inputs, one output for the STL scheme and two outputs for the MTL scheme, and 12 nodes in the hidden layer. The learning parameters are the following. The base learning rate is dynamic, it initial value is 0.05 with an increment of 1.1 or a decrement of 0.5 depending on the performance of the validation set. Random initial weight values are selected in the range −0.1 to 0.1 for all runs. This experiment is repeated ten times, with 20.000 epochs for each one of them.

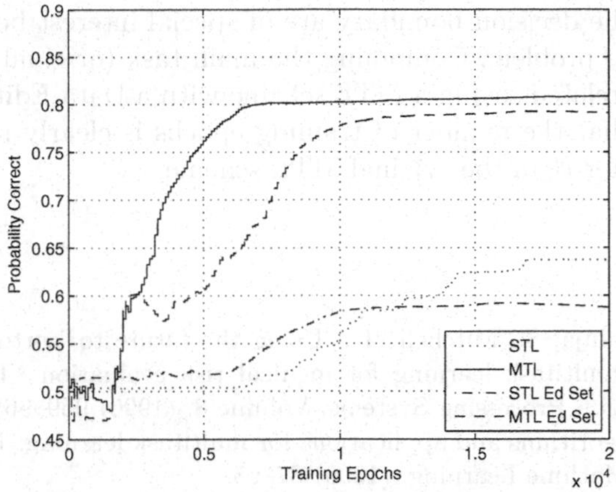

Fig. 4. Training epochs vs. probability correct (measured over the test set) from the different tested schemes

Figure 4 illustrates the evolution of the probability of correct classification over the test set using the different tested procedures. This figure shows that the MTL with Data Editing scheme improve the performance of the STL scheme using Data Editing, and it maintains the advantage of quick performance.

In Table 2, we show the obtained results of the different training schemes. These results, measured over the test set, have been selected using a validation set. MTL approach with edited sets gets good results working with a small samples set, reducing the training epochs and the algorithm complexity. The performance is very fast due to the small samples set.

Table 2. Correct probability average (over the test set) and standard deviation from the different tested schemes

Scheme	*%Correct*	*std*	*Epochs*	*Best*
STL	0.637	0.001	19990	0.640
MTL	0.808	0.003	16742	0.814
STL Ed-Set	0.588	0.004	6762	0.594
MTL Ed-Set	0.761	0.025	7635	0.782

5 Conclusions

Multitask Learning (MTL) is a procedure for training a neural network that learns several related tasks simultaneously considering one of them as main task and the others as secondary tasks. In this paper, a Data Editing procedure is used to reduce the algorithm complexity. In particular, Data Editing has been implemented by means of {K,L}-NN and Condensed-NN algorithms. Data Editing is a sample selection procedure that gets a reduced set of critical samples. Samples close to the decision boundary are of special interest because they are critical to solve the problem. Combining the main task (original data set) with a secondary task (related set) in a MTL scheme with a Data Editing procedure, we have checked that the number of training epochs is clearly reduced and its performance is faster than the original MTL scheme.

References

1. Caruana, R., Baluja, S., Mitchell, T.: Using the future to 'sortout' the present: Rankprop and multitask learning for medical risk evaluation. In: Advanced in Neural Information Processing Systems. Volume 8. (1996) 959–965
2. Caruana, R.: Algorithms and applications for multitask learning. In: International Conference on Machine Learning. (1996) 87–95
3. Caruana, R.: Multitask learning. Machine Learning **28** (1997) 41–75
4. Caruana, R.: Learning many related tasks at the same time with backpropagation. In: Advanced in Neural Information Processing Systems. (1995) 656–664
5. Ghosn, J., Bengio, Y.: Bias learning, knowledge sharing. IEEE Transactions on Neural Networks **14** (2003) 87–95
6. Silver, D.L., Mercer, R.E.: Selective functional transfer: Inductive bias from related tasks. In: Proceedings of the IASTED International Conference on Artificial Intelligence and Soft Computing, Cancun, Mexico, ACTA Press (2001) 182–191
7. Ripley, B.D.: Pattern Recognition and Neural Networks. Cambridge University Press, New York, USA (1996)
8. Rockett, P., Choi, S.: The training of neural classifiers with condensed datasets. IEEE Transactions on Systems, Man, and Cybernetics - part B: Cybernetics **32** (2002) 202–206
9. Hart, P.E.: The condensed nearest neighbour rule. IEEE Transactions on Information Theory **14** (1968) 515–516
10. Hand, B.G., Batchelor, D.J.: An edited condensed nearest neighbour rule. Information Sciences **14** (1978) 171–180

Efficient BP Algorithms for General Feedforward Neural Networks*

S. España-Boquera[1], F. Zamora-Martínez[2],
M.J. Castro-Bleda[1], and J. Gorbe-Moya[1]

[1] DSIC, Universidad Politécnica de Valencia, Valencia Spain
{sespana,mcastro,jgorbe}@dsic.upv.es
[2] LSI, Universitat Jaume I, Castellón Spain
fzamora@guest.uji.es

Abstract. The goal of this work is to present an efficient implementation of the Backpropagation (BP) algorithm to train Artificial Neural Networks with general feedforward topology. This will lead us to the "consecutive retrieval problem" that studies how to arrange efficiently sets into a sequence so that every set appears contiguously in the sequence. The BP implementation is analyzed, comparing efficiency results with another similar tool. Together with the BP implementation, the data description and manipulation features of our toolkit facilitates the development of experiments in numerous fields.

1 Introduction and Motivation

The Backpropagation (BP) algorithm [11] is the most widely used supervised learning technique to train feedforward Artificial Neural Networks (ANNs). In this work, we present an efficient implementation of the BP algorithm to train general feedforward ANNs, that is, networks that have no feedback.

There are many variations of the BP algorithm. The simplest implementation of BP learning updates the network weights and biases in the direction in which the performance function decreases most rapidly, the negative of the gradient. There are two different ways in which this gradient descent algorithm can be implemented: incremental mode and batch mode. In incremental mode, the gradient is computed and the weights are updated after each input is applied to the network. In batch mode, all the inputs are applied to the network before the weights are updated. Incremental training is usually significantly faster than batch training. On the other hand, adding a momentum term [10] is a standard technique that often provides faster convergence and maintain generalization performance. Momentum allows a network to respond not only to the local gradient, but also to recent trends in the error surface: without momentum a network can get stuck in a shallow local minimum and with momentum a network can slide through such a minimum.

By the above reasons, we have decided to implement the incremental BP algorithm with momentum with some characteristics that most of the available BP implementations lack. Section 2 describes in more detail the most important features of our BP

* This work has been partially supported by the Spanish Government under contract TIN2006-12767 and by the Generalitat Valenciana under contract GVA06/302.

J. Mira and J.R. Álvarez (Eds.): IWINAC 2007, Part I, LNCS 4527, pp. 327–336, 2007.

implementation regards to efficiency. This will lead us to the "consecutive retrieval problem" [7,4] that studies how to arrange efficiently a list of sets of neurons in a vector so that every set appears contiguously in the vector. Next, Section 3 lists some additional characteristics of the BP implementation. Section 4 describes the data facility mechanisms and presents an example of use of the application. Finally, some conclusions and future work are drawn in Section 5.

2 An Efficient BP Algorithm for General Feedforward ANNs

2.1 Preprocessing the ANN: The Consecutive Retrieval Problem

The bottleneck in the simulation of a neural network, at least for big networks with many connections, is the scalar product of vectors: the input vector $\overrightarrow{x}$ and the weight vector $\overrightarrow{w}$ of each neuron, that must be executed once by input sample and unit. We must take into account that the *forward* pass and the *backward* pass present different memory access patterns, so a representation that improves the locality in a pass will not do it in the other one. When the networks to be trained are big enough, the weights cannot be stored entirely in cache memory during a complete forward or backward pass, and this causes an important speed reduction due to the inevitable cache misses.

There seems to be a tradeoff between efficiency and flexibility: On the one hand, some implementations are specialized in certain restricted topologies like the layered feedforward ANNs with all-to-all connections between layers. For these specialized topologies, storing the connection weights in matrices improves the calculation of the scalar product by favoring data locality and simplifying data access. On the other hand, algorithms prepared for general feedforward topologies usually represent the networks by means of a list of neurons arranged in a topological order. Each neuron needs information of its predecessors and this list representation simplifies the algorithm, but the cost of traversing the data and their locality are worse than with the matrix representation.

Our proposed implementation is able to achieve the speed of specialized BP algorithms for general feedforward topologies. The activation values of neurons are arranged in a vector as consecutively as possible by means of a preprocessing of the network topology. In this way, given a neuron, the activation values of its predecessors is a consecutive subvector which can be efficiently traversed. The weights of the neuron are stored in the same order so the scalar product of vectors $\overrightarrow{x}$ and $\overrightarrow{w}$ of each neuron, as well as the backpropagation of the error, are improved. In order to assure this property, some neuron values may need to be duplicated. The problem of finding the *optimal* arrangement is known in the literature as "consecutive retrieval problem" [7,4]:

> Let $X = \{1, 2, \ldots, N\}$ be a set, having P subsets denominated $C_1, C_2, \ldots, C_P$, not necessarily disjoint. The goal is to obtain a sequence $A = a_1, \ldots, a_k$ of elements of X so that every C_i appears in A as a contiguous subsequence, while keeping the length of A as small as possible.

This problem is related to our neural network arrangement problem as follows: X represents the set of neurons that have one or more outputs connected to another neuron. P

is the number of neurons which are not inputs. Each set C_i represents the set of inputs of each neuron. Finally, A is the desired vector of activation values of the neurons. The advantage of this representation is twofold. On the one hand, iterating over a vector is cheaper than over a linked list. On the other hand, weights are packed consecutively in memory and the spatial cost of pointers of linked list nodes is avoided, therefore reducing the number of cache misses. The cost of copying duplicated activation values is very cheap so the obtained benefits exceed the cost of the replication.

Since the consecutive retrieval problem is proven to be NP-Complete [7,3], we need an efficient method to obtain good approximations to the optimal solution in order to make practical the packing of big ANNs. We have designed a greedy algorithm [13] which achieves packing rates very superior to a "naive" consecutive arrangement.

2.2 Tests of Efficiency

The BP algorithm described above is a part of the `April` (A Pattern Recognizer in Lua) toolkit which has been implemented in `C++` and can be extended using the `Lua` [6] scripting language. `April` provides a homogeneous environment to perform pattern recognition tasks (ANNs, hidden Markov models, dynamic time warping, clustering, and others). Neural network experiments can be easily performed in `April` as described in Section 4.

We are going to compare empirically the temporal cost of `April` BP implementation with other BP implementation: the Stuttgart Neural Network Simulator (`SNNS`) [14], that is one of the most well-known tools for training ANNs. `SNNS` has a great descriptive capacity and allows to train networks with any topology. In order to compare both BP implementations, an example of OCR classification problem is used. The task consists of classifying handwritten digits standardized to 16×16 black and white pixels.

A total of 1 000 image are stored in a unique image in PNG format (see Figure 1). Since the ANN receives a digit, the input has 256 input values (the size of a image). There is an output neuron per digit class.

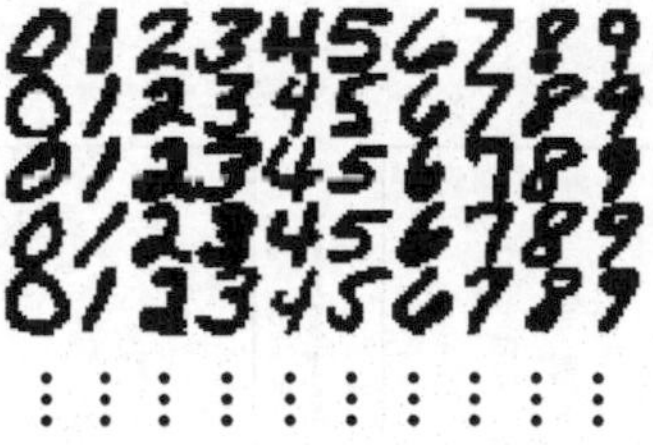

Fig. 1. The OCR corpus: $1\,600 \times 160$ pixels image containing a matrix of 100×10 image digits

Execution time of a complete training with validation epoch has been measured for both tools in an AMD Athlon (1 333 MHz.) with 384 MB of RAM under Linux. Similar results have been obtained in other architectures.

Efficiency with layered feedforward topologies. Layered feedforward networks with one and two hidden layers with all-to-all connections between layers have been used (see Figure 2), where the first hidden layer has from 10 to 200 neurons and the second layer, if used, has fewer neurons than the first one. Sigmoid activation functions have been used, and each network has been trained for 10 epochs. The networks have been generated and initialized with `April` and later have been exported to `SNNS` format in order to have the same initial weights.

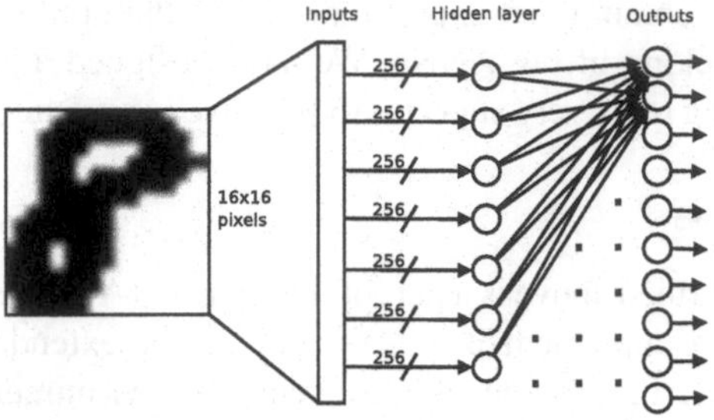

Fig. 2. Layered feedforward ANN used for efficiency tests. 1 and 2 hidden layers have been used.

Figure 3 shows the temporal cost per epoch for `April` and `SNNS` and the ratio between both costs. The first graph shows that the running time is always smaller in `April` than in `SNNS`. In the second graph, we can see that the speed-up achieved by `April` over `SNNS` varies between 4 and 16. For small networks, with less than 10 000 connections, `April` is 4 times faster. For bigger networks, `April` is up to 16 times faster (when the number of connections is between 20 000 and 30 000). With more than 30 000 connections, the difference between both tools decreases progressively until, at 50 000 connections, it becomes stable around `SNNS` being 4 times slower than `April`.

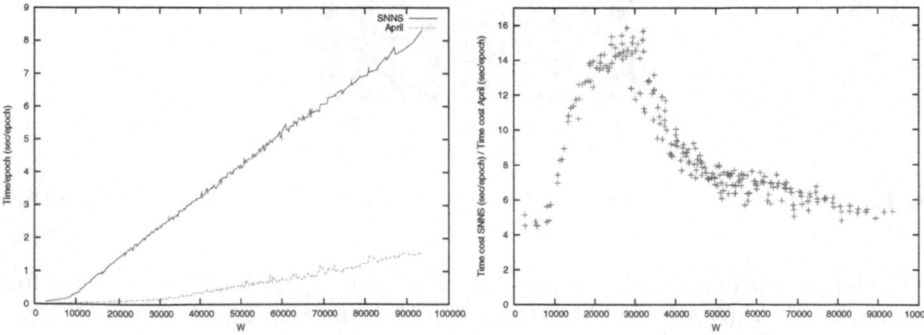

Fig. 3. Temporal cost by epoch (sec./epoch) of `April` and `SNNS` (left) and ratio between both costs (right)

Table 1. Cache misses in training networks with different number of weights W. "L1 misses" is the percentage of data accesses which result in L1 cache misses, and "L1&L2 misses" shows the percentage of data accesses which miss at both L1(fastest) and L2(slightly slower) cache, resulting in a slow access to main memory.

	April			SNNS		
W	# Accesses	L1 misses	L1&L2 misses	# Accesses	L1 misses	L1&L2 misses
2 790	3.10×10^8	0.14%	0.10%	4.45×10^8	4.46%	0.05%
25 120	1.70×10^9	1.71%	0.03%	2.43×10^9	7.39%	4.81%
62 710	4.25×10^9	1.85%	1.11%	6.00×10^9	7.18%	5.63%

Repercussion of data locality in efficiency. One of the most influential factors in these results is the locality of the data in memory. In order to corroborate this influence we have used the `valgrind`[1] [9] analysis tool, which executes a program in a simulated CPU and provides quantitative data on the number of memory accesses and cache performance (see Table 1). We have trained three layered feedforward networks, each one belonging to the three intervals that we have distinguished in the previous experiment: a network with less than 10 000 connections, between 20 000 and 30 000 connections, and with more than 50 000 connections. The results shown in Table 1 can be interpreted in terms of the network size:

Small networks. ($W = 2\,790$): the number of L1 cache misses in `April` is an order of magnitude lower (0.14% as opposed to 4.46% of SNNS). In the case of L2 cache, `April` has many more misses, but it also needs many fewer accesses. The better data locality in `April` reveals why `April` is 4 times faster than SNNS.

Medium-sized networks. ($W = 25\,120$): `April` displays a very small index of cache misses: 1.71% of data accesses are L1 misses, and only 0.03% result in main memory accesses, whereas SNNS shows a 7.39% of L1 misses and a 4.81% of combined cache misses, with more total accesses.This great difference is the reason why SNNS is 16 times slower than `April`.

Big networks. ($W = 62\,710$): SNNS obtains miss ratios which are similar to the ones in medium-sized networks, and `April` presents a combined cache miss ratio of 1.11%. This great increase is caused by the enormous growth in data size, but the cache-related behaviour of `April` is still better for both L1 and L2 caches, which allows `April` to be approximately 4 times faster than SNNS. One of the reasons of this smaller increase of cache misses in `April` is the use of single-precision values instead of the double-precision ones used by SNNS.

Efficiency with general feedforward topologies. We have repeated the experiment with networks which have each layer connected to all the previous ones (see Figure 4a), with one and two hidden layers. We have also tested networks in which connections

[1] `Valgrind` is a debugging and analyzing tool for the x86 architecture which detects programming errors and studies the efficiency at different levels (cache, bottlenecks, etc.).

are no longer all-to-all. We have divided the 16×16 pixel input in four 8×8 pixel fragments (see Figure 4b). Thus, the hidden neurons are distributed in four groups, connecting each group to a different part of the image. The rest of neurons display all-to-all connections. Also we have generated networks with one and two hidden layers, with the restriction that the number of hidden neurons must be a multiple of four. Training these topologies with `April` and SNNS has given analoguous time results as before. Thus, we have verified that `April` is able to train efficiently general feedforward topologies as well as specific feedforward topologies.

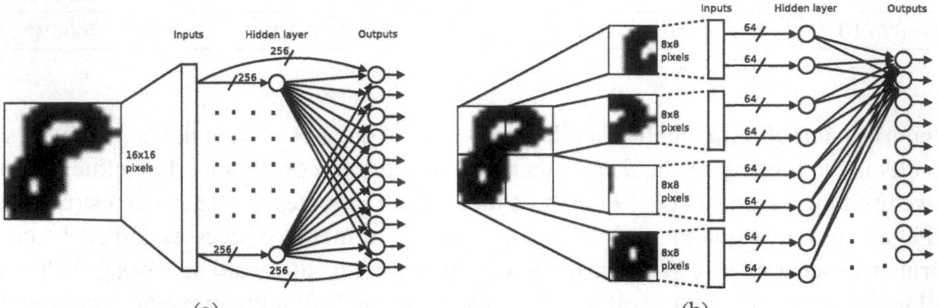

Fig. 4. (a) Feedforward network with shortcuts between the layers: a neuron is connected to neurons of all the previous layers. **(b)** Feedforward network with segmented input: the whole 16×16 image is divided in four 8×8 fragments forming four groups of hidden neurons.

3 Additional Features of the BP implementation

In addition to the usability facilities which will be described in Section 4, the following features deserve mention:

Softmax activation function. Softmax outputs suffer from numerical instability problems and many implementations do not deal properly with them. If the outputs of a ANN are required to be interpretable as posterior probabilities, it is needed for those outputs to lie between zero and one and to sum to one. The purpose of the softmax activation function [2] is to enforce these constraints on the outputs by defining the output of each neuron as $o_i = \exp(\gamma_i + c)/(\sum_{j=1}^{n} \exp(\gamma_j + c))$, for $i = 1, \ldots, n$, where γ_i are the potentials of the neurons. The value c is arbitrary and disappears when working out the equation. Some implementations set $c = \min \gamma_i$ in order to avoid problems of numerical instability, but numerical errors occasionally continue taking place if $\exp(\max \gamma_i - \min \gamma_i)$ overflows. For this reason, we use $c = \min \gamma_i$ if $\max \gamma_i - \min \gamma_i < c2$ but use $c = \max \gamma_i - c2$ in other case, where $c2$ is a constant that depends on the floating-point data type used, avoiding this problem. Other very useful feature of our implementation is the possibility of grouping output neurons so that the softmax calculations are performed independently in each group. This feature, which we have not found in other similar tools, can be useful if we want to use a given network to estimate different probabilities simultaneously on a same input dataset.

Tied and constant weights. It is useful to be able to indicate that certain weights of the network are tied and have the same value. Marking certain weights of the networks as constant, so that they do not undergo modifications during the training, is also possible.

Weight decay. Weight decay was introduced by Werbos [12] and consists of decreasing the weights during the BP training. In addition to each update of a weight, the weight is decreased by a part of its old value. The effect is similar to a pruning algorithm [1]: weights are driven to zero unless reinforced by BP. In particular, if we have a weight w_{ij} that connects the output of neuron i with neuron j, and δ_{BP} is the increment incorporated to that weight by the BP algorithm, $\triangle w_{ij}(t+1) = \delta_{bp} - \lambda w_{ij}(t)$, being λ the value of the weight decay parameter.

Value representation. Many ANN tools use double precision floating-point data. However, single precision (*float*) is enough to represent data and ANNs weights for most applications. Moreover, using ANNs in embedded systems (PDAs, cell phones, etc.) whose architectures do not usually feature a FPU requires a fixed-point representation in order to accelerate the calculations enormously. We have implemented a fixed-point version of the forward pass because a network is usually trained in a conventional computer and is only used to calculate outputs in the embedded applications.

Reproducibility of experiments. Another important characteristic included in our BP implementation is the ability to stop and resume experiments, what can be useful to recover from system failures, migrate processes, perform *grid* computing, etc. Therefore, we control all parameters of the networks and the experimentation. This includes the pseudo-random generators used to obtain the initial weights and to shuffle the training samples. The weight values of the network can be saved without loss of accuracy, and also the weight values from the previous iteration can be saved to calculate the momentum step after a recovering.

4 Usability Issues

There are many factors to take into account when designing a training experiment besides the speed of the BP algorithm. It is very important the possibility of describing easily the network topology and the training, validation and test corpora. Other important issue is the space occupied by the training data, which can be very huge, making impractical some experimentations even with various gigabytes of main memory.

Some BP implementations offer just a library which can be used to incorporate neural networks in any application, but which leaves the user with the responsibility of preparing the set of input and output pairs needed for training and validation. Other toolkits also offer a complete (graphical) environment where it is very easy to design the neural topology, to inspect the datasets and to perform the training experiments, but these environments are not useful for automatically testing combinations of parameters and topologies or when very sophisticated stopping conditions are used. Our BP C++

implementation can be used as a library, but is intended to be used with `April`, which provides a homogeneous environment to perform pattern recognition tasks.

4.1 Data Description and Manipulation Facilities

Neural network experiments can be easily described in `April` using the `Lua` [6] scripting language. `Lua` is used to describe and to save general network topologies and in general to describe experiments. `Lua` is an extensible procedural embedded programming language especially designed for extending and customizing applications with powerful data description facilities. Besides the `Lua` description facilities, `April` adds the *matrix* and *dataset* classes which allow the definition and manipulation of possibly huge sets of samples easier and more flexibly than simply enumerating the pairs of inputs and outputs.

The *matrix* class represents conventional n-dimensional matrices and can store images, sequences of words, etc. The neural network expects one or several *datasets* which represent an ordered set of patterns of a given size. Thus, the method for training a network during an epoch requires two *datasets* whose number of patterns coincide and whose pattern size is, respectively, the number of input and output neurons of the net.

A *dataset* can be defined from a n-dimensional matrix: The set of patterns is obtained by displacing a submatrix window over the original matrix. It is possible to specify the relative displacement between each subpattern, the starting position, the traversal order, whether the matrix is circullary closed in some dimensions, the default values when the windows is partially outside the matrix, etc. improving a similar feature found in `SNNS`.

Other types of *datasets* can also be defined by combining previous ones. For instance, an *indexed dataset* receives a *dataset* whose patterns are interpreted as indexes of a second *dataset* interpreted as a dictionary in order to obtain the set of patterns. This *dataset* is very useful to code the input and output values of an ANN which do not need to be explicitly stored because all *datasets* use lazy evaluation to compute the desired patterns only when needed. In this way, the memory usage is practically limited to the underlying matrices.

Also, the possibility of *splitting* and *joining dataset* allows the combination of patterns from different sets and the automatic partition of a big *dataset* in training, validation and test *datasets*. These *datasets* together with others for selecting, filtering and combining subpatterns offer a great deal of flexibility in the description of corpora.

4.2 Example of Use of the Application

We present in detail a very simple example for the task of handwritten digit classification described in Section 2. First, the image of Figure 1 is loaded in a *matrix* and later a *dataset* containing the 10×100 samples of size 16×16 pixel values is generated from it. Later, the *matrix* $[1\ 0\ 0\ 0\ 0\ 0\ 0\ 0\ 0\ 0]$ is iterated cicularly in order to obtain the *dataset* for the associated desired output.

The corresponding training, validation and test input and output *datasets* are obtained by slicing the former *datasets*. Although more complex ANN can be described, for layered ANNs it is possible to give a simple description like "`256 inputs 30 logistic 10 linear`".

```
samples = matrix.loadImage("digits.png")  -- loads the corpus pixel matrix
input_data = dataset.matrix(samples, { -- create a dataset from matrix
  patternSize = {16,16}, -- sample size
  offset      = {0,0},    -- initial window position
  numSteps    = {100,10}, -- number of steps in each direction
  stepSize    = {16,16},  -- step size
  orderStep   = {1,0}     -- step direction
})

-- matrix used for computing the output values
m2 = matrix(10,{1,0,0,0,0,0,0,0,0,0})
output_data = dataset.matrix(m2, {
  patternSize = {10},
  offset      = {0},
  numSteps    = {input_data:numPatterns()},
  circular    = {true}, -- circular dataset
  stepSize    = {-1}    -- the window moves backwards, "1" moves forwards
})

-- datasets created by slicing the previous input and output data
train_input       = dataset.slice(input_data  ,  1, 600) -- first 60% of data
train_output      = dataset.slice(output_data,   1, 600)
validation_input  = dataset.slice(input_data  ,601, 800) -- next  20% of data
validation_output= dataset.slice(output_data,601, 800)
test_input        = dataset.slice(input_data  ,801,1000) -- next  20% of data
test_output       = dataset.slice(output_data,801,1000)

-- Layered feedforward MLP generation, a general description is also possible
the_net=mlp.generate("256 inputs 30 logistic 10 linear", rnd, -0.7, 0.7)
rnd = random(1234) -- pseudo-random generator object used for shuffling

for i=1,100 do     -- Training and validation for 100 cycles,
                   -- more complex stopping criteria are also possible
  mse_train = the_net:train {
    learning_rate  = 0.2,
    momentum       = 0.2,
    input_dataset  = train_input,  -- input patterns
    output_dataset = train_output, -- desired output patterns
    shuffle        = rnd           -- used in sample shuffling
  }
  mse_val = the_net:validate {
    input_dataset  = validation_input,  -- input patterns
    output_dataset = validation_output  -- desired output patterns
  }
  printf ("Cycle %3d MSE %f %f\n", i, mse_train, mse_val)
end

mse_test = the_net:validate {
  input_dataset  = test_input,  -- input patterns
  output_dataset = test_output  -- desired output patterns
}
printf ("MSE of the test set: %f\n", mse_test)
```

5 Conclusions and Future Work

`April` toolkit is up to 16 times faster than SNNS. In addition, its capacity to train general feedforward networks does not decrease its efficiency due to the use of data structures with a great memory locality instead of linked lists (as SNNS). An approximation algorithm for the NP-Complete *consecutive retrieval problem* has been designed in order to mantain this efficiency.

At the present moment, the following extensions are being considered:

- **Pruning algorithms.** To include *pruning algorithms* such as *optimal cerebral damage* [8] or *optimal cerebral surgery* [5].
- **Recurrent networks.** This type of networks has demonstrated to be very useful in diverse fields, like in Natural Language Processing.
- **Graphical interface.** Adding a graphical interface could orient the application towards a didactic use.
- **Automatization of experiments.** To be able to automatically adjust some parameters of the training phase.
- **Grid computing.** In order to distribute an experiment in different machines, so that each one trains a network with different topologies or different parameters.

To conclude, we want to emphasize that `April` includes other utilities for pattern recognition tasks such as hybrid ANN/HMM training (which uses the proposed BP implementation), finite state automata parsing, voice parametrization, dynamic time warping alignment, image preprocessing, etc.

References

1. C. M. Bishop. *Neural networks for pattern recognition.* Oxford University Press, 1996.
2. J. Bridle. *Neuro-computing: Algorithms, Architectures, and Applications.*, chapter Probabilistic interpretation of feedforward classification network outputs, with relationships to statistical pattern recognition, pages 227–236. Springer-Verlag, 1989.
3. Michael R. Garey and David S. Johnson. *Computers and Intractability. A Guide to the Theory of NP-Completeness.* W. H. Freeman and Company, 1979.
4. Sakti P. Ghosh. File organization: the consecutive retrieval property. *Commun. ACM*, 15(9):802–808, 1972.
5. Babak Hassibi and David G. Stork. Second order derivatives for network pruning: Optimal brain surgeon. In *Advances in NIPS*, volume 5, pages 164–171. Morgan Kaufmann, 1993.
6. Roberto Ierusalimschy. *Programming in Lua.* Published by Lua.org, December 2003.
7. L. T. Kou. Polynomial complete consecutive information retrieval problems. *SIAM J. Comput.*, 6:67–75, 1977.
8. Y. LeCun, J. Denker, S. Solla, R. E. Howard, and L. D. Jackel. Optimal brain damage. In *Advances in NIPS II.* Morgan Kauffman, 1990.
9. Nicholas Nethercote and Jeremy Fitzhardinge. Valgrind: A Program Supervision Framework. *Electronic Notes in Theoretical Computer Science*, 89(2), 2003.
10. D. Plaut, S. Nowlan, and G. Hinton. Experiment on learning by back propagation. Technical Report CMU-CS-86-126, Department of Computer Science, Carnegie Mellon University, 1986.
11. D. E. Rumelhart, G. E. Hinton, and R. J. Williams. *PDP: Computational models of cognition and perception, I*, chapter Learning internal representations by error propagation, pages 319–362. MIT Press, 1986.
12. P. Werbos. Backpropagation: Past and future. In *Proceedings of the IEEE International Conference on Neural Networks*, pages 343–353. IEEE Press, 1988.
13. Francisco Zamora Martínez. Implementación eficiente del algoritmo de retropropagación del error con *momentum* para redes hacia delante generales, 2005. Proyecto Final de Carrera.
14. Andreas Zell, Niels Mache, Ralf Huebner, Michael Schmalzl, Tilman Sommer, and Thomas Korb. SNNS: Stuttgart Neural Network Simulator. User manual Version 4.1. Technical report, Stuttgart, 1995.

Coefficient Structure of Kernel Perceptrons and Support Vector Reduction

Daniel García, Ana González, and José R. Dorronsoro*

Dpto. de Ingeniería Informática and Instituto de Ingeniería del Conocimiento
Universidad Autónoma de Madrid, 28049 Madrid, Spain

Abstract. Support Vector Machines (SVMs) with few support vectors are quite desirable, as they have a fast application to new, unseen patterns. In this work we shall study the coefficient structure of the dual representation of SVMs constructed for nonlinearly separable problems through kernel perceptron training. We shall relate them with the margin of their support vectors (SVs) and also with the number of iterations in which these SVs take part. These considerations will lead to a remove–and–retrain procedure for building SVMs with a small number of SVs where both suitably small and large coefficient SVs will be taken out from the training sample. Besides providing a significant SV reduction, our method's computational cost is comparable to that of a single SVM training.

1 Introduction

It is well known that one of the potential drawbacks of support vector machines (SVMs) is a large number of support vectors (SVs), as this implies long computing times when an SVM is applied to new patterns not seen before. This has led to several proposals for controlling the final number of SVs. A natural approach [4,8] is to limit beforehand the number of SV allowable and maintain that number during training. However, choosing the right number is not easy and the test performance of the final classifier may not be good enough. Another common approach [9,10] is to "grow" the final SV set, carefully selecting new SVs to be added to an already trained SVM. This guarantees a small final number of SVs, although probably after a somewhat involved and costly growing procedure. Other approaches start from a "full" SV set obtained by SVM construction and either try to approximate a reduced SV set [2] or incorporate an SV sparseness constrain to the SVM convex minimization problem [16].

We will explore here an alternative approach, where we start from a fully trained SVM and progressively try to reduce its SV set by removing appropriately selected SVs and constructing new SVMs with less SVs while also maintaining the accuracy of the final classifier. One clear drawback of any such approach would be the cost of the several retrainings needed. This makes mandatory the use, first, of simple training procedures, but also to ensure that successive retrainings are significantly less costly than the initial one. We also need a criterion

* All authors have been partially supported by Spain's TIN 2004–07676.

J. Mira and J.R. Álvarez (Eds.): IWINAC 2007, Part I, LNCS 4527, pp. 337–345, 2007.

to decide which SVs to take out of the final SV set. To deal with these issues we shall use the Schlesinger–Kozinec (SK) algorithm, a classical method [12] for perceptron training that constructs a maximal margin linear classifier for linearly separable samples and that has been recently extended [5] to a kernel setting and non-linearly separable problems. This requires the introduction of a slack quantity on the SV margins to allow for patterns not correctly classified. After briefly reviewing the SK algorithm in section 2, we shall show in section 3 that there are two ways of looking at the coefficients α_i of the final SVs that are relevant to this work. First, it turns out that the α_i essentially coincide with the margin slack variables. More precisely, a small α_i is associated with an SV correctly classified with a large margin. On the other hand, if α_i is sufficiently large, its SV will require a large slack value and, hence, a large negative margin, and will not be correctly classified. Moreover, many final SVs are a consequence of the training process, that uses them at some point but later concentrates in more relevant SVs. We shall argue also that the α_i coefficients somehow measure the number of times their associated SV have appeared during training. In particular, large α_i SVs imply larger training costs.

Our SV reduction procedure takes advantage of the above considerations. In fact, small α_i SVs represent "safe" patterns that could be removed without much affecting the final classifier performance (the same can also be said of those sample vectors for which $\alpha_i = 0$). On the other hand, SVs with quite large α_i correspond to hard to classify patterns that, moreover, require longer training efforts. Thus, we will successively apply a "remove and retrain" SV selection method, where at each step we will take out from the training sample patterns with either zero or small α_i coefficients and, also, all the large α_i SVs that are not correctly classified and retrain then a new SVM. To decide when to stop these iterations, we will use the first type of removed SVs as a validation set, stopping the procedure when the error over this set is no longer zero. As we shall illustrate in section 4, this will lead to SVMs with a small number of SVs and good test set performances, at a cost which is comparable to that of the first perceptron training (we point out that, to begin with, the SK algorithm is computationally competitive [5] with state of the art SVM training procedures such as those in [7,11]).

The rest of the paper is organized as follows. In section 2 we will quickly review SVM construction and the SK algorithm for nonlinearly separable problems with quadratic penalties, and will analyze the above mentioned interpretation of the SV coefficients in section 3, that also contains a detailed description of our proposed remove and retrain procedure. Section 4 illustrates the procedure over ten datasets and the paper ends with a short discussion section.

2 Kernel Perceptron Training

Assume we have a linearly separable training sample $\mathcal{S} = \{(X_i, y_i) : i = 1, \ldots, N\}$, where $y_i = \pm 1$. We shall work in a homogeneous setting, where we consider patterns of the form $X_i' = (X_i, 1)$ and separating hyperplanes defined by

$W' = (W, b)$. Then $W' \cdot X'_i = W \cdot X_i + b$, and we have to solve the minimization problem:

$$\min \frac{1}{2}\|W'\|^2 \text{ subject to } y_i W' X'_i \geq 1. \tag{1}$$

Those W' verifying the constrains in (1) are said to be in canonical form. While not strictly equivalent to the convex minimization problem that is standard in SVM construction [13,15], it can be argued that the solution of (1) is a good approximation of the optimum SVM separating hyperplane, particularly for high dimension problems. In what follows we shall drop the prime superscript, writing from now on W and X.

Problem (1) is usually rewritten in dual form and then solved by either relatively ad–hoc quadratic programming methods or, simply, by gradient ascent in its dual form. Alternatively, an equivalent way [1] of obtaining an optimal solution of (1) is to solve the following problem

$$W^* = \arg\min\{\|W\|^2 \ : W \in C(\tilde{\mathcal{S}})\}, \tag{2}$$

where $\tilde{\mathcal{S}} = \{y_i X_i : X_i \in \mathcal{S}\}$ and $C(\mathcal{A})$ denotes the convex hull of a set $\mathcal{A}$, that is, the set of all linear combinations $\sum \alpha_i Z_i$ with $Z_i \in \mathcal{A}$ and $\sum \alpha_i = 1$. Moreover, the maximum margin m^o verifies $m^o = \|W^*\|$. Notice that W^* is not in canonical form, as we have $y_i W^* \cdot X_i \geq m^o = \|W^*\|$. The canonical vector is then $W^o = W^*/\|W^*\|$. The Schlesinger–Kozinec (SK) algorithm [12] is a classical method of solving (2) using an update rule very close to the classical delta rule of Frank Rosenblatt's perceptrons. More precisely, at each step the current weight W_{t-1} is updated as

$$W_t = (1 - \lambda^t)W_{t-1} + \lambda^t y_{l(t)} X_{l(t)} \tag{3}$$

where the convex factor λ^t is given by $\lambda^t = \arg\min_\lambda\{\|(1-\lambda)W_{t-1} + \lambda X_{l(t)}\|\}$, and the updating pattern $X_{l(t)}$ at step t is chosen as

$$l = \arg\min_i\{y_i W_t \cdot X_i\}; \tag{4}$$

that is, $X_{l(t)}$ is the pattern that determines the margin of W_{t-1}. Notice that these updates keep the successive W_t in $C(\tilde{\mathcal{S}})$ and also that $\|W_t\| \leq \|W_{t-1}\|$. To provide a stopping criterion for the iterations in (3), observe that for any $W \in C(\tilde{\mathcal{S}})$, it follows from (2) that $\|W\| \geq \|W^*\| = m^o \geq m(W)$, where $m(W)$ denotes the margin of W. Defining $g(W) = \|W\| - m(W)$, we have $0 = g(W^*) \leq g(W)$ for all $W \in C(\tilde{\mathcal{S}}$. We shall use g as a criterion function to stop perceptron training when $g(W_t)$ becomes smaller than a specified small value. The choice of λ^t guarantees that $\|W_t\| \leq \|W_{t-1}\|$. Moreover, we have $y_{l(t)} W_t \cdot X_{l(t)} = y_{l(t)} W_{t-1} \cdot X_{l(t)} + \lambda^t \left(\|X_{l(t)}\|^2 - y_{l(t)} W_{t-1} \cdot X_{l(t)}\right)$. Since it can be shown that if $\lambda_t < 1$,

$$\|X_{l(t)}\|^2 - y_l W_{t-1} \cdot X_l = (1 - \lambda^t)\|W_{t-1} - y_l X_l\|^2 > 0,$$

it follows that $y_{l(t)} W_t \cdot X_{l(t)} \geq y_{l(t)} W_{t-1} \cdot X_{l(t)}$; that is, W_t has a larger margin at $X_{l(t)}$. Therefore, we may expect that $g(W_t) \leq g(W_{t-1})$ and, indeed, it can be verified experimentally that $g(W_t)$ decreases. Thus, the SK algorithm achieves a maximal margin separating hyperplane. In fact, the same is true of standard perceptrons trained under Rosenblatt's delta rule. Notice that the delta rule builds a weight vector W_t' of the form $W_t' = \sum_1^t y_j X_j$, which can be brought into $C(\tilde{\mathcal{S}})$ by setting $W_t = W_t'/t$. A simple consequence of this is the update formula

$$W_t = (1 - \frac{1}{t})W_{t-1} + \frac{1}{t} y_{l(t)} X_{l(t)}, \tag{5}$$

which means that using in (3) the values $1/t$ instead of λ^t results in a convex form of the delta rule. It can also be seen experimentally [6] that the resulting W_t also converge to a maximal margin weight vector.

The extension of the SK algorithm (and of the convex delta rule) to a kernel setting is fairly simple. First, writing $W_t = \sum_j \alpha_j^t y_j X_j$, it follows from (3) that $\alpha_j^t = (1-\lambda^t)\alpha_j^{t-1} + \lambda^t \delta_{j,l(t)}$. Moreover the selection of the minimum margin vector $X_{l(t)}$ and the computation of λ^t require the update of the quantities $\|W_t\|^2$ and $D_j^t = y_j W_t \cdot X_j$, $j = 1, \ldots, N$. For this we have

$$\begin{aligned} \|W_t\|^2 &= (1-\lambda^t)^2 \|W_{t-1}\|^2 + 2(1-\lambda^t)\lambda^t D_l^{t-1} + (\lambda^t)^2 X_l \cdot X_l), \\ D_j^t &= (1-\lambda^t) D_j^{t-1} + \lambda^t y_l y_j X_l \cdot X_j, \end{aligned} \tag{6}$$

and all of them can be expressed in terms of dot products. Thus, if $X = \phi(x)$ is the non–linear transformation associated by Mercer's theorem to a positive definite kernel, we can replace the above dot products $X_l \cdot X_j$ by their kernel counterparts $K(x_l, x_j)$, which allows us to work on the extended space vectors X_j from their lower dimensional sample counterparts x_j. Moreover, it follows from (6) that the cost of an update is $O(N)$ kernel operations and that the cost of a T iteration kernel perceptron training is $O(T \times N)$ such operations.

Turning our attention to non–separable problems, the standard approach in SVM training is to relax the margin restrictions by introducing slack variables ξ_i for which we allow $y_i W \cdot X_i \geq 1 - \xi_i$, and to add a penalty $C \sum \xi_i^k$ to the criterion function in (1). While $k = 1$ is usually taken, we shall work here with a quadratic penalty choosing $k = 2$, the reason being that it is then straightforward to extend the point of view in (2) to the non–linear setting. More precisely, we shall consider extended weights $\tilde{W}$ and patterns $\tilde{X}_i$ defined as

$$\tilde{W} = \left(W, \sqrt{C}\xi_1, \ldots, \sqrt{C}\xi_N\right); \ \tilde{X}_i = \left(X_i, 0, \ldots, \frac{y_i}{\sqrt{C}}, \ldots, 0\right).$$

It is then easy to check that the quadratic penalty criterion function $J(W, \xi) = \|W\|^2 + C \sum \xi_i^2$ verifies $J(W, \xi) = \|\tilde{W}\|^2$; moreover, the slack margin conditions $y_i W \cdot X_i \geq 1 - \xi_i$ are equivalent to $1 \leq y_i W \cdot X_i + \xi_i = y_i \tilde{W} \cdot \tilde{X}_i$. This allows the direct application of (2) to the extended weights and vectors. In particular, we can write the optimum $\tilde{W}^*$ as $\tilde{W}^* = \sum \alpha_i^* y_i \tilde{X}_i$ with $\sum \alpha_i^* = 1$. This yields the extended vector equalities

$$(W^*, \sqrt{C}\xi_1^*, \ldots, \sqrt{C}\xi_N^*) = \tilde{W}^* = \sum \alpha_i^* y_i \tilde{X}_i = (W^*, \alpha_1^*/\sqrt{C}, \ldots, \alpha_N^*/\sqrt{C}),$$

which imply that $\sqrt{C}\xi_i^* = \alpha_i^*/\sqrt{C}$, that is, $\alpha_i^* = C\xi_i^*$. Here again we can write $\tilde{W}^*$ in canonical form as $\tilde{W}^o = \tilde{W}^*/\|W^*\|^2$ and the margin slack values are then $\xi_i^o = \xi_i^*/\|\tilde{W}^*\|^2 = \alpha_i^*/C\|\tilde{W}^*\|^2$. The previous kernel version of the SK method can be extended to this setting simply by working with the kernel $K'(x, x') = K(x, x') + 1/C$. We examine next how to exploit these facts for the removal of support vectors of kernel perceptrons.

3 Support Vector Removal and Retraining

We start by observing two possible interpretations of the final α_i^* coefficients. Recall that after putting the optimal $\tilde{W}^*$ in its canonical form W^o, the slack margin of a SV X_i is $\xi_i^* = \alpha_i^*/C\|\tilde{W}^*\|^2$. Let us write $\Lambda^* = C\|\tilde{W}^*\|^2 = C\|W^*\|^2 + \sum(\alpha_i^*)^2$. We then have for any SV X_i

$$y_i W^o \cdot X_i = \frac{y_i}{\|\tilde{W}^*\|^2}\left(\tilde{W}^* \cdot X_i - \xi_i^*\right) \geq 1 - \frac{\xi_i^*}{\|\tilde{W}^*\|^2} = 1 - \frac{\alpha_i^*}{C\|\tilde{W}^*\|^2} = 1 - \frac{\alpha_i^*}{\Lambda^*}.$$

Therefore, small α_i^* patterns can be considered "safe", as they have positive margins close to 1; moreover, they have little weight influence on the optimal W^*. Thus, their removal from the training set should not greatly affect the final classifier obtained. This reduction in sample patterns should make subsequent trainings less expensive, as training costs depend directly on the sample size. On the other hand, large α_i^* patterns also require large slacks; in fact, their classification will be wrong if $\alpha_i^* > \Lambda^*$.

We can look to the α_i^* from another point of view. We have mentioned that the SK and the convex delta rule updates lead to the same final SVs and to essentially the same coefficients for them. For the convex delta rule, the coefficient of a given SV represents the number of times it has been selected as the updating vector. Notice that if T training iterations have been performed and each SV X_i has appeared T_i times as the updating pattern, the final convex separating vector obtained through the convex delta rule has the form

$$W^* = \frac{1}{T}\sum_1^T y_t X_t = \sum_1^N \frac{T_i}{T}\, y_i X_i \simeq \sum \alpha_i^* y_i X_i;$$

thus, a large α_i^* means that the corresponding SV intervenes in a sizeable part of the training iterations. Hence, while these SV have a considerable weight in W^*, they also take up a large part of SVM training. Moreover, since they will not be correctly classified, it is conceivable that a classifier constructed after their removal could still have a good performance.

Based on the above considerations, we will apply a remove and retrain procedure that iteratively constructs a series of SVM classifiers $\mathcal{C}_t$. More precisely, after a new reduced sample SVM has been trained, we will remove all patterns with zero coefficients or such that $\alpha_i^* > \Lambda^*$. We will also remove small coefficient SVs, namely those X_i such that $\alpha_i^* \leq \alpha_{Min} + \rho(\Lambda^* - \alpha_{Min})$, where $\alpha_{Min} = \min\{\alpha_i\}$; we shall take $\rho = 0.02$ in our experiments. To decide when to stop, we will use at step t a "validation" subset $\mathcal{V}_t$ initially empty and that will grow at each step with the addition of the zero or small coefficient patterns that we remove. When the error of $\mathcal{C}_t$ over $\mathcal{V}_t$ is not 0, the procedure stops and outputs the previous classifier $\mathcal{C}_{t-1}$. In order to achieve faster trainings, each new $\mathcal{C}_t$ will be built starting from initial α_i values close of those defining $\mathcal{C}_{t-1}$. Assume that we want to remove a certain X_l; we have to change its weight α_l^* to 0 and to reassign the other weights as $\alpha_i' = \alpha_i^*/(1-\alpha_l^*)$ so that we still have a convex combination. If we choose $\eta = \alpha_l^r/(\alpha_l^r - 1)$, We can write this as an update

$$\alpha_j' = (1-\eta)\alpha_j^* + \eta\delta_{j,l}.$$

Notice that the α_j' updates have again the form used in the SK algorithm; as a consequence, D_j' and $\|W'\|^2$ can also be updated according to (6). Therefore, the cost of these margin and norm updates is of O(N) kernel operations for each pattern to be removed. This is much smaller than the cost of the overall training; moreover, as we shall illustrate numerically, it makes the second and successive retrainings much faster than the first one.

4 Numerical Experiments

In this section we shall explore the evolution of the previous support vector removal procedure over 10 datasets (see table 1) taken from the UCI problem database [14] or from the LIBSVM repository [3]. Some of these data sets, namely, those of the heart disease, Wisconsin breast cancer or thyroid disease, originally contain data from more than two classes. We have reduced them to 2–class problems considering their patterns as coming from either healthy or sick patients. The fourclass problem has originally 4 classes; we use the 2 class version in [3]. We shall work with the gaussian kernel $k(x,x') = \exp\left(-\|x-x'\|^2/\sigma^2\right)$, and normalize the original patterns to componentwise 0 mean and 1 variance. We shall also use common values $\sigma^2 = 25$ and $C = 20$ for all datasets; the test accuracies reported here are comparable with those achieved by other methods.

For each dataset we have performed a 10×10 cross validation procedure, dividing 10 times the full datasets randomly into 10 subsets with approximately equal sizes and have used 9 of these subsets for training and the remaining one for test. Table 1 gives the original training sample size, the number of SVs obtained after the first training pass and the final SV number, which is clearly smaller in all cases. Our procedure achieves quite large SV reductions for the fourclass problem (94.2%), Ripley (93.5%), Wisconsin breast cancer (86.7%) and Pima (85.8%). The smallest reduction is that of the German credit problem (15.4%); in all other cases the SV set reduction is at least of a 36% of the SVs obtained after just one training. On the other hand, the test set accuracies after the initial and final trainings remain essentially the same. Table 2 shows them for all datasets together with their standard deviations. The table also shows the significance $P(t)$ of a Student's t–test for different means; while the test's hypotheses may not hold, the significances are quite large in all cases except in the Ripley dataset, for which a null hypothesis of equal test accuracies before

Table 1. Starting sample size, SV # after the initial and final trainings and percentage of SV reduction

Dataset	sample s.	# SV Ini.	# SV Fin.	% SV red.
Wisc. Br. C.	629	120.6 ± 10.2	16.1 ± 3.0	86.7
Heart disease	240	165.2 ± 6.4	69.1 ± 16.1	58.1
Ionosphere	284	143.0 ± 5.1	74.5 ± 8.3	47.9
Pima	622	554.5 ± 7.9	78.9 ± 35.9	85.8
Ripley	1125	576.0 ± 11.5	37.5 ± 6.4	93.5
Sonar	168	136.7 ± 2.5	81.9 ± 9.2	40.1
Thyroid	6480	1173.5 ± 42.7	744.9 ± 244.6	36.5
Austral. cred.	690	351.3 ± 11.4	134.3 ± 61.6	61.8
German cred.	1000	652.2 ± 8.3	551.5 ± 119.3	15.4
Fourclass	862	602.8 ± 8.4	34.8 ± 7.6	94.2

and after SV removal should be rejected. On the other hand, equality of means cannot be rejected for the other datasets after SV removal. Even in the Ripley problem, the average accuracy only falls slightly, passing from an average of 89.62% to an 88.53%.

The extra effort over a single training pass is also shown in table 2. In all retrainings we have iterated the SK convex update procedure until the criterion function $g(W)$ is < 0.001. The table's sixth column gives for the entire data sets a comparison between the number of kernel operations of a single training pass and of the full remove and retrain procedure. As it can be seen, the extra work is quite modest for all problems, the largest number of extra kernel operations being made in the heart disease and Pima problems, where the SV removal and retraining costs are about 75% and 67% more than the first full sample training. This extra effort is essentially below 50% for all the other problems; we recall

Table 2. Test accuracies after the first training (Ini.) and after SV removal (Fin.) and probability values of equal means t-test. Values shown represent 10 × 10 cross validation averages and standard deviations. Columns 5 and 6 show the average number of retrainings and the training time increases for full dataset samples.

Dataset	Acc. Ini.	Acc. Fin.	$P(t)$	# iters.	over. cost
WBCancer	96.88 ± 1.85	96.65 ± 2.03	0.40	10.8 ± 2.1	115.17
Heartdis	80.34 ± 7.34	81.97 ± 6.70	0.11	10.8 ± 3.6	175.47
Ionosphere	91.50 ± 4.03	90.94 ± 3.92	0.32	5.9 ± 1.1	129.72
Pima	77.11 ± 4.17	76.83 ± 4.09	0.64	16.4 ± 4.2	167.10
Ripley	89.62 ± 3.00	88.53 ± 2.91	0.01	32.0 ± 3.6	151.50
Sonar	86.95 ± 6.55	87.35 ± 6.65	0.67	11.2 ± 2.1	137.27
Thyroid	97.79 ± 0.92	97.63 ± 0.88	0.22	3.5 ± 0.8	119.22
Austral. cred.	86.18 ± 2.82	85.82 ± 3.16	0.41	6.6 ± 1.9	141.50
German cred.	74.31 ± 3.72	74.28 ± 3.68	0.95	4.0 ± 2.7	149.52
Fourclass	79.18 ± 4.41	79.33 ± 3.50	0.79	27.2 ± 4.6	152.90

[5] that the cost of the SK method is competitive with that of other state of the art SVM construction algorithms. As mentioned above, key reasons for this small extra complexity are the removal of those SVs that make training costlier and the restart of subsequent trainings from α values close to the optimum ones obtained in the previous training.

5 Conclusions

A common problem of kernel classifier construction methods is the high number of final support vectors they must use, which results in costly classification of new patterns. This affects support vector machines as well as kernel perceptrons. In this paper we have shown for quadratic penalties in non separable problems how the coefficients of a kernel perceptron are related to the SV margins and also to the impact of a given SV during training. In turn, this coefficient interpretation suggests which SVs to remove iteratively from an already trained classifier so that successive trainings are faster and the final classifiers have a good test performance. Although very simple, the procedure may lead to quite large SV reduction while only requiring a modest extra effort. There are several points of further research. For instance, linear penalties are likely to result in fewer SVs than quadratic ones. Thus, a question of interest is to adapt the present analysis to these linear penalties. On the other hand, it is clear, particularly for large sample problems, that it is preferable to maintain a reduced number of SVs at all times, to which the coefficient interpretations given here may be applied. These and other related questions are presently being considered.

References

1. K. Bennett, E. Bredensteiner. Geometry in learning. In **Geometry at Work**, C. Gorini, E. Hart, W. Meyer, T. Phillips (eds.), Mathematical Association of America, 1997.
2. C. J. C. Burges. Simplified support vector decision rules. In **Proc. 13th International Conference on Machine Learning**, L. Saitta (editor), 71-77. Morgan Kaufmann, 1996.
3. Ch. Chang, Ch. Lin, LIBSVM: a library for support vector machines, 2001. http://www.csie.ntu.edu.tw/ cjlin/LIBSVM.
4. O. Dekel, S. Shalev-Shwartz, Y. Singer. The Forgetron: A Kernel-Based Perceptron on a Fixed Budget. **Advances in Neural Processing Systems** 18 (2005), 259–266.
5. V. Franc, V. Hlavac. An iterative algorithm learning the maximal margin classier. **Pattern Recognition** 36 (2003), 1985-1996.
6. D. García, A. González, J.R. Dorronsoro. Convex Perceptrons. In **Proceedings of the 7th International Conference on Intelligent Data Engineering and Automated Learning - IDEAL 2006**. Lecture Notes in Computer Science 4224, 578-585, Springer Verlag 2006
7. T. Joachims. Making Large-Scale Support Vector Machine Learning Practical. In **Advances in Kernel methods**, B. Schlkopf, C. Burges, A. Smola (eds.). MIT Press, 1999, 169-184.

8. Y. Lee and O. L. Mangasarian. RSVM: reduced support vector machines. In **CD Proceedings of the First SIAM International Conference on Data Mining**, Chicago, 2001.
9. S. Keerthi, O. Chapelle, D. de Coste. Building support vector machines with reduced complexity. **Journal of Machine Learning Research** 7 (2006) 1493–1515.
10. E. Parrado-Hernández, I. Mora-Jiménez, J. Arenas-García, A.R. Figueiras-Vidal, A. Navia-Vázquez. Growing support vector classifiers with controlled complexity. **Pattern Recognition** 36 ((2003), 1479–1488.
11. J.C. Platt. Fast training of support vector machines using sequential minimal optimization. In **Advances in Kernel methods**, B. Schlkopf, C. Burges, A. Smola (eds.). MIT Press, 1999, 185–208.
12. M. Schlesinger, V. Hlavac. **Ten Lectures on Statistical and Structural Pattern Recognition.** Kluwer Academic Pub., 2002.
13. B. Schölkopf, A. J. Smola. **Learning with Kernels**. MIT Press, 2001.
14. UCI-benchmark repository of machine learning data sets. University of California Irvine. http://www.ics.uci.edu.
15. V.N. Vapnik. **The Nature of Statistical Learning Theory**. Springer, Berlin, 1995.
16. M. Wu, B. Scholkopf, G. Bakir. A Direct Method for Building Sparse Kernel Learning Algorithms. **Journal of Machine Learning Research** 7 (2006), 603–624.

The Max-Relevance and Min-Redundancy Greedy Bayesian Network Learning Algorithm

Feng Liu and QiLiang Zhu

Department of Computer Science and Technology, Beijing University of Posts and Telecommunications
lliufeng@hotmail.com, zhuqiliang@tom.com

Abstract. Existing algorithms for learning Bayesian network require a lot of computation on high dimensional itemsets which affects reliability, robustness and accuracy of these algorithms and takes up a large amount of time. To address the above problem, we propose a new Bayesian network learning algorithm MRMRG, Max Relevance-Min Redundancy Greedy. MRMRG algorithm is a variant of K2 which is a well-known BN learning algorithm. We also analyze the time complexity of MRMRG. The experimental results show that MRMRG algorithm has much better efficiency. It is also shown that MRMRG algorithm has better accuracy than most of existing learning algorithms for limited sample datasets.

Keywords: Bayesian network; Max Relevance; Min Redundancy; Greedy search.

1 Introduction

There are many problems in fields as diverse as medical diagnosis, weather forecast, fault diagnosis, where there is a need for models that allow us to reason under uncertainty and take decisions, even when our knowledge is limited. To model this type of problems, AI community has proposed Bayesian network which allows us to reason under uncertainty.[1] During the last two decades, many Bayesian network learning algorithms have been proposed. But, the recent explosion of high dimensional data sets in the biomedical realm and other domains has induced a serious challenge to these BN learning algorithms. The existing algorithms must face higher dimensional and limited sample datasets.

In general, BN learning algorithms take one of two approaches: constraint-based methods [2], [3] and search & score methods [4], [5], [6]. The constraint-based approach estimates from the data whether certain conditional independencies hold between variables. Typically, this estimation is performed using statistical or information theoretical measures. The search & score approach attempts to find a graph that maximizes the selected score. Score function is usually defined as a measure of fitness between the graph and the data. These algorithms use a score function in combination with a search method in order to measure the goodness of each explored structure from the space of feasible solutions. During the exploration process, the score function is applied in order to evaluate the fitness of each candidate structure to the data.

J. Mira and J.R. Álvarez (Eds.): IWINAC 2007, Part I, LNCS 4527, pp. 346–356, 2007.

Although encouraging results have been reported, the two approaches both suffer some difficulties for high dimensional data. With the constraint-based approach, a statistical or information theoretical measure may become unreliable for high dimensional and limited sample datasets. If the measure would return incorrect independence statements, errors could arise in graphical structure. The search & score approach suffers from the exponential search space. The search space is so vast for high dimension that the category algorithms have to use heuristic methods to find approximately optimal Bayesian network.

The K2 algorithm [3] is a typical search & score method. Although it has already been presented for 15 years, K2 is still one of the most effective BN learning algorithms. Moreover, K2 is often used by other BN learning algorithms in order to improve performance. So, it is practically valuable to improve K2 algorithm.

In this paper, we propose MRMRG algorithm that improve K2 algorithm for high dimensional and limited sample datasets. MRMRG imports Max Relevance - Min Redundancy feature selection technology as an efficient score function to obtain better reliability, robustness and efficiency, even accuracy. MRMRG proposes Local Bayesian Increment function to terminate program in order to avoid overfitting and improve accuracy.

This paper is organized as follows. Section 2 provides a brief review of some basic concepts and theorems. Section 3 describes K2 algorithm. In Section 4, we propose Local Bayesian Increment function. Section 5 represents the details of MRMRG algorithm. At the same time, we also analyze the correctness, robustness and time complexity of MRMRG. Section 6 shows an experimental comparison among K2 and MRMRG. Finally, we conclude and present future work.

2 Concepts and Theorems

2.1 Bayesian Network

A Bayesian network is defined as a pair $B = \{G, \Theta\}$, where G is a directed acyclic graph $G = \{V(G), A(G)\}$, with a set of nodes $V(G) = \{V_1, \ldots, V_n\}$, representing a set of stochastic variables and a set of arcs $A(G) \subseteq V(G) \times V(G)$, representing independent relationships that exist between variables. Θ represents the set of parameters that quantifies the network. It contains a parameter $\theta = P(x_i \mid \pi_i)$ for each possible value x_i of X_i, and $\phi_i[j]$ of π_i. Here π_i denotes the set of parents of X_i in G and π_i is a particular instantiation of the parents.

The network structure G encodes all assertions of conditional independence among variables. The joint probability distribution of any particular instantiation of all n variables in the BN is given as $P(x_1, x_2, \ldots, x_n) = \prod_{i=1}^{n} \theta_{x_i|\pi_i}$, where x_i represents the instantiation of the variable X_i and π_i represents the instantiation of parents of X_i.[7]

2.2 Mutual Information and Relative Entropy

Lemma 1 (Mutual Information and Relative Entropy)
The relative entropy $D_{KL}(P\|Q)$ between the true distribution P and any BN model distribution Q is a monotonically decreasing function of the sum of mutual information between every node X_i and it's parents $Pa_Q(X_i)$ in distribution Q.

$$\sum_{i=1, Pa_Q(X_i)\neq\phi}^{n} MI(X_i, Pa_Q(X_i)), \textit{where} \quad (1)$$

$$MI(X_i, Pa_Q(X_i)) = \sum_{X_i, Pa_Q(X_i)\neq\phi} P(X_i, Pa_Q(X_i)) \log_2\left(\frac{P(X_i, Pa_Q(X_i))}{P(X_i)P(Pa_Q(X_i))}\right).$$

Note that proof of Lemma 1 refer to Appendix in [8].

2.3 MRMR and Mutual Information

In feature selection, Maximum Mutual Information (MMI) [9] is the method which is to find a feature set S_m with m features $\{x_i\}$, which jointly have the largest mutual information on the target class c.

In feature selection, Max Relevance-Min Redundancy (MRMR) [9] is that suppose we already have the feature set with $m-1$ features S_{m-1}, if we want to select one feature from the set $\{X - S_{m-1}\}$, we can select the feature x_m that maximizes the following formula:

$$x_m = \underset{x_j \in X - S_{m-1}}{\arg} \max \left[I(x_j; c) - \frac{1}{m-1} \sum_{x_i \in S_{m-1}} I(x_j; x_i) \right]. \quad (2)$$

Lemma 2 (MRMR and Maximum Mutual Information)
If one feature is added at one time, MRMR is equivalent to MMI.

Note that proof of Lemma 2 refer to Section 2.3 in [9].

3 K2 Algorithm

Given a complete dataset D, K2 searches for the Bayesian network G^* with maximal $P(G, D)$.

Let D be a dataset of m cases, where each case contains a value for each variable in V. D is sufficiently large. Let V be a set of n discrete variables, where x_i in V has r_i possible values $(v_{i1}, v_{i2}, \ldots, v_{ir_i})$. Let G denote a Bayesian network structure containing just the variables in V. Each variable x_i in G has the parents set π_i. Let $\phi_i[j]$ denote the jth unique instantiation of π_i relative to D. Suppose there are q_i such unique instantiation of π_i. Define $N_{ijk}$to be the number of cases in D in which variable x_i is instantiated as v_{ik} and π_i is instantiated as $\phi_i[j]$. Let $N_{ij} = \sum_{k=1}^{r_i} N_{ijk}$.

Given a Bayesian network model, cases occur independently. Bayesian network prior distribution is uniform. It follows that

$$P(G, D) = p(G) \prod_{i=1}^{n} \prod_{j=1}^{q_i} \frac{(r_i - 1)!}{(N_{ij} + r_i - 1)!} \prod_{k=1}^{r_i} N_{ijk}! \ . \tag{3}$$

Given a Bayesian network model, cases occur independently. Bayesian network prior distribution is uniform. It follows that

$$g(i, \pi_i) = \prod_{j=1}^{q_i} \frac{(r_i - 1)!}{(N_{ij} + r_i - 1)!} \prod_{k=1}^{r_i} N_{ijk}! \ . \tag{4}$$

It starts by assuming that a node has no parents, and then in every step it adds incrementally the node which can most increase the probability of the resulting BN, to the parents set. K2 stops adding nodes to parents set when the addition cannot increase the probability of the BN given the data.

K2 algorithm

Input: A set of n nodes $V = \{x_1, x_2, \ldots, x_n\}$, an ordering on the nodes, an upper bound u_{max} on the number of parents a node may have, Pre_idenotes the set of nodes that precede x_i, and a dataset $D = \{d_1, \ldots, d_m\}$ containing m cases. The dataset D is sufficiently large.

Output: For each node x_i, a printout of the parents set π_i of the node.

Procedure:

```
for i=1 to n do
begin
  π_i = NULL;
  P_old = g(i, π_i);
  OK=TRUE;
  while OK and (| π_i |< u_max) do
  begin
    y =      arg       max [g(i, π_i ∪ x_j)] ;
         x_j∈Pre_i−π_i
    P_new = g(i, π_i ∪ x_j);
    if P_new > P_old then
    begin
      P_old = P_new;
      π_i = π_i ∪ {y};
    else
      OK=FALSE;
    end if
  end while
  Output(x_i, π_i);
end for
```

4 Local Bayesian Increment Function

Let X and Y be two discrete variables, **Z** be a set of discrete variables, and z be an instantiation for **Z**. X,Y$\notin$**Z**.

According to Moore's recommendation [10] about chi-squared test, we assume that the dataset D satisfying at least one of the following two conditions is "sufficiently large" for {X$\cup$Y}:

1. The number of cases of the dataset D is much larger than the number of the values of {X$\cup$Y}, such as $\|D\| > 5 \times (\|X\| \times \|Y\|)$;
2. All cells of {X$\cup$Y} in the contingency table have expected value greater than 1, and at least 80% of the cells in the contingency table about {X$\cup$Y} have expected value greater than 5.

According to Moore's recommendation [10] about chi-squared test, we assume that the sub-dataset $D_{\mathbf{Z}=z}$ satisfying at least one of the following two conditions is "locally sufficiently large" for {X$\cup$Y} given **Z**=z:

1. The number of cases of the sub-dataset $D_{\mathbf{Z}=z}$ is much larger than the number of the values of {X$\cup$Y}, such as $\|D_{\mathbf{Z}=z}\| > 5 \times (\|X\| \times \|Y\|)$;
2. All cells of {X$\cup$Y} in the contingency table conditioned on **Z**=z have expected value greater than 1, and at least 80% of the cells in the contingency table about {X$\cup$Y} on **Z**=z have expected value greater than 5.

Let D be a dataset of m cases, where each case contains a value for each variable. Let V be a set of n discrete variables, where x_i in V has r_i possible values $(v_{i1}, v_{i2}, \ldots, v_{ir_i})$. B_P and B_S denote Bayesian network structures containing just the variables in V. B_S exactly has one more edge $y \rightarrow x_i$ than B_P has. Variable x_i in B_P has the parents set π_i. Variable x_i in B_S has the parents set $\pi_i \cup y$. Let $\phi_i[j]$ denote the unique instantiation of π_i relative to D. Suppose there are q_i such unique instantiations of π_i. Define $N_{ijk}$to be the number of cases in D in which variable x_i is instantiated as v_{ijk} and π_i is instantiated as $\phi_i[j]$.

Let $N_{ijk} = \sum\limits_{y} N_{i,\{j \cup y\},k}, N_{ij} = \sum\limits_{k=1}^{r_i} N_{ijk}$.

$\theta_{ijk}^{B_P}$ denotes $P(x_i = v_{ik} \mid \pi_i = \phi_i[j])$, $\theta_{ijk}^{B_P} > 0$, $\sum\limits_{k=1}^{r_i} \theta_{ijk}^{B_P} = 1, \theta_{ij}^{B_P} = \cup_{j=1}^{q_i}\{\theta_{ijk}^{B_P}\}$, $\theta_i^{B_P} = \cup_{j=1}^{q_i}\{\theta_{ij}^{B_P}\}, \theta^{B_P} = \cup_{i=1}^{n}\{\theta_i^{B_P}\}$. $\hat{\theta}_{ijk}^{B_P}, \hat{\theta}_{ij}^{B_P}, \hat{\theta}_i^{B_P}, \hat{\theta}^{B_P}$ denote maximum likelihoods of $\theta_{ijk}^{B_P}, \theta_{ij}^{B_P}, \theta_i^{B_P}, \theta^{B_P}$.

Given a Bayesian network model, cases occur independently. Bayesian network prior distribution is uniform. Given Bayesian network B_S , there exist two properties: Parameter Independence and Parameter Modularity. [5]

In order to control the complexity of BN model, we apply BIC (Bayesian Information Criterion) approximation formula [11] :

$$BIC(B_S) = \left[\log\left(L\left(\hat{\Theta}\right)\right) - \frac{1}{2}\log(m)dim\left(\hat{\Theta}\right)\right] \approx \log(P(D \mid B_S)) \quad (5)$$

to $\log(P(B_S, D)/P(B_P, D))$. BIC adds the penalty of BN structure complexity to Local Bayesian Increment function in order to avoid overfitting.

For high dimensional and limited sample datasets, with the dimension increase of current parents set π_i, the maximum likelihood estimations $\hat{\Theta}$ of parameters Θ in Bayesian network will be more and more unreliable because the dataset D does not satisfy either condition of "sufficiently large" for π_i. But we find that for some specific instantiations of π_i, the sub-datasets $D_{\pi_i=\phi_i[j]}$ are "locally sufficiently large" for {X∪Y} given $\pi_i = \phi_i[j]$.

Definition 1 (Local Bayesian Increment Function)

$$\begin{aligned}
&Lbi(y, i, \pi_i) \\
&= \log(P(B_S, D)/P(B_P, D)) \\
&= \log(P(D \mid B_S) \times P(B_S)) - \log(P(D \mid B_P) \times P(B_P)) \\
&\approx BIC(B_S) - BIC(B_P) \\
&= \left(\frac{\log\left(L\left(\hat{\Theta}^{B_S}\right)\right)}{\log\left(L\left(\hat{\Theta}^{B_P}\right)\right)} \right) - \frac{1}{2}\log(m)\left[dim\left(\hat{\Theta}^{B_S}\right) - dim\left(\hat{\Theta}^{B_P}\right)\right]
\end{aligned} \tag{6}$$

$$\begin{aligned}
&\log\left(L\left(\hat{\Theta}^{B_S}\right)\right) - \log\left(L\left(\hat{\Theta}^{B_P}\right)\right) \\
&= \log\left(P\left(D \mid \Theta^{B_S}\right)\right) - \log\left(P\left(D \mid \Theta^{B_P}\right)\right) \\
&= \sum_{l=1}^{m} \left(\log\left(P\left(d_l \mid \Theta^{B_S}\right)\right) - \log\left(P\left(d_l \mid \Theta^{B_P}\right)\right)\right) \\
&= \sum_{i=1}^{n} \left(\log\left(P\left(x_i^l \mid \hat{\Theta}_i^{B_S}, \pi_i^l\right)\right) - \log\left(P\left(x_i^l \mid \hat{\Theta}_i^{B_P}, \pi_i^l\right)\right)\right)
\end{aligned} \tag{7}$$

According to the definition of overfitting(the likelihood of the training dataset is larger with the overfitting), we assume that the log-likelihood of x^l does not change for $d_l \in D_{\pi_i=\phi_i[*]}$ in the sub-datasets $D_{\pi_i=\phi_i[*]}$ which are not "locally sufficiently large(lsl)" for {X∪Y},

$$\log\left(P\left(x^l \mid \hat{\Theta}^{B_S}, \pi_l \cup y\right)\right) = \log\left(P\left(x^l \mid \hat{\Theta}^{B_P}, \pi_l\right)\right) . \tag{8}$$

According to (7) and (8), we infer the following results:

$$\begin{aligned}
&\log\left(L\left(\hat{\Theta}^{B_S}\right)\right) - \log\left(L\left(\hat{\Theta}^{B_P}\right)\right) \\
&= \sum_{j}\sum_{k=1}^{r_i}\sum_{y} N_{i,\{y\cup j\},k} \log\left(\frac{N_{i,\{y\cup j\},k} N_{ij}}{N_{i,\{y\cup j\}} N_{ijk}}\right), \text{for } j, D_{\pi_i=\phi_i[j]} \text{ is "locally sufficiently large"} \\
&= \sum_{j}\sum_{k=1}^{r_i}\sum_{y} N \times I_j(\mathrm{X,Y}), \text{for } j, D_{\pi_i=\phi_i[j]} \text{ is "locally sufficiently large"}
\end{aligned} \tag{9}$$

$$\dim\left(\hat{\Theta}^{B_S}\right) - \dim\left(\hat{\Theta}^{B_P}\right) = (r_y - 1)(r_i - 1)q_i \tag{10}$$

$$\begin{aligned} & Lbi(y, i, \pi_i) \\ & = \log\left(\frac{P(B_S, D)}{P(B_P, D)}\right) \\ & = \sum_j \sum_{k=1}^{r_i} \sum_y N \times I_j(\text{X,Y}) - \frac{1}{2}(r_y - 1)(r_i - 1)q_i \log(m), \\ & \qquad \text{for } j, D_{\pi_i=\phi_i[j]} \text{ is "locally sufficiently large"}. \end{aligned} \tag{11}$$

Note: I_j(X,Y) is the mutual information between variable X and variable Y in the sub-dataset $D_{\pi_i=\phi_i[j]}$.

5 MRMRG Algorithm

In this section, we discuss MRMRG algorithm, Max Relevance-Min Redundancy Greedy algorithm. For high dimensional and limited sample datasets, MRMRG algorithm improves K2 in two ways.

Firstly, for high dimensional dataset, the computation of $g(i, \pi_i)$ has less and less reliability and robustness with the dimension increase of π_i, so that formula

$$y = \underset{x_j \in Pre_i - \pi_i}{\arg} \max[g(i, \pi_i \cup x_j)]$$

cannot accurately obtain node y which has the maximal mutual information increment between node i and $\pi_i \cup y$ in K2 algorithm. According to Lemma 2, MRMRG algorithm replaces

$$y = \underset{x_j \in Pre_i - \pi_i}{\arg} \max[g(i, \pi_i \cup x_j)]$$

by

$$y = \underset{x_j \in Pre_i - \pi_i}{\arg} \max\left[I(x_i, x_j) - \frac{1}{\mid \pi_i \mid} \sum_{x \in \pi_i} I(x, x_j)\right].$$

MRMRG reduces the dimension of the computation by 2 dimension. Therefore, MRMRG algorithm greatly improves reliability and robustness of previous BN learning algorithms in high dimensional and limited sample datasets.

Secondly, When the dataset D is "sufficiently large" for {X∪Y∪π_i}, MRMRG algorithm uses Local Bayesian Increment function to control the complexity of BN and to avoid overfitting. When dataset D is not "sufficiently large" for {X∪Y∪π_i}, but there exist sub-datasets $D_{\pi_i=\phi_i[j]}$ are "locally sufficiently large" for {X∪Y} given $\pi_i = \phi_i[j]$, MRMRG algorithm can apply Local Bayesian Increment function to improve accuracy and avoid overfitting. The technique also makes it unnecessary to set the maximum parents number u_{max} of a node.

5.1 MRMRG Algorithm

Input: A set of n nodes , an ordering on the nodes, the mutual information between any two nodes is saved in 2-dimension array MI(n, n), prei denotes the set of nodes that precede xi, the parents set of node xi, a dataset containing m cases.

Output: For each node xi, a printout of the parents set of the node.

Assume to satisfy all conditions and assumptions in section 4.

Procedure:

```
* First Part *
  /* get mutual information between any two nodes */
  initialize MI(n,n) to 0;
  for i=1 to n do
    for j=i+1 to n do
    begin
      MI[i,j] = MI[j,i] = I(x_i, x_j);
    end for
  end for
* Second Part *
  /* obtain a set of parents of each node, and output results */
  for i=1 to n do
  begin
    π_i=NULL;
    OK=TRUE;
    while OK do
    begin
      y = arg max_{x_j ∈ Pre_i − π_i} [ I(x_i, x_j) − 1/|π_i| Σ_{x∈π_i} I(x, x_j) ];
      if Lbi(y, i, π_i) > 0 then
        π_i = π_i ∪ {y};
      else
        OK=FALSE;
      end if
    end while
    Output(x_i, π_i);
  end for
```

where

$$y = \arg\max_{x_j \in Pre_i - \pi_i} \left[I(x_i, x_j) - \frac{1}{|\pi_i|} \sum_{x \in \pi_i} I(x, x_j) \right];$$

MRMRG algorithm has two parts:

The first part includes the computation of mutual information between any two nodes.

In the second part, MRMRG algorithm initializes the parents set of given node to NULL, and then add one by one the node, which can get the maximal increment of mutual information, to the parents set, until the computation result of Local Bayesian Increment function is no more than 0. Repeating the above steps for every node, we can obtain an approximately optimal Bayesian network.

5.2 The Time Complexity of MRMRG

The complexity of first part of MRMRG is $O(n^2)$.

The second part of MRMRG: The complexity of computing

$$y = \underset{x_j \in Pre_i - \pi_i}{\arg \max} \left[I(x_i, x_j) - \frac{1}{|\pi_i|} \sum_{x \in \pi_i} I(x, x_j) \right]$$

is $O(n)$. The complexity of computing $Lbi(y, i, \pi_i)$ is $O(mnr)$ (Note: Let $r = max(r_i) i = 1, \ldots, n$). The other steps in the while statement require $O(1)$ time. The while statement loops at most $O(n)$ times, each time it is entered. The for statement loops n times. In the worst case, the complexity of second part in MRMRG is $(O(n)+O(mnr))*O(n)*n=O(mn^3r)$.

The overall time complexity of MRMRG is $O(n^2)+O(mn^3r)=O(mn^3r)$ at worst. On the other hand, the worst-case time complexity of K2 is $O(mn^4r)$.

6 Experimental Results

We implemented MRMRG algorithm, K2 algorithm, TPDA algorithm and presented the experimental comparison results of the three implementations.

Tests were run on a PC with Pentium4 1.5GHz and 1GB RAM. The operating system was Windows 2000. The program was developed under Matlab 7.0. 3 Bayesian networks were used. Table 1 shows the networks characteristics. From each of the networks we randomly sampled 3 datasets with 500, 2000, 10000 training cases each.

Table 1. Bayesian networks

BNs	Vars Num	Arcs Num	Max In/Out Degree	Domain Range
Insur	27	52	3/7	2-5
Alarm	37	46	4/5	2-4
Munin	189	282	3/15	1-21

6.1 Comparison of Runtime Among Algorithms

A summary of the time results of the execution of all three algorithms is in Table 2. We normalized the times reported by dividing by the corresponding running time of MRMRG on the same datasets and reported the averages over sample sizes. Thus, a normalized running time of greater than 1 implies a slower algorithm than MRMRG on the same learning task. A normalized running time of lower than 1 implies a faster algorithm than MRMRG.

From the results, we can see that MRMRG has better performance than other two algorithms K2 and TPDA. For small sample sizes (500, 2000), MRMRG runs several times faster than K2 and TPDA. For larger sample sizes (10000), MRMRG performs nearly one magnitude faster than K2.

Table 2. Normalized Runtime of Algorithms

Size	MRMRG	K2	TPDA
500	1.0	1.58	3.79
2000	1.0	2.39	3.26
10000	1.0	6.98	2.18

6.2 Comparison of Accuracy Among Algorithms

We compared the accuracy of bayesian networks learned by three algorithms according to Bayesian score of every learned Bayesian netowrk. Table 3,4,5 reports the Bayesian scores(BDeu).

Table 3. Bayesian Score BDeu(Insur)

Size	MRMRG	K2	TPDA
500	-19.0322	-19.5833	-28.7857
2000	-18.0233	-18.2722	-24.7111
10000	-17.8519	-17.8498	-17.8571

Table 4. Bayesian Score BDeu(Alarm)

Size	MRMRG	K2	TPDA
500	-13.8577	-13.9500	-18.0973
2000	-13.1732	-13.3824	-14.5286
10000	-12.9641	-12.9644	-12.9640

Table 5. Bayesian Score BDeu(Munin)

Size	MRMRG	K2	TPDA
500	-62.4827	-63.2500	-125.6250
2000	-61.1359	-61.3175	-143.4762
10000	-59.8601	-59.8576	-110.3646

From the results, we can see that MRMRG can learn more accurately than K2 for limited datasets relative to the learned Bayesian network, such as Insur(500,2000), Alarm(500,2000), Munin(500,2000). For sufficiently large datasets for K2 algorithm and MRMRG algorithm, the accuracy of MRMRG and K2 is almost same, such as Insur(10000), Alarm(10000), Munin(10000).

7 Conclusion

Accuracy, reliability, robustness and efficiency are main indices in evaluating algorithms for learning Bayesian network. This paper proposes MRMRG algorithm which applies Max Relevance-Min Redundancy technology, Local Bayesian Increment function. This algorithm greatly reduces the number of high dimensional computations and improves scalability of learning. From experimental results, MRMRG has better performance on reliability, robustness and efficiency. Moreover, for high dimensional and limited sample datasets, MRMRG also has better accuracy than most of existing algorithms.

References

1. Heckerman, D.: Bayesian Networks for Data Mining. Microsoft Press, Redmond, USA (1997).
2. Spirtes, P., Glymour, C., Scheines, R.: Causation, Prediction and Search. MIT Press, Massachusetts, USA (2000).
3. Cheng, J., Greiner, R., Kelly, J., Bell, D., Liu, W.: Learning Belief Networks form Data: An Information Theory Based Approach. Artificial Intelligence, Vol.137, No.1-2, (2002) 43-90.
4. Cooper, G., Herskovits, E.: A Bayesian Method for the Induction of Probabilistic Networks from Data. Machine Learning, Vol.9, No.4, (1992) 309-347.
5. Heckerman, D., Geiger, D., Chickering, D.M.: Learning Bayesian Networks: the Combination of Knowledge and Statistical Data. Machine Learning, Vol.20, No.3, (1995) 197-243.
6. Suzuki, J.: Learning Bayesian Belief Networks Based on the MDL Principle: An Efficient Algorithm Using the Branch and Bound Technique. In: Proceedings of the Thirteenth International Conference on Machine Learning ICML'1996, Morgan Kaufmann, (1996) 462-470.
7. Pearl, J.: Probabilistic Reasoning in Intelligent Systems: Networks of Plausible Inference. Morgan Kaufman, San Francisco, USA (1998).
8. Wai, Lam., Fahiem, B.: Learning Bayesian Belief Networks An approach based on the MDL Principle. Computational Intelligence. Vol.10, No.4 (1994) 269-293.
9. HanChuan, P., Fuhui, L., Chris, Ding.: Feature Selection Based on Mutual Information: Criteria of Max-Dependency, Max-Relevance,and Min-Redundancy. IEEE Transactions on Pattern Analysis and Machine Intelligence. Vol.27, No.8 (2005) 1226-1238.
10. Moore, D.S.: Goodness-of-Fit Techniques. Marcel Dekker, New York, USA (1986).
11. Kass, R.E., Raftery, A.E.: Bayes factors. Journal of the American Statistical Association. Vol.90, No.430 (1995) 773-796.

On Affect and Self-adaptation: Potential Benefits of Valence-Controlled Action-Selection

Joost Broekens, Walter A. Kosters, and Fons J. Verbeek

Leiden Institute of Advanced Computer Science, Leiden University, P.O. Box 9500, 2300 RA Leiden, The Netherlands

Abstract. Psychological studies have shown that emotion and affect influence learning. We employ these findings in a machine-learning meta-parameter context, and dynamically couple an adaptive agent's artificial affect to its action-selection mechanism (Boltzmann β). The agent's performance on two important learning problems is measured. The first consists of learning to cope with two alternating goals. The second consists of learning to prefer a later larger reward (global optimum) for an earlier smaller one (local optimum). Results show that, compared to several control conditions, coupling positive affect to exploitation and negative affect to exploration has several important benefits. In the alternating-goal task, it significantly reduces the agent's goal-switch search peak. The agent finds its new goal faster. In the second task, artificial affect facilitates convergence to a global instead of a local optimum, while permitting to exploit that local optimum. We conclude that affect-controlled action-selection has adaptation benefits.

1 Introduction

Affect influences thought and behavior in many ways [1,2,3,4]. While affective states can be complex and composed of multiple components, in this paper we use the term *affect* to refer to valence: the positiveness versus negativeness of an agent's affective state (in our case, the agent's mood: a long term, low intensity affective state) [3]. Valence can be seen as a further undifferentiated component of an affective state that defines an agent's situation as good versus bad [5].

We focus on the influence of affect on *learning*. Numerous psychological studies support the idea that enhanced learning is related to positive affect [6], while others show that enhanced learning is related to negative affect [7], or to both [8]. Currently it is not yet clear how affect influences learning. Computational modeling might give insights into the possible underlying mechanisms.

From a machine learning point of view the influence of affect on learning suggests that adaptive agents can benefit from artificial affect, once we know how to (1) simulate affect in a way useful for learning, (2) know what parts of the adaptive agent's architecture can be influenced by artificial affect, and (3) know how to *connect* artificial affect to the appropriate parts of that architecture.

We investigate the relation between affect and learning with a self-adaptive agent in a simulated gridworld. Our agent autonomously influences its action-selection mechanism. It uses artificial affect to control its amount of exploration.

J. Mira and J.R. Álvarez (Eds.): IWINAC 2007, Part I, LNCS 4527, pp. 357–366, 2007.

Our agent uses a standard form of model-based reinforcement learning (see Section 4). We present results based on two different learning tasks. In the first the agent has to cope with two different alternating goals in a two-armed grid-world. We call this the "alternating goal task". This is an important task to be able to learn. An agent with a changing set of goals that has to cope with a dynamic environment has to learn to modify its behavior in order to reflect a change in the set of goals; it has to be flexible enough to give up on an old goal and it has to be persistent enough to continue trying an active goal. In other words the agent has to decide when to explore versus exploit its knowledge, a.k.a. the exploration-exploitation problem [9].

The second task consists of learning to prefer a later larger reward (global optimum) for an earlier smaller one (local optimum). We call this task the "Candy task"; candy represents the local optimum being closest to the agents starting position, while food represents the global optimum being farther away from its starting position. The ability to learn this task is important as it enables survival with the knowledge an agent has, while trying to find better alternatives. Failure to do so results in getting stuck in local optima or slow convergence.

2 The Influence of Affect on Learning

In this section we review some of the evidence that affect influences natural information processing and learning. Some studies find that negative affect enhances learning [7]. Babies aged 7 to 9 months were measured on an attention and learning task. The main result is that negative affect correlates with faster learning. Attention was found to mediate this influence. Negative affect related to more diverse attention, i.e., the babies' attention was "exploratory", and both negative affect and diverse attention related to faster learning. Positive affect had the opposite effect as negative affect (i.e., slower learning and "less exploratory" attention). This relation suggests that positive affect relates to exploitation, while negative affect relates to exploration.

Other studies suggest an inverse relation [6], and find that mild increases in positive affect related to more flexible attention but also to more distractible attention. So it seems that in this study positive affect facilitated a form of exploration, positive affect removes attention bias towards solving the old task.

Of course, attention is not equivalent to learning. It is, however, strongly related to exploration: an important precursor to learning. Flexible distribution of attentional resources favors processing of a wide range of external stimuli. So, in the study by Dreisbach and Goschke [6] positive affect facilitated exploration, as it helped to remove bias towards solving the old task thereby enabling the subject to faster adapt to the new task. In the study by Rose et al. [7] negative affect facilitated exploration as it related to defocused attention.

Other studies, e.g., [8] show that both negative and positive affect can relate to faster learning. The authors found that both flow (a positive state characterized by a learner receiving the right amount of new material at the right speed [10]) and confusion related to better learning.

Combined, these results suggest that positive and negative affective states can help learning at different phases in the process, a point explicitly made in [8]. Our paper investigates this in an adaptive agent context.

3 Simulated Affect Influences Action Selection

To model the influence of affect on learning, we simulate affect as follows. Our agent learns based on reinforcement learning (RL), so at every time step it receives a reward r. Simulated affect is based on this r:

$$e_p = (r_{star} - (r_{ltar} - f\sigma_{ltar}))/2f\sigma_{ltar} \tag{1}$$

Here, e_p is the measure for positive affect, where e_p ranges from 0 to 1, modeling negative affect versus positive affect respectively. The short-term running average reinforcement signal, r_{star}, has a parameter *star* defining the window-size (in steps) of that running average. At every step of the agent, r_{star} is used as input to calculate a long-term running average reinforcement signal, r_{ltar}, with *ltar* a parameter again defining the window-size. The standard deviation of r_{star} over that same long-term period is denoted by σ_{ltar}, and f is a multiplication factor defining the sensibility of the measure. The standard deviation is included as a measure to normalize the affect signal based on the natural variance of r_{star}. Artificial affect measures "how well the agent is doing compared to what it is used to".

Two issues regarding natural affect are important. First, in studies that measure the influence of affect on cognition, affect relates more to long-term mood than to short-term emotion. Affect is usually induced before or during the experiment aiming at a continued, moderate effect instead of short-lived intense emotion-like effect [6,3,7]. This is reflected by the fact that e_p is based on reinforcement signal *averages*, not on r itself.

Second, affect induction (the method used in psychological experiments to investigate the influence of affect on information processing) is compatible with the administration of reward in reinforcement learning. Affect is usually induced by giving subjects small *unanticipated* rewards [1,11]. The reinforcement signal in RL only exists if there is a difference between predicted and received reward. Predicted rewards thus have the same effect as no reward. It seems that both reward and positive affect follow the same rule: if it's predicted it isn't important. This is reflected in our measure. It compares a short-term estimate r_{star} with a long-term estimate r_{ltar}. As the first, short-term average reacts quicker to changes in the reward signal than the second, long-term average, a comparison between the two yields a measure for how well the agent is doing compared to what it is used to. If the environment and the agent's behavior in that environment do not change, e_p converges to a neutral value of 0.5. This reflects the fact that anticipated rewards do not influence affect much.

Our agent uses a Boltzmann distribution to select actions:

$$p(a) = \frac{e^{V_t(s,a)\cdot\beta}}{\sum_{b=1}^{n} e^{V_t(s,b)\cdot\beta}} \tag{2}$$

Here, $p(a)$ is the probability that a certain action a out of n possible ones is chosen, and $V_t(s,a)$ is the value of action a in state s at time t. The inverse temperature parameter β determines the randomness of the distribution. High β's result in a greedy selection strategy (low temperature, small randomness). If β is zero the distribution function results in a random selection strategy, regardless of the predicted reward values (high temperature, high randomness).

In our experiments artificial affect e_p controls an agent's β parameter and thereby exploration versus exploitation. This is compatible with Doya's approach [12], who proposes that emotion is also a system for meta-learning.

3.1 Type-A: Positive Affect Relates to Exploitation

To investigate the influence of artificial affect on exploration, we study two types of relations. Type-A models that positive affect increases exploitation [7]:

$$\beta = e_p \cdot (\beta_{max} - \beta_{min}) + \beta_{min} \tag{3}$$

If e_p increases to 1, β increases towards β_{max}. As e_p decreases to 0, β decreases towards β_{min}. So positive affect results in more exploitation, while negative affect results in more exploration. In essence, our agent is autonomously adapting how selective its attention process is: "when happy, be greedy in the actions to consider, when sad: consider all possible actions equally".

3.2 Type-B: Negative Affect Relates to Exploitation

Type-B models the inverse of the previous relation (as suggested in [6]):

$$\beta = (1 - e_p) \cdot (\beta_{max} - \beta_{min}) + \beta_{min} \tag{4}$$

As affect e_p increases to 1, β decreases towards β_{min} and as ep decreases to 0, β consequently increases towards β_{max}. So positive affect results in more exploration, while negative affect results in more exploitation. In this case our agent uses a different way to adapt attention: "when sad, be greedy in the actions to consider, when happy: consider all possible actions equally".

4 Experiment and Method

Our experiments are performed in two different simulated gridworlds (Fig. 1). The first is a two-armed maze with a potential goal at the end of each arm. This maze is used for the Alternating-Goal (AG) task, i.e., coping with two alternating goals, find food or find water (only one goal is active during an individual trial, goal reward $r = +2.0$). The second maze has two active goal locations. The nearest goal location is the location of the candy (i.e., a location with a reward $r = +0.25$), while the farthest goal location is the food location ($r = +1.0$). This maze is used for the Candy task. The walls in the mazes are *lava* patches, on which the agent can walk, but is discouraged to do so by a negative

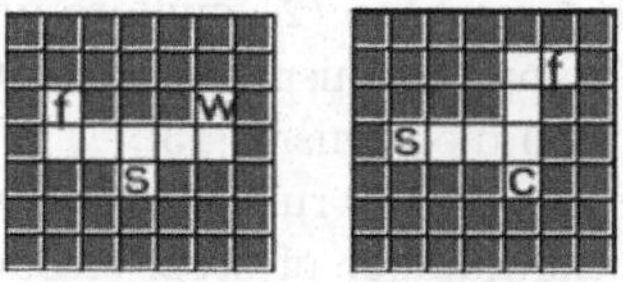

Fig. 1. The task-mazes used. Left maze is the Alternating-Goal task. Right maze is the Candy task. s denotes agent's starting position, f is food, c is candy and w is water.

reinforcement ($r = -1.0$). The agent learns by acting in the maze and perceiving its direct environment using an 8-neighbor and center metric (i.e., it senses its eight neighbors and the location it is at). An agent that arrives at a goal location is replaced to its starting location. Agents start with an empty world model and construct a Markov Decision Process (MDP) as usual (a perceived stimulus is a state s in the MDP, and an action a leading from state s_1 to s_2 is an edge in the MDP). The agent counts state occurrences, $N(s)$, and uses this count in a standard weighting mechanism. Values of states are updated using as follows:

$$R(s) \leftarrow R(s) + \alpha \cdot (r - R(s)) \tag{5}$$

$$V(s) \leftarrow R(s) + \gamma \cdot \sum_i V(s_{a_i}) \frac{N(s_{a_i})}{\sum_j N(s_{a_j})} \tag{6}$$

So, a state s has a learned reward $R(s)$ and a value $V(s)$ that incorporates predicted future reward. $R(s)$ converges to the reward for state s with a speed proportional to the learning rate α. $V(s)$ is updated based on $R(s)$ and the weighted values of the next states reachable by action $a_{1\ldots i}$ (with a discount factor of γ). So, we use a standard model-based RL approach [9]. In the Alternating-Goal task the learning rate α and discount factor γ are respectively 1.0 and 0.7, and in the Candy task respectively 1.0 and 0.8. We have fixed the artificial affect parameters $ltar$, $star$ and f to 400, 50 and 1, respectively.

4.1 Learning Tasks

To analyze the difference in learning behavior of agents that use affect control of type-A, B and a control condition using static β values we did the following. In the Alternating-Goal task agents first have to learn goal one (food). After 200 trials the reinforcement for food is set to $r = 0.0$, while the reinforcement for water is set to $r = +2.0$. The water is now the active goal location (so an agent is only reset at its starting location if it reaches the water). This reflects a change in goals, of which the agent is initially unaware. It has to search for the new goal location. After 200 trails, the situation is set back; i.e., food becomes the active goal. This is repeated 2 times, resulting in 5 learning phases (phases 0 to 4 referring to learning of food, then water, food, water, and finally food). This represents 1 run, and we repeated these runs.

The setup of the Candy task is simpler. The agent has to learn to optimize reward in the Candy maze. The problem for the agent is to (1) exploit the local reward (candy), but at the same time (2) explore and then exploit the global reward (food). This relates to opportunism, an important ability that should be provided by an action-selection mechanism [13].

All AG task results are based on 800 runs, while Candy task results are based on 400 runs. We averaged the number of steps needed to get to the goal over these runs, resulting in an average learning curve. The same was done for β (the *exploration-exploitation* curve), and the agent's quality of life (QOL) (measured as the sum of the rewards received during 1 trial). In all plots the trials are on the x-axis, while β, steps or quality of life on the y-axis.

4.2 Experiment 1: Alternating-Goal Task

Our main finding is that type-A (positive affect relates to exploitation, negative to exploration) has by far the lowest switch cost between different goals, as measured by the number of steps taken at the trial in which the goal-switch is made (Fig. 3b). This is an important adaptation benefit. All goal-switch peaks (phases 1–4) of the 4 variations of type-A (i.e., dotted lines labeled "AG dyn 3–6, 3–7, 3–9, and 2–8") are smaller than the peaks of the controls (straight lines labeled "AG static 3,4,5,6, and 7") and type-B (i.e., striped lines labeled "AG dyn inv 3–6, 3–7, 3–9 and 4–9"). Initial learning (phase 0) is marginally influenced by affective feedback (peaks at phase 0 in Fig. 3b), as can be expected: no goal switch occured before the initial learning phase. Closer investigation of the first goal switch (trial 200, phase 1) shows that the trials just after the goal switch also benefit considerably from type-A (Fig. 2b). When we computed for all settings an average peak for trial 200, 201 and 202 together, and compared these averages statistically, we found that type-A performs significantly better ($p < 0.001$ for all comparisons, Mann-Whitney, $n = 800$). Analysis of the peak at trial 800 (phase 4), reveals about the same picture. The trial in which the goal is switched benefits significantly from type-A ($p < 0.01$) for all comparisons.

All other comparisons between peaks revealed significantly ($p < 0.001$) smaller peaks for type-A (Fig. 3b). This effect is most clearly shown for the peaks of phase 3 and 4, where the relative peak-height difference between type-A peaks and static peaks ranges between 1.25 and 2. This means that using positive affect to control action-selection in the way described in type-A can result in up to a 2 fold decrease of search investment needed to find a new goal. As expected, the smallest difference between control and type-A is when β is small (3 or 4) in the control condition (small β = exploration = less tied to old goal). However, small β's have a classical downside: less convergence due to less exploitation (Fig. 3a). In contrast, type-A curves in Fig. 3a show that the agent does converge to the minimum number of steps needed to get to the goal (i.e., 4).

For completeness we show the β curves for the complete phase 1 of the control group and one type-A and one type-B (Fig. 2a). These curves confirm the expected β dynamics. For type-A, the goal-switch induces high exploration (β near β_{min}) for type-A due to the lack of reinforcement (it is going worse than expected), after which β quickly moves up to β_{max}, and then decays to average. For type-B this is exactly the inverse.

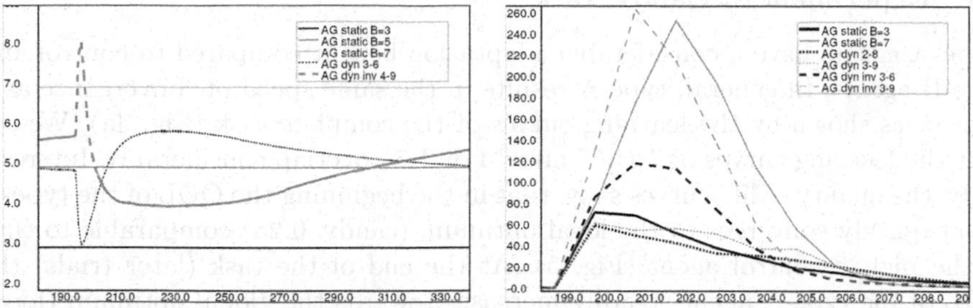

Fig. 2. (a) AG mean β for phase 1. (b) AG mean steps, detail phase 1.

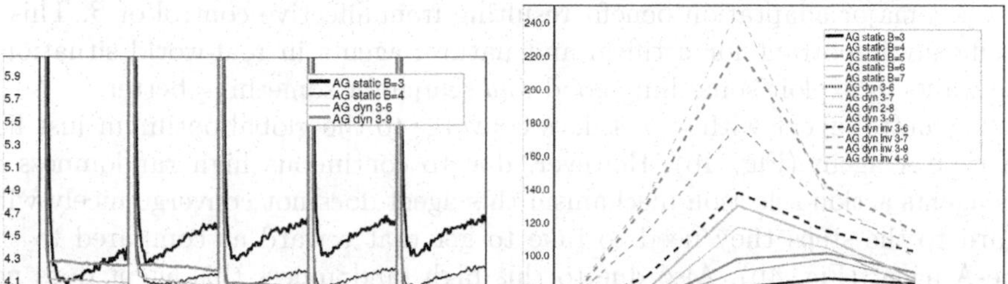

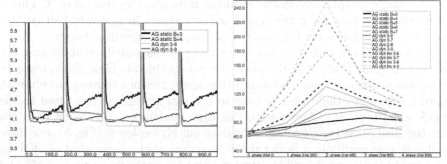

Fig. 3. (a) AG all phases. (b) AG peaks at phases 0–4 (steps in trial 0, 200, 400, etc.).

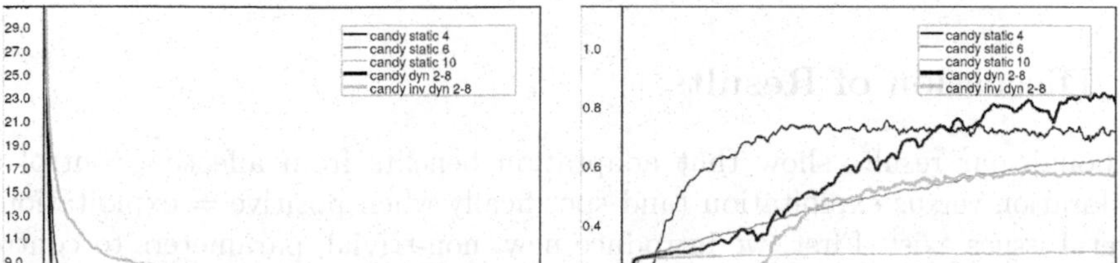

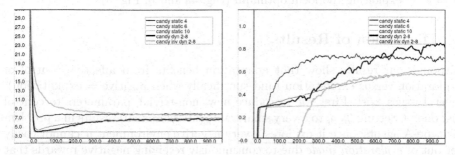

Fig. 4. (a) Candy task complete, mean steps. (b) Candy task complete, mean QOL.

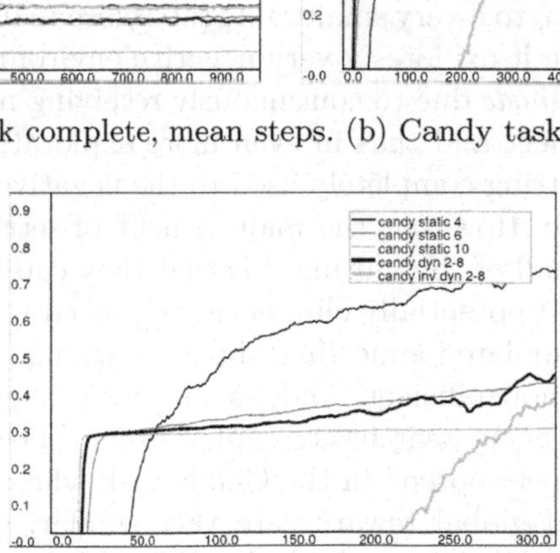

Fig. 5. Candy task starts learning, mean QOL

4.3 Experiment 2: Candy Task

Type-A agents have a considerable adaptation benefit compared to control and type-B agents. In general, type-A results in the same speed of convergence as a high β, as shown by the learning curves of the complete task (Fig. 4a). We see that the learning curves of "β=6" and "dyn 2–8" overlap considerably. Interestingly, the quality of life curves show that in the beginning the QOL of the type-A agent quickly converges to the local optimum (candy, 0.25) comparable to that of the high β control agent (Fig. 5). At the end of the task (later trials) the QOL of the type-A agent steadily increases towards the global optimum (food, +1.0; Fig. 4b). This shows that type-A affective feedback helps to first exploit a local optimum, while at a later stage explore for and exploit a global optimum. This is a major adaptation benefit resulting from affective control of β. This is specifically important for artifical and natural agents in real-world situations. One wants to exploit something good *and* search for something better.

The control agent with $\beta = 4$ does converge to the global optimum just like the type-A agent (Fig. 4b). However, due to continuous high randomness in this agents action-selection mechanism this agent does not converge nicely with regard to the steps they need to take to get that reward as compared to the type-A agent (Fig. 4b). Also due to this high randomness this agent does not learn the local reward consistently enough to quickly exploit it (Fig. 5). For these smaller βs this results in a major delay in arrival at the same level of QOL as compared to the larger βs and the type-A agent (compare the curves "β=5" and "dyn 2-8" in Fig. 5). The type-B agent does not perform well at converging or at quickly exploiting the local optimum (Fig. 4a and b, Fig. 5).

5 Discussion of Results

Although our results show that adaptation benefits from affective control of exploration versus exploitation (and specifically when positive = exploitation), several issues exist. First, we introduce new, non-trivial, parameters to control just one, β. Setting β_{min} to a very small value (i.e., close to 0) results in a problem for type-A agents: once it explores a very negative environment it cannot easily get out of *exploration mode* due to continuously receiving negative rewards that result in even lower affect and thus in even more exploration. The only way to get out of this is by getting completely used to the negative environment, which might take a long time. However, the main benefit of setting meta-parameters such as *ltar*, *star* instead of configuring β is that they can be set for a complete (set of) learning task(s) potentially eliminating the need to adapt β using other mechanism (such as simulated annealing) during learning. Therefore, configuration of these values is more efficient. And, as our results show, the β is controlled in such a way that the agent switches to exploration in the alternating-goal task when needed, and it "gets bored" in the Candy task when needed.

Second, if local and global rewards are very similar, even a type-A agent cannot learn to prefer global reward, as the difference becomes very small. So, the candy and food reward have to be significantly different, such that the agent can exploit this difference once both options have been found. You don't walk a long way for a little gain. The discount factor is related to this issue, as a small γ results in discarding rewards in the future and therefore the agent is more prone to fall for the nearer local optimum. So γ should be set such that the agent is at least theoretically able to prefer a larger later reward for a smaller earlier one (which is $\gamma = 0.8$ in the Candy task, as compared to 0.7 in the AG task).

5.1 Related Work

Our work relates to computational modeling of emotion and motivation based control/action-selection. It explicitly defines a role for emotion in biasing behavior-selection (e.g, [14]). The main difference is that we have explicitly experimented with a psychologically plausible model of affect as a way to directly, and continuously, control the randomness of action-selection.

Although affective control of exploration is promising for adaptive behavior, our learning model is specific, and our claims hard to generalize. Other learning architectures, such as Soar or ACT-R, should be used to further investigate the mechanisms introduced here. Belavkin [15] has shown using ACT-R that affect can be used to control the search through the solution space, which resulted in better problem-solving performance. He used an information-theoretic approach towards modeling affect that is related to the rule-state of the ACT-R agent. A key difference is thus that our measure for affect is based on a comparison of reinforcement signal averages. Further we explicitly model affect based on different theoretical views on the relation between affect and information processing and compared these views experimentally. The *Salt* model [16] relates to Belavkin's approach in the sense that the agent's effort to search for a solution in its memory depends on, among other parameters, the agent's mood valence.

Schweighofer and Doya [17] used a similar measure for "how well the agent is doing compared to what it is used to"; however, they use it differently. Instead of directly controlling β, affect is used as basis for a search-based method. If a random change to β results in positive affect (agent is doing better), the new β is kept, and vice versa. Recently we have extended this work by comparing these methods on the same tasks in a different learning environment (Soar-RL) [18].

6 Conclusions

We have defined a measure for affect for adaptive agents, and used it to control action-selection. Based on experimental results with learning agents in simulated gridworlds, we conclude that coupling positive affect to exploitation and negative affect to exploration has several important adaptation benefits, at least in the tasks we have experimented with. First, it significantly reduces the agent's *goal-switch search peak* when the agent learns to adapt to a new goal: the agent finds this new goal faster. Second, artificial affect facilitates convergence to a global instead of a local optimum, while exploiting that local optimum. However, additional experiments are needed to verify the generality of our results, e.g., in continuous problem spaces, and other learning architectures (see [18]).

References

1. Ashby, F.G., Isen, A.M., Turken, U.: A neuropsychological theory of positive affect and its influence on cognition. Psychological Review **106** (1999) 529–550
2. Damasio, A.R.: Descartes' error: Emotion, reason, and the human brain. Gosset/Putnam Press, New York (1994)
3. Forgas, J.P.: Feeling is believing? The role of processing strategies in mediating affective influences in beliefs. In: Frijda, N. et al. (Eds.). Emotions and Beliefs, Cambridge, UK: Cambridge University Press (2000) 108–143
4. Phaf, R.H., Rotteveel, M.: Affective modulation of recognition bias. Emotion **15** (2005) 309–318
5. Russell, J.A.: Core affect and the psychological construction of emotion. Psychological Review **110** (2003) 145–172
6. Dreisbach, G., Goschke, T.: How positive affect modulates cognitive control: Reduced perseveration at the cost of increased distractibility. Journal of Experimental Psychology **30** (2004) 343–353
7. Rose, S.A., Futterweit, L.R., Jankowski, J.J.: The relation of affect to attention and learning in infancy. Child Development **70** (1999) 549–559
8. Craig, S.D., Graesser, A.C., Sullins, J., Gholson, B.: Affect and learning: An exploratory look into the role of affect in learning with Autotutor. Journal of Educational Media **29** (2004) 241–250
9. Sutton, R., Barto, A.: Reinforcement learning, An introduction. MIT Press, Cambridge, Massachusetts (1998)
10. Csikszentmihalyi, M.: Flow: The psychology of optimal experience. New York: Harper Row (1990)
11. Custers, R., Aarts, H.: Positive affect as implicit motivator: On the nonconscious operation of behavioral goals. Journal of Personality and Social Psychology **89** (2005) 129–142
12. Doya, K.: Metalearning and neuromodulation. Neural Networks **15** (2002) 495–506
13. Tyrrell, T.: Computational mechanisms for action selection, PhD thesis. University of Edinburgh (1993)
14. Avila-Garcia, O., Cañamero, L.: Using hormonal feedback to modulate action selection in a competitive scenario. In: Proceedings of the 8th Intl. Conf. on Simulation of Adaptive Behavior. (2004) 243–252
15. Belavkin, R.V.: On relation between emotion and entropy. In: Proceedings of the AISB'04 Symposium on Emotion, Cognition and Affective Computing, Leeds, UK (2004) 1–8
16. Botelho, L.M., Coelho, H.: Information processing, motivation and decision making. In: Proc. 4th International Workshop on Artificial Intelligence in Economics and Management. (1998)
17. Schweighofer, N., Doya, K.: Meta-learning in reinforcement learning. Neural Networks **16** (2003) 5–9
18. Hogewoning, E., Broekens, J., Eggermont, J., Bovenkamp, E.G.P.: Strategies for affect-controlled action-selection in Soar-RL, submitted to IWINAC'2007. (2007)

Detecting Anomalous Traffic Using Statistical Discriminator and Neural Decisional Motor

Paola Baldassarri, Anna Montesanto, and Paolo Puliti

Dipartimento di Elettronica Intelligenza Artificiale e Telecomunicazioni
Università Politecnica delle Marche, Ancona, Italy
{p.baldassarri,a.montesanto,p.puliti}@univpm.it

Abstract. One of the main challenges in the information security concerns the introduction of systems able to identify intrusions. In this ambit this work takes place describing a new Intrusion Detection System based on anomaly approach. We realized a system with a hybrid solution between host-based and network-based approaches, and it consisted of two subsystems: a statistical system and a neural one. The features extracted from the network traffic belong only to the IP Header and their trend allows us detecting through a simple visual inspection if an attack occurred. Really the two-tier neural system has to indicate the status of the system. It classifies the traffic of the monitored host, distinguishing the background traffic from the anomalous one. Besides, a very important aspect is that the system is able to classify different instances of the same attack in the same class, establishing which attack occurs.

1 Introduction

Intrusion Detection Systems (IDS) are being designed to protect the availability, confidentiality and integrity of critical networked information systems [5]. Existing intrusion detection techniques fall into two major categories: misuse detection [6,11] and anomaly detection [1]. Misuse detection techniques try a match between activities in an information system and signatures of known intrusions and signal intrusions. Anomaly detection techniques establish a profile of the normal behaviour of a user and/or a system. They compare the observed behaviour with the normal profile, and they signal intrusions when the observed behaviour deviates significantly from its normal profile. Statistical modelling followed with classical or neural network classification has been utilized in some anomaly intrusion detection systems [13]. The work of Ghosh et al. [1] studied the employment of neural network classifiers to detect anomalous and unknown intrusions against a software system. Horeis [3] observed that Neural Networks are tolerant of imprecise data and uncertain information. So, with their ability to generalize from learned data they seem to be an appropriate approach to IDS. Current research in the area of Intrusion Detection based on Neural Networks shows encouraging results. Besides Neural Networks were specifically proposed to learn the typical characteristics of system users and identify statistically significant variations from their established behaviour [10]. The Self-Organizing

J. Mira and J.R. Álvarez (Eds.): IWINAC 2007, Part I, LNCS 4527, pp. 367–376, 2007.

Maps (SOM) developed by Kohonen [4] automatically categorizes the varieties of input presented during the training and can then express how well new input fit the patterns it has discerned. However the use of the SOM approach is a relatively new choice for anomaly detection. Rhodes et al. [10] developed a system that uses multiple SOM for intrusion detection. They suggest that each neural network become a kind of specialist trained to recognize the normal activity of a single protocol and ready to raise an alarm when a significant deviation is detected. The paper [7] describes an anomaly approach that characterizes each connection based on some features. Using these features, the SOM classifies the network traffic into normal or suspicious. In [9] the authors characterize each network connection based on a different set of features and build SOM for each individual network service of interest.

In this work, the SOM has to classify the network traffic from and towards a victim host: such as a web server, a mail server or a remote authentication server. From this point of view, we realized a system similar to an IDS "host based", but our original hybrid solution derives from the fact that the IDS does not reside on the monitored host, but in another computer exclusively dedicated to this purpose. So, the system does not depend on the operating system of the monitored host, guarantying integrity and robustness, also in the case of attacked and compromised system.

2 The System Overview

The proposed intrusion detection system can be classified as a hybrid IDS based on "anomaly detection". The system observes the network traffic of a host, probably a server that provides network services. Under this point of view the system is similar to a "host based" IDS, but the hybrid solution depends on the fact that the IDS is not implemented in the monitored host but in a different dedicated computer. Besides, the proposed anomaly detection system represents an approach for the development of an "anomaly detection" IDS based on the integration of two systems: a statistical system and a neural one. The neural system which receives in input the results obtained by the previous statistical processing, does not directly process the raw data of the network traffic (as IP addresses, port, TCP flag, and so on), but it works with more homogeneous and processed data.

2.1 The Statistical System

The statistical system plays the role of "discriminator" and its output has to allow through a simple visual inspection the distinction between a normal behaviour and an anomalous one. The system uses fixed temporal windows, in which the network traffic is observed and it memorizes some characteristic parameters. We considered a temporal window of 60 seconds (corresponding to 1 period). During each period the packets in transit in the network are acquired thanks to a network board in promiscuous mode and to the wincap library [12], which allows us capturing packets in the lower levels of the ISO/OSI structure. From the available features, 8 were selected for use in the system: 4 for the input packets (the first four), and 4 for the output packets (the last four):

1. Source IP Address (for TCP, UDP and ICMP protocols);
2. Source Port (for TCP and UDP protocols);
3. Source IP Address and Source Port, IP:Port (for TCP and UDP protocols);
4. Total number of packets (for TCP, UDP and ICMP protocols);
5. Destination IP Address (for TCP, UDP and ICMP protocols);
6. Source Port (for TCP and UDP protocols);
7. Destination IP Address and Source Port, IP:Port (for TCP and UDP protocols);
8. Total number of packets (for TCP, UDP and ICMP protocols);

In order to elaborate a continuous flow of data, the statistical system uses sliding windows including 5 periods and each window is processed as follows. In the first period, the occurrences have to be ordered according to a decreasing sorting. This choice derives from the fact that in this way, in our opinion, we have a common basis to detect possible variations in the traffic observed in the following four periods. After the sorting, the graphs are updated using the data obtained from the four successive periods. For example in the case of a packet coming from an IP address yet observed, in the graph the occurrence related to the IP address is updated (increased of 1 units). Instead, in the case of an unseen IP address this value and its occurrence are added to the queue, as the last value of the graph. Moreover, the trend of these features was modelled using the "first momentum". As final result, we obtain the values of the "first momentum" averaged on the five periods. The data obtained determining the "first momentum" on the windows consisting of 5 periods undergoes a successive processing. This processing intends to make discrete the results of the statistical system in order to obtain more significant data for the successive neural system.

2.2 The Neural System

The neural system based on an unsupervised learning, is the "decisional motor": it has to distinguish the "normal" traffic from the anomalous network traffic. Besides, in order to evaluate the performances of a self-organising network model, the architecture has been realized using the SOM. We evaluate the ability of the SOM network to automatically and spontaneously indicates if an attack occurred.

The figure 1 shows the two-tier neural architecture [7]. As regards the first level the figure 1 shows the 4 maps, each of which was trained on 2 of the 8 features: one concerning the processing of the data related to the "IP:Port" (named IP:Port network), the second concerning the processing of the data related to the "Port" (named Port network), the third concerning the processing of the data related to the "IP" (named IP network), and finally the fourth network classifies the data concerning the number of packets observed both in input and

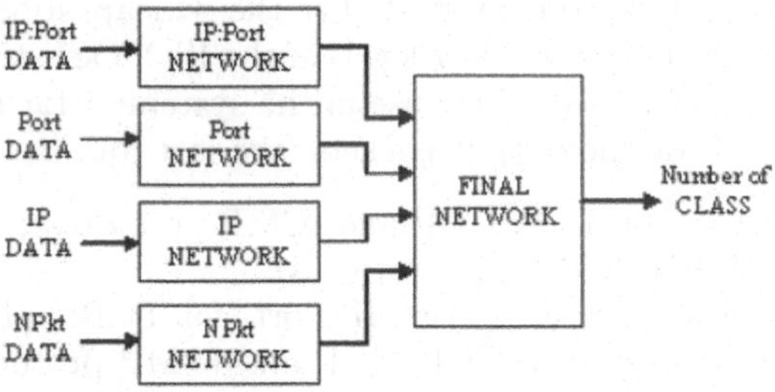

Fig. 1. Architecture of the neural system

in output of the monitored server (named NPkt network). Each network has in output the winner unit vector: the weight vectors of the winner unit according to the Euclidian metric, for each input presented to each network. After the training the nodes of each network in the first layer represent a class. For each network, the traffic recognized as normal would belong to the same class (node), while in the case of anomalous behaviour, different classes would identify different attacks. As shown in figure 1, each of the four networks of the first layer receives in input a vector of two elements; this two-dimensional vector represents the output of the statistical algorithm. The first element of the vector contains the value concerning the input connections, while the second element contains the value concerning the output connections. Each network considers only one information: for example, the network that classifies the IP:Port data will receive the two values (one for the input communications and the other for the output communications) related to the IP:Port, and so on.

The second layer consisted of a single neural network (named Final network). It receives as input the weight vectors of the four winner units, one for each network of the first layer, and so simultaneously processes the classifications of the first layer. The second layer must promptly communicate the status of the network that is if the network traffic has a normal behaviour or there is an anomalous situation.

3 The Experimental Results

In the development of an experimental project it is necessary a complete and a wide dataset. We chose the 1999 DARPA/MIT Lincoln Laboratory intrusion detection evaluation dataset (IDEVAL)[2] which is the only available in Internet. It was specifically created to effect benchmark and experimentations on the IDS, and it has the following characteristics:

- **1 Week** (**From** 1/3/1999, 8:00 am **To** 6/3/1999, 6:00 am): Background traffic;
- **2 Week** (**From** 8/3/1999, 8:00 am **To** 13/3/1999, 6:00 am): Background traffic with a low rate of attacks;

- **3 Week** (**From** 15/3/1999, 8:00 am **To** 20/3/1999, 6:00 am): Background traffic;
- **4 Week** (**From** 29/3/1999, 8:00 am **To** 3/4/1999, 6:00 am): Background traffic with a high rate of attacks;
- **5 Week** (**From** 5/4/1999, 8:00 am **To** 10/4/1999, 6:00 am): Background traffic with a high rate of attacks;

This dataset was used in many works of current literature referring to the development of IDS [8]. We used the "entire dataset" in order to test in depth the performances of our system. The dataset is produced in order to recreate the background traffic and a given number of attacks in a purposely dedicated network. The background traffic was generated considering both the characteristics of the dataflow observed near to the Hanscom Air Force Base military American base, and the statistics on the traffic reaped from other basis. The attacks and the hacking code are extracted from Internet or autonomously developed. Besides, the taxonomy of attacks used for DARPA evaluation is characterized as follows:

- Denial of Service, DoS (Mailbomb, Smurf, Udpstorm, SshProcessTable, Neptune): the intruder attempts to reduce the performance of a host, possibly going as far as making the host unavailable;
- Surveillance/Probing, PROBE (Portsweep, Mscan): the intruder attempts to gather information about the host;
- User to Root, U2R: the intruder tries to access the superuser account by using buffer overflow;
- Remote to Local User, R2L: the intruder attempts to gain unauthorized access from a remote machine.

3.1 Experimental Results of the Statistical System

As explained during the description of the system (paragraph 2), through the packets in input and in output, we observed the behaviour of one host in the network. We chose *Pascal.eyrie.af.mil* with IP address: 172.16.112.50. This is a victim host with Solaris 2.5 operating system on an UltraOne system. So, we intend to determine if an attack or an anomaly occurs, when the statistical distribution of the packets has a significant variation as regards the normal situation. The statistical system was tested observing the network traffic of Pascal with the data related to the five weeks, five days for each week.

Results related to the background traffic. In this paragraph we show a sequence of graphs resulting from the processing of the statistical system related to a day of normal traffic: the second day of the first week. In this way we could observe the typical trend of a day containing background traffic only, pointing out some characteristics of these trends.

The graphs in figure 2 are related to the input connections and show that the maximum value of the averaged "first momentum" is always nearly low. This behaviour is also reflected in the graphs of the other day of background traffic.

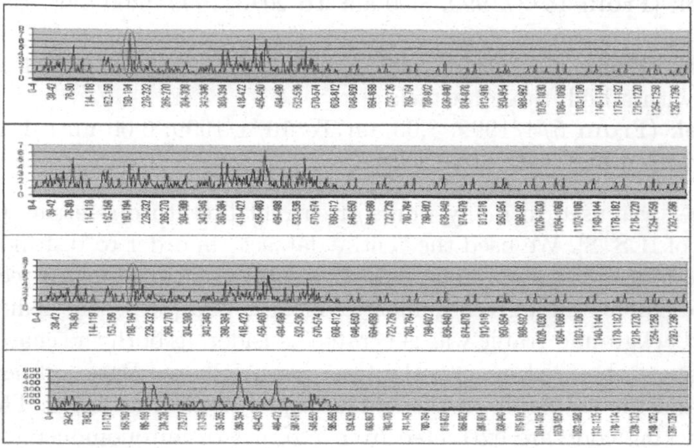

Fig. 2. Graphs related to the background traffic observed for a period in the input packets: (a) "IP:Port graph"; (b) "Port graph"; (c) "IP graph"; (d) "NPkt graph"

Moreover, as shown in the last graph of the figure 2, the number of packets achieves the maximum value near 600. In other day of only background traffic the maximum number of packets can be near 1400. In the first and in the third graphs of the figure 2, the maximum value corresponds to the 5 consecutive periods: from the 194 to the 198 period. Analyzing the content of the data file related to these temporal windows, we can note that the period 194 has a different behaviour from the others: this is due to the fact that all the IP addresses are communicating on the same TCP 25 Port. Summarizing, the "IP:Port graph" has rather high values where a consistent number of IP addresses are observed or however different Ports are considered. The "IP graph" has a similar behaviour only in the case of a high number of IP addresses, on the contrary when an only IP address communicates with a different number of Port, there are not high values in the "IP graph". The "Port graph" shows this aspect.

Results related to the attacks. An attack will be pointed out by some observed parameters. For example a kind of attack could be detected by the "IP:Port graph" and the "IP graph", while another kind of attack could be detected by the "IP:Port graph" and "Port graph". So, this depends on the implicated parameters and perhaps on the typology of attack. The statistical system has to show through a simple visual inspection if an attack occurs, but really the neural system has to indicate the status of the system.

As an example we showed the graphs related to a particular DoS attack (Mailbomb attack) and so only the graphs related to the statistical characteristics that pointed out one instance of the attack will be shown.

"Mailbomb" attack

The "Mailbomb" attack belongs to DoS category and is based on the mailing of a high number of e-mail against a mail server in order to crash the system.

A typical "Mailbomb" attack occurs through the mailing of 10000 messages from some users (10 Mbyte of data for each user).

In IDEVAL there are three different instances of this attack against our monitored server Pascal. Our statistical system was able to identify all the three instances of this specific attack. The three graphs of the figure 3 evidence the trend which identifies the first instance (in the second day of the second week) of the "Mailbomb" attack.

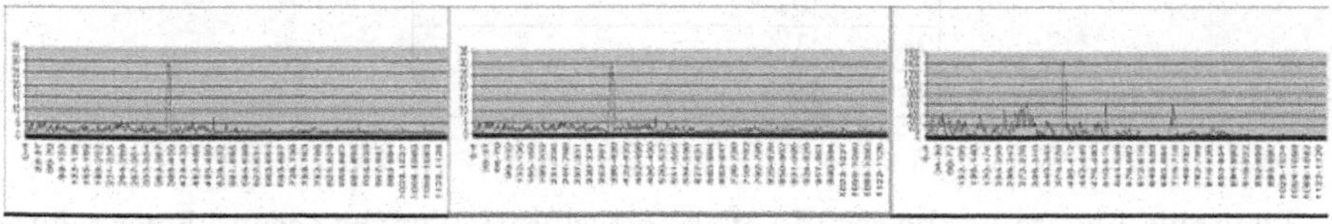

Fig. 3. Graphs related to traffic with attacks observed in a period in the input packets: (a) "IP:Port Graph"; (b)"Port Graph";(c) "NPkt Graph"

As we can see in figure 3 the peak of attack occurred corresponding to the 390-394 temporal window. The peak that identifies the attack has a value much higher than the values shown for the background traffic. In particular, concerning the input connection for the IP:Port and Port graphs the value corresponding to the peak is 28; while for the number of observed packets the value corresponding to the peak is 1440.

3.2 Experimental Results of the Neural System

As we said, the neural system consisted of a self-organizing neural network: the SOM. In this paragraph, the performances of the SOM are analyzed, pointing out the choice regarding the learning, and the choice concerning the topology that is "a priori" fixed. The inputs refer to temporal windows of 5 consecutive periods. The output represents the class in which each neural network classifies the input, both for the four networks of the first layer and for the only one network of the second layer.

The aim of our system is to classify the background traffic in a class, while the attacks would be classified in other classes. The best situation is that different instances of the same attack are classified in the same class, and moreover this class would include only one typology of attack.

The four networks of the first layer have a two-dimensional lattice of neurons (5×5 dimension), so each network has 25 different classes in which to classify the input pattern. The network of the second layer is a one-dimensional lattice, with 9 neural units, which are used to distinguish the 8 different attacks and the background traffic. Before the learning the nodes of the networks are initialized using some values related to the windows of 5 periods extracted from the dataset. During the learning the inputs of the neural system are presented in order to allow the neural networks observing the trend of the statistical results refer to a complete day. For the testing, in the first experiment we only considered the weeks 1 and 3, in order to evaluate the classification of the only background traffic on side of the neural system.

Table 1. Results related to the background traffic

Class	Background traffic
0	4%
1	92%
2	1%
3	3%

The table 1 contains the results of the classification of the second layer related to the only background traffic. We remember that in this first experiment, we are processing the background traffic, this implicates that the classes with a zero rate would be considered as death units or probably they will be used to classify the attacks.

Again for the testing, in the second experiment we considered all the five weeks that compose the dataset. Concerning the results we expect that the classes (4,5,6,7,8) not used for the background traffic are reserved for the attacks.

Table 2. Results of the test related to the all dataset

	Attacks							
Class	Mailb	Smurf	Portsweep	UdpStorm	Ssh	Neptune	Mscan	Secret
0								
1								
2			50%					
3		67%			100%		100%	100%
4				100%				
5		33%						
6								
7						100%		
8	100%		50%					

The table 2 shows the results of the classification of the attacks. First, we analyze the 2 and the 3 classes used for the classification of a limited number of attacks. These two classes are also used for the classification of the background traffic, even if for a lower rate than the 1 class that it mainly characterizes the normal traffic (see table 1). We could consider the 2 and 3 classes of the Final Network as pre-alarm classes. That is, if some temporal windows are classified in 2 and/or 3 classes, but all the SOM of the first layer identify the same input as normal traffic, then even if the input is slightly different from the classical normal trend, but there is not attack. On the contrary, if the networks of the first layer detect an anomalous situation, the input has a behaviour that can be the symptom of a recognized attack. This interpretation is supported by the fact that in several cases the temporal windows these immediately precede and follow an attack are classified in 2 and/or 3 classes of the Final Network.

In some cases the Final Network recognized a particular attack classifying in an only class all the instances of this attack. This happened for the different Mailbomb instances (classified in 8 class); for the different UdpStorm instances (classified in 4 class), for the different Neptune instances (classified in 7 class) and finally for the two of three instances of the Smurf attack (classified in 3 class). Moreover the Final Network reserved some classes for the classification of only one typology of attack; this means that the patterns classified in these classes are representative of a particular attack.

4 Conclusion

In this work we proposed a new model of IDS. For this purpose we realized a system based on an "anomaly detection" technique that combines the "host based" and a "network based" approaches, in order to exploit the advantages of both the models. The proposed architecture consisted of two different systems. The first is the statistical system, which is the "discriminator", since it allows a first distinction between a normal behaviour and an anomalous one on the basis of some statistical characteristics of the network traffic. The second system is the neural system that receives as input, the output of the first system and it is defined "decisional motor". This last one will communicate the status of the network: if there is a normal situation or an attack (or anomaly) occurs. In order to demonstrate the performances of the proposed system we showed some experimental results, by using the 1999 DARPA dataset. The statistical system is able to point out a 100% rate of attacks belonging to DoS category. Moreover, a 67% rate of instances of attacks belonging to Probe category is pointed out. The decisional motor consisted of a neural system that was able to correctly classify the background traffic. Some classes are exclusively reserved for the classification of the only background traffic. In fact the second layer of the neural network classified the data related to the background traffic in 1 class with a rate of 92%. The capability of the system to classify instances of attacks in different classes from the background traffic evidences the ability of the system to distinguish the normal behaviour from the malicious one. Besides the neural system gives significant results, classifying different instances of the same attack in the same class, and reserving this class for a specific attack. So we have not limited our research to distinguish only the background traffic from the attacks, but to distinguish also different attacks. This characteristic is very important for the intervention mode.

Our analysis is based on a single source of network traffic due to the lack of other available data. Obviously, every environment is different; we plan to confirm our results using other sources of real traffic.

References

1. Ghosh, A.K., Wanken, J., Charron, F.: Detection Anomalous and Unknown Intrusions Against Programs, In Proceedings of IEEE 14th Annual Computer Security Applications Conference, (1998) 259–267.
2. Haines, J.W., Lippmann, R.P., Fried, D.J., Tran, E., Boswell, S., Zissman, M.A.: 1999 DARPA Intrusion Detection System Evaluation: Design and Procedures, MIT Lincoln Laboratory Technical Report (1999).
3. Horeis, T.: Intrusion Detection with Neural Networks - Combination of Self-Organizing Maps and Radial Basis Function Networks for Human Expert Integration, Computational Intelligence Society Student Research Grants (2003).
4. Kohonen, T.: Self-Organizing Maps, 3rd edition, Springer-Verlag, Berlino (2001).
5. Labib, K., Vemuri, V.R.: Detecting and Visualizing Denial-of-Service and Network Probe Attacks Using Principal Component Analysis, In the 3rd Conference on Security and Network Architectures, La Londe, Cote d'Azur (France), (2004).
6. Lee, W., Stolfo, S.J., Mok, K.: A Data Mining Framework for Building Intrusion Detection Models, In Proceedings of 1999 IEEE Symposium of Security and Privacy, (1999) 120–132.
7. Lichodzijewski, P., Zincir-Heywood, A.N., Heywood, M.I.: Dynamic Intrusion Detection Using Self-Organizing Maps, In the 14th Annual Canadian Information Technology Security Symposium (2002).
8. Mahoney, M.V., Chan, P.K.: An Analysis of the 1999 DARPA/Lincoln Laboratory Evaluation Data for Network Anomaly Detection, In Proceeding of the 6th International Symposium on Recent Advances in Intrusion Detection (RAID)-2003, Pittsburgh, Pennsylvania, Lecture Notes in Computer Science **2820**, (2003) 220–237.
9. Ramadas, M., Ostermann, S., Tjaden, B.: Detecting anomalous network traffic with self-organizing maps, In Proceeding of the 6th International Symposium on Recent Advances in Intrusion Detection (RAID)-2003, Pittsburgh, Pennsylvania, 8-10 September 2003, Lecture Notes in Computer Science **2820**, (2003) 36–54.
10. Rhodes, B.C., Mahaffey, J.A., Cannady, J.D.: Multiple Self-Organizing Maps for Intrusion Detection, In Proceedings of the 23rd National Information Systems Security Conference, Baltimore, MD (2000).
11. Vigna, G., Kemmerer, R.A.: NetSTAT a network-based Intrusion Detection Approach, In Proceedings of 14th Annual Computer Security Applications Conference, Scottsdale, AZ, USA, (1998), 25–34.
12. wincap http://winpcap.polito.it/
13. Zhang, Z., Li, J., Manikopoulos, C.N., Jorgenson, J., Ucles J.: HIDE: a Hierarchical Network Intrusion Detection System Using Statistical Preprocessing and Neural Network Classification, In Proceeding of the 2001 IEEE Workshop on Information Assurance and Security, United States Military Academy, West Point, NY, (2001), 85–90.

A Learning Based Widrow-Hoff Delta Algorithm for Noise Reduction in Biomedical Signals

Jorge Mateo Sotos[1], César Sánchez Meléndez[1], Carlos Vayá Salort[1], Raquel Cervigon Abad[1], and José Joaquín Rieta Ibáñez[2]

[1] Innovation in Bioengineering Research Group
Castilla la Mancha University, Cuenca, Spain
jorge.mateo@uclm.es
[2] Biomedical Synergy, Valencia University of Technology, Spain

Abstract. This work presents a noise cancellation system suitable for different biomedical signals based on a multilayer artifical neural network(ANN). The proposed method consists of a simple structure similar to the MADALINE neuronal network (Multiple ADAptive LINear Element). This network is a grown artificial neuronal network which allows to optimize the number of nodes of one hidden layer and coefficients of several matrixes. These coefficients matrixes are optimized using the Widrow-Hoff Delta algorithm which requires smaller computational cost than the required by the back-propagation algorithm.

The method's performance has been obtained by computing the cross correlation between the input and the output signals to the system. In addition, the signal to interference ratio (SIR) has also been computed. Making use of the aforementioned indexes it has been possible to compare the different classical methods (Filter FIR, biorthogonal Wavelet 6,8, Filtered Adaptive LMS) and the proposed system based on neural multilayer networks . The comparison shows that the ANN-based method is able to better preserve the signal waveform at system output with an improved noise reduction in comparison with traditional techniques. Moreover, the ANN technique is able to reduce a great variety of noise signals present in biomedical recordings, like high frequency noise, white noise, movement artifacts and muscular noise.

Keywords: Biomedical Signals, Noise Reduction, Electrocardiogram, Neural Networks.

1 Introduction

The Electrocardiogram (ECG) is a graphical representation of the electrical activity of the heart that offers information about the state of the cardiac muscle. This representation consists of a base line and several deflections and waves (the so called P–QRS–T).

However, due to the low amplitude of the ECG signal (1mV approx.), it has been a traditional biomedical problem to remove as much noise as possible from

J. Mira and J.R. Álvarez (Eds.): IWINAC 2007, Part I, LNCS 4527, pp. 377–386, 2007.

the ECG before any extra analysis or processing. The objective of this filtering procedure is both to reduce the noise level in the signal and to prevent the waveform distortion. This last issue is very important. In this way, it would be possible to improve the diagnosis of some heart diseases and diverse pathologies[1].

The signals have been contaminated with noise without correlation with ECG's signals (myoelectric, thermal, etc.) which approximate to white noise. To these signals are added other types of noise to verify the effectiveness of the methods, as muscular and artefacts noise.

Nowadays, there is not a only one method to cancel out noise in biomedical recordings [2]. For example: Adaptive filtering LMS [3,4], FIR filters [5] and filters in the transform domain [6]; are several systems to cancel noise's effects. Besides, there are two problems to work with these filters: convergence problems (LMS) and computational cost. Other applications based on Kalman filters [7], neural networks [8] and Wavelets [9,10,11], are also able to reduce the ECG's noise but they can distort the signal and use to be computationally expensive.

The proposed system is based on a grown ANN, which optimizes the number of neurons in the hidden layer. This method has not been applied in the cancellation of muscular and artifacts noise in ECG signals yet. The system has important advantages: to reduce the processing time, to provoke low signal distortion, to reduce diverse noise types and, in addition, it can be applied to a wide range of biomedical signals.

2 Materials

The electrocardiography treated signals validation requires a set of signals which will have to cover the pathologies, leads, etc in real situations. For this study, two types of signals were used: real recordings from the PhysioNet Database[12], and synthetic signals.

550 recordings with different pathologies have been obtained from PhysioNet with different types of QRS morphologies. These recordings were sampled at a frequency of 360 Hz and, later, they were upsampled to obtain a frequency of 1kHz.

Table 1. Signals used for the study

	N° of Registers	Time (seg)
Synthetic	200	1049
Real	565	106
Real+noise	550	106

Synthetic signals with different noises have been generated making use of the ECGSyn software [12]. White noise, muscular and artefacts are included in these registers. The sampling frequency used is 1kHz.

3 Methods

3.1 Neural Networks

The neuronal perceptron multilayer network(MLP) using the algorithm of back-propagation has been applied to diverse practical problems [13]. Perceptron multilayers method consists of at least three layers: A hidden entrance layer, one or more hidden layers and an exit layer.

A way to consider the optimal number of nodes in the hidden layer is to stop the training after a certain number of iterations and to determine how many signals were filtered with the present number of neurons used in the hidden layer. If the result of this test is not satisfactory it will add one or more neurons in the hidden layer to improve the performance of the network. In these cases the network must be completely trained [13].

An alternative that seems more attractive is the development of increasing networks in which nodes are added in the hidden layer in systematic form during the learning process. With this idea, diverse structures have been proposed such as the increasing network cascade-correlation [14], as well as neural networks [15,16,17]. These networks have been applied in the solution of diverse problems [18].

Proposed System. The proposed system consists initially in a structure similar to the neural network[19] ADALINE (ADAptive LINear Element) as it's shown in the Figure 1 and 2, which is used like initial structure because it is simple and easy to optimize using the algorithm of square minimums average, LMS [15]. It is had initially an input layer layer, one hidden layers(with four neurons) and an exit layer, where they will be added neurons in the intermediate layer.

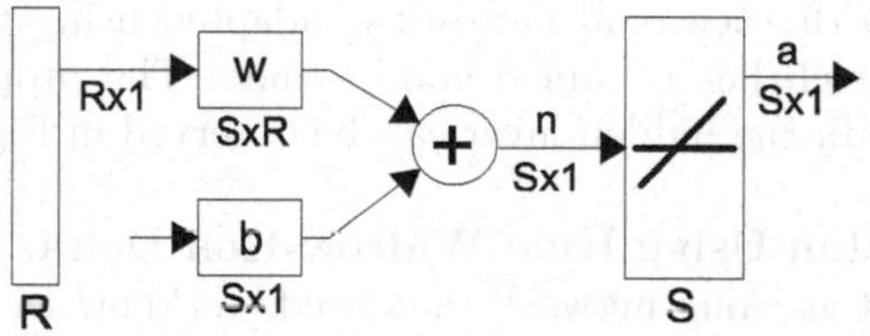

Fig. 1. Structure of a network Adaline, S neurons

Once the network has converged, if the operation obtained by the system is not the required one, a neuron is added in the hidden layer. In this case, the weights that connect the input layer with the nodes of the intermediate layer are congealed, these are been previously trained. The gains that connect the hidden layer with the exit layer are adapted, as well as the weights that connect the layer of entrance with the neuron added in the intermediate layer, as it's shown in 3.

Once trained the network, it returns to evaluate its operation. If it is not correct, a new neuron will be added in the hidden layer. Next, the gains that connect the hidden layer with the exit, as well as the weights connect the layer

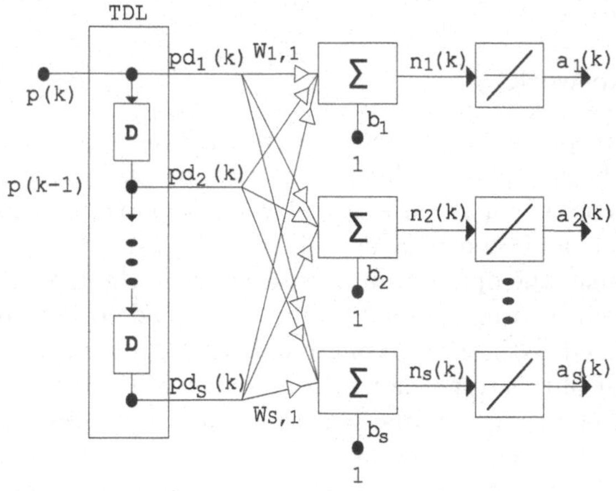

Fig. 2. Initial stage of the proposed neural network

of entrance with the new neuron added in the hidden layer are trained.This procedure is repeated until obtaining the wished operation, becoming a neural network of the type adaline multilayers (Madeline).

This new structure has got a special characteristic: it grows while it learns. It means that the added neurons in the intermediate layer adapted weights and gains whereas the weights of the input layer conserves the learning of the obtained network. This mechanism, although sometimes could produce neural networks with an sub-optimal number of neurons in the hidden layer, allows to consider the size of the network. Thus, the added neurons allow a good operation in the network.

In all the stages the neuronal network is adapted using the Widrow - Hoff Delta algorithm which has obtained good results. The proposed system with two neurons added in the hidden layer can be observed in Figure 3.

Learning Algorithm Using Rule Widrow-Hoff Delta. The network Adaline is a supervised learning network that needs to know the associated values in each input. The pairs of input/output are:

$$\{p_1, t_1\}, \{p_2, t_2\}, ..., \{p_Q, t_Q\} \tag{1}$$

Where $p_Q$is the input to the network and t_Q is its corresponding wished exit, when a input p is presented to the network, the exit of the network is compared with the value of t (hoped exit) that is associated to him.

Adaptive LMS algorithm derives from the Widrow-Hoff rule Delta [20], A network Adaline, is deduced of the following way, according to the procedure described in Widrow [21,22].

$$W(k+1) = W(K) + \alpha \frac{e(k)p(k)}{|p(k)|^2} \tag{2}$$

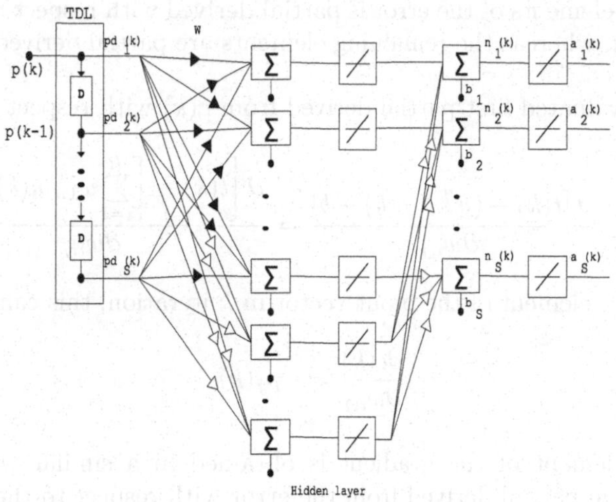

Fig. 3. Proposed Neuronal Network with two neurons in the hidden layer. The black coefficients are constants.

In which k shows the present iteration of the update process, $W(k+1)$ is the following value that will take the vector from weights and $W(k)$ is the present weights vector's value. Present error $e(k)$ is defined as the difference between wished answer $t(k)$ and the exit of network $a(k) = W^T(k)p(k)$ before the update:

$$e(k) = t(k) - W^T(k)p(k) \tag{3}$$

The variation of the error in each iteration is represented by:

$$\Delta e(k) = \Delta(t(k) - W^T(k)p(k)) = -p^T(k) * W(k) \tag{4}$$

The main characteristic of LMS algorithm is that safes the error and it reduces the average quadratic error. In order to explain the quadratic mean error it will be considered a network Adaline and will be used an algorithm of approximated steps, like Widrow and Hoff ; with this algorithm the function for the quadratic mean error is:

$$e^2(k) = (t(k) - a(k))^2 \tag{5}$$

In equation 5, $t(k)$ shows the wanted exit in iteration k and $a(k)$ shows the exit of the network; the quadratic error has been replaced in iteration k, therefore in each iteration is had a gradient of the error of the following way:

$$\left[\nabla e^2(k)\right]_j = \frac{\partial e^2(k)}{\partial w_{i,j}} = 2e(k)\frac{\partial e(k)}{\partial w_{i,j}} \; para \; j = 1, 2, ...R \tag{6}$$

and

$$\left[\nabla e^2(k)\right]_{R+1} = \frac{\partial e^2(k)}{\partial b} = 2e(k)\frac{\partial e(k)}{\partial b} \tag{7}$$

The first R elements of the error is partial derived with respect to the weights of the network, whereas the remaining elements are partial derived with respect to the gains.

It will be evaluated first partial derived from $e(k)$ with respect to $w_{i,j}$:

$$\frac{\partial e(k)}{\partial w_{i,j}} = \frac{\partial \left[t(k) - (w^T * p(k) + b\right]}{\partial w_{i,j}} = \frac{\partial \left[t(k) - \left[\sum_{i=1}^{R} w_{1,i} p_i(k) + b\right]\right]}{\partial w_{i,j}} \tag{8}$$

where $p_i(k)$ is i element of the input vector in k iteration, this can be simplified thus:

$$\frac{\partial e(k)}{\partial w_{i,j}} = -p_j(k) \tag{9}$$

The final element of the gradient is obtained in a similar way, it can be expressed as the partial derived from the error with respect to the gain:

$$\frac{\partial e(k)}{\partial b} = -1 \tag{10}$$

In this equation it's possible to see the simplification advantages of the quadratic mean error, it can be calculated by means of the error in iteration k.

The approach of $\nabla e(k)$ found in the equation 6 is replaced in the equation 2 that defines the process of update of weights for LMS algorithm; after to have evaluated partial derived the update process can be expressed as it follows:

$$w(k+1) = w(k) + 2\alpha e(k) p(k) \tag{11}$$

$$b(k+1) = b(k) + 2\alpha e(k) \tag{12}$$

Now $t(k)$ and $w(k)$ are independent terms. The equations 11 and 12 show the rule used by a network Adaline, the rate of learning α is constant during the algorithm's process.

The weights and gains algorithm for the network Adaline is expressed:

$$W(k+1) = W(k) + 2\alpha e(k) p(k) \tag{13}$$

$$b(k+1) = b(k) + 2\alpha e(k) \tag{14}$$

4 Results

For further research of the Neural Networks filter, it was simulated a clean pulse signal and corrupted with different muscular and artifacts noises. It was compared the performance of the neural network approach with the standard filtering techniques. The Butterworth high pass digital filter was nonlinear phase filter, so the pulse waveform should be distorted. LMS presented convergence's problems. The proposed system was evaluated using several types of noise in biomedical signals as: high frequency, artefacts and muscular noise, obtaining,

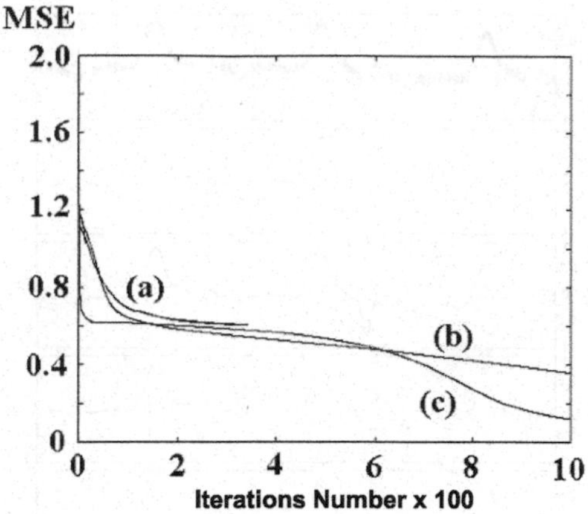

Fig. 4. General operation to reduce the muscular noise with (a) 10 hidden neurons, (b) 11 hidden neurons and (c) 12 hidden neurons

a suitable operation in all cases. Besides, the neural network was evaluated in order to obtain the smaller possible number of hidden neurons.

Figure 4 shows the neural network increasing proposal with 10, 11 and 12 hidden neurons respectively. In this way the weights of the neuronal network were initialized with random uniformly distributed numbers in the interval $[-0.5, 0.5]$. With 12 neurons in the hidden layer is obtained the best results, more neurons are not improved the results.

The filtered percentage can be increased optimizing the network weights by means of a training the weights W an the gains b in the matrix.

The method's performance has been obtained by computing the cross correlation between the input and the output signals to the system. In addition, the signal to interference ratio (SIR) has also been computed. Equation 15, shows SIR expression where x_{in} shows the input signal to the system, x_{out} the output and x the original recording without noise. This parameter has been used to evaluate synthetic signals (without noise). In table 2 the obtained results are the following.

$$SIR = 20 \log \left(\sqrt{\frac{E\{||x_{in} - x||^2\}}{E\{||x_{out} - x||^2\}}} \right) \tag{15}$$

Table 2 shows the cross average correlation values in synthetic signals and real signals. Nevertheless the table 2 shows the average SIR values calculated for synthetic recordings, contaminated with muscular noise and movement artifacts.

With high frequency noise added to the muscular noise more differences were observed between the different tested techniques (Figure 5), the Neural Networks obtains better cross correlation. The second method that approaches the data is Wavelet. The adaptive methods as LMS depend on the ECG and therefore its result is more variable. Methods FIR obtain intermediate values of noise cancellation.

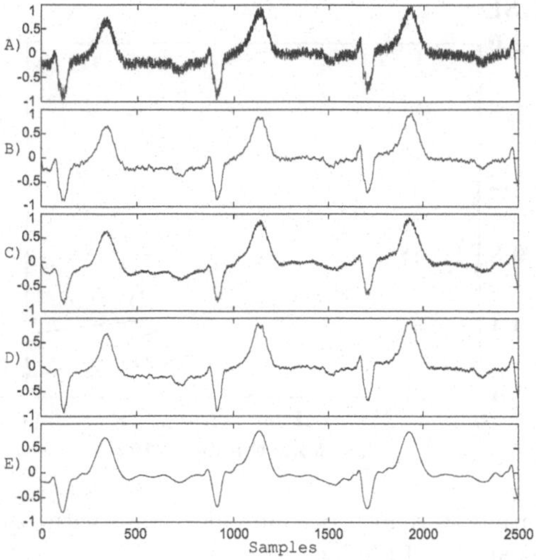

Fig. 5. A. Input Signal. B. FIR Filter C. LMS Filter D. Wavelet Filter E. Neural Filter.

Table 2. Obtained results of the cross correlation of noise of artifacts and muscular, average values. And Obtained results of the SIR for synthetics recordings, average values

	Artefacts		Muscular		SIR	
Methods	Synthetic	Real	Synthetic	Real	Muscular	Artefacts
FIR	$0,91 \pm 0.04$	$0,91 \pm 0.04$	$0,93 \pm 0.03$	$0,92 \pm 0.03$	13.2 ± 0.3	12.2 ± 0.4
LMS	$0,64 \pm 0.34$	$0,58 \pm 0.36$	$0,61 \pm 0.32$	$0,59 \pm 0.35$	5.8 ± 2.23	10.8 ± 2.2
Wavelets	$0,93 \pm 0.03$	$0,92 \pm 0.03$	$0,95 \pm 0.02$	$0,94 \pm 0.02$	15.2 ± 0.3	14.4 ± 0.3
Neural Networks	$0,97 \pm 0.02$	$0,96 \pm 0.03$	$0,97 \pm 0.02$	$0,96 \pm 0.02$	18.2 ± 0.3	17.4 ± 0.3

5 Conclusions

The present work shows an alternative and powerful tool to reduce noise in biomedical signals, as the electrocardiogram, and therefore is able to facilitate the later clinical analysis and study.

The system is based on Neuronal Networks and has been proved as the better method to reduce muscular and high frequency noise. It is possible to emphasize as well, that this new methodology is suitable for high processing speed applications, it is of easy hardware implementation and it needs minimum memory requirement. Finally, the neuronal network-based approach obtains better signal reduction and low distoriton results in comparison with systems based on FIR, LMS and Wavelets (biortogonal 6.8) and it can be applied to different types of signals changing little network parameters.

Acknowledgements

This work was partly funded by the project PAC-05-008-1 from Consejería de Educación de la Junta de Comunidades de Castilla-La Mancha, GV06/299 from Consellería de Empresa, Universidad y Ciencia de la Generalitat Valenciana and TEC2007–64884 from the Spanish Ministry of Science and Education.

References

1. Sörnmo, L., Laguna, P.: Bioelectrical Signal Processing in Cardiac an Neurological Applications. Elsevier Academic Press (2005)
2. Hamilton, P.S.: A comparison of adaptive and nonadaptive filters for reduction of power line interference in the ECG. IEEE Trans. Biomed. Eng. **43** (1996) 105–109
3. Olmos, S., Laguna, P.: Steady-state MSE convergence of LMS adaptative filters with deterministic reference inputs with applications to biomedical signals. IEEE Trans. Sign. Proces. **48** (2000) 2229–2241
4. Olmos, S., Laguna, P.: Block LMS adaptive filter with deterministic reference inputs for event-related signals. Proccedings of the 23rd Annual EMBS International Conference (2001) 1828–1831
5. Lian, Y., Ho, P.C.: ECG noise reduction using multiplier-free FIR digital filters. Proceedings of 2004 International Conference on Signal Processing (2004) 2198–2201
6. Paul, J.S., Reddy, M.R., Kumar, V.J.: A transform domain SVD filter for suppression of muscle noise artefacts in exercise ECG's. IEEE Trans. Biomed. Eng. **47** (2000) 654–663
7. Sameni, R., Shamsollahi, M., Jutten, C., Babaie-Zadeh, M.: Filtering noisy ECG signals using the extended kalman filter based on a modified dynamic ECG model. In: Computers in Cardiology. (2005) 1017–1020
8. Reaz, M.B.I., Wei, L.S.: Adaptive linear neural network filter for fetal ECG extraction. In: The International Conference on Intelligent Sensing and Information Processing. (2004) 321–324
9. Nibhanupudi, S.: Signal Denoising Using Wavelets. PhD thesis, University of Cincinnati (2003)
10. Chmelka, L., Kozumplík, J.: Wavelet-based wiener filter for electrocardiogram signal denoising. Computers in Cardiology **32** (2005) 771–774
11. Xu, L., Zhang, D., Wang, K.: Wavelet-based cascaded adaptive filter for removing baseline drift in pulse waveforms. IEEE Trans. Biomed. Eng. **53** (2005) 1973–1975
12. Goldberger, A.L., Amaral, L.A.N., Glass, L., Hausdorff, J.M., Ivanov, P.C., Mark, R.G., Mietus, J.E., Moody, G.B., Peng, C.K., Stanley, H.E.: PhysioBank, PhysioToolkit, and PhysioNet: Components of a new research resource for complex physiologic signals. Circulation **101** (2000 (June 13)) e215–e220 Circulation Electronic Pages: http://circ.ahajournals.org/cgi/content/full/101/23/e215.
13. Haykin, S.: Neural networks: A comprehensive approach. In: IEEE Computer Society Press. (1994) Piscataway, USA.
14. Lehtokangas, M.: Fast initialization for cascade-correlation learning. IEEE Trans. on Neural Networks **10** (1999) 410–414
15. Sanchez, G., Toscano, K., Nakano, M., Perez, H.: A growing cell neural network structure with backpropagation learning algorithm. Telecommunications and Radio Engineering **56** (2001) 37–45

16. Hodge, V.: Hierarchical growing cell structures, trees GCS. IEEE Trans. on Knowledge and Engineering **13** (2001) 207–218
17. Schetinin, V.: A learning algorithm for evolving cascade neural networks. Neural Letters **17** (2003) 21–31
18. Toscano, K., Sánchez, G., Nakano, M., Perez, H.: Of-line signatur recognition and verification using multiple growing cell neural network structure. Científica **6** (2002) 175–184
19. Hush, R., Horne, B.G.: Progress in supervised neural networks. IEEE Signal Processing Magazine (1993) 8–39
20. Hui, S., Zak, S.H.: The widrow-hoff algorithm for McCulloch-pits type neurons. IEEE Trans. on Neural Networks **5** (1994) 924–929
21. Widrow, B., Lehr, M.A.: 30 years of adaptive neural networks: Perceptrons, madeline and backpropagation. Proc. of IEEE **78** (1990) 1415–1442
22. Wang, Z.Q., Manry, M.T., Schiano, J.L.: LMS learning algorithms: Misconceptions and new results on convergence. IEEE Trans. on Neural Networks **11** (2000) 47–56

Hopfield Neural Network and Boltzmann Machine Applied to Hardware Resource Distribution on Chips

F. Javier Sánchez Jurado and Matilde Santos Peñas

Departamento de Arquitectura de Computadores y Automática
Facultad de Informática, Universidad Complutense de Madrid
C/ Prof. José García Santesmases s/n, 28040 Madrid, Spain
fjsjurado@fdi.ucm.es, msantos@dacya.ucm.es

Abstract. On chip resource distribution is a problem that, due to its complexity, is susceptible to be solved by using artificial intelligence optimization procedures. In this paper, a Hopfield recurrent neural network and a Boltzmann machine are proposed for searching good solutions.

The main challenge of this approach is proposing an energy function to be minimized so it mixes all the problem-related restrictions.

Experimental data shows that we can get good enough solutions in a reasonable time using Hopfield nets or close to the global minimum solutions using Boltzmann machines.

1 Introduction

One of the most complex tasks to be done when designing an integrated circuit is how to distribute all its different elements among the available space. Usually there are some restrictions to be accomplished when choosing a particular module placement, all of them mainly related with power consumption, heat dissipation and wire length issues.

More over, if instead of developing an integrated circuit, a bitstream is being built for a FPGA device [4], the complexity of this task increases due to its limited resources for wiring.

Taking both design models into consideration, one can conclude that the minimization of distances between elements to be placed is an important problem to overcome, not only for reducing latency, but also for decreasing power consumption related to communication busses.

An exhaustive searching procedure to achieve the best solution is too expensive to be done. In this paper, an artificial neural network based algorithm is proposed in order to find good enough solutions in a considerably short time.

The application of such techniques to this kind of problems is not something new. There are previous works that propose their utilization in work fields like improving CISC processors microcode [2]. This article, based in these researches, enounces a general procedure for solving the placement problem, previously formulated.

J. Mira and J.R. Álvarez (Eds.): IWINAC 2007, Part I, LNCS 4527, pp. 387–396, 2007.

In the next section the problem to be solved will be presented. The third and fourth sections describe two different methodologies for solving optimization problems: Hopfield neural nets and Boltzmann machines. After that, the fifth section describes a procedure for applying both algorithms to the previously defined problem. Simulations results are shown in sixth section. Finally, conclusions and further work ends this paper.

2 Problem Overview

The task to be accomplished is placing n distinct modules in n free slots inside a chip. Power dissipation of each module is supposed not to be important when choosing a particular distribution. Communications are made through single non-directional links, just for getting simplicity. The persecuted objective is finding a distribution which allows placing all the modules in such a way that the mean link distance is as short as possible.

The main idea is representing module placement with the binary square matrix **M**, in which if the ith module is in the jth position then the value of element m_{ij} will be 1; otherwise, this value must be 0.

Also, the module dependencies graph is expressed with the binary symmetric matrix **C**. In this matrix, the element c_{ij} will be 1 if the ith module is related to the jth one and it will be 0 if this condition is not satisfied.

The distance between slots is stored into the symmetric matrix **D**, where the value of element d_{ij} is the square length of the link between the ith slot and the jth slot. The square function is used in order to get a better behaviour of the algorithm, punishing long distances and rewarding short ones. This improvement increases the difference between the local minima in the energy function.

3 Hopfield Artificial Neural Networks

Hopfield neural networks were proposed by the physicist John J. Hopfield in 1982 [1]. These are a type of recurrent artificial neural networks in which the input layer and the output one are the same. Also there are no hidden layers.

$$U_{hidden} = \emptyset \wedge U_{input} = U_{output} \ . \tag{1}$$

Each neuron receives in its input all the output values from the rest of them, without considering its own output value.

$$C = U \times U - \{(u, u) \mid u \in U\} \ . \tag{2}$$

Connection weights must be symmetric.

$$\forall u, v \in U : \quad w_{uv} = w_{vu} \ . \tag{3}$$

The next graphic shows an example of Hopfield neural network with 3 nodes. Notice that weights, thresholds and activation state have no specific value.

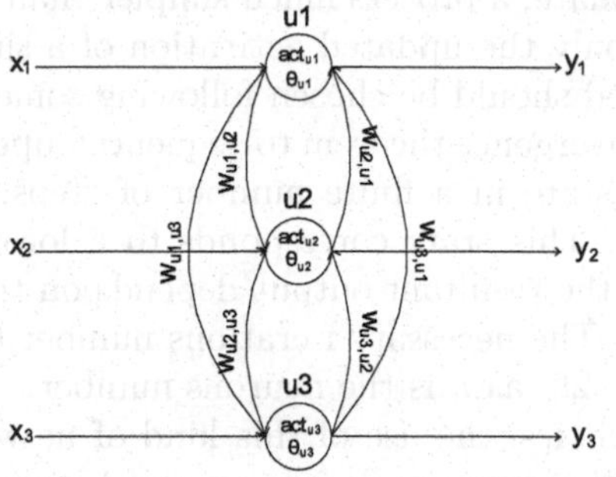

Fig. 1. Hopfield net example

Then, the value of the neuron input network function can be calculated as the weighted addition of all the outputs.

$$\forall u \in U: \quad f_{net}(u) = \overrightarrow{w} \cdot \overrightarrow{f}_{input}(u) \quad . \tag{4}$$

$$\forall u \in U: \quad f_{net}(u) = \sum_{v \in U - \{u\}} w_{uv} \cdot f_{output}(v) \quad . \tag{5}$$

The activation function is usually expressed with some threshold value.

$$\forall u \in U: \quad f_{activation}(u) = \begin{cases} 1 & \text{if } f_{net} > \theta_u \\ act_u & \text{if } f_{net} = \theta_u \\ 0 & \text{if } f_{net} < \theta_u \end{cases} \quad . \tag{6}$$

Also, the output function is the identity.

$$\forall u \in U: \quad f_{output}(u) = act_u \quad . \tag{7}$$

Known all the activation states of the network neurons and all the weights values and thresholds, the network energy function is defined as a Lyapunov function [3] like this:

$$E = -\frac{1}{2} \sum_{u,v \in U | u \neq v} w_{u,v} \cdot act_u \cdot act_v + \sum_{u \in U} \theta_u \cdot act_u \quad . \tag{8}$$

In other terms, transition between states can be calculated with two different procedures: parallel relaxing or sequential update of neurons.

The first mechanism consists on changing all the neuron states at once, depending on the current activation of them. Despite the fact that it is a very intuitive way, it cannot be warranted that the network will achieve a stable state. Because of this, the taken output depends on when the recursive process is stopped.

About sequential update, a process much simpler than the previous one, consisting on calculating only the updated activation of a single neuron at a time, the neuron to be revised should be chosen following some predefined schedule.

According to the convergence theorem to sequential update [3], every Hopfield net tends to a stable state in a finite number of steps, independently of the activation initial state. This state corresponds to a local minimum of the net energy function. Then the resultant output depends on the initial state and the neuron selection order. The necessary iterations number to achieve convergence is always lower than $n \cdot 2^n$, as n is the neurons number.

Due to its characteristics, the use of this kind of networks focuses on applications such as associative memory design, patterns recognition or optimization problem resolution.

4 Boltzmann Machine

The Boltzmann machine is a type of simulated annealing stochastic recurrent neural network proposed by Geoffry Hinton and Terry Sejnowski [6]. It can be seen as an improvement of the Hopfield net model.

Energy function, input function and output function are the same as in the Hopfield net. Also weights matrix has the restrictions of the previous model.

Then, the energy function increment that results from a single unit i being 0 instead of 1 is:

$$\Delta E_i = \sum_{j=1}^{n} w_{ij} \cdot act_j - \theta_i \ . \tag{9}$$

Thus the probability p of a neuron i to be activated is defined by this expression:

$$p_i = \frac{1}{1 + e^{-\frac{\Delta E_i}{T}}} \ . \tag{10}$$

A possible way for calculating transitions between states is random sequential update. This mechanism consists on updating the activation of a randomly selected neuron at a time. The new value for the neuron depends on the probability calculated with the expression (n). The Boltzmann machine must be run at the same temperature time enough for its reaching the "thermal equilibrium", equivalent to the Hopfield net stable state when the machine temperature is very low.

Following a predefined schedule, the temperature decreases whenever the machine is at "thermal equilibrium": Starting with a high temperature and gradually reducing it, normally multiplying it per a less-than-one factor, the convergence at low temperature to a distribution where the energy fluctuates around a global minimum is guarantied.

For a given problem, algorithm behaviour should be better if right values are chosen to the initial temperature and scaling factor. If the initial temperature is too high, the net will simply change its state randomly, independently of the energy function value. But if this temperature is too low, the net will act just as a Hopfield net.

The scaling factor must also be carefully chosen. The closer this number is to 1, the more steps must be done. Even more, the lower this value is, the less the probability is to find the global minimum.

5 Proposed Algorithm

This method is based in the application of the general procedure to solve optimization problems by using Hopfield networks. The main challenge is how to formulate correctly the energy function, which must represent all necessary restrictions for getting a correct solution.

First step in this general procedure is converting the function to be optimized into a function to be minimized. After that, the function has to be transformed so that it responds to a Hopfield net energy function form. Then, the values of the weights matrix and thresholds, which are needed to implement the net, can be calculated. Finally, several net simulations must be done with different random initial activation states, taking itself the best from all obtained solutions as the result.

Therefore, considering the previously defined matrix **M**, **C** and **D**, the objective is to find the values for **M** which makes the links length addition minimum. Besides, there cannot be in **M** more than a single 1 in each row (11) or column (12).

$$\forall j \in \{1, \ldots, n\} \quad \sum_{i=1}^{n} m_{ij} = 1 \ . \tag{11}$$

$$\forall i \in \{1, \ldots, n\} \quad \sum_{j=1}^{n} m_{ij} = 1 \ . \tag{12}$$

Considering that **C** and **D** matrix are symmetric, the first restriction can be expressed with the energy function (13).

$$E_1 = -\frac{1}{2} \sum_{\substack{(i_1,j_1)\in\{1,\ldots,n\}^2 \\ (i_2,j_2)\in\{1,\ldots,n\}^2}} c_{j_1 j_2} \cdot d_{i_1 i_2} \cdot m_{i_1 j_1} \cdot m_{i_2 j_2} \ . \tag{13}$$

The restriction reflected in (11) can be expressed through the function to be minimized (14). Analogously, restriction (12) becomes (15).

$$E_2 = \sum_{\substack{(i_1,j_1)\in\{1,\ldots,n\}^2 \\ (i_2,j_2)\in\{1,\ldots,n\}^2}} \delta\left(j_1, j_1\right) \cdot m_{i_1 j_1} \cdot m_{i_2 j_2} - \sum_{(i,j)\in\{1,\ldots,n\}^2} m_{ij} \ . \tag{14}$$

$$E_3 = \sum_{\substack{(i_1,j_1)\in\{1,\ldots,n\}^2 \\ (i_2,j_2)\in\{1,\ldots,n\}^2}} \delta\left(i_1, i_1\right) \cdot m_{i_1 j_1} \cdot m_{i_2 j_2} - \sum_{(i,j)\in\{1,\ldots,n\}^2} m_{ij} \ . \tag{15}$$

Being the δ function:

$$\delta(a,b) = \begin{cases} 1 & \text{if } a = b \\ 0 & \text{otherwise} \end{cases} . \tag{16}$$

In order to make the Hopfield net using these restrictions, necessary operations must be done so the energy functions looks like a Lyapunov function (8).

$$E_1 = -\frac{1}{2} \sum_{\substack{(i_1,j_1)\in\{1,\dots,n\}^2 \\ (i_2,j_2)\in\{1,\dots,n\}^2}} -c_{j_1 j_2} \cdot m_{i_1 j_1} \cdot m_{i_2 j_2} . \tag{17}$$

$$E_2 = -\frac{1}{2} \sum_{\substack{(i_1,j_1)\in\{1,\dots,n\}^2 \\ (i_2,j_2)\in\{1,\dots,n\}^2}} -2 \cdot \delta(j_1, j_2) \cdot m_{i_1 j_1} \cdot m_{i_2 j_2} + \sum_{(i,j)\in\{1,\dots,n\}^2} -m_{ij} . \tag{18}$$

$$E_3 = -\frac{1}{2} \sum_{\substack{(i_1,j_1)\in\{1,\dots,n\}^2 \\ (i_2,j_2)\in\{1,\dots,n\}^2}} -2 \cdot \delta(i_1, i_2) \cdot m_{i_1 j_1} \cdot m_{i_2 j_2} + \sum_{(i,j)\in\{1,\dots,n\}^2} -m_{ij} . \tag{19}$$

Once the conditions are set, the resultant net energy function is obtained by linear combining the energy functions associated with each restriction(20).

$$E = a \cdot E_1 + b \cdot E_2 + c \cdot E_3 . \tag{20}$$

Choosing a, b and c parameters determinates how important are each condition in the global result. E_2 and E_1 weights must be higher than E_3 so the net application resultant solution would be valid. A possible relation between these parameters could be the one defined in the next expression.

$$\frac{b}{a} = \frac{c}{a} > 2 \cdot \max(d_{ij}) . \tag{21}$$

If a takes 1 as its value, then b and c are the same. In this case, weights (22) and threshold (23) values can be:

$$w_{(i_1,j_1)(i_2,j_2)} = -c_{j_1 j_2} \cdot d_{i_1 i_2} - 2 \cdot b \cdot (\delta(i_1, i_2) + \delta(j_1, j_2)) . \tag{22}$$

$$\theta_{(i,j)} = -2 \cdot b . \tag{23}$$

This energy function is applied to a certain problem by giving right values for c and d in the expressions (22) and (23). Then a Hopfield net can be build with the obtained $\mathbf{W}$ and θ. The sequential update method must be started with a random initial neuron activation state. The process is repeated until it reaches a stable state. If the result is not acceptable, all must be repeated with a new initial activation state.

The Hopfield net only find certain local minimum solutions. If finding a solution very close to the global minimum is more important than finding a good enough solution in a short time, the Boltzmann machine should be used. The particular values for initial temperature and scaling factor must be calculated by making some test to the algorithm so its behaviour is acceptable.

6 Experimental Results

Tests have been made using a 16 modules dependency graph, with 16 possible slots to be filled. The scheme shown in figure 2 represents this graph. As it can be observed, the structure to be optimized is composed by a loop, which encloses the 8 firs modules, and a hierarchical tree. This allows to watch the algorithm behaviour when processing a mixed topology configuration.

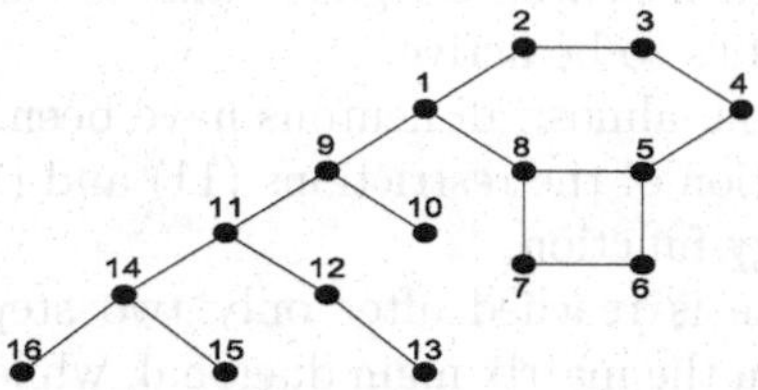

Fig. 2. Test dependency graph

All the slots, where modules are placed, have the same size and are uniformly distributed inside a square. They are identified by numbers like this:

1	2	3	4
5	6	7	8
9	10	11	12
13	14	15	16

Fig. 3. Slots numbering

The distance between slots is calculated by adding the distance measured in rows and columns. Then the matrix of square distances is calculated. Due to the fact that the maximum value of this matrix is 36, b can take the value 72. If so, weights and threshold are:

$$w_{(i_1,j_1)(i_2,j_2)} = -c_{j_1 j_2} \cdot d_{i_1 i_2} - 144 \cdot (\delta (i_1, i_2) - \delta (j_1, j_2)) \ . \tag{24}$$

$$\theta_{(i,j)} = -144 \ . \tag{25}$$

Figure 4 shows the activation state in the distinct steps of the sequential update method for the Hopfield net. Each black box indicates a 1 state and each white one indicates a 0 state. Columns represent possible positions and rows modules.

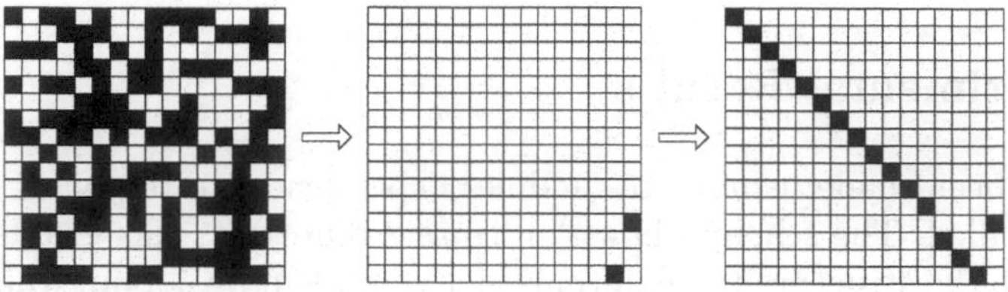

Fig. 4. Hopfield net test

At the initial state, each neuron assigned value is calculated randomly considering a 50% possibilities to be active.

After the first iteration, almost all neurons have been deactivated. This happens due to the application of the restrictions (11) and (12), which have a great importance in the energy function.

The final stable state is reached after only two steps. At this state, loop modules are placed down the matrix main diagonal, while those which are in the hierarchical tree are distributed in the remaining free slots of the matrix.

This simulation has been done in a Pentium III 700MHz with 128MB ram. The execution total time was less than 2 seconds for this test.

Table 1. Energy function values

Step	Net energy
Initial	124639
1	-288
Stable	-2232

It can be observed in table 1 that in each step there have only been made those changes which produce an immediate reduction in the energy function value. Therefore, the stable state must be an energy function local minimum. The net convergence to right solutions demonstrates that the energy function has been correctly formulated.

The resultant solution is interpreted as in figure 5. The boxes are the free slots and each number is the module which is supposed to be in it.

It can be concluded, taking a close look at the solution in figure 5, that the loop part has not been fully optimized. It was noticed in successive tests that this behaviour keeps on whatever is the initial activation state. The reason for always assigning these modules to main diagonal slots is that, when doing sequential update, the matrix is always processed following a fixed order: from left to right and then from up to down.

1	2	3	4
5	6	7	8
9	10	11	12
14	15	16	13

Fig. 5. Hopfield net obtained solution

The figure 6 shows the normalized values for the energy function and temperature for each step in a Boltzmann machine solving the same problem. It can be observed that if the temperature is high, the energy function variation is very high. But also, the lower the temperature is, the faster the energy value tends to a minimum value.

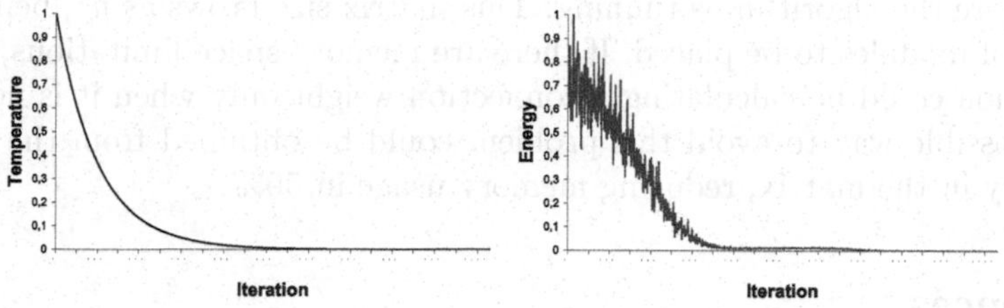

Fig. 6. Boltzmann machine test

After having run 20 seconds the algorithm in the same machine as with the Hopfield net, the resultant activation state for this simulation, wich has an energy value of -2240, is shown in figure 7.

1	8	16	12
4	2	3	7
5	6	13	14
9	15	11	10

Fig. 7. Boltzmann machine obtained solution

7 Conclussion

Once the energy function is formulated correctly, the application of a Hopfield net or a Boltzmann machine to a specific problem is immediate.

On the other hand, for a given number of modules, dependency graph complexity does not affect the algorithm yield.

Although good solutions in a reasonable time are found, even with such limited hardware resources, the Hopfield model behaviour is not good enough for optimizing loop distributed modules. The problem is that this kind of neural net is susceptible to fall into local minimums, in which a change of a module position does not led to an energy function decrement.

Unlike Hopfield net, the Boltzmann machine should find solutions surrounding a global minima. Despite the fact that it is a good way to expand the solution search field, it spends a lot of time on reaching thermal equilibrium and convergence to the solution. Because of that, this procedure must only be used if the solution improvement compensates the cost in time and memory.

Other possibility to expand the searching field is combining the proposed approximations with a genetic algorithm procedure, so all neural network solutions will be used for generating the most part of the initial population [5].

Finally, the high memory usage due to the weights matrix could be a very important problem, depending on how many resources are available in the machine where the algorithm is running. This matrix size grows as n^4, being n the number of modules to be placed. If there are memory space limitations, a possible solution could be calculating a connection weight only when it is necessary. Other possible way to avoid this problem could be obtained from the existing symmetry in the matrix, reducing memory usage in 50%.

References

1. Hopfield, J. J., (1982) "Neural Networks and Physical Systems with Emergent Collective Computational Abilites". Proceedings of the National Academy of Sciences of the USA, vol. 79 no. 8 pp. 2554-2558
2. Bharitkar, S. et al, (1999) "Microcode Optimization with Neural Networks", IEEE-NN, vol. 10 no. 3 pp. 698
3. Krusne, R., (2006) "Chapter 8: Hopfield Networks", available at `http://fuzzy.cs.uni-magdeburg.de/studium/nn/txt/folien060615.pdf`
4. Xilinx Inc., (2002) "Virtex 2.5V Field Programmable Gate Arrays", available at `http://direct.xilinx.com/bvdocs/publications/ds003-2.pdf`
5. Xin Yao, (1992) "A Review of Evolutionary Artificial Neural Networks", Commonwealth Scientific and Industrial Research Organization, Australia
6. Hinton, G. E., Sejnowski, T. J., (1985) "A learning Algorithm for Boltzmann Machines", Cognitive Science, vol. 9 pp. 147-169

A New Rough Set Reduct Algorithm Based on Particle Swarm Optimization

Benxian Yue[1,2], Weihong Yao[1], Ajith Abraham[3,4,⋆], Hongfei Teng[1,2], and Hongbo Liu[1,4]

[1] Department of Computer, Dalian University of Technology, Dalian 116023, China
yuebenxian@vip.sina.com, ywh@dl.cn, tenghf@dlut.edu.cn, lhb@dlut.edu.cn
[2] School of Mechanical and Engineeing, Dalian Universityof Technology, Dalian 116023, China
[3] Centre for Quantifiable Quality of Service in Communication Systems, Norwegian University of Science and Technology, Trondheim, Norway
ajith.abraham@ieee.org
http://www.softcomputing.net/
[4] School of Computer Science, Dalian Maritime University, Dalian 116026, China

Abstract. Finding appropriate features is one of the key problems in the increasing applications of rough set theory, which is also one of the bottlenecks of the rough set methodology. Particle Swarm Optimization (PSO) is particularly attractive for this challenging problem. In this paper, we attempt to solve the problem using a particle swarm optimization approach. The proposed approach discover the best feature combinations in an efficient way to observe the change of positive region as the particles proceed through the search space. We evaluate the performance of the proposed PSO algorithm with Genetic Algorithm (GA). Empirical results indicate that the proposed algorithm could be an ideal approach for solving the feature reduction problem when other algorithms failed to give a better solution.

1 Introduction

Rough set theory [1,2,3] provides a mathematical tool that can be used for both feature selection and knowledge discovery. It helps us to find out the minimal attribute sets called '*reducts*' to classify objects without deterioration of classification quality and induce minimal length decision rules inherent in a given information system. The idea of reducts has encouraged many researchers in studying the effectiveness of rough set theory in a number of real world domains, including medicine, pharmacology, control systems, fault-diagnosis, text categorization, social sciences, switching circuits, economic/financial prediction, image processing, and so on [4,5,6,7,8,9,10].

Usually real world objects are the corresponding tuple in some decision tables. They store a huge quantity of data, which is hard to manage from a computational point of view. Finding reducts in a large information system is still an

⋆ Corresponding author.

J. Mira and J.R. Álvarez (Eds.): IWINAC 2007, Part I, LNCS 4527, pp. 397–406, 2007.

NP-hard problem [11,15]. The high complexity of this problem has motivated investigators to apply various approximation techniques to find near-optimal solutions. Many approaches have been proposed for finding reducts, e.g., discernibility matrices, dynamic reducts, and others [12,13]. The heuristic algorithm is a better choice. Hu [14] proposed a heuristic algorithm using discernibility matrix. The approach provided a weighting mechanism to rank attributes. Zhong [15] presented a wrapper approach using rough sets theory with greedy heuristics for feature subset selection. The aim of feature subset selection is to find out a minimum set of relevant attributes that describe the dataset as well as the original all attributes do. So finding reduct is similar to feature selection. Zhong's algorithm employed the number of consistent instances as heuristics. Banerjee [16] presented various attempts of using Genetic Algorithms (GA) in order to obtain reducts. Although several variants of reduct algorithms are reported in the literature, at the moment, there is no accredited best heuristic reduct algorithm. So far, it's still an open research area in rough sets theory.

Particle swarm algorithm is inspired by social behavior patterns of organisms that live and interact within large groups. In particular, it incorporates swarming behaviors observed in flocks of birds, schools of fish, or swarms of bees, and even human social behavior, from which the Swarm Intelligence (SI) paradigm has emerged [17]. The swarm intelligent model helps to find optimal regions of complex search spaces through interaction of individuals in a population of particles [18,19]. As an algorithm, its main strength is its fast convergence, which compares favorably with many other global optimization algorithms [20,21]. It has exhibited good performance across a wide range of applications [22,23,24,25,26]. The particle swarm algorithm is particularly attractive for feature selection as there seems to be no heuristic that can guide search to the optimal minimal feature subset. Additionally, it can be the case that particles discover the best feature combinations as they proceed throughout the search space. This paper investigates how particle swarm optimization algorithm may be applied to the difficult problem of finding optimal reducts.

The rest of the paper is organized as follows. Some related terms and theorems on rough set theory are explained briefly in Section 2. The proposed approach based on particle swarm algorithm is presented in Section 3. In Section 4, experiment results and discussions are provided in detail. Finally conclusions are made in Section 5.

2 Rough Set Reduction

The basic concepts of rough set theory and its philosophy are presented and illustrated with examples in [1,2,3,15,27,28]. Here, we illustrate only the relevant basic ideas of rough sets that are relevant to the present work.

In rough set theory, an information system is denoted in 4-tuple by $S = (U, A, V, f)$, where U is the universe of discourse, a non-empty finite set of N

objects $\{x_1, x_2, \cdots, x_N\}$. A is a non-empty finite set of attributes such that $a : U \to V_a$ for every $a \in A$ (V_a is the value set of the attribute a).

$$V = \bigcup_{a \in A} V_a$$

$f : U \times A \to V$ is the total decision function (also called the information function) such that $f(x, a) \in V_a$ for every $a \in A$, $x \in U$. The information system can also be defined as a decision table by $S = (U, C, D, V, f)$. For the decision table, C and D are two subsets of attributes. $A = \{C \cup D\}$, $C \cap D = \emptyset$, where C is the set of input features and D is the set of class indices. They are also called condition and decision attributes, respectively.

Let $a \in C \cup D$, $P \subseteq C \cup D$. A binary relation $IND(P)$, called an equivalence (indiscernibility) relation, is defined as follows:

$$IND(P) = \{(x, y) \in U \times U \quad | \quad \forall a \in P, f(x, a) = f(y, a)\} \tag{1}$$

The equivalence relation $IND(P)$ partitions the set U into disjoint subsets. Let $U/IND(P)$ denote the family of all equivalence classes of the relation $IND(P)$. For simplicity of notation, U/P will be written instead of $U/IND(P)$. Such a partition of the universe is denoted by $U/P = \{P_1, P_2, \cdots, P_i, \cdots\}$, where P_i is an equivalence class of P, which is denoted $[x_i]_P$. Equivalence classes U/C and U/D will be called condition and decision classes, respectively.
Lower Approximation: Given a decision table $T = (U, C, D, V, f)$. Let $R \subseteq C \cup D$, $X \subseteq U$ and $U/R = \{R_1, R_2, \cdots, R_i, \cdots\}$. The R-lower approximation set of X is the set of all elements of U which can be with certainty classified as elements of X, assuming knowledge R. It can be presented formally as

$$R_{-}(X) = \bigcup \{R_i \quad | \quad R_i \in U/R, R_i \subseteq X\} \tag{2}$$

Positive Region: Given a decision table $T = (U, C, D, V, f)$. Let $B \subseteq C$, $U/D = \{D_1, D_2, \cdots, D_i, \cdots\}$ and $U/B = \{B_1, B_2, \cdots, B_i, \cdots\}$. The B-positive region of D is the set of all objects from the universe U which can be classified with certainty to classes of U/D employing features from B, i.e.,

$$POS_B(D) = \bigcup_{D_i \in U/D} B_{-}(D_i) \tag{3}$$

Reduct: Given a decision table $T = (U, C, D, V, f)$. The attribute $a \in B \subseteq C$ is $D - dispensable$ in B, if $POS_B(D) = POS_{(B-\{a\})}(D)$; otherwise the attribute a is $D - indispensable$ in B. If all attributes $a \in B$ are $D - indispensable$ in B, then B will be called $D - independent$. A subset of attributes $B \subseteq C$ is a $D - reduct$ of C, iff $POS_B(D) = POS_C(D)$ and B is $D - independent$. It means that a reduct is the minimal subset of attributes that enables the same classification of elements of the universe as the whole set of attributes. In other words, attributes that do not belong to a reduct are superfluous with regard to classification of elements of the universe.

Reduced Positive Universe and *Reduced Positive Region*: Given a decision table $T = (U, C, D, V, f)$. Let $U/C = \{[u_1^{'}]_C, [u_2^{'}]_C, \cdots, [u_m^{'}]_C\}$, Reduced Positive Universe $U^{'}$ can be written as:

$$U^{'} = \{u_1^{'}, u_2^{'}, \cdots, u_m^{'}\}. \tag{4}$$

and

$$POS_C(D) = [u_{i_1}^{'}]_C \cup [u_{i_2}^{'}]_C \cup \cdots \cup [u_{i_t}^{'}]_C. \tag{5}$$

Where $\forall u_{i_s}^{'} \in U^{'}$ and $|[u_{i_s}^{'}]_C/D| = 1 (s = 1, 2, \cdots, t)$. Reduced positive universe can be written as:

$$U_{pos}^{'} = \{u_{i_1}^{'}, u_{i_2}^{'}, \cdots, u_{i_t}^{'}\}. \tag{6}$$

and $\forall B \subseteq C$, reduced positive region

$$POS_B^{'}(D) = \bigcup_{X \in U'/B \wedge X \subseteq U_{pos}^{'} \wedge |X/D|=1} X \tag{7}$$

where $|X/D|$ represents the cardinality of the set X/D. $\forall B \subseteq C$, $POS_B(D) = POS_C(D)$ if $POS_B^{'} = U_{pos}^{'}$ [28]. It is to be noted that $U^{'}$ is the reduced universe, which usually would reduce significantly the scale of datasets. It provides a more efficient method to observe the change of positive region when we search the reducts. We didn't have to calculate U/C, U/D, U/B, $POS_C(D)$, $POS_B(D)$ and then compare $POS_B(D)$ with $POS_C(D)$ to determine whether they are equal to each other or not. We only calculate U/C, $U^{'}$, $U_{pos}^{'}$, $POS_B^{'}$ and then compare $POS_B^{'}$ with $U_{pos}^{'}$.

3 Particle Swarm Approach for Reduction

Given a decision table $T = (U, C, D, V, f)$, the set of condition attributes, C, consist of m attributes. We set up a search space of m dimension for the reduction problem. Accordingly each particle's position is represented as a binary bit string of length m. Each dimension of the particle's position maps one condition attribute. The domain for each dimension is limited to 0 or 1. The value '1' means the corresponding attribute is selected while '0' not selected. Each position can be "decoded" to a potential reduction solution, an subset of C. The particle's position is a series of priority levels of the attributes. The sequence of the attribute will not be changed during the iteration. But after updating the velocity and position of the particles, the particle's position may appear real values such as 0.4, etc. It is meaningless for the reduction. Therefore, we introduce a discrete particle swarm optimization for this combinatorial problem.

During the search procedure, each individual is evaluated using the fitness. According to the definition of rough set reduct, the reduction solution must ensure the decision ability is the same as the primary decision table and the number of attributes in the feasible solution is kept as low as possible. In our algorithm, we first evaluate whether the potential reduction solution satisfies

$POS'_E = U'_{pos}$ or not (E is the subset of attributes represented by the potential reduction solution). If it is a feasible solution, we calculate the number of '1' in it. The solution with the lowest number of '1' would be selected. For the particle swarm, the lower number of '1' in its position, the better the fitness of the individual is. $POS'_E = U'_{pos}$ is used as the criterion of the solution validity.

As a summary, the particle swarm model consists of a swarm of particles, which are initialized with a population of random candidate solutions. They move iteratively through the d-dimension problem space to search the new solutions, where the fitness f can be measured by calculating the number of condition attributes in the potential reduction solution. Each particle has a position represented by a position-vector $\boldsymbol{p}_i$ (i is the index of the particle), and a velocity represented by a velocity-vector $\boldsymbol{v}_i$. Each particle remembers its own best position so far in a vector $\boldsymbol{p}_i^{\#}$, and its j-th dimensional value is $p_{ij}^{\#}$. The best position-vector among the swarm so far is then stored in a vector $\boldsymbol{p}^*$, and its j-th dimensional value is p_j^*. When the particle moves in a state space restricted to zero and one on each dimension, the change of probability with time steps is defined as follows:

$$P(p_{ij}(t) = 1) = f(p_{ij}(t-1), v_{ij}(t-1), p_{ij}^{\#}(t-1), p_j^*(t-1)). \tag{8}$$

where the probability function is

$$sig(v_{ij}(t)) = \frac{1}{1 + e^{-v_{ij}(t)}}. \tag{9}$$

At each time step, each particle updates its velocity and moves to a new position according to Eqs.(10) and (11):

$$v_{ij}(t) = wv_{ij}(t-1) + c_1 r_1 (p_{ij}^{\#}(t-1) - p_{ij}(t-1)) + c_2 r_2 (p_j^*(t-1) - p_{ij}(t-1)). \tag{10}$$

$$p_{ij}(t) = \begin{cases} 1 & \text{if } \rho < sig(v_{ij}(t)); \\ 0 & \text{otherwise.} \end{cases} \tag{11}$$

Where c_1 is a positive constant, called as coefficient of the self-recognition component, c_2 is a positive constant, called as coefficient of the social component. r_1 and r_2 are the random numbers in the interval [0,1]. The variable w is called as the inertia factor, which value is typically setup to vary linearly from 1 to near 0 during the iterated processing. ρ is random number in the closed interval [0, 1]. From Eq.(10), a particle decides where to move next, considering its current state, its own experience, which is the memory of its best past position, and the experience of its most successful particle in the swarm. The pseudo-code for the particle swarm search method is illustrated in Algorithm 1..

4 Experiment Settings, Results and Discussions

In this experiment, Genetic algorithm (GA) was used to compare the performance with PSO. The two algorithms share many similarities [29,30]. Both

Algorithm 1. A Rough Set Reduct Algorithm Based on Particle Swarm

01.Calculate U', U'_{pos} using Eqs.(4) and (6).
02.Initialize the size of the particle swarm n, and other parameters.
03.Initialize the positions and the velocities for all the particles randomly.
04.While (the end criterion is not met) do
05. $t = t + 1$;
06. Calculate the fitness value of each particle,
06. if $POS'_E = U'_{pos}$, the fitness is punished
06. as the total number of the condition attributes,
06. else the fitness is the number of '1' in the position.
07. $\boldsymbol{p}^* = argmin_{i=1}^{n}(f(\boldsymbol{p}^*(t-1)), f(\boldsymbol{p}_1(t)), f(\boldsymbol{p}_2(t)), \cdots, f(\boldsymbol{p}_i(t)), \cdots, f(\boldsymbol{p}_n(t)))$;
08. For i= 1 to n
09. $\boldsymbol{p}_i^{\#}(t) = argmin_{i=1}^{n}(f(\boldsymbol{p}_i^{\#}(t-1)), f(\boldsymbol{p}_i(t))$;
10. For j = 1 to d
11. Update the j-th dimension value of $\boldsymbol{p}_i$ and $\boldsymbol{v}_i$
11. according to Eqs.(10) and (11);
12. Next j
13. Next i
14.End While.

methods are valid and efficient methods in numeric programming and have been employed in various fields due to their strong convergence properties. In GA, the probability of crossover is set to 0.8 and the probability of mutation is set to 0.08. In PSO, self coefficient c_1 and social coefficient c_2 both are 1.49, and the inertia weight w is decreasing linearly from 0.9 to 0.1. The size of the population in GA and the swarm size in PSO both are set to $(int)(10+2*sqrt(D))$, where D is the dimension of the position, i.e., the number of condition attributes. In each trial, the maximum number of iterations is $(int)(0.1 * recnum + 10 * (nfields - 1))$, where $recnum$ is the number of records/rows and $nfields$ is the number of condition attributes. Each experiment (for each algorithm) was repeated 3 times with different random seeds. If the standard deviation is larger than 20%, the times of trials would be set to larger, 10 or 20. We consider the datasets in Table 1 from AFS[1], AiLab[2] and UCI[3].

Figs. 1, 2 and 3 illustrate the performance of the algorithms for lung-cancer, lymphography and mofn-3-7-10 datasets, respectively. For lung-cancer dataset, the results (the number of reduced attributes) for 3 GA runs all were 10: {1, 3, 9, 12, 33, 41, 44, 47, 54, 56} (The number before the colon is the number of condition attributes, the numbers in brackets are attribute index, which represents a reduction solution). The results of 3 PSO runs were 9: { 3, 8, 9, 12, 15, 35, 47, 54, 55}, 10: {2, 3, 12, 19, 25, 27, 30, 32, 40, 56}, 8: {11, 14, 24, 30, 42, 44, 45, 50}. For lymphography datasets, the results of 3 GA runs all were 7: {2, 6, 10,

[1] http://sra.itc.it/research/afs/
[2] http://www.ailab.si/orange/datasets.asp
[3] http://www.datalab.uci.edu/data/mldb-sgi/data/

13, 14, 17, 18}, the results of 3 PSO runs were 6: {2, 13, 14, 15, 16,18}, 7: {1, 2, 13, 14, 15, 17,18}, 7: {2, 10, 12, 13, 14, 15, 18}. For mofn-3-7-10 datasets, the results of 3 GA runs all were 7: {3, 4, 5, 6, 7, 8, 9}, the results of 3 PSO runs all were 7: {3, 4, 5, 6, 7, 8, 9}. Other results are shown in Table 1. PSO usually can obtain a better result than GA, specially for a large scale problem. although GA and PSO both got the same results, PSO usually uses only very few iterations, as illustrated in Fig. 2.

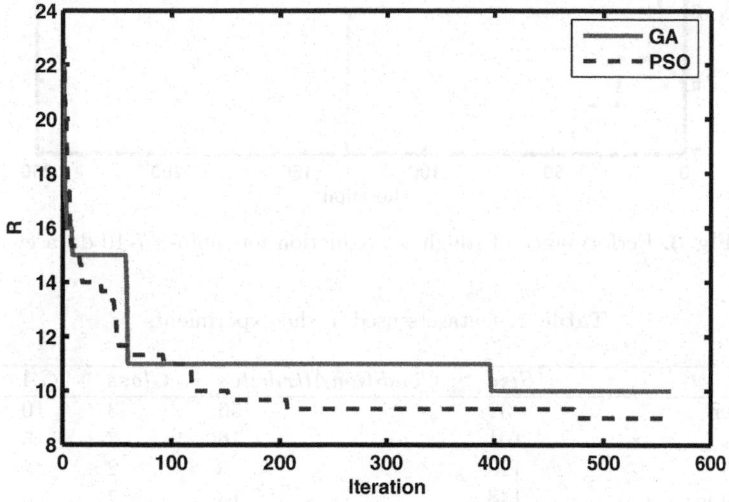

Fig. 1. Performance of rough set reduction for lung-cancer dataset

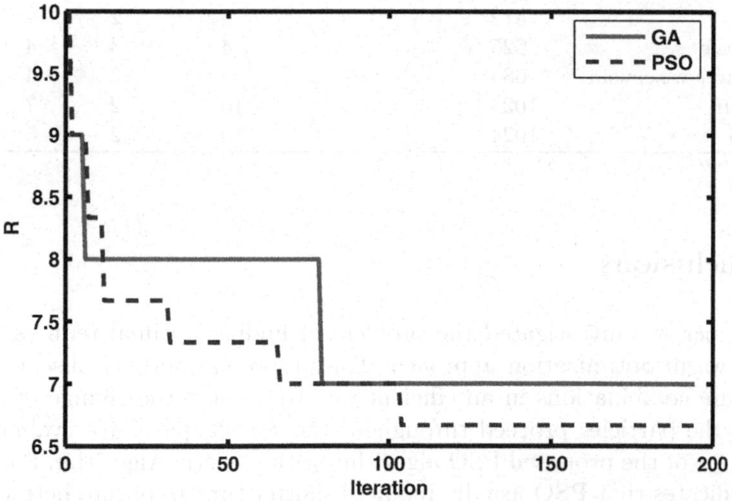

Fig. 2. Performance of rough set reduction for lymphography dataset

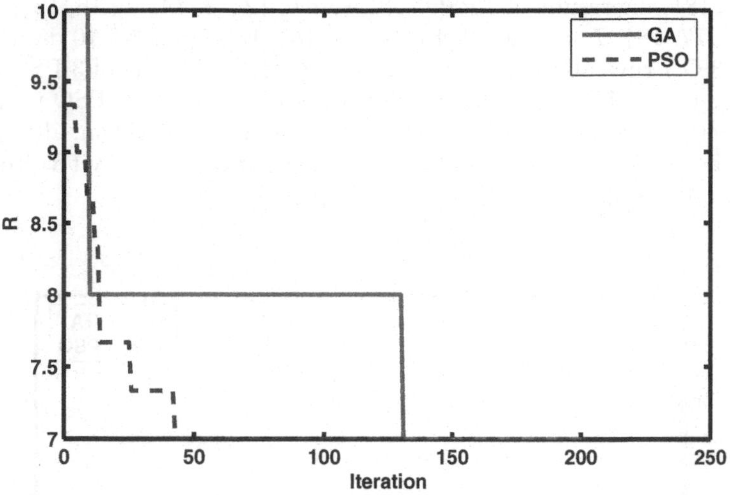

Fig. 3. Performance of rough set reduction for mofn-3-7-10 dataset

Table 1. Datasets used in the experiments

Dataset	*Size*	*ConditionAttributes*	*Class*	*GA*	*PSO*
lung-cancer	32	56	3	10	8
zoo	101	16	7	5	5
corral	128	6	2	4	4
lymphography	148	18	7	6	3
hayes-roth	160	4	3	3	3
shuttle-landing-control	253	6	2	6	6
monks	432	6	2	3	3
xd6-test	512	9	2	9	9
balance-scale	625	4	3	4	4
breast-cancer-wisconsin	683	9	2	4	4
mofn-3-7-10	1024	10	2	7	7
parity5+5	1024	10	2	5	5

5 Conclusions

In this paper, we investigated the problem of finding optimal reducts using a particle swarm optimization approach. The proposed approach discovered the best feature combinations in an efficient way to observe the change of positive region as the particles proceed throughout the search space. We evaluated the performance of the proposed PSO algorithm with Genetic Algorithm (GA). The results indicates that PSO usually required shorter time to obtain better results than GA, specially for large scale problems, although its stability need to be improved in further research. The proposed algorithm could be an ideal approach for solving the reduction problem when other algorithms failed to give a better solution.

Acknowledgments

This work is supported by NSFC (60573087), MOST (2005CB321904), and MOE (KP0302). *Ajith Abraham* is supported by the Centre for Quantifiable Quality of Service in Communication Systems, Centre of Excellence, appointed by The Research Council of Norway, and funded by the Research Council, NTNU and UNINETT.

References

1. Pawlak, Z.: Rough Sets. International Journal of Computer and Information Sciences 11 (1982) 341–356
2. Pawlak, Z.: Rough Sets: Present State and The Future. Foundations of Computing and Decision Sciences 18 (1993) 157–166
3. Pawlak, Z.: Rough Sets and Intelligent Data Analysis. Information Sciences 147 (2002) 1–12
4. Kusiak, A.: Rough Set Theory: A Data Mining Tool for Semiconductor Manufacturing. IEEE Transactions on Electronics Packaging Manufacturing 24 (2001) 44–50
5. Shang, C., Shen, Q.: Rough Feature Selection for Neural NetworkBased Image Classifiction. International Journal of Image and Graphics 2 (2002) 541–555
6. Francis E.H. Tay, Shen, L.: Economic And Financial Prediction Using Rough Sets Model. European Journal of Operational Research 141 (2002) 641–659
7. Świniarski, R. W., Skowron, A.: Rough Set Methods in Feature Selection and Recognition. Pattern Recognition Letters 24 (2003) 833–849
8. Beaubouef, T., Ladner, R., Petry, F.: Rough Set Spatial Data Modeling for Data Mining. International Journal of Intelligent Systems, 19 (2004) 567–584
9. Shen, L., Francis E. H. Tay: Applying Rough Sets to Market Timing Decisions. Decision Support Systems 37 (2004) 583–597
10. Gupta, K. M., Moore ,P. G., Aha, D. W., Pal, S. K.: Rough Set Feature Selection Methods for Case-Based Categorization of Text Documents. Lecture Notes in Computer Science 3776 (2005) 792–798
11. Boussouf, M.: A Hybrid Approach to Feature Selection. Lecture Notes in Artificial Intelligence 1510 (1998) 231–238
12. Skowron, A., Rauszer, C.: The Discernibility Matrices and Functions in Information Systems. In: Świniarski, R. W.(ed.), Handbook of Applications and Advances of the Rough Set Theory, Kluwer Academic Publishers (1992) 331–362
13. Zhang, J., Wang, J., Li, D., He, H., Sun, J.: A New Heuristic Reduct Algorithm Base on Rough Sets Theory. Lecture Notes in Artificial Intelligence 2762 (2003) 247–253
14. Hu, K., Diao, l., Shi, C.: A Heuristic Optimal Reduct Algorithm. Lecture Notes in Computer Science, 1983 (2000) 139–144
15. Zhong, N., Dong, J.: Using Rough Sets with Heuristics for Feature Selection. Journal of Intelligent Information Systems 16 (2001) 199–214

16. Banerjee, M., Mitra, S., Anand, A.: Feature Selection Using Rough Sets. Studies in Computational Intelligence, Springer 16 (2006) 3–20
17. Kennedy, J., Eberhart, R.: Swarm Intelligence, Morgan Kaufmann Publishers, San Francisco, CA (2001)
18. Clerc, M., Kennedy, J.: The Particle Swarm-Explosion, Stability, and Convergence in a Multidimensional Complex Space. IEEE Transactions on Evolutionary Computation 6 (2002) 58–73
19. Clerc, M.: Particle Swarm Optimization. ISTE Publishing Company, London (2006)
20. Parsopoulos, K. E., Vrahatis, M. N.: Recent Approaches to Global Optimization Problems Through Particle Swarm Optimization. Natural Computing 1 (2002) 235–306
21. Abraham, A., Guo, H., Liu, H.: Swarm Intelligence: Foundations, Perspectives and Applications. In: Nedjah, N., Mourelle, L. (eds.), Swarm Intelligent Systems, Studies in Computational Intelligence, Chapter 1, Springer (2006) 3–25
22. Salman, A., Ahmad, I., Al-Madani, S.: Particle Swarm Optimization for Task Assignment Problem. Microprocessors and Microsystems 26 (2002) 363–371
23. Sousa, T., Silva A., Neves A.: Particle Swarm Based Data Mining Algorithms for Classification Tasks. Parallel Computing 30 (2004) 767–783
24. Liu, B., Wang, L., Jin, Y., Tang, F., Huang D.: Improved Particle Swarm Optimization Combined With Chaos. Chaos, Solitons and Fractals 25 (2005) 1261–1271
25. Schute, J. F., Groenwold, A. A.: A Study of Global Optimization Using Particle Swarms. Journal of Global Optimization 31 (2005) 93–108
26. Liu, H., Abraham, A., Choi, O., Moon, S. H.: Variable Neighborhood Particle Swarm Optimization for Multi-Objective Flexible Job-Shop Scheduling Problems. Lecture Notes in Computer Science 4247 (2006) 197–204
27. Wang, G.: Rough Reduction in Algebra View and Information View. International Journal of Intelligent Systems 18 (2003) 679–688
28. Xu, Z., Liu, Z., Yang, B., Song, W.: A Quick Attibute Reduction Algorithm with Complexity of Max$(O(|C||U|), O(|C|^2|U/C|))$. Chinese Journal of Computers 29 (2006) 391–399
29. Cantú-Paz, E.: Efficient and Accurate Parallel Genetic Algorithms, Kluwer Academic publishers, Netherland (2000)
30. Abraham A., Evolutionary Computation, Handbook for Measurement Systems Design, Peter Sydenham and Richard Thorn (Eds.), John Wiley and Sons Ltd., London, ISBN 0-470-02143-8, (2005) pp. 920-931

Use of Kohonen Maps as Feature Selector for Selective Attention Brain-Computer Interfaces

Miguel Angel Lopez, Hector Pomares, Miguel Damas, Alberto Prieto, and Eva Maria de la Plaza Hernandez

Department of Computer Architecture and Computer Technology
University of Granada
{malopez,hpomares,mdamas}@atc.ugr.es, aprieto@ugr.es

Abstract. Selective attention to visual-spatial stimuli causes decrements of power in alpha band and increments in beta. For steady-state visual evoked potentials (SSVEP) selective attention affects electroencephalogram (EEG) recordings, modulating the power in the range 8-27 Hz. The same behaviour can be seen for auditory stimuli as well, although for auditory steady-state response (ASSR), it is not fully confirmed yet. The design of selective attention based braincomputer interfaces (BCIs) has two major advantages: First, no much training is needed. Second, if properly designed, a steady-state response corresponding to spectral peaks can be elicited, easy to filter and classify. In this paper we study the behaviour of Kohonen Maps as feature selector for a selective attention to auditory stimuli based BCI system.

Keywords: Brain-computer interfaces, Artificial Neural Networks, Self-Organizing Maps (SOMs), Selective attention, Auditory Steady-state Response.

1 Introduction

Many types of BCIs have been developed based on the classification of different features extracted from EEG recordings. For example, BCIs based on Event-related brain potentials (ERPs) are one of the most popular. ERPs are indicators of brain activities that occur in preparation for, or in response to, discrete events [1]. The P300 is an ERP with a typical latency exceeding 300 ms that shows up after the stimulus is presented and a cognitive task, typically counting target stimuli, is performed. One of the reasons for using the P300 in BCI systems is because it is a large ERP with maximum amplitude in the range of units of microvolts, big enough to be detected even in single-trial experiments [2]. Other BCIs are based on the voluntary modulation by the subject of spectral bands, such as alpha (8-13 Hz), beta (14-20) Hz or theta (5-8 Hz). One of the first BCIs used the spectral power of alpha band as a feature to extract and classify, based on the assumption that human beings can easily modify it. Recently, BCIs based on selective attention to visual stimuli that elicit SSVEP have been developed [3].

J. Mira and J.R. Álvarez (Eds.): IWINAC 2007, Part I, LNCS 4527, pp. 407–415, 2007.

The SSVEP is a periodic response elicited by the repetitive presentation of a visual stimulus, at a rate of more than 6 Hz. SSVEP power extends over an extremely narrow bandwidth as the periodicity of the response matches that of the stimulus [4]. The SSVEP amplitude is substantially increased when attention is focused upon the location of the flickering stimulus and it is more pronounced in recordings over the posterior scalp contralateral to the visual field of stimulation in the range 8.6-28 Hz [5]. SSVEP based BCIs measure the spectral power at flicker frequency in order to discriminate whether the stimuli is attended or ignored.

BCIs based on selective attention to auditory stimuli that elicit ASSRs have not been reported yet. ASSRs are composed of a train of superimposed auditory brainstem responses, that added in phase, conforms an averaged response with most of the energy located around the frequency of repetition [6] (see Fig. 1). Treatment of ASSRs signals have two major drawbacks: On the one hand the low amplitude, typically in the range of hundreds of nanovolts, and on the other hand it is not clear yet the influence of selective attention on signals as auditory brainstem responses not generated in the cortex. Peripheral effects of selective attention would only occur if the auditory system is "obliged" to do so adapting for the most efficient result at the lowest energetic cost [7]. That could happen in a very noisy environment with a very weak auditory stimulus.

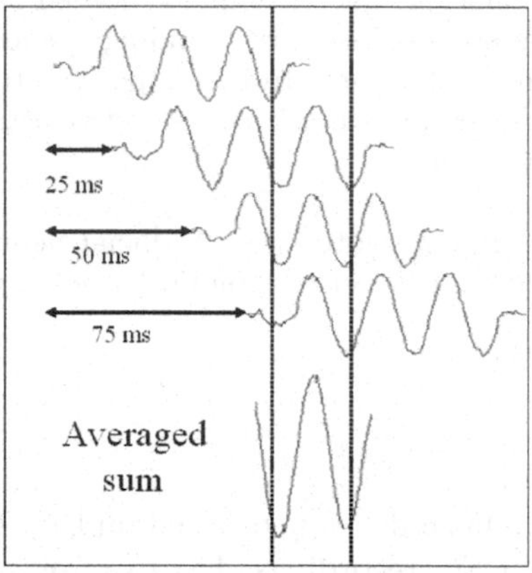

Fig. 1. Rows one to four are simulated potentials evoked during the first 100 msec after four auditory click stimuli, delayed 25 ms interstimulus, were applied. Last row shows the averaged sum. ASSRs are generated, in a similar way, as an averaged sum of single ABRs. This figure has been adapted from [6].

BCIs based on classification of features extracted from EEG recordings have some problems in common. First, the target features are immerse in low SNR. That is a weakness as the classification and extraction of the target features is difficult and not always successful. This issue can be minimized either by using high energy features with amplitudes in the range of microvolts, or by grand-averaging many trials as the SNR increases with the number of trials according to equation 1, where N is the number of trials averaged.

$$\text{SNRnew(dB)} = \text{SNRoriginal(dB)} + 10\log10(\text{N}) \quad . \tag{1}$$

Another issue is the low average transfer rate. Currently, a throughput of 27.15 bits/min has been reported for SSVEP based BCIs [8]. In order to improve the transfer rate a classification based on several features extracted simultaneously from EEG, or the use of contextual information, when available, have been proposed [9]. A third problem is that EEG signals are not considered to be stationary and the design of experiments have to bear in mind that the same experiment on the same subject could produce different results. In order to avoid this problem, adaptive systems such as ANNs, can be used.

A BCI based on the simultaneous extraction of several high energy features and its classification by means of an adaptive system seems to be the basis to enhance performance. This paper shows the use of a Kohonen map as a feature selector. The features under analysis are spectral power in alpha and beta bands and frequencies of an ASSRs and the possible use in BCIs systems.

The organization of the rest of this paper is as follows: Section 2 describes the stimuli and the experimental design. Section 3 presents and discuss the results obtained and finally, in section 4, some final conclusions are stated.

2 Methodology

2.1 Recordings and Stimulation

One male subject, 30 years old, with university studies and normal hearing participated in the experiment. The subject remained comfortably sat down in a quiet testing room, isolated from noise and external disruptions. The subject was encouraged to relax and close his eyes in order to reduce the background noise level when the EEG was being recorded.

The system used for recording was the Geodesic EEG System 200, by Electrical Geodesic. Each electrode was amplified by 1000. Data was collected at a sampling rate of 250 Hz, filtered with a low-pass filter of 100 Hz bandwidth and digitized using 16 bits per sample.

The dense-array Geodesic Sensor Net with 128 channels was inmersed in a container of electrolyte and impedance was reported below 5 kOhms. Despite the reference, commonly named channel 129, is located in the vertex for symmetry reasons, it was changed during analysis off-line to the left mastoid (sensor 57). Once applied, a test for electrolyte bridge detection was also performed. Fifteen electrodes were used at positions 6, 13, 31, 38, 54, 62, 80, 88, 106, 113, 7, 32, 55, 81, 107 of Geodesic Sensor Net. Ten of those electrodes match the positions FCz, FC1, C1, CP1, Pz, CP2, C2, FC2, CPz and Cz of standard 10-20 whereas five of them do not have equivalent positions (see Fig. 2).

The auditory stimuli were presented simultaneously to both ears through insert earphones at comfortable level, between 50 to 60 dB. Each stimulus consisted of a carrier, 1kHz for left ear and 2.5kHz for right, 100 per cent amplitude-modulated (AM) by a pure tone, 38 Hz for left ear and 42Hz for right, applied during 42 to 46 seconds. This kind of stimulus elicits an ASSR with spectral peaks around the

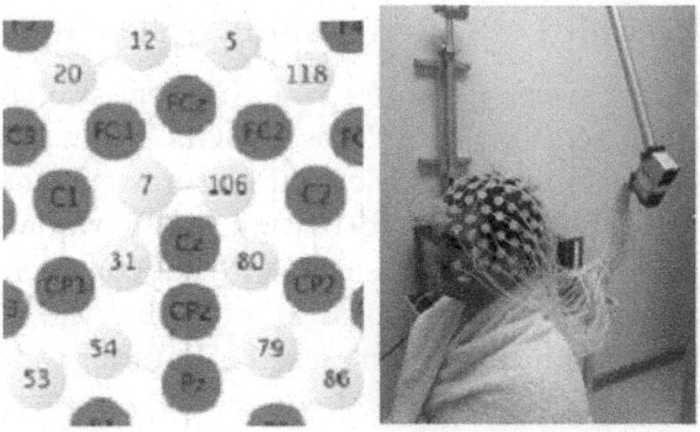

Fig. 2. On the left: Central top view of electrodes position. In dark circles the electrodes labelled according to the standard 10-20. In grey the specific electrodes for Geodesic Sensor Net. The electrodes used for training and evaluation were FCz, FC1 C1, CP1, 54, Pz, 79, CP2, C2, FC2, 31, CPz, Cz, 7, 106. On the rigth: Geodesic Sensor Net properly positioned and adjusted. A towel was used to prevent the leak of electrolyte to disturb the subject.

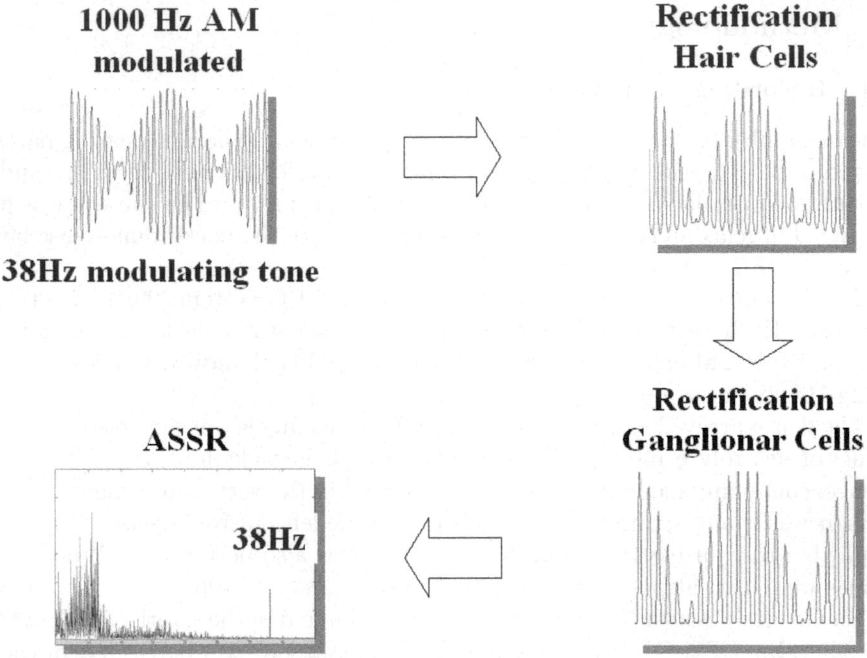

Fig. 3. Rectification of an AM modulated stimulus in the internal auditory system. It illustrates that the response to an AM modulated stimulus is a spectral peak at modulating frequency. This figure has been adapted from [11].

frequencies of the modulating tones [10]. That is the expected behaviour as the auditory system acts like an envelope detector, similar to an AM demodulator (see Fig. 3)

In order to facilitate selective attention, the AM modulated stimuli were used to codify different meaningful, but unknown by the subject, Morse messages. The subject received some training in Morse before the test was executed.

2.2 Experimental Design

The experiment was performed in 2 sessions with 10 minutes intersession rest and 10 trials per session. Each trial consists of one question with binary answer (Yes/No). Each question was displayed on the screen, in front of the subject, for 10 seconds at a comfortable distance and height. Afterwards, two auditory stimuli as described before were presented simultaneously to both ears. The subject was instructed to focus attention to stimulus from left ear if the answer was Yes and ignore the stimulus from right ear. If the answer was No, the subject had to ignore both stimuli. Due to the design of the stimuli and the experiment, it is expected to cause two effects during the attended condition: First, an increase of spectral power in alpha band and a decrease in beta band. Second, enhancement of spectral power of AM modulating frequency for the left ear (38 Hz), although the second effect is not truly confirmed yet. Fig. 4 shows data in the frequency domain collected during a trial in electrode Cz. On the left we see the EEG spectrum up to 45 Hz whereas on the right we see the ASSRs with two peaks at both AM modulating frequencies, 38 and 42 Hz for left and right ears respectively.

As selective attention is an inherent feature of human beings, the subject did not experiment much difficulty to focus attention to the target stimulus and to ignore the other one. Only a little training was needed for Morse code. Despite the subject was told to decode the Morse message each trial, the real purpose of the message was help the subject to focus attention. Correct decoding of the message was irrelevant to this experiment.

3 Results

The FFT was computed to measure spectral power in alpha and beta bands and for the ASSR at frequencies 38 and 42 Hz. Noise levels in weak ASSRs recordings were reduced increasing the duration of the trials up to 44 seconds. However in order to avoid start-stop problems due to the nature of selective attention, only the central data of each trial was submitted to Fourier analysis. All collected data during both sessions were used to train and evaluate the SOM. Trials with amplitude bigger than 50 microvolts were rejected, to avoid muscular artifacts.

We used a Kohonen SOM to classify the features. An array of 16-by-16 neurons was arranged. For training we used data from fifteen electrodes collected along twenty trials in both sessions. That makes 300 input vectors. Each 4-dimensions vector is composed of: The spectral power of alpha and beta bands and the two ASSRs at the frequencies of the AM modulating tones (38 and 42 Hz). Once the

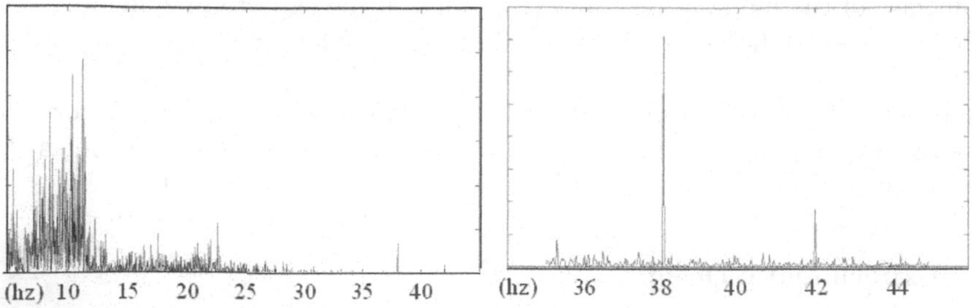

Fig. 4. On the left: EEG recording in the frequenciy domain where most of the power is located around the alpha band. On the rigth: EEG amplitude spectrum of two ASSRs AM modulated at both 38 and 42 Hz. The amplitude of peak at 38 Hz might be modulated due to selective attention in the attended condition. The compared asymmetrical amplitude of both peaks is not significant as it could correspond to the maximum efficiency of ASSRs around 40Hz but closer to the first peak.

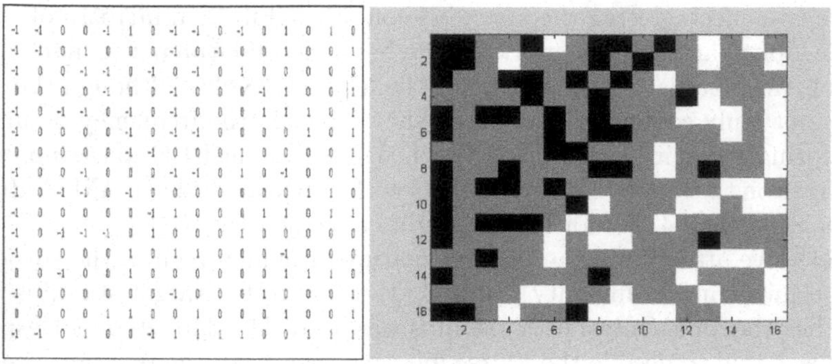

Fig. 5. On the left, neurons showing negative values are activated by an input vector associated to attended stimulus. Positive values correspond to trials where the stimuli were ignored. Neurons showing zero value are not activated by any input vector. On the right the values are represented in grey scale with zero as neutral grey.

net was trained, 150 input vectors randomly picked from both sessions (75 for attended condition and 75 for ignored) were presented to the network for their classification. In this way only one neuron is activated for each input vector. As the number of neurons is greater than the number of vectors, some neurons never are activated, whereas some other could be activated for more than one input vector. The idea behind this is to gather input vectors in clusters related to selective attention and subsequently analyze the values of their components. Fig. 5 shows the array of 16-by-16 neurons. Neurons in black correspond with neurons activated by input vectors related to attended stimuli, whereas neurons in white correspond to neurons activated by input vectors related to ignored stimuli. Neurons in grey correspond to neurons not activated by any input vector.

In order to facilitate visualization of clusters, a simple algorithm was executed to make grey neurons become white or black making the boundaries clearer (see Fig. 6). The SOM shows two clusters of approximately the same area, in black and white, that correspond to the attended/ignored condition respectively. The size of the areas matches the ratio of attended/ignored trials (50% each). The analysis of the components of the neurons along the diagonal shows a relation between the power in alpha band and selective attention according to [14]. Beta seems to have the same behaviour as Alpha and that is not the expected behaviour. For spectral power at 38 Hz we see selective attention to enhance level of attended stimulus. For 42 Hz no clear relations can be assured.

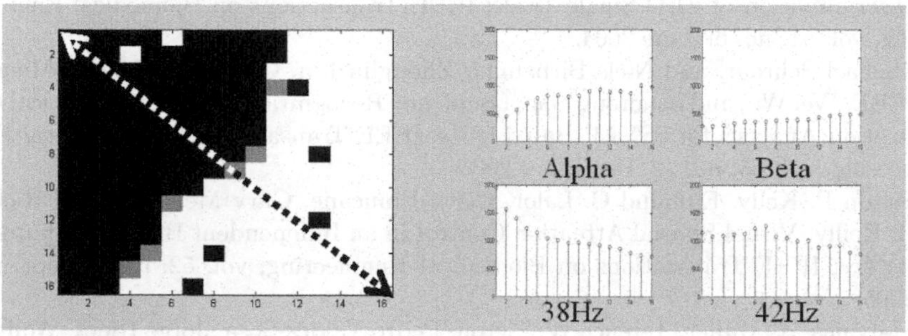

Fig. 6. On the left, the SOM determines two clusters: In black for stimuli attended and white for stimuli ignored. According to the topology of the SOM, the diagonal represents the direction of maximum variation in the attended/ignored condition. On the right the values of the four components along the 16 neurons of the diagonal.

4 Conclusions

In this paper we have presented a study of the behaviour of a Kohonen map used as feature selection for a selective attention to auditory stimuli based BCI system. Four different features extracted from EEG were submitted to analysis: The spectral power of alpha and beta bands and the two ASSRs at the frequencies of the AM modulating tones (38 and 42 Hz).

As it has been reported in previous papers, we have seen evidence of modulation of spectral power in alpha according to [13,14], beta bands and attended ASSR by selective attention. That is an advantage for BCI systems as selective attention hardly needs the subject to be trained. However we have to keep in mind that this experiment has been executed only in one subject, hence no closed conclusions can be stated. We plan to execute the experiment in a more significant number of subjects.

In further studies we will add more EEG features such as the ERPs N100 and P300 to the four features studied in this paper as they are expected to be highly influenced by selective attention [15].

Acknowledgments. The authors would like to thank the Department of Experimental Psychology and Physiology of Behavior of the University of Granada for the support in the design and execution of the experiment.

References

1. J. Cacioppo, L. Tassinary G. Berntson (Eds); "Handbook of Psychophysiology", 2nd Edition. Cambridge University Press 2000.
2. Benjamin Blankertz, Klaus-Robert Mller, Gabriel Curio, Theresa M. Vaughan, Gerwin Schalk, Member, IEEE, Thilo Hinterberger, Michael Schrder, and Niels Birbaumer; The BCI Competition 2003: Progress and Perspectives in Detection and Discrimination of EEG Single Trials; IEEE Transactions on Biomedical Engineering, vol. 51, no. 6, June 2004.
3. Michael Schrder, and Niels Birbaumer Zhonglin Lin, Changshui Zhang, Member, IEEE, Wei Wu, and Xiaorong Gao; Frequency Recognition Based on Canonical Correlation Analysis for SSVEP-Based BCIs; IEEE Transactions on Biomedical Engineering, vol. 53, no. 12, December 2006.
4. Simon P. Kelly, Edmund C. Lalor, Ciarn Finucane, Gary McDarby, and Richard B. Reilly; Visual Spatial Attention Control in an Independent Brain-Computer Interface; IEEE Transactions on Biomedical Engineering, vol. 52, no. 9, September 2005.
5. Matthias M. Muller, Terence W. Picton, Pedro Valdes-Sosa, Jorge Riera, Wolfgang A. Teder-Salejarvi, Steven A. Hillyard; Effects of spatial selective attention on the steady-state visual evoked potential in the 2028 Hz range; Cognitive Brain Research 6 1998 249261.
6. R. Galambos, S. Makeig, P. J. Talmachoff; A 40-Hz auditory potential recorded from the human scalp; Proc. Nati. Acad. Sci. USA Vol. 78, No. 4, pp. 2643-2647, April 1981 Psychology.
7. Marie-Helene Giard, Alexandra Fort, Yolande Mouchetant-Rostaing, Jacques Pernier; Neurophysiological mechanisms of auditory selective attention in humans; Frontiers in Bioscience, 5, d84-94, January 1, 2000.
8. Ming Cheng, Xiaorong Gao, Shangkai Gao, Senior Member, IEEE, and Dingfeng Xu; Design and Implementation of a Brain-Computer Interface With High Transfer Rates; IEEE Transactions on Biomedical Engineering, vol. 49, no. 10, October 2002.
9. Jessica D. Bayliss and Dana H. Ballard; A Virtual Reality Testbed for BrainComputer Interface Research; IEEE Transactions on Rehabilitation Engineering, vol. 8, no. 2, June 2000.
10. T.W. Picton, A. Dimitrijevic, M.Sasha John; Muliple Auditory Steady-state Responses; The Annals of Otology, Thinology and Laryngology; May 2002;111,5; Health and Medical Complete.
11. Maria Cecilia Perez Abalo Alejandro Torres Fortuny Guillermo Savio Lopez E. Eimil Suarez; Los potenciales evocados auditivos de estado estable a mltiples frecuencias y su valor en la evaluacin objetiva de la audicion; Centro de Neurociencias de Cuba. La Habana. Cuba, Auditio: Revista Electrnica de Audiologa, Vol. 2, 2003.
12. Simon Haykin; Neural Networks a comprehensive foundation;2ND Edition; Prentice Hall.

13. C. Gomez, M.Vazquez, E. Vaquero, D. Lopez-Mendoza, M.J. Cardoso; Frequency Anlisis of the EEG during spatial selective attention; Internacional Journal of Neuroscience, vol. 95, 17-32 (1998).
14. M.Vazquez, E. Vaquero, , M.J. Cardoso , C. Gomez; Temporal Evolution of Alpha and Beta bands during visual spatial attention; Cognitive Brain Research, vol 12, n2, 315-320 (2001).
15. Sara Maatta, Ari Paakkonen, Pia Saavalainen, Juhani Partanen; Selective attention event-related potential effects from auditory novel stimuli in children and adults; Clinical Neurophysiology 116 (2005) 129141; Ed. Elsevier.

Nature-Inspired Congestion Control: Using a Realistic Predator-Prey Model

Morteza Analoui and Shahram Jamali

Iran University of Science & Technology,Tehran, Iran
{analoui,jamali}@iust.ac.ir

Abstract. Nature has been a continuous source of inspiration for many successful techniques, algorithms and computational metaphors. We outline such a inspiration here, in the context of bio-inspired congestion control (BICC) algorithms. In this paper a realistic predator-prey model is mapped to the Internet congestion control mechanism. This mapping leads to a bio-inspired congestion control scheme. Dynamic and equilibrium properties of developed algorithm are good enough according to the simulation results.

Keywords: Communication Networks, Congestion Control, Biology, Predator-Prey.

1 Introduction

Biologically inspired approaches have already proved successful in achieving major breakthroughs in a wide variety of problems in information technology (IT). A more recent trend is to explore the applicability of bio-inspired approaches to the development of self-organizing, evolving, adaptive and autonomous information technologies. The central aim of this paper is to obtain methods on how to engineer congestion control algorithms, which have similar high stability and efficiency as biological entities often have. Several examples are available in the area of computer technology. The most known examples are swarm intelligence, evolutionary or genetic algorithms, and the artificial immune system. The adapted mechanisms find application in computer networking for example in the areas of network security [1,2], pervasive computing, and sensor networks [3].

Previous congestion control research has been heavily based on measurements and simulations, which have intrinsic limitations. There are also some theoretical frameworks and especially mathematical models that can greatly help us understand the advantages and shortcomings of current Internet technologies and guide us to design new protocols for identified problems and future networks [4,5,6,7,8,9,10,11,12].

In our previous work [13], a multidisciplinary conceptual framework has been proposed that provides principles for designing and analyzing bio-inspired congestion control algorithms. We proposed that the biological population control approaches such as predator-prey is susceptible for mapping to the congestion

J. Mira and J.R. Álvarez (Eds.): IWINAC 2007, Part I, LNCS 4527, pp. 416–426, 2007.

control problem in the Internet. In [14,15] inspiring by predator-prey interaction we developed a bio-inspired congestion control algorithm (BICC). We discussed on how a skillful parameters setting can help us to achieve good equilibrium properties such as fairness and performance. One of the weak points of proposed algorithm was its dynamic properties such as stability and speed of convergence. This paper attempts to design another bio-inspired congestion control algorithm that has better dynamic performance in compare to BICC. This algorithm that is called RBICC uses principals of a more realistic model of predator-prey interaction. RBICC not only inherent some intrinsic characteristics of biology such as stability and robustness, but also provides us a theoretic and mathematical framework that we can benefits from its facilities in analysis and development of the model.

Section 2, briefly explains analogy between the biological environment and the communication networks. Section 3 presents a methodology for applying the realistic models of predator-prey mathematical model to the Internet congestion control scheme. Section 4 presents an illustrative example for the proposed algorithm. The implementation issues for the proposed algorithm will be discussed in section 5 and we conclude in section 6 with future works.

2 Internet as an Ecosystem Analogy

Consider a network with a set of k source nodes and a set of k destination nodes. We denote $S = \{S_1, S_2, ..., S_k\}$ as the set of source nodes with identical round-trip propagation delay (RTT), and $D = \{D_1, D_2, ..., D_k\}$ as the set of destination nodes. Our network model consists of a bottleneck link, with capacity of B packet per RTT, from LAN to WAN as shown in Fig. 1 (left) and uses a window-based algorithm for congestion control.

We propose that, this network can be imagined as an ecosystem that connects a wide variety of habitats such as routers, hosts, and etc. We consider this network from congestion control viewpoint and assume that there is some species in these habitats such as *Congestion Window (W)*, *Packet Drop (P)*, *Queue (q)* and *link utilization (u)*. The size of these network elements refers to their population size in Internet ecosystem. Fig. 1 (right) shows the typology of Internet ecosystem from congestion control perspective. In this ecosystem the species are interacting and hence, the population size of each species is affected. Let the population of W in source S_i be W_i (congestion window size of connection i). It is clear that if the population size of this species is increased, then the number of sent packet would be inflated. Hence, in order to control the congestion in the communication networks the population size of W (all of the W_is) must be controlled. This means that *the population control problem in the nature can be mapped to the congestion control problem in the communication networks.* We can use the natural population control tactics for this purpose. Nature uses many tactics such as predation, competition, parasites and etc to control the population size of species. In this paper a methodology is proposed to use the predation tactic to control the population size of W species.

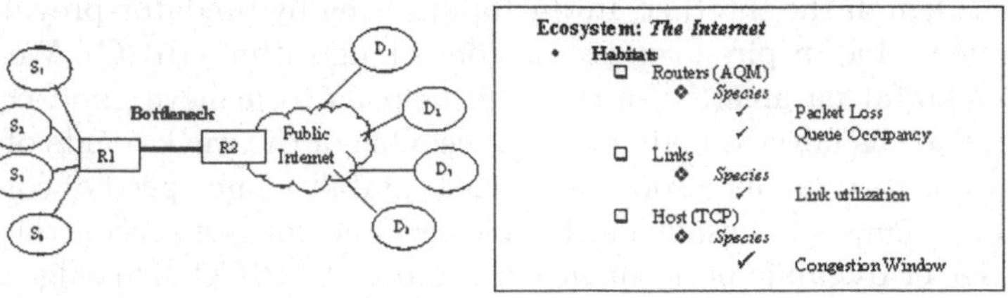

Fig. 1. Test network *(left)* and its ecological analogy *(right)*

Predator-Prey Interaction. This interaction refers to classical predators that kill individuals and eat them: (1) In the absence of predators, prey would grow exponentially. (2) The effect of predation is to reduce the prey's growth rate. (3) In the absence of prey, predators will decline exponentially. (4) The prey's contribution to the predator's growth rate is proportional to the available prey as well as to the size of the predator population. (5) A prey carrying capacity puts a ceiling on the prey population. If r and f represents the number of rabbits and foxes, then the Lotka-Volterra model [16,17] is:

$$\frac{dr}{dt} = ar - brf \tag{1}$$

$$\frac{df}{dt} = crf - hf \tag{2}$$

Where the parameters are defined by: a is the natural growth rate of rabbits. b is the death rate per encounter of rabbits due to predation. c is the efficiency of turning predated rabbits into foxes. h is the natural death rate of foxes in the absence of food.

One of the unrealistic assumptions in the Lotka-Volterra model (1)-(2) is that the prey growth is unbounded in the absence of predation. As a reasonable first step we might expect the prey to satisfy a logistic growth [16,17], say, in the absence of any predators, has some maximum carrying capacity:

$$\frac{dr}{dt} = a\left(1 - \frac{r}{K}\right)r - brf \tag{3}$$

In which K is the constant carrying capacity for the prey (r) when $f = 0$.

3 Congestion Control by Realistic Predator-Prey Model

In order to clarify the similarity between TCP/AQM congestion control mechanism and predator-prey interaction, we explain the TCP/AQM running on a network: (1) In the absence of packet drop (P), congestion window (W) would grow. (2) On the occurrence of a packet drop, congestion window size would decline. (3) Incoming packet rate contribution to packet drop rate growth is proportional to available traffic intensity, as well as, the packet drop rate itself. (4) In the absence of a packet stream, packet drop rate will decline. (5) Bottleneck bandwidth is a limit for packet rate (carrying capacity). We see that this behavior is so similar to the predator-prey interaction. A similar discussion can also be carried out to show that the interaction of W and q (queue size) is also similar to the predator-prey interaction. These similarities motivate us to use Predator-Prey model to design congestion control scheme: For this purpose we assume that there are two species P and q that prey the individuals of W species to control its population size. Note that P and q species are not exactly the packet drop probability and the queue size. Later we will discuss about their interpretation in the network context. Since according to the Fig. 1, W contains k species $(W_1, W_2, ..., W_k)$, so accordingly suppose that P have also k species $(P_1, P_2, ..., P_k)$ in the congested router. To specify a congestion control system, we can in general postulate a dynamic model of the form

$$\dot{W}_i = F(W_j, Price_j) \qquad i, j = 1, 2, ..., k$$

$$\dot{Price}_i = G(W_j, Price_j) \quad i, j = 1, 2, ..., k$$

Since we adopted predator-prey interaction for population control of W, hence, the generalized version of realistic predator-prey model of (2) and (3) is used to drive F and G. In this model $k+1$ species $P_1, P_2, ..., P_k$ and q predate and control the population size of other k species $W_1, W_2, ..., W_k$. This deliberation leads to the following Bio-inspired distributed congestion control algorithms:

$$\frac{dW_i}{dt} = W_i\left(h_i\left(1 - \frac{W_i}{B_n}\right) - \sum_{j=1}^{k} b_{ij}P_j - r_i q\right) \qquad i = 1, 2, ..., k \tag{4}$$

$$\frac{dP_i}{dt} = P_i\left(\sum_{j=1}^{k} c_{ij}W_j - d_i\right) \qquad i = 1, 2, ..., k \tag{5}$$

$$\frac{dq}{dt} = \left(\sum_{j=1}^{k} e_{ij}W_j - m\right) \qquad m = Min(B, q + \sum_{j=1}^{k} W_j) \tag{6}$$

In which h_i is the growth rate of W_i in the absence of P and q. b_{ji} is the decrement rate per encounter of W_i due to P_j. r_i is the decrement rate per encounter of W_i due to q. d_i is the decrement rate of P_i in the absence of W. c_{ij} and e_j are the efficiency of turning predated W_j into P_i and q respectively. We set $B_n = 0.9B$, this means that in the absence of any P_i and q the congestion window i can grow up to 90 percent of bottleneck bandwidth. Interpret the first term of equation (4) as follows: when the window size is small then $(h_i(1 - W_i/B_n))$ will be large, pushing up window growth rate. When the window size becomes large then $(h_i(1 - W_i/B_n))$ will be small, pushing down source rates.

3.1 Discussion on Behavior and Parameter Setting

RBICC is a unified source/AQM in which the equation (4) refers to source algorithm and the equations (5)-(6) describe the AQM algorithm. Any source that implements RBICC maintains a variable that we call it "*Price*", as a congestion measure. This variable is used to adjust sending rate and are updated at the sources by using the congestion information that is sent by the routers:

$$Price_i = \sum_{j=1}^{k} b_{ij} P_j + r_i q \tag{7}$$

P_is and q are calculated using the equations (5)-(6). The *Price* is incremented if the weighted sum of P_is and q is positive and decremented otherwise. When the number of users increases, P_is and q grow, pushing up *Price* and hence marking probability. This sends a stronger congestion signal to the sources which then reduce their rates. When the source rates are too small, the P_is and q will be small, pushing down *Price* and raising source rates, until eventually, the *Price* is driven to equilibrium P_is and q. In equilibrium the queue size, the window size and the P_is should be stabilized and hence, the following conditions must hold for i=1, 2, ..., k:

$$(a) \sum_{j=1}^{k} b_{ij} P_j + r_i q = h_i \left(1 - \frac{W_i}{0.9B}\right) \quad (b) \sum_{j=1}^{k} c_{ij} W_j = d_i \quad (c) \sum_{j=1}^{k} e_{ij} W_j = m \tag{8}$$

The equations (8) depict the fact that the steady state behavior of the RBICC is significantly affected by the choice of the control parameters such as b_{ij}s, c_{ij}s, e_js, h_is and d_is. Other features such as stability, scalability and convergence time are also affected by these parameters. In what follows, based on these features we discuss how a good parameter configuration can help us for achieving better results. We use the following settings for the parameters of proposed model:

$$c_{ij} = b_{ij}, h_i = 0.25, d_i = d = 0.9B/k, r = 0.02, e_i = 1 where i, j = 1, 2, ..., k \tag{9}$$

$$c_{ii} = c_1 = 0.9, c_{ij} = c_2 = (1 - c_1)/k \quad \Rightarrow \sum_{j=1}^{k} c_{ij} = 1 \;\; and \;\; c_1 \gg c_2 \tag{10}$$

Where k is the number of sources sharing congested link and can be estimated by some approaches such as in [18]. These settings lead to the following results:

- According to the (8.b) and (10), due to symmetry of c_{ij} coefficients, all of the W_is will be equal to the d in equilibrium (fairness).
- By summing up the equations (8.b) aggregate traffic on the bottleneck link can be computed as follows:

$$\sum_{i=1}^{k} \sum_{j=1}^{k} c_{ij} W_j = k d_i = 0.9B \tag{11}$$

The equation (11) shows that in equilibrium the bottleneck link is utilized around 0.9. Since in the steady state the input rate is less than true link capacity and hence, q approaches to zero.

- The ratio of c_{ij}/c_{ii} indicates the degree of the influence of the connection j that share the same bottleneck link with connection i. To converge window size to a positive value despite its share of bandwidth d_i, it is necessary to satisfy the condition $0 < c_{ij}/c_{ii} < 1$. Furthermore, smaller c_{ij}/c_{ii} leads to faster convergence. Hence, if any c_{ii} be several times larger than other c_{ij}s then equations (4)-(6) will be stable. With these discussions any c_{ii} must be several times larger than any c_{ij}. These conditions can be hold by (10).
- According to the formal definition of queue dynamics any e_i must be set to 1. Degree of influence of queue length on $Price_i$ is indicated by r. We set r to 0.02.
- According to (5) P_i refers to rate mismatch i.e. difference between input rate and target capacity (d_i). P_i can also be an unfairness measure for source i i.e. P_i is incremented if source i uses more than its fair share of bandwidth (d_i) and is decremented otherwise. q refers to queue mismatch and is positive if there is any waiting packet in queue. Hence the $Price$ is positive when the input rate exceeds the link capacity or there is excess backlog to be cleared and is zero otherwise. This is so similar to REM [19] that measures congestion by incoming traffic rate and queue length.
- It is desirable that in equilibrium $Price$ has a small value that can be realized by small P_is and q. As mentioned q approaches to zero in equilibrium. So according to the equation (8.a) P_i will be equal to αh_i ($\alpha l1$) in equilibrium. This means that h_i must be set to a small non-zero value.

3.2 Scalability

When the number of connections is increased, the load on the router is increased too. Hence this scheme cannot be scalable by number of connections. In order to solve this problem we outline an approach in which the congested router only use aggregated and local information to measure congestion and send it to the sources:

By using the equation (10) we can rewrite the equation (5) as follows.

$$\frac{dP_i}{dt} = P_i \left((c_1 - c_2)W_i + \frac{(1-c_1)\sum_j W_j - 0.9B}{k} \right) \tag{12}$$

We assume that k is large enough to regard $c_1 \gg c_2$ and hence, rewrite the equation (12) as follows:

$$\frac{dP_i}{dt} = P_i \left(c_1 W_i + \frac{(1-c_1)\sum_j W_j - 0.9B}{k} \right) \tag{13}$$

In order to enable the source i to compute P_i the congested router applies (14) and computes the P_g by using only local and aggregated information:

$$P_g = \frac{(1-c_1)\sum_j W_j - 0.9B}{k} \tag{14}$$

Note that P_g refers to share of any source from rate mismatch in the congested router. The router also monitors the queue length (q) of the congested link and then disseminates and q to all of the sources that share the congested link. Any source i that receives P_g computes new P_i by using the equation (13). Since $c_1 \gg c_2$, then by an acceptable estimation (if we assume that all of the P_is are equal) the equation (4) can be rewritten as follows:

$$\frac{dW_i}{dt} = W_i(h_i - P_i - rq) \tag{15}$$

Regarding these considerations RBICC uses only local and aggregated information in the router and in the sources, therefore it is scalable by complexity.

4 Illustrative Example

To study RBICC we apply it to the network of Fig. 1 (left) and consider a four-connection network that has a single bottleneck link of capacity 50 pkts/RTT (for example if RTT=20 ms and packet size=1500 Byte then this capacity will refer to capacity of 30 Mbps). Other links of this network have bandwidth of 100 pkt/RTT. We suppose that all flows are long-lived, have the same end-to-end propagation delay and always are active. To simulate this network behavior the parameters of the proposed congestion control system of (4)-(6) are set according to the (9)-(10) and leads to the following equations:

$$dW_1/dt = W_1(0.33(1 - W_1/45) - 0.9P_1 - 0.033P_2 - 0.033P_3 - 0.033P_4 - 0.02q)$$

$$dW_2/dt = W_1(0.33(1 - W_2/45) - 0.033P_1 - 0.9P_2 - 0.033P_3 - 0.033P_4 - 0.02q)$$

$$dW_3/dt = W_1(0.33(1 - W_3/45) - 0.033P_1 - 0.033P_2 - 0.9P_3 - 0.033P_4 - 0.02q)$$

$$dW_4/dt = W_1(0.33(1 - W_4/45) - 0.033P_1 - 0.033P_2 - 0.033P_3 - 0.9P_4 - 0.02q)$$

$$dP_1/dt = P_1(0.9W_1 + 0.033W_2 + 0.033W_3 + 0.033W_4 - 11.25)$$

$$dP_2/dt = P_2(0.033W_1 + 0.93W_2 + 0.033W_3 + 0.033W_4 - 11.25)$$

$$dP_3/dt = P_3(0.033W_1 + 0.033W_2 + 0.9W_3 + 0.033W_4 - 11.25)$$

$$dP_4/dt = P_4(0.033W_1 + 0.033W_2 + 0.033W_3 + 0.0W_4 - 11.25)$$

$$dq/dt = (W_1 + W_2 + W_3 + W_4 - Min(q + W_1 + W_2 + W_3 + W_4, 50))$$

Considering the following initial state we use Matlab 7.1 to solve this system. $P1(0) = P2(0) = P3(0) = P4(0) = 0.1, q(0) = 1, W1(0) = 1, W2(0) = 2, W3(0) = 4, W4(0) = 6$

The simulation results of this congestion control system are shown in figures (2)-(5). In order to reference to the results of these figures, we note that: 1. *Utilization:* According to the Fig. 3, after the startup transient of the sources, utilization of bottleneck link remains always over the 90 percent that is good enough.

Table 1. Average of W_is and P_is

Traffic sources	Source 1	Source 2	Source 3	Source 4
Mean of Wi	11.26	11.26	11.26	11.26
Mean of Pi	0.231	0.231	0.231	0.231

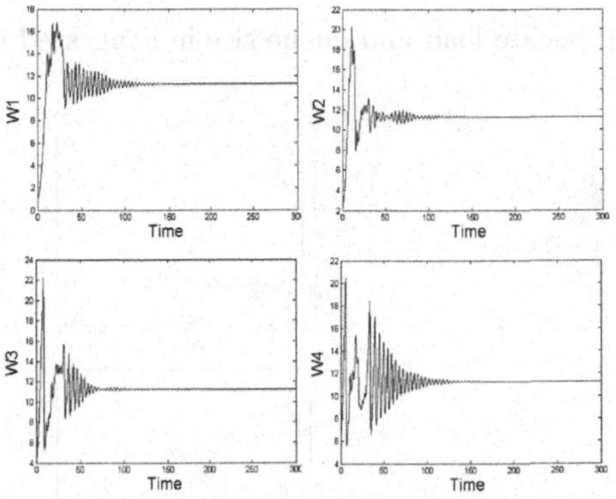

Fig. 2. Evolution of congestion windows size ($W_i s$)

2. *Fairness:* At each equilibrium stage, the bandwidth is shared equally among sources despite their heterogeneous initial state (max-min fairness [20,21]). Table 1 shows this equality.

3. *Queue evolution:* As can be found in Fig. 3 the queue size is zero in equilibrium and around 3 packets in transient. This means that queuing delay and jitter is negligible. This figure shows that if we set the queue size of the congested router around 10 packets then there won't be any packet loss.

4. *Stability and speed of convergence:* As we can see in Fig. 2, Fig. 3, Fig. 4, and Fig.5, the source rates, queue size and the P_is has decreasing oscillation level and track stable behavior. In steady state, there is no oscillation, and the speed of convergence to this steady state is more than BICC algorithm. Part of the results of BICC that had been developed based on the equation (1)-(2) is shown in Fig. 6. Comparison of results shows that RBICC have better dynamic performance.

Simulation results show that speed of convergence for aggregate rate is more than any individual source. This is due to global coordination between the sending sources.

5 Implementation Issues

The following algorithm summarizes the implementation process and addresses the implementation issues of RBICC.

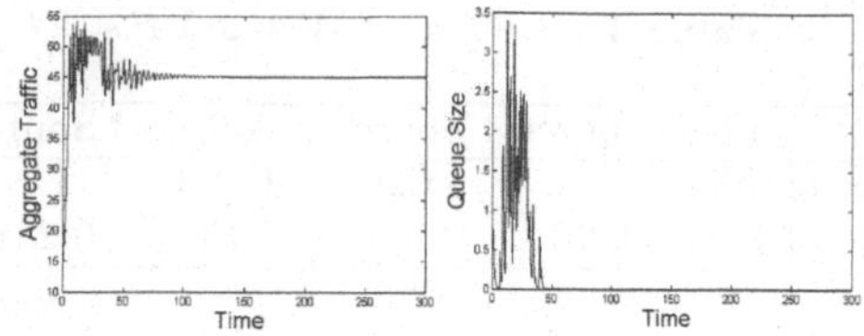

Fig. 3. Aggregate load and queue size in congested router

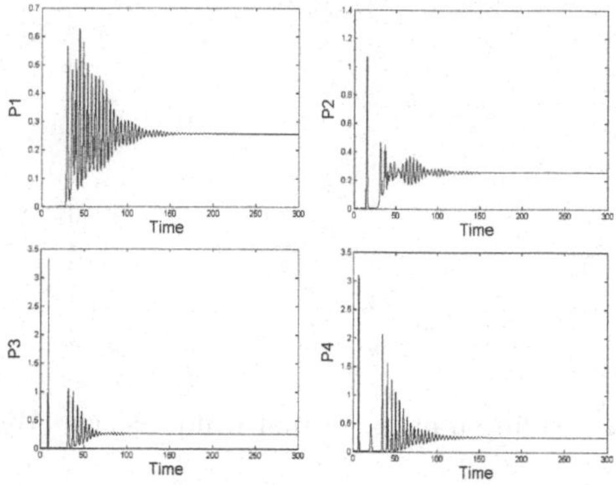

Fig. 4. Evolution of P_is

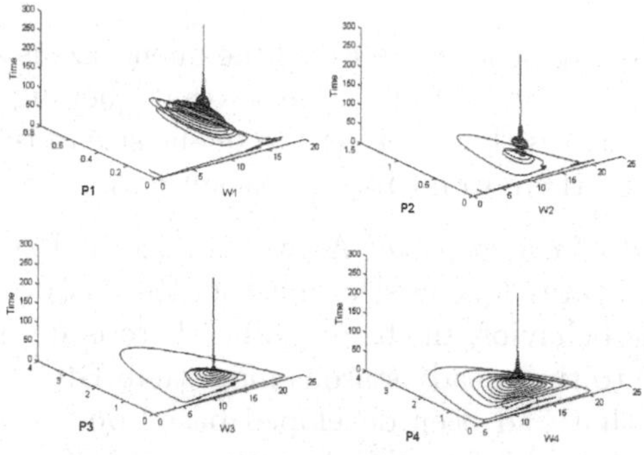

Fig. 5. Phase-plane trajectories for stability analysis

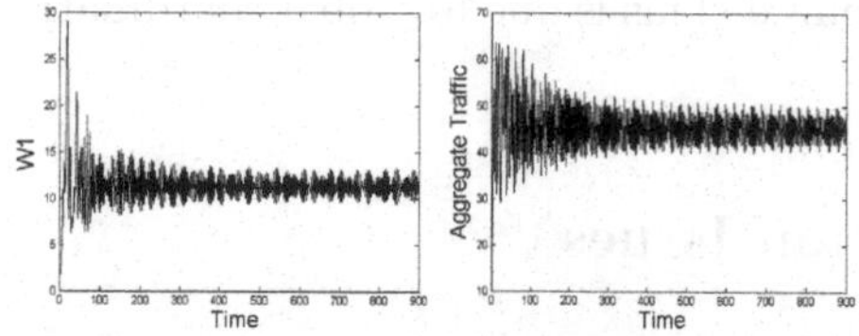

Fig. 6. Evolution of W_1 and aggregate load in BICC

Congested router's algorithm:
At time $RTT, 2RTT, 3RTT, ...$ congested router:
1. Receives W_s packet from all of the sources that goes through bottle link l.
2. Computes the new P_g and q for all of the sources that use link l:

$$P_g = \frac{(1-c_1)\sum_j W_j - 0.9B}{k} \quad and \quad \frac{dq}{dt} = (\sum_{j=1}^{k} e_{ij}W_j - m)$$

3- Computes marking probability p_{p_g} and p_q through

$$p_q = 1 - \Phi^{-q} \quad and \quad p_{p_g} = 1 - \Phi^{p_g} \ (\Phi > 1)$$

Note that according to (14), P_g is non positive but q is positive always, so $0 \leq p_{pg}$ and $p_q \leq 1$.
4- Uses two ECN bits [22] to communicate new P_g and q to all of the sources that use link.

Source i's algorithm:
At time $RTT, 2RTT, 3RTT, ...$ source i:
1- Receives from the congested router marked packets and computes p_{p_g} and p_q.
2- Extract from marking statistics the P_g and the q then computes P_i.
$p_g = log_\phi(1 - p_{p_g})$, $q = -log_\phi(1 - p_q)$, $\frac{dP_i}{dt} = P_i(c_1 W_i + p_g)$
3- Choose a new window size for the next period:
$\frac{dW_i}{dt} = W_i(h_i - P_i - rq)$

The exponential form of the marking probability is critical in a large network. When and only when, individual link marking probability is exponential in its link price, this end-to-end marking probability will be exponentially increasing in the sum of the link prices at all the congested links in its path [19].

6 Conclusion Remarks

In this paper we have used a model based on predator-prey interaction to design a congestion control mechanism in communication network. We used a realistic predator-prey model and have seen that with some consideration on parameters, this model leads to a stable, fair and high performance congestion control algorithm. The dynamic performance of this algorithm is better than the previous works. A number of avenues for future extensions remain. First this work needs some analytical foundations. Second, with mathematical characterization of network objective such as fairness, stability and etc. we can use mathematical rules for definition of parameters of purposed model to achieve well-designed communication network.

References

1. D'haeseleer, P. Forrest, S.: An Immunological Approach to Change Detection: Algorithms, Analysis and Implications. IEEE. (1996)
2. Hofmeyer, S.: An Immunological Model of Distributed Detection and Its Application to Computer Security. University of New Mexico. (1999)

3. Dressler, F.: Efficient and Scalable Communication in Autonomous Networking using Bio-inspired Mechanisms - An Overview. informatica 29. (2005)
4. Wang, J.: A Theoretical Study of Internet Congestion Control: Equilibrium and Dynamics. PhD thesis, university of Caltech. (2005)
5. Floyd, S.: Connections with multiple congested gateways in packet-switched networks part 1: One-way traffic. Computer Communications Review. (1991)
6. Handley, M. Floyd, S. Padhye, J. Widmer, J.: TCP Friendly Rate Control (TFRC): Protocol specification. RFC 3168, Internet Engineering Task Force. (2003)
7. Hollot, C. Misra, V.: D. Towsley, and W. Gong, A control theoretic analysis of RED. IEEE Infocom. (2001)
8. Low, S. Paganini, F. Wang, J. Doyle, J.: Linear stability of TCP/RED and a scalable control. Computer Networks Journal. (2003)
9. Low, S. Peterson, L. Wang, L.: Understanding Vegas: a duality model. Journal of ACM. (2002)
10. Kelly, F.: Charging and rate control for elastic traffic. European Transactions on Telecommunications. (1997)
11. Kelly, F. Maulloo, A. Tan, D.: Rate,control for communication networks: Shadow prices, proportional fairness and stability. Journal of Operations Research Society. (1998)
12. Low, S. Lapsley, D.: Optimization flow control I: basic algorithm and convergence. IEEE/ACM Transactions on Networking. (1999)
13. Analoui, M. Jamali, Sh.: A Conceptual Framework for Bio-Inspired Congestion Control In Communication Networks. IEEE/ACM BIONETICS. (2006)
14. Analoui, M. Jamali, Sh.: Inspiring by predator-prey interaction for Congestion Control In Communication Networks. CSICC2007, Iran, Tehran. (2007)
15. Analoui, M. Jamali, Sh.: Bio-Inspired Congestion Control: Conceptual Framework, Case Study and Discussion. Book chapter, Springer CI series. (To appear)
16. Elizabeth, S. Rhodes, A.: Mathematical Models in Biology : An Introduction. Cambridge press. (2003)
17. Murray, J.: Mathematical Biology: I. an Introduction. Third Edition, Springer press. (2002)
18. Ohsaki, H. Mera, Y. Murata, M. Miyahara, H.: Steady state analysis of the RED gateway: stability, transient behavior, and parameter setting. ACM SIGMETRICS. (2000)
19. Athuraliya, S. Li, V. Low, S. Yin, Q.: REM: active queue management. IEEE Network. (2001)
20. Kelly, F.: Mathematical Modeling of the Internet. Mathematics Unlimited-2001 and Beyond, Springer-Verlag, Berlin. (2001)
21. Analoui, M. Jamali, Sh.: TCP Fairness Enhancement Through a parametric Mathematical Model. CCSP2005,IEEE International Conference. (2005)
22. Ramakrishna, K. Floyd, S. Black, D.: The addition of explicit congestion notification (ECN) to IP. RFC 3168.

EDNA: Estimation of Dependency Networks Algorithm

José A. Gámez, Juan L. Mateo, and José M. Puerta

Computing Systems Department
Intelligent Systems and Data Mining Group - $i^3\mathcal{A}$
University of Castilla-La Mancha
Albacete, 02071, Spain

Abstract. In this work we present a new proposal in order to model the probability distribution in the estimation of distribution algorithms. This approach is based on using dependency networks [1] instead of Bayesian networks or simpler models in which structure is limited. Dependency networks are probabilistic graphical models similar to Bayesian networks, but with a significant difference: they allow directed cycles in the graph. This difference can be an important advantage because of two main reasons. First, in some real problems cyclic relationships appear between variables an this fact cannot be represented in a Bayesian network. Secondly, dependency networks can be built easily due to the fact that there is no need to check the existence of cycles as in a Bayesian network.

In this paper we propose to use a general (multivariate) model in order to deal with a richer representation, however, in this initial approach to the problem we also propose to constraint the construction phase in order to use only bivariate statistics. The algorithm is compared with classical approaches with the same complexity order, i.e. bivariate models as chains and trees.

1 Introduction

Estimation of distribution algorithms (EDAs) [2] are, as genetic algorithms [3], a kind of evolutionary metaheuristics. Instead of using crossover and mutation operators EDAs are based on learning a probability distribution over the variables in the problem representation[1] and sampling that distribution in order to get the next population. Then the key point in an EDA is to estimate the joint probability distribution $p_l(\mathbf{x})$ in each iteration l. Obviously this is a NP-hard problem as soon as the number of variables increase, so we need to simplify this problem by factorising the distribution according with the (in)dependencies observed in the variables or genes codified in the current population. Depending on the type of dependencies allowed we have models of different complexity. The simplest EDA is perhaps the UMDA algorithm [4] (Univariate Marginal

[1] Although EDAs can deal with different representations in this paper we only used the binary case.

J. Mira and J.R. Álvarez (Eds.): IWINAC 2007, Part I, LNCS 4527, pp. 427–436, 2007.

Distribution Algorithm) which does not consider any kind of dependencies between the variables, i.e, the joint distribution is codified as the product of the marginal distributions. More complex proposals only consider pair-wise dependencies, e.g. MIMIC [5] (Mutual Information Maximization for Input Clustering) and COMIT [6] (Combining Optimizer with Mutual Information Trees). Finally the richest models are those in which multivariate dependencies are allowed. In that case the usual choice is to use a Bayesian network [7] to factorise the joint distribution. A representative algorithm from this group is EBNA [2] (Estimation of Bayesian Network Algorithm). The disadvantage of multivariate models is the computational complexity of the learning phase. Besides, in some problems, simpler models performs similar to multivariate models.

In this work we propose the use of a multivariate probabilistic model in the sense that the number of dependencies are not limited a priori, but we build it as an approximation, based on an heuristic expression, in which only second order statistics are used (as in bivariate models), thus we maintain quadratic complexity. As we will discuss later the use of dependency networks makes possible this approximation in an easy way, on the contrary as it happens with Bayesian networks.

This paper is organized as follows. In the next section dependency networks are described, and in section 3 we expose our proposal for building the multivariate probabilistic model with a dependency network. In section 4 we preset some experiments and their results in order to evaluate the proposed method attending to convergence and scalability criteria. Finally, in section 5 we conclude with some remarks and the outline of future research in this topic.

2 Probabilistic Graphical Models: Dependency Networks

Dependency networks (DNs) were proposed by Heckerman et al. [1] as an alternative to Bayesian networks (BN). They can be defined as a tuple $(\mathbb{G}, \mathbf{P})$ over a domain $\mathbf{X}$ where $\mathbb{G}$ is a directed graph (not necessarily acyclic) and $\mathbf{P}$ is a set of conditional probability distributions, one for each variable in $\mathbf{X}$. Every $P \in \mathbf{P}$ must be such that $P(X_i|\mathbf{Pa}_i) = P(X_i|X_1, \ldots, X_{i-1}, X_{i+1}, \ldots, X_n) = P(X_i|\mathbf{X} \setminus X_i)$. This means that the set of parents $Pa(X_i)$ for every variable X_i is its Markov blanket $MB(X_i)$.

For example, if we consider a domain with three variables (Age, Gender and Income) and they are represented by the Bayesian network shown in the figure 1(a), then the DN for the same domain must be the one in figure 1(b).

This definition requires consistency in the sense that the joint probability distribution for $\mathbf{X}$ can be exactly recovered from $\mathbf{P}$. This is a very restrictive condition for automatic learning so in [1] the authors defined *general dependency networks* in order to relax the factorization: $P(\mathbf{X}) \approx \prod P(X_i|Pa_i)$.

It can be observed that a DN can be learned from data by independently learning the parent set for each variable, which quickly lead to the design of parallel learning algorithms. The fact of allowing directed cycles, although enlarges the representation issue has the disadvantage of avoiding the use of traditional

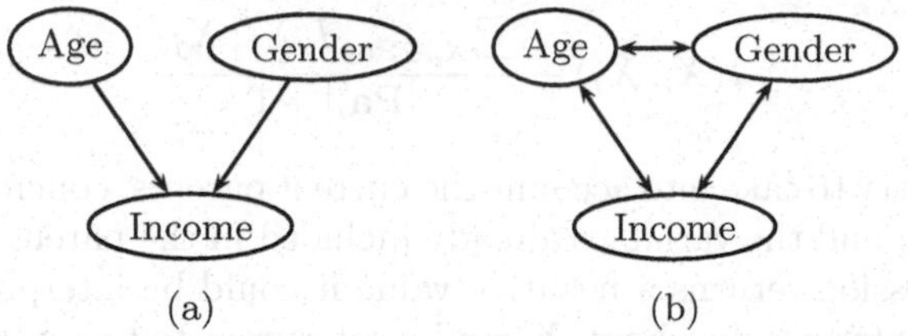

Fig. 1. Example of a Bayesian and a dependency network for the same domain

BNs exact inference algorithms. In this case, Heckerman et al. in [1] propose to use approximate inference carried out by using Gibbs sampling [1]. Sampling is usually carried out in graphical models based EDAs by using probabilistic logic sampling. In the case of using DNs this method cannot be applied because it needs a topological ordering of the network variables and such an ordering can not be assured in a DN. On the other hand, Gibbs sampling does not require such an ordering so this will be the sampling method used in our proposal.

3 EDNA: Estimation of Dependency Networks Algorithm

In this section we present EDNA, our estimation of distribution algorithm which is based in the use of a dependency network to model the relation among the variables. As mentioned before, consistence condition is too strict and so we propose the use of general DNs, because although they introduce a new level of approximation, the resulting model is enough expressive for our goals.

EDNA fits perfectly in the canonical EDAs framework, so we only specify the two main steps related with the probabilistic model: learning and simulation. In the learning phase EDNA runs over all the variables and looks for the parent set of the considered variable. This process can be done in parallel and consists in the identification of the more relevant variables with respect to the considered one. In our proposal the degree of dependence between variables is measured by using mutual information (MI), which is an usual measure. As an initial approach we can think about including as parents of X_i all those variables $\{X_j\}$ such that $I(X_i, X_j)$ is greater than a given threshold. However, this procedure could lead to overfitting because it is possible that a variable X_k selected as parent for X_i makes unnecessary the inclusion of some other variables in the parent set even if they pass the threshold. Because of this we propose an alternative approach that consists in looking for the parent set of X_i in an iterative way. Thus, in each step the algorithm looks for the variable X_j with highest MI with respect to X_i, but having a small degree of relation with the current parents of X_i, that is, $I(X_i; X_k|\mathbf{Pa}_i)$ is small. However, computing this multivariate statistics can be computationally expensive, and given the philosophy of this initial approach, we propose to measure this relationship between variables by using the following (heuristic) approximation:

$$I(X_i;X_j) - \frac{\sum_{X_p \in \mathbf{Pa}_i} I(X_p;X_j)}{|\mathbf{Pa}_i|+1}.$$

In this way we try to take into account the current parents, concretely the average MI between X_j and the variables already included in the parent set. Thus, if the previous expression returns a negative value it could be interpreted as X_j does not give new information about X_i and so it is rejected as a candidate parent for X_i. For those candidate parents such that the previous expression returns a positive value we act greedily and the one with greater value is selected.

In this way we try to make an approximation of a multivariate model by using only second order statistics. Notice that this bivariate approximation can not be directly applied to Bayesian networks because it could introduce cycles.

Figure 2 shows the pseudo-code of the described process:

```
For each variable X_i ∈ X
  Pa_i = ∅ // Parents of variable X_i
  cand_i = X − {X_i} // Candidate parents for X_i
  While cand_i ≠ ∅
    For each X_j ∈ cand_i assess it by
      I(X_i; X_j) − (Σ_{X_p ∈ Pa_i} I(X_p; X_j)) / (|Pa_i| + 1)
    Let X_max the variable with best assessment
    If X_max has positive assessment Then
      Pa_i = Pa_i ∪ X_max
      cand_i = cand_i − X_max
      For each X_c in cand_i with negative assessment
        cand_i = cand_i − X_c
    Else cand_i = ∅

p_l(x) ~ ∏_{i=1}^{n} p_l(x_i|Pa_i)
```

Fig. 2. Pseudo-code for estimating the population's joint probability distribution by using a dependency network-based factorization

With respect to the simulation phase in order to generate the new population we use Gibbs sampling because of the reasons commented above. As it is well known Gibbs sampling need positive conditional probability distributions to assure convergence, so Laplace correction is used during probabilities estimation. Also, there is two more parameters related with Gibbs sampling, both related with the idea of avoiding the correlation or dependence between samples: the burning time, i.e. the number of samples that are discarded at the beginning of the sampling process; and the latency, that indicate the number of samples discarded between each two valid samples.

Concerning implementation, this algorithm has been coded by using LiO[2], a metaheuristics library developed in our group [8]. Also the rest of algorithms used in this paper for comparison have been coded using LiO, which have similar complexity and they are an univariate algorithm (UMDA) and two bivariate algorithms whose structure is respectively a chain (MIMIC) and a tree (COMIT).

4 Results

In this section we present our experiments that are focused in the analysis of two aspects: (1) convergence reliability, i.e. the number of times in which the algorithm achieves the optimum; and (2) scalability, defined as the ability of maintaining the algorithm performance as the dimension of problem grows.

In order to be able to drawn significant conclusions we have perform 50 independents runs for each problem, and in the i-th run the same initial population was used for all the algorithms considered. As stopping criterion we set a maximum of 10^5 evaluations. Population size is set to $10n$ individuals, where n is the problem size or dimension, and only the (best) half population is used to learn the model. In the simulation phase (Gibbs sampling) we set the burning samples as the double of the number of variables and the latency is set to 1. After the 50 runs, and in order to assess our conclusions we carry out a parametric statistic test. We chose a paired test because the results obtained in each generation by each algorithm came from the same initial population.

With respect to scalability, the dependence between the performance of the algorithm and the dimensionality of the problem is measured as the number of iterations needed to reach the optimum value of the considered problem. The experimental settings is the same as for the convergence analysis, but in this case the population size is set to 100, we only perform 30 runs and five different problem dimesion for each case are considered. For each model and each setting (problem plus dimesion) we only consider those runs in which the optimum fitness is reached, then the 10 best results (less iterations) are selected in order to compute the average number of required iterations.

Regarding the test suite we have selected some problems that have been extensively used in the literature: OneMax, Plateau, CheckerBoard, $Fc4$ and $Fc5$ [2], and random decomposable problems [9], which are described as follows:

4.1 Test Functions

Function OneMax. This function is well known and implements a simple linear problem. It can be written as

$$F_{OneMax}(\mathbf{x}) = u(\mathbf{x}) = \sum_{i=1}^{n} x_i.$$

[2] `http://www.dsi.uclm.es/simd/SOFTWARE/LIO/`

The target is maximize the function F_{OneMax} with $x_i \in \{0,1\}$. The global optimum is therefore equal to n, the problem size, and it is represented by an individual in which all components are one $(1,1,\ldots,1)$

Function Plateau. This problem was proposed in [10]. Here each individual is a vector with size n, such that n is multiple of 3, $n = m \times 3$. In order to define this function is needed to use another auxiliary:

$$g(x_1,x_2,x_3) = \begin{cases} 1 \text{ if } x_1 = 1 \text{ y } x_2 = 1 \text{ y } x_3 = 1 \\ 0 \text{ otherwise} \end{cases}$$

Then the Plateau function is defined as:

$$F_{Plateau}(\mathbf{x}) = \sum_{i=1}^{m} g(\mathbf{s}_i)$$

where $\mathbf{s}_i = (x_{3i-2}, x_{3i-1}, x_{3i})$. The target is maximize this function too, that has the same best individual than the former with value m.

CheckerBoard function. This problem [11] defines a matrix of $s \times s$ positions with values 0 or 1. The target is make in this matrix a chess table in which white and black cells are substituted by ones and zeros. In each position where there was a 1 it must be surrounded by zeros in the four basic directions, and vice versa. The positions in the borders are no taken into account. The optimum value is $4(s-2)^2$, and the problem size is $n = s^2$. Considering the matrix as $\mathbf{x} = [x_{ij}]_{i,j=1\ldots,s}$, and $\delta(a,b)$ as the delta Kronecker function, the Checkerboard function can be written as:

$$F_{Checkerboard}(\mathbf{x}) = 4(s-2)^2 - \sum_{i=2}^{s-1}\sum_{j=2}^{s-1}\{\delta(x_{ij}, x_{i-1j}) + \\ + \delta(x_{ij}, x_{i+1j}) + \delta(x_{ij}, x_{ij-1}) + \delta(x_{ij}, x_{ij+1})\}$$

Deceptive functions. Here are presented the functions $Fc4$ y $Fc5$ which are two examples of deceptive functions proposed in [12]. They are decomposable and have overlapping beetwen adjacent components. The problem size, in both cases, is n, which is defined based on the number of subproblems m of size 5. Before it is needed to define some auxiliary functions.

$$F^3_{cuban1}(\mathbf{x}) = \begin{cases} 0.595 \text{ if } \mathbf{x} = (0,0,0) \\ 0.200 \text{ if } \mathbf{x} = (0,0,1) \\ 0.595 \text{ if } \mathbf{x} = (0,1,0) \\ 0.100 \text{ if } \mathbf{x} = (0,1,1) \\ 1.000 \text{ if } \mathbf{x} = (1,0,0) \\ 0.050 \text{ if } \mathbf{x} = (1,0,1) \\ 0.090 \text{ if } \mathbf{x} = (1,1,0) \\ 0.150 \text{ if } \mathbf{x} = (1,1,1) \end{cases}$$

$$F^5_{cuban1}(\mathbf{x}) = \begin{cases} 4F^3_{cuban1}(x_1,x_2,x_3) \text{ if } x_2 = x_4 \ \& \ x_3 = x_5 \\ 0 \quad \text{otherwise} \end{cases}$$

$$F^5_{cuban2}(\mathbf{x}) = \begin{cases} u(\mathbf{x}) & \text{if } x_5 = 0 \\ 0 & \text{if } x_1 = 0 \ \& \ x_5 = 1 \\ u(\mathbf{x}) - 2 & \text{if } x_1 = 1 \ \& \ x_5 = 1 \end{cases}$$

Fc4 function

$$Fc4(\mathbf{x}) = \sum_{j=1}^{m} F^5_{cuban1}(\mathbf{s}_j)$$

where $\mathbf{s}_j = (x_{4j-3}, x_{4j-2}, x_{4j-1}, x_{4j}, x_{4j+1})$, and the problem size is $n = 4m + 1$.

Fc5 function

$$Fc5(\mathbf{x}) = F^5_{cuban1}(\mathbf{s}_1) + \sum_{j=1}^{(m-1)/2} (F^5_{cuban2}(\mathbf{s}_{2j}) + F^5_{cuban1}(\mathbf{s}_{2j+1}))$$

where $\mathbf{s}_j = (x_{4j-3}, x_{4j-2}, x_{4j-1}, x_{4j}, x_{4j+1})$, and the problem size is $n = 4m + 1$.

Random decomposable problems. We use this kind of problems defined in [9]. In this work it is presented an algorithm in order to generate instances for decomposable problems, i.e. they can be written as sum of subfunctions, and with the possibility that they have overlapping beetwen subchains. We use some pre-generated instances which can be got, as well as the source code for the generator, as is shown in [13]. This instances are identified by three numbers. The first tells the problem size (parameter n), the second tells the long of each subchain (parameter k), and the third tells the quantity of overlap beetwen subchains (parameter o). If m is the number of subchains,then the relation among these parameters is:

$$n = k + (m - 1) * (k - o)$$

4.2 Convergence

In order to get a good analysis of convergence we have chosen three complex problems: $Fc4$, $Fc5$ and some instances of the random decomposable problems. For $Fc4$ and $Fc5$ we have considered three configurations with 5, 7 and 10 subproblems, and for random decomposable problem we have chosen 6 instances with different size and overlap. So, we test the algorithms over 12 cases.

The results obtained for these experiments are shown in table 1, where the best value for each configuration is highlighted in bold face and a • symbol is used to mark those results having no statistical difference with respect to the best one according to the Student's paired t-test used ($\alpha = 0.05$).

In these tables we can see that, even for only one problem, the best model is not always the same. As it is logical, the univariate model shows the worst results because it is the simpler one. Nonetheless the other three are quite similar, in almost all cases we can say the three model are statistically similar, although we can say also that the tree model is slightly better because shows the best values in 8 of 12 considered cases. In any case, with these results we can say that our model is at least as good as the other models analyzed, and despite trying to be more complete by considering multivariate dependencies, it does not introduce more computational complexity.

Table 1. Convergence results for the problems *FC*4 (a), *FC*5 (b) and DP (c)

	m=5	m=7	m=10
Univariate	13.56	18.33	25.48
Tree	•13.59	**18.39**	**25.57**
Chain	•13.59	•18.38	•25.57
DN	**13.59**	•18.38	25.56

(a)

	m=5	m=7	m=10
Univariate	•14.77	20.18	28.44
Tree	•14.79	**20.18**	**28.58**
Chain	**14.80**	•20.18	•28.57
DN	•14.80	•20.18	•28.57

(b)

	22_4_2	31_4_1	40_4_0	42_4_2	61_4_1	80_4_0
Univariate	8.16	8.57	8.93	16.51	17.62	18.23
Tree	**8.26**	**8.71**	9.06	**16.91**	**17.83**	•18.46
Chain	•8.26	•8.71	**9.07**	•16.84	•17.82	**18.46**
DN	•8.25	•8.70	9.06	16.77	•17.83	•18.46

(c)

4.3 Scalability

With respect to scalability we have selected a different set of problems: CheckerBoard, OneMax and Plateau. These problems are simpler than those selected in the previous section, but in this way the algorithms will be able to reach the optimum value and it can be assessed the effort in terms of the number of required iterations. The results obtained are shown graphically in the figure 3.

From these results we have to stand out that the slope of the line representing our model is always very low, specially for the more complex problem: Checkerboard. The univariate model shows a good scalability in OneMax and Plateau, that are easier, but its performance heavily decreases for CheckerBoard. The tree model shows the opposite behavior, i.e., it is good for CheckerBoard but not as good as our proposal. Finally, the chain model is comparable to the model with DN for OneMax, but is worse in the others.

5 Conclusions and Future Work

In this work we introduce dependency networks as a probabilistic model valid for the estimation of distribution algorithms. At a first glance, DNs main advantage is their ability for modeling cyclic relationships between the individual components or variables. Furthermore the construction of models based on dependency networks can be simpler, specially compared with model based on Bayesian networks. These two advantages make interesting the use of dependency networks, nonetheless we have tried to show its usefulness with some empirical results.

Here we have presented a multivariate-model EDA but with limited computational complexity so only second statistics are used to approximate the multivariate model. Our proposal is compared with other known models with similar complexity. Attending the experiments carried out we can conclude that EDNA has similar convergence propierties than UMDA, COMIT and MIMIC

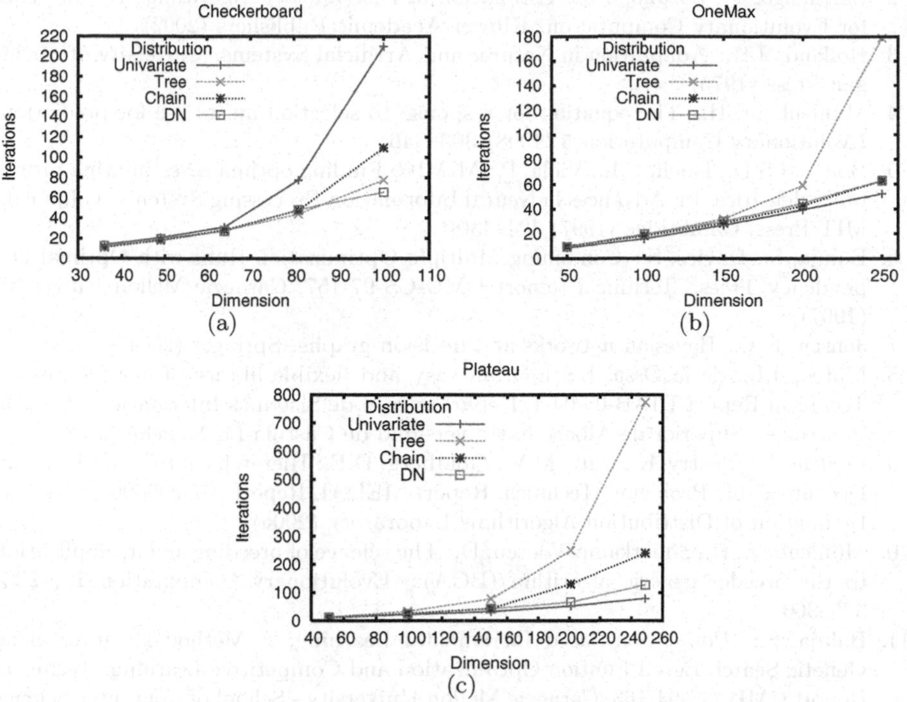

Fig. 3. Evolution of the needed iterations to reach the optimum value by each model for the problems Checkerboard (a), OneMax(b) and Plateau(c)

but behaves better with respect to scalability. So, from a global analysis we can say that this model based on DN should be take into account as a good choice.

As this work constitutes a starting point in this research line, we can expect huge room for a future work. As an example of future research a way to improve the results keeping the complexity of the model is to study alternative ways to the heuristic used here, e.g. by using the approximation presented in [14]. Other option could be to build the model without complexity restrictions and then compare its results with algorithms like EBNA.

Acknowledgments

This work has been partially supported by Spanish Ministerio de Ciencia y Tecnología, Junta de Comunidades de Castilla-La Mancha and European Social Fund under projects TIN2004-06204-C03-03 and PBI-05-022.

References

1. Heckerman, D., Chickering, D.M., Meek, C.: Dependency networks for inference, collaborative filtering and data visualization. Journal of Machine Learning Research **1** (2000) 49–75

2. Larrañaga, P., Lozano, J.A.: Estimation of Distribution Algorithms: A New Tool for Evolutionary Computation. Kluwer Academic Publishers (2002)
3. Holland, J.H.: Adaptation in Natural and Artificial Systems. University of Michigan Press (1975)
4. Mühlenbein, H.: The equation for response to selection an its use for prediction. Evolutionary Computation **5** (1998) 303–346
5. Bonet, J.S.D., Isbell, C.L., Viola, P.: MIMIC: Finding optima by estimating probability densities. In: Advances in Neural Information Processing Systems. Volume 9., MIT Press, Cambridge (1997) 424–430
6. Baluja, S., Davies, S.: Combining Multiple Optimization Runs with Optimal Dependency Trees. Technical Report CMU-CS-97-157, Carnegie Mellon University (1997)
7. Jensen, F.V.: Bayesian networks and decision graphs. Springer (2001)
8. Mateo, J.L., de la Ossa, L.: LiO: an easy and flexible library of metaheuristics. Technical Report DIAB-06-04-1, Departamento de Sistemas Informáticos, Escuela Politécnica Superior de Albacete, Universidad de Castilla-La Mancha (2006)
9. Pelikan, M., Sastry, K., Butz, M.V., Goldberg, D.E.: Hierarchical BOA on Random Decomposable Problems. Technical Report MEDAL Report No. 2006001, Missouri Estimation of Distribution Algorithms Laboratory (2006)
10. Mühlenbein, H., Shlierkamp-Voosen, D.: The science of breeding and its application to the breeder generic algorithm (BGA). Evolutionary Computation **1** (1993) 335–360
11. Baluja, S.: Population-Based Incremental Learning: A Method for Integrating Genetic Search Based Funtion Optimization and Competitive Learning. Technical Report CMU-CS-94-163, Carnegie Mellon University - School of Computer Science (1994)
12. Mühlenbein, H., Mahnig, T., Ochoa-Rodriguez, A.: Schemata, Distributions and Graphical Models in Evolutionary Optimization. Journal of Heuristics **5** (1999) 215–247
13. Pelikan, M., Sastry, K., Butz, M.V., Goldberg, D.E.: Generator and Interface for Random Decomposable Problems in C. Technical Report MEDAL Report No. 2006003, Missouri Estimation of Distribution Algorithms Laboratory (2006)
14. Roure Alcobé, J.: An approximated MDL score for learning Bayesian networks. In: Proceedings of the Second European Workshop on Probabilistic Graphical Models 2004 (PGM'04). (2004) 185–192

Grammar-Guided Neural Architecture Evolution

Jorge Couchet, Daniel Manrique, and Luis Porras

Departamento de Inteligencia Artificial, Facultad de Informática,
Universidad Politécnica de Madrid, 28660 Boadilla del Monte, Madrid, Spain
{jcouchet,dmanrique,lporras}@fi.upm.es

Abstract. This article proposes a context-free grammar to be used in grammar-guided genetic programming systems to automatically design feed-forward neural architectures. This grammar has three important features. The sentences that belong to the grammar are binary strings that directly encode all the valid neural architectures only. This rules out the appearance of illegal points in the search space. Second, the grammar has the property of being ambiguous and semantically redundant. Therefore, there are alternative ways of reaching the optimum. Third, the grammar starts by generating small networks. This way it can efficiently adapt to the complexity of the problem to be solved. From the results, it is clear that these three properties are beneficial to the convergence process of the grammar-guided genetic programming system.

Keywords: context-free grammar, grammar-guided genetic programming, feed-forward neural architecture, ambiguity, semantic redundancy.

1 Introduction

Evolutionary computation (EC) is the study of computational systems that borrow ideas from and are inspired by natural evolution and adaptation. EC's primary aims are to understand the mechanism of such computational systems and to design highly robust, flexible and efficient algorithms for solving real-world problems that are generally very difficult for standard methods to deal with. EC could be divided into four branches [1]: evolution strategy, evolutionary programming, genetic algorithm and genetic programming. All the approaches used in EC employ a population-based search with reproduction, mutation and selection to find better solutions.

Grammar-guided genetic programming (GGGP) is an extension of traditional genetic programming (GP) systems whose goals are to simplify the search space and solve the closure problem [2]. This problem involves always generating valid individuals (points or possible solutions that belong to the search space). To solve the closure problem, GGGP employs a context-free grammar (CFG), which establishes a formal definition of the syntactical restrictions of the problem to be solved and its possible solutions. Each of the individuals handled by GGGP is a derivation tree that generates and represents a sentence (solution) belonging to the language defined by the CFG.

J. Mira and J.R. Álvarez (Eds.): IWINAC 2007, Part I, LNCS 4527, pp. 437–446, 2007.

One of the main fields of artificial intelligence research is the development of self-adaptive systems that are capable of transforming in order to solve different problem types. The properties of learning by examples, generalizing to unseen data, and noise filtering make feed-forward neural networks applicable to a wide variety of real-world tasks [3]. Despite the advantages of these networks, the design of the neural architecture for a particular application is typically governed by heuristics mainly due to the size and complexity of the available design space.

Feed-forward neural architectures can be evolved using evolutionary computation. The two most commonly used techniques are: genetic programming [4] and genetic algorithms [5]. In the field of genetic programming, there have been several approaches to efficiently codify neural networks as genotypes. An attribute grammar was proposed to encode the neural design principles, as well as structural and behavioural elements [6]. This approach provides mechanisms to automatically translate the grammar parse trees into complete neural network specifications in an XML-based format termed the Generic Neural Markup Language. The main disadvantage of this approach is that it is too complex. It can be inefficient for complex problems with an extensive search space.

As regards genetic algorithms, there are several research works on the evolution of network architectures. A new technique has been presented for incorporating human-generated advice in real time in the neural evolution of a network [7]. The advice is given in a formal language, and a recursive algorithm converts this into its equivalent neural network structure. The rtNEAT neuroevolution method [8] then incorporates the advice into existing networks by evolving network weights and topology. The rtNEAT encodes the network in a genome that includes a list of connection genes, each of which specifies two connected node genes and the weight on that connection. Other techniques explicitly separate design and training, introducing two interconnected genetic algorithms that work in parallel [9]. In this case, the design of the neural architectures uses the basic architectures codification method, while the training process employs a real-valued codification. The main drawback of this approach is the fixed length of the codification, which is inadequate when the optimal solution is a small-sized neural architecture.

This paper presents an ambiguous context-free grammar with semantic redundancies as an efficient alternative for encoding all the possible valid one hidden-layer feed-forward neural networks with any number of input and output neurons and direct connections from the input layer to the output layer. Encoding only one-hidden layer is not a limitation because this type of neural networks can approximate any function with an arbitrary error [10]. The proposed grammar is used in a grammar-guided genetic programming system with the genetic operator grammar-based crossover (GBX) [11] to automatically generate neural architectures. The performance of such a system can be improved by using ambiguous grammars with semantic redundancies [12]. Semantic redundancy also leads to a many-to-one mapping between the genotype and phenotype. It is suggested that the Non Free Lunch Theorem [13] is not valid due to this non-uniform many-to-one mapping between the description of an object and the actual object [14]. Experimental lab tests have been run and compared to the attribute grammar and basic architectures encoding method, showing that the proposed approach performs better. The system has also been successfully applied to a real-world problem: breast cancer diagnosis. Compared to the prognosis given by expert radiologists in the subject, accuracy is good.

2 The Proposed Context-Free Grammar

A context-free grammar G is a 4-tuple $G = (S, \Sigma_{\mathrm{N}}, \Sigma_{\mathrm{T}}, P)$, where $S \in \Sigma_{\mathrm{N}}$ is the start variable or axiom, Σ_{N} is a finite set called the variables or non-terminal symbols, Σ_{T} is a finite set, disjoint from Σ_{N}, called the terminals and P is a finite set of production rules, with each rule being a non-terminal and a string of non-terminals and terminals. If u, v, and w are strings of non-terminals and terminals, and $A \rightarrow w$ is a grammar rule, it is said that uAv yields uwv, written $uAv \Rightarrow uwv$. Say that u derives v, written $u \Rightarrow^* v$, if $u = v$ or if a sequence $u_1, u_2, \cdots, u_k$ exists for $k \geq 0$ and $u \Rightarrow u_1 \Rightarrow \cdots \Rightarrow u_k \Rightarrow v$. The language of the grammar is $\{w \in \Sigma_T^* | S \Rightarrow^* w\}$.

Equation 1 shows the proposed ambiguous context-free grammar capable of generating one-hidden layer feed-forward neural networks architectures with I input neurons, $i_1, \cdots i_p, \cdots, i_I$, O output neurons, $o_1, \cdots o_r, \cdots, o_O$, an indeterminate number of hidden neurons D, $d_1, \cdots d_q, \cdots, d_D$, and direct connections from the input to the output layers. The sentences that belong to the grammar are binary strings that encode the neural architectures.

$$
\begin{aligned}
G_{\mathrm{I,O}} &= (S, \Sigma_{\mathrm{N}}, \Sigma_{\mathrm{T}}, P_{\mathrm{I,O}}) \text{ with:} \\
\Sigma_{\mathrm{N}} &= \{S, L, R, H, A, Z, B\} \\
\Sigma_{\mathrm{T}} &= \{0, 1\} \\
P_{\mathrm{I,O}} &= \{S \rightarrow LR|R \ , \ R \rightarrow B_{I \cdot O} \ , \ L \rightarrow LH|H \ , \ H \rightarrow AZ \ , \ B \rightarrow 0|1 \ , \\
&\qquad A \rightarrow 1B_{I-1}|B1B_{I-2}|\cdots|B_j1B_{I-1-j}|\cdots|B_{I-2}1B|B_{I-1}1 \ , \\
&\qquad Z \rightarrow 1B_{O-1}|B1B_{O-2}|\cdots|B_j1B_{O-1-j}|\cdots|B_{O-2}1B|B_{O-1}1\} \ .
\end{aligned}
\tag{1}
$$

Where the production rule $\alpha \rightarrow \beta$ means that α yields a string composed of s symbols β.

Architectures are encoded directly: each connection is represented explicitly as a binary digit. The codification schemes differ depending on whether the connections are with a hidden layer or direct connections between input and output. Given a neuron d_i of the hidden layer, this can have up to a maximum of I connections with the input layer and O with the output layer. Therefore, a string of $I + O$ bits represents the state of the connections of d_i, where 1 or 0 respectively indicate the presence or absence of a connection.

The connections of the hidden neurons are generated by the rule $L \rightarrow LH|H$. The recursiveness of the rule can generate strings of non-terminal symbols H of indeterminate length D, which match architectures of D hidden neurons.

The non-terminal H placed in the qth position generates a binary substring that represents the state of the connections of the hidden layer neuron d_q to the input and output layer neurons. The rule $H \rightarrow AZ$ produces this state of connections. The non-terminal A generates all the possible connections I to the input neurons through the rule $A \rightarrow 1B_{I-1}|B1B_{I-2}|\cdots|B_{I-1}1$, where each non-terminal B yields 1 or 0 through rule $B \rightarrow 1|0$. The bit placed in position p represents the existence or otherwise of a connection between d_q and i_p. Similarly, the non-terminal Z encodes the possible O connections of the hidden neuron d_q with the output neurons $o_1, \cdots, o_r, \cdots, o_O$ through rule $Z \rightarrow 1B_{O-1}|B1B_{O-2}|\cdots|B_{O-1}1$.

For an architecture to be valid, a hidden neuron should be connected at the same time to an input layer neuron and output layer neuron or not be connected to any other. More formally, if there exists a connection between i_p and d_k, then there also exists another connection between d_k and o_r. The proposed grammar assures this behaviour by entering terminal 1 in the right-hand side of all productions whose left-hand sides are the non-terminals A and Z.

The direct connections between the input and output layers are generated by the rule $D \rightarrow B_{I \cdot O}$. Given a neuron or of the output layer, there are a maximum of I connections with I input neurons, where a string of I bits is necessary to represent its state of connections. Generalizing for O output neurons, a string of $I \cdot O$ bits is needed to represent the state of all its possible connections. Each string is the one generated with the $I \cdot O$ non-terminal symbols B of the production rule $D \rightarrow B_{I \cdot O}$. The rth substring of length I bits represents the state of connections of o_r. So, the pth position within the rth substring encodes the connection between o_r and i_p. This production is also employed to generate the null architecture, which is encoded as a string of $I \cdot O$ zeros $(S \rightarrow D; D \rightarrow B_{I \cdot O}; B \rightarrow 0)$.

$$
\begin{aligned}
&G_{3,2} = (S, \Sigma_{\mathrm{N}}, \Sigma_{\mathrm{T}}, P_{3,2}) \text{ with:} \\
&\Sigma_{\mathrm{N}} \;= \{S, L, R, H, A, Z, B\} \\
&\Sigma_{\mathrm{T}} \;= \{0, 1\} \\
&P_{3,2} = \{S \rightarrow LR|R \text{ , } R \rightarrow BBBBBB \text{ , } L \rightarrow LH|H \text{ , } H \rightarrow AZ \text{ ,} \\
&\qquad B \rightarrow 0|1 \text{ , } A \rightarrow 1BB|B1B|BB1 \text{ , } Z \rightarrow 1B|B1\} \text{ .}
\end{aligned}
\tag{2}
$$

Equation 2 is an example of an instance of the proposed grammar for generating architectures with any number of neurons in the hidden layer (D), three inputs (I=3) and two outputs (O=2). On the left-hand side of Fig. 1 is one of the possible sentences generated by the grammar defined earlier that encodes the neural architecture shown on the right-hand side.

The substring 11011 at the start of the sentence encodes the hidden neuron connections d_1 with the input and output layers. The first three bits 110, labeled as Input, encode the connections with the input neurons: the first 1 indicates that d_1 is connected with i_1, the second that d_1 and i_2 are connected and the third digit set to zero that there is no connection between d_1 and i_3. Similarly, the following substring 11, labeled as Output, represents the connections of d_1 with the output neurons o_1 and o_2.

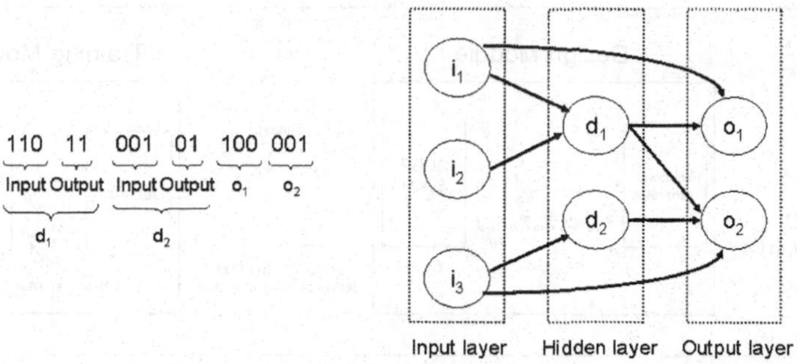

Fig. 1. Neural architecture codification with $I = 3$, $D = 2$ and $O = 2$

The grammar is ambiguous because it generates sentences that have more than one derivation tree. Figure 2 shows two possible derivation trees for the sentence in Fig. 1.

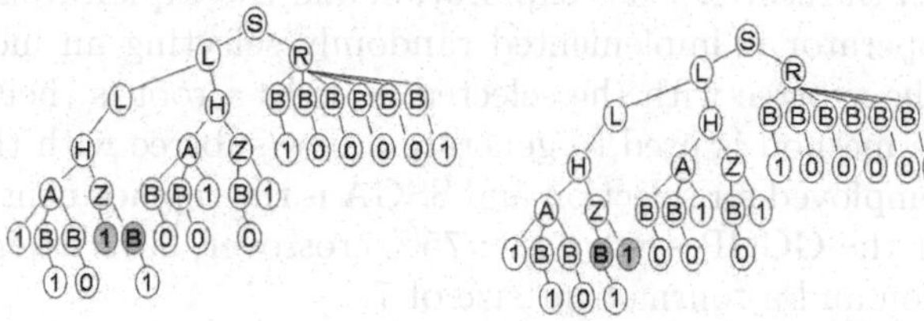

Fig. 2. Two of the possible derivation trees that parse the sentence 1101100101100001

The proposed grammar is not only structurally ambiguous, but it is also semantically redundant: different sentences (binary strings) encode the same architecture. For example, the sentence 0010111011100001 generates a neural network whose architecture is equivalent to the one shown in Fig. 1. The difference between them is that the position of the two hidden neurons d_1 and d_2 has been swapped.

3 Neuroevolution System

The neuroevolution system responsible for automatically building the generalized feed-forward neural networks is composed of two modules as shown in Fig. 3: design module and training module. The design module works on the set of all possible valid neural architectures and makes use of a GGGP system to find the best. The input for the module is the number of input and output neurons needed to solve a given problem. These parameters are used to instantiate (1) and generate the grammar then used by the GGGP system. The initial population in the GGGP is generated randomly. The initialization method starts from the

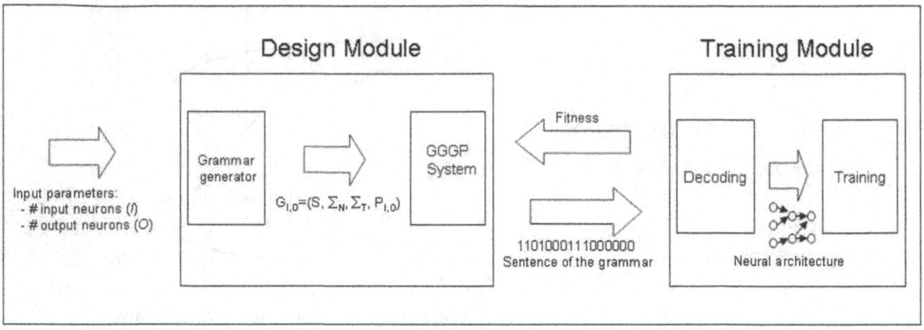

Fig. 3. Neuroevolution system

grammar axiom and randomly continues with the non-terminal symbols of the right-hand sides of the productions executed until no more non-terminal symbols are obtained. The crossover operator is the GBX, which assures an offspring composed of valid individuals, prevents code bloat and provides a satisfactory trade-off between the search space exploration and the exploitation capabilities. The mutation operator is implemented randomly selecting an individual node for mutation. The subtree with the selected node as a root is then deleted, and the initialization method is used to generate a new subtree with the same root. Tournament is employed for selection and SSGA is the replacement method. The settings used for the GGGP system are: 75% crossover, 20% straight reproduction, 5% mutation and a tournament size of 7.

The fitness function used in the GGGP system is shown in (3). This function combines two criteria for evaluating an individual: its effectiveness, expressed as its mean square error (MSE), and its efficiency, expressed as a function of the number of network connections (C_{A}). Both are weighted by a factor $\alpha \in [0, 1]$ which can be used to adjust the level of importance.

$$F = \alpha \cdot MSE + (1 - \alpha) \cdot C_{\mathrm{A}} \ . \quad (3)$$

The training module adapts the weights of the connections for a given network by means of the enhanced back-propagation (EBP) method with a learning factor of 0.1 and 0.85 of momentum. Its input is a sentence (binary string) generated by the GGGP system. It decodes this input in its respective neural architecture and trains it using EBP, returning its fitness to the GGGP system.

4 Results

Two different experiments were carried out to test the neuroevolution system. First, the encoder/decoder laboratory problem with eight inputs and outputs. This problem illustrates the performance of the proposed grammar compared with a GGGP system using the attribute grammar encoding method and a genetic algorithm using the basic architectures codification method. Second, the problem of prognosing breast cancer is used to test the accuracy of the proposed system with a real-world problem. Each experiment was run 100 times and average results are shown.

The objective for the encoder/decoder experiment is to find an optimal neural architecture that is capable of returning an output vector that is equal to its input. This experiment was carried out with a maximum of 120 generations of 30 individuals in the population. Figure 4 shows the average convergence speed of the three different evolutionary algorithms with their respective codification methods.

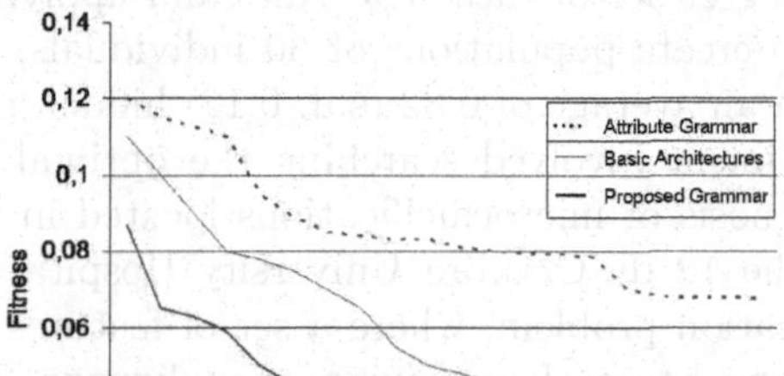

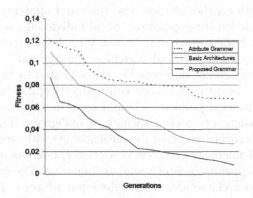

Fig. 4. Average fitness evolution for the encoder/decoder problem

The optimum architecture that solves this problem has no hidden neurons and each of the eight input neurons d_i, with $i \in [1, 8]$, is directly connected to another eight output neurons o_i. Therefore, we have I=8, D=0 and O=8. Table 1 shows the relative frequency of the target architectures for the different encoding methods used.

This table shows that the use of the proposed grammar provides clearly better results than the other two approaches. Specifically, the neuroevolution system achieves the optimum solution (I=8, D=0, O=8) in 87% of cases, taking an average of 82.49 (s.d. 5.2) generations to converge. The second-best evolutionary system in terms of results is the one that employs the basic architectures codification method. This system achieves the optimum solution in 62% of cases,

Table 1. Architectures output by the different encoding types

Solutions	Basic architectures	Attribute grammar	Proposed
I=8, D=0, O=8	62%	55%	87%
I=8, D=4, O=8	25%	23%	9%
I=8, D=5, O=8	9%	13%	4%
I=8, D=6, O=8	4%	9%	0%

taking an average of 95.2 (s.d 4.7) generations to converge. The neuroevolution system achieves such good results because the proposed grammar is able to take advantage of structural ambiguity and semantic redundancy to explore all possible paths that lead to the optimal architecture. An additional advantage is the grammar's capability to start with solutions with few hidden neurons to which it can add depending on the problem complexity. This property is explained by the rules $L \rightarrow LH|H$. If the probability of selecting rule $L \rightarrow H$ is equal to or greater than the probability of selecting the recursive rule $L \rightarrow LH$, then this rule cannot be executed very often in the initialization method. This creates an initial population of individuals with few hidden neurons. Using a 0.5 probability of selecting either of these two rules and applying the initialization method 100 times to create populations of 30 individuals, we output artificial neural networks with an average of 0.82 (s.d. 0.12) hidden neurons.

The second experiment involved searching the optimal neural architecture that can give a prognosis of microcalcifications located in the breast tissue of real patients from the 12 de Octubre University Hospital in Madrid. This is a non-trivial classification problem, where a set of features describing a breast lesion has to be assigned to a class: benign or malignant. There are eight features describing microcalcifications: the patient's age, site (region, side, depth), size and number of microcalcifications in a cluster, how microcalcifications are grouped and their appearance after visual examination. The grammar employed can generate neural architectures with eight input neurons (I=8) for each feature, and one output neuron (O=1) for prognosis. The neuroevolution system using this grammar was run 100 times, each with a maximum of 150 generations and a population size of 100 individuals. A set of 184 lesions, of which half matched malignant cases, were employed to train (80%) and test (20%) the artificial neural networks built by the system. The most commonly output smallest feed-forward neural architecture to provide the best results (Table 2) has seven fully connected hidden neurons with no direct connections from the input to the output. The criterion employed to compare the neural architectures output by the neuroevolution system is based on the success and the prognosis failure rate of lesions in the test set. It is usual in the field of medicine to use three criteria, defined in (4) to report statistical results: accuracy (Ac), specificity (Sp) and sensitivity (Se).

$$Ac = \frac{T_{\mathrm{P}} + T_{\mathrm{N}}}{\#\ \text{instances}}; \; Sp = \frac{T_{\mathrm{N}}}{T_{\mathrm{N}} + F_{\mathrm{P}}}; \; Se = \frac{T_{\mathrm{P}}}{T_{\mathrm{P}} + F_{\mathrm{N}}} \; . \tag{4}$$

Where T_{P} (true positives) are the correctly classified malignant cases, T_{N} (true negative) are the correctly classified benign cases, F_{P} are the incorrectly classified benign cases and F_{N} are the incorrectly classified malignant cases. Table 2 shows the results in terms of Ac, Sp and Se of the prognosis given by the network solution compared with the expert radiologists' diagnosis. The artificial neural network output by the proposed system outperforms the radiologists as regards Sp and Ac, although it is worse as regards Se. This is because of the method used by the radiologist to diagnose breast cancer: the lesion is only classified as benign when he or she is really sure. If there is any doubt, then the lesion is

Table 2. Prognosis results for the breast cancer problem

	Accuracy	Specificity	Sensitivity
Proposed grammar	71%	77%	68%
Radiologist A	56%	39%	78%
Radiologist B	49%	28%	77%

classified as malignant. The system has no such prejudices, which means that it provides better results than radiologists for true negatives, but also worse results for false negatives, with the resulting risk for patients with cancer.

5 Conclusions

In this paper, we propose a grammar able to evolve neural architectures through a GGGP system to find solutions to any problem that can be solved with a feed-forward neural network. This grammar has three remarkable properties: the sentences generated by the grammar are binary strings that belong to the space of valid neural architectures. This reduces the GGGP system's search space complexity. Second, the grammar is structurally ambiguous and semantically redundant. Ambiguity enables one and the same neural architecture to be generated from different derivation trees, whereas semantic redundancy allows different sentences generated by the grammar to codify the same architecture. Finally, the probability of the rule $L \rightarrow H$ being selected can control the tree sizes of the initial population. This way, it is possible to start with composite architectures with few hidden neurons. Then it will be the GGGP system's evolution process that gradually increases their size depending on the complexity of the problem to be solved.

The results section shows that the GGGP system with the proposed grammar improves the convergence speed. It does this thanks to the reduction of the search space, ambiguity and semantic redundancy as shown in previous empirical results. Additionally, if the Non Free Lunch Theorem does not hold because of the proposed grammar's semantic redundancy, then it is not necessarily true that for every problem that the GGGP system does well on, there exists another problem on which it does poorly. Similarly, the proposed system provides smaller neural architectures than output by the other methods with which they were compared because of the possibility of adapting the tree sizes to problem complexity. As a result, a system is produced that is capable of providing high accuracy in more complex problems like breast cancer prognosis.

We are now investigating how to define a measure of the ambiguity of a context-free grammar. Empirical tests run have shown that if the grammar is too ambiguous, the search space increases considerably. This causes a loss in system performance. Therefore, the goal is to find a formal method that can establish the exact grammar ambiguity and achieve the maximum convergence speed.

Acknowledgments. This research is being funded by the Spanish Ministry of Science and Education under project no. DEP2005-00232-C03-03 .

References

1. Raidl, G.: Evolutionary computation: An overview and recent trends. OGAI Journal **2** (2005) 2–7
2. Whigham, P.: Grammatically-based genetic programming. In: Proc. of the Workshop on Genetic Programming: From Theory to Real-World Apps. (1995) 33–41
3. Haykin, S.: Neural Networks: A Comprehensive Foundation. Prentice-Hall, New Jersey (1999) 2nd ed.
4. Koza, J., Rice, J.: Genetic generation of both the weight and architecture for a neural network. In: Proceedings of the International Joint Conference on Neural Networks, Vol. II. (1991) 397–404
5. Manrique, D., Rios, J., Rodriguez-Paton, A.: Evolutionary system for automatically constructing and adapting radial basis function networks. Neurocomputing **69** (2006) 2268–2283
6. Hussain, T.: Attribute grammar encoding based upon a generic neural markup language: Facilitating the design of theoretical neural network models. In: Proc. of the International Joint Conference on Neural Networks , Vol. I. (2004) 25–29
7. Yong, C., Stanley, K., Miikkulainen, R., Karpov, I.: Incorporating advice into neuroevolution of adaptive agents. In: Proceedings of the Artificial Intelligence and Interactive Digital Entertainment Conference. (2006)
8. Stanley, K., Bryant, B., Miikkulainen, R.: Real-time neuroevolution in the NERO video game. IEEE Transactions on Evolutionary Computation **9** (2005) 653–668
9. Barrios, D., Carrascal, A., Manrique, D., Ros, J.: Cooperative binary-real coded genetic algorithms for generating and adapting artificial neural networks. Neural Computing and Applications **12** (2003) 49–60
10. Funahashi, K.: On the approximate realization of continuous mappings by neural networks. Neural Networks **2** (1989) 183–192
11. Manrique, D., Marquez, F., Rios, J., Rodrguez-Paton, A.: Grammar based crossover operator in genetic programming. In Mira, J., Alvarez, J., eds.: Artificial Intelligence and Knowledge Engineering Applications: A Bioinspired Approach. Lec. Notes in Computer Science. Volume 3562., Springer Verlag, Berlin Heidelberg New York (2005) 252–261
12. Hoai, N., Shan, Y., McKay, R., D, E.: Is ambiguity useful or problematic for grammar guided genetic programming? a case study. 4th Asia-Pacific Conference on Simulated Evolution and Learning (2002) 449–454
13. Woodward, J., Neil, J.: No free lunch, program induction and combinatorial problems. In: Proceedings of Genetic Programming. (2003) 475–484
14. Woodward, J.: GA or GP? that is not the question. In: Proceedings of the Congress on Evolutionary Computation. (2003) 1056–1063

Evolutionary Combining of Basis Function Neural Networks for Classification

César Hervás[1,*], Francisco Martínez[2], Mariano Carbonero[2], Cristóbal Romero[1], and Juan Carlos Fernández[1]

[1] Department of Computing and Numerical Analysis University of Córdoba, Spain
{chervas,cromero,i82fecaj}@uco.es

[2] Department of Management and Quantitative Methods, ETEA, Spain
{fjmestud,mcarbonero}@etea.com

Abstract. The paper describes a methodology for constructing a possible combination of different basis functions (sigmoidal and product) for the hidden layer of a feed forward neural network, where the architecture, weights and node typology are learned based on evolutionary programming. This methodology is tested using simulated Gaussian data set classification problems with different linear correlations between input variables and different variances. It was found that combined basis functions are the more accurate for classification than pure sigmoidal or product-unit models. Combined basis functions present competitive results which are obtained using linear discriminant analysis, the best classification methodology for Gaussian data sets.

1 Introduction

There are a number of neural network approaches for statistical pattern recognition. One of the oldest is the perceptron network [1], in which the transfer function is a hard threshold. This function is not differentiable at its point of transition, and it had to be replaced by sigmoidal basis function units to construct the multilayer perceptron (MLP) model. Other well-known transfer functions are the linear function, often used for output units that must be capable of producing a wide range of values, the Gaussian transfer functions used by radial basis function architecture (RBF).

Since there are different types of basis functions which determine net typology (MLP, RBF, PUNN, etc.) [2,3,4], both the structure and the learning capacity of different nets depend on the problem at hand.

Sigmoidal and product unit basis functions have global support, and they produce values significantly different from zero over an infinite domain of input values. Other architectures in neural networks employ functions with local support, using transfer functions that yield values that are not close to zero for only a small domain of input values. These networks use Gaussian or Gaussian-like functions. There is research comparing the functionality of sigmoidal versus

* Corresponding author.

J. Mira and J.R. Álvarez (Eds.): IWINAC 2007, Part I, LNCS 4527, pp. 447–456, 2007.

Gaussian functions, but the question about whether or not commonly used function types are in some sense optimal remains largely unanswered. Taking into account the almost infinite number of possibilities, one can easily see that an exhaustive search for function space is a NP-hard problem and so this paper considers two specific types of basis functions, sigmoidal and product.

Except for nonlinear ability, what can not be overlooked is that MLP, RBF, even DBF [5] neural networks all have some flaw points in local optimization problems, needing bigger memory space, and a slower convergence speed, all of which can hopefully ultimately be resolved by variable nonlinear function. A combination of nonlinear transfer function neurons could implement the random nonlinear mapping relationship between input layer and output layer, which could allow variable transfer function neurons to have a much wider application in pattern recognition.

The combination of sigmoidal, Gaussian or product unit activation functions can have several advantages, if we consider a classification task which has areas which are separate in general, but where it is difficult to situate the exact location of the border because the best discriminant functions for each class can be very different. The above-mentioned family of parameterized transfer functions can provide flexible decision borders. The idea behind the combination of several functions is to increase the good performance of a traditional sigmoidal network by adding targeted Gaussian and/or product basis functions which will cover the weak parts of the pure sigmoidal network.

From a theoretical point of view in this context, the work of Donoho [6] demonstrates that any continuous function can be decomposed into two mutually exclusive functions, such as radial and crest (based on the projection). But although theoretically this decomposition is justified, in practice it is difficult to separate the different locations of a function. To the best of our knowledge, no theoretical result has been found to show that some continuous function can be decomposed into two mutually exclusive parts associated to other projection typologies and in particular related to the hybridation of the sigmoidal/product unit basis functions in neural networks.

The objectives proposed in this paper are various. First, we want to show hybrid-unit neural network models for multi-class classification problems based on a combination of the basis functions (sigmoidal and product). In that classification problem, measurements x_i, $i = 1, 2, ..., k$, are taken on a single individual (or object), and the individuals are to be classified into one of the J classes based on these measurements. Secondly, we have analyzed the accuracy of the different basis functions in classification problems where the generated measurements have Gaussian distributions with equal covariance matrices for all classes, though in several experiments there are different variances and correlations. It is necessary to remember that the best classifier for that classification structure is a linear function because of the minimum Bayes error produced.

In this way we present the first studies on the use of hybrid models that are associated with two specific types of functions used as functional projection approximators: product (PU) and sigmoidal (SU) basis functions, yielding the

SPU neural network as a combination of both basis functions. Evolutionary algorithms are employed for the optimization of the parameters and the architecture of the model as an alternative to the classic choice based on gradient methods, due to the high complexity of the error functions.

2 Sigmoidal Versus Product Basis Functions for Classification

Product-unit basis functions used in the hybridation process and their representation by means of a neural network structure is a class of high-order neural network [7], having only unidirectional interlayer connections. They are also a generalization of sigma-pi networks [8], which are multilayer perceptrons having high-order connections, and functional link networks [9]. Product units enable a neural network to form higher-order combinations of inputs, with the advantages of increased information capacity and smaller network architectures when we have interaction between the input variables. Thus neurons with multiplicative responses can act as powerful computational elements in real neural networks [10].

In this work, several types of basis functions have been used, namely PU function

$$B_k(\mathbf{x}, \mathbf{w}_k) = \prod_{i=1}^{p} x_i^{w_{ki}} \qquad k = 1, ..., m_2 \tag{1}$$

sigmoidal-unit basis functions, SU, in the form

$$B_j(\mathbf{x}, \mathbf{u}_j) = \frac{1}{1 + \exp(-u_{j0} - \sum_{i=1}^{p} u_{ji} x_i)} \qquad j = 1, ..., m_1 \tag{2}$$

and a linear combination of both provided the hybrid function (SPU):

$$f(\mathbf{x}) = \sum_{j=1}^{m_1} \alpha_j B_j(\mathbf{x}, \mathbf{u}_j) + \sum_{k=1}^{m_2} \beta_k B_k(\mathbf{x}, \mathbf{w}_k) \tag{3}$$

The method involves finding a sufficient number of basis functions (its architecture) to permit an approach that can estimate the classification function, in such a way, and taking into account that both types of basis functions are universal approximators, and that for every $\epsilon > 0$ it should be possible to find a value of m_1 and m_2 as well as the estimators of the parameters $\hat{\alpha}_j$, $\hat{\beta}_k$, $\hat{\mathbf{u}}_j$ and $\hat{\mathbf{w}}_k$ for $j = 1, ..., m_1$ and $k = 1, ..., m_2$ assuming that

$$\left\| f(\mathbf{x}) - \left(\sum_{j=1}^{m_1} \hat{\alpha}_j B_j(\mathbf{x}, \hat{\mathbf{u}}_j) + \sum_{k=1}^{m_2} \hat{\beta}_k B_k(\mathbf{x}, \hat{\mathbf{w}}_k) \right) \right\| < \epsilon \tag{4}$$

This optimization problem is similar to that involved in the "projection pursuit" regression model with the special feature that the "ridge functions" are

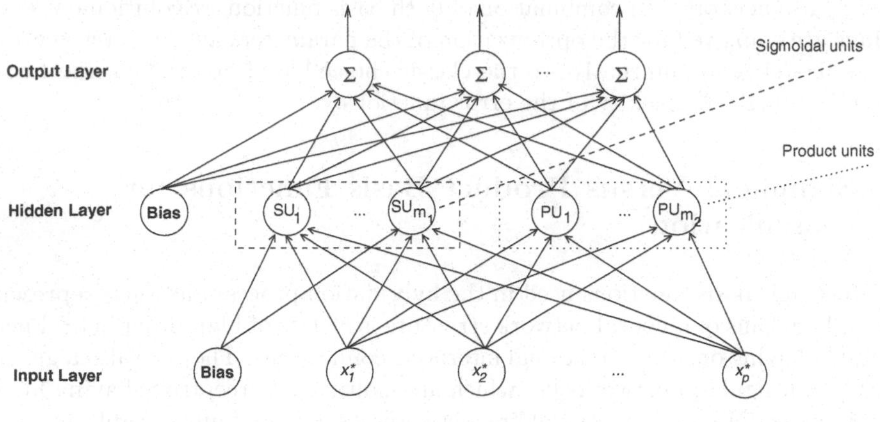

Fig. 1. Hybrid neural network for classification

exclusively of two types. An evolutionary algorithm (EA) similar to those reported by Yao and Liu [11] and García et al. [12] and have been used in this work to obtain the ANN architecture and for estimating the coefficients of the model. The kind of function typology proposed in (3) where we use hybrid transfer functions can be represented by an ANN architecture, as shown in Figure 1, where we consider the softmax activation function given by

$$g_l(\mathbf{x}, \boldsymbol{\theta}_l) = \frac{\exp f_l(\mathbf{x}, \boldsymbol{\theta}_l)}{\sum_{l=1}^{J} \exp f_l(\mathbf{x}, \boldsymbol{\theta}_l)} \qquad l = 1, 2, ..., J \tag{5}$$

3 Evolutionary Algorithm

The optimization of the ANN topology consisted of the search for the structure of the sigmoid and product unit base functions that best fit the data of the training set, by determining the values of m_1 and m_2 parameters related to the optimum number of base functions of each type involved. On the other hand, the estimation of the weights of the network is based on the evolution of the $\mathbf{u}_j$ and $\mathbf{w}_j$ vectors, which determine the coefficients in each base function, as well as with the α_j and β_k coefficients involved in the linear combination of the base functions. The population of the neural networks for classification is subject to the operations of replication and structural and parametric mutations. The general structure of the evolutionary algorithm (EA) is given in Algorithm 1..

Firs we calculate the fitness of a neural network of the population in the form:

If we use the training data set $D = \{(\mathbf{x}_n, \mathbf{y}_n)\}$ where $x_{in} > 0, \forall i, n$ then the cross-entropy error function (J-class multinomial deviance) for those observations is:

Algorithm 1. Evolutionary Programming Algorithm

1: Generate a random population of size NP
2: **repeat**
3: Calculate the fitness of every individual in the population.
4: Rank the individuals with respect to their fitness.
5: Copy best individual into the new population.
6: The best 10% of population individuals are replicated and substitute the worst 10% of individuals.
7: Apply parametric mutation to the best 10% of individuals.
8: Apply structural mutation to the remaining 90% of individuals.
9: **until** stopping criterion is fulfilled
10: Select the best individual of the population in the last generation and return it as the final solution.

$$
\begin{aligned}
l(\boldsymbol{\theta}) &= -\frac{1}{N}\sum_{n=1}^{N}\sum_{l=1}^{J} y_n^{(l)} \log g_l(\mathbf{x}_n, \boldsymbol{\theta}_l) \\
&= \frac{1}{N}\sum_{n=1}^{N}\left[-\sum_{l=1}^{J} y_n^{(l)} f_l(\mathbf{x}_n, \boldsymbol{\theta}_l) + \log \sum_{l=1}^{J} \exp f_l(\mathbf{x}_n, \boldsymbol{\theta}_l)\right]
\end{aligned} \tag{6}
$$

where $\boldsymbol{\theta} = (\boldsymbol{\theta}_1, \boldsymbol{\theta}_2, ..., \boldsymbol{\theta}_J)$, $f(\mathbf{x})$ is calculated in Equation 3. Let $l(\boldsymbol{\theta})$ to be the error function of an individual g of the population. Observe that g, obtained in Equation 5 is a SU, PU or SPU neural network and can be seen as the multivaluated function $g(x, \boldsymbol{\theta}) = (g_1(x, \boldsymbol{\theta}_1), ..., g_l(x, \boldsymbol{\theta}_l))$. The fitness measure is a strictly decreasing transformation of the error function $l(\boldsymbol{\theta})$ given by $A(g) = \frac{1}{1+l(\boldsymbol{\theta})}$. Over the intermediate population the adjustement of both weights and structure of the SU, PU or SPU networks is performed by the complementary action of two mutation operators: parametric and structural mutation. Parametric mutation implies a Gaussian modification in the coefficients α_j, β_k, and the $\mathbf{u}_j$ and $\mathbf{w}_j$ vectors of the model, using a self-adaptive simulated annealing algorithm.

Parametric mutation is accomplished for each coefficient α_j, β_k, $\mathbf{u}_i j$ and $\mathbf{w}_j k$ of the model with Gaussian noise. Once the mutation is performed, the fitness of the individual is recalculated and the usual simulated annealing algorithm [13] is applied. Thus, if ΔA is the difference in the fitness function after and preceding the random step, the criterion is: if $\Delta A \geq 0$ the step is accepted, if $\Delta A < 0$, the step is accepted with a probability $\exp(\Delta A/T(g))$, where the temperature $T(g)$ of an individual g is given by $T(g) = 1 - A(g), 0 \leq T(g) < 1$.

Structural mutation modifies the topology of the neural nets, helping the algorithm to avoid local minima and increasing the diversity of the trained individuals. Five structural mutations are applied sequentially to each network: node deletion, connection deletion, node addition, connection addition and node fusion. In this hybrid implementation of the basis functions, when a new node should be added to the networks of size m, it is necessary to estimate the probability of adding sigmoidal or product units base functions. We define the corresponding probabilities $p_{m,S}$ and $p_{m,P} = 1 - p_{m,S}$ of adding a basis sigmoidal or product function. These probabilities will be determined with the same value throughout the whole evolutionary process. Finally, the stop criterion is reached whenever one of the following two conditions is fulfilled: (i) the algorithm achieves a given number of generations; (ii) there is no improvement for a number of generations either in the average performance of the best 20% of the population or in the fitness of the best individual.

The parameters used in the evolutionary algorithm for learning the ANN models are common for all experiments: Components of the $\mathbf{u}_j$ and $\mathbf{w}_j$ vectors and the coefficients α_j and β_k, are initialized in the [-5, 5] interval; the maximum number of hidden nodes is $M = 6$; initially $m_1 = [M/2]$, where [.] is the entire part function, and $m_2 = M - m_1$; and the size of the population is $N_R = 1000$. The number of nodes that can be added or removed in a structural mutation is within the [1, 2] interval, and for node addition $p_{m,S} = 0.5$, whereas the number of connections that can be added or removed in a structural mutation is within the [1, 6] interval. Finally, the accuracy of each model is assessed in terms of the CCR for the results obtained for both data sets, that is, CCR_T for the training set, and CCR_G for the generalization set. We define the corrected classified rate by $CCR = \frac{1}{N}\sum_{n=1}^{N} I(C(\mathbf{x}_n) = \mathbf{y}_n)$, where $I(.)$ is the zero-one loss function. More details about the EA can be found in [2].

4 Experiments and Results

In order to gain an understanding of the accuracy of the different transfer functions and hybrid transfer function in neural networks for multi-class classifications problems, we have carried out five experiments where we use a three-dimensional, three-class synthetic data set, and where the three populations have normal density functions $\mathrm{N}(\boldsymbol{\mu}_i, \boldsymbol{\Sigma})$. The experimental design is conducted using a holdout cross-validation procedure where the size of the training set is approximately $3n/4$ and $n/4$ for the generalization set, where n is the size of the data set; for that procedure we use the NtRand 2.01 software [14]. Then, we randomly generate 300 samples for the training set and 90 for the generalization set. Denoting by $\boldsymbol{\mu}_i$ for $i = 1, 2, 3$ the means of the generating samples for class i, the centres are located at μ_1 =(1, 0, 0), μ_2 =(0, 1, 0), μ_3 =(0, 0, 1). Covariance matrices, $\boldsymbol{\Sigma}$, for all the populations are isotropic and different for each experiment, E_j, that is:

- E_1) Corr(x_1, x_2)= 0.1, Corr(x_1, x_3)= 0.5, Corr(x_2, x_3)= 0.9, Var(x_i)= 0.3.
- E_2) Corr(x_1, x_2)= 0.1, Corr(x_1, x_3)= 0.5, Corr(x_2, x_3)= 0.9, Var(x_i)= 0.6
- E_3) Corr(x_i, x_j)= 0, Var(x_i)= 0.9
- E_4) Corr(x_1, x_2)= 0.5, Corr(x_1, x_3)= 0.7, Corr(x_2, x_3)= 0.9, Var(x_i)= 0.9
- E_5) Corr(x_i, x_j)= 0.9, Var(x_i)= 0.9, Σ is singular.

Corr denotes the linear correlations between the input variables, and Var represents the variance of each variable.

Initially, we use a linear discriminant analysis, LDA [15] for classification. It is well known that if we assume that the cost of misclassification are equals, then the regions of classification R_i, that minimize the expected cost are defined by

$$R_i : u_{ij}(x) = \left[\mathbf{x} - \frac{1}{2}(\boldsymbol{\mu}_i + \boldsymbol{\mu}_j)\right]' \boldsymbol{\Sigma}^{-1}(\boldsymbol{\mu}_i - \boldsymbol{\mu}_j) > 0 \qquad i = 1, 2, 3, j \neq i \quad (7)$$

In this three class classification problem we first compute the coefficients of $\boldsymbol{\Sigma}^{-1}(\boldsymbol{\mu}_1 - \boldsymbol{\mu}_2)$, $\boldsymbol{\Sigma}^{-1}(\boldsymbol{\mu}_1 - \boldsymbol{\mu}_3)$ and then $\boldsymbol{\Sigma}^{-1}(\boldsymbol{\mu}_2 - \boldsymbol{\mu}_3) = \boldsymbol{\Sigma}^{-1}(\boldsymbol{\mu}_1 - \boldsymbol{\mu}_3) - \boldsymbol{\Sigma}^{-1}(\boldsymbol{\mu}_1 - \boldsymbol{\mu}_2)$. If we calculate $\frac{1}{2}(\boldsymbol{\mu}_i + \boldsymbol{\mu}_j)\boldsymbol{\Sigma}^{-1}(\boldsymbol{\mu}_i - \boldsymbol{\mu}_j)$ we obtain the linear discriminant functions for the experiments.

Table 1. Accuracy and statistical results of CCR provided by different basis functions used for a three class problem using synthetic data obtained in five experiments for thirty runs of the EA

Experiments	CCR_T				CCR_G				#conn	
Basis functions	Mean	SD	Best	Worst	Mean	SD	Best	Worst	Mean	SD
LDA	100				100				12	
SU	95.43	2.19	98.66	92.33	95.33	1.36	97.77	93.33	21.00	3.33
PU	97.83	1.03	99.33	96.33	98.33	1.07	100.00	96.66	13.30	2.58
SPU	98.00	0.64	99.00	97.00	**98.33**	**1.07**	100.00	96.66	20.30	2.49
LDA	98.66				96.66				12	
SU	89.00	2.61	93.66	86.00	88.66	2.55	92.22	84.44	21.50	4.45
PU	93.83	2.43	97.66	91.00	93.22	1.84	96.66	90.00	13.90	2.18
SPU	95.13	1.72	97.33	91.66	**94.33**	**1.69**	97.77	92.22	20.80	3.96
LDA	63.66				64.44				12	
SU	64.06	0.82	65.33	62.66	63.77	2.92	67.77	57.77	19.40	4.85
PU	63.70	1.19	65.33	62.00	**63.88**	2.41	67.77	60.00	13.10	2.18
SPU	64.90	1.14	67.33	63.00	62.44	**1.63**	65.55	60.00	19.20	3.70
LDA	92.00				91.11				12	
SU	85.73	1.64	87.66	83.00	83.33	1.95	86.66	81.11	20.60	3.06
PU	89.86	0.84	91.00	88.66	88.22	1.67	92.22	86.66	15.20	2.20
SPU	90.60	1.35	92.00	87.33	**88.66**	**1.02**	90.00	86.66	21.80	3.93
LDA	Singular Matriz									
SU	95.40	1.33	97.00	92.66	95.33	2.27	97.77	90.00	22.40	4.29
PU	97.30	0.67	98.33	96.33	96.77	1.61	98.88	93.33	14.80	1.87
SPU	97.83	0.52	98.33	96.66	**97.00**	**1.28**	100.00	95.55	22.60	3.23

Bold face: Best performance; SD: Standard deviation; #conn: Number of connections.

Since a goal of this work is to evaluate different ANN models supported on three different transfer functions, namely SU, PU and SPU, we use the experiments to analyze the accuracy of these models to be compared with LDA, because the classification rule (Equation 7) produces the best Bayes error over the training set.

To start processing data, each of the input variables were scaled in the ranks [0.1, 0.9]. The lower bound is chosen to avoid inputs values near to 0 that can produce very large values of the function for negative exponents. The upper bound is chosen near 1 to avoid dramatic changes in the outputs of the network when there are weights with large values (especially in the exponents). The new scaled variables are designed by x^*.

Table 2. CCR (correct classification rate) obtained for the best models for the E4 design of the synthetic data set with LDA, SU, PU and SPU models

LDA,SU,PU,SPU	Training			Generalization		
Target/Pre.	G=1	G=2	G=3	G=1	G=2	G=3
G= 1	90,84,93,90	8,12,6,7	2,4,1,3	24,26,26,25	5,4,4,4	1,0,0,1
G= 2	12,15,17,15	88,84,83,85	0,1,0,0	2,2,2,3	28,28,28,27	0,0,0,0
G= 3	1,17,2,0	1,1,1,2	98,82,97,98	0,6,1,0	0,0,0,1	30,24,29,29
CCR	92.00,83.33,91.00,91.00			91.11,86.66,92.22,90.00		

Table 3. Quantification equations and accuracy provided by LDA and the optimized SPU network topologies for the E4 and E5 designs

Equations LDA	$u_{12}(x) = 2.223 + 0.833x_1 - 5.278x_2 + 4.167x_3$
	$u_{13}(x) = -3.889 + 6.111x_1 + 9.444x_2 - 13.889x_3$
	$u_{23}(x) = -1.667 + 5.277x_1 + 14.722x_2 - 18.056x_3$
Equations SPU	$f_1(\mathbf{x}) = -2.84 + 2.88B_1 + 3.48B_2 - 4.38B_3 + 3.96B_4$
	$f_2(\mathbf{x}) = -4.98 + 5.06B_2$
Basis functions	$B_1 = \frac{1}{1+\exp-(-0.57+0.35x_2^*)}$
	$B_2 = (x_2^*)^{5.67}(x_3^*)^{-4.75}$
	$B_3 = (x_1^*)^{-1.39}(x_3^*)^{1.90}$
	$B_4 = (x_1^*)^{2.93}(x_2^*)^{-0.38}(x_3^*)^{-0.93}$
	Architecture=3:4:2SPU; Effective links= 14; CCR_T=91; CCR_G= 90
Equations SPU	$f_1(\mathbf{x}) = 3.95 + 5.98B_1 - 2.98B_2 - 10.06B_4$
	$f_2(\mathbf{x}) = 4.50 + 2.35B_1 - 8.85B_3 - 1.82B_5$
Basis functions	$B_1 = \frac{1}{1+\exp-(-5.27-2.43x_1^*)}$
	$B_2 = \frac{1}{1+\exp-(-7.23-3.63x_3^*)}$
	$B_3 = (x_2^*)^{-5.50}(x_3^*)^{-4.78}$
	$B_4 = (x_1^*)^{-5.03}(x_2^*)^{1.40}(x_3^*)^{3.47}$
	$B_5 = (x_1^*)^{3.72}(x_2^*)^{-4.68}(x_3^*)^{3.69}$
	Architecture=3:5:2SPU; Effective links= 20; CCR_T=96.66; CCR_G= 100

In Table 1 we present statistical results of CCR for training and generalization sets using SU, PU and SPU models for five experiments in 10 runs of the EA, where in general, the SPU design produces the best basis function models. Table 2 presents the confusion matrix generated by the best models for each basis function of these ANN and using LDA for the E_4 experiment where we can see the best CCR for Class 3 in function of its higher correlation with the other two classes.

Table 3 shows the best discriminant linear classifier and the best functions for constructing the softmax classification functions using a SPU basis function neural network for the E_4 and E_5 experiment. We can see that the best model using SPU nets in experiment 4 does not have a much higher number of coefficients, 14, than the model using LDA, 12. On the other hand the best SPU model for experiment 5 presents a CCR_G of the 100, classifying well all patterns of the generalization set, while it is not possible to construct the discriminating functions for LDA because the matrix is singular.

5 Conclusions

We propose a classification method that combines two different types of basis functions for feed-forward neural networks, and an evolutionary algorithm that finds the optimal structure of the model and estimates its corresponding parameters. To the best of our knowledge, this is the first study that applies a combination of basis function neural networks using evolutionary algorithm to solve a range of synthetic classification problems evolving both structure and weights. Our method uses softmax transformation and the cross-entropy error function. From a statistical point of view, the approach can be seen as nonlinear multinomial logistic regression where we use evolutionary computation to optimize log-likelihood. The designed combined basis function models provided accurate results, more robust models, and they present results comparable to those obtained by linear discriminant analysis, the methodology that obtains the best Bayes error throughout the training set. On the other hand, the experimental results show that the sigmoidal-product unit model could be an interesting alternative to resolve a classification task when the variance-covariance matrix is singular and thus does not allow the application of the classic LDA technique.

Acknowledgments

This work has been financed in part by the TIN2005-08386-C05-02 project of the Spanish Inter-Ministerial Commission of Science and Technology (CICYT) and FEDER funds.

References

1. Minsky, M., Papert, S.: Perceptrons. MIT Press, Cambridge, MA (1969)
2. Martínez-Estudillo, A., Martínez-Estudillo, F.J., Hervás-Martnez, C., García-Pedrajas, N.: Evolutionary product unit based neural networks for regression. Neural Networks **19** (2006) 477–486
3. Cybenko, G.: Aproximation by superpositions of a sigmoidal function. Mathematics of Control, Signals and Systems **2** (1989) 302–366
4. Leshno, M., Lin, V., Pinkus, A., Schocken, S.: Multilayer feedforward networks with a nonpolynomical activation can approximate any function. Neural Networks **6** (1993) 861–867

5. Sankar, A., Mammone, R.: Growing and pruning neural tree networks. IEEE Trans. on Computers **3** (1993) 291–299
6. Donoho, D.: Projection based in approximation and a duality with kernel methods. Annals Statistics **17** (1989) 58–106
7. Lee, Y., Doolen, G., Chen, H., Sun, G., Maxwell, T., Lee, H., Giles, C.: Machine learning using a higher order correlation network. Physica D: Nonlinear Phenomena **22** (1986.) 276–306
8. Rumelhart, D., McClelland, J., the PDP Research Group: Parallel Distributed Processing: Explorations in the Microstructure of Cognition, vol. 1: Foundations. MIT Press., Cambridge, MA (1986)
9. Pao, Y.: Adaptive Pattern Recognition and Neural Networks. Addison-Wesley, Reading, MA (1989)
10. Salinas, E., Abbott, L.: A model of multiplicative neural responses in parietal cortex. In: Proc. Natl Acad. Sci. 93. (1996) 11956–11961
11. Yao, X., Liu, Y.: A new evolutionary system for evolving artificial neural networks. IEEE Trans. Neural Networks **8** (1997) 694713
12. García, N., C.Hervás, Munoz, J.: Covnet: Cooperative coevolution of neural networks. IEEE Transactions on Neural Networks **14** (2003) 575596
13. Otten, R.H.J.M., van Ginneken, L.P.P.P.: The annealing algorithm. Ed. Kluwer, Boston, MA (1989)
14. Numerical Technologies Incorporated Jingumae, Shibuya-ku, Tokio: NtRand Version 2.01. (2003)
15. Duda, R.O., Hart, P.E., Stork, D.G.: Pattern Classification. 2nd edn. Wiley-Interscience, New York (2001)

Non-linear Robust Identification: Application to a Thermal Process*

J.M. Herrero, X. Blasco, M. Martínez, and J.V. Salcedo

Predictive Control and Heuristic Optimization Group
Department of Systems Engineering and Control,
Polytechnic University of Valencia Spain
Tel.: +34-96-3879571; Fax: +34-96-3879579
{juaherdu,xblasco,mmiranzo,jsalcedo}@isa.upv.es

Abstract. In this article, a methodology to obtain the Feasible Parameter Set (FPS) and a nominal model in a non-linear robust identification problem is presented. Several norms are taken into account simultaneously to define the FPS which improves the model quality but, as counterpart, it increases the optimization problem complexity. To determine the FPS a multimodal optimization problem with an infinite number of minima, which constitute the FPS, is presented and a special evolutionary algorithm ($\epsilon-$GA) is used to characterize it. Finally, an application to a thermal process identification, where $||\cdot||_\infty$ and $||\cdot||_1$ norms have been considered simultaneously, is presented to illustrate the technique.

1 Introduction

In Robust Process Control all designs are based on a nominal process model and a reliable estimate of the uncertainty associated to this nominal model through robust identification (RI). Uncertainty can be caused mainly by measurement noise and model error [7] (e.g. dynamics not captured by model). Although uncertainty can have different sources, it always will appear as an identification error (IE) that means the difference between model and process outputs for a specific experiment.

Two different approaches are possible in RI: stochastic or deterministic. In the first one, the IE is assumed to be modelled as a random variable with several statistical properties. Under this approach, it is possible to use classical techniques of identification [6] to obtain the nominal model and its uncertainty which is related to the covariance matrix of the estimated parameters. When these assumptions do not work, the deterministic approach can be more appropriated [8], where the IE, although unknown, is assumed to be bounded.

The objective of the deterministic approach consists of the obtaining of the nominal model and its uncertainty or directly the (FPS), i.e. the parameter set which keeps the IE bounded for certain IE function or norms and their bounds.

* Partially supported by MEC (Spanish government) and FEDER funds: projects DPI2005-07835, DPI2004-8383-C03-02 and GVA-026.

J. Mira and J.R. Álvarez (Eds.): IWINAC 2007, Part I, LNCS 4527, pp. 457–466, 2007.

When the model is linear in its parameters, the FPS is a convex polytope. However, when the model is non-linear, the FPS can be a non-convex even disjoint polytope, and it can be approximated by orthotopes [1], ellipsoids [3], parallelotopics [2], which results, in a more conservative obtaining of the FPS.

There exist techniques such as interval computation [9], support vector machine [5] and others which although do not approximate the FPS, they have either limitations (the type of function for bounding the IE, the inability to characterize a non-convex or disjoint FPS) or their utilization is complicated when the model is complex (non differentiable respect to its parameters, etc.).

As a consequence of these handicaps, in this work a more flexible and powerful methodology to characterize the FPS is presented. The proposed methodology is based on the optimization of a function which is built from IE norms and bounds, and whose global minima will characterize the FPS. It will be a multimodal function, which can be non-convex and/or present local minima, and therefore classical optimizers (e.g. sequential quadratic programming) can be inappropriate. Therefore the ϵ-GA algorithm will be used.

The paper is organized as follows. The ϵ-GA algorithm is described in section II. The RI methodology is given in section III. An experimental example to illustrate the theory is shown in section IV. Finally, some concluding remarks are reported in section V.

2 ϵ-GA Evolutionary Algorithm

ϵ-GA [4] is an evolutionary algorithm (EA) designed to optimize multimodal monoobjective functions which even have an infinite number of global optima.

2.1 Related Concepts of the ϵ-GA

The optimization problem consists of:

Definition 1. *(Global minimum set) Given a finite domain $D \subseteq \mathcal{R}^L$, $D \neq \emptyset$ and a function to optimize $J : D \to \mathcal{R}$, the set Θ^* will be the global minimum set of J if and only if Θ^* contains all the global optima of J.*

$$\Theta^* := \{\theta \in D : J(\theta) = J^*\},$$

being J^* a global minimum of J for the searching space D.

From this definition, Θ^* is assumed to be a unique set and the best that can be made is to obtain a discretized approximation to Θ^* in the solution space D, that means, a finite set Θ^*_ϵ. For that, the solution space is divided by a grid into boxes of width ϵ_i for each dimension $i \in [1 \ldots L]$ and the algorithm is forced to produce just one solution at the same box. So, thanks to the grid, the solutions in Θ^*_ϵ are forced to be well distributed and to characterize Θ^*.

Concepts such as approximation and discretization must be specified to obtain Θ^*_ϵ, so quasi-global minimum and box representation definitions are shown.

Definition 2. (Quasi-global minimum) Given a finite domain $D \neq \emptyset$ and a function to optimize $J : D \rightarrow \mathcal{R}$, the solution θ is considered as a quasi-global minimum of J if and only if

$$J(\theta) \leq J^* + \delta,$$

being $\delta > 0$ and J^* a global minimum of J.

So, a global minimum solution is also a quasi-global minimum solution.

Definition 3. (Quasi-global minimum set) Given a finite domain $D \neq \emptyset$ and a function to optimize $J : D \rightarrow \mathcal{R}$, the set Θ^{**} will be the quasi-global minimum set of J if and only if Θ^{**} contains all the quasi-global minimum solutions of J.

$$\Theta^{**} := \{\theta \in D : J(\theta) \leq J^* + \delta\},$$

being J^* a global minimum of J for the searching space D and $\delta > 0$.

From this definition, $\Theta^* \subseteq \Theta^{**}$. Besides if $\delta \rightarrow 0$ then $\Theta^{**} \rightarrow \Theta^*$.

Definition 4. (Box) Given a vector θ in the solution space $D \subseteq \mathcal{R}^L$, its box for $\epsilon_i > 0$ is defined as the vector $\mathbf{box}(\theta) = [box_1(\theta) \ldots box_L(\theta)]$ where:

$$box_i(\theta) = \left\lfloor \frac{\theta_i - \theta_i^{min}}{\epsilon_i} \right\rfloor \forall i \in [1 \ldots L].$$

So $box_i(\theta) \in [0 \ldots (n\_box_i - 1)]$, being n_box_i the number of divisions of the grid in the dimension i

$$n\_box_i = \left\lceil \frac{\theta_i^{max} - \theta_i^{min}}{\epsilon_i} \right\rceil, \; (\theta_i^{max} - \theta_i^{min}) \geq \epsilon_i$$

where θ_i^{max} and θ_i^{min} determine the limits of the solution space D.

Definition 5. (Box-representation) Given two vectors $\theta^1, \theta^2 \in D$, whose images in the space of the function J are $J(\theta^1)$ and $J(\theta^2)$ respectively, it is said that θ^1 *box*-represents θ^2 (denoted by $\theta^1 \preceq \theta^2$) for a certain $\epsilon_i > 0$ if

$$\mathbf{box}(\theta^1) = \mathbf{box}(\theta^2) \wedge J(\theta^1) \leq J(\theta^2).$$

Therefore, Θ_ϵ^* can be defined as:

Definition 6. (ϵ-global minimum set) Given a solution set Θ in the solution space, the set $\Theta_\epsilon^* \subseteq \Theta$ will be an ϵ-global minimum set of Θ if and only if

1. It only contains quasi-global minimum solutions of Θ

$$\Theta_\epsilon^* \subseteq (\Theta \cap \Theta^{**}).$$

2. Any vector in $\Theta \cap \Theta^{**}$ has a *box*-representation in Θ_ϵ^*, that is:

$$\forall \theta \in \Theta \cap \Theta^{**}, \; \exists \theta^* \in \Theta_\epsilon^* : \theta^* \preceq \theta.$$

Therefore, given a set Θ, Θ^*_ϵ has not to be a unique set, because global minimum solutions of Θ which share the same box can *box*-represent each other.

Definition 7. ($\Phi_\epsilon(\Theta)$ set) The set of all the ϵ-global minimum sets of Θ will be called as $\Phi_\epsilon(\Theta)$.

With these definitions, it is possible to establish the procedure to manage the contents of the archive $A(t)$ (where the optimization problem solution Θ^*_ϵ will be stored). So it is necessary to know the global minimum J^*, although it is not always possible. The best approximation to J^* which the algorithm can provide will be J^{min}_Θ, the approximation whose value of the function J is the smallest.

$$J^{min}_\Theta = \min_{\theta \in \Theta} J(\theta).$$

Definition 8. (Inclusion of θ in $A(t)$) Given a vector in the solution space θ and the archive $A(t)$, θ will be included in the archive if and only if

$$J(\theta) \leq J^{min}_{A(t)} + \delta \tag{1}$$
$$\wedge$$
$$\neg\exists\theta^* \in A(t) : J(\theta^*) \preceq J(\theta). \tag{2}$$

At the same time, the inclusion of θ in the archive could modify $J^{min}_{A(t)}$ (the best solution included in $A(t)$), and therefore, all the solutions θ^* satisfying this condition

$$J(\theta^*) > J^{min}_{A(t)} + \delta \tag{3}$$
$$\vee$$
$$J(\theta) \preceq J(\theta^*). \tag{4}$$

will be removed from $A(t)$.

Due to the inclusion procedure of the definition 8, the contents of archive $A(t)$ converge towards an ϵ-global minimum set (the first of the proposed objectives) as long as $J^{min}_{A(t)}$ converges towards the global minimum J^*. This is possible since the algorithm can provide any solution in D with a probability greater than zero (this is achieved by the crossover and mutation operator of the ϵ-GA) and it is an elitist algorithm since the best solution will not be lost.

Coefficients ϵ_i show the desired discretization degree to apply to Θ^*_ϵ and they are directly related to the physical meaning of the parameters. The lower ϵ_i is, the higher n_box_i and the solutions number $|\Theta^*_\epsilon|$ are.

$$|\Theta^*_\epsilon| \leq \prod_{i=1}^{L} n\_box_i. \tag{5}$$

2.2 ϵ-GA Description

The objective of the ϵ-GA algorithm is to provide an ϵ-global minimum set, Θ^*_ϵ. ϵ-GA uses the populations $P(t)$, $A(t)$ y $G(t)$:

1. $P(t)$ is the main population and it explores the searching space D. The population size is $Nind_P$.
2. $A(t)$ is the archive where Θ^*_ϵ is stored. Its size $Nind_A$ is variable but bounded (equation (5)).
3. $G(t)$ is an auxiliary population which is used to store the new individuals generated at each iteration by the algorithm. The population size is $Nind_G$.

The pseudocode of the ϵ-GA algorithm is given by:

```
1. t:=0,A(t):=∅
2. P(t):=ini_random(D)
3. eval(P(t))
4. A(t):=store(P(t),A(t))
5. mode:=exploration
6. while t<t_max do
7.       G(t):=create(P(t),A(t))
8.       eval(G(t))
9.       A(t+1):=store(G(t),A(t))
10.      P(t+1):=update(G(t),P(t))
11.      mode:=determinemode(P(t))
12.      t:=t+1
13. end while
```

The main steps of the above algorithm are detailed:

Step 2. Population $P(0)$ is initialized with $Nind_P$ individuals, created inside the searching space D.

Steps 3 and 8. Function *eval* calculates the value of the fitness function $J(\theta)$ for every individual θ from $P(t)$ (step 3) or $G(t)$ (step 8).

Step 11. The function *determinemode* selects the algorithm operation mode between the exploration and exploitation modes. These modes affect the way of creating new individuals (function create). When the population $P(t)$ has converged, the exploitation mode must be selected, by using the difference between the best value $J^{min}_{P(t)} = \min_{\theta \in P(t)} J(\theta)$ and the worst one $J^{max}_{P(t)} = \max_{\theta \in P(t)} J(\theta)$ at iteration t. If $J^{max}_{P(t)} - J^{min}_{P(t)} < \delta$ the *exploitation* mode [1] will be selected, on the contrary it will be selected the *exploitation* one.

Step 4 and 9. Function *store* analyzes whether every individual of $P(t)$ (step 4) or $G(t)$ (step 9) must be included in archive $A(t)$. So the individual will have to satisfy the inclusion condition (definition 8), and according to this definition other individuals will be removed. When including a new individual θ^1 in the

[1] If $J^{min}_{P(t)} = J^*$ all the individuals in $P(t)$ will be quasi-global minimum solutions.

archive, its box ($\mathbf{box}(\theta^1)$) is occupied by other individual θ^2 from the archive, that is $\mathbf{box}(\theta^1) = \mathbf{box}(\theta^2)$, and $J(\theta^1) = J(\theta^2)$, finally that individual which is nearest to the centre of the box they occupy, will be included. So, a better distribution of the solutions inside the archive is achieved.

Step 7. Function *create* creates new individuals and stores them in population $G(t)$ using the following procedure until $G(t)$ is full [4]:

1. Two individuals are randomly selected, θ^{p1} from $P(t)$, and θ^{p2} from $A(t)$.
2. If the algorithm operates in *exploration* mode θ^{p2} is not altered whereas if the mode is *exploitation*, the individual is mutated.
3. A random number $u \in [0 \ldots 1]$ is selected. If $u > P_{c/m}$ (crossover-mutation probability) step 3 (crossover) is taken, otherwise step 4 (mutation).
4. θ^{p1} and θ^{p2} are crossed over by the extended linear recombination technique.
5. θ^{p1} and θ^{p2} are mutated by random mutation with gaussian distribution.

Step 9. Function *update* updates $P(t)$ with individuals from $G(t)$. One individual θ^G from $G(t)$ will be inserted in $P(t)$ and it will replace θ^p, if $J(\theta^G) < J(\theta^p)$ being $\theta^p = \arg\max_{\theta \in P(t)} J(\theta)$ so, the contents of $P(t)$ is converging.

Finally, when $t = t_{max}$, the individuals included in the archive $A(t)$ will be the solution Θ^*_ϵ to the multimodal optimization problem.

3 RI Problem Statement

Assuming the following model structure:

$$\dot{\mathbf{x}}(t) = f(\mathbf{x}(t), \mathbf{u}(t), \theta), \quad \hat{\mathbf{y}}(t, \theta) = g(\mathbf{x}(t), \mathbf{u}(t), \theta) \tag{6}$$

where: $f(.), g(.)$ are the non-linear functions of the model; $\theta \in D \subset R^L$ is the vector[2] of unknown model parameters and $\mathbf{x}(t) \in R^n$, $\mathbf{u}(t) \in R^m$ and $\hat{\mathbf{y}}(t, \theta) \in R^l$ are the vectors of model states, inputs and outputs respectively.

The objective is that the model behaviour (obtained by simulation) will be the most similar possible to the real process one (obtained by experiments). This objective is achieved by a minimization of a function which penalizes the IE.

Definition 9. (Identification Error) The identification error $\mathbf{e}_j(\theta)$ for the output $j \in [1 \ldots l]$ is stated:

$$\mathbf{e}_j(\theta) = \mathbf{y}_j - \hat{\mathbf{y}}_j(\theta),$$

where: $\mathbf{y}_j = [y_j(t_1), y_j(t_2) \ldots y_j(t_N)]$ are the process output j measurements[3] when the inputs $\mathbf{U} = [\mathbf{u}(t_1), \mathbf{u}(t_2) \ldots \mathbf{u}(t_N)]$ are applied to the model and $\hat{\mathbf{y}}_j(\theta) = [\hat{y}_j(t_1, \theta), \hat{y}_j(t_2, \theta) \ldots \hat{y}_j(t_N, \theta)]$ are the simulated model output j when the same inputs $\mathbf{U}$ are applied to the model[4].

[2] θ, $\mathbf{x}(t)$, $\mathbf{u}(t)$ e $\hat{\mathbf{y}}(t, \theta)$ are all column vectors.

[3] $\mathbf{y}(t) \in \mathcal{R}^l$ is the column vector of process outputs.

[4] N is the measurements number of each output and input. The interval between measurements is constant $t_i = i \cdot T_s$, being T_s the sample time.

It is assumed that the IE can be bounded by several norms simultaneously.

Definition 10. (IE norm) Let N_i denote the p-norm of the identification error vector for the output j as:

$$N_i(\theta) = \|\mathbf{e}_j(\theta)\|_p, \quad i \in A := [1, 2, \ldots, s],$$

where s is the number of norms.

Therefore, there exists an FPS_i consistent with each N_i and η_i bound

$$FPS_i := \{\theta \in D : N_i(\theta) \leq \eta_i, \eta_i > 0\}.$$

And therefore the FPS for all the norms simultaneously is stated as:

$$FPS := \{\bigcap_{i \in A} FPS_i\} = \{\theta \in D : \forall i \in A, N_i(\theta) \leq \eta_i, \eta_i > 0\}.$$

and its boundary

$$\partial FPS := \{\theta \in D : \exists i | N_i(\theta) = \eta_i \wedge N_j(\theta) \leq \eta_j,\ \eta_i, \eta_j > 0,\ i, j \in A\}. \quad (7)$$

To characterize the FPS and especially its boundary ∂FPS a function $J(\theta)$ is stated in such a way that its global minima constitute the ∂FPS and the FPS constitutes quasi-global minimum solutions.

$$J(\theta) := \begin{cases} \sum_B J_i & if\ B(\theta) \neq \emptyset \\ \min(\delta, \prod_A J_i)\ if\ B(\theta) = \emptyset \end{cases}$$

where: $B(\theta) := \{i \in A : N_i(\theta) > \eta_i\}$, and $J_i(\theta) = |N_i(\theta) - \eta_i|$.

In order to select the bounds η_i on N_i, a priori process knowledge (for instance, non-modelled dynamics) and noise characteristics must be taken into account. However, this can be a difficult task, and many times the bounds may be selected according to the desired performance for the model predictions.

Once the feasible parameter set FPS is determined using the identification data $\Omega_{ide} = \{\mathbf{Y}_{ide}, \mathbf{U}_{ide}\}$, it must be validated by using different validation data $\Omega_{val} = \{\mathbf{Y}_{val}, \mathbf{U}_{val}\}$. The validation consists of checking whether the FPS contains models which are consistent with data Ω_{val}, the norms and their bounds.

4 Robust Identification. Experimental Results

A scale furnace with a resistance placed inside is considered. A fan continuously introduces air from outside (air circulation) while energy is supplied by an actuator controlled by voltage. Using a data acquisition system, both resistance and air temperatures are measured when voltage is applied to the process. The dynamics of the resistance temperature can be modelled by

$$\dot{x}(t) = \frac{\left(\theta_1 u(t)^2 - \theta_2\left(x(t) - T_a(t)\right) - \frac{\theta_3(273 + x(t))^4}{100^4}\right)}{1000}, \quad \hat{y}(t) = x(t), \quad (8)$$

where: $\dot{x}(t)$ is the model state; $u(t)$ is the input voltage with rank 0 - 100 (%); $\hat{y}(t)$ is the resistance temperature (^oC) (model output); $T_a(t)$ is the air temperature (^oC) and $\theta = [\theta_1, \theta_2, \theta_3]^T$ are the model parameters.

Both ∞-norm $N_1(\theta)$ and absolute norm $N_2(\theta)$ are simultaneously used to determine the FPS. Fig. 1 shows identification and validation data.

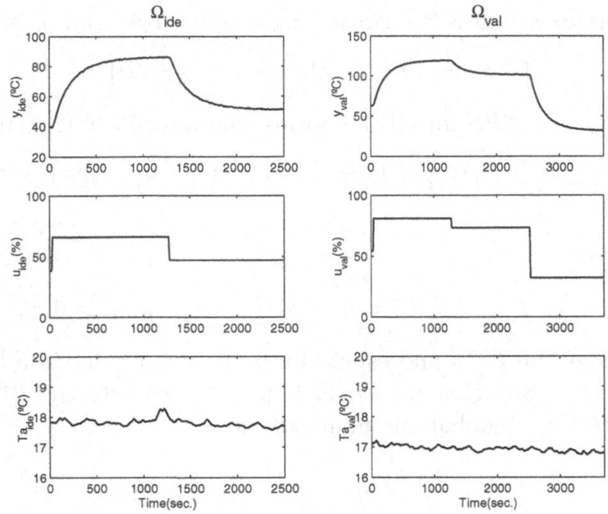

Fig. 1. On the left, identification data Ω_{ide}. On the right, validation data Ω_{val}.

Bounds $\eta_1 = 2$ and $\eta_2 = 0.8$ are selected in order to hold the FPS models predictions errors not greater than 2^oC and their average values not greater than 0.8^oC.

The FPS is determined next by ϵ-GA with the following parameters: Searching space $\theta_1 \in [0.01 \ldots 0.15]$, $\theta_2 \in [2 \ldots 10.0]$, $\theta_3 \in [0 \ldots 0.8]$; $t_{max} = 9975$ and $\epsilon = [0.0028, 0.16, 0.016]$ so the grid contains 50 divisions per dimension; $Nind_P = 100$, $Nind_G = 4$, $P_{c/m} = 0.1$ and the parameter $\delta(t)$ is tuned as $\delta(t) = \delta'(t) \cdot \bar{J}$, in order to be useful for other optimization problems, where $\bar{J}$ is the J average for all the individuals inserted in the population $P(t)$ during the optimization process. $\delta'(t)$ is determined by:

$$\delta'(t) = \frac{\delta_{ini}}{\sqrt{1 + \left(\left(\frac{\delta_{ini}}{\delta_{fin}}\right)^2 - 1\right)\frac{t}{(t_{max}-1)}}}, \qquad \delta_{ini} = 0.1, \delta_{fin} = 0.01$$

Fig. 2 shows the ϵ-GA optimization process result, i.e. FPS^*. The FPS has been characterized by 304 models from which 38 are consistent with Ω_{val} therefore the FPS^* is validated. The $J(\partial FPS^*)$ average is 0.021, which shows the

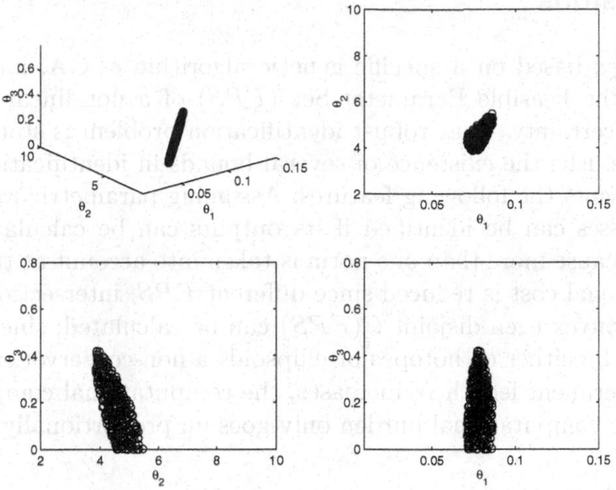

Fig. 2. On the left top the FPS^* models, inside searching space D, marked with $\circ$. The others figures are the FPS^* models projections.

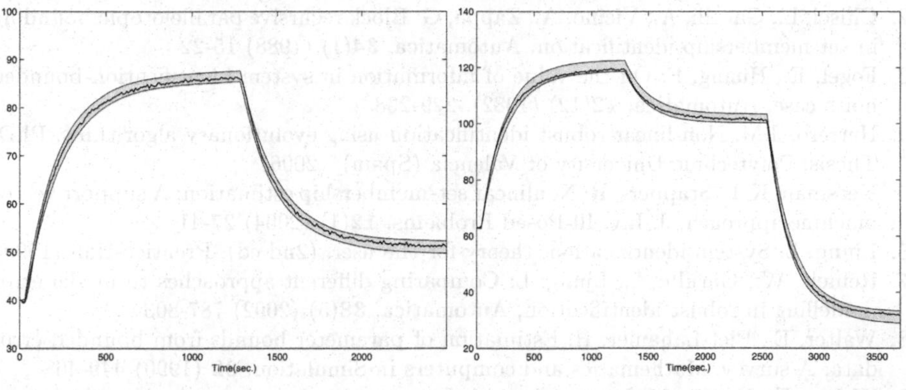

Fig. 3. On the left: $y_{ide}(t)$ and the FPS^* models envelop. On the right: $y_{val}(t)$ and the FPS^* models envelop.

good algorithm convergence (the ideal $J(\partial FPS^*)$ average would be 0). The number of function $J(\theta)$ evaluations results in 40000 which is approximately the third part of the computational cost if the cost function had been evaluated in every grid box. Although the accuracy obtained with the ϵ-GA algorithm is higher than the exhaustive search one.

Fig. 3 shows the $\mathbf{Y}_{ide}$ data, and the envelop generated by the FPS^*, as well as the $\mathbf{Y}_{val}$ data and the envelop from FPS^*. These envelops show both how the process dynamics has been modelled satisfactorily and the bounds have not been conservatively selected.

5 Conclusions

A methodology, based on a specific genetic algorithm ϵ−GA, has been developed to find the Feasible Parameter Set (FPS) of a non-linear model under parametric uncertainty. That robust identification problem is stated by assuming, simultaneously, the existence of several bounds in identification error. The algorithm presents the following features: Assuming parametric uncertainty, all kind of processes can be identified if its outputs can be calculated by model simulation; because more than one norm is taken into account at the same time, the computational cost is reduced since different FPS_i intersection is done implicitly; non-convex even disjoint $C(FPS)$ can be calculated; since FPS is not approximated by either orthotopes or ellipsoids a non-conservatism is provided and if the experiment length N increases, the computational complexity is preserved and the computational burden only goes up proportionally to N.

References

1. Belforte, G., Bona, B., Cerone, V: Parameter estimation algoritms for a set-membership description of uncertainty, Automatica, **26**(5) (1990) 887-898
2. Chisci, L., Garulli, A., Vicino, A., Zappa, G: Block recursive parallelotopic bounding in set membership identification, Automatica, **34**(1) (1988) 15-22
3. Fogel, E., Huang, F: On the value of information in system identification-bounded noise case, Automatica, **12**(12) (1982), 229-238
4. Herrero, J.M: Non-linear robust identification using evolutionary algorithms, PhD. Thesis. Polytechnic University of Valencia (Spain), (2006)
5. Keesman, K.J., Stappers, R: Nonlinear set-membership estimation: A support vector machine approach, J. Inv. Ill-Posed Problems, **12**(1) (2004) 27-41
6. Ljung, L: System identification, theory for the user, (2nd ed), Prentice-Hall (1999)
7. Reinelt, W., Garulli, A., Ljung, L: Comparing different approaches to model error modelling in robust identification, Automatica, **38**(5) (2002) 787-803
8. Walter, E., Piet-Lahanier, H: Estimation of parameter bounds from bounded-error data: A survey, Mathematics and computers in Simulation, **32** (1990) 449-468
9. Walter, E., Kieffer, M: Interval analysis for guaranteed nonlinear parameter estimation, In Proc. of the 13th IFAC Symposium on System Identification (2003)

Gaining Insights into Laser Pulse Shaping by Evolution Strategies

Ofer M. Shir[1], Joost N. Kok[1], Thomas Bäck[1,⋆], and Marc J.J. Vrakking[2]

[1] Natural Computing Group, Leiden University
Niels Bohrweg 1, 2333 CA Leiden, The Netherlands
{oshir,joost,baeck}@liacs.nl
http://natcomp.liacs.nl

[2] Amolf-FOM, Institute for Atomic and Molecular Physics
Kruislaan 407, 1098 SJ Amsterdam, The Netherlands
vrakking@amolf.nl

Abstract. We consider the numerical *evolutionary* optimization of dynamic molecular alignment by shaped femtosecond laser pulses. We study a simplified model of this *quantum control* problem, which allows the full physical investigation of the optimal solutions. By using specific variants of Derandomized Evolution Strategies, subject to parameterizations which are known to be superior for this problem, the numerical results reveal different *conceptual physics structures* for the different optimization procedures. These results are strong both from the algorithmic as well as from the physics perspectives. This shows that *Natural Computing* techniques can be used to derive new insights into *Physics*.

1 Introduction

To investigate and, more importantly, to control the motion of atoms or molecules by irradiating them with laser light, one has to provide laser pulses with durations on the same time scale as the motion of the particles. Recent technological development has made lasers with pulse lengths on the order of femtoseconds (1 fs=$10^{-15}s$) routinely available. Moreover, the time profile of these laser pulses can be shaped to a great extent. By applying a self-learning loop using an *evolutionary* mechanism, the interaction between the system under study and the laser field can be steered, and optimal pulse shapes for a given optimization target can be found. To this end, the role of the experimental feedback in the self-learning loop is played by a **numerical simulation** [1]. We plan to transfer, later on, these techniques to the laboratory for experimental optimization [2].

Control of molecular motion with shaped laser pulses is subject of intense current theoretical and experimental efforts. The success of femtosecond laser pulse shaping can considerably contribute even to the field of *computation*, as its application to *Molecular Quantum Computing* has been suggested (see, e.g., [3]). More specifically, *molecular alignment* is of considerable interest in this context because of its many practical consequences: a multitude of chemical and physical

⋆ NuTech Solutions, Martin-Schmeisser-Weg 15, 44227 Dortmund, Germany.

J. Mira and J.R. Álvarez (Eds.): IWINAC 2007, Part I, LNCS 4527, pp. 467–477, 2007.

processes ranging from bimolecular reactions [4] to high harmonic generation [5] are influenced by the angular distribution of the molecular sample. Furthermore, in many fundamental molecular dissociation or ionization experiments the interpretation of the collected data becomes much easier when the molecules are known to be aligned with respect to a certain axis. Hence, techniques to generate molecular alignment are much needed. For a review, see [6].

A recent study presented a survey of modern evolutionary approaches to the problem, and showed that it payed off to use more elaborated optimization schemes, and in particular Derandomized Evolution Strategies (DES), for such a high-dimensional optimization task [7]. We rely on that study in our choice of two DES variants.

In this study we focus in the simplified variant of the original problem, at zero temperature and with only a single rotational level in the initial distribution. The motivation for this simplification is to allow studying the physics nature of the optimal solutions, which would not have been possible for the general case, for reasons that will be explained in section 4.4 of this paper. The optimization procedure is subject here to two different parameterizations, which have been proposed for the laser problem [8].

The remainder of this paper is organized as follows. In section 2 we provide the reader with the details concerning the optimization routines in use. This is followed in section 3 with the introduction of the *dynamic molecular alignment* problem. In section 4 we present the experimental procedure, and section 5 outlines the conclusions which were drawn from this study.

2 Algorithms

Based on previous experience with this laser pulse shaping problem, and due to experimental results that showed that certain variants of Derandomized Evolution Strategies perform well with respect to other Evolutionary Algorithms on those problems [7], we restrict our study to these state-of-the-art algorithms. Our goal here is not to compare performance of algorithms, but rather to show that *natural computing* techniques can be used to derive new insights in *physics*. In this section we provide a short background of the specific variants in use.

2.1 Evolution Strategies

Evolution Strategies [9] are a *canonical evolutionary algorithm* for real-valued function optimization, due to their straightforward real-valued encoding, their specific variation operators, as well as to their high performance in this domain in comparison with other methods on benchmark problems. The higher the dimensionality of the search space, the more suitable a task becomes for an ES.

2.2 Derandomized Evolution Strategies

Mutative step-size control tends to work well for the adaptation of a global step-size, but tends to fail when it comes to the individual step size. This is due to several disruptive effects [10] as well as to the fact that the selection of the *strategy parameters* setting is indirect. The so-called *derandomized mutative step-size control* aims to tackle those disruptive effects. It is important to note that the different variants of *derandomized-ES* hold different numbers of strategy parameters to be adapted, and this is a factor in the speed of the optimization course: it is either a linear or quadratic order in terms of the dimensionality of the search problem n, and there seems to be a trade-off between the number of strategy parameters and the time needed for the adaptation/learning process of the step sizes.

The $(1,\lambda)$-DR2 Algorithm

The **DR2** Algorithm [11] is considered to be the second generation of the derandomized Evolution Strategies. This variant uses a linear number in n of strategy parameters, and it aims to accumulate information about the correlation or anti-correlation of past mutation vectors in order to adapt the step size:

$$\boldsymbol{x}^{g+1} = \boldsymbol{x}^{g} + \delta^{g} \boldsymbol{\delta}^{g}_{scal} \boldsymbol{Z}^{k} \qquad \boldsymbol{Z}^{k} = \boldsymbol{\mathcal{N}}(0,1) \tag{1}$$

$$\boldsymbol{Z}^{g} = c\boldsymbol{Z}_{sel} + (1-c)\,\boldsymbol{Z}^{g-1} \tag{2}$$

$$\delta^{g+1} = \delta^{g} \cdot \left(\exp\left(\frac{|\boldsymbol{Z}^{g}|}{\sqrt{n}\sqrt{\frac{c}{2-c}}} - 1 + \frac{1}{5n} \right) \right)^{\beta} \tag{3}$$

$$\boldsymbol{\delta}^{g+1}_{scal} = \boldsymbol{\delta}^{g}_{scal} \cdot \left(\frac{|\boldsymbol{Z}^{g}|}{\sqrt{\frac{c}{2-c}}} + 0.35 \right)^{\beta_{scal}} \tag{4}$$

where $\boldsymbol{\xi}_{scal} = \boldsymbol{\mathcal{N}}(0,1)^{+}$, $\boldsymbol{Z} \in \{-1,+1\}^{n}$, and β, β_{scal}, b and ξ^{k} are constants.

The (μ_W, λ) Covariance Matrix Adaptation ES

The (μ_W, λ)**-CMA-ES** algorithm [10] is known as the state-of-the-art among of the derandomized ES variants (could also be considered as DR4). It has been successful for treating correlations among object variables, where it applies *principal component analysis* (PCA) to the *selected* mutations during the evolution, also referred to as *"the evolution path"*, for the adaptation of the covariance matrix of the distribution. The concept of *weighted recombination* is introduced: applying intermediate multi-recombination on the best μ out of λ with given weights $\{w_i\}_{i=1}^{\mu}$. The result is denoted with $\langle \boldsymbol{x} \rangle_W$. Furthermore, $\boldsymbol{p}_{\sigma}^{(g)} \in \mathbb{R}^n$ is the so-called evolution path, $\boldsymbol{p}_{c}^{(g)} \in \mathbb{R}^n$, sum of weighted differences of points $\langle \boldsymbol{x} \rangle_W$, $\mathbf{C}^{(g)} \in \mathbb{R}^{n \times n}$, the covariance matrix of the mutation distribution $\left(\mathbf{C}^{(g)} = \mathbf{B}^{(g)}\mathbf{D}^{(g)} \left(\mathbf{B}^{(g)}\mathbf{D}^{(g)}\right)^{T}\right)$:

$$\boldsymbol{x}^{g+1} = \langle \boldsymbol{x} \rangle_W + \sigma_g \mathbf{B}^{g} \mathbf{D}^{g} \boldsymbol{z}_{k}^{g+1} \tag{5}$$

$$\boldsymbol{p}_c^{g+1} = (1 - c_c) \cdot \boldsymbol{p}_c^g + c_c^u \cdot c_W \mathbf{B}^g \mathbf{D}^g \langle \boldsymbol{z} \rangle_W^{g+1} \tag{6}$$

$$\mathbf{C}^{g+1} = (1 - c_{cov}) \cdot \mathbf{C}^g + c_{cov} \cdot \boldsymbol{p}_c^{g+1} \left(\boldsymbol{p}_c^{g+1}\right)^T \tag{7}$$

$$\boldsymbol{p}_\sigma^{g+1} = (1 - c_\sigma) \cdot \boldsymbol{p}_\sigma^g + c_\sigma^u \cdot c_W \mathbf{B}^g \langle \boldsymbol{z} \rangle_W^{g+1} \tag{8}$$

$$\sigma^{g+1} = \sigma^g \cdot exp\left(\frac{1}{d_\sigma} \cdot \frac{\left\|\boldsymbol{p}_\sigma^{g+1} - \hat{\chi}_n\right\|}{\hat{\chi}_n}\right) \tag{9}$$

where $\hat{\chi}_n$ is the expected length of $\boldsymbol{p}_\sigma$. c_c, c_{cov}, c_σ and d_σ are learning/adaptation rates, $\{w_i\}_{i=1}^{\mu}$ are the recombination weights, and $c_c^u := \sqrt{c_c(2-c_c)}$, $c_W := \frac{\sum w_i{}_{i=1}^{\mu}}{\sqrt{\sum w_i^2{}_{i=1}^{\mu}}}$ and $c_\sigma^u := \sqrt{c_\sigma(2-c_\sigma)}$ are derived respectively.

All weighting variables and learning rates were applied as suggested in the given citations, and particular in [10].

3 Dynamic Molecular Alignment

In this section we describe the optimization problem under investigation. The reader who wants to abstract from the physics details can view the problem as a single-criterion 80-dimensional optimization task, subject to maximization, with a punishment term for handling a physics constraint.

3.1 Quantum Control: Physics Background

The interaction of a generic linear molecule with a laser field is described within the framework introduced in [1]. We calculate the time evolution of a thermal ensemble of molecules quantum mechanically, by considering a single initial rotational level, characterized by the rotational quantum number $J_{initial} = 0$ (and the projection of the angular momentum on the laser polarization axis $M_{initial}$, respectively). We take the molecule to be a rigid rotor, which allows a description of its wavefunction solely in terms of the rotational wave functions $|JKM\rangle$ (where $K = 0$ for a diatomic molecule). Two electronic states are taken into account, the ground state denoted by X and an off-resonant excited state denoted by A. Hence the wavefunction for a given M is expanded as

$$\Psi_M(t) = \sum_{J=M}^{J_{max}} \alpha_{XJM}(t)\psi_{XJM} + \alpha_{AJM}(t)\psi_{AJM} \tag{10}$$

The time dependence of the molecular wave function is given by

$$i\frac{\partial \Psi_M}{\partial t} = (H_0 + V)\Psi_M(t) \tag{11}$$

The Hamiltonian consists of a molecular part H_0 and the interaction with the laser field, given by

$$V = \mu \cdot \mathbf{E}(\mathbf{t}) \cos(\omega t) \tag{12}$$

The Eigenenergies of H_0 are given by

$$E(J) = hcBJ(J+1) \tag{13}$$

where B is the *rotational constant* of the molecule. The laser field induces transitions between the rotational states which, in the off-resonant case, occur via subsequent Raman processes. The transitions between X and A were assumed to proceed via the selection rules $\Delta J = \pm 1, \Delta M = 0$.

The envelope of the laser field (which completely determines the dynamics after the transition to the rotating frame has been performed) is described by

$$E(t) = \int A(\omega) \exp(i\phi(\omega)t) \exp(i\omega t)\, d\omega. \tag{14}$$

The spectral function $A(\omega)$ is taken to be a fixed Gaussian. The control function is the phase function $\phi(\omega)$, which defines the phase at a set of n frequencies that are equally distributed across the spectrum of the pulse. These parameters are taken to be the decision parameters of the evolutionary search; the search space is therefore an n-dimensional hypercube spanning a length of 2π in each dimension.

3.2 Optimization

We consider the goal of optimizing the alignment of a sample of generic diatomic molecules undergoing irradiation by a shaped femtosecond laser. We have used the maximum $\langle \cos^2(\theta) \rangle$ that occurs under field free conditions after the laser pulse, where θ is the angle between the molecular and the laser polarization axis, as a measure of the alignment. The temperature of the ensemble was $T = 0K$ and the rotational constant was chosen to be $B = 5\text{cm}^{-1}$. The peak Rabi frequency between the two electronic states X and A, that determines the interaction strength, was $\Omega_{XA} = 1.6 \cdot 10^{14}\ \text{s}^{-1}$.

Since we want to achieve a high degree of alignment with a peak intensity as low as possible, an additional constraint was introduced as a punishment for pulses that are too intense. We have used

$$I_p = \int E^2(t) \Theta(E^2(t) - I_{thr})\, dt \tag{15}$$

(where $\Theta(x)$ is the Heaviside step function) for this purpose, so that the fitness function assigned to a pulse shape was

$$F = \max_{\mathrm{E(t)=0}} \langle \cos^2(\theta) \rangle - \beta \mathrm{I_p}. \tag{16}$$

By choosing β large enough, I_{thr} was shown [7] to effectively operate the evolutionary algorithms only on a subset of pulses whose maximum peak intensity approaches the threshold intensity from below. We have typically used $\beta = 1$; unless otherwise specified, I_{thr} was $0.36 \cdot\ I_{FTL}$.

4 Experimental Procedure

Given the optimization routines which were introduced in section 2, we describe here the different parameterizations in use, present the numerical results, and finally apply an analytical tool for the investigation of the optimal solutions.

4.1 Parameterization

As introduced earlier, the phase function $\phi(\omega)$ is the target function to be calibrated. Here, two parameterizations of $\phi(\omega)$ are considered.

The traditional approach was to interpolate $\phi(\omega)$ at n frequencies $\{\omega_i\}_{i=1}^{n}$; the n values $\{\phi(\omega_i)\}_{i=1}^{n}$ are the decision parameters to be optimized. In order to achieve a good trade-off between high resolution and optimization efficiency, the value of $n = 80$ turned to be a good compromise. We define this calibration of $\phi(\omega)$, i.e. learning $n = 80$ function values and interpolating, as the so-called *'plain-parameterization'* optimization:

$$\phi_P(\omega) = (\phi(\omega_1), \phi(\omega_2), ..., \phi(\omega_n)) \tag{17}$$

An alternative parameterization for the phase function, which has been presented at [8], considers the *Hermite* polynomials,

$$H_k(x) = (-1)^k \exp\left\{x^2\right\} \frac{d^k}{dx^k}\left(\exp\left\{-x^2\right\}\right),\ k = 0, 1, ... \tag{18}$$

as building blocks for the phase function, and aims to learn their coefficients in order to form the phase function:

$$\phi_H(\omega) = \sum_{k=0}^{K_{max}} c_k \cdot H_k(\omega) \tag{19}$$

Note that the *Hermite polynomials* form a complete set of functions over the infinite interval $-\infty < x < \infty$ with respect to the weight function $\exp\left\{-\frac{1}{2}x^2\right\}$. We define this as the *'Hermite-parameterization'* optimization. $K_{max} = 40$ turned out to perform best.

4.2 Setup

Some technical details concerning the experimental setup and our *modus operandi* are outlined:

- Every function evaluation has the duration of 7 seconds on a P4 2.6GHz.
- Based on past experience, we choose the $(1, 10)$ strategy for the DR2; $(7, 15)$-CMA for $n = 40$ and $(8, 17)$-CMA for $n = 80$.
- **Each run is limited to** $20,000$ **function evaluations.**
- Implementation was done in Fortran for the numerical simulation (in order to stay close to the systems used by the physics researchers), and in MATLAB 7.0 for the optimization routines.

4.3 The Optimization: Numerical Results

Table 1 presents the mean values and the standard deviations of the cosine-squared alignment, obtained after 20 runs of 20,000 function evaluations. As can be observed, the DR2 routine clearly outperforms the CMA in the *plain* parameterization, in consistency with previous results on the general problem $J_{max}^{initial} = 7$ (see, e.g., [7]). For the *Hermite* parameterization, however, the picture is different. The DR2 does not seem to deliver, and fails to obtain high-quality solutions. The CMA does succeed in this task, with highly-satisfying results. Essentially, this suggests that the *Hermite* parameterization introduces strongly correlated decision parameters, whereas the *plain* parameterization can be tackled successfully by a strategy which does not consider the correlations between the decision parameters. It is nevertheless surprising to observe the low performance of the CMA on the latter. We may conclude that the *Hermite*

Table 1. Maximizing the cosine-squared alignment over 20 runs with 20,000 function evaluations per run; *mean* and *std*; the *maximal* value obtained is in brackets

$\max_{E(t)=0} < \cos^2(\theta) >$	**DR2**	**CMA**
Plain Parameterization	0.9559 ± 0.0071 (0.9622)	0.9413 ± 0.0058 (0.9508)
Hermite Parameterization	0.9501 ± 0.0043 (0.9570)	0.9583 ± 0.0026 (0.9618)

parameterization is slightly better, but not dramatically superior on this simplified variant, especially due to the fact that it requires the CMA routine in order to obtain optimal solutions, in comparison to the DR2 with the *plain* parameterization.

4.4 Investigation of Optimal Solutions

An optimal solution is represented by its phase function $\phi(\omega)$, and the electric field respectively, but one can also examine the revival structure. **Only due to our simplified variant, i.e. $J_{initial}^{max} = 0$ at initialization, it is possible to study the *population*[1] of the rotational levels as a function of time**. Otherwise, in the general case, all levels are initially populated, and a thermal averaging is applied. Explicitly, the wavefunction can be expressed as a superposition of those levels,

$$\psi = \sum_j a_j^{(t)} \cdot |j\rangle \cdot e^{-i\frac{E_j t}{\hbar}} \tag{20}$$

the expectation of the cosine-squared alignment (the objective function) is calculated directly from these complex amplitudes $a_j^{(t)}$, whereas the *population* of the rotational levels is $\left|a_j^{(t)}\right|^2$. This population of rotational levels can be analyzed

[1] The careful reader should note that *'population'* is used here exclusively in the context of *quantum mechanics*, e.g., populating quantum levels.

in a fairly simple technique, known as the *Sliding Window Fourier Transform* (SWFT), which provides us with a **powerful visual tool**. Given the revival structure of an optimal solution, a sliding time window is Fourier transformed, to produce the frequency picture through the alignment process. This windowing creates a transformation which is localized in time. Due to the *quantization* of the rotational levels, only certain frequencies (or *energy* levels, respectively) are expected to appear.

Analysis Results. We applied the SWFT routine to the optimal solutions which were found in the various runs. Figures 3, 4, 5 and 6 visualize the typical population process of the rotational levels for four typical solutions of the different optimization procedures (2 parameterizations *times* 2 DES variants). The quantum energy levels are indeed observed as expected from theory.

The results reveal two different conceptual physical structures, which correspond to optimal and sub-optimal solutions in terms of the success-rate, i.e., the cosine-squared alignment. The Plain-DR2 as well as the Hermite-CMA procedures obtain the best solutions, which share the same structure - they are characterized by the dominant population of the 4th quantum energy level in the SWFT picture. On the other hand, the Plain-CMA and Hermite-DR2 procedures obtain inferior solutions, which are characterized by a gradually increasing population of the energy levels.

The original *revival structures* for two optimal solutions, representing the two conceptual structures, are given in Figures 1 and 2. The *optimal* family of solutions (Fig. 1) presents a dramatic revival structure, with a typical strong pulse in the train which lies on the boundary of the punished regime ($I \approx 0.36$). This pulse seems to be essential in giving the molecules the right 'kick', and

Alignment and Revival Structure of two obtained solutions (Fig. 1 and 2). Thin red line: alignment; thick black line: intensity of the laser pulse.

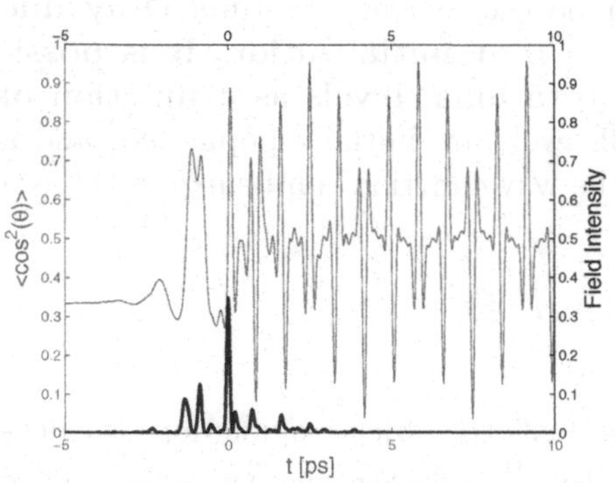

Fig. 1. Optimal solution type: no governing structure for the revival structure is observed; $< \cos^2(\theta) >= 0.9622$

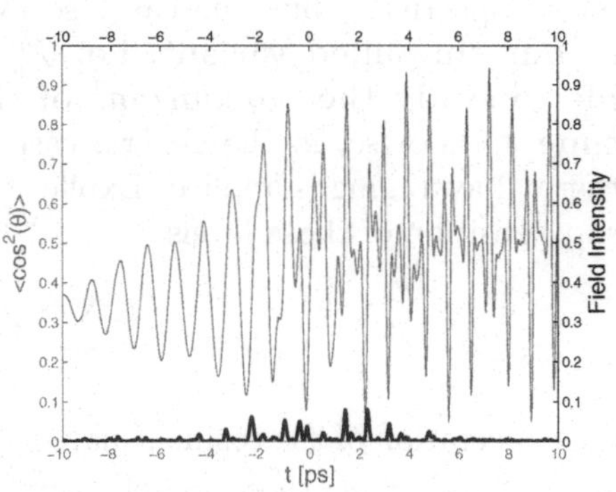

Fig. 2. Suboptimal solution type: a smooth exponential envelope on top of the revival structure; $< \cos^2(\theta) >= 0.9505$

Each of the SWFT figures (Fig. 3 - 6) represents a *Fourier transform* applied to the revival structures of the optimal solutions (the thin-red alignment curves of Fig. 1-2). The values are log-scaled, and represent how high the rotational levels of the molecules are populated as a function of time. Thus, an exponential envelope (Fig. 2) is represented by a gradual building-up of frequencies (Fig. 4). Note that the quality of the laser pulse cannot be measured in those plots.

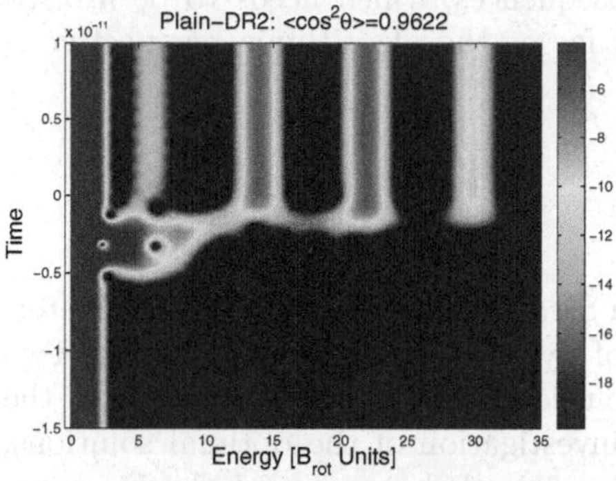

Fig. 3. 4th rotational level is mostly populated after the interaction

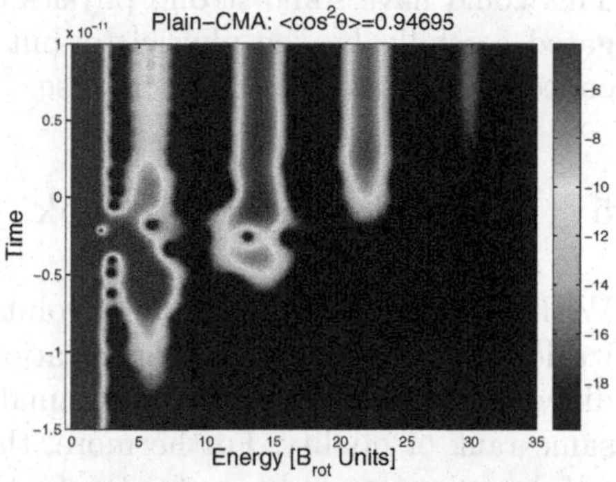

Fig. 4. All five first rotational levels are populated gradually after the interaction

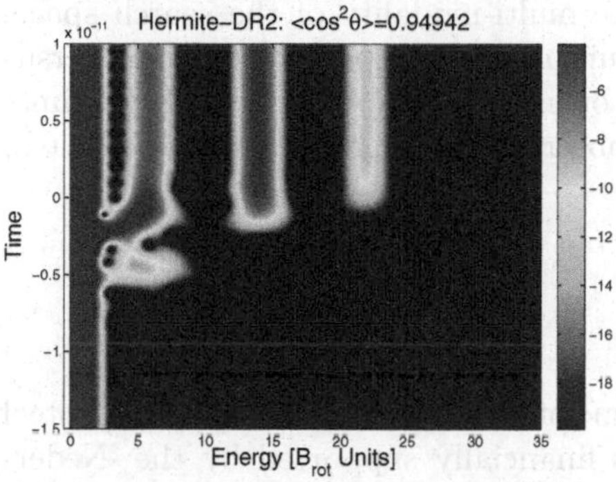

Fig. 5. The four first rotational levels are poppopulated gradually after the interaction

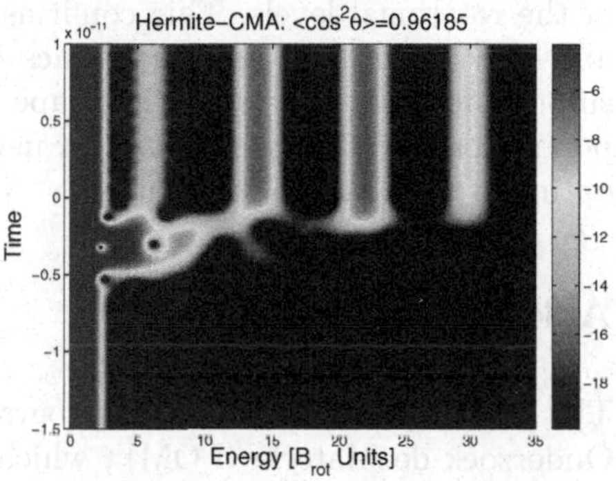

Fig. 6. 4th rotational level is mostly populated after the interaction

most likely responsible for the dominant population of the 4th quantum energy level in the SWFT picture. The *sub-optimal* family of solutions (Fig. 2) yields a revival structure with a smooth exponential envelope, and thus has a gradual building-up of the quantum energy levels in the SWFT picture, respectively. It typically contains a train of medium pulses and lacks a dominant one.

We would like to emphasize the fact that we obtained the same family of optimal solutions, with the same structure, from two different optimization approaches - the one learning a vector of successful variations of interpolated points of the target function, whereas the other learning a covariance matrix of coefficients of Hermite polynomials that span the target function. Based on our experience with the alignment problem, and due to the new results, we claim that this might suggest that the regime of the global optimum has been reached. This could have some strong physics consequences, which needs to be investigated carefully by the physicists, but as far as the algorithmic perspective is concerned, this seems to be the case.

5 Discussion and Outlook

We have applied derandomized Evolution Strategies, subject to two parameterizations, to the numerical optimization of dynamic molecular alignment. Two different approaches obtained optimal solutions with numerical results in the same rank of quality. Furthermore, the investigation of the optimal solutions, which was performed here for the first time, revealed two typical physics structures, through the time-evolution of the populated quantum energy levels. The first structure typically contains a strong pulse, which is followed by the population of a specific quantum rotational level for the molecules, whereas the second structure is a train of medium pulses with an outcome of a gradual population of the rotational levels. This confirms the multi-modality of the search space, and provides us with strong physics intuition with respect to optimal versus sub-optimal solutions, e.g., the shape of optimal pulses as well as the optimal population of rotational levels. Our new observation suggests that the regime of optimal solutions has been found.

Acknowledgments

This work is part of the research programme of the 'Stichting voor Fundamenteel Onderzoek de Materie (FOM)', which is financially supported by the 'Nederlandse Organisatie voor Wetenschappelijk Onderzoek (NWO)'.

References

1. Rosca-Pruna, F., Vrakking, M.: Revival structures in picosecond laser-induced alignment of i2 molecules. Journal of Chemical Physics **116**(15) (2002) 6579–6588
2. Zamith, S. Eur. Phys. J. D **12**(255) (2000)
3. Tesch, C.M., Kurtz, L., de Vivie-Riedle, R. Chem. Phys. Lett. **343** (2001)
4. Friedrich, B., Herschbach, D. Phys. Chem. Chem. Phys. **2**(419) (2000)
5. Hay, N. Phys. Rev. **A 65**(053805) (2000)
6. Stapelfeldt, H. Rev. Mod. Phys. **75**(543) (2003)

7. Shir, O.M., Siedschlag, C., Bäck, T., Vrakking, M.J.: Evolutionary algorithms in the optimization of dynamic molecular alignment. In: 2006 IEEE World Congress on Computational Intelligence, IEEE Computational Intelligence Society (2006) 9817–9824
8. Shir, O.M., Siedschlag, C., Bäck, T., Vrakking, M.J.: The complete-basis-functions parameterization in es and its application to laser pulse shaping. In: GECCO '06: Proceedings of the 8th annual conference on Genetic and evolutionary computation, New York, NY, USA, ACM Press (2006) 1769–1776
9. Beyer, H.G., Schwefel, H.P.: Evolution strategies a comprehensive introduction. Natural Computing: an international journal **1**(1) (2002) 3–52
10. Hansen, N., Ostermeier, A.: Completely derandomized self-adaptation in evolution strategies. Evolutionary Computation **9**(2) (2001) 159–195
11. Ostermeier, A., Gawelczyk, A., Hansen, N.: Step-size adaptation based on non-local use of selection information. In: Parallel Problem Solving from Nature (PPSN3). (1994)

Simulated Evolution of the Adaptability of the Genetic Code Using Genetic Algorithms

Ángel Monteagudo and José Santos

Departamento de Computación
Facultade de Informática. Universidade da Coruña.
Campus de Elviña, 15071 A Coruña Spain
santos@udc.es

Abstract. In this work we use simulated evolution to corroborate the adaptability of the natural genetic code. An adapted genetic algorithm searches for optimal hypothetical codes. The adaptability is measured as the average variation of the hydrophobicity that experiment the encoded amino acids when errors or mutations are presented in the codons of the hypothetical codes. Different types of mutations and base position mutation probabilities are considered in this study.

1 Introduction and Previous Work

In this work we use a genetic algorithm as a method to corroborate the adaptability of the natural genetic code. Although this code is not universal (for example, mitochondrial DNA has variations), it is the one that practically is present in the vast majority of complex genomes. The genetic code, with the four nitrogenated bases (A, T, G and C) encodes, when grouped in genes, mainly the amino acids that are linked to determine the proteins. That code is redundant because three bases are needed to establish a codon that codifies each one of the 20 amino acids that are present in proteins, plus a “stop translation” signal found at the end of every gene. That manner, in the code, most of the amino acids are specified by more than one codon, what implies that redundancy of the code.

Nevertheless, as there are 64 possible codons to encode the 21 meanings, a huge number of hypothetical “genetic codes” could be defined, with different associations than the ones of the natural genetic code. That number of possible codes is $1.4 \cdot 10^{70}$, as Yockey [3] has calculated, taking into account the encoding of each each amino acid with a maximum of 6 codons (as in the natural code).

The establishment of the genetic code is still in discussion. It is not definitively clear if it was a random assignment or it was an adaptive process by means of natural evolution. The second case means that those codes with less harmful effects in the possible errors of the protein synthesis machinery and in the final proteins, have an evolutionary advantage against those codes that present a greater number of harmful effects. An argument in favour of this possibility is the fact that the natural genetic code appeared to be arranged such that the amino acids with similar chemical properties are coded by similar codons. For

J. Mira and J.R. Álvarez (Eds.): IWINAC 2007, Part I, LNCS 4527, pp. 478–487, 2007.

example, the codons that share two of the three bases tend to correspond to amino acids that have a similar hydrophobicity.

In a first computational experiment, Haig and Hurst [2] have corroborated it by means of a simple simulation. They found that of 10.000 randomly generated codes, only 2 performed better at minimizing the effects of error, when polar requirement was taken as the amino acid property. They thus estimated that the chance that a code as conservative as the natural code arose by chance was 0.0002, and, therefore, concluded that the natural code was a product of natural selection for load minimization. To quantify the efficiency of each one of the possible codes they used a measure that considers the changes in a basic property of the amino acids when all the possible mutations are considered in a generated code (section 2.2 explains the measure in better detail). The property used by the authors was the polar requirement, which may be considered a measure of hydrophobicity, as the one that gave the most significant evidence of load minimization from an array of four amino acid properties (hydropathy, molecular volume, isoelectric point and the one chosen).

Freeland and Hurst [7] refined the previous estimate, with a greater sample of 1.000.000 possible codes. The authors found 114 better codes (a proportion of 0.000114), indicating, according to the author's results, a refinement of the previous estimate for relative code efficiency such that the code was even more conservative. In addition, they extended the work to investigate the effect of weighting transition errors differently from transversion errors and the effect of weighting each base differently, depending on reported mistranslation biases (section 3.2 explains these type of mutations). When they employed weightings to allow for biases in translation, they found that only 1 in every million random alternative codes generated was more efficient than the natural code.

Gilis et al. [1] extended Freeland and Hurst's work taking into account the frequency at which different amino acids occur in proteins. Their results indicate that the fraction of random codes that beat the natural code decreases. In addition, they use a new function of error measurement that evaluates *in silico* the change in folding free energy caused by all possible point mutations in a set of protein structures, being a measure of protein stability. With that function the authors estimated that around two random codes in a billion (10^9) are fitter than the natural code.

Opposite to that brute-force search of possible codes that are better adapted, we have used simulated evolution, by means of a genetic algorithm, to have a guided search of better adapted hypothetical codes and to have a method to guess the progression and the difficulty to find such alternative codes, which could serve as a method of "traceability" in the evolution of the codes in their fight for survival in the evolution in the "RNA World".

2 Adapted Genetic Algorithm to the Problem

Each individual of the genetic population must encode a hypothetical code. In our solution, each individual has 64 positions, that correspond to the 64 codons, and each position encodes the particular amino acid associated with the codon. A basic procedure ensures that the individuals of the initial population encode, at least in one position, the 20 amino acids. As in Haig and Hurst [2], a fixed number of three codons are used for the STOP label (as in the natural code), that determines the end in the production of the protein. The genetic operators must ensure that a given individual always encode the 21 labels.

2.1 Genetic Operators

We have used a mutation operator and a swap operator. A mutation changes the amino acid encoded in each one of the 64 positions, according to a mutation probability, with another different one. The mutation does not operate if the amino acid to mutate is the unique instance in all the code. Those mutations simulate the possible errors in the transcription process form DNA to RNA and in the translation process when incorrect transfer RNAs join a given codon of the messenger RNA. From our application point of view, it is the operator that varies the number of codons associated with a particular amino acid.

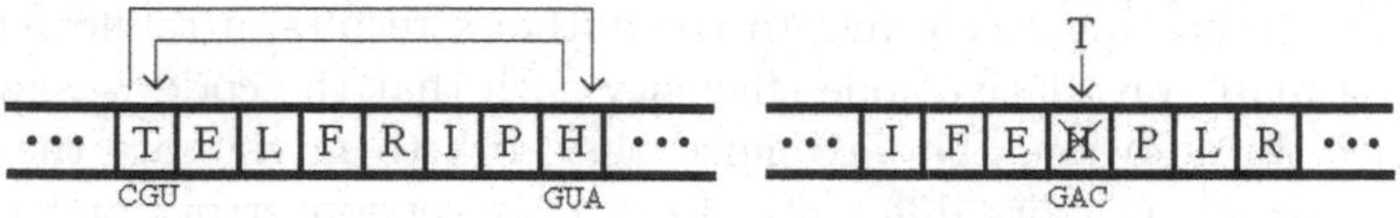

Fig. 1. Swap operator (left) and mutation operator (right)

The other operator is the swap operator, hardly ever used in GA applications, although here is perfectly suited. The operator interchanges the contents of two genes (codons), that is, once two genes are randomly selected, the amino acids codified by the two respective codons are swapped. Figure 1 shows the functioning of these genetic operators. The two operators guarantee that the 20 amino acids are always represented in the individuals. Other operators, like the classical crossover operator, do not guarantee that important restriction. Finally, as selection operator we have used a tournament selection.

2.2 Fitness Function

We have used as fitness function the measure used by Haig and Hurst [2] and Freeland and Hurst [7] to quantify the relative efficiency of any given code. The measure calculates the mean squared (MS) change in an amino acid property resulting from all possible changes to each base of all the codons within a given code. Any one change is calculated as the squared difference between the property value of the amino acid coded for by the original codon and the value of the amino acid coded for by the new (mutated) codon. The changes from and to "stop" codons are ignored, while synonymous changes (the mutated codon encodes the same amino acid) are included in the calculation. Figure 2 summarizes the error calculation, when the first base of the codon UUU is mutated, taking

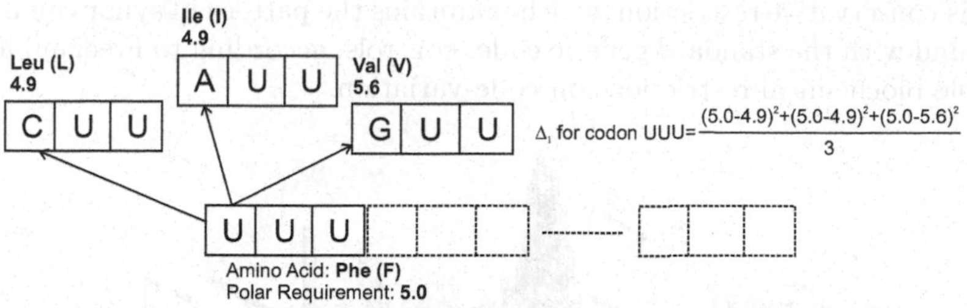

Fig. 2. Calculation of the error value of a code to define the fitness function

into account the new values of the polar requirement of the new coded amino acids. The final error is an average of the effects of all those substitutions over the whole code.

Many other alternative types of weighting are imaginable, being hard to know what would be the best model relating chemical distance to code fitness, as commented by the previously cited authors. In the MS measure we can consider the MS1, MS2 and MS3 values that correspond to all single-base substitutions in the first, second and third codon positions, respectively, of all the codons in a given genetic code. The MS value (or any of the components) defines the fitness value of a given code and the evolutionary algorithm will try to minimize it.

3 Results

3.1 Equal Transition/Transversion Bias

In this first analysis, we test the capability of simulated evolution to find better codes than the natural one, taking into account the MS value previously commented, and with an equal probability of mutations in all the three bases and the two types of mutations. As we have outlined, Haig and Hurst [2] only found 2 alternative codes with lower MS than the natural code in a set of 10.000 randomly generated codes, and Freeland and Hurst [7] refined the probability with a greater sample of 1.000.000, where they found only 114 better codes. These authors, when hypothetical codes were generated, have taken into account two restrictions:

1. The codon space (the 64 codons) was divided into the 21 nonoverlapping sets of codons observed in the natural code, each set comprising all codons specifying a particular amino acid in the natural code. Twenty sets correspond to the amino acids and one set for the 3 stop codons.
2. Each alternative code is formed by randomly assigning each of the 20 amino acids to one of these sets. All the three stop codons remain invariant in position for all the alternative codes. In addition, these three codons are the same as the ones of the natural genetic code (UAA, UAG and UGA).

This conservative restriction, which maintains the pattern of synonymous coding found with the standard genetic code, controls, according to Freeland [6], for possible biochemical restrictions on code variation.

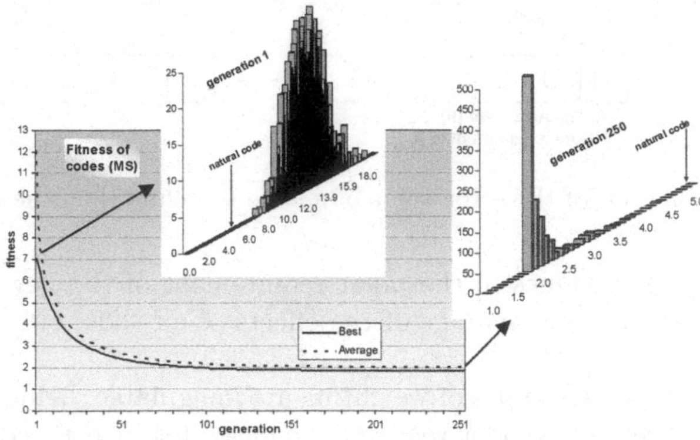

Fig. 3. Evolution of the MS without restrictions (except the number of stop codons), and histograms of the population at the beginning and end of the evolutionary process.

We have tested the evolution with these restrictions, and with a more free evolution with a unique restriction: we only impose three codons for the STOP signal. Our results clearly change the analysis, as the GA easily finds better alternative codes. Figure 3 shows the evolution of the MS across 250 generations of the genetic algorithm, in that case without the 2 previously commented restrictions. The best individual and the average qualities of the population are the result of an average of 10 evolutions with different initial populations. The initial population was 1000 individuals in each one of the different tests. Larger populations do not improve the results. The other evolutionary parameters were a mutation probability of 0.01 and a swap probability of 0.5, although the results show little variations respect these parameters. As selection we have used tournament with low selective pressure (tournament window of 3% of the population). The best (minimum) MS found in one of the evolutions was 1.784, really better than the best value found by Freeland and Hurst (~ 4.7) and than the value of the natural genetic code (5.19).

The MS measures of each sample of codes in each generation form a probability distribution against which the natural code MS value may be compared. Figure 3 also shows the "histograms" of the initial population and at the end of the evolution in generation number 250. In the histograms, the X axis gives a particular range of categories of MS values whereas the Y coordinate indicates the number of individuals with an MS in that category. The histogram of the initial population presents a similar distribution as the ones of the cited authors, as the population is random. There is not any better code (than the natural one) by chance in that initial population. At the end of the evolutionary process the situation was changed radically, where all the individuals have a better MS than the one of the natural genetic code.

The left part of Figure 4 shows the same analysis of evolution, in that case with all the restrictions commented. Even with the restrictions, evolution finds in few generations better codes than the natural one, although the minimum MS values (around 3.48) are not as good as with the more free evolution. The right part of Figure 4 shows the assignments of the amino acids to the codons in the natural genetic code as well as in the best obtained code with those restrictions. The polar requirement values are also showed, associated with their amino acids of that best evolved code.

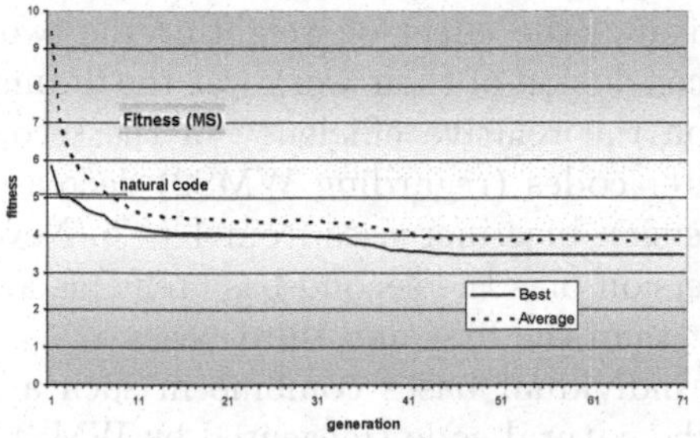

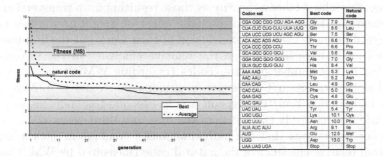

Codon set	Best code		Natural code
CGA CGC CGG CGU AGA AGG	Gly	7.9	Arg
CUA CUC CUG CUU UUA UUG	Gln	8.6	Leu
UCA UCC UCG UCU AGC AGU	Ser	7.5	Ser
ACA ACC ACG ACU	Pro	6.6	Thr
CCA CCC CCG CCU	Thr	6.6	Pro
GCA GCC GCG GCU	Val	5.6	Ala
GGA GGC GGG GGU	Ala	7.0	Gly
GUA GUC GUG GUU	His	8.4	Val
AAA AAG	Met	5.3	Lys
AAC AAU	Trp	5.2	Asn
CAA CAG	Leu	4.9	Gln
CAC CAU	Phe	5.0	His
GAA GAG	Cys	4.8	Glu
GAC GAU	Ile	4.9	Asp
UAC UAU	Tyr	5.4	Tyr
UGC UGU	Lys	10.1	Cys
UUC UUU	Asn	10.0	Phe
AUA AUC AUU	Arg	9.1	Ile
AUG	Glu	12.5	Met
UGG	Asp	13.0	Trp
UAA UAG UGA	Stop		Stop

Fig. 4. Evolution of the MS with Freeland and Hurst's restrictions (left), and best obtained code with those restrictions (right)

An analysis of the amino acids encoded in the codons of a variety of the best codes indicates two considerations: there is a great variety of better codes, with very different assignments; and there are not clear coincidences between the best codes with the assignments of the natural code. When evolution works without restrictions (except for the stop signal), 5 amino acids incorporate the majority of codes: alanine (Ala), asparagine (Asn), proline (Pro), serine (Ser) and threonine (Thr), while the rest are codified by only one or two codons. These amino acids are codified in the natural code with 6 or 4 codons (except Asn), although the other two amino acids codified by 6 codons in the natural code (Arg and Leu) are codified by only one codon in the majority of the best evolved codes.

If restrictions are taken into account, there are not practically any coincidences in the amino acids assignments in the 20 sets of codons between the natural code and the best obtained codes, such as the one showed in the right part of Figure 4. This last result coincides with the observations of Freeland and Hurst [7], although their better codes correspond to those obtained by chance in a great number of samples. Nevertheless, the best evolves codes, like the one of figure 4, have the same property of the natural code: amino acids that share the two first bases have similar values of polar requirement.

3.2 Introducing Transition/Transversion Bias

In nature, transition errors tend to occur more frequently than transversion mutations, because the unequal chemical similarity of the 4 nucleotides to one another [6]. A transition error is the substitution of a purine base (A, G) into another purine, or a pyrimidine (C, U/T) into another pyrimidine (i.e., $C \leftrightarrow T$ and $A \leftrightarrow G$), whereas a transversion interchanges pyrimidines and purines (i.e., $C, U \leftrightarrow A, G$).

We can use the MS values for each code calculated at different weightings of transition/transversion bias, and turning the MS measures into WMS measures. That manner, at a weighting of 1, all possible mutations are equally when calculating the MS values for each position of each codon. And, for example, at a weighting of 2, the differences in amino acid attribute resulting from transition errors were weighted twice as heavily as those resulting from transversion errors.

Freeland and Hurst [7] investigated the effect of weighting the two types of mutations differently. The main conclusion of their work was the dramatic effect of transition/transversion bias on the relative efficiency of the second codon base; that is, the number of better codes (regarding WMS2) decreases almost six fold as the transition/transversion bias increases from 1 to 5. Nevertheless, even at higher transition/transversion bias the second base remains an order of magnitude less relatively efficient than the first and third bases.

Another aspect was that the individual bases combine in such a way that the overall relative efficiency of the natural code (measured by WMS) increases with increasing transition/transversion bias up to a bias of approximately 3. Moreover, this effect is clearer with WMS1. As commented by the authors, that observation coincides quite well with typical empirical data, which reveal general transition/transversion biases of between 1.7 and 5. In addition, as the causes of that natural biases are physiochemical, basically size and shape of purines and pyrimidines, "it seems reasonable to suppose that the biases observed now were present to a similar extent during the early evolution of life" [7].

Figure 5 shows the evolution of the three individual components of the MS. That means, for example, that the evolution with MS1 uses a fitness that only considers errors in that first base. All the evolutions in that figure were with a population of 1000 individuals and with the commented Freeland and Hurst's restrictions. The upper figures correspond to a weight of 1 and the figures at the bottom part were obtained with a weight of 3. The figures also include the improvement obtained in fitness by the best evolved individual, measured as the final fitness respect to the corresponding MS value of the natural code.

If we consider Freeland and Hurst's analysis, regarding the random individuals of the first generation, the second base is clearly the worst adapted in the natural code, since in the first generation there are random codes with better MS2, and additionally, in less than five generations all the individuals overcome the MS2 value of the natural code (10.56). The first base seems better adapted in the natural code, as the simulated evolution has more difficulty to have all the population with less MS1 than the natural MS1 value (4.88). Finally, the third base is the best adapted in the natural code, as there are not any random codes with lower values than the MS3 value of the natural code (0.14) and the genetic algorithm requires a few generations to obtain better individuals, and it is not able that the whole population gets a lower value. However, the improvements

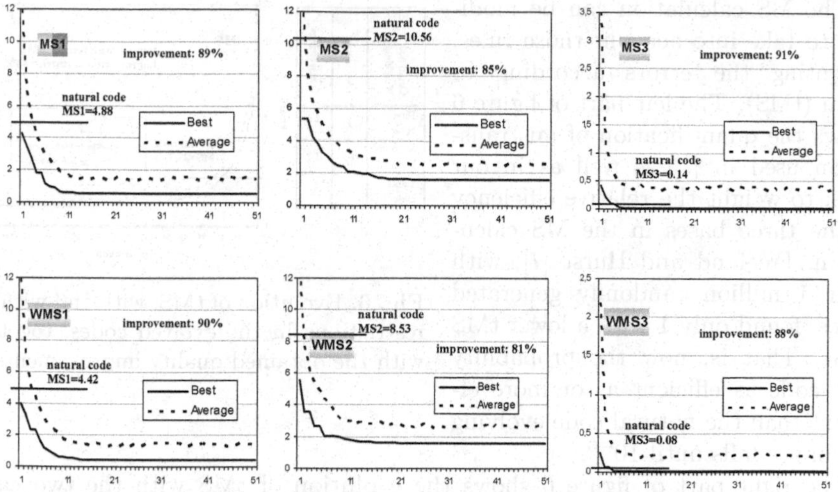

Fig. 5. Evolution of the individual components of MS with restrictions, with equal transition/transversion bias (upper figures) and with a bias of 3 (bottom figures)

that can be obtained by simulated evolution are not in accordance with the previous analysis, as the three bases present similar levels of improvement, with the second base being the one with the worst result. Finally, with a bias of 3, there are not any appreciable differences, with also contradicts the analysis of the previous authors, as we cannot infer that the first base is better adapted with that bias of 3.

3.3 Different Codon Position Errors

The previous works assumed that mistakes are equally likely to be made at any of the three codon positions. That assumption is correct when we consider point mutations in the DNA sequence and that are accurately translated via mRNA into an erroneous amino acid. However, the assumption must be reconsidered if we take into account mistranslation of mRNA. The translation machinery acts upon mRNA reading bases in triplets (codons), and that translation accuracy varies according to the base position of the codon. We have used the same rules from [7], that were used to consider the empirical data, summarized as:

- A Mistranslation of the second base is much less frequent than the other two positions, and mistranslation of the first base is less frequent than the third base position.
- B The mistranslations at the second base appear to be almost-exclusively transitional in nature.
- C At the first base, mistranslations appear to be fairly heavily biased toward transitional errors.
- D At the third codon position, there is very little transition bias.

The MS calculation can be modified to take into account those rules, weighting the errors according to them (tMS). The left part of figure 6 shows the quantification of mistranslation used in [7] as well as in our work to weight the relative efficiency of the three bases in the MS calculation. Freeland and Hurst [7], with their 1 million randomly generated codes, found only 1 with a lower tMS value. That is, now the probability of a code as efficient as or more efficient than the natural code evolving by chance falls until 10^{-6}.

Combined Weighting	Transitions	Transversions
First Base	1	0.5
Second Base	0.5	0.1
Third Base	1	1

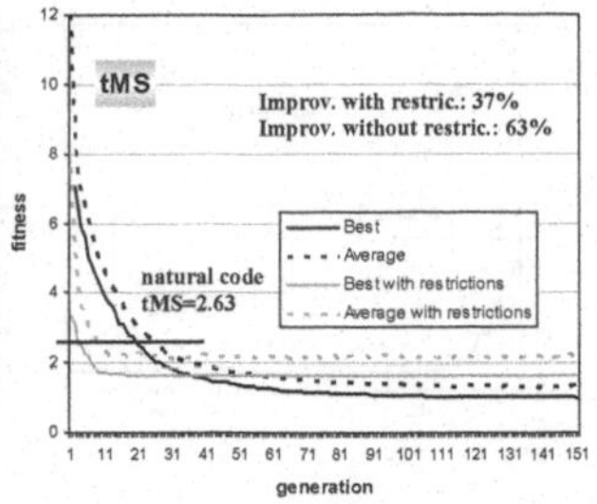

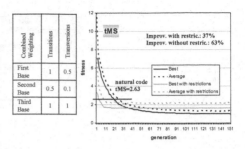

Fig. 6. Evolution of tMS, with and without restrictions in the evolved codes, together with the obtained quality improvements

The right part of figure 6 shows the evolution of tMS with the two cases previously considered, with and without the introduction of Freeland and Hurst's restrictions. In both cases the improvement in fitness quality is similar respect to an equal probability of errors in the three bases: the improvement in MS (decrease of fitness of the best individual respect to the natural case) is 32% with restrictions and 64.3% without restrictions. That is, both cases (MS and tMS) provide evolution with the same level of possible optimization, although there is more difficulty to obtain better codes by change with tMS.

4 Discussion and Conclusions

Giulio [4], in a review of the theories about the origin of the genetic code, distinguishes two basic alternatives about the evolution of the genetic code. The stereochemical theory claims that the origin of the genetic code must lie in the stereochemical interactions between anticodons or codons and amino acids. On the other hand, the physicochemical theory claims that the force defining the origin of the genetic code structure was the one that tended to reduce the deleterious effects of physicochemical distances between amino acids codified by codons differing in one base. The commented work of Freeland and Hurst [7] is obviously in this line.

Yockey [3] is critic with the idea of evolution of the genetic code in the sense of minimization of the effects of mutations. As argument he exposes that the $1.4 \cdot 10^{70}$ possible codes couldn't have been checked in the $8 \cdot 10^{8}$ years between the event of Earth formation ($4.6 \cdot 10^{9}$ years ago) and the origin of life in it ($3.8 \cdot 10^{9}$ years ago). We do not consider this statement as a correct argument, as neither natural evolution has to check all possible codes to minimize the deleterious effects of mutations, nor simulated evolution checks all the possibilities in the search space.

Nevertheless, the fact that the GA easily finds better codes than the natural genetic code has not imply that there was not any adaptive evolution of the genetic code. Two considerations must be taken in this regard. Firstly, we only have considered one (important) property. In Knight et al. [5] words "the average effect of amino acid changes in proteins is unlikely to be perfectly recaptured by a single linear scale of physical properties". Secondly, we agree with the same authors that the code could be trapped in a local, rather than global, optimum. Again, with words from the authors "The fact that the code is not the best of all possible codes on a particular hydrophobicity scale does not mean that it has not evolved to minimize changes in hydrophobicity under point misreading". However, our results indicate that the local optimum obtained in the natural code is not as optimal as supposed by Freeland and Hurst [7], as the simulation of the evolution can easily find a great variety of better codes.

It has been proposed, as Knight et al. [5] pointed out, that the standard genetic code evolved from a simpler ancestral form encoding fewer amino acids. Our work will follow in this line, trying to find out the possible routes that evolution could have followed once a codon with two letters coded the possible 15 amino acids (plus the stop signal), although there is not consensus about which amino acids entered the code and in what order.

References

1. Gilis D., Massar S., Cerf N.J., and Rooman M. Optimality of the genetic code with respect to protein stability and amino-acid frequencies. *Genome Biology*, 2(11), 2001.
2. Haig D. and Hurst L.D. A quantitative measure of error minimization in the genetic code. *Journal of Molecular Evolution*, 33:412–417, 1991.
3. Yockey H.P. *Information Theory, Evolution, and the Origin of Life.* Cambridge University Press, NY, 2005.
4. Giulio M.D. The origin of the genetic code: theories and their relationship, a review. *Biosystems*, 80:175–184, 2005.
5. Knight R.D., Freeland S.J., and Landweber L.F. Adaptive evolution of the genetic code. *The Genetic Code and the Origin of Life*, 80:175–184, 2004.
6. Freeland S.J. The darwinian genetic code: an adaptation for adapting? *Genetic Programming and Evolvable Machines, Kluwer Academic Publishers*, 3:113–127, 2002.
7. Freeland S.J. and Hurst L.D. The genetic code is one in a million. *Journal of Molecular Evolution*, 47(3):238–248, 1998.

GCS with Real-Valued Input

Lukasz Cielecki and Olgierd Unold

The Institute of Computer Engineering, Control and Robotics (CECR),
Wroclaw University of Technology, Wyb. Wyspianskiego 27, 50-370 Wroclaw, Poland
lukasz.cielecki@pwr.wroc.pl, olgierd.unold@pwr.wroc.pl

Abstract. The new learning classifier system is described to classify real-valued data. The approach applies the continous-valued context-free grammar based system GCS. In order to handle effectively, the terminal rules have been replaced by the so-called environment probing rules. The rGCS model has been tested on the checkerboard problem.

1 Introduction

GCS is a Learning Classifer System LCS [2] introduced by Unold [4] in which a knowledge about the solved problem is represented by context-free grammar (CFG) in Chomsky Normal Form productions. The GCS is described in detail in [5][6]. An integer-value representation (in fact the set of letters a-z) is used with GCS. However, many real-world problems are not expressed in terms of a simple non-continuous representation and several alternate representations have been suggested to allow LCSs to handle these problems more readily (for references see [3]). In this paper we introduce an extension to GCS that allows the representation of continuous-valued inputs.

The remainder of this paper is organized as follows. Section 2 describes the GCS - original Grammar Classifier System that works with sentence strings as an input data. Section 3 introduces the new rGCS - extended system prepared to work with real-valued input. Section 4 illustrates our experiments with the checkerboard problem and the last one summarizes the paper and tells about some of our future plans.

2 The GCS

The GCS operates similar to the classic LCS but differs from them in (i) representation of classifiers population, (ii) scheme of classifiers' matching to the environmental state, (iii) methods of exploring new classifiers.

Population of classifiers has a form of a context-free grammar rule set in a Chomsky Normal Form (CNF). This is not a limitation actually because every context-free grammar can be transformed into equivalent CNF. Chomsky Normal Form allows only production rules in the form of $A \rightarrow a$ or $A \rightarrow BC$, where A, B, C are the non-terminal symbols and a is a terminal symbol. The first rule is an instance of *terminal rewriting rule.* These ones are not affected by the GA,

J. Mira and J.R. Álvarez (Eds.): IWINAC 2007, Part I, LNCS 4527, pp. 488–497, 2007.

and are generated automatically as the system meets unknown (new) terminal symbol. Left hand side of the rule plays a role of classifier's action while the right side a classifier's condition. System evolves only one grammar according to the so-called Michigan approach. In this approach each individual classifier – or grammar rule in GCS – is subject of the genetic algorithm's operations. All classifiers (rules) form a population of evolving individuals. In each cycle a fitness calculating algorithm evaluates a value (an adaptation) of each classifier and a discovery component operates only on a single classifier.

Automatic learning context-free grammar is realized with so-called grammatical inference from text [1]. According to this technique system learns using a training set that in this case consists of sentences both syntactically correct and incorrect. Grammar which accepts correct sentences and rejects incorrect ones is able to classify unseen so far sentences from a test set. Cocke-Younger-Kasami (CYK) parser, which operates in $\Theta(n^3)$ time [8], is used to parse sentences from the sets.

Environment of classifier system is substituted by an array of CYK parser. Classifier system matches the rules according to the current environmental state (state of parsing) and generates an action (or set of actions in GCS) pushing the parsing process toward the complete derivation of the sentence analyzed.

The discovery component in GCS is extended in comparison with standard LCS. In some cases a "covering" procedure may occur, adding some useful rules to the system. It adds productions that allow continuing of parsing in the current state of the system. This feature utilizes for instance the fact that accepting 2-length sentences requires separate, designated rule in grammar in CNF.

Apart from the "covering" a GA also explores the space searching for new, better rules. First GCS implementation used a simple rule fitness calculation algorithm which appreciated the ones commonly used in correct recognitions. Later implementations introduced the "fertility" technique, which made the rule fitness dependant on the amount of the descendant rules (in the sentence derivation tree) [5] [6]. This approach is particularly useful since in GCS population individuals must cooperate to parse sentences successfully. Appreciating linked rules we help to preserve the structure of evolved grammar. In both techniques classifiers used in parsing positive examples gain highest fitness values, unused classifiers are placed in the middle while the classifiers that parse negative examples gain lowest possible fitness values.

3 rGCS

3.1 Overview of the rGCS

rGCS exploits the main idea of the classic GCS system. The CYK table is the environment and the area where the rGCS operates. The learning process is divided into the cycles. During every cycle evolved grammar is tested against every example of the train set, then new rules are evolved or existing ones are modified and another cycle begins (Fig. 1).

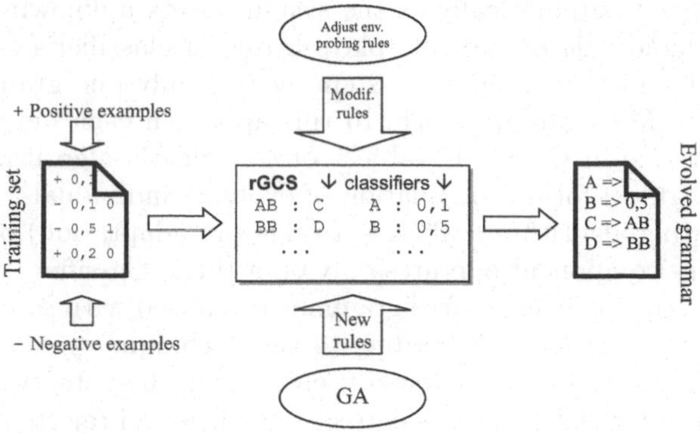

Fig. 1. rGCS block diagram

3.2 Environment Probing Rules

Structure of the rules that are used in the very first row of the CYK table during parsing is the main difference between the classic GCS and rGCS. Since their role is to sense the input data and then to launch the CYK process we called them the environment probing rules. In classic GCS system the environment situation consisted of input string terminals so these rules were called very accurately - the terminal rewriting rules. Now in rGCS the input data (environmental situation) is formed by the vector of real numbers that may describe various kinds of data (Fig. 2). Each rule has the form of

$$A \rightarrow f \tag{1}$$

where A is the non-terminal symbol, f is the real number value.

f value is used during the matching process and the non-terminal A is to put into the first row of the CYK table. Additionally a special environment probing rule may be used - the most general one that accepts every single input value. This one is called the wildcard (don't care symbol) and has the form of

$$F \rightarrow * \tag{2}$$

where F is the non-terminal symbol chosen to play the role if wildcard (every time it appears in the CYK table it means: put any value here) and $*$ mean that any real number is accepted here.

3.3 Regular Grammar Rules

These rules are identical to the ones used in the classic GCS. They are used in the CYK parsing process and the GA phase. They are in the form of

$$A \rightarrow BC \tag{3}$$

where A, B and C are non-terminal symbol.

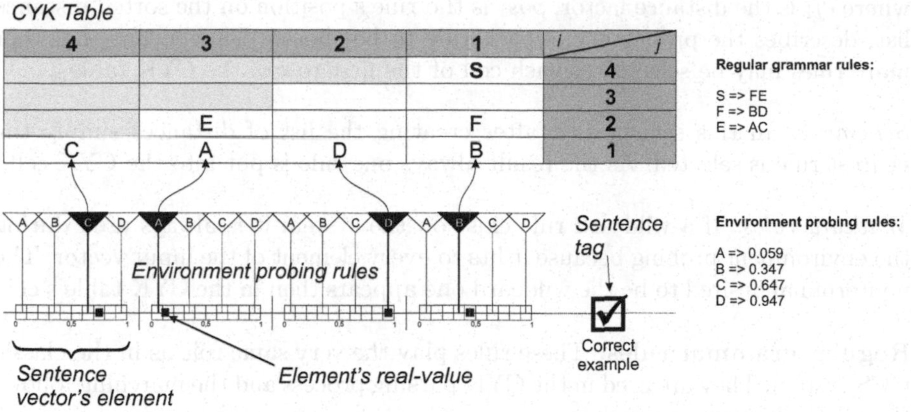

Fig. 2. The main idea of rGCS. Environment probing rules and their cooperation with CYK parsing procedure.

3.4 Generating the Rules

Every time the new experiment is started a set of random rules is generated. It contains the specified number of environment probing rules and the regular grammar rules. All non-terminals in the set are from the range limited by the system parameter. In the environment probing rules system keeps the equal number of various non-terminal symbols in the rules. Real number values may be from the range determined by the minimum and the maximum value of the input data to speed up the learning process but this is not necessary.

3.5 Matching Phase

Environment Probing Rules

Scheme 1. First a list of distances between the element of the input vector (real number value) and the each rule's real number value is created. Then all values from the list are scaled using the equation:

$$df_i = 1 - \left(\frac{dist_i}{maxdist} \right) \tag{4}$$

where df_i is the factor calculated for rule i (distance factor), $dist_i$ is the distance of rule i and maxdist - is the maximum distance value (distance value of the most distant rule).

The most distant rule receives factor 0 and rule that is located exactly at the input vector's value receives 1. In the next step the rules are sorted from the nearest one to the most distant one. Finally the equation:

$$p_i = \frac{df_i}{pos_i} \tag{5}$$

where df_i is the distance factor, pos_i is the rule's position on the sorted distance list, describes the probability of each rule to be chosen. That means that 0 or more rules may be selected to each cell of the first row of the CYK table.

Scheme 2. In this scheme just after creating the list of distances simply the nearest rule is selected. As the result always one rule is put into the CYK cell.

Wildcard rules. If a wildcard rule exist on the system it is always used during the environment probing because it fits to every element of the input vector. The nonterminal desired to be the wildcard one appears then in the CYK table's cell.

Regular grammar rules. These rules play the very same role as in the classic GCS system. They are used in the CYK parsing process and the matching follows the CYK algorithm procedure.

3.6 Adjusting Environment Probing Rules

As the environment probing rules match the input vector some data about the environment is collected. Every single rule of this kind keeps the copy of its real number factor. At the beginning of the each learning cycle it is set to the same value as the factor itself. Just after matching phase if the rule was used this copy is modified according to the equation:

$$vn_i = vc_i + wsp * g * ch \tag{6}$$

where vn_i is the copy of the factor value of the i-th classifier in the population, vc_i is the current factor's value of the i-th classifier in the population, wsp is the learning factor that is learning cycle dependant (see below), g is the neighborhood function dependant on the rule's position on the list sorted by the rule's distance from the environmental situation (see below), ch is the distance from the rule's factor to the environmental situation value, calculated according to the situation:

$$ch = ve - vc_i \tag{7}$$

where ve is the input vector element's value.

Learning factor is calculated according to the rule:

$$wsp = pMaxLearningRate * \left(\frac{pMinLearningRate}{pMaxLearningRate}\right)^{\left(\frac{cycle}{pCycles}\right)} \tag{8}$$

where $pMinLearningRate$ is the minimal learning factor value (parameter), $pMaxLearningRate$ is the maximal learning factor value (parameter), $cycle$ is the current learning cycle, $pCycles$ is the desired learning cycles values (parameter).

Neighborhood function is calculated according to the equation:

$$g = e^{\frac{-pos_i}{ps}} \tag{9}$$

where pos_i is the rule's position on the list sorted by the rule's distance from the environmental situation, ps is the neighborhood radius calculated according to the equation:

$$ps = pMaxNeigbourhoodRadius * \left(\frac{pMinNeigbourhoodRadius}{pMaxNeigbourhoodRadius}\right)^{\left(\frac{cycle}{pCycles}\right)} \quad (10)$$

where $pMinNeigbourhoodRadius$ is the minimal radius value (parameter) and $pMaxNeigbourhoodRadius$ is the maximal radius value (parameter).

It is important to work on the copy of the real number factor since we want the system to classify each example in the learning set using the same rules. This enables us to estimate correct competence of the current grammar evolved by the rGCS. As soon as the cycle finishes the copy of the factor replaces the old one moving the factor towards the values the rules accepts frequently. The change is more significant during the initial cycles of the learning process. As the induction goes on only small adjustments of the factors take place.

3.7 Evolving Regular Grammar Rules

Regular grammar rules are evolved just like in the classic GCS system during the evolutionary process. Genetic algorithm is then launched at the end of the learning cycle. Fitness evaluation uses the fertility measurement technique (see [7] for discussion), for the rules that were present in any complete parsing tree generated during the cycle:

$$f_i = FTrim + (tf_i * FertSig) \quad (11)$$

where f_i is a fitness measure of i-th classifier in the population, $FTrim$ is a fitness trim parameter - a base value given to unused classifier, tf_i is a pure fertility measure (see below) of i-th classifier in the population, $FertSig$ is a fertility significance parameter.

Pure fertility parameter is calculated according to the equation:

$$tf_i = \frac{FertPos_i - FertNeg_i}{FertPos_i + FertNeg_i} \quad (12)$$

where $FertPos_i$ is the number of positive fertility points of i-th classifier in the population, $FertNeg_i$ is the number of negative fertility points of i-th classifier in the population.

Rules that were unused in the complete parsing trees but still appeared in the CYK table take the following fitness measure:

$$f_i = FTrim + (tn_i * FSig) \quad (13)$$

where f_i is a fitness measure of i-th classifier in the population, $FTrim$ is a fitness trim parameter - a base value given to unused classifier, tn_i is a pure fitness measure (see below) of i-th classifier in the population, $FSig$ is a fitness significance parameter.

Pure fitness parameter is calculated according to the equation:

$$tn_i = \frac{PosPoints_i - NegPoints_i}{PosPoints_i + NegPoints_i} \tag{14}$$

where $PosPoints_i$ is the number of positive usage points of i-th classifier in the population, $NegPoints_i$ is the number of negative usage points of i-th classifier in the population.

It is important that the fitness of these rules is downgraded by the fitness significance parameter. Complete trees parsing rules are more valuable for the system as they cooperate with the others. Finally - unused rules take the constant fitness trim value (parameter - usually 0.5).

GA in the rGCS chooses parents using the roulette-wheel or random selection, then crossover and mutation operators are applied to the offspring with the probability given by the system parameters. The crowding technique is (discussed in) replaces rules in the population with the offspring.

4 The Checkerboard Problem

The checkerboard problem was proposed as a benchmark in [3]. It divides the n-dimensional space into hypercubes of two colors (i.e. black and white). Each hypercube has the same size and is surrounded by the others with alternate color. This means that for two dimensional space it looks like a chess or checkers board (so that's the source of the problem's name). There are two parameters describing the problem's complexity. First - mentioned above - is the number of space's dimension (n). Second one is the number of divisions of each dimension of space (n_d). In the following paragraphs we use the sets of checkerboard problem examples with $n = 3$ and $n_d = 3$. We evolve grammars telling whether the point in solution space of given coordinates is inside black or white hypercube. Single CNF grammar can only tell us if the given example is positive (belongs to the grammar's language) or not. We have to choose if we evolve the grammar related to white or black hypercubes. There are only two classes of hypercubes so example rejected by one class's grammar is assumed to be from the another. Every example in the set consist of three real numbers which are coordinates and the example's tag telling whether example is positive or negative - depending on the color of hypercubes we want the grammar to be evolved for. Example sets consisted of 100 examples - 50 positive and 50 negative ones.

Parameters seeking experiments presented below helped us to choose optimal parameters for the checkerboard problem. For $N = 3$ and $N_d = 3$ this include:

- Number of non-terminals = 6,
- Number of environment probing rules = 3,
- Number of regular grammar rules = 60,
- Crowding factor = 35,
- Crowding subpopulation = 5,
- Desired learning cycles = 50000.

rGCS system was able to evolve perfect grammar, accepting all positive and rejecting all negative sentences, at every single run after on average 24506 learning cycles (mean from 10 runs, min. 19898 cycles, max. 31893 cycles, see Fig. 3). Every run ended with 100% grammar competence factor.

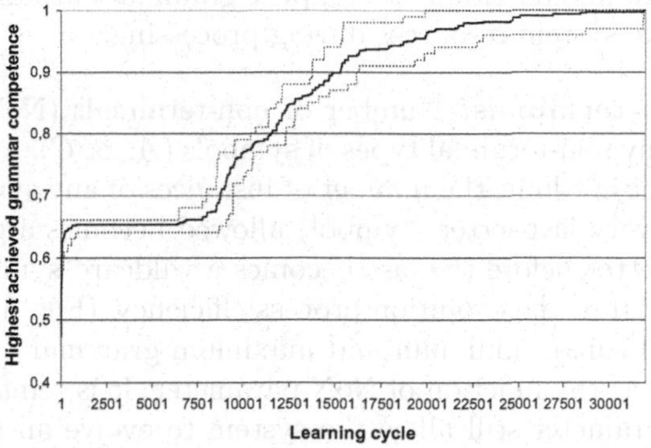

Fig. 3. Minimum, mean and maximum highest grammar competence achieved during the learning cycles

4.1 Parameter Seeking Experiments

rGCS tends to inherit a GCS property - it is quite sensitive to the system parameters settings. It looks like kind of disadvantage, however correct settings can results with rapid grammar evolution. Our previous work discussed many of these settings (see [5][7] for details). In this article we explain only rGCS-specific ones. In the next paragraphs we investigate the best parameters settings for checkerboard problem $N = 3$, $N_d = 3$. Single experiment tests a range of settings for examined parameter while the other (constant) parameters are set to the random values.

Number of environment probing rules. Some preliminary experiments confirmed our first guess that we should generate at least as many rules as the number of classes the input vectors' elements are divided into. That means that checkerboard problem requires at least N_d environment probing rules. (Fig. 4a) shows the mean (averaged over 10 runs), minimum and maximum grammar competence of the grammar evolved as the function of environment probing rules. Setting the number of environment probing rules to value lower than N_d (3 in this case) doesn't allow the system to evolve an efficient grammar. However higher values ($> Nd$) do not have any significant affect on the grammar evolution, moreover it slightly decreases the maximum competence of the grammar developed by the system.

Number of regular grammar rules. Number of regular grammar rules defines the size of the grammar evolved by the system. This parameter is used at the beginning of the learning process when random rules are generated. Since the rules evolved by GA replace the old ones, the number of regular grammar rules stays constant. (Fig. 4b) shows the mean (averaged over 10 runs), minimum and maximum grammar competence of the grammar evolved as the function of number of regular rules parameter. More regular rules enable the system to evolve highly efficient grammars. However complex grammars are difficult to analyze and consume more system resources during processing.

Number of non-terminals. Number of non-terminals (NoN) parameter determines how many non-terminal types of symbols (A, B, C...) are allowed in the grammar. This doesn't limit the number of instances of any symbol. In rGCS we assume that the very last letter (symbol) allowed becomes a grammar starting symbol and the letter before the last becomes a wildcard symbol. Actually this parameter is crucial to the evolution process efficiency. (Fig. 5) shows the mean (averaged over 10 runs), minimum and maximum grammar competence of the grammar evolved as the function of NoN parameter. It is remarkable that lower values of NoN parameter still allow the system to evolve an efficient grammar

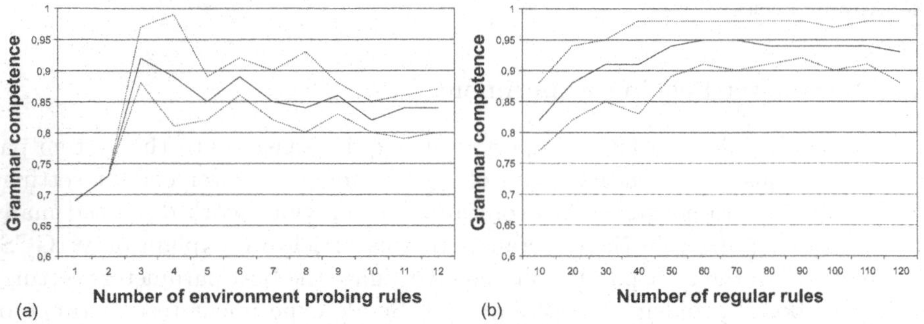

Fig. 4. Minimum, mean and maximum grammar competence as the function of the number of environment probing rules (a) and regular grammar rules (b)

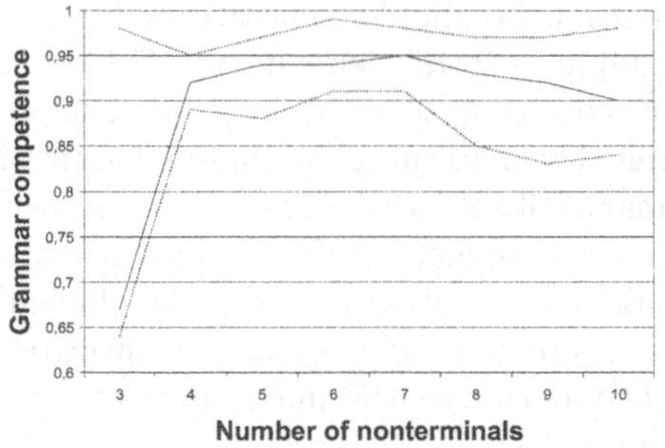

Fig. 5. Minimum, mean and maximum grammar competence as the function of the number of non-terminals

(high maximum competence values). On the other hand worse mean competence values prove that it is easier for the system to operate on larger grammars (6, 7 non-terminals for this problem).

5 Conclusions and Future Plans

We proved that it is possible to use a mutation of Grammar Classifier System to solve problems represented by vectors of real numbers. So a "real GCS" – rGCS – hybrid system was created, that allowed us to introduce grammar classification to a brand new set of problems. Our future plans include tuning the system, improving some mechanisms (i.e. new GA operators). We also plan to use rGCS on some noisy data sets.

References

1. E. Gold. Language identification in the limit. Information Control, 10, 447-474, 1967.
2. J. Holland. Adaptation. In Progress in theoretical biology, New York, 1976. Plenum.
3. C. Stone and L. Bull. For real! XCS with continuous-valued inputs. Evolutionary Computation, 11(3), 299-336, 2003.
4. O. Unold. Context-free grammar induction with grammar-based classifier system, Archives of Control Science, Vol. 15 (LI), 4, 681-690, 2005.
5. O. Unold. Playing a toy-grammar with GCS, In: Mira J., lvarez J.R. (eds.) IWINAC 2005, LNCS 3562, 300–309, 2005.
6. O. Unold and L. Cielecki. Grammar-based Classifier System, In: Hryniewicz O. at all (eds.) Issues in Intelligent Systems: Paradigms, EXIT, Warszawa, 273-286, 2005.
7. O. Unold and L. Cielecki. How to use crowding selection in Grammar-based Classifier System, In: Kwasnicka H., Paprzycki M. (eds.) Proc. of 5th International Conference on Intelligent Systems Design and Applications, Los Alamitos, IEEE Computer Society Press, 126–129, 2005.
8. D. Younger. Recognition and parsing of context-free languages in time n3. Technical report, University of Hawaii, Department of Computer Science, 1967.

A Study on Genetic Algorithms for the DARP Problem

Claudio Cubillos, Nibaldo Rodriguez, and Broderick Crawford

Pontificia Universidad Católica de Valparaíso, Escuela de Ingeniería Informática,
Av. Brasil 2241, Valparaíso, Chile
{claudio.cubillos,nibaldo.rodriguez,broderick.crawford}@ucv.cl

Abstract. This work presents the results on applying a genetic approach for solving the Dial-A-Ride Problem (DARP). The problem consists of assigning and scheduling a set of user transport requests to a fleet of available vehicles in the most efficient way according to a given objective function. The literature offers different heuristics for solving DARP, a well known NP-hard problem, which range from traditional insertion and clustering algorithms to soft computing techniques. On the other hand, the approach through Genetic Algorithms (GA) has been experienced in problems of combinatorial optimization. We present our experience and results of a study to develop and test different GAs in the aim of finding an appropriate encoding and configuration, specifically for the DARP problem with time windows.

1 Introduction

Genetic Algorithms (GA) have been successfully applied to diverse optimization problems in a wide range of domains. This well-known evolutionary computation technique inspires on nature and its evolutive mechanism to better adapt the different species to their environment. In nature, the most adapted organisms from the species have a higher expectancy of life and thus a higher possibility to reproduce and leave descendants before they die.

This simple mechanism is mimicked artificially through genetic algorithms as an effective optimization technique. By creating a specie (population of individuals) that "evolves" through selection-reproduction-mutation cycles allow us to search in the state space of the problem under optimization. This process lasts until the specie converges to certain characteristics, that is, a certain genomic sequence which represents a near-optimal solution for the problem.

This bio-inspired optimization technique has been applied to a variety of optimization problems including scenarios that require routing and scheduling, as the problem matter of this work.

The DARP problem consists in finding a minimal fleet of vehicles with limited capacity that must transport a set of clients from an initial pickup point to a final delivery location. Normally "Time Windows" constraints are added, which specify the time intervals within which each client must be picked-up and

J. Mira and J.R. Álvarez (Eds.): IWINAC 2007, Part I, LNCS 4527, pp. 498–507, 2007.

delivered, generating the DARPTW (or PDPTW) problem, which is known to be NP-hard.

Traditional approaches in Dial-a-Ride service planning are usually implemented as heuristic procedures that extend basic graph search algorithms, acting over large collections of data that describe the entities of the domain problem (vehicles, service requests, schedules). The most commercially used transport planning algorithms correspond to extensions of the solomon's heuristic for the VRPTW [16]. An example can be found in our past research [4], where we implemented a version of Jaw's Advanced Dial-A-Ride with Time Windows (ADARTW) algorithm [8].

Regarding GA-based solutions for DARP, it is hard to find works in this field and at the best of our knowledge, no relevant work could be found. Some works can be found on a similar but simpler problem, the VRP (Vehicle Routing Problem) from which DARP derivates. In these cases the proposed solutions make use of the GA together with other post-optimization techniques for improving the final solution or use the GA as a clusterization technique for assigning clients to vehicles but not for the scheduling of the trips. This can be explained because VRP (and hence DARP) corresponds to a deceptive problem, so traditional Bit-String based GAs do not perform well.

In Addition, DARPTW imposes further requirements of precedence (pickup precedes the delivery) and time-windows that make it harder to find feasible solutions in the problem's state space.

For all the above, the aim of this work is to provide some lights and experience on applying genetic algorithms to the dial-a-ride problem. We pursue the objective of developing a specific GA encoding for the DARPTW problem and obtain non deceptive results. In contrast to other research available in literature, our approach uses the genetic algorithm for the whole DARPTW problem, that is, the assignment of clients to vehicles and their scheduling, without post-optimization procedures.

To achieve our intentions diverse chromosome's encoding and evolutionary operators (selection, crossover and mutation) were implemented on a GA framework. Through it, diverse GAs with different configurations of genetic operators, chromosome's encoding and parameters were tested under diverse request scenarios.

2 The DARP Problem

The problem we are treating consists of a set C of geographically distributed transportation requests, coming from customers that should be served by a set of identical vehicles V.

The service can be defined as picking-up the client from the origin node ns_i, and conducting it to a destination node nd_i, where $\{ns_i, nd_i\} \in N$, with N the entire set of nodes that represent the network.

The service must be executed considering a time window for delivery constraint defined for each customer, expressed in terms of an earliest delivery time and a latest delivery time, the pair (edt_i, ldt_i) with $edt_i < ldt_i \forall i \in C$. In this way, a vehicle serving the customer i must reach nd_i, neither not before the edt_i time, nor after the ldt_i time.

Two functions, $DRT(N \times N) \in \Re$ and $MRT(N \times N) \in \Re$, define the direct ride time (optimistic time) and the maximum ride time (pessimistic time) required to reach the node nd_j from ns_i $\forall i \neq j$. Delivery times define a time window for pick-up, the pair (ept_i, lpt_i), where $ept_i = edt_i - MRT(ns_i, nd_i)$ and $lpt_i = edt_i - DRT(ns_i, nd_i)$ (see Figure 1).

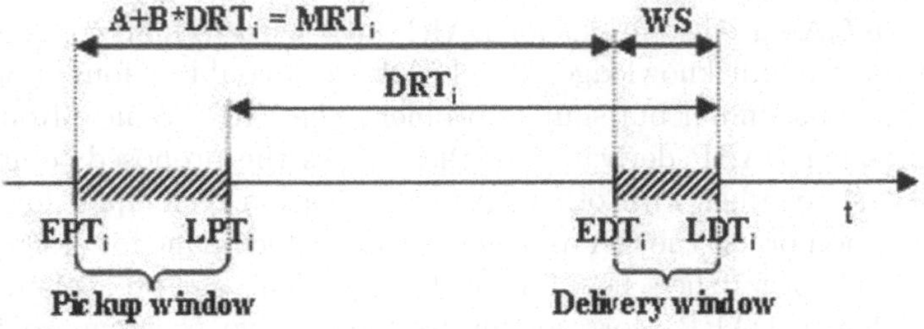

Fig. 1. Users Time Windows

In practical terms the customer is supposed to attend the vehicle at the pick-up point not after the time ept_i. Vehicles are not allowed to wait for a client, so they have to be scheduled to reach the point ns_i for serving the request i not before the time ept_i. On the other hand, the passenger has to be picked-up not after the time lpt_i otherwise his request (theoretically) will not be satisfied.

Service requests have to be assigned to Vehicles and scheduled according to the time restrictions. There exists no restriction about the minimum number of passengers to serve, but the maximum capacity of the vehicles must never be exceeded.

In our model we consider the possibility of a multi-depot scenario, that is, the vehicle i starts from the depot DS_i and after serving their last clients they turn back to the depot DF_i, where $\{DS_i, DF_i\} \in N$.

Finally, the objective function pursues the minimization of a disutility function that considers the weighted sum of performance measures coming from the fleet operator – number of vehicles required, total bus travel time and slack time – and from the served users – effective waiting time and excess ride time.

3 Related Work

Research in the field of passenger transport planning systems has received an increasing attention during last years, due to the congestion and contamination problems, and the number of accidents generated by an always increasing number of vehicles in our cities. As response, new alternatives to satisfy transport demands of citizens are being conceived [1].

In literature, the passenger transportation problem can be found under different names. It is a sub-type of the Travel Salesman Problem (TSP) and more specifically the Vehicle Routing Problem (VRP) and the Pickup and Delivery Problem (PDP) devoted to goods transport. Under the name of dial-a-ride problem (DARP) were developed the first passenger-transport planning systems, usually based on greedy insertion heuristics (see [8] [14]).

The VRP has been studied more deeply as it treats a simpler planning problem; assumes a central depot for all the vehicles, considers only the picking-up of goods (not the pickup & delivery), and therefore no disutility function exists for the transported entities. More recent research in the field tries to include newer techniques to improve the quality of the obtained solutions. In [11], Li presented a meta-heuristic for the PDPTW. A tabu-search can be found in [15], implemented for real-life problems including time-window constraints. Kohout and Erol [10] showed an agent-based implementation that uses stochastic improvement in the final solutions.

As mentioned before, under some domains traditional GAs do not perform well [6]. These are the so called deceptive problems in which the GA is not able to converge to a near optimal solution mainly due to a bad linking of the building-blocks. Much research can be found about this linking problem ([5][6][7]) and its possible solutions.

On the other hand, Genetic Algorithms (GA) have been successfully used as optimization technique in different research domains, including simple route optimization as in [3]. In the VRP field [2] reports an hybrid GA for multiple vehicles, that combines the genetic algorithm together with a greedy constructive heuristic. In [12] Maeda uses GA alone for the VRP problem but with time deadlines. Thangiah [17] uses a genetic approach for clustering initial routes. In [9] an hybrid approach uses GA together with dynamic programming and [13] combines GA with tabu-search.

4 GA Implementation

We have developed a GA framework for the assignment and scheduling of customers in the Dial-a-Ride problem. Such framework was implemented as a parameterized C++ program able to execute different runs with different treatments for problem parameters.

The GA tackles the whole DARPTW problem, so each individual of the population corresponds to an entire solution of the problem, that is, a set of vehicles and their routes that fulfill all customer requests. In our model the chromosomes evolve minimizing the fitness function in a continuous process that applies the genetic operators (selection, reproduction, and mutation) over the population. The process finishes after a pre-specified number of generations.

As mentioned earlier, in the GA framework were considered different kinds of genetic operators and chromosome's encoding. In particular the following ones were implemented.

Genotype. For the application of GA to a problem, is required to model a representation of the solution in a chromosome. As first encoding we choose a representation in which the chromosome is made up of a bus-passenger list, where each bus-passenger pair corresponds to a gene as shows the Figure 2.

A simple parser was provided to decode the chromosome, which consists in interpreting the first passenger's occurrence always as a pick-up, and the second one, always as a delivery. In this way the chromosome must have exactly two genes for each passenger, which determines the length of the chromosome.

Note that these 2 genes associated to a customer (pickup/delivery) can specify different buses. Therefore, the bus finally assigned to the customer is the one specified by the pick-up. The bus information in the delivery gene is used as a sort of recessive genetic material useful upon recombination.

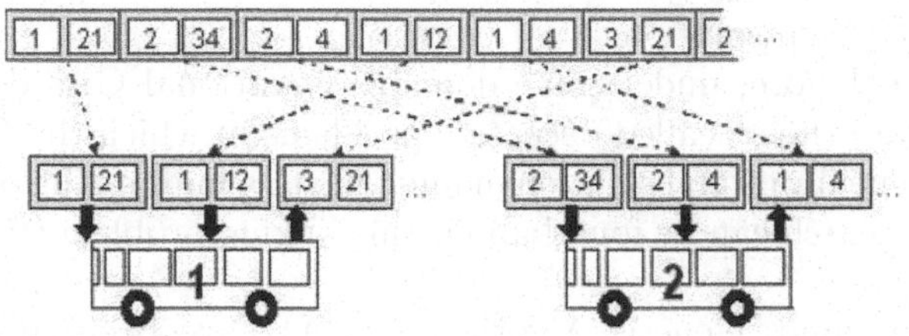

Fig. 2. The bus-passenger encoding scheme for the chromosome

Other encodings are considered to be implemented, like the locus-bus-passenger gene, that will allow putting the information about the relative position of the gene in the chromosome inside the gene itself, enabling the adoption of a linkage learning model.

Phenotype. In order to evaluate the "quality" of a chromosome and to decide its reproduction probability, a fitness function was defined. In the case of the DARP problem we adopted as fitness function the disutility function already defined by the end of Section 2.

Initial Population. The generation of an initial population that covers the whole problem's state-space is crucial for the final performance and convergence of a given Genetic algorithm. In this sense, we have implemented two alternative procedures for the population initialization: the first is the common random generation of the chromosome's genes at bit-level.

The second initialization procedure is based on a basic insertion heuristic for scheduling the passengers into the available vehicles. The heuristic picks one by one the clients from a list and tries to schedule them in the first possible vehicle. Different solutions (population's individuals) are generated by applying a random ordering to the list of clients and vehicles each time the insertion heuristic is to be used.

Selection. Firstly, a tournament selection has been used to choose the individuals(mates) for reproduction. It works by choosing a number of ts (tournament size) individuals from the population in a random way. Then the individual with the lowest fitness value is selected (the fitness function in our case is a cost function, so lower values are better than higher values). The number of tournaments applied in a generation depends on the number of individuals required to reproduce.

Crossover. The reproduction is the operation that produces a new individual pertaining to a new generation, due to the combination of the genetic material of its parents. For the framework we have already implemented the basic bit-level one-point crossover and a modified version of the Partial Match Crossover (PMX) which has been used in other scheduling problems with good results.

Mutation. The mutation allows an individual to slightly change the inherited genetic material. We have already implemented a bit-level mutation and the 2-opt operator.

5 Tests and Results

Firstly, we have implemented a traditional bit-level GA model with the following characteristics: bit-string chromosome using the bus-passenger gene, tournament selection, and bit-level population initialization, crossover and mutation. As expected, this GA model was unable to solve small problems (no feasible solution found) for 10 requests. This is because the DARP behaves in a deceptive way, specially with traditional bit-string GAs.

A second implementation changed the bit-string by an integer chromosome representation. The crossover and mutation operators together with the population initialization procedure were modified to operate over the integer representation. This GA behaved well for small request sizes, up to 25 requests. From 30 requests on, the GA was not able to arrive to any feasible solution. This happened because the used population initialization step (random integer generation) produced too much chromosomes with unfeasible solutions. This happened with population sizes of 50, 100 and 200 chromosomes.

A third GA implementation was done, which included an insertion heuristic for the population initialization procedure together with the PMX crossover and 2-opt operator. The introduction of an insertion heuristic for the population initialization procedure, allowed starting the GA with a population with some individuals having feasible solutions. This allowed us to use the GA with request sizes bigger than 50 but lower than 100 and to obtain better results.

With this last GA implementation different tests were carried out varying some of the parameters. At first view, the GA seems to work more efficiently and quickly with small sizes of requests (Under 100). For that reason we decided to use a small number of requests (5 scenarios of 25 trip requests). In particular we have performed an initial test of 640 runs, considering the combination of the following parameters:

- Scenarios: 5 demand scenarios, each with 25 trip requests distributed uniformly in a two hours horizon.
- Maximum number of available buses: 25
- Population size: 50 - 100 - 150 - 200 individuals.
- Number of generations: 1.000 and 11.000 generations
- Crossover probability: 0,7 - 0,8 - 0,9 - 1.0
- Mutation probability: 0,0015 and 0,0095
- Tournament size: 50% and 60% of the entire population.

Table 1 summarizes the best solutions obtained by the GA model in terms of the fitness value, and in terms of number of vehicles. A general conclusion from the analysis of this first study is that the ADARTW heuristic provides much better results than the implemented GA model, in terms of the fitness function value. However, in terms of the number of required vehicles the GA model has produced better results than the heuristic in 4 of the five scenarios with 25 trip requests.

Table 1. Comparison of best solutions obtained

Scenario	Genetic Algorithm						ADARTW			Notes
	Best solution in terms of minimum fitness value			Best solution in terms of minimum number of vehicles			Fitness	Vehicles	Time [sec]	
	Fitness	Vehicles	Time [sec]	Fitness	Vehicles	Time [sec]				
1	74.7	11	876	103.7	8	393	56.2	9	1	(*)
2	92.6	11	234	125.2	9	1220	90.0	11	1	(*)
3	75.7	9	1013	77.0	7	717	39.8	8	1	(*)
4	85.1	9	930	91.7	8	945	27.8	8	1	
5	88.0	9	461	92.5	8	681	42.1	9	1	(*)

*Better results in (minimum) number of vehicles given by the GA program.

In the following, an analysis on the most relevant GA parameters is presented through bar graphs and their results are analyzed and discussed.

The graph of Figure 3 (left) shows the tendency observed on the fitness function when the population size increases. Better results are given for bigger population sizes either with 1000 or 11000 generations. This is because a bigger population size allows examining a larger portion of the solutions space on each generation. It is also shown that a bigger number of generations allows for better fitness function results, independently from the considered population sizes.

The graph of Figure 3 (right) corroborates the importance of the population size by showing how steadily lowers the fitness mean when raising the population size. This phenomenon is independent from the crossover probability as it is appreciable on all its four scenarios, ranging from 70% to 100% of crossover.

The graphs of Figure 4 show the influence of the mutation probability in the value obtained for the fitness function. The left graphic shows that independently from the evaluated crossover probability values, the higher mutation probability

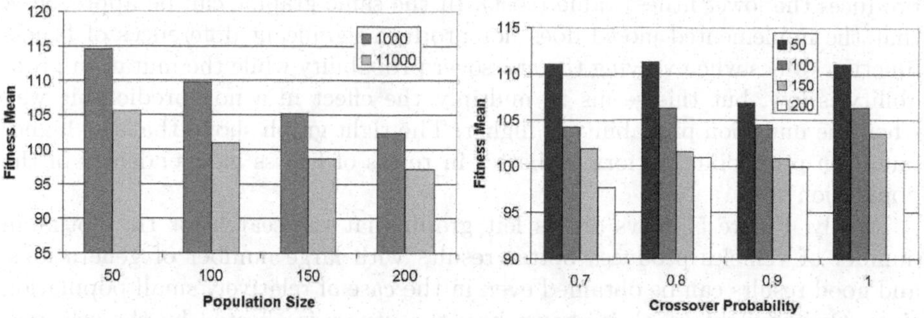

Fig. 3. (left) Fitness mean versus population size for 1000 and 11000 generation sizes. (right) Fitness mean versus crossover probability for population sizes of 50, 100, 150 and 200 individuals.

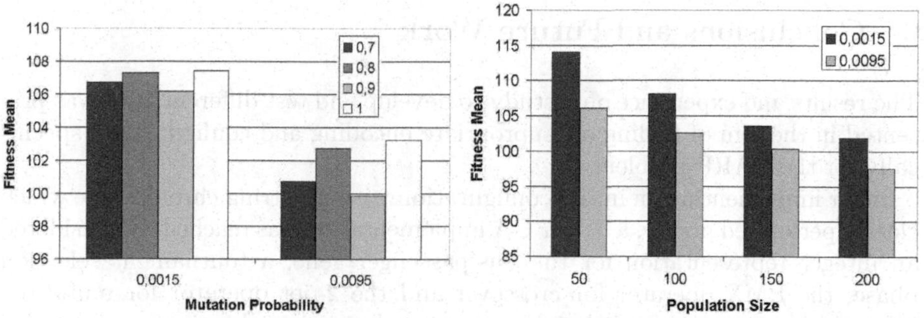

Fig. 4. (left) Fitness mean versus mutation probability for crossover probabilities 0,7 - 0,8 - 0,9 - 1,0.(right) Fitness mean versus population size for mutation probabilities of 0,0015 and 0,0095

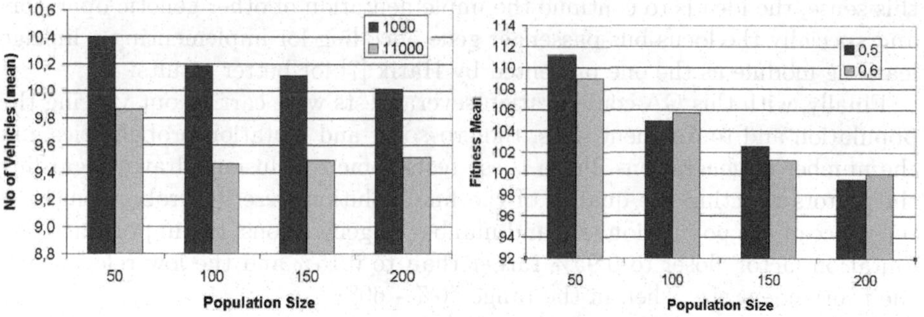

Fig. 5. (left) Number of vehicles versus population size for 1.000 and 11.000 generations. (right) Fitness mean versus population size for 50% and 60% of tournament size.

produces the lower fitness value (cost). In the same graphic can be appreciated that the implemented model does not produce significant differences of fitness function values when varying the crossover probability while the mutation probability is low, but this seems to multiply the effect in a non predictable way when the mutation probability is higher. The right graph shows that the higher mutation probability performed better in terms of fitness independently of the population size.

Finally, Figure 5 shows on its left graph that the search for the optimum number of vehicles produces better results with large number of generations, and good results can be obtained even in the case of relatively small population sizes. On the right graph is shown how the fitness is affected by the selection tournament sizes of 50% and 60%. In fact, from the graphic is possible to see that no trend exists and that one tournament size is not better not worse that the other in a consistent way varying the population size.

6 Conclusions and Future Work

The results and experience on a study to develop and test different GAs was presented in the aim of finding an appropriate encoding and configuration, specifically for the DARP problem.

After implementing an initial configuration with bit-string chromosome which clearly performed poorly, a better GA implementation was reached. It considered an integer representation for the bus-passenger gene, a tournament selection phase, the PMX operator for crossover and the 2-opt operator for mutation. Afterward an insertion heuristic for the population initialization was included resulting in even better behavior and results.

These solutions were comparable to the ones obtained using the ADARTW algorithm implementation [8] regarding the number of vehicles used but not with respect to the fitness value, aspect that can be improved in the future. In this sense, the idea is to continue the implementation of other genetic operators and specially the locus-bus-passenger gene encoding for implementing a linkage learning module as the one presented by Harik [7] for better results.

Finally, with this GA configuration several tests were carried out varying the population and tournament sizes, the crossover and mutation probabilities and the number of generations. From these tests some conclusions drawn regarding the factors affecting the quality of the final solution were: (corroborating) the relevance of the population size and number of generations, the importance of a mutation factor closer to 0.95% rather than to 0.15% and the low relevance of the tournament size when in the range 50% - 60%.

References

1. Ambrosino, G. et al: EBusiness Applications to Flexible Transport and Mobility Services. (2001). Available online at: http://citeseer.nj.nec.com/ambrosino01ebusiness.html.

2. Blanton, J; Wainwright, R.: Multiple Vehicle Routing with Time and Capacity constraints Using Genetic Algorithms. In Proc. of the 5th International Conference on Genetic Algorithms. **1** (1993) 452–459
3. Chang Wook, A.; Ramakrishna, R.S. A Genetic Algorithm For Shortest Path Routing Problem And The Sizing Of Populations. In IEEE Transactions on Evolutionary Computation, Vol. 6, No. 6, 2002, pp. 566–579.
4. Cubillos, C. et al. On user requirements and operator purposes in Dial-a-Ride services. Proceedings of the 9th Meeting of the EURO Working Group on Transportation. 2002, pp. 677–684.
5. Goldberg, D. et al. Toward a Better understanding of Mixing in Genetic Algorithms. J. Soc. Instrument and Control Engineers. Vol. 32, No 1, pp. 10–16, 1993.
6. Harik, G. Learning Gene linkage to Efficiently solve Problems of Bounded Difficulty Using Genetic Algorithms. PhD Thesis. University of Michigan. 1997.
7. Harik, G; Goldberg, D.E. Learning Linkage. In foundations of Genetic Algorithms. Vol. 4, 1997, pp. 247–262.
8. Jaw, J. et al. A heuristic algorithm for the multiple-vehicle advance request dial-a-ride problem with time windows. Transportation Research. Vol. 20B, No 3, 1986, pp. 243–257.
9. Jih, W; Hsu, J. Dynamic Vehicle Routing using Hybrid Genetic Algorithms. In Proc. Of the IEEE Int. Conf. on robotics & Automation. Detroit, May, 1999.
10. Kohout, R; Erol, K. Robert C. In-Time Agent-Based Vehicle Routing with a Stochastic Improvement Heuristic. In Proc. Of the AAAI/IAAI Int. Conf. Orlando, Florida, 1999, pp. 864–869.
11. Li, H; Lim, A. A Metaheuristic for the Pickup and Delivery problem with Time Windows. In 13th IEEE International Conference on Tools with Artificial Intelligence (ICTAI'01). Texas, November, 2001.
12. Maeda, O. et al. A Genetic Algorithm Approach to vehicle Routing Problem with Time Deadlines in Geographical Information Systems. In IEEE International Conference on Systems, Man, and Cybernetics. Vol II, Tokyo, Oct. 1999, pp. 595–600.
13. Ombuki, O. et al. A Hybrid Search Based on Genetic Algorithms and Tabu Search for Vehicle Routing. In the 6th IASTED International Conference on Artificial Intelligence and Soft Computing. Banff, Canada, July 2002, pp. 176–181.
14. Psaraftis, N.H. Dynamic vehicle routing: Status and prospects. Annals of Operation Research. No 61, 1995, pp. 143–164.
15. Rochat, Y; Semet, F. A Tabu Search Approach for delivering Pet food and flour in Switzerland. Journal of Operation Research Society. 1:1233–1246, 1994.
16. Solomon, M. Algorithms For The Vehicle Routing And Scheduling Problems With Time Window Constraints. Operations Research, No. 35. 1987, pp. 254–265.
17. Thangiah, S. Vehicle Routing with Time Windows using Genetic Algorithms. Application Handbook of Genetic Algorithm: New Frontier, 2:253–277, 1995.

Optimization of the Compression Parameters of a Phonocardiographic Telediagnosis System Using Genetic Algorithms

Juan Martínez-Alajarín, Javier Garrigós-Guerrero, and Ramón Ruiz-Merino

Universidad Politécnica de Cartagena, Cartagena 30202, Spain
juanc.martinez@upct.es

Abstract. Auscultation of the heart is a medical technique that still today is used to provide a fast diagnosis about the heart condition. Compression of the heart sounds (or phonocardiogram) is very convenient to reduce bandwidth in telediagnosis systems that aid the physician in the evaluation of the cardiovascular state. The compression algorithm used depends on several parameters, which can take a diversity of values. Genetic algorithms have been used to obtain the optimal set of values for the compression parameters that optimize the performance of the compression for a test set of cardiac recordings. The optimized results were obtained very quickly, and optimal values agreed for (almost) all the test records.

1 Introduction

Many efforts are nowadays directed towards the development of systems for cardiac diagnosis or monitoring using the recording of heart sounds or phonocardiograms (PCG). To improve bandwidth and storage efficiency, the heart sounds audio signal can be transmitted compressed using a lossy specific method based in the wavelet transform and in additional methods to decrease the size of the resulting signal. ASEPTIC (Aided System for Event-based Phonocardiographic Telediagnosis with Integrated Compression) [1] is a system that analyzes remote PCG recordings and provides the diagnosis of the cardiovascular state. The input signal to ASEPTIC is usually a compressed PCG record, in order to reduce bandwidth. Thus, a compression/decompression stage is included in ASEPTIC, together with the main analysis stage.

The PCG compression algorithm [2] depends on a number of parameters, that have great influence on the performance of the compression. These parameters can be defined over a wide range of values, so deciding the optimal set of the parameters values is not an easy task. Brute force assures to find the optimal solution, but its high computational time to explore permutations of many parameters and values makes it not a practical solution.

In this paper we propose to use genetic algorithms (GA) to find the optimal set of values for the compression parameters of PCG. This technique has been already applied to compression successfully to optimize the compression parameters of bi-level images [3], the length of the block for image compression [4],

J. Mira and J.R. Álvarez (Eds.): IWINAC 2007, Part I, LNCS 4527, pp. 508–517, 2007.

or to analyze sound files to determine the chunks that are most likely to contain irrelevant signals in audio compression [5]. GA performs an stochastic random search over the entire space of search, which provides the optimal solution with less computation time than other approaches.

The paper is organized as follows: section 2 provides a brief description of the PCG compression algorithm and parameters. Section 3 describes the procedure for GA optimization adapted to PCG compression, and section 4 presents the results obtained with this technique. Finally, section 5 remarks conclusions obtained.

2 PCG Compression

In this section, a brief description of the PCG compression algorithm and the parameters that take part is presented, together with the expressions used to measure the performance of the compression.

2.1 The PCG Compression Algorithm

The compression/decompression stage in ASEPTIC [1] is based in a specific algorithm for PCG lossy compression [2] (Figure 1). Previously to the compression, the PCG signal is divided in blocks of Lb samples (the last block is zero-padded if needed), and then each block (or compression window) is compressed independently of the others in two basic steps:

1. wavelet compression, which consists of:
   - decomposition of PCG signal using the wavelet transform (WT) or the wavelet packets transform (WPT),
   - to assign value 0 to those wavelet coefficients that are below a threshold, and
   - to remove 0-coefficients from the wavelet coefficient vector, and the last block of 0's from the significance map vector (binary string),
2. additional compression, formed by:
   - compression of the wavelet coefficient vector using lineal quantization, and
   - compression of the significance map using Run-Length-Encoding (RLE) and Huffman coding.

2.2 Performance of the Compression

The performance of lossy compression methods is usually measured with two parameters, considered jointly [6]: the compression rate and the distortion error.

The compression rate (CR) of a compressed signal $\mathbf{C}$ with respect to the original signal $\mathbf{X}$ is defined as:

$$CR = \frac{L_{\mathbf{X}}}{L_{\mathbf{C}}} = \frac{N_B \cdot N}{L_{\mathbf{C}}} \tag{1}$$

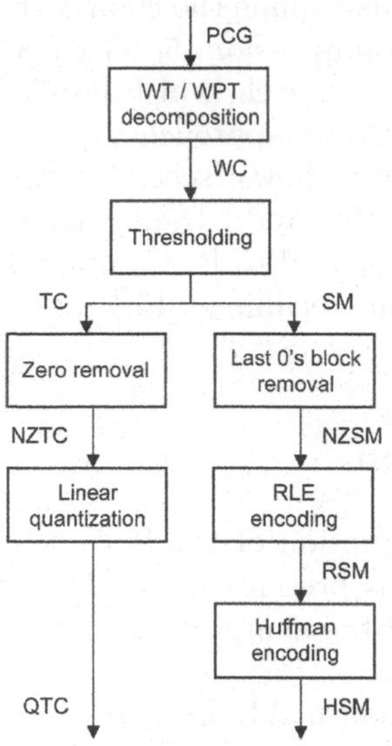

Fig. 1. Block diagram of the PCG compression algorithm

where $L_{(\cdot)}$ is the length in bits of the considered signal, N_B is the number of bits used to represent each sample of $\mathbf{X}$, and N is the number of samples of $\mathbf{X}$.

On the other side, a signal that is compressed and decompressed using lossy techniques is not exactly equal to the original signal. This difference (distortion error) is usually measured by the percent root-mean-square difference (PRD), modified to be independent of the offset of the signal [7], defined as :

$$PRD = \sqrt{\frac{\sum_{i=1}^{N}(x_i - \hat{x}_i)^2}{\sum_{i=1}^{N}(x_i - \mu_x)^2}} \times 100 \tag{2}$$

where μ_x is the mean value of $\mathbf{X}$, and x_i and $\hat{x}_i$ are samples of the original signal and the decompressed (reconstructed) signal, respectively.

Since CR and PRD are related (as PRD increases, CR increases, and viceversa), it is necessary to found a trade-off between them, according to the performance of the compression algorithm and the needs of the application.

2.3 Compression Parameters

The algorithm used to compress the PCG signal depends on the following parameters:

- type of wavelet transform (WT o WPT), Tt,
- wavelet mother function or wavelet filter, wmf,
- wavelet decomposition level or deep, M,

– number of bits used to quantize the wavelet coefficients, N_b,
– number of samples of each block (compression window) in which the original PCG signal is divided, Lb,
– compression method, Cm, that reflects the combination between the quantized thresholded wavelet coefficients (**QTC**) and the compressed signals of the significance map (**NZSM**, **RSM** and **HSM**): **QTC+NZSM** (1), **QTC+RSM** (2), and **QTC+HSM** (3).

It is also necessary to set the maximum value of the coefficients that will be zeroed in the thresholding. The threshold thr can be set manually by the user, although the lack of a direct relation between its value and the losses of the compression due to thresholding makes manually setting not suitable. Another possibility is to set the minimum percentage fraction of the coefficients energy of the original signal that must be retained after thresholding ($RtEn$). Finally, it is also possible to set one of the parameters used to measure the performance of the compression, CR or PRD, as the target value.

For our application, the main objective for the transmission and storing of PCG signals is to reduce the size of the recordings as much as possible but assuring a minimum quality level (or a maximum error value) that guarantees that those signals are still suitable to diagnose the cardiovascular condition after decompression.

The algorithm described in [8] allows achieving compressions with a specific PRD error, by adjusting iteratively the initial threshold that decides which wavelet coefficients will be zeroed. For each compression window Lb_i, the initial threshold is set as , $thr_i^0 = 2 \cdot C_i^{\max}$, where $C_i^{\max}$ is the maximum coefficient for the compression window Lb_i. Compression is executed using thr_i^0, and the resulting PRD is computed. If PRD is in the range $[0.95 \cdot PRD_{\text{target}}, 1.05 \cdot PRD_{\text{target}}]$, compression ends; if not, thr_i is adjusted iteratively as described in [8] until PRD reaches the $\pm 5\%$ tolerance band centered at PRD_{target}. After a maximum number of iterations (25), if PRD_{target} has not been reached, the threshold that provides the best result until that moment is chosen.

Although that iterative algorithm has been proposed for ECG compression, its use for PCG compression is straightforward. For our case, that algorithm has been adapted so not only PRD, but also CR, can be used as target values. Another modification made to the original algorithm concerns the method to set the initial value of the threshold for each compression window. In [8], it has been said that the double of the maximum coefficient is used to set thr_i^0. However, thr_i often takes similar values for consecutive compression windows for a target value of PRD or CR. This can be exploited to reduce the number of iterations. In our case, from the second to the last compression window, the initial threshold is computed as the double of the final threshold for the previous window: $thr_i^0 = 2 \cdot thr_{i-1}$. Thus, the threshold for the target value is computed with fewer iterations, so the computation time is reduced. If the threshold does not decrease for 5 iterations, then its value is reset to $2 \cdot C_i^{\max}$, and the search for the correct threshold begins from the start.

3 Optimization of Compression Parameters

Due to the large number of parameters that take part in the compression of PCG, and to the diversity of values that each of them can achieve, a systematic approach based in permutations of the different parameter values to establish the optimal set (brute force) is not efficient. Instead, an approach based in optimization of a cost function using GA has been used.

GA [9] are numeric optimization algorithms based in randomization and in concepts like natural selection and evolution. GA has proven to be very robust and effective for optimization problems too complex for the traditional optimization techniques. The advantage is less computational time than for a systematic search of the optimal set of parameters in the full search space defined by the ranges of the different compression parameters.

The block diagram of the GA used for the determination of the optimal parameters for PCG compression is shown in Figure 2. Previously, each of the parameters to be optimized has been encoded as a gene. A chain of genes is called a chromosome, and it contains the set of parameters to be optimized. Each individual of the population is characterized by its chromosome (genotype). The GA performs the following tasks:

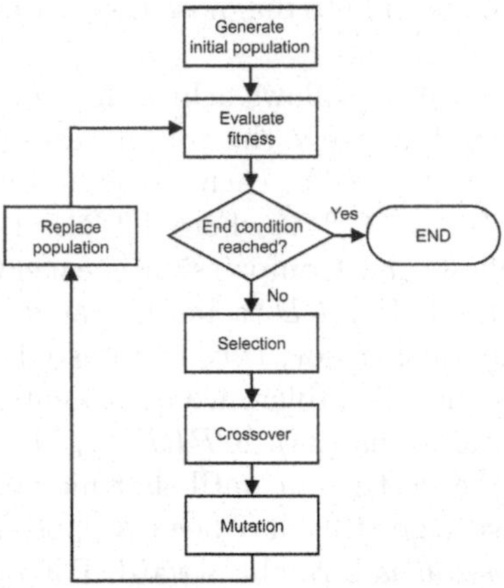

Fig. 2. Optimization of parameters for PCG compression using genetic algorithms

1. A new population is generated assigning random values for the chromosomes of each individual.
2. Each individual in the initial population is evaluated for its *fitness* or goodness in solving the problem.
3. If the stop criterium is reached (at least one individual in the present population gets the target fitness, or the maximum number of iterations is reached), the GA ends; if not, it continues with step 4.
4. Individuals are selected for reproduction for the next generation according to their fitness.

5. Crossover and mutation are applied to the selected individuals (*parents*) in order to create a new population (*children*).
6. The present population is replaced with the new chromosomes, and it continues with step 2.

The GA was executed for 50 iterations, and with a population of $N = 50$ individuals. Each chromosome consisted of 5 genes (each gene represents a compression parameter), and each gene had a different length in bits, according to the range it can take. Table 1 shows the genes, their possible values, and the number of bits used for each gene.

The fitness expression has been defined in function of CR, PRD and TPS (time need to compress 1 second of PCG signal):

$$fitness = \left(\frac{1}{1 + K_1 \cdot PRD}\right)^{\left(\frac{1}{1+K_2 \cdot CR}\right)} \cdot \left(\frac{1}{1 + K_3 \cdot TPS}\right)^{K_4} \tag{3}$$

In this expression, the first exponential term takes values in the range [0,1], approaching to 1 for high values of CR and low values of PRD (ideal situation). The second exponential term reflects the influence of the compression time for 1 second of signal. One of the parameters that have greater influence in this term is the wavelet filter. There are several filters that achieve similar values of CR and PRD, although the higher order filters usually takes more computation time. The second term gives priority to those filters that for similar values of CR and PRD have lower order and so, use less computational time. Constants K_1, K_2, K_3 and K_4 are used to modify the relative importance of the parameters. Their values were assigned experimentally to the following: $K_1 = K_3 = 5$, $K_2 = 0.2$, and $K_4 = 0.1$.

Table 1. Genes of the chromosome, possible values for each gene, and number of bits used for each one

Gene	Values	Num. of bits
Transform	WT, WPT	1
Wavelet filter	Db1–16, Sym4–16, Coif2–4	5
Decomposition level	2–9	3
Num. of quantization bits	1–16	4
$\log_2$ of the compression window length	9–12	2

The tournament selection algorithm [10] has been used to select the individuals. In it, a subset of n individuals (with $2 \leq n \leq 6$) are chosen among the population randomly. These individuals compete, and that with the highest fitness wins the tournament and is selected for reproduction. This process is repeated N times until all the parents of the new population are determined. In this algorithm, the probability that the winner of the tournament is the individual with the highest fitness is P_t. In our case, the values used for n and P_t have been 5 and 1, respectively.

The recombination of the selected individuals is done using the crossover and mutation operators. Their probabilities have been $P_c = 0.7$ and $P_m = 0.05$, respectively. The crossover algorithm uses two points, so each individual is divided in three parts using two crossing points in bits 4 and 7. Each pair of parents generates a pair of children exactly like the parents if $P_{\text{parents}} \geq P_c$ (with the probability of crossing the parents, P_{parents}, determined randomly). However, if $P_{\text{parents}} < P_c$, the chromosome of the first child will be formed by the parts 1 and 3 of the first parent and the part 2 of the second parent, whereas the second child will be formed by the parts 1 and 3 of the second parent and the part 2 of the first parent.

The mutation operator explores areas of the space of search that are not included in the present population. All the bits of each individual remains without changes for the next generation if $P_{\text{bit}} > P_m$; in other case, they are changed. The probability of mutation of each bit, P_{bit}, is determined randomly.

4 Experimental Results

To obtain the results, two data sets have been used: set 1 is formed by 6 PCG records, and set 2 consists of 50 real PCG records from [11], which includes those of set 1. Set 1 includes 1 normal record, 1 prosthetic valve record, and 4 murmur records of different pathologies, and it has been used for the optimization of the compression parameters with GA. Set 2 includes a variety of records of many pathologies, and it was used to determine the compression method, Cm. All these records were stored in WAV format, with 8000 Hz sampling frequency, 1 channel, and 16 bits.

The optimization of the compression parameters with GA applied to the first 5 seconds of the six records of set 1 to give generality to the results. The target value was to retain the $RtEn = 99.9\%$ of the energy of the wavelet coefficients of the original record. Results obtained are detailed in Table 2. This table shows that for the six records, compression was optimized using WT with a compression window length of $2^{12} = 4096$ samples (approximately 0.5 seconds), which corresponds to the maximum compression window length allowed. The number of quantization bits and the level of decomposition were 7 and 4 for five of the six records; for the remaining case, their values were 6 and 3, very similar to the former. As for the wavelet filter, there were 3 records with Db9 as optimal value and 3 records with Db10.

The best compression method was determined using the records of set 2. The first 5 seconds of each record of set 2 were compressed with a target value of $PRD = 3\%$ and with the parameters obtained from the previous GA optimization: $M = 4$, $\log_2(Lb) = 12$, $N_b = 7$ and $Tt = \text{WT}$. Table 3 shows the three compression methods evaluated and the number of records for which the highest CR was obtained (for the same PRD error), together with the mean CR and its standard deviation obtained for the compression of the 50 records with each of the three compression methods. The result of this analysis showed that the best compression method is 3, which combines the quantized signal from the wavelet coefficients (**QTC**) with the significance map compressed with Huffman coding (**HSM**).

Table 2. PCG compression parameters obtained for each record for the optimization with GA using (3) as fitness function

Record	Transform	Wavelet filter	Level of decomposition	Num. of quantization bits	$\log_2$ of compression window length
dm009	WT	Db9	4	7	12
dm037	WT	Db10	4	7	12
dm084	WT	Db9	4	7	12
dm086	WT	Db10	4	7	12
dm113	WT	Db9	4	7	12
dm127	WT	Db10	3	6	12

Table 3. Number of records for which the indicated compression method obtained the highest CR (for the same PRD error). Mean CR and standard deviation are also shown for the compression of the first 5 seconds of the 50 records of set 2 with each of the three compression methods.

Compression method	Number of records	Mean CR
QTC+NZSM (1)	2	9.1806 ± 7.3160
QTC+RSM (2)	10	9.5085 ± 7.2191
QTC+HSM (3)	38	9.6673 ± 7.3894

Table 4. Number of records for which the indicated wavelet filter achieved the highest CR (for the same PRD error). Mean CR and standard deviation are also shown for the compression of the first 5 seconds of the 50 records of set 2 with each of the two wavelet filters.

Wavelet filter	Number of records	Mean CR
Daubechies 9 (Db9)	25	9,1632 ± 7,0399
Daubechies 10 (Db10)	25	9,6673 ± 7,3894

To decide which wavelet filter (Db9 or Db10) provided the best results, set 2 was used again. Compression was made with the best parameter values achieved. Table 4 shows the number of records for which the indicated wavelet filter achieved the highest CR (for the same PRD error), together with the mean CR and standard deviation. Again, the two filters obtained the same number of records for the best results. Finally, Db10 was selected, since its CR was slightly higher than for Db9.

Finally, Table 5 sums up the optimal set of values achieved for the different compression parameters using (3).

4.1 Computation Time

The main reason to use GA to obtain the optimal compression values is the high computation time that a systematic exploration of the full space of search

Table 5. Optimal values obtained for the compression parameters using genetic algorithms and the cost function defined by (3)

Parameter	Value
Transform (Tt)	WT
Wavelet filter (wmf)	Db10
Decomposition level (M)	4
Num. of quantization bits for the wavelet coefficients (N_b)	7
$\log_2$ of the length of the compression window ($\log_2(Lb)$)	12 (4096 samples)
Compression method (Cm)	3 (**QTC+HSM**)

(defined by the parameters and values in Table 1) would require. For example, for the record *dm009*, it has been roughly estimated that the systematic determination of the optimal set of values using a target of $RtEn = 99.99\%$ would require 21 days (in a PC laptop with Intel Pentium IV 2.8 GHz). On the other side, the 50 iterations of the GA only took 2 hours and 37 minutes, which is the 0.52% of the time needed for the systematic exploration. The GA reached the optimal values at the iterations 23, 14, 27, 4, 35 and 33, for records *dm009*, *dm037*, *dm084*, *dm086*, *dm113*, and *dm127*, respectively.

Although the systematic search could be quicker if a previous rough analysis was done to determine the optimal region, and then a thorough analysis was perform considering only a restricted subset of all the possible values of the compression parameters, the computation time needed would be still too high in comparison to what the GA needed to find the optimal solution.

The excessive computation time needed for the systematic exploration is due to 1) the high number of evaluations needed for the compression, and 2) the high computation time needed for some parameter values, like high values of M with the WPT, low values of $\log_2(Lb)$, and some of the wavelet filter values (Db1 and Sym12–16).

5 Conclusions

In this paper, an optimization procedure for parameter values of PCG compression has been presented. The optimal set of values obtained is that which performs best in average for a set of representative test records, and it has been obtained in a very reduced computational time, with respect to a full exploration of the search space. The optimization criterium (fitness function) has considered three parameters: compression rate, PRD error, and the time elapsed in compressing 1 second of signal. This fitness function provides a reasonable criterium for optimization, but it could be changed depending on the interest of the application.

The analysis of the optimal values suggests that the length of the compression block should be the maximum allowed, although this can have practical limits, depending on the platform of the implementation of the compression algorithm (especially in hardware). Finally, it could be also possible to define several profiles, or optimal sets of values, according on the properties of the signal (ambient noise, normal or pathological records, etc.).

Acknowledgments

This work has been supported by Ministerio de Ciencia y Tecnología of Spain under grants TIC2003-09400-C04-02 and TIN2006-15460-C04-04.

References

1. Martínez-Alajarín, J., López-Candel, J., Ruiz-Merino, R.: ASEPTIC: Aided system for event-based phonocardiographic telediagnosis with integrated compression. Computers in Cardiology **33** (2006) 537–540
2. Martínez-Alajarín, J., Ruiz-Merino, R.: Wavelet and wavelet packet compression of phonocardiograms. Electronics Letters **40** (2004) 1040–1041
3. Sakanashi, H., Iwata, M., Higuchi, T.: Evolvable hardware for lossless compression of very high resolution bi-level images. IEE Proceedings–Computers and Digital Techniques **151** (2004) 277–286
4. Sakihama, H., Nakao, Z., Ali, F.E.A.F., Chen, Y.W.: Evolutionary block design for video image compression. In: IEEE SMC '99 Conference Proceedings. Volume 4., Tokyo (1999) 928–930
5. Chen, H., Yu, T.L.: Comparison of psychoacoustic principles and genetic algorithms in audio compression. In: 18th International Conference on Systems Engineering. (2005) 270–275
6. Blanco-Velasco, M., Cruz-Roldán, F., Godino-Llorente, J.I., Blanco-Velasco, J., Armiens-Aparicio, C., López-Ferreras, F.: On the use of PRD and CR parameters for ECG compression. Medical Engineering & Physics **27** (2005) 798–802
7. Alshamali, A., Al-Fahoum, A.S.: Comments on "An efficient coding algorithm for the compression of ECG signals using the wavelet transform". IEEE Transactions on Biomedical Engineering **50** (2003) 1034–1037
8. Blanco-Velasco, M., Cruz-Roldán, F., Godino-Llorente, J.I., Barner, K.E.: ECG compression with retrieved quality guaranteed. Electronics Letters **40** (2004) 1466–1467
9. Coley, D.A.: An Introduction to genetic algorithms for scientists and engineers. World Sci., New York (1999)
10. Goldberg, D.E., Deb, K.: A comparative analysis of selection schemes used in genetic algorithms. In: Foundations of Genetic Algorithms. Morgan Kaufman (1991) 69–93
11. Mason, D.: Listening to the heart. F. A. Davis Co. (2000)

An Integrated Resolution of Joint Production and Maintenance Scheduling Problem in Hybrid Flowshop

Fatima Benbouzid-Sitayeb, Mourad Tirchi, and Abid Mahloul

Institut National de formation en Informatique, BP 68M, 16270, Oued Smar, Algeria
f_sitayeb@ini.dz, m_tirchi@ini.dz, a_mahloul@ini.dz
http://www.ini.dz

Abstract. The article presents an integrated resolution of the joint production and maintenance scheduling problem in hybrid flowshop. Two resolution methods are used on the basis of a new coding to represent a joint production and maintenance scheduling: Taboo search where we proposed an algorithm for the generation of a joint initial solution and neighbourhood, and GA where we proposed new joint operators for crossover and mutation. Computational experiments are conducted on a large set of instances and the resulting genetic algorithm gives the best results so far.

Keywords: Maintenance, Production, Joint scheduling, Genetic Algorithms, Taboo Search, Hybrid Flowshop.

1 Introduction

Maintenance and production are two functions, which act on the same resources. However the scheduling of their respective activities is independent, and does not take into account this constraint. The interaction between maintenance and production, particularly their joint scheduling, is relatively little studied and rather recent in the literature. One account in the literature two joint scheduling strategies which aim is to solve conflicts between production and maintenance [1]. The sequential one consists of two steps: First the scheduling of the production jobs then the insertion of the maintenance tasks, taking the production scheduling as a strong constraint. The integrated one consists of a scheduling of the production and maintenance tasks at the same time, on the basis of a joint representation of production and maintenance data. The joint production and maintenance scheduling is a complex problem because of the scheduling of two different activities: production and maintenance. The problem as defined is a multicriteria none. On one hand, one schedules the production respecting deadlines, cost and products quality. On the other hand, one plans maintenance under the constraints of equipment reliability which ensures the perenniality of the production equipment. The uncertainties related in particular to the data and target makes us think that an approach by exact methods is not possible. Therefore we propose to use heuristic methods which solve, even partially, this type of problems.

J. Mira and J.R. Álvarez (Eds.): IWINAC 2007, Part I, LNCS 4527, pp. 518–527, 2007.

In this paper, we investigate an integrated resolution of the joint production and maintenance scheduling problem in hybrid flowshop. Thus, the resolution of the joint production and maintenance scheduling will be done by Taboo search (TS) and GA. The aim being to optimize a common objective function which takes into account the criteria of maintenance and production into same. In this paper, we study the general hybrid flowshop scheduling problem with parallel and identical machines at all stages. A set of n jobs J = 1, 2, . . . , n has to be sequenced in a flowshop environment with k stages. For each stage i a set Mi = 1, 2, . . . ,mi of identical machine is considered. A job consists in a sequence of k tasks, one task denoted by Tij for each stage. Each task within a job requires one machine. The processing time of task Tij will be denoted by pij . it has been proved that the problem is NP-complete even if there are two stages and there is only one machine at one of the two stages [2], [3]; also, heuristics have been proposed in the case that there is one machine at stage one and parallel machines at stage two [4]. The maintenance used is a systematic preventive one [Barlow60]. The tasks are periodic interventions occurring every T*j periods (T*ij indicates the ideal maintenance period of task i on the machine j). Each preventive maintenance task is characterized by a range of maintenance tasks pre-established by the maintenance department or the manufacturer of the considered equipment. It consists of a succession of elementary operations which duration p'ij is evaluated with more or less certainty. Moreover, the periodicity T* of these tasks is authorized to vary in a tolerance interval noted [Tminij,Tmaxij]. This interval represents a compromise between the maintenance cost and the machine unavailability risk. The respect of the maintenance periods influences the constraints of the production system. The choice of systematic preventive maintenance, with specificities such as the periodicity of the maintenance tasks, for this study, is a consequence of its planned aspect which makes it the most adapted for the maintenance scheduling. The rest of the paper is organized as follows. Section 2 is devoted to the resolution methods proposed to solve the joint preventive maintenance and production scheduling in a hybrid flowshop. We use two heuristics on the basis of a new coding to represent a joint production and maintenance scheduling: Taboo search where we proposed a new technique for the generation of neighbourhood, and GA where we proposed new joint operators for crossover and mutation. The goal is to optimize a global objective function which takes into account maintenance and production criterion at the same time. The third section present the various results obtained for the different methods. Finally, we will conclude with some perspectives and extensions for this work.

2 Resolution Methods

The integrated strategy aims to optimize a global objective function which takes into account production and maintenance data. The resolution by the integrated strategy is based on a joint representation of production and maintenance data before the resolution itself. We will, first, present in this section the proposed representation of a joint production and maintenance scheduling and the global objective function to optimise. Then for each heuristic (Taboo search and GA) the adaptations we proposed. We will not explain here the operating mode of GA and Taboo search. However, the reader can refer to Goldberg's [6] and Glover [7] work for details.

2.1 Representation of a Joint Solution

We propose to represent each joint solution as a structure with tree fields:
- the first one is a sequence S that represents the execution order of the production jobs in the first stage;
- the second one is a matrix R called "order matrix" that represents the execution order of the production tasks on each machine. Each line of this matrix relates to a machine;
- the last one is a matrix M that represents the sites of the maintenance tasks insertion. The element M[i,j] represents the insertion of the jth maintenance task of the ith machine, in the sequence of the tasks relating to the ith line of the order matrix R.

Example: let us consider a HFS with two stages; compose of 3 identical machines in the first stage (M0, M1, M2) and 2 identical machines in the second one (M3, M4).

Sequence S	7	5	2	6	4	1	3	0

$$\textbf{Matrix R}\quad \begin{matrix} M_0 \\ M_1 \\ M_2 \\ M_3 \\ M_4 \end{matrix} \begin{pmatrix} 7 & 4 & 3 & & \\ 5 & 1 & & & \\ 2 & 6 & 0 & & \\ 2 & 5 & 4 & 3 & 1 \\ 7 & 6 & 0 & & \end{pmatrix} \qquad \textbf{Matrix M}\quad \begin{matrix} M_0 \\ M_1 \\ M_2 \\ M_3 \\ M_4 \end{matrix} \begin{pmatrix} 1 & 3 & \\ 2 & & \\ 0 & 2 & \\ 1 & 3 & 4 \\ 0 & 3 & \end{pmatrix}$$

Fig. 1. Example

S= (7; 5; 2; 6; 4; 1; 3; 0)
M [4,2] = 3 means that the second maintenance task on the first machine (of the second stage M3) is inserted in position 3, after the production job 4 corresponding to the line 4 of the order matrix R.

The execution of the tasks on the five machines according of the preceding example is the following:
M0 : P7 , M0 1, P4 , P3 , M02
M1 : P5 , P1 , M11
M2 : M21 , P2 , P6 , M22 , P0
M3 : P2 , M3 1, P5 , P4 , M32 , P3 , M33, P1
M4 : M41, P7 , P6 , P0, M42

2.2 Global Objective Function

The goal of joint scheduling is to propose a method that provides a common planning for the production jobs and maintenance tasks. Thus, the objective of optimization must be a compromise between the target objective maintenance and production functions. The constraints imposed by the customers to their suppliers are often expressed in term of time, which lead us naturally the minimization of the makespan Cmax, i.e., the completion time of the last job at the last stage. One will note f1 the production objective function:
f1 = Cmax = Max(Cij)[1]

From the point of view of the supplier, the respect of the maintenance periods influences the constraints of the production system. One will note f2 the maintenance objective function
$f2 = \sum_{j=1}^{m} \sum_{k=1}^{kj} E'_{jk} + L'_{jk}$

With
- E'_{jk} : Advance of the kth occurrence of the maintenance task Mj ;
- L'_{jk} : Delay of the kth occurrence of the maintenance task Mj ;
- k_j : Effective occurrence number of the maintenance task Mj.

To optimize the two criteria, we take into account the following common global objectif function: f = f1 + f2

2.3 An Integrated Taboo Search

In this section, we present the new parameters of the Taboo search. We proposed an algorithm for the generation of the initial solution and the neighbourhood.

Initial solution. Taboo search can start the search with a random solution, or a provided one. For us a solution can be complete or partial. A complete solution represents a joint production and maintenance scheduling solution. In this case, a complete initial solution is obtained after the insertion of maintenance tasks on the partial one, which represents the production scheduling, according to one of heuristics developed by Benbbouzid & al. [8] Algorithm 1 present the generation of a complete random solution.

Random complete solution
Begin
Generate a random production sequence S
Generate the order matrix R associated to the sequence S
Generate the maintenance matrix M by inserting maintenance tasks on this sequence with ones of maintenance insertion heuristics [8]
End.
Algorithm 1: Generation of a random complete solution.

[1] Cij: completation time of the task i on the machine j.

Neighbourhood. The move from a solution to another in the neighbourhood can be done by shifts in the production sequence, the maintenance sequence, or on both at the same time. We defined two types of shifts allowing generating neighbour solutions from the current solution. The first concerns the production tasks and the second the maintenance tasks. The choice of the neighbour solution is done according to one of the following strategies: Best move, First Improve or Randomly. - Maintenance tasks shifting A maintenance task can have several possible sites in its tolerance interval. It is thus interesting to define a solution neighbourhood as being the whole of the possible sites for the insertion of one or more maintenance tasks, on one or more machines. Algorithm 2 presents the principal of this neigbourhood .

The neighbourhood with maintenance tasks shifting

```
Local variables.
rj [0..effective occurrence number of maintenance tasks on the machineMj];
Begin
  For each machine Mj
    If There are late tasks
    Then Advance the task which has the greatest delay
    Else Generate a random number t
      If t is less than ¡rj (the Mj machine is selected)
      Then Select randomly a maintenance task k on machine Mj
        Shift this task on the right (Delay it)
      EndIf
    EndIf
  EndFor
End.
```

Algorithm 2. Generation of neighbourhood with maintenance shifting.

- Production tasks shifting The goal of this operation is to create new individuals by changing the execution order of the production tasks, while keeping the initial maintenance tasks site. The set of possible moves is defined by a neighbourhood of the current sequence. Most often the following neighbourhood structures are used:

(i) right shifting of job at ith position to (i+1)th position, i=1,,n-1; (right-shifting-moves);

(ii) swaps of two neighbouring jobs at position i and i+1, i=1,,n-1 (swap-moves);

(iii) exchanges of jobs placed at the ith and the jth position, i≠ j (interchange-moves);

(iv) remove the job at the ith position and insert it in the jth position (insertion-moves).

Local search based on swap-moves is very fast, as only a low number of possible moves have to be inspected, yet the obtained solution quality is rather low and we do not consider it further. In [12] it was shown that the neighbourhood based on insertion-moves can be evaluated more efficiently than the one based on right shifting moves and additionally gives at least the same solution quality. The number of production tasks executed on a given machine can change after mutations on the initial production sequence because of the type of our workshop (HFS). For that we proposed a technique to improve the muted solution. The principle of this improvement is presented in algorithm 3.

Improvement of muted solution

```
Begin
   For each machine Mj
      If the number of production tasks increase after mutation
      Then Add randomly a site in the maintenance matrix M
      Else Remove randomly a site in the maintenance matrix M
      End If
      Calculate the new cost.
      If the cost is improved
      Then Maintain this improvement
      Else Reject it.
      End If
   EndFor
End.
```

Algorithm 3. Improvement of muted solution

Stopping criterion. The method tries to improve the current solution during a certain number of iterations.

2.4 An Integrated Genetic Algorithm

We introduce in this section the new GA operators which we propose. These operators have the particularity of working on a joint production and maintenance sequence.

Reproduction Operators. A valid individual will be generated from two parents. This individual will inherit its information on the production and the maintenance of its parents. This leads us to define the following crossover operators:

1- The crossover on Production only. We use SJOX, SJ2OX, SBOX and SB2OX crossover [9]

2- The crossover on Maintenance only.

According to the type of our workshop (HFS) where the number of the production jobs executed on each machine is not necessarily the same one. A crossover on the maintenance matrix M only, without taking into account the number of production jobs executed on each machine, according to the order matrix R, is not enough significant, because of unauthorized site for certain maintenance tasks. The site of maintenance tasks on a machine depends on the number of production jobs executed on this machine. A site is called unauthorized maintenance site on a machine if this one is higher than the number of production jobs which will be executed on this machine.

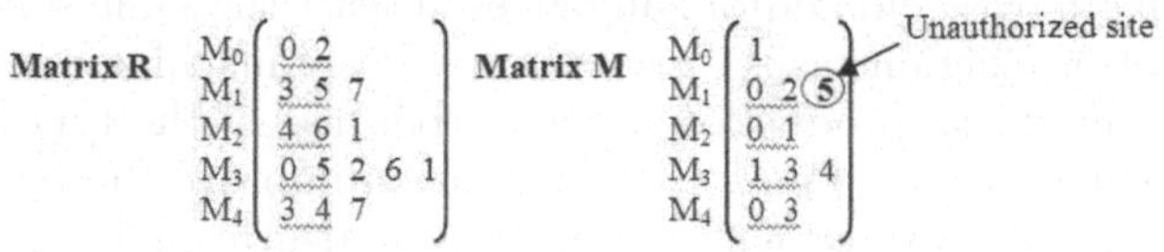

Fig. 2. Example

The third maintenance task on machine M1 has to be inserted after the fifth production job on the same machine. However there are only tree production jobs on machine M1: P3, P5 and P7. This is why this insertion site is called unauthorized site.

Within this framework, we proposed five new crossovers knowing the bound which exists between the number of production tasks and the insertion site of maintenance tasks on the same machine.
- K-points crossover on maintenance. It consists on permuting k lines, selected randomly, in the maintenance matrix of the two parents. For each permutation, we must check the unauthorized sites according to the order matrix, in order to remove them.
- Even (Odd) crossover on stages. The sites of the maintenance tasks on the machines which are in even stages of the first parent (respectively the second one) are recopied in the first child (respectively the second 2). And the sites of the maintenance tasks on the machines which are in the odd stages of the first child (respectively the second) are recopied from the odd stages of the second parent (respectively the first one). Lastly, a checking on the child maintenance matrix according to the order matrix is done, in order to remove the unauthorized sites.
- Even (Odd) crossover inside stages. It consists in generating k stages randomly (k $\in$ [0, number of stages - 1]).

Then for each generated stage, the sites of the maintenance tasks, on the even machines, of the first parent (respectively the second) are recopied in the first child (respectively the second), and the sites of the maintenance tasks, on the odd machines, of the child 1 (respectively the second) are recopied from the odd machines of the second parent (respectively the first). The sites of the maintenance tasks on the machines of the stages which are not generated are recopied directly. Lastly, a checking on the child maintenance matrix according to the order matrix is done, in order to remove the unauthorized sites.

The mutation. The mutation can also be done on the production or maintenance. Four operators of mutation are proposed:
- Random mutation on production. It consists in using right shifting moves, swaps-moves or interchange-moves on production sequence.
- Random mutation on maintenance. It consists in selecting randomly two machines in the maintenance matrix, then permuting the insertion sites of the maintenance tasks of these two machines, keeping the same sites for the other

machines. Lastly, a checking on this matrix is done, in order to remove the unauthorized sites.
- Add/Remove mutation. It consists first to done mutation on the production sequence and generating the corresponding order matrix. Then to copy the same maintenance matrix with removing the unauthorized sites.

3 Computational Experiments

In this section, we present the evaluation of the results obtained by Taboo search and GA in an integrated resolution of the joint production and maintenance scheduling problem in HFS. We generate the benchmarks used for evaluating the proposed heuristics; because the benchmarks proposed by Vignier [10] and Nron [11] are not exploitable in our case as the tested objective function is the computing time. The form of the proposed benchmarks is Tx Cy, where Tx represent the number of the production tasks, and Cy the number of stages. The number of machines on each stage is variable, with the possibility of existence of neck stages, in different positions. These problems have different sizes, but do not include maintenance data. For that, we developed a generator of random maintenance tasks to generate benchmarks in systematic preventive maintenance. The used parameters are: the number of machines, lower and higher bounds for each maintenance task parameter (T*, Tmin and Tmax). Each one of these three parameters itself is limited by two minimal and maximum values, to avoid having identical values. To carry out our tests, we generated only one task of maintenance per machine for each problem. Moreover, the processing time of a maintenance task is identical for all its occurrences. The objective functions are the minimization of makespan for the production, the minimization of the sum of the delays and advances for maintenance (2.2). Taboo search was executed with the following parameters: Number of generations: 100; neighbourhood size: 50; Taboo list size: 10; stagnation: 20. For each benchmark, the

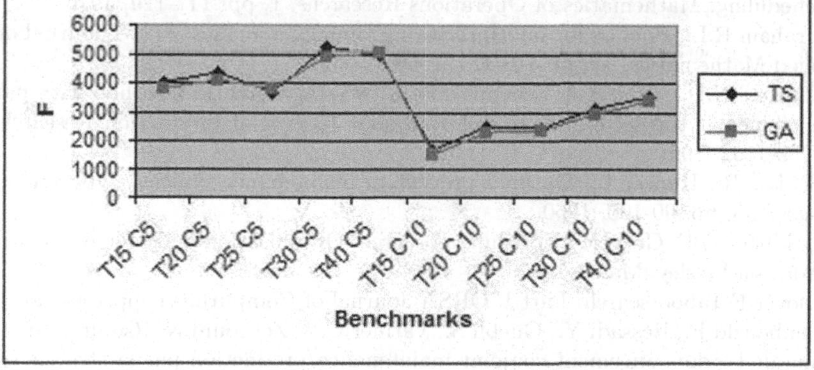

Fig. 3. Integrated joint production and maintenance scheduling

best result is retained. The results of GA are obtained after 100 executions of the method. The best result is saved, as well as the associated parameters. The following parameters are the same for all the executions of the genetic algorithms: crossover rate: 0.8; mutation rate: 0.1; renewal strategy: Nworst, the replacement is done between the selected population and the mute one. Population size: between 30 and 100. We use all the proposed operators. Figure 2 presents the computational results obtained on HFS instances with t TS and GA. GA proves its superiority over TS, since the values performed by the integrated GA represent more than 80.

4 Conclusion

In this paper we propose an integrated Taboo search and genetic algorithm to solve the joint production and maintenance scheduling problem in HFS. We proposed first, a new structure with tree fields to represent a joint production and maintenance scheduling. Then algorithms to generate an initial solution and neighbourhood in the case of integrated TS, and new operators in the case of integrated GA. The instance of HFS used for testing our approach proves than integrated GA performs better than integrated TS. These preliminary experiments are encouraging and prove that the approach is consistent. Despite these results, it is necessary to improve our approach with a preliminary study which allows deducing the best operators for each tested method. Future research will be to investigate a bi-criterion approach to solve this problem.

References

1. Kaabi J., Contribution l'ordonnancement des activits de maintenance dans les systmes de production. Thse de doctorat soutenue l'Universit de Franche Comt, France, 2004.
2. Garey M.R, Johnson D.S., and Sethi R. the complexity of flowshop and jobshop scheduling. Mathematics of Operations Researchs, 1, pp. 117-129, 1976.
3. Graham R.L., Bounds for multiprocessing timing anomalies. SIAM Journal of Applied Mathematics, 17, pp.416-429, 1969.
4. Gupta J.N.D., Tunc E.A. Schedules for a two-stages hybrid flowshop with parallel machines at the second stage. International Journal of production Research, 29, 1489-1502, 1991.
5. Barlow R., Hunter L., Optimal preventive maintenance policies. Operations Research, 8, pp. 90-100, 1960.
6. Goldberg D.E. Genetic algorithms in search, Optimisation and Machine Learning. Addison-Wesley, Mass., 1989.
7. Glover F.Taboo search: Part I. ORSA Journal of Computing,1, pp.190-206, 1989.
8. Benbouzid F., Bessadi Y., Guebli S., Varnier C. & Zerhouni N. Rsolution du problme de l'ordonnancement conjoint maintenance/production par la stratgie squentielle. MOSIM'03, 2, pp. 627-634, Toulouse (France), 2003.
9. Ruiz R., Maroto C. A genetic algorithm for hybrid flowshops with sequence dependent setup times and machines eligibility. EJORS, June 2004.

10. Nron E., Du flowshop hybride au problme cumulatif. Thse de doctorat soutenue l'Universit de Technologie de Compigne, France, 1999.
11. Vignier A. contribution la rsolution des problmes d'ordonnancement de type monogamme, multimachines (flowshop hybride). Thse de doctorat soutenue l'universit de Tours, France, 1997.
12. Taillard E. Some Efficient Heuristic Methods for the Flow Shop Sequencing Problem. European Journal of Operational Research, 47, 65-74, 1990.

Improving Cutting-Stock Plans with Multi-objective Genetic Algorithms*

César Muñoz, María Sierra, Jorge Puente, Camino R. Vela, and Ramiro Varela

University of Oviedo, Dep. of Computer Science, Artificial Intelligence Center
Campus de Viesques, 33271 Gijón, Spain
{mariasierra,puente,camino,ramiro}@aic.uniovi.es
http://www.aic.uniovi.es/Tc

Abstract. In this paper, we confront a variant of the cutting-stock problem with multiple objectives. The starting point is a solution calculated by a heuristic algorithm, termed *SHRP*, that aims to optimize the two main objectives, i.e. the number of cuts and the number of different patterns. Here, we propose a multi-objective genetic algorithm to optimize other secondary objectives such as changeovers, completion times of orders pondered by priorities and open stacks. We report experimental results showing that the multi-objective genetic algorithm is able to improve the solutions obtained by *SHRP* on the secondary objectives.

1 Introduction

This paper deals with a real Cutting-Stock Problem (*CSP*) in manufacturing plastic rolls. The problem is a variant of the classic *CSP*, as it is usually considered in the literature, with additional constraints and objective functions. We have solved this problem in [1,2] by means of a *GRASP* algorithm [3] termed Sequential Heuristic Randomized Procedure (*SHRP*), which is similar to other approaches such as the *SVC* algorithm proposed in [4]. Even though *SHRP* tries to optimize all objective functions, in practice it is mainly effective in optimizing the main two ones: the number of cuts and the number of patterns. It is due to *SHRP* considering all objective functions in a hierarchical way that it pays much more attention to the first two ones than to the remaining. In this work we propose a Multi-Objective Genetic Algorithm (*MOGA*) that starts from a solution computed by *SHRP* algorithm and tries to improve it regarding three secondary objectives: the cost due to changeovers or setups, the orders' completion time weighted by priorities and the maximum number of open stacks. As we will see, the *MOGA* is able to improve the solutions obtained by *SHRP*. The paper is organized as follows. Next section is devoted to problem formulation. In section 3, we describe the main characteristics of the proposed *MOGA*. In section 4, we report results from a small experimental study. Finally, in section 5, we summarize the main conclusions and some ideas for future work.

* This work has been partially supported by the Principality of Asturias Government under Research Contract FC-06-BP04-021.

J. Mira and J.R. Álvarez (Eds.): IWINAC 2007, Part I, LNCS 4527, pp. 528–537, 2007.

2 The Production Process

Figure 1 shows the schema of the cutting machine. A number of rolls are cut at the same time from a big roll according to a cutting pattern. Each roll is supported by a set of cutting knives and a pressure roller of the appropriate size. At each of the borders, a small amount of product should be discarded, so that there is a maximum width that can be used from the big roll, there is also a minimum width due to the limited capability of the machine to manage trim loss. Also, there is a maximum number of rolls that can be cut at the same time. When the next cut requires a different cutting pattern, the process incurs in a setup cost due to changing cutting knives and pressure rollers. The problem has also a number of constraints and optimization objectives that make it different from conventional formulations. For example underproduction is not allowed and the only possibility for overproduction is a stock declared by the expert. Once a cut is completed, the rolls are packed into stacks. The stack size is fixed for each roll width, so a given order is composed by a number of stacks, maybe the last one being uncompleted. Naturally, only when a stack is completed it is taken away from the proximity of the cutting machine. So, minimizing the number of open stacks is also convenient in order to facilitate the production process. Moreover, some orders have more priority than others. Consequently the delivery time of orders pondered by the client priorities is an important criterion as well.

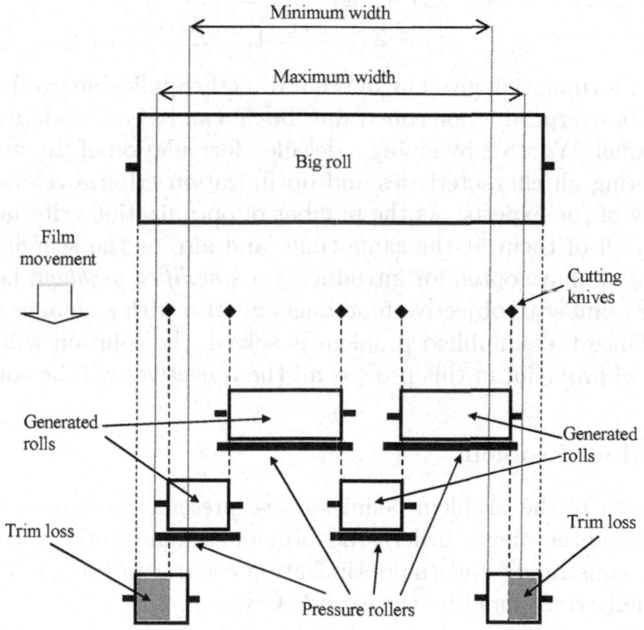

Fig. 1. Working schema of the cutting machine

3 Problem Formulation

The problem is a variant of the One Dimensional Cutting-Stock Problem, also denoted $1D - CSP$. In [5] Gilmore and Gomory proposed the first model for this problem. It is defined by the following data: $(m, L, l = (l_1, ..., l_m), b = (b_1, ..., b_m))$, where L denotes the length of each stock piece (here the width of the big roll), m denotes the number of piece types (orders) and for each type $i = 1, ..., m$, l_i is the piece length (roll width), and b_i is the order demand. A cutting pattern describes how many items of each type are cut from a stock length. Let column vectors $A^j = (a_{1j}, ..., a_{mj}) \in Z_+^m$, $j = 1, ..., n$, represent all possible valid cutting patterns, i.e. those satisfying

$$\sum_{i=1,\ldots,m} a_{ij} l_i \leq L$$

where a_{ij} is the number of pieces of order i that are generated by one application of the cutting pattern A^j. Let $x_j, j = 1, ..., n$, be the frequencies, i.e. the number of times each pattern is applied in the solution. The model of Gilmore and Gomory aims at minimizing the number of stock pieces, or equivalently minimizing the trim-loss, and is stated as follows

$$Z^{1D-CSP} = \min \textstyle\sum_{j=1,\ldots,n} X_j$$

subject to:

$$\textstyle\sum_{j=1,\ldots,n} a_{ij} X_j \geq b_i, \quad i = 1, ..., m$$
$$x_j \in Z_+, \quad j = 1, ..., n$$

The classic formulation given in [5] is not directly applicable to our case mainly due to the non-overproduction constraint, but it can be easily adapted as we will see in the sequel. We start by giving a detailed formulation of the *main problem*; that considering all characteristics and optimization criteria relevant from the point of view of the experts. As the number of optimization criteria is too large to deal with all of them at the same time, and also as the search space could be very large, we have opted for introducing a *simplified problem*; i.e. a problem with a lower number of objective functions and also with a smaller search space in general. Once the simplified problem is solved, the solution will be adapted to the original problem; in this process all the objectives will be considered.

3.1 The Main Problem

In order to clarify the problem definition, we present the data of the machine environment and the clients' orders, the form and semantic of a problem solution, the problem constraints and the optimization criteria in the hierarchical order they are usually considered by the expert. Given

- The set of parameters of the cutting machine: the maximum width of a cut L_{max}, the minimum width of a cut L_{min}, the maximum number of rolls that can be generated in a cut C_{max}, the minimum and the maximum width of a single roll, W_{min} and W_{max} respectively, and the increment of width $\triangle W$ between two consecutive permitted roll widths.

- The setup costs. There is an elementary setup cost SC and some rules given by the expert that allows calculating the total setup cost from a configuration of the cutting machine to the next one. The setup cost is due to roller and cutter changes as follows. The cost of putting in or taking off a pressure-roller is SC; the cost of putting in an additional cutting knife is $3SC$, and the cost of dismounting a cutting knife is $2SC$.
- The types of pressure-rollers $PR = PR_1, ..., PR_p$ and the mapping F_{PR} from roll widths to pressure-rollers.
- The mapping F_{ST} from roll widths to stack sizes or number of rolls in each stack unit.
- The orders description, given by ($M = 1, ..., m, b = (b_1, ..., b_m), l = (l_1, ..., l_m)$, $p = (p_1, ..., p_m)$) where for each order $i = 1, ..., m$, b_i denotes the number of rolls, l_i denotes the width of the rolls and p_i the order priority.
- The stock allowed for overproduction, given by ($S = \{m+1, ..., m+s\}, bs = (b_{m+1}, ..., b_{m+s}), ls = (l_{m+1}, ..., l_{m+s})$) where for each $i = 1, ..., s$, b_{m+i} denotes the number of rolls of type $m+i$ allowed for overproduction and l_{m+i} denotes the width of these rolls.
- The set of feasible cutting patterns, for the orders and stock given, $\mathbf{A}$ where each $A^j \in \mathbf{A}$ is $A^j = (a_{1j}, ..., a_{mj}, a_{(m+1)j}, ..., a_{(m+s)j}) \in Z_+^{m+s}$ and denotes that, for each $i = 1, ..., m+s$, a_{ij} rolls of order i are cut each time the cutting pattern A^j is applied. A cutting pattern A^j is feasible if and only if both of the following conditions hold

$$L_{min} \leq L_j = \sum_{i \in M \bigcup S} a_{ij} l_i \leq L_{max}, C_j = \sum_{i \in M \bigcup S} a_{ij} \leq C_{max}$$

  being L_j and C_j the total width and the number of rolls of pattern A^j respectively. $D_j = L_{max} - L_j$ denotes the trim-loss of the cutting pattern.

The objective is to obtain a *cutting plan* (Π, x), where $\Pi = (A^1, ..., A^{|\Pi|}) \in \mathbf{A}^{\Pi}$ and $x = (x1, ..., x_{|\Pi|}) \in Z_+^{|\Pi|}$ denotes the pattern frequencies. The cutting patterns of Π are applied sequentially, each one the number of times indicated by its frequency. A_l^j, $0 \leq j \leq |\Pi|$, $0 \leq l \leq x_j$, denotes the lth cut corresponding to pattern A^j and $CI(A_l^j)$ is the cut index defined as

$$CI(A_l^j) = \sum_{k=1,...j-1} x_k + l$$

Given an order $i \in M$ its first roll is generated in cut A_l^j such that A^j is the first pattern of Π with $a_{ij} \neq 0$, this cut is denoted $CU_{start}(i)$. Analogously, the last roll of order i is generated in cut A_{xk}^k so that A^k is the last pattern of Π with $a_{ik} \neq 0$, this cut is denoted $CU_{end}(i)$.

As we have considered feasible cutting patterns, the only constraint that should be required to a solution is the following

- The set of rolls generated by the application of the cutting plan (Π, x) should be composed by all rolls from the orders and, eventually, by a number of rolls from the stock. That is, let s_i be the number of rolls of stock $i \in S$ in the solution

$$\forall i \in S, s_i = \sum_{A^j \in \Pi} a_{ij} x_j$$

Then, the constraint can be expressed as follows:

$$\forall i \in M, \sum_{A^j \in \Pi} a_{ij} x_j = b_i$$

$$\forall i \in S, 0 \leq s_i \leq b_i$$

Regarding objective functions, as we have remarked, we consider two main functions

1. Minimize the number of cuts, given by $\sum_{j=1,\ldots,|\Pi|} x_j$. The optimum value is denoted z^{1D-CSP}.
2. Minimize the setup cost, given by $\sum_{j=1,\ldots,|\Pi|} SU(A^{j-1}, A^j)$; $SU(A^{j-1}, A^j)$ being the setup cost from pattern A^{j-1} to pattern A^j calculated as it is indicated above. Configuration A^0 refers to the situation of the cutting machine previous to the first cut.

And three secondary functions

3. Minimize the completion times of orders weighted by their priorities given by

$$\sum_{i \in M} CI(CU_{end}(i)) p_i$$

4. Minimize the maximum number of open stacks along the cut sequence. Let $R(i, A_l^j)$ denote the number of rolls of order i generated from the beginning up to completion of cut A_l^j

$$R(i, A_l^j) = \sum_{k=1,\ldots,j-1} a_{ik} x_k + a_{ij} l$$

and let $OS(i, A_l^j)$ be 1 if after cut A_l^j there is an open stack of order i and 0 otherwise. Then, the maximum number of open stacks along the cut sequence is given by

$$\max_{j=1,\ldots,|\Pi|, l=0,\ldots,x_j} \sum_{i \in M} OS(i, A_l^j)$$

3.2 The Simplified Problem

As the main problem has too many objectives to deal with all of them at the same time, we have developed a two step procedure in which a simplified version is solved and then a solution to this problem is transformed into a solution to the main problem. This simplification consists in merging all orders with the same width into one only, so as objectives 3 and 4 can not longer be considered. Moreover, objective 2 is simplified so as we only consider the number of different patterns. To solve this simplified problem, in [1,2] we have proposed a *GRASP* algorithm. Then, the solution given by this algorithm is transformed into a solution to the main problem by a greedy algorithm that assigns items to actual orders so as it tries to optimize objectives 2, 3, and 4 in hierarchical order, while keeping the values of the first one. To be more precise, we clarify how a simplified solution in transformed into an actual solution by means of an example.

The problem data and final results are displayed as in the application program. Figure 2 shows an instance and the corresponding simplified problem. A real instance is given by a set of orders, each one defined by a client name, a order identification number, the number of rolls, the width of the rolls and the order priority. Additionally, the maximum and minimum allowed width of a cut should be given, in this case 5500 and 5700 respectively, and also a stock description to choose a number of rolls from if it is necessary in order to obtain valid cutting patterns. In this example up to 10 rolls of each width 1100, 450 and 1150 could be included in the cutting plan. Furthermore, some other parameters (not shown in Figures) are necessary, for instance, two additional data should be given to evaluate the number of open stacks and setup cost: the number of rolls that fit in a stack (mapping F_{ST}) and the correspondence between the size of pressure rollers and the width of the supported rolls (mapping F_{PR}). Here we have supposed that every stack contains 4 rolls and that the correspondence between types of pressure rollers and roll widths is the following: type 1 $(0 - 645)$, type 2 $(650 - 1045)$, type 3 $(1050 - 1345)$, type 4 $(1350 - 1695)$. All the allowed widths are multiples of 5 and the minimum width of a roll is 250 while the maximum is 1500. Finally, the maximum number of rolls in a pattern is 10.

As we can observe in Figure 2, the main instance with 10 orders is reduced to a simplified instance with only 6 orders. This simplified instance is actually a conventional $1D - CSP$ instance with two additional constraints: the

Problem Data (Main Problem)							(Simplified Problem)	
ROLLS	WIDTH	ORDER	CLIENT	PRIORITY	Max. width	5700	ROLLS	WIDTH
20	600	20001	Client 1	2	Min. width	5500	30	600
10	600	20002	Client 2	2			28	850
15	850	20003	Client 3	1	Stock		15	950
13	850	20004	Client 4	1	1150	10	14	1350
15	950	20005	Client 5	1	550	10	20	550
14	1350	20006	Client 6	1	1500	10	33	900
20	550	20007	Client 7	1				
18	900	20008	Client 8	2				
15	900	20009	Client 9	1				

Fig. 2. An example of problem data (main and simplified instance)

Cutting Plan for the Simplified Problem

FREQUENCY	14	3	3	1	
P	600	550	950	900	
A	900	950	950	600	
T	1350	950	900	600	
T	850	550	900	**1500**	
E	600	900	950	**1500**	
R	850	900	900	**550**	
N	550	900	-	-	
PATTERN WIDTH	5700	5700	5550	5650	
TRIM LOSS	0	0	150	50	500
				NUMBER OF PATTERNS	4
				NUMBER OF CUTS	21

Fig. 3. A solution to the simplified instance of Figure 2. Bold face values are stock rolls.

Cutting Plan for the Main Problem (roll widths and evaluation functions)

FREQUENCY	8	2	4	3	3	1	
P	600	600	600	550	950	900	
A	900	900	900	950	950	600	
T	1350	1350	1350	950	900	600	
T	850	850	850	550	900	**1500**	
E	600	600	600	900	950	**1500**	
R	850	850	850	900	900	**550**	
N	550	550	550	900			
PATTERN WIDTH	5700	5700	5700	5700	5550	5650	
TRIM LOSS	0	0	0	0	150	50	500
CHANGEOVERS	28	0	0	4	5	10	47
OPEN STACKS	5-3-5-0-5-3-5-1	5-3	5-1-4-3	2-3-2	1-1-0	0	5
						WEIGHTED TIME	188
						NUMBER OF CUTS	21

Cutting Plan for the Main Problem (order identifiers)

FREQUENCY	8	2	4	3	3	1
P	20001	20001	20002	20007	20005	20009
A	20008	20008	20008	20005	20005	20002
T	20006	20006	20006	20005	20009	20002
T	20003	20004	20004	20007	20009	0
E	20001	20001	20002	20008	20005	0
R	20003	20004	20004	20008	20009	0
N	20007	20007	20007	20008		

Fig. 4. Solution to the instance (main problem) of Figure 3 obtained from the solution of Figure 3

maximum number of rolls in a pattern and the minimum width of a pattern. Figure 3 shows a solution to the simplified problem with 21 cuts and 4 different patterns, where 3 stock rolls have been included in order that the last pattern to be valid. Figure 4 shows the final solution to the main problem. The figure shows the order identifiers, where 0 represents the stock. A solution is a sequence of cutting patterns, where each pattern represents not only a set of roll widths, but also the particular order the roll belongs to. The actual solution is obtained from a simplified solution by means of a greedy algorithm that firstly considers the whole set of individual cuts as they are expressed in the simplified solution. Then it assigns a customer order to each one of the roll widths in the simplified cuts, and finally considers all different actual patterns maintaining the order derived from the simplified solution. The *MOGA* proposed in this paper starts from this solution and tries to improve it by considering different arrangements of cuts so giving rise to different sequences of patterns and frequencies.

The changeover of each pattern refers to the cost of put in and out cutting knives and pressure rollers from the previous pattern to the current one. As we can observe the first pattern has a changeover cost of 28 because it is assumed that it is necessary to put in all the 7 cutting knives and 7 pressure rollers before this pattern. In practice, this is not often the case as a number of cutting knives and pressure rollers remain in the machine from previous cuts. Regarding open stacks, each column shows the number of them that remain incomplete in the proximity of the machine from a cut to the next one, i.e. when a stack gets full after a cut, or it is the last stack of an order, it is not considered.

4 The Multi-Objective Genetic Algorithm

According to previous section, the encoding schema is a permutation of the set of single cuts comprising a solution. So, each chromosome is a direct representation of a solution, which is alternative to the initial solution produced by the greedy algorithm. The starting solution is the one of figure 4 which is codified by chromosome (1 2 3 4 5 6 7 8 . . . 21), i.e. each gene represents a single cut. As objectives 2, 3 and 4 depends on the relative ordering of cuts and also on their absolute position in the chromosome sequence, we have used genetic order based operators that maintains these characteristics from parents to offsprings.

The algorithm structure is quite similar to a conventional single *GA*: it uses generational replacement and roulette wheel selection. The main differences are due to its multi-objective nature. The *MOGA* maintains, apart from the current population, a set of non-dominated chromosomes. This set is updated after each generation, so as at the completion it contains an approximation of the pareto frontier for the problem instance.

In order to assign a single fitness to each chromosome, the whole population is organized into dominant groups as it follows. The first group is comprised by the non-dominated chromosomes. The second group is comprised by the non-dominated chromosomes in the remaining population and so on. The individual fitness is assigned so that a chromosome in a group has a larger value than any chromosome in the subsequent groups and, inside each group, the fitness of a chromosome is adjusted by taking into account the number of chromosomes in its neighborhood in the space defined by the three objective functions. The chromosomes' neighbors are those that are in the chromosome's *niche count.* The evaluation algorithm is as follows.

Step 0. Set F to a value sufficiently large.

Step 1. Determine all non-dominated chromosomes Pc from the current population and assign F to their fitness.

Step 2. Calculate each individual's *niche count* m_j:

$$m_j = \sum_{k \in P_c} sh(d_{jk})$$

where

$$sh(d_{jk}) = \begin{cases} 1 - (d_{jk}/\sigma_{share})^2 & \text{if } d_{jk} < \sigma_{share} \\ 0 & \text{otherwise} \end{cases}$$

and d_{jk} is the phenotypic distance between two individuals j and k in P_c and σ_{share} is the maximum phenotypic distance allowed between any two chromosomes of P_c to become members of a niche.

Step 3. Calculate the shared fitness value of each chromosome by dividing it fitness value by its *niche count*.

Step 4. Create the next non-dominated group with the chromosomes of P_c, remove these chromosomes from the current population, set F to a value lower than the lowest fitness in P_c, go to step 1 and continue the process until the entire population is all sorted.

This evaluation algorithm is taken from [6]. In their paper, G. Zhou and M. Gen propose a *MOGA* for the Multi-Criteria Minimum Spanning Tree (*MCMSP*). In the experimental study they consider only two criteria.

In order to compute d_{jk} and σ_{share} values we have normalized distances in each one of the three dimensions to take values in [0, 1]; this requires calculating lower and upper bounds for each objective. The details of these calculations are given in [7]. As we will see in the experimental study, we have determined empirically that $\sigma_{share} = 0,5$ is a good choice.

5 Experimental Study

In this section we neither present results form an exhaustive experimental study nor compare with other methods. We only give results from 3 runs, of a prototype implemented in [7], for the problem instance commented in Figures 2, 3 and 4. The *MOGA* starts from the solution of Figure 4 and searches for the pareto

Table 1. Summary of results from four runs of *MOGA* starting from the solution of Figure 4 for the problem of Figure 2. Parameters of *MOGA* refers to /Population size/Number of generation/Crossover probability/Mutation probability/σ_{share}/. Each cell shows the cost of /changeovers/weighed times/maximum open stacks.

Run	1	2	3
Parameters	/200/200/0,9/0,1/0,5/	/500/500/0,9/0,1/0,5/	/700/700/0,9/0,1/0,5/
Time (s.)	37	649	1930
Pareto frontier reached	49/188/6 49/186/7 **44/196/5**	47/176/6 **47/184/5** 39/179/7 45/184/6	**39/172/6** **47/184/5**

Values in **bold** represent solutions non-dominated for any other reached in all three runs.

frontier of solutions to the problem. Table 1 summarizes the values of the three objective functions (change-overs, weighed time and maximum open stacks) for each of the solutions in the pareto frontiers obtained in three runs with different parameters. The target machine was Pentium 4 at 3′2GHz with HT and 1GB of RAM. As we can observe, the quality of the solutions is in direct ratio with the processing time given to the *MOGA*. The values of objective functions for the starting solution of Figure 4 are 47/188/5, which is dominated by some of the solutions of Table 1. Hence, it is clear that it is possible to improve the secondary objectives in solutions obtained by procedure *SHRP*.

6 Concluding Remarks

In this paper we have proposed a multi-objective genetic algorithm (*MOGA*) which aims to improve solutions to a real cutting stock problem obtained previously by another heuristic algorithm. This heuristic algorithm, termed *SHRP*, focuses on two objectives and considers them hierarchically. Then, the *MOGA* tries to improve other three secondary objectives at the same time, while keeping the value of the main objective. We have presented some preliminary results over a small real problem instance showing that the proposed *MOGA* is able to improve the secondary objective functions with respect to the initial solution, and that it offers the expert a variety of non-dominated solutions. As future work, we plan reconsidering the *MOGA* strategy in order to make it more efficient and more flexible so that it considers the preferences of the experts with respect to each one of the objectives.

References

1. Puente, J., Sierra, M., González, I., Vela, C., Alonso, C., Varela, R.: An actual problem in optimizing plastic rolls cutting. In: Workshop on Planning, Scheuling and Temporal Reasoning (CAEPIA). (2005) 21–30
2. Varela, R., Puente, J., Sierra, M., González, I., Vela, C.: An effective solution for an actual cutting stock problem in manufacturing plastic rolls. In: Proceedings APMOD. (2006)
3. Resende, M., Ribeiro, G. In: Hadbook of Metaheuristics.Greedy randomized adaptive search procedures. Kluwer Academic Publishers (2002) 219–249
4. Belov, G., Scheithauer, G.: Setup and open stacks minimization in one-dimensional stock cutting. INFORMS Journal of Computing **19**(1) (2007) XX–XX
5. Gilmore, P.C., Gomory, R.E.: A linear programming approach to the cutting stock problem. Operations Research **9** (1961) 849–859
6. Zhou, G., Gen, M.: Genetic algorithm approach on multi-criteria minimum spanning tree problem. European Journal of Operational Research **114** (1999) 141–152
7. Muñoz, C.: A Multiobjective Evolutionary Algorithm to Compute Cutting Plans for Plastic Rolls. Technical Report. University of Oviedo, Gijón School of Computting (2006)

Sensitivity Analysis for the Job Shop Problem with Uncertain Durations and Flexible Due Dates

Inés González-Rodríguez[1], Jorge Puente[2], and Camino R. Vela[2]

[1] Department of Mathematics, Statistics and Computing,
University of Cantabria, Spain
ines.gonzalez@unican.es
[2] A.I. Centre and Department of Computer Science,
University of Oviedo, Spain
{puente,crvela}@uniovi.es
http://www.aic.uniovi.es/Tc

Abstract. We consider the *fuzzy job shop problem*, a job shop scheduling problem with uncertain task durations and flexible due dates, with different objective functions and a GA as solving method. We propose a method to generate benchmark problems with variable uncertainty and analyse the performance of the objective functions in terms of the objective values and the sensitivity to variations in the uncertainty.

1 Introduction

In the last decades, scheduling problems have been subject to intensive research due to their multiple applications in areas of industry, finance and science [1]. To enhance the scope of applications, fuzzy scheduling has tried to model the uncertainty and vagueness pervading real-life situations, with a great variety of approaches, from representing incomplete or vague states of information to using fuzzy priority rules with linguistic qualifiers or preference modelling [2],[3].

Incorporating uncertainty to scheduling usually requires a significant reformulation of the problem and solving methods, in order that the problem can be precisely stated and solved efficiently and effectively. Furthermore, in classical scheduling the complexity of problems such as open shop and job shop means that practical approaches to solving them usually involve heuristic strategies, for instance, genetic algorithms, local search etc [1]. Some attempts have been made to extend these heuristic methods to fuzzy scheduling problems. For instance, 6-point fuzzy numbers and simulated annealing are used for single objective job shop problem in [4], while triangular fuzzy numbers and genetic algorithms are considered for multiobjective job shop problems in [5] and [6]. Flow shop problems with triangular fuzzy numbers are solved using an adapted Johnson's algorithm and evolutionary algorithms in [7], [8] and [3]. A study of critical paths in activity networks with triangular fuzzy numbers as uncertain durations can be found in [9] and a semantics for job shop with uncertainty is proposed in [10].

J. Mira and J.R. Álvarez (Eds.): IWINAC 2007, Part I, LNCS 4527, pp. 538–547, 2007.

In the sequel, we describe a fuzzy job shop problem with uncertain durations and flexible due dates, solved by means of a genetic algorithm. We analyse the performance of the GA using different objective functions with respect to the objective values and the effect in the obtained schedule of variable uncertainty in the problem data.

2 Description of the Problem

The *job shop scheduling problem*, also denoted *JSSP*, consists in scheduling a set of jobs $\{J_1, \ldots, J_n\}$ on a set of physical resources or machines $\{M_1, \ldots, M_m\}$, subject to a set of constraints. There are *precedence constraints*, so each job J_i, $i = 1, \ldots, n$, consists of m tasks $\{\theta_{i1}, \ldots, \theta_{im}\}$ to be sequentially scheduled. Also, there are *capacity constraints*, whereby each task θ_{ij} requires the uninterrupted and exclusive use of one of the machines for its whole processing time. In addition, we may consider *due-date constraints*, where each job has a maximum completion time and all its tasks must be scheduled to finish before this time. The goal is to find a *feasible* schedule, so that all constraints hold, which is *optimal*, in the sense that its *makespan* (i.e., the time it takes to finish all jobs) is minimal.

2.1 Uncertain Processing Times and Flexible Constraints

In real-life applications, it is often the case that the exact duration of a task is not known in advance. For instance, in ship-building processes, some tasks related to piece cutting and welding are performed by a worker and, depending on his/her level of expertise, the task will take a different time to be processed. However, based on previous experience, an expert may have some knowledge about the duration, thus being able to estimate, for instance, an interval for the possible processing time or its most typical value. In the literature, it is common to use fuzzy numbers to represent such processing times, as an alternative to probability distributions, which require a deeper knowledge of the problem and usually yield a complex calculus.

When there is little knowledge available, the crudest representation for uncertain processing times would be a human-originated confidence interval. If some values appear to be more plausible than others, a natural extension is a a fuzzy interval or a fuzzy number. The simplest model of fuzzy interval is a *triangular fuzzy number* or *TFN*, using only an interval $[a^1, a^3]$ of possible values and a single plausible value a^2 in it. For a TFN A, denoted $A = (a^1, a^2, a^3)$, the membership function takes a triangular shape completely determined by the three real numbers, $a^1 \leq a^2 \leq a^3$ as follows:

$$\mu_A(x) = \begin{cases} \frac{x-a^1}{a^2-a^1} & : a^1 \leq x \leq a^2 \\ \frac{x-a^3}{a^2-a^3} & : a^2 < x \leq a^3 \\ 0 & : x < a^1 \text{ or } a^3 < x \end{cases} \quad (1)$$

To compute the completion time of a given task, it is necessary to add the task's duration to its starting time. This can be done using *fuzzy number addition*, which in the case of TFNs $A = (a^1, a^2, a^3)$ and $B = (b^1, b^2, b^3)$ is reduced to adding three pairs of real numbers so $A + B = (a^1 + b^1, a^2 + b^2, a^3 + b^3)$. A consequence of this operation is that completion times are TFNs as well.

The starting time for a given task θ is calculated as the maximum between two TFNs, the completion time of the task preceding θ in its job J and that preceding θ in its resource M. For two TFNs $A = (a^1, a^2, a^3)$ and $B = (b^1, b^2, b^3)$, the *maximum* $A \vee B$ is obtained by extending the lattice operation max on real numbers using the Extension Principle. However, computing the membership function is not trivial and the result is not guaranteed to be a TFN. For these reasons, we approximate $A \vee B$ by a TFN, $A \sqcup B = (a^1 \vee b^1, a^2 \vee b^2, a^3 \vee b^3)$. This approximation was first proposed in [4] for 6-point fuzzy numbers, a particular case of which are TFNs. It artificially increases the value of $A \vee B$ while both sets $A \vee B$ and $A \sqcup B$ have identical support and the modal point in $A \sqcup B$ also has full membership in $A \vee B$.

Using the addition and the maximum $\sqcup$, it is possible to find the completion time for each job. The fuzzy makespan C_{max} would then correspond to the greatest of these TFNs. Unfortunately, neither the maximum $\vee$ nor its approximation $\sqcup$ can be used to find such TFN, because they do not define a total ordering in the set of TFNs. Instead, it is necessary to use a method for *fuzzy number ranking* [11]. The chosen method consists in obtaining from each TFN A three real numbers $C_1(A) = \frac{a^1+2a^2+a^3}{4}$, $C_2(A) = a^2, C_3(A) = a^3 - a^1$ and then use real number comparisons to establish a total ordering. First, the TFNs are ordered according to the value of C_1. Ties are then broken with the value C_2 and, should any ties persist, C_3 is finally used. This corresponds to a lexicographical ordering of TFNs.

To measure the non-specificity of durations and completion times, including the makespan C_{max}, we use a standard uncertainty measure for fuzzy sets, the *U-uncertainty* [12]. It is a generalisation of Hartley's measure for classical sets which, in the particular case of a TFN A, is given by $U(A) = 0$ if $a^1 = a^3$ and otherwise by the formula:

$$U(A) = \frac{1 + a^3 - a^1}{a^3 - a^1} \ln(1 + a^3 - a^1) - 1 \tag{2}$$

In practice, if due-date constraints exist, they are often flexible. For instance, customers may have a preferred delivery date d^1, but some delay will be allowed until a later date d^2, after which the order will be cancelled. They would be completely satisfied if the job finishes before d^1 and after this time their level of satisfaction would decrease, until the job surpasses the later date d^2, after which date they will be clearly dissatisfied. The satisfaction of a due-date constraint becomes a matter of degree, our degree of satisfaction that a job is finished on a certain date. A common approach to modelling such satisfaction levels is to use a fuzzy set D with linear decreasing membership function:

$$\mu_D(x) = \begin{cases} 1 & : x \le d^1 \\ \frac{x-d^2}{d^1-d^2} & : d^1 < x \le d^2 \\ 0 & : d^2 < x \end{cases} \tag{3}$$

Such membership function expresses a flexible threshold "less than", representing the satisfaction level $sat(t) = \mu_D(t)$ for the ending date t of the job [2]. However, when dealing with uncertain task durations, the job's completion time is no longer a real number t, but a TFN C. In this case, the degree to which a completion time C satisfies the due-date constraint D may be measured using the following *agreement index* [8],[5]:

$$AI(C, D) = \frac{area(D \cap C)}{area(C)} \tag{4}$$

2.2 Definition of the Objective Function

Let us assume that we have a schedule s such that resource and precedence constraints hold (otherwise, the schedule is unfeasible and hence is not a solution). A fuzzy makespan C_{max} may be obtained from the completion times of all jobs C_i, $i = 1, \ldots, n$, and, in the case that a due date D_i exists for job J_i, the agreement index $AI_i = AI_i(C_i, D_i)$ measures to what degree the due date is satisfied. Based on this information, it is necessary to decide on the quality of this schedule.

If flexible due-date constraints exist, the degree of overall due-date satisfaction for schedule s may be obtained by combining the satisfaction degrees AI_i, $i = 1, \ldots, n$. We may expect due dates to be satisfied in average or, being more restrictive, expect that all due dates be satisfied. The degree to which schedule s satisfies due dates is then given, respectively, by the following:

$$AI_{av} = \frac{1}{n}\sum_{i=1}^{n} AI_i, \quad AI_{min} = \min_{i=1,\ldots,n} AI_i \tag{5}$$

Clearly, both AI_{av} and AI_{min} should be maximised. Notice however that the two measures model different requirements and encourage different behaviours.

Regarding makespan, if we consider the total ordering defined by the ranking method, minimising C_{max} translates into minimising $C_1(C_{max})$. Curiously, for TFNs this coincides with minimising the expected value for the makespan, as defined in [13].

Given both measures of feasibility and the makespan and depending on the final goal of the job-shop scheduling problem, the following objective functions were defined in [14]:

$$f_1 = \frac{1}{C_1(C_{max})}, f_2 = AI_{av}, f_3 = AI_{min}, f_4 = \frac{AI_{av}}{C_1(C_{max})}, f_5 = \frac{AI_{min}}{C_1(C_{max})} \tag{6}$$

f_1 corresponds to the case when no due-date constraint is considered and the only goal is to find a schedule with minimum makespan. Similarly, f_2 and f_3 correspond to the case when due-date constraints are present and makespan is not relevant. Finally, f_4 and f_5 are obtained when both due-date constraints and makespan are considered.

An alternative definition of the objective function combining feasibility and makespan can be given in the framework of fuzzy decision making [10],[5]:

$$f_6 = \min\left(\mu_1(AI_{av}), \mu_2(AI_{min}), \mu_3(C_1(C_{max}))\right) \tag{7}$$

Here μ_i, $i = 1, 2, 3$, must be defined by an expert and represent the expert's satisfaction degrees with respect to due date constraints and makespan. This objective function is heavily parameterised, assuming a deeper knowledge of the problem. In the absence of an expert with such knowledge, a heuristic method with high computational cost can be used to define the satisfaction degrees [10].

3 Using Genetic Algorithms to Solve FJSSP

In classical JSSP, the search for an optimal schedule is usually limited to the space of active schedules. *G&T Algorithm* [15] is widely used to find active schedules, allowing to use complementary techniques to reduce the search space [16]. It can also be used as a basis for efficient genetic algorithms (GA). We describe a possible extension of G&T for the FJSSP (see Alg. 1) and a GA to solve the FJSSP based on such extension (see Alg. 2). Both have been successfully used in for the above objective functions [14],[10] and were inspired in the work from [5].

$A = \{\theta_{i1},\ i = 1, \ldots, n\}$; /**first task of each job**/
while $A \neq \emptyset$ **do**
  Find the task $\theta' \in A$ with minimum earliest completion time /*$CT(\theta)^1$*/;
  Let M' be the machine required by θ' and B the subset of tasks in A requiring machine M';
  Delete from B any task that cannot overlap with θ'; /*$ST(\theta)^1 > CT(\theta')^3$*/
  Select $\theta^* \in B$ (according to some criteria) to be scheduled;
  Remove θ^* from A and, if θ^* is not the last task of its job, insert in A the task following θ^* in the job;

Alg. 1: Fuzzy G&T

Regarding the GA, *chromosomes* are a direct codification of schedules. For n jobs and m machines, each individual will be a $n \times m$ matrix, where element (i, j) is the completion time for the task in job J_i requiring resource M_j. Therefore, each row is the schedule of a job's tasks over the corresponding resources. Each chromosome in the *initial population* for the GA can be generated with fuzzy G&T algorithm, choosing a task at random from the conflict set B. To introduce diversity and prevent premature convergence, a new individual will only be incorporated to the population if its similarity to other members of the population is below a given threshold σ. Let $Pr_I(\theta)$ be the set of tasks preceding θ in its machine according to the ordering induced by individual I and let $Su_I(\theta)$ be the set of tasks following θ in its machine w.r.t. the same ordering. Then, the *similarity* between two individuals I_1 and I_2 is defined using phenotype distance as follows:

$$S(I_1, I_2) = \frac{\sum_{i=1}^{n} \sum_{j=1}^{m} (|Pr_{I_1}(\theta_{ij}) \cap Pr_{I_2}(\theta_{ij})| + |Su_{I_1}(\theta_{ij}) \cap Su_{I_2}(\theta_{ij})|)}{n \cdot m \cdot (m-1)} \quad (8)$$

The value of the *fitness function* for a chromosome is simply the value of the objective function for the corresponding schedule. The *crossover operator*, applied with probability p_m, consists in performing the fuzzy G&T algorithm and solve non-determinism situations using the information from the parents. Every time the conflict set B has more than one element, the selected task is that with earliest completion time in the parents, according to the ranking algorithm. The *mutation operator* is embedded in the crossover operator, so that, with a given probability pm, the task from the conflict set is selected at random.

```
generate initial population divided in k groups P_1,...,P_k with K individuals each;
while terminating condition T_1 is not satisfied do
  for i = 1; i ≤ k; i++ do
    repeat
      select 2 parents at random from P_i;
      obtain 3 children by crossover and mutation;
      select the best of 3 children and the best of remaining children and parents
      for the new population NP_i;
    until a new population NP_i is complete
    replace the worst individual in NP_i with the best of P_i;
merge P_1,...,P_k into a single population P;
while terminating condition T_2 is not satisfied do
  obtain a new population from P following the scheme above;
```

Alg. 2: Genetic Algorithm for FJSSP

The *general scheme* of the GA, in Alg. 2, is designed to avoid premature convergence by using a niche-based system. The population is initially divided in k sub-populations, containing K individuals each. Each sub-population evolves separately for I_{min} generations. At this stage, these sub-populations are merged into a single population of N individuals, which will again evolve until a total of I_{max} generations is reached.

4 Generation of New Problems for Sensitivity Analysis

Unfortunately, benchmark examples of FJSSP in the literature are scarce, clearly a problem for any thorough experimentation, where a sufficiently large and diverse set of problems is needed. It is also interesting to have problems with variable uncertainty, to see the consequences on the resulting schedule. For these reasons, we propose a novel heuristic method to generate an adequate sample of problems that allow to test the different objective functions introduced in Section 2 and perform sensitivity analysis with respect to the problem's uncertainty. The underlying idea is to define new problems by "fuzzifying" benchmark crisp problems from the literature, as done in [4] for a different fuzzy job shop.

Given a crisp job-shop problem, capacity constraints are already defined and, for each task θ, we take the crisp duration to be the most typical value a^2. Then, given $\alpha \in [0,1]$, the least and greatest possible durations a^1 and a^3 are random values from $[\max\{0, int(a^2(1-\alpha))\}, a^2]$ and $[a^2, int(a^2(1+\alpha))]$ respectively, where $int(x)$ denotes the closest integer to a real number x. Different values of α may be used to introduce more or less uncertainty in the durations, with a maximum spread of $2\alpha a^2$; in any case, crisp durations may be generated.

Due-date values are the most difficult to define. If they are too strict, the problem will have no solution and if they are too lenient, due-date constraints will always be satisfied, which is equivalent to having no constraints at all. For a given job J_i, let $\iota_i = \sum_{j=1}^{m} a_{i,j}^2$ be the sum of most typical durations across all its tasks. Also, for a given task $\theta_{i,j}$ let $\rho_{i,j}$ be the sum of most typical durations of all other tasks requiring the same machine as $\theta_{i,j}$, $\rho_{i,j} = \sum_{\theta_{i,j} \neq \theta : M(\theta)=M(\theta_{i,j})} a^2(\theta)$, where $M(\theta)$ denotes the machine required by task θ and $a^2(\theta)$ denotes its most typical duration. Finally, let $\rho_i = \max_{j=1,\ldots,m} \rho_{i,j}$ be the maximum of such values across all tasks in job J_i. Then, the earlier due date d^1 is taken as a random value from $[d_m, d_M]$, where $d_m = \iota_i + 0.5\rho_i$ and $d_M = \iota_i + \rho_i$ and the later due date d^2 is a random value from $[d^1, int(d^1(1+\delta))]$, where $\delta \in [0,1]$ is a parameter that allows to have more or less flexible due dates.

5 Experimental Results

The results shown hereafter correspond to 50 problems generated from FT10, a well-known crisp benchmark problem of size 10×10, using the method proposed in Section 4. These 50 problems are subdivided in five families of 10 instances each, generated with different values for the uncertainty and flexibility parameters α and δ: 0.05, 0.10, 0.15, 0.20 and 0.25. This will allow to analyse the performance of the different objective functions with an increasing level of uncertainty and flexibility in the problems. We consider the six objective functions described in Section 2, where the additional parameters for f_6 are obtained with the method from [10]. For each problem and objective function, the GA from Section 3 is executed 20 times with the parameter values used in [10] for other problems of size 10×10 ($p_m/p_c/\sigma/N/I_{min}/I_{max}$ equal to 0.03/0.9/0.8/200/100/200).

A summary of the results can be seen in Table 1. It shows the average value of AI_{av}, AI_{min}, $C_1(C_{max})$ as well as $U(C_{max})$ across the problems of each family when the six objective functions are used. The average values of U-uncertainty of the problems' processing times, U_p, is also given for completeness.

The minimum values for $U(C_{max})$ are always obtained with f_1, i.e., when the only objective is to minimise the makespan. However, differences among objective functions are relatively low (between 0.22% and 8.70%) and they tend to reduce as the problem's U-uncertainty increases. Indeed, for those problems generated with $\alpha = \delta = 0.25$, the uncertainty values are almost identical for all six objective functions. The soving method seems to be robust to changes in the problem's uncertainty, with the non-specificity of the makespan depending

Table 1. Average values of AI_{av}, AI_{min}, $C_1(C_{max})$ and $U(C_{max})$

		f_1	f_2	f_3	f_4	f_5	f_6
$\alpha, \delta = 0.05$ $U_p = 0.67$	AI_{av}	0.69	0.90	0.17	0.89	0.17	0.17
	AI_{min}	0.01	0.16	0.00	0.16	0.00	0.00
	$C_1(C_{max})$	993.18	1146.31	1313.63	1048.75	1313.63	1313.63
	$U(C_{max})$	2.99	3.10	3.25	3.02	3.25	3.25
$\alpha, \delta = 0.10$ $U_p = 1.04$	AI_{av}	0.73	0.93	0.30	0.92	0.30	0.28
	AI_{min}	0.03	0.35	0.09	0.33	0.09	0.05
	$C_1(C_{max})$	997.08	1118.45	1279.84	1044.33	1278.89	1282.94
	$U(C_{max})$	3.67	3.76	3.90	3.70	3.90	3.90
$\alpha, \delta = 0.15$ $U_p = 1.29$	AI_{av}	0.76	0.94	0.64	0.92	0.64	0.52
	AI_{min}	0.06	0.48	0.43	0.37	0.43	0.25
	$C_1(C_{max})$	997.31	1111.76	1189.82	1037.39	1179.93	1212.70
	$U(C_{max})$	4.03	4.14	4.21	4.07	4.20	4.23
$\alpha, \delta = 0.20$ $U_p = 1.47$	AI_{av}	0.78	0.94	0.85	0.92	0.85	0.81
	AI_{min}	0.09	0.49	0.64	0.40	0.62	0.52
	$C_1(C_{max})$	996.37	1107.74	1122.94	1041.04	1105.20	1096.87
	$U(C_{max})$	4.30	4.41	4.42	4.34	4.40	4.40
$\alpha, \delta = 0.25$ $U_p = 1.65$	AI_{av}	0.83	0.95	0.91	0.94	0.92	0.90
	AI_{min}	0.23	0.63	0.76	0.61	0.75	0.67
	$C_1(C_{max})$	994.28	1092.34	1079.62	1020.64	1052.37	1038.91
	$U(C_{max})$	4.57	4.64	4.64	4.58	4.61	4.60

mainly on the way uncertainty is propagated using TFN operations and with little influence of the objective function used.

In Table 1, values of AI_{min} show that the heuristic method has generated due dates very difficult to satisfy, specially for the first two families of problems (α and δ equal to 0.05 or 0.10). In consequence, AI_{min} values are too low for the objective functions f_3 and f_5 to guide the GA's search procedure. That is, the GA "evolves" almost randomly, which explains the bad performance of the obtained solutions, compared to other objective functions.

As expected when the objective functions were introduced, the results indicate that, when only the productivity goal of minimising C_{max} is considered, f_1 should be used. Notice however that the lowest values of C_{max} (and those with least uncertainty) are obtained with schedules for which at least a due-date constraint is not satisfied at all.

Regarding due-date satisfaction, f_3 cannot be used because, as mentioned above, it is too strict when constraints are "tight" (especially for the first three families of problems). We may feel tempted to conclude that, if the goal is to respect delivery dates, f_2 should be used. This would certainly correspond to the motivation for its definition in Section 2. However, a more careful look shows that, if we use f_4 instead of f_2, values of AI_{av} and AI_{min} are very similar but $C_1(C_{max})$ decreases considerably. Indeed, comparing both functions, f_2 yields slightly better values of AI_{av}, with a relative improvement w.r.t. f_4 between 0.65% and 2.28%, but at the cost of an increase in makespan ranging from

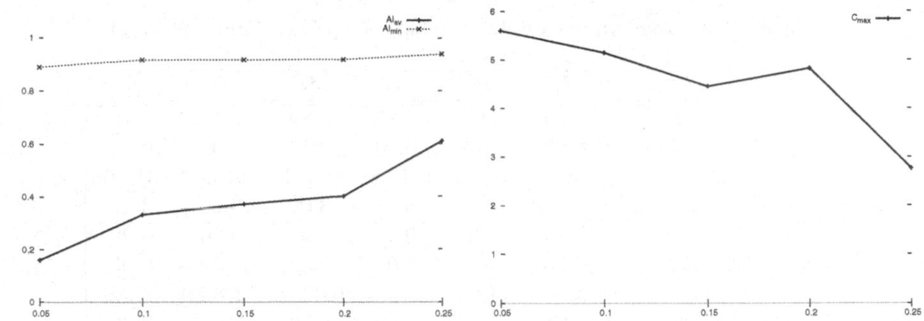

a) Overall Agreement Index b) Relative error of Makespan

Fig. 1. Performance of f_4 w.r.t. increasing of uncertainty

6.41% to 9.3%. That is, with f_4 due date satisfaction is similar and there is the added benefit of considerably reducing the makespan. Therefore, even if the goal is to satisfy due dates, it seems preferable to use f_4 instead of just using f_2, as first proposed.

Finally, if both optimisation of due date constraints and makespan is considered, the objective function should be f_4. Clearly, it yields better results than the other multi-objective functions f_5 and f_6. Additionally, the parameter setting of f_6 is much more complex and involves some heavy computation. In fact, the results for f_4 are overall the best. We have already seen that f_4 is comparable to f_2 regarding delivery dates. Compared to f_1, $C_1(C_max)$ increases from 2.65% to 5.6%, being $U(C_{max})$ only between 0.22% and 1% worse, whilst AI_{av} improves between 13.01% and 29.52%. Therefore, f_4 may be preferable to f_1, in the sense that the increase in makespan seems to be compensated by a considerable improvement in due-date satisfaction. In fact, notice that f_4 is the most "robust" function, as it yields the second best solutions regarding the four performance measures AI_{av}, AI_{min}, $C_1(C_{max})$ and $U(C_{max})$ in almost all cases. Also, its performance seems to improve with an increasing uncertainty in the input data, as can be seen in Fig. 1.

6 Conclusions and Future Work

We have considered the FJSSP, a version of JSSP that tries to model the imprecise nature of data in real-world problems, using fuzzy sets to represent uncertain processing times and flexible due dates. Different objective functions are described, considering the goals of optimising productivity or respecting delivery dates. In order to solve the FJSSP, a GA has been described. Finally, a method for generating test problems has been proposed and used to obtain a sample of problem instances that allow to analyse the results of the GA with the different objective functions in terms of the schedule's performance and its sensitivity to the uncertainty in the data. Results suggest that the makespan uncertainty depends mainly on the fuzzy arithmetic and not on the objective function. Overall, it seems preferable to use the multi-objective function f_4, which obtains good results both for makespan and due dates and has the further advantage that its definition involves no additional parameters.

References

1. Brucker, P., Knust, S.: Complex Scheduling. Springer (2006)
2. Dubois, D., Fargier, H., Fortemps, P.: Fuzzy scheduling: Modelling flexible constraints vs. coping with incomplete knowledge. European Journal of Operational Research **147** (2003) 231–252
3. Słowiński, R., Hapke, M., eds.: Scheduling Under Fuzziness. Volume 37 of Studies in Fuzziness and Soft Computing. Physica-Verlag (2000)
4. Fortemps, P.: Jobshop scheduling with imprecise durations: a fuzzy approach. IEEE Transactions of Fuzzy Systems **7** (1997) 557–569
5. Sakawa, M., Kubota, R.: Fuzzy programming for multiobjective job shop scheduling with fuzzy processing time and fuzzy duedate through genetic algorithms. European Journal of Operational Research **120** (2000) 393–407
6. Fayad, C., Petrovic, S.: A fuzzy genetic algorithm for real-world job-shop scheduling. In: Innovations in Applied Artificial Intelligence. Volume 3533 of Lecture Notes in Computer Science., Springer (2005) 524–533
7. Petrovic, S., Song, X.: A new approach to two-machine flow shop problem with uncertain processing times. Optimization and Engineering **7** (2006) 329–342
8. Celano, G., Costa, A., Fichera, S.: An evolutionary algorithm for pure fuzzy flowshop scheduling problems. International Journal of Uncertainty, Fuzziness and Knowledge-Based Systems **11** (2003) 655–669
9. Chanas, S., Dubois, D., Zieliński, P.: On the sure criticality of tasks in activity networks with imprecise durations. IEEE Transactions on Systems, Man and Cybernetics–Part B: Cybernetics **32** (2002) 393–407
10. González Rodríguez, I., Puente, J., Vela, C.R., Varela, R.: Evaluation methodology of schedules for the fuzzy job-shop problem. IEEE Transactions on Systems, Man and Cybernetics, Part A **X** (2007) XX–XX Accepted for publication.
11. Bortolan, G., Degani, R.: A review of some methods for ranking fuzzy subsets. In Dubois, D., Prade, H., Yager, R., eds.: Readings in Fuzzy Sets for Intelligence Systems. Morgan Kaufmann, Amsterdam (NL) (1993) 149–158
12. Klir, G.J., Smith, R.M.: On measuring uncertainty and uncertainty-based information: Recent developments. Annals of Mathematics and Artificial Intelligence **32** (2001) 5–33
13. Liu, B., Liu, Y.K.: Expected value of fuzzy variable and fuzzy expected value models. IEEE Transactions on Fuzzy Systems **10** (2002) 445–450
14. González Rodríguez, I., Vela, C.R., Puente, J.: Study of objective functions in fuzzy job-shop problem. ICAISC 2006, Lecture Notes in Artificial Intelligence **4029** (2006) 360–369
15. Giffler, B., Thomson, G.L.: Algorithms for solving production scheduling problems. Operations Research **8** (1960) 487–503
16. Varela, R., Vela, C.R., Puente, J., Gómez, A.: A knowledge-based evolutionary strategy for scheduling problems with bottlenecks. European Journal of Operational Research **145** (2003) 57–71

Comparative Study of Meta-heuristics for Solving Flow Shop Scheduling Problem Under Fuzziness

Noelia González, Camino R. Vela, and Inés González-Rodríguez

Centro de Inteligencia Artificial
Universidad de Oviedo
Campus de Viesques, E-33271, Gijón, Spain
{gonzaleznoelia,crvela}@uniovi.es, ines.gonzalez@unican.es
http://www.aic.uniovi.es/Tc

Abstract. In this paper we propose a hybrid method, combining heuristics and local search, to solve flow shop scheduling problems under uncertainty. This method is compared with a genetic algorithm from the literature, enhanced with three new multi-objective functions. Both single objective and multi-objective approaches are taken for two optimisation goals: minimisation of completion time and fulfilment of due date constraints. We present results for newly generated examples that illustrate the effectiveness of each method.

Keywords: fuzzy sets, flow shop scheduling, meta-heuristics, evolutive algorithms.

1 Introduction

Scheduling problems appear profusely in areas such as industry, finance and science, thus proving of great interest to researchers [1]. In particular, flow shop scheduling problems (*FSP*) are specially suitable to manage certain manufacturing systems. Due to the complexity of *FSP*, many heuristic strategies have been proposed in literature to solve them, for instance, genetic algorithms, local search etc [2]. Usually these methods assume that processing times and due dates can be modelled as deterministic values; however in real production environments, it is often necessary to deal with uncertainty or flexibility in the process data. Incorporating uncertainty and vagueness to scheduling usually requires a significant reformulation of the problem and solving methods, in order that the problem can be precisely stated and efficiently and effectively solved.

Some authors have proposed the use of fuzzy set theory to represent uncertain processing times or flexible constraints in scheduling problems [3], [4], [5]. For instance, job shop problems with fuzzy durations (6-point and triangular fuzzy numbers and λ-sets) are considered in [6] and [7], and also in [8], [9] and [10] combined with fuzzy due dates. The methods used to solve the resulting problems range from simulated annealing, defuzzification together with Johnson's algorithm and single or multi-objective genetic algorithms.

J. Mira and J.R. Álvarez (Eds.): IWINAC 2007, Part I, LNCS 4527, pp. 548–557, 2007.

The flow shop problem in presence of uncertainty has been considered by a number of authors with different definitions of the objective function and methods to solve it. In [11], the two-machine fuzzy flow shop problem is solved using a method based on Johnson's algorithm, adapted for triangular fuzzy processing times using ranking techniques for the comparison of fuzzy numbers. In [5] triangular fuzzy numbers and an evolutionary algorithm are used for a generic fuzzy flow shop problem, with a comparison of different objective functions. In [4] and [12], two GAs are described to solve a fuzzy flow shop problem, the second one hybridised with neighbourhood search but considering fuzzy sets to model only the due dates of the jobs. In [13] a method to maximise the satisfaction degree of a flow shop problem with fuzzy constraints is presented.

Among heuristic methods, locally improved constructive methods have shown to be effective and fast when dealing with scheduling problems. In this paper we propose a new meta-heuristic method of this type based on the one presented in [14] for the crisp flowshop scheduling problem. We compare results of this method with those obtained with a genetic algorithm as described in [5], enhanced with three new multi-objective functions that take into account not only the completion time minimisation but also the due date constraints fulfilment.

2 Description of the Problem and the Genetic Algorithm

The *FSP*, denoted as $n/m/P/C_{max}$ in $\alpha|\beta|\gamma$ notation, consists in scheduling a set of jobs $\{J_1, \ldots, J_n\}$ on a set of physical resources or machines $\{M_1, \ldots, M_m\}$, subject to a set of constraints. There are *precedence constraints*, so each job J_i, $i = 1, \ldots, n$, consists of m tasks $\{\theta_{i1}, \ldots, \theta_{im}\}$ to be sequentially scheduled. Also, there are *capacity constraints*, whereby each task θ_{ij} requires the uninterrupted and exclusive use of machine M_j for its whole processing time or duration du_{ij}. In addition, we may consider *due-date constraints*, where each job has a maximum completion time and all its tasks must be scheduled to finish before this time. The goal is twofold: we need to find a *feasible* schedule, so that all constraints hold and then we want this schedule to be *optimal*, in the sense that its *makespan* (i.e., the time it takes to finish all the jobs) is minimal.

Formally, the makespan, usually denoted by C_{max}, is the completion time C of the last scheduled job n, on the last machine, m, that is, $C_{max} = C_{n,m}$. The completion time of each scheduled job i is calculated recursively as $C_i = max\{C_{i,m-1}; C_{i-1,m}\} + du_{i,m}$, $i = 1, \ldots, n$.

When a fuzzy approach is taken, job processing times and due dates are expressed as fuzzy numbers. One of the simplest models for fuzzy processing times is a triangular fuzzy number (TFN) denoted $A = (a, b, c)$ with the following membership function μ_A:

$$\mu_A(x) = \begin{cases} \frac{x-a}{b-a} & : a \leq x \leq b \\ \frac{c-x}{c-b} & : b \leq x \leq c \\ 0 & : \text{otherwise} \end{cases} \tag{1}$$

where b is the modal value ($\mu_A(b) = 1$) and $(c - a)$ is the spread of the fuzzy number A, which is associated with its level of uncertainty.

In practice, if due-date constraints exist, they are often flexible: we are completely satisfied if the job finishes before a delivery date d^1 and after this time our level of satisfaction decreases until a later date d^2, after which we are clearly dissatisfied. A common approach to modelling such satisfaction levels is to use a fuzzy set D with linear decreasing membership function as follows:

$$\mu_D(x) = \begin{cases} 1 & : x < d^1 \\ \frac{x-d^2}{d^1-d^2} & : d^1 < x \leq d^2 \\ 0 & : d^2 < x \end{cases} \tag{2}$$

According to Dubois *et al* [3], the above membership function expresses a flexible threshold "less than" and expresses the satisfaction level $sat(t) = \mu_D(t)$ for the ending date t of the job.

The *fuzzy sum* and the *fuzzy maximum* are the two main arithmetical operations needed to calculate the objective function in scheduling problems. The fuzzy sum is a linear shape conservative operation, while the fuzzy maximum does not verify this property. The sum between two triangular fuzzy numbers (*TFNs*) $A = (a_1, b_1, c_1)$ and $B = (a_2, b_2, c_2)$ is equal to $A + B = (a_1 + a_2, b_1 + b_2, c_1 + c_2)$. The fuzzy maximum $A \vee B$ between the same pair of fuzzy numbers is defined using the Extension Principle but computing the membership function is not trivial and the result is not guaranteed to be a TFN. For these reasons, we use an approximation proposed in [6] that consists on replacing $A \vee B$ by the TFN $A \sqcup B = (a_1 \vee b_1, a_2 \vee b_2, a_3 \vee b_3)$. Both sets have identical support and the modal point in $A \sqcup B$ also has full membership in $A \vee B$.

Unfortunately neither the maximum $\vee$ nor its approximation $\sqcup$ define a total ordering in the set of TFNs, so it is not trivial to optimise a schedule in terms of fuzzy makespan. In the literature, this problem is tackled using some ranking method for fuzzy numbers, lexicographical orderings, comparisons based on λ-cuts or defuzzification methods. Here two different criteria are used to compare TFNs taking into account the defuzzification of the makespan: the *Area Compensation criterion* proposed in [15]:

$$F_{AC}(A) = \frac{a + 2b + c}{4} \tag{3}$$

and the *Intersections Average criterion* proposed in [5]:

$$F_{IA}(A) = \frac{1}{3}\frac{b^3 - a^3 - c^3 + 7abc - 5a^2b - 5c^2b + 2ac^2 + 2ca^2}{b^2 - 2a^2 - 2c^2 - ab - bc + 5ac} \tag{4}$$

When due dates are considered, the parameter to evaluate the performance of a processing order of jobs is the *Agreement Index (AI)* [8], [5]. The AI of a job J_i measures the degree to which the fuzzy completion time of J_i is contained in the job's due date. When due date constraints are present we can find feasible schedules maximizing the minimum or the average value of AI for the n scheduled jobs (denoted AI_{min} and AI_{ave} respectively).

A scheduling problem in a fuzzy environment differs substantially from a traditional crisp model, due to the more complex evaluation and the performance indexes of the sequence. However, from a computational perspective, the complexity of the problem remains unchanged. Analogously to the crisp model described in [14], a feasible sequence of scheduled jobs may be represented by a permutation in the number of jobs. Such permutations are the individuals of the population for the evolutionary algorithm. This algorithm starts from an initial population formed by N_s chromosomes randomly generated, which evolve for a fixed number of iterations, N_{it}, through the application of the genetic operators.

As in [5], two different crossover operations are proposed: position based crossover (PBC) and two points crossover (TPC); and the mutation operators used are gene swapping operator and block swapping operator. The fitness function is a performance index of the sequence; its expression depends on the selected objective and defuzzification criterion. When the minimisation of fuzzy completion times is considered as the only aim, the fitness coincides with the crisp defuzzified makespan calculated through one of two ranking criterion described above. When fuzzy due dates are considered as a production goal, the fitness function to be maximised is AI_{min} or AI_{ave}.

3 Introduction of Heuristic Methods for Solving *FFSP*

According to [16], traditional construction of a heuristic method involves three different phases: development of an index function, construction of an initial solution and improvement of this initial solution. The heuristic method proposed in [14] for the *FSP* consists in combining the index and constructive phase of heuristic as proposed by Liu and Reeves in [17] with an improvement (or local search) phase inspired in that proposed by Ho in [18]. We have adapted this heuristic, designed for the crisp problem with the criterion of minimising total flow time (C_{sum}), in order to cope with the new fuzzy features of the *FFSP* and new optimisations criteria.

3.1 Constructive Phase

The designed index function combines two parts, one considering the machine idle time and another considering the effects on the completion times of later jobs. We now present the two parts of the index function first and then combine them together.

The Weighted Total Machine Idle Time. Consider any stage in the heuristic procedure where a partial sequence S with k jobs has been constructed and a decision is to be made for selecting one job to append to S from the set of unscheduled jobs, U. Let $C_{i,j}$ be the completion time on machine j of a job i in U if it is chosen to be appended to S as the $(k+1)$th job. Then, we define the weighted total machine idle time between the processing of the kth job and job i as follows:

$$IT_{ik} = \sum_{j=2}^{m} w_{jk} \max\{C_{i,j-1} - C_{[k],j}, 0\}, \tag{5}$$

where the weights are

$$w_{jk} = \frac{m}{j + \frac{k(m-j)}{(n-2)}} \tag{6}$$

and $\max\{C_{i,j-1} - C_{[k],j}, 0\}$ is the idle time of machine j. The form of the weight functions is important, since the weights need first to emphasise the idle time of the early stage machines, as they may have a greater effect on the starting time of the jobs in U, but, as the length of S increases, the number of jobs in U becomes small. In general, the value of the weights should decrease as either j or k increases. It is clear that the above choice of weights does accomplish this.

The Artificial Total Flow Time. While the weighted total machine idle time focuses more on the matching between the last job in S and the job to be appended, artificial total flow time tries to consider the effect of this choice on the remaining jobs in U. To consider this, the second part of the index function is defined in the following way. When we consider appending job i to S with k jobs, we calculate the *average* processing time of all the other jobs in U on each machine and take these average processing times to be the processing times of an artificial job p:

$$du_{pj} = \frac{\sum_{\substack{q \in U \\ q \neq p}} du_{qj}}{(n-k-1)} \tag{7}$$

We then calculate the completion time of job i and the completion time of this artificial job p as if it were scheduled after job i:

$$\begin{aligned} C_{i,1} &= C_{[k],1} + du_{i1} \\ C_{i,j} &= \max\{C_{[k],j}, C_{i,j-1}\} + du_{ij}, \quad j = 2, \ldots, m \\ C_{p,1} &= C_{i,1} + du_{p1} \\ C_{p,j} &= \max\{C_{i,j}, C_{p,j-1}\} + du_{pj}, \quad j = 2, \ldots, m \end{aligned} \tag{8}$$

The total artificial time AT_{ik} is defined here as the sum of completion times of these two jobs:

$$AT_{ik} = C_{i,m} + C_{p,m} \tag{9}$$

When scheduling the second last job, obviously this artificial total flow time is actually the *real* total flow time of the last job.

The Combined Index Function for *FFSP*. Combining the above two parts, the index function for choosing a job i in U to append to S with k jobs is defined as follows.

$$f_{ik} = (n-k-2)IT_{ik} + AT_{ik} \tag{10}$$

The job with the *minimum* value of this index function is selected. Notice that the result of this index function will be a TFN and not a crisp value, since the durations of the tasks are TFN, and the result of a TFNs sum is another TFN.

In order to obtain a ranking for the eligible jobs according to the described index function, we must use any of the two presented comparison criterions (F_{AC} and F_{IA}) for defuzzificating TFNs. Once we have obtained this real number we will be able to obtain an ordered sequence of the predictably most suitable jobs.

3.2 Improvement Phase

For the improvement phase, we propose a possible adaptation to the fuzzy domain of the local search schema given by Ho in [18] for the *FSP*, that combines two kinds of sorting methods: exchange sort and insertion sort. The essence of exchange sort consists in systematically exchanging two items that are out of order until no more such pairs exist. Two types of exchange sort, adjacent pairwise interchange (also known as bubblesort) and non-adjacent pairwise interchange, are proposed to improve the initial solutions. The bubblesort exchanges an adjacent pair of jobs, $J_{[h]}$ and $J_{[h+1]}$, when a reduction in total flow time is obtained, where $h = 1, 2, \ldots, n-1$. The bubblesort terminates when no more improvements can be made in a pass. The merit of bubblesort is that it guarantees to provide a locally optimal sequence. Therefore, any non-locally optimal sequence can definitely be improved by the bubblesort. Unlike sorting a set of numbers, a local optimum does not guarantee the global optimum.

Hence, two additional sorting procedures are introduced to improve a locally optimal solution obtained by the bubblesort. The first one is non-adjacent pairwise interchange, belonging to the exchange sort. This procedure first initialises $h = 1$. It systematically considers every non-adjacent pair of jobs, i.e. $J_{[h]}$ and $J_{[h+2]}$ and $J_{[h+3]}$, ..., and $J_{[h]}$ and $J_{[n]}$, and they are swapped if a reduction in total flow time is obtained. If $h < n-2$, then it increment h by one and repeats the process, otherwise the procedure terminates. The second procedure is based on the idea of insertion sort method. This procedure works as follows: it first initialises $h = 1$, and then systematically considers inserting $J_{[h]}$ into the other $n-1$ positions, that is, from position 1 through position n except for position h. The insertion is made when an improvement is obtained. If $h < n$, then it increments h by one and repeats the process; otherwise, the procedure stops. The total number of insertions made is $n^2 - n$.

Regarding the stopping rule, existing heuristics have consistently shown that as the number of jobs increases, the quality of heuristic solutions deteriorates rapidly, though their quality is not very sensitive to the number of machines. Hence, the number of improvement iterations is set as a function of n, concretely $\lambda = n - 4$, where λ is the number of iterations. The proposed heuristic is given in Alg. 1.

As in the constructive phase of the heuristic, here it must be considered that all the obtained solutions, from the initial to the definitive one, are TFNs. In order to compare the different solutions obtained we must use any of the two presented comparison criterions (F_{AC}, and F_{IA}) for defuzzificating TFNs. Once we have obtained this real numbers we will be able to decide whether a locally improved solution is better than the previous one or not.

initialise $\lambda = n - 4$ and $i = 1$;
obtain an initial solution using a constructive method;
sort the initial solution using the bubblesort method (call it *current solution*);
repeat
 set Z_1 = *current solution*;
 sort the *current solution* by the insertion sort;
 sort the *current solution* using the bubblesort method;
 set Z_2 = Solution from previous step;
 set $i = i + 1$;
until $i > \lambda$ **or** $Z_1 \neq Z_2$
current solution becomes the final solution;

Algorithm 1: 1: Heuristic method for *FFSP*

4 Experimental Results

Following as closely as possible the method from [5], we have generated a set of 40 problems with a number of processed parts n equal to 10 or 20 and a number of workstations m equal to 5 or 10. For the fuzzy triangular processing times, the modal value is obtained at random following a discrete uniform distribution in the interval [1,B] and then, the left $(b - a)$ and the right $(c - b)$ spreads are obtained at random following a discrete uniform distribution in $[1, S]$. Here, we set $B = 99$ and $S = 20$, as done in [5]. Additionally, we propose to generate fuzzy trapezoidal due dates for each a job J_i as follows. First, the earliest value (d_1) is taken at random from $[d_m, d_M]$, where $d_m = 0.5B(m + (i - 1)))$ and $d_M = B(m + (i - 1))$. Then, the latest value (d_2) is generated by adding to d_1 a value selected at random from $[1, S]$ according to a discrete uniform distribution.

Regarding genetic parameters, the number of chromosomes (N_s) equals the number of jobs n, that is, 10 or 20; the mutation probability mutation (M_p) is 0.18, and the number of iterations (N_{it}) is 10000 for problems 10×5 and 10×10 and 40000 for the 20×5 and 20×10 problems. Notice that our GA is a Steady-state GA, so, in each iteration, a pair of new individuals are generated at most.

We shall use the following notation for the different solving methods considered in this section:

- GA-MAC and GA-MIA: GAs for makespan optimisation with Area Compensation AC and Intersections Average IA ranking methods respectively.
- GA-AID and GA-AIM: GAs for due date constraints fulfilment using AI_{ave} and AI_{min} respectively.
- H-MAC, H-MIA: Heuristic methods for makespan optimisation with AC and IA ranking methods respectively.
- MOGA-DAC, MOGA-MIA and MOGA-MAC: multi-objective GAs as proposed in [5], with the new fitness functions proposed herein, the ratio between AI_{ave} and F_{AC}, the ratio between AI_{min} and F_{IA} and the ratio between AI_{min} and F_{AC} respectively.

For each of these methods, Table 1 presents the average values of productivity (F_{AC} and F_{IA}), constraint fulfilment (AI_{min} and AI_{ave}) and CPU time across the 40 problems. For comparison purposes, the table also presents the results obtained with MOGA-DIA, the multi-objective GA with the ratio between the Average Agreement Index AI_{ave} and the defuzzified makespan using F_{IA} as fitness function proposed in [5].

Table 1. Average values of F_{AC},F_{IA}, AI_{min}, AI_{ave} and time in seconds obtained for the four families of problems

	F_{AC}	F_{IA}	AI_m	AI_d	$t(s)$	F_{AC}	F_{IA}	AI_m	AI_d	$t(s)$
					10×5					*10×10*
GA-MAC	730.3	728.7	0.47	0.90	0.17	1108.7	1106.8	0.33	0.76	0.20
GA-MIA	735.8	734.3	0.52	0.90	0.27	1103.1	1100.8	0.31	0.84	0.33
GA-AIM	800.3	799.4	0.78	0.95	0.16	1235.5	1235.1	0.44	0.84	0.18
GA-AID	804.7	803.9	0.72	0.96	0.18	1189.4	1188.3	0.41	0.85	0.20
MOGA-MAC	761.1	760.01	0.78	0.96	0.17	1193.3	1191.9	0.44	0.84	0.19
MOGA-MIA	763.2	762.1	0.78	0.96	0.17	1195.6	1194.9	0.44	0.84	0.20
MOGA-DAC	746.6	745.5	0.69	0.95	0.17	1122.0	1119.8	0.40	0.89	0.21
MOGA-DIA	742.6	741.2	0.72	0.93	0.18	1123.7	1121.6	0.44	0.82	0.21
H-MAC	711.4	710.1	0.70	0.93	0.03	1065.6	1063.0	0.25	0.86	0.05
H-MIA	748.4	747.1	0.59	0.93	0.00	1114.7	1112.5	0.30	0.88	0.00
					20×5					*20×10*
GA-MAC	1247.3	1246.1	0.46	0.90	1.23	1574.8	1572.6	0.23	0.87	1.58
GA-MIA	1250.0	1248.6	0.44	0.90	1.76	1569.9	1566.9	0.20	0.86	2.38
GA-AIM	1393.7	1393.3	0.84	0.97	0.99	1726.7	1725.1	0.64	0.90	1.54
GA-AID	1396.5	1395.6	0.87	0.94	1.13	1722.6	1721.3	0.60	0.94	1.58
MOGA-MAC	1310.8	1309.7	0.85	0.97	1.20	1685.9	1683.8	0.73	0.81	1.53
MOGA-MIA	1312.8	1311.3	0.85	0.97	1.21	1683.2	1681.1	0.63	0.90	1.60
MOGA-DAC	1270.7	1296.1	0.79	0.98	1.23	1602.4	1599.8	0.48	0.93	1.56
MOGA-DIA	1265.7	1264.1	0.87	0.90	1.27	1607.0	1604.2	0.51	0.93	1.60
H-MAC	1215.9	1214.2	0.53	0.92	1.07	1513.5	1509.7	0.30	0.90	2.16
H-MIA	1304.4	1303.5	0.65	0.94	0.03	1631.8	1628.7	0.33	0.88	0.04

Let us first analyse the performance of the different fitness functions of the GA. As could be expected, the results indicate that, when only the productivity goal of minimising C_{max} is considered, GA-MAC or GA-MIA should be used. Notice that the lowest values of C_{max} are obtained with schedules for which at least a due-date constraint is not satisfied at all, in fact these solutions are, in average, a 45% and a 46% worst than the best value of AI_{min} found. More surprising are the results with respect to due-date satisfaction. At first, we may feel tempted to conclude that, if the goal is to respect due dates GA-AIM or GA-AID should be used. However, a more careful look shows that if we use multi-objective versions instead, due dates are satisfied to almost the same degree and there is the added benefit of reducing the makespan and improving in productivity with a very similar computational effort, specially with the MOGA-MAC proposed in this work.

If we also consider the heuristic approach, H-MAC obtains the best results when minimising makespan. However, notice that the heuristic methods are more sensitive to the defuzzification criterion than the GA. Indeed, the GA is quite indifferent to criterion used and, in general, relative differences between both criteria are less than 1%. The only exception is that AI_{min} values with MOGA-DAC are worse than with MOGA-DIA while AI_{ave} values are better. On the other hand, when H-MAC is used, the values for F_{AC} e F_{IA} are the best but, with H-MIA, they are about 6% worse in average and they are even worse than the values obtained with some multiobjective versions of the GA.

Regarding due-date satisfaction, AI_{min} values for the heuristic approach are much worse than for the GA versions where feasibility is incorporated to the objective function and slightly better than the GA versions using only makespan. However, results for AI_{ave} are similar to those obtained with a multiobjective GA.

We may conclude that, even if there is not a clearly better method, the most promising ones seem to be the heuristic approaches proposed herein. The final decision as to what method should be used may depend on the time available to obtain a solution: if the method must be fast, we recommend a heuristic method, otherwise, a multiobjective GA might be used.

5 Conclusions

We have considered the *FFSP*, a version of *FSP* that tries to model the imprecise nature of data in real-world problems. Using a fuzzy set representation, we have modelled uncertain processing times and flexible due-date constrains. We have proposed a heuristic method that combines a greedy algorithm and a local search procedure. Also, we have proposed three new multi-objective functions to enhance the genetic algorithm proposed by [5]. We have conducted an experimental study across a set of problem instances generated by ourselves to compare our approaches with this genetic algorithm. From this study, we can conclude that the heuristic method is the best option to optimize the makespan only. However, in order to optimize all three objectives considered at the same time, the genetic algorithm with one of our proposed objective functions is clearly the best option.

In the future, we plan developing new heuristic methods to solve the FFSP. In particular, we will try to improve the greedy algorithm strategy by means of problem specific knowledge, in order to fulfill the due-dates requirements, but keeping the computational cost within reasonable bounds.

References

1. Brucker, P.: Scheduling Algorithms. 4th edn. Springer (2004)
2. Brucker, P., Knust, S.: Complex Scheduling. Springer (2006)
3. Dubois, D., Fargier, H., Fortemps, P.: Fuzzy scheduling: Modelling flexible constraints vs. coping with incomplete knowledge. European Journal of Operational Research **147** (2003) 231–252

4. Słowiński, R., Hapke, M., eds.: Scheduling Under Fuzziness. Volume 37 of Studies in Fuzziness and Soft Computing. Physica-Verlag (2000)
5. Celano, G., Costa, A., Fichera, S.: An evolutionary algorithm for pure fuzzy flowshop scheduling problems. International Journal of Uncertainty, Fuzziness and Knowledge-Based Systems **11** (2003) 655–669
6. Fortemps, P.: Jobshop scheduling with imprecise durations: a fuzzy approach. IEEE Transactions of Fuzzy Systems **7** (1997) 557–569
7. Lin, F.T.: Fuzzy job-shop scheduling based on ranking level $(\lambda, 1)$ interval-valued fuzzy numbers. IEEE Transactions on Fuzzy Systems **10**(4) (2000) 510–522
8. Sakawa, M., Kubota, R.: Fuzzy programming for multiobjective job shop scheduling with fuzzy processing time and fuzzy duedate through genetic algorithms. European Journal of Operational Research **120** (2000) 393–407
9. Fayad, C., Petrovic, S.: A fuzzy genetic algorithm for real-world job-shop scheduling. Innovations in Applied Artificial Intelligence, Lecture Notes in Computer Science **3533** (2005) 524–533
10. González Rodríguez, I., Puente, J., Vela, C.R., Varela, R.: Evaluation methodology of schedules for the fuzzy job-shop problem. IEEE Transactions on Systems, Man and Cybernetics, Part A. Accepted for publication.
11. Petrovic, S., Song, X.: A new approach to two-machine flow shop problem with uncertain processing times. Optimization and Engineering **7** (2006) 329–342
12. Ishibuchi, H., Yamamoto, N., Murata, T., Tanaka, H.: Genetic algorithms and neighborhood search algorithms for fuzzy flowshop scheduling problems. Fuzzy Sets and Systems **67** (1994) 81–100
13. Song, X., Petrovic, S.: Handling fuzzy constraints in flow shop problem. In De Baets, B., Fodor, J., Radojevic, D., eds.: Proceedings of the 9th Meeting of the EURO Working Group on Fuzzy Sets, EUROFUSE2005. (2005) 235–244
14. González, N., Vela, C.R.: Un algoritmo metaheurístico para la resolución de problemas $p//\sum c_i$. In: Actas del IV Congreso Espaol sobre Metaheurísticas, Algoritmos Evolutivos y Bioinspirados (MAEB'2005). (2005) 449–456
15. Bortolan, G., Degani, R.: A review of some methods for ranking fuzzy subsets. In Dubois, D., Prade, H., Yager, R., eds.: Readings in Fuzzy Sets for Intelligence Systems. Morgan Kaufmann, Amsterdam (NL) (1993) 149–158
16. Framinan, J., Gupta, J., Leisten, R.: A review and classification heuristics for permutation flow shop scheduling with makespan objective. Journal of the Operational Research Society **55** (2004) 1243–1255
17. Liu, J., Reeves, C.: Constructive and composite heuristic solutions to the $p//\sum c_i$ scheduling problem. European Journal of Operational Research **132** (2001) 439–452
18. Ho, J.: Theory and methodology, flowshop sequencing with mean flowtime objective. European Journal of Operational Research **81** (1995) 571–578

Fusion of Neural Gas*

Sebastián Moreno[1], Héctor Allende[1], Rodrigo Salas[2], and Carolina Saavedra[1]

[1] Universidad Técnica Federico Santa María; Dept. de Informática; Chile
smoreno@inf.utfsm.cl, hallende@inf.utfsm.cl, saavedra@inf.utfsm.cl
[2] Universidad de Valparaíso; Departamento de Ingeniería Biomédica; Chile
rodrigo.salas@uv.cl

Abstract. One of the most important feature of the Neural Gas is its ability to preserve the topology in the projection of highly dimensional input spaces to lower dimensions vector quantizations. For this reason, the Neural Gas has proven to be a valuable tool in data mining applications.

In this paper an incremental ensemble method for the combination of various Neural Gas models is proposed. Several models are trained with bootstrap samples of the data, the "codebooks" with similar Voronoi polygons are merged in one fused node and neighborhood relations are established by linking similar fused nodes. The aim of combining the Neural Gas is to improve the quality and robustness of the topological representation of the single model. We have called this model *Fusion-NG*.

Computational experiments show that the *Fusion-NG* model effectively preserves the topology of the input space and improves the representation of the single Neural Gas model. Furthermore, the *Fusion-NG* explicitly shows the neighborhood relations of it prototypes. We report the performance results using synthetic and real datasets, the latter obtained from a benchmark site.

Keywords: Machine ensembles, Neural Gas, Machine Fusion.

1 Introduction

Vector quantizations techniques have been successfully applied in several areas as pattern recognition and data mining (see [11]). In particular the Neural Gas (NG), introduced by Martinetz et. al. [7], is a variant of the Self Organizing Map (SOM) where the neighborhoods are adaptively defined during training through the ranking order of the distance of prototypes from the given training sample.

The success of these methodologies are due to their special property of effectively creating spatially organized internal representations of various features of input signals and their abstractions [4]. Tools for visualizing high dimensional data are crucial in discovering patterns in the data, due to the complicated relationships in real world data are difficult to perceive.

* This work was supported by Research Grant Fondecyt 1061201, 1070220 and DGIP-UTFSM.

J. Mira and J.R. Álvarez (Eds.): IWINAC 2007, Part I, LNCS 4527, pp. 558–567, 2007.

Despite their great success, there is a need to improve the quality and performance of the solution of these models. Recently, several techniques of machines learning ensembles have been proposed (see for example [2], [10], [6]). The machine ensemble consists of a set of weak learners where the models decisions of the phenomenon under study are combined to obtain a global decision. The aim of making a machines ensemble is to improve the individual answers.

In this paper we propose an incremental ensemble method consisting of fusing a collection of Neural Gas beginning with two models and successively increasing the number until some stopping criterion is met. The fusion process is accomplished by merging similar nodes based on the information of the data modeled by each prototype. Furthermore, the *Fusion-NG* connects fused nodes that are similar in order to improve the topological representation and visualization. We empirically show that the resulting *Fusion-NG* model will improve the performance and it will be more stable than a single model even under the presence of small quantity of additive outliers in the data.

The remainder or this paper is organized as follows. In the next section we briefly introduce the Machines Ensemble framework. In section 3 the Neural Gas model is introduced. Our proposal of the *Fusion-NG* model is stated in section 4. In section 5 we provide some simulation results on synthetic and real data sets. Conclusions and further work are given in the last section.

2 Machines Ensembles

An ensemble of machines is a set of learners whose individual decisions are combined in some way to produce more accurate results. The aim of combining the decisions is to improve the quality and robustness of the results (see [6], [8]).

The task of constructing ensembles of learners consists in two parts. The first part is achieved by creating and training a diverse set of base learners, where the diversity of the machines can be accomplished in several ways, for example, using different training sets [2], different training parameters or different machines. The second part consists in combining the decisions of the individual learners, for example voting rules are used if the learners outputs are regarded as simple classification "labels".

The combination of the models can be accomplished at any of these three levels: (1) In the input space, process known as data fusion; (2) In the architecture of the machines, process that we call Fusion; and (3) In the output space, process known as aggregation. According to [8] there are generally two types of combinations: machine selection and machine fusion. In machine selection, each model is trained to become an expert in some local area of the total feature space, and the output is aggregated or selected according to their performance. In machine fusion all the learners are trained with samples of the entire feature space, the combination process involves merging the individual machine designs to obtain a single (stronger) expert of superior performance. For example, in [5] a combination of classifier selection and fusion is presented.

3 Neural Gas

In this section we briefly introduce the Neural Gas (NG) model, for further details please refer to [7]. The "Neural-Gas" model consists of an ordered set $\mathcal{W} = \{\mathbf{w}_1, ..., \mathbf{w}_q\}$ of q "codebooks" (prototypes or neurons) vectors $\mathbf{w}_r \in \mathbb{R}^d$, $r = 1, .., q$ arranged according to a neighborhood ranking relation between the units.

When the data vector $\mathbf{x} \in \mathcal{X} \subseteq \mathbb{R}^d$ is presented to the NG model, it is projected to a neuron position by searching the best matching unit (*bmu*), i.e., the prototype that is closest to the input, and it is obtained as

$$c(\mathbf{x}) = \arg \min_{r=1..q} \{\|\mathbf{x} - \mathbf{w}_r\|\} \tag{1}$$

where $\|\cdot\|$ is the classical Euclidean norm. This procedure divides the manifold $\mathcal{X}$ into a number of subregions $\mathcal{V}_r, r = 1..q$, as follows

$$\mathcal{V}_r = \{\mathbf{x} \in \mathcal{X} | \ \|\mathbf{x} - \mathbf{w}_r\| \leq \|\mathbf{x} - \mathbf{w}_i\| \, \forall i\} \tag{2}$$

these subregions are called Voronoi polygons or Voronoi polyhedra, where each data vector $\mathbf{x}$ is described by its corresponding reference vector $\mathbf{w}_r$.

The neighborhood relation of the prototypes in the NG model is defined by the ranking order $(\mathbf{w}_{i_0}, \mathbf{w}_{i_1}, ..., \mathbf{w}_{i_{q-1}})$ of the distance of the codebooks to the given sample $\mathbf{x}$, with $\mathbf{w}_{i_0}$ being the closest to $\mathbf{x}$, $\mathbf{w}_{i_1}$ being second closest to $\mathbf{x}$, and $\mathbf{w}_{i_k}, k = 0, .., q-1$, being the reference vector for which there are k vectors $\mathbf{w}_r$ with $\|\mathbf{x} - \mathbf{w}_r\| < \|\mathbf{x} - \mathbf{w}_{i_k}\|$. If $k_r(\mathbf{x}, \mathcal{W})$ denotes the number k associated with each vector $\mathbf{w}_r$, which depends on $\mathbf{x}$ and the whole set $\mathcal{W}$ of reference vectors, then the adaptation step for adjusting the $\mathbf{w}_r$'s is given by:

$$\mathbf{w}_r(t+1) = \mathbf{w}_r(t) + \alpha h_\lambda(k_r(\mathbf{x}, \mathcal{W}))(\mathbf{x} - \mathbf{w}_r) \qquad r = 1, .., q \tag{3}$$

where both the learning parameter function $\alpha = \alpha(t) \in [0, 1]$ and the characteristic decay function $\lambda = \lambda(t)$ are monotonically decreasing functions with respect to time. For example for α the function could be linear $\alpha(t) = \alpha_0 + (\alpha_f - \alpha_0)t/t_\alpha$ or exponential $\alpha(t) = \alpha_0(\alpha_f/\alpha_0)^{t/t_\alpha}$, where α_0 is the initial learning rate (< 1.0), α_f is the final rate (≈ 0.01) and t_α is the maximum number of iteration steps to arrive α_f. Analogously for λ (See [12] for further details).

The neighborhood kernel $h_\lambda(k_r(\mathbf{x}, \mathcal{W}))$ is unity for $k_r = 0$ and decays to zero for increasing k_r. In this paper we use $h_\lambda(k_i(\mathbf{x}, \mathcal{W})) = \exp^{k_i(\mathbf{x}, \mathcal{W})/\lambda}$. Note that if $\lambda \to 0$ then (3) is the K-means adaptations rule, whereas for $\lambda \neq 0$ not only the "winner" (bmu) $\mathbf{w}_{i_0}$ but the second closest reference vector $\mathbf{w}_{i_1}$, third closest vector $\mathbf{w}_{i_2}$, etc., are also updated.

To evaluate the quality of adaptation to the data the *mean square quantization error* is used:

$$MSQE = \frac{1}{|\mathcal{D}|} \sum_{\mathbf{x}_i \in \mathcal{D}} \left\|\mathbf{x}_i - \mathbf{w}_{c(\mathbf{x}_i)}\right\|^2 \tag{4}$$

where $|\mathcal{D}|$ is the number of data that belongs to the input set $\mathcal{D} = \{\mathbf{x}_1, ..., \mathbf{x}_n\}$, and $\mathbf{w}_{c(\mathbf{x}_i)}$ is the best matching unit to the data $\mathbf{x}_i$ defined in equation (1).

4 Fusion of Neural Gas

The *Fusion-NG* model is an ensemble of Neural Gas models that are combined by fusing prototypes that are modeling similar Voronoi polygons (partitions) and with the capability of create a lattice by connecting similar fused codebooks. The aim of combining the NG is to improve the quality and robustness of the results of a single model.

The effectiveness of the ensemble methods relies on creating a collection of diverse yet accurate learning models [6]. The diversity is created by using different training sets. Unfortunately, we have access to one training set $\mathcal{D} = \{\mathbf{x}_1, ..., \mathbf{x}_n\}$, and we have to imitate the process of random generation of T training sets. To create a new training set of length n we apply the Efron's Bootstrap sampling technique [3], where we sample with replacement from the original training set. Indeed, for the construction of the base NG we use the Breiman's Bagging algorithm [2] with different number of prototypes for the training phase. All the models are trained independently with the learning rule given in equation (3).

The construction of the *Fusion-NG* is accomplished by incrementally combining trained neural gas models $\mathcal{W}_t$, $t = 1, 2, ...$, starting with two models and successively adding one more NG model until a desired stopping criterion is met. In the T-th iteration a fuse model $\mathcal{M}_T$ is obtained by combining the first T Neural Gas models.

Let consider the Voronoi polygon $\mathcal{V}_r^{(t)}$, defined in equation (2), of the r-th codebook $\mathbf{w}_r^{(t)}$ of the neural gas $\mathcal{W}_t$ and let $|\mathcal{V}_r^{(t)}|$ be the number of samples that belong to the Voronoi polygon $\mathcal{V}_r^{(t)}$.

Let $q_t = |\mathcal{W}_t|$ the number of units of the model $\mathcal{W}_t$. For each codebook $\mathbf{w}_r^{(t)}, r = 1..q_t, t = 1..T$, compute the mean square quantization error of the prototype as

$$msqe_r^{(t)} = \frac{1}{|\mathcal{V}_r^{(t)}|} \sum_{\mathbf{x}_i \in \mathcal{V}_r^{(t)}} \left\| \mathbf{x}_i - \mathbf{w}_r^{(t)} \right\|^2 \tag{5}$$

We identify all the nodes with low usage, i.e., prototypes whose Voronoi polygon is almost empty and we proceed to delete all the nodes $\mathbf{w}_r^{(t)}$ such that $|\mathcal{V}_r^{(t)}| < \theta_u$, where θ_u is the usage threshold.

We order all the non-deleted codebooks according to their $msqe_r^{(t)}$ performance from the smallest value to greatest in a vector $[\mathbf{w}_{(1)}, ..., \mathbf{w}_{(M_T)}]$, where $M_T = \sum_{t=1}^{T} q_t$ is the total number of codebooks of all the T neural gas models. The map $j(\mathbf{w}_r^{(t)}) = \mathbf{w}_{(j)}$ indexes the performance position of the units, i.e., the prototype $\mathbf{w}_r^{(t)}$ has the j-th best performance.

The co-association matrix between the data and the codebooks is the matrix $C = \{C_{ij}\}$ of size $n \times M_T$ where each component of the co-association matrix takes value 1 if the data $\mathbf{x}_i$ belongs to the Voronoi polygon $\mathcal{V}_{(j)}$ of the prototype $\mathbf{w}_{(j)}$ and 0 otherwise, i.e.,

$$C_{ij} = \begin{cases} 1 & \text{if } \mathbf{x}_i \in \mathcal{V}_{(j)} \\ 0 & otherwise \end{cases} \tag{6}$$

where the i-th row corresponds to the i-th sample of $\mathcal{D} = \{\mathbf{x}_1, ..., \mathbf{x}_n\}$ and the j-th column corresponds to the Voronoi polygon $V_{(j)}$ of the codebook $\mathbf{w}_{(j)}$.

Let define the dissimilarity measure as:

$$ds(C_{\cdot r}, C_{\cdot p}) = \frac{\sum_{i=1}^{n} XOR(C_{ir}, C_{ip})}{\sum_{l=1}^{n} OR(C_{lr}, C_{lp})} \tag{7}$$

where $C_{\cdot r}$ corresponds to the r-th column of the matrix C. With this measure we construct the incidence matrix $\mathcal{I} = \{\mathcal{I}_{rp}\}$ of size $M_T \times M_T$ whose elements are given by

$$\mathcal{I}_{rp} = \begin{cases} 1 & \text{if } ds(C_{\cdot r}, C_{\cdot p}) < \theta_f \\ 0 & otherwise \end{cases} \tag{8}$$

where θ_f is the fusion threshold.

Now we consider the incidence matrix starting from the first row corresponding to the unit $\mathbf{w}_{(1)}$ and we add it to the set W_1 together with all the similar codebooks, i.e., add to the set W_1 all the codebooks $\mathbf{w}_{(i)}$ such that $\mathcal{I}_{1i} = 1, i = 1..M_T$. Let s be the number of sets that have been already created. Then, with the rests of the units, $j = 2..M_T$, if the prototype $\mathbf{w}_{(j)}$ was not previously included in any set, increment in one the number s and create the set W_s with the prototype $\mathbf{w}_{(j)}$ and with all its similar codebooks $\mathbf{w}_{(i)}$ such that $\mathcal{I}_{ji} = 1$, $i = j+1..M_T$. Otherwise, if the prototype $\mathbf{w}_{(j)}$ is in any previous set, all its similar nodes $\mathbf{w}_{(i)}$ that satisfy $\mathcal{I}_{ji} = 1, i > j$, are included in all the sets W_l where $\mathbf{w}_{(j)} \in W_l$. Suppose that the final number of sets created is λ. The fused prototypes $\overline{\mathbf{w}}_s$, $s = 1..\lambda$ will correspond to the centroid of their respective sets W_s, computed as:

$$\overline{\mathbf{w}}_s = \frac{1}{|W_s|} \sum_{\mathbf{w}_{(j)} \in W_s} \mathbf{w}_{(j)} \qquad s = 1..\lambda \tag{9}$$

The lattice of the *Fusion-NG* is obtained by pairwise connecting the fused nodes. The link between the fused prototypes $\overline{\mathbf{w}}_k$ and $\overline{\mathbf{w}}_q$ is established if the dissimilarity between the sets W_k and W_q are less than the connection threshold θ_c, i.e.,

$$\min_{\mathbf{w}_r \in W_k, \mathbf{w}_q \in W_q} ds(\nu(\mathbf{w}_r), \nu(\mathbf{w}_q)) < \theta_c \tag{10}$$

All the established links $\{\overline{\mathbf{w}}_k, \overline{\mathbf{w}}_q\}$ are included in the set of connections $\mathcal{N}^T$. The resulting *Fusion-NG* model $\mathcal{M}^T = \mathcal{M}(\mathcal{W}^T, \mathcal{N}^T)$ obtained at iteration T is composed with the set of fused prototypes $\mathcal{W}^T = \{\overline{\mathbf{w}}_1, ..., \overline{\mathbf{w}}_\lambda\}$ together with the set of connections $\mathcal{N}^T$. If the stopping criterion is not met, we repeat the process by incrementing in one the number of base models T, otherwise the process is finalized.

Finally, algorithm 1 shows the Fusion process for the Neural Gas models explained earlier.

Algorithm 1. The *Fusion-NG* Algorithm

Given is a training data set $\mathcal{D}$ with n elements.
Initialize the parameters. Pick the usage threshold θ_u, the fusion threshold θ_f and the connection threshold θ_c, $0 \leq \theta_f < \theta_c \leq 1$. Select a stopping criterion. Let $T = 0$ and $s = 1$.
repeat
  Increment T by one.
  Take a bootstrap sample $\mathcal{D}_T$ from $\mathcal{D}$.
  Create and train the NG model $\mathcal{W}_T$ using $\mathcal{D}_T$ as the training set.
  Compute the $msqe_r^{(t)}$ to all the prototypes $\mathbf{w}_r^{(t)}, r = 1..q_t, t = 1..T$ with equation (5).
  Order all the codebooks according to their $msqe_r^{(t)}$ performance from the smallest value to greatest in a vector $[\mathbf{w}_{(1)}, ..., \mathbf{w}_{(M_T)}]$.
  Eliminate all the nodes with low usage, i.e, delete the node $\mathbf{w}_r^{(t)}$ if $|\mathcal{V}_r^{(t)}| < \theta_u$.
  Compute the co-association matrix C with equation (6)
  Compute the incidence matrix $\mathcal{I}$ with equation (8)
  Create the set W_1 with the unit $\mathbf{w}_{(1)}$ together with all the codebooks $\mathbf{w}_{(i)}$ such that $\mathcal{I}_{1i} = 1, i = 1..M_T$.
  for $j = 2..M_T$ **do**
    if the prototype $\mathbf{w}_{(j)}$ was not previously included in any set **then**
      Increment in one the number s. Create the set W_s with the prototype $\mathbf{w}_{(j)}$ and all the codebooks $\mathbf{w}_{(i)}$ that satisfy $\mathcal{I}_{ji} = 1$, $i = j + 1..M_T$.
    else
      All the nodes $\mathbf{w}_{(i)}$ that satisfy $\mathcal{I}_{ji} = 1, i > j$ are included in all the sets W_l where $\mathbf{w}_{(j)} \in W_l$.
    end if
  end for
  Compute the location of the fused prototype $\overline{\mathbf{w}}_s$ as the centroid of the set $W_s, s = 1..\lambda$ with equation (9). Create the set of fused prototypes $\mathcal{W}^T = \{\overline{\mathbf{w}}_1, ..., \overline{\mathbf{w}}_\lambda\}$
  Connect all fused nodes of the set $\mathcal{W}^T$ that satisfy the equation (10). Add the connections to the set $\mathcal{N}^T$.
  The resulting Fusion-NG model is $\mathcal{M}^T = \mathcal{M}(\mathcal{W}^T, \mathcal{N}^T)$.
until The stopping criterion is met.
Output: The Fusion-NG model $\mathcal{M}^T$

5 Simulation Results

In this section we empirically show the capabilities of our *Fusion-NG* model proposal compared to a single Neural Gas (NG) model in both Synthetic and Real data sets, the latter was obtained from a benchmark site.

For the synthetic experiment we used three types of synthetic data sets. The first two data sets are the well known *"Doughnut"* and *"Spiral"* data sets (see the first and second columns of figures 1 and 2 respectively). The third data set, that we call the *"Three Objects"* consists of three types of clusters: a bi-dimensional gaussian, a noisy sinusoidal curve and the bi-dimensional uniform random samples (see the third column of figures 1 and 2). The models executed

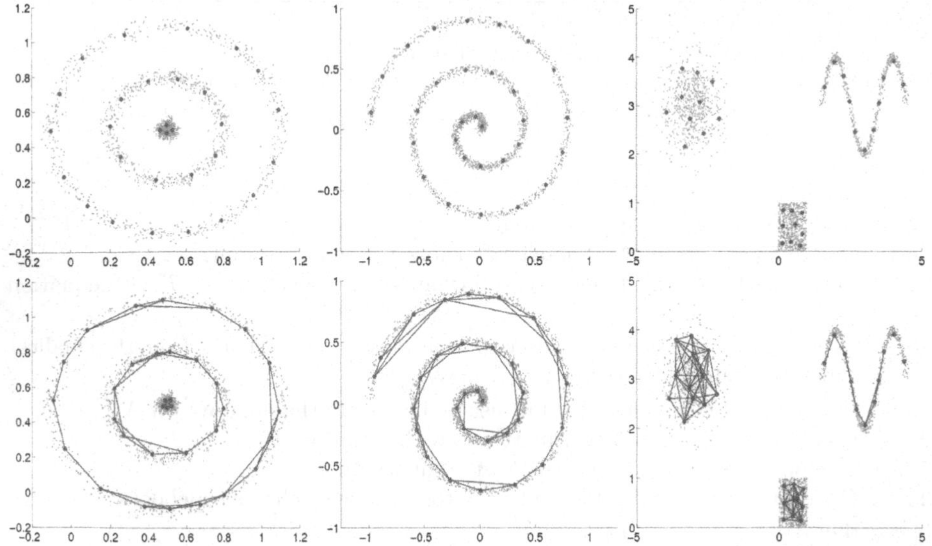

Fig. 1. Simulation Results for the Synthetic data sets without outliers: The figures show the topology approximation of the NG (first row), the *Fusion-NG* (second row) to the Doughnut (first column), Spiral (second column) and Three Objects (third column) data sets

were the NG model (see first row of figures 1 and 2) and the *Fusion-NG* model (see second row of figures 1 and 2) .

First we experiment with the synthetic data set free of outlying observations. Figure 1 shows the topology approximation of the NG and *Fusion-NG* models. Note that the two models were able to effectively learn the topologies. Nevertheless the *Fusion-NG* shows the topology relations between its neurons favoring a better visualization.

We experiment with the synthetic data, but in this time we introduce 1% of additive outliers. Figure 2 shows the topology approximation results. Note that in all three cases, some prototypes of the NG were located far from the bulk of data because they learn these outliers. However, the *Fusion-NG* was more stable and resistant to the outliers in all three cases and obtained a better topological representation of the data.

The quality of adaptation of the models to the data were computed with equation (4). Let $MSQE_T$ be the MSQE of the *Fusion-NG* composed by the first T trained NG, $T = 1..20$. The left and the middle graphs of figure 3 shows the ratio between the $MSQE_T$ and the $MSQE_1$ times 100% for data without outliers and with 1% outliers respectively, where $MSQE_1$ is the MSQE evaluation of the first and single NG model. The left side of figure 3 shows that for the case without outliers, the performance of the *Fusion-NG* improves the performance of the single model between 15% to 24%. However, for the case with the presence of outliers (middle graph of figure 3), the *Fusion-NG* obtained worse MSQE values

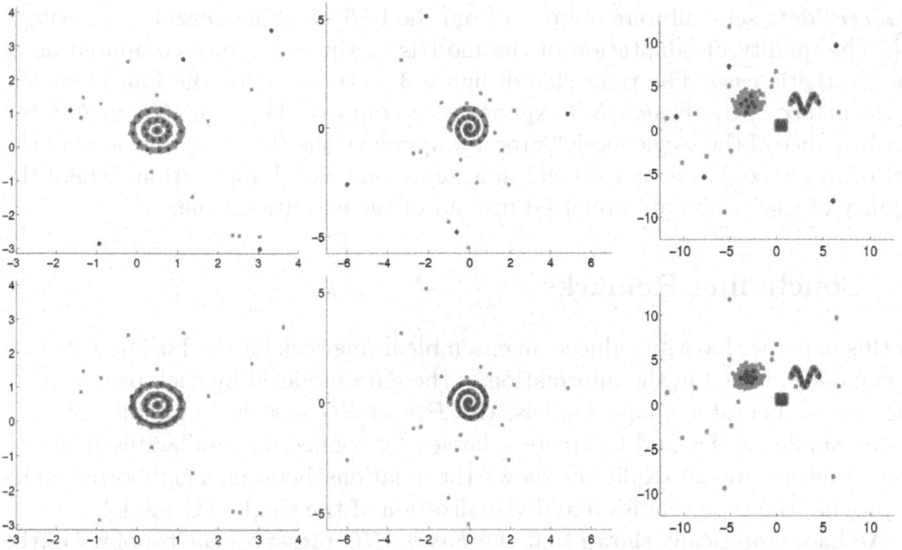

Fig. 2. Simulation Results for the Synthetic Datasets with 1% outliers: The figures show the topology approximation with outliers of the NG (first row) *Fusion-NG* (second row) to the Doughnut (first column), Spiral (second column) and Three Objects (third column) datasets

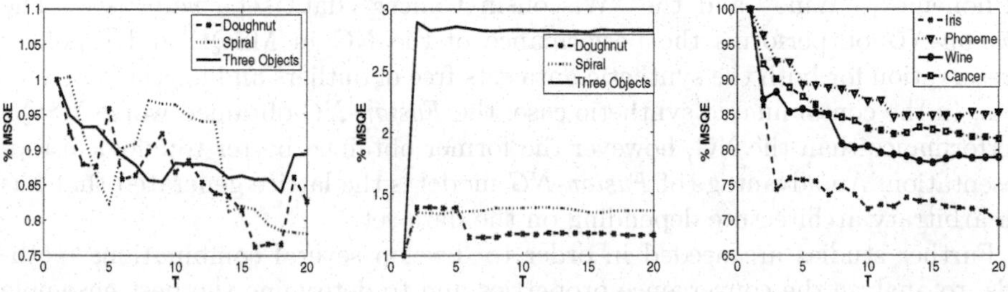

Fig. 3. Performance Evaluation The figures shows the graphs of the ratio between the MSQE of the *Fusion-NG* obtained with T models and the MSQE of the first single NG times 100% evaluated for the "Doughnut", "Spiral" and "Three objects" data sets without outliers **(left)**, with outliers **(middle)** and **(right)** the "Iris", the "Phoneme", the "Wine" and the "Wisconsin Cancer" data sets

than the single model, where the MSQE for the three cases were incremented between 115% to 290%. This increments to the MSQE is explained by the fact that the NG is modeling the outliers implying in a lower MSQE value. This is caused because the MSQE is not a good performance metric of the adaptation quality under contaminated data [9].

In the experiment with real data we test the algorithm with the following benchmarks datasets the *"Iris"*, the *"Phoneme"*, the *"Wine"* and the *"Wisconsin*

Cancer" data sets, all were obtained from the UCI Machine Learning repository [1]. The quality of adaptation of the models to the data were computed as in the synthetic case. The right side of figure 3 shows that for the four cases the performance of the *Fusion-NG* exponentially improves between 15% to 30% the performance of the single model after few aggregations. This improvement of the performance could be very crucial in a many real world applications where the quality of the results are preferred instead of the execution time.

6 Concluding Remarks

In this paper we have introduced an ensemble framework for the Fusion of Neural Gas models. Based in the information of the data modeled by each prototype of the several neural gas base models, the *Fusion-NG* has the capability of both fusing similar nodes and to create a lattice by connecting similar fused nodes. The resulting model explicitly shows the relations between neighboring units improving the representation and visualization of the single NG model.

We have empirically shown that the *Fusion-NG* preserves the topology of the input space by effectively locating the prototypes improving the performance of the single model. Furthermore the *Fusion-NG* is more stable and resistant to the presence of outlying observations, obtaining a better topological representation than the NG.

For the simulation study with synthetic data the "Doughnut", "Spiral" and "Three objects" data sets were used, while for the real experiment, the "Iris", "Phoneme", "Wine" and the "Wisconsin Cancer" data sets were used. The *Fusion-NG* outperforms the performance of the NG in MSQE and topology preservation for both the synthetic data sets free of outliers and for the real data sets. In the contaminated synthetic case, the *Fusion-NG* obtained worse MSQE performance than the NG, however the former obtained better topological representation. An advantage of *Fusion-NG* model is the lattice generated that has an arbitrary architecture depending on the data set.

Further studies are needed in order to develop several combinations methods, to analyze the convergence properties and to determine the best ensemble parameters. Possible interesting applications of the *Fusion-NG* could be Health sciences, where the quality of the results is of paramount importance.

References

1. C.L. Blake and C.J. Merz, *UCI repository of machine learning databases*, 1998.
2. L. Breiman, *Bagging predictors*, Machine Learning **24** (1996), no. 2, 123–140.
3. B. Efron, *Bootstrap methods: another look at the jacknife*, The Annals of Statistics **7** (1979), 1–26.
4. T. Kohonen, *Self-Organizing Maps*, Springer Series in Information Sciences, vol. 30, Springer Verlag, Berlin, Heidelberg, 2001, Third Extended Edition 2001.
5. L. Kuncheva, *Switching between selection and fusion in combining classifiers: An experiment*, IEEE Trans. on System Man, And Cybernetics – Part B **32** (2002), no. 2, 146–156.

6. ———, *Combining pattern classifiers: Methods and algorithms*, Wiley, 2004.
7. T. Martinetz, S. Berkovich, and K. Schulten, *Neural-gas network for vector quantization and its application to time-series prediction*, IEEE Trans. on Neural Networks **4** (1993), no. 4, 558–568.
8. R. Polikar, *Ensemble based systems in decision making*, IEEE Circuits and Systems Magazine **6** (2006), no. 3, 21–45.
9. C. Saavedra, S. Moreno, R. Salas, and H. Allende, *Robustness analysis of the neural gas learning algorithm*, LNCS **4225** (2006), 559–568.
10. R. Schapire, *The boosting approach to machine learning: An overview*, 2001.
11. U. Seiffert and L. Jain (eds.), *Self-organizing neural networks: Recent advances and applications*, Studies in Fuzziness and Soft Computing, vol. 78, Springer Verlag, 2002.
12. M. Su and H. Chang, *Fast self-organizing feature map algorithm*, IEEE Trans. on Neural Networks **11** (2000), no. 3, 721–733.

Decision Making Graphical Tool for Multiobjective Optimization Problems

X. Blasco, J.M. Herrero, J. Sanchis, and M. Martínez

Department of Systems Engineering and Control
Polytechnic University of Valencia
xblasco@isa.upv.es
http://ctl-predictivo.upv.es

Abstract. Multiobjective optimization problems have become an important issue at many engineering problems. A tradeoff between several design criteria is required and important efforts are made for the development of Multiobjective Optimization Techniques and, in particular, Evolutionary Multiobjective Optimization. Usually these algorithms produce a set of optimum solutions in Pareto sense, there is not a unique solution. The designer (Decision Maker) has to finally select one solution for each particular problem, then he has to select from a set of Pareto solutions, the most adequate solution according to his preferences. It is widely accepted that visualization tools are valuable tools to provide the Decision Maker (DM) with a meaningful way to analyze Pareto set and then to help to select an adequate solution. This work describes a new graphical way to represent high dimensional and large sets of Pareto solutions, allowing an easier analysis, and helping the DM to select an adequate solution.

1 Introduction

In numerous engineering areas the obtention of suitable designs is turned into a multiobjective (or multicriteria) problem. That means that it is necessary to look for a solution in the design space that satisfies several specifications (objectives) in the performance space. In general, these specifications are in conflict with each other, that is, there is no optimal solution for all of them simultaneously. In this context, the solution is not unique, and there is instead a set of possible solutions none of which is the best for all objectives. This set of optimal solutions in the design space is called Pareto set. The region defined by the performances (value of all objectives) for all Pareto set points is called Pareto front.

The first step to solve a multiobjective optimization problem could be to obtain the Pareto set points (and the Pareto front). This is an open research field where numerous techniques have already been developed [8,7]. An alternative and very active research line is Multiobjective Evolutionary Algorithms [4,5]. In general, these algorithms supply a reasonable solution for the Pareto set and front. The following step for the designer is to select one or several solutions inside the Pareto set. The final solution is often selected including designer preferences and following a handmade procedure based on designer experience.

J. Mira and J.R. Álvarez (Eds.): IWINAC 2007, Part I, LNCS 4527, pp. 568–577, 2007.

Decision-Making methodology is a field in constant development with very interesting solutions. It is widely accepted that visualization tools are valuable to provide the Decision Maker a meaningful way to analyze Pareto set and select good solutions. But these tools lose potentiality for high dimensional problems with a large Pareto set. For a 2-Dimensional problem (and sometimes for 3-Dimensional) it is normally easy to make an accurate graphical analysis of the Pareto set points, but for higher dimensions it becomes more difficult. There are several interesting alternatives offering graphical representation [1,6,9].

This work contributes a new alternative, called *Layer Graph*, that allows an easier analysis of Pareto set and front becoming, a useful tool for the Decision Maker. Following sections describe the proposed graphical representation and show simple examples. Subsequently this representation is used in a more complex problem, in order to choose an adequate solution in a multiobjective problem with four dimensions in the performances space and three dimensions in the parameter space.

2 Layer Graph for Pareto Front

A multiobjective problem can be formalized as follows:

$$\theta = [\theta_1, \ldots, \theta_l] \in \mathcal{D} \subseteq \mathcal{R}^l \; ; \; \mathbf{J}(\theta) = [J_1(\theta), \ldots, J_s(\theta)] \; ; \; \min_{\theta \in \mathcal{D}} \mathbf{J}(\theta)$$

Without loss of generality, it consists of a simultaneous minimization of all objectives $J_i(\theta)$. In general, there is no single solution, but a set of solutions none of which is better than the others. Using the definition of *dominance*, the Pareto set $\mathbf{\Theta}_p$ is the set which contains all non-dominated solutions.

It is said that a solution θ^1 dominates another solution θ^2, denoted by $\theta_1 \prec \theta_2$, iff

$$\forall i \in B, J_i(\theta^1) \leq J_i(\theta^2) \;\wedge\; \exists k \in B : J_k(\theta^1) < J_k(\theta^2) \, .$$

Therefore the Pareto optimal set $\mathbf{\Theta}_P$, is given by

$$\mathbf{\Theta}_P = \{\theta \in D \mid \nexists \; \tilde{\theta} \in D \; : \; \tilde{\theta} \prec \theta\} \, . \tag{1}$$

$\mathbf{\Theta}_P$ is unique and normally includes infinite solutions. Hence a set $\mathbf{\Theta}_P^*$, with a finite number of elements from $\mathbf{\Theta}_P$, should be obtained ($\mathbf{\Theta}_P^*$ is not unique).

At this point, the Decision Maker has a set as $\mathbf{\Theta}_p^* \subset \mathcal{R}^l$, that constitutes the Pareto set and an associated set of objective values for every point that constitutes a description of Pareto front $\mathbf{J}(\mathbf{\Theta}_p^*) \subset \mathcal{R}^s$.

The *Layer Graph* tool is based on the classification by sorted layers according to the proximity to the ideal point[1] at the Pareto front points $\mathbf{J}(\mathbf{\Theta}_p^*)$.

For the definition of a layer, every objective ($J_i(\theta)$, $i = 1 \ldots s$) is normalized relative to its minimum and maximum values at the Pareto front, $\bar{J}_i(\theta)$:

$$J_i^M = \max_{\theta \in \mathbf{\Theta}_p^*} J_i(\theta); \; J_i^m = \min_{\theta \in \mathbf{\Theta}_p^*} J_i(\theta); \; i = 1 \ldots s \tag{2}$$

[1] Ideal point is a point with the minimum value of the Pareto front at each objective.

$$\bar{J}_i(\theta) = \frac{(J_i(\theta) - J_i^m)}{J_i^M - J_i^m} \tag{3}$$

Assuming that nl is the number of layers required for the classification, the normalized range of variation for each objective is divided in nl intervals of the same length, and each one is numbered in ascending order.

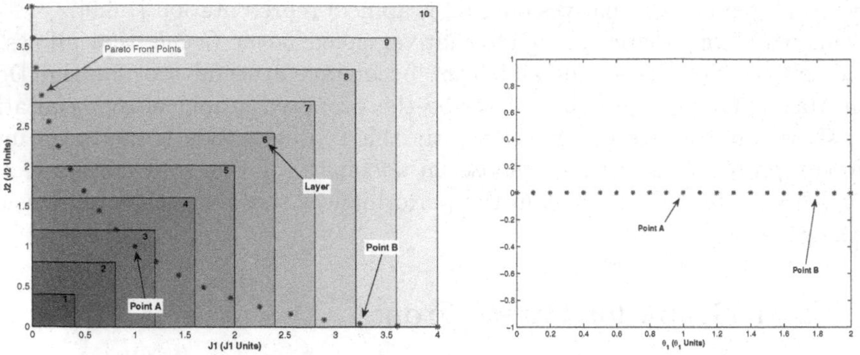

Fig. 1. Pareto front classical 2D representation and interpretation of layer for $nl = 10$

Then it is possible to find the number of the interval where each objective of a point θ is situated ($lo_i(\theta)$, $i = 1 \ldots s$):

$$lo_i(\theta) = \lceil \bar{J}_i(\theta) \cdot nl \rceil \tag{4}$$

The concept of a point belonging to a layer implies that all points of a layer have the objectives with the highest number of interval in this layer.

Then, the number of layer ($L^{nl}_{\mathbf{\Theta}^*_p}(\theta)$) that is assigned to a point $\theta \in \mathbf{\Theta}^*_p$ for a a classification in nl layers is obtained as follows:

$$L^{nl}_{\mathbf{\Theta}^*_p}(\theta) = \max_{i=1\ldots s} (lo_i(\theta)) \tag{5}$$

For instance, normalizing every objective $\bar{J}_i(\theta) \in [0, 1]$ and establishing $nl =$ 10 layers, the ranges of values ($\bar{J}_i(\theta)$) for each layer are: layer 1 have the worst objectives in the range $]0, 0.1]$, layer 2 in the range $]0.1, 0.2]$, ..., layer 10 in the range $]0.9, 1]$. A point that belongs to layer 5 has, at least, one objective value in the range $]0.4, 0.5]$ and the rest of its objectives have values below 0.5.

Once every point is classified, the graphical representation of Pareto front and Pareto set is performed with the following methodology. Each objective (J_i) and decision variable (θ_j) have its own graphical representation. The vertical axis of all graphs corresponds to layers, which means that all graphs are synchronized relative to this axis. The horizontal axis corresponds to the value of the objective or decision variable in physical units. Layers are shadowed in darker grey for lower layers and lighter grey for higher layers.

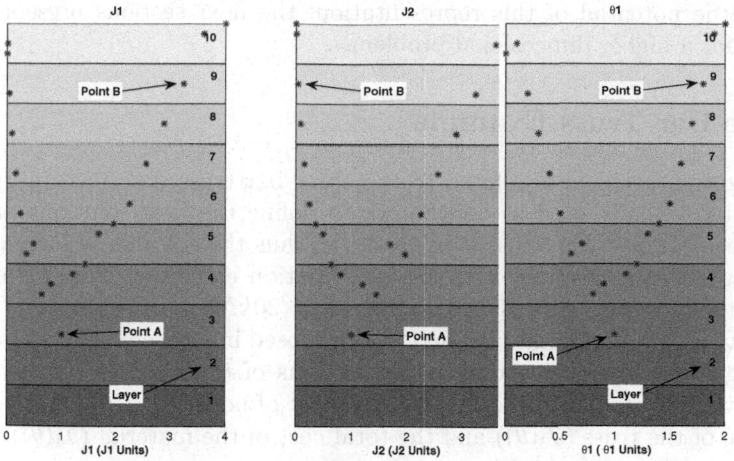

Fig. 2. *Layer Graph* representation for MOP1 problem (for $nl = 10$)

Points in higher layers correspond to points with a high value in one or more objectives, but, obviously they must be good in at least one objective. However, the level in the graphical representation inside a layer does not established any classification of proximity to the ideal point. Therefore, points represented higher inside a layer do not have to be interpreted as worse than lower ones. For an adequate interpretation of *Layer Graph*, it is important to remark that each objective and component of a point is represented at the same level position for all graphs, which means all information of a single point is drawn at the same position at vertical axis for all graphs J_i and θ_j.

The next simple example of *Layer Graph* is shown to clarify this new alternative. A classical 2D problem (MOP1) is selected, although *Layer Graph* is not necessary for a 2D problem (classical representation is enough). Characteristics of MOP1 are:

$$J_1(\theta) = \theta^2 \; ; \; J_2(\theta) = (\theta - 2)^2 \; ; \quad \min_{10^5 \leq \theta \leq 10^5} [J_1(\theta), J_2(\theta)]$$

Figures 1 and 2 show a 2D classical representation and *Layer Graph*, respectively, for a discrete set of Pareto points of a MOP1 problem. Figures 1 shows the most common type of representation of a 2D Pareto front, and also the zones corresponding to each layer are shown shadowed as on *Layer Graph*.

Each point of Pareto front $\mathbf{J}(\mathbf{\Theta}_p^*)$ corresponds to a point in each graph (J_1 and J_2) on *Layer Graph* (see figure 2). For instance, point A on figure 1 corresponds to points A on both graphs (J_1 and J_2) of figure 2, etc.

Pareto set $\mathbf{\Theta}_p^*$ is drawn in a similar way. The classification of Pareto front points by layers is maintained, then for each component of a Pareto set point there is an associated graph. A point is drawn at the same level (vertical coordinate) on each graph and, this level is the same for the associated Pareto front point (see graph θ_1 at figure 2).

To show the potential of this representation, the next sections present the application for a higher dimensional problem.

3 Three-Bar Truss Example

The optimization problem is related to the three-bar truss described in figure 3. It is a truss broadly used as benchmark to define the best solutions based on some specifications. The truss is hyperstatic; thus the solution of balance of forces has to be supplemented with the deformation equations. For this case, the parameters $l = 1m$, $\beta = 45^o$, $\alpha = 30^o$, $F = 20kN$, $\rho = 7580K_g/m^3$ and $K = 15\, euro/K_g$ (the material cost per K_g) proposed in [7] were selected.

The design variables correspond to the sections of the bars $\theta = [\theta_1, \theta_2, \theta_3]$. The objectives correspond to the the displacement of node P ($J_1(\theta)$, $J_2(\theta)$), the total volume of the truss ($J_3(\theta)$) and the total cost of the material ($J_4(\theta)$).

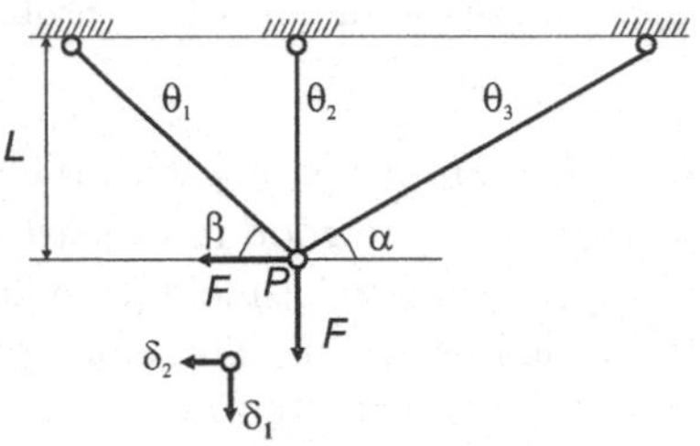

Fig. 3. Three-bar truss problem $\beta = 45^o$ y $\alpha = 30^o$

The problem can be formulated as follows:

$$\min \mathbf{J}(\theta) = [J_1(\theta), J_2(\theta), J_3(\theta), J_4(\theta)] \tag{6}$$

$$s.t. : 0.1 \cdot 10^{-4} m^2 \leq \theta_i \leq 2 \cdot 10^{-4} m^2,\ i = 1 \ldots 3,$$

where:

$$J_1(\theta) = \delta_1,\ \ J_2(\theta) = \delta_2,\ \ J_3(\theta) = l(\frac{\theta_1}{\sin\beta} + \theta_2 + \frac{\theta_3}{\sin\alpha}),$$

$$J_4(\theta) = K\rho l(\frac{\theta_1}{\sin\beta} f_1 + \theta_2 f_2 + \frac{\theta_3}{\sin\alpha} f_3).$$

f_1, f_2 y f_3 are coefficients related to the manufacture cost which depend on the section area of each bar. Theses ones can be calculated as

$$f_i = \begin{cases} 1 & if\ 0.9e^{-4} \leq \theta_i \leq 1.5e^{-4} \\ 1.4 & otherwise \end{cases}$$

Deformations δ_1 and δ_2 are calculated as [2]:

$$\begin{bmatrix} \delta_1 \\ \delta_2 \end{bmatrix} = \frac{l}{E} \begin{bmatrix} \gamma_{11} & -\gamma_{12} \\ \gamma_{12} & \gamma_{22} \end{bmatrix}^{-1} \begin{bmatrix} F \\ -F \end{bmatrix},\ \ \begin{aligned} \gamma_{11} &= \theta_2 + \theta_1 \sin^3\beta + \theta_3 \sin^3\alpha, \\ \gamma_{12} &= \theta_1 \sin^2\beta \cos\beta - \theta_3 \sin^2\alpha \cos\alpha, \\ \gamma_{22} &= -\theta_1 \sin\beta \cos^2\beta - \theta_3 \sin\alpha \cos^2\alpha. \end{aligned} \tag{7}$$

where $E = 200GPa.$ is the *Young* module. Besides, the problem is subject to three constraints ($\sigma = 200MPa.$) related to the reaction forces in each bar N_i. These reaction forces are calculated according to the following expressions [2]:

$$\frac{|N_1|}{\theta_1} = \frac{E}{l}(\delta_1 \sin\beta - \delta_2 \cos\beta)\sin\beta \leq \sigma, \ \frac{|N_2|}{\theta_2} = \frac{E}{l}\delta_1 \leq \sigma, \tag{8}$$

$$\frac{|N_3|}{\theta_2} = \frac{E}{l}(\delta_1 \sin\alpha + \delta_2 \cos\alpha)\sin\alpha \leq \sigma. \tag{9}$$

The constraints (8)(9) will be taken into account through static penalty functions [3]. Therefore the objective functions results in:

$$J_1(\theta) = \delta_1 + C(\theta), \ J_2(\theta) = \delta_2 + C(\theta), \ J_3(\theta) = l(\frac{\theta_1}{\sin\beta} + \theta_2 + \frac{\theta_3}{\sin\alpha}) + C(\theta),$$

$$J_4(\theta) = K\rho l(\frac{\theta_1}{\sin\beta}f_1 + \theta_2 f_2 + \frac{\theta_3}{\sin\alpha}f_3) + C(\theta)\ , \ C(\theta) = \sum_{i=1}^{3} max\left[0, \ \frac{|N_i|}{\theta_i} - \sigma\right].$$

4 Pareto Front Analysis with Layer Graph

Pareto front has been obtained applying a ϵ-MOGA algorithm[2] [5]. For the graphical representation the number of layers is adjusted to 25. Increasing the number of layers in the graphical representation shows the classification by layers more exactly. Figure 4 shows the layer graph representation for the set of solutions supplied by ϵ-MOGA (1084 points).

The representation shows, at objective graphs, a ∇ layout. This has to be the typical shape of a Pareto front. I must be remembered that all points are non-dominated, so if one objective has lower values, then at least one of the other objectives has to be increased. Graphically that means that points on higher layers have to tend to the extremes of the objective range.

An advantage of this graphical representation is that it is easy to see what is the range of objective values for points nearest to the ideal point (in this example layer 9): $J_1 \in [0.04, 0.05]\,cm$, $J_2 \in [0.12, 0.13]\,cm$, $J_3 \in [420, 500]\,cm^3$ and $J_4 \in [60, 75]\,euros$. For this layer the values of design parameters are: $\theta_1 = 10^{-5}\,m^2$ or $0.5 \cdot 10^{-4}\,m^2$, $\theta_2 \in [0.9 \cdot 10^{-4}, 1.4 \cdot 10^{-4}]\,m^2$ and $\theta_3 \in [1.5 \cdot 10^{-4}, 1.9 \cdot 10^{-4}]\,m^2$, all this values are in physical units.

Other important characteristics can be extracted: the extremes of Pareto front supply the worst case for all objectives ($J_1 \approx 0.1\,cm$ $J_2 \approx 0.22\,cm$ $J_3 \approx 900\,cm^3$ and $J_4 \approx 145\,euros$). A cost (J_4) under $80\,euros$ can be easily obtained, there are Pareto points for layers 9 to 25 (all layers with points). Displacement J_2 is higher than J_1 (between 2 and 4 times higher).

[2] The parameters of the algorithm were set to: $Nind_G = 4$, $Nind_P = 100$, $t_{max} = 20000$, resulting in 80100 evaluations of J_1 and J_2. $P_{c/m} = 0.1$, $d_{ini} = 0.25$, $d_{fin} = 0.1$, $\beta_{ini} = 10.0$, $\beta_{fin} = 0.1$, $n\_box_1 = n\_box_2 = n\_box_3 = n\_box_4 = 50$.

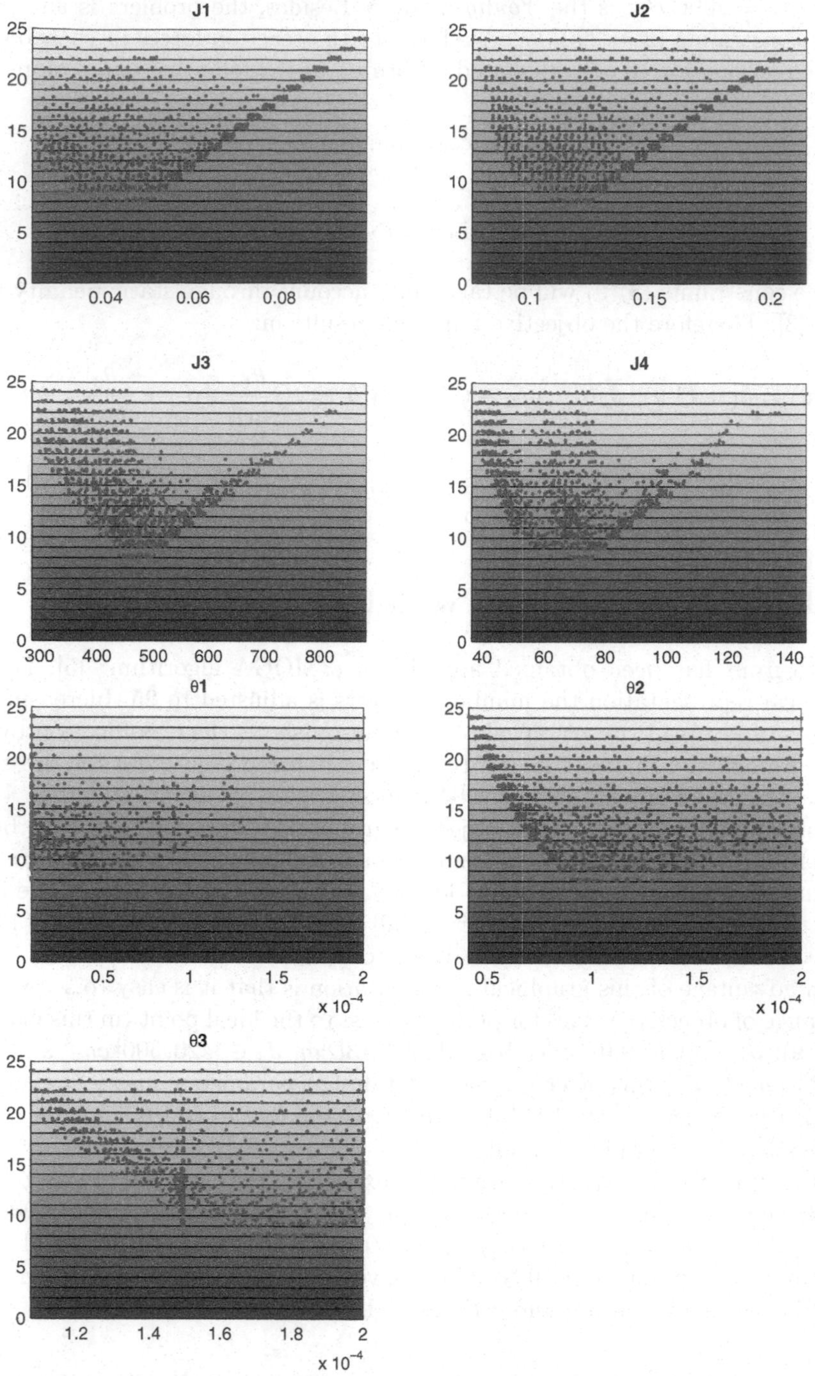

Fig. 4. *Layer Graph* representation of Pareto front and set for Three-bar truss problem

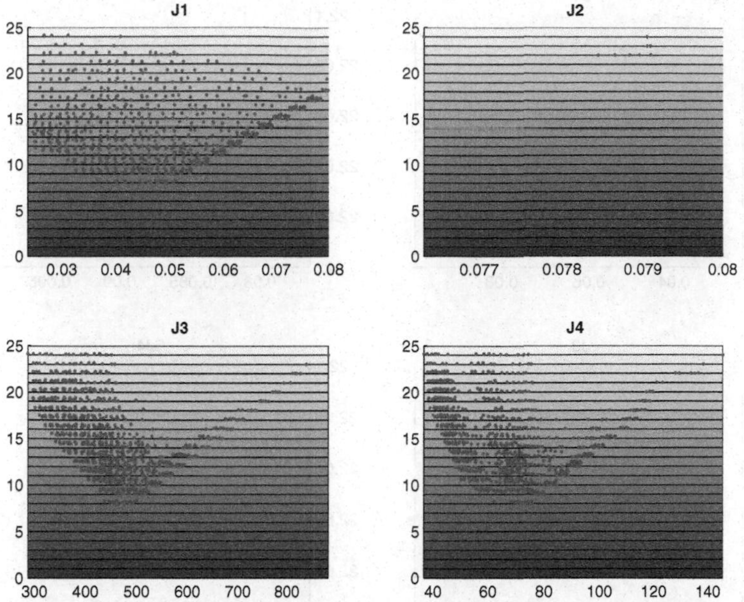

Fig. 5. *Layer Graph* representation of Pareto front and set for Three-bar truss problem. Zoom of J_1 and J_2 between minimum value and 0.08 cm.

Looking at parameter design it is possible to evaluate qualitatively the Pareto solutions. It can be seen that a good choice for θ_1 is $10^{-5}\,cm^2$ (the lower limit). θ_3 is always over 10^{-4} and it has multiple solutions for $1.5 \cdot 10^{-4}$ (the economic limit) and for $2 \cdot 10^{-4}$ (maximum allowed value). θ_2 has no clear pattern but the range $[0.9 \cdot 10^{-4}, 1.4 \cdot 10^{-4}]\,m^2$ covers solutions for nearly all layers.

All this type of information is valuable for an adequate choice of a final solution, but in order to determine a unique solution, the DM has to introduce his preferences. For this example, it will be assumed that preferences are established on displacement: J_1 and $J_2 \leq 0.08\,cm$. Visualization of possible solutions is shown at figure 5 where only the preferred range of value for J_1 and J_2 is shown. For J_1 it is possible to satisfy constraints even for the nearest value to ideal the point, but the limitation is in J_2. The only solutions are available at layers 23 to 25. That means the final solution has to be near to the extreme value of the front. To select the solution, a closer analysis of layer 23 (solutions nearest to the ideal point) is performed. Figure 6 shows the relevant information for layer 23. An adequate solution for this problem is one of the 5 lower points shown at this figure: $J_1 \approx 0.05\,cm$, $J_2 \approx 0.079\,cm$, $J_3 \approx 850\,cm^3$ and $J_4 \approx 110\,euros$, Design parameters for these solutions are: $\theta_1 \approx 1.8 \cdot 10^{-4}\,m^2$, $\theta_2 \approx 1.5 \cdot 10^{-4}\,m^2$ and $\theta_3 \approx 2 \cdot 10^{-4}\,m^2$.

It is easy to see that if the constraints are relaxed to J_1 and J_2 below $0.09\,cm$ the cost and volume of the truss are clearly reduced: $J_3 \approx 450\,cm^3$ and $J_4 \approx 75\,euros$.

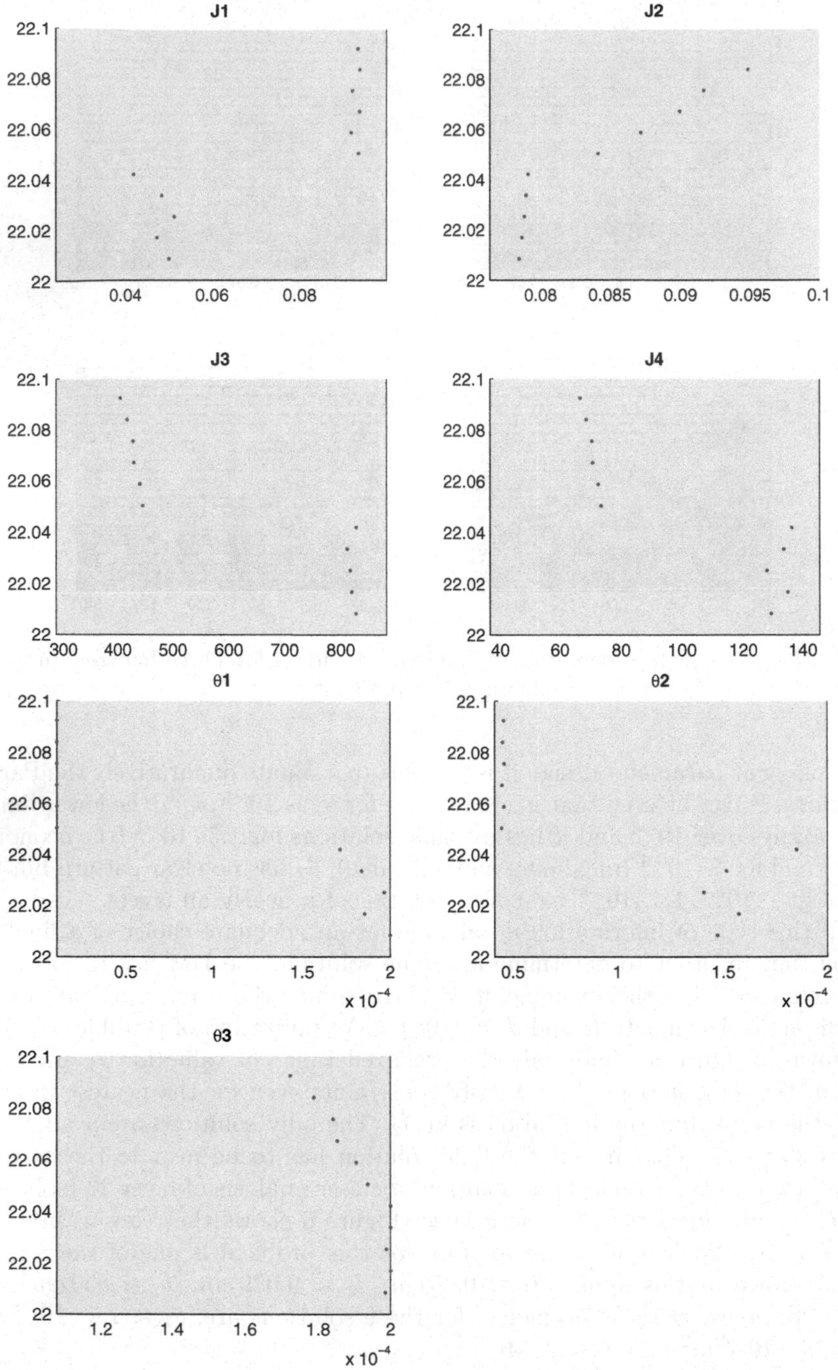

Fig. 6. Partial representation for layer 23 for Three-bar truss problem. Zoom of J_1 and J_2 between minimum value and 0.08 cm.

5 Conclusions

A new visualization methodology for Pareto front representation, called *Layer Graph*, is presented. It allows the analysis of large high-dimensional Pareto fronts and sets. The fundamental idea of this alternative is the classification in layers and the synchronous representation of all objectives and parameters. It is shown that this representation permits a good analysis of Pareto front and provides an excellent tool to help Decision Making. In this article only some of the Pareto front characteristics have been evaluated (closeness to ideal point, ranges of attainable values) but it already gives valuable information and seems to be open to others evaluations of different characteristics. New possibilities for incorporating designer preferences to this representation are being developed and will contribute to improve the decision-making tools for multiobjective problems.

All developments have been made with Matlab©. A beta version for basic *Layer Graph* function and additional tools are available online at: `http://ctl-predictivo.upv.es/programas.htm`

Acknowledgments

Partially supported by MEC (Spanish government) and FEDER funds: projects DPI2005-07835, DPI2004-8383-C03-02 and GVA-026.

References

1. G. Agrawal, C.L. Bloebaum, and K. Lewis. Intuitive design selection using visualized n-dimensional pareto frontier. In *46th AIAA/ASME/ASCE/AHS/ASC Structures, Structural Dynamics and Materials Conference*, Austin, Texas, 2005.
2. Batill, S.M.: Course: ME/AE 446. Finite Element Methods in Structural Analysis, Planar truss applications. www.nd.edu, (1995)
3. C. Coello. Theoretical and numerical constraint-handling techniques used with evolutionary algorithms: a survey of the state of the art. *Computer Methods in applied Mechanics and Engineering*, 191:1245–1287, 2001.
4. K. Deb. *Multi-Objective Optimization using Evolutionary Algorithms.* Wiley, NY, 2001.
5. Juan Manuel Herrero. *Non-linear robust identification using evolutionary algorithms.* PhD thesis, Universidad Politécnica de Valencia, Valencia, Spain, 2006.
6. Chen-Hung Huang and C.L. Bloebaum. Visualization as a solution aid for multi-objective concurrent subspace optimization in a multidisciplinary design environment. In *10th AIAA/ISSMO Multidisciplinary Analysis and Optimization Conference*, Albany, New York, 2004.
7. M. Martínez, J. Sanchis, and X. Blasco. Global and well-distributed pareto frontier by modified normalized normal constraint methods for bicriterion problems. *Structural and Multidisciplinary Optimization*, doi:10.1007/s00158-006-0071-5, 2007.
8. Kaisa M. Miettinen. *Nonlinear multiobjective optimization.* International series in operation research and management science. Kluwer Academic Publisher, 1998.
9. Yee Swian Tan and Niall M. Fraser. The modified star graph and the petal diagram: Two new visual aids for discrete alternative multicriteria decision making. *Journal of Multi-Criteria Decision Analysis*, 7:20–33, 1998.

Electromagnetic Interference Reduction in Electronic Systems Cabinets by Means of Genetic Algorithms Design

Antonio José Lozano-Guerrero[1], Alejandro Díaz-Morcillo[1], and Juan Vicente Balbastre-Tejedor[2]

[1] Depto. Tecnologías de la Información y las Comunicaciones, Universidad Politécnica de Cartagena, Cuartel de Antiguones, 30202 Cartagena, Spain
antonio.lozano@upct.es
[2] Instituto Itaca, Universidad Politécnica de Valencia, Camino de Vera s/n, 46022 Valencia, Spain

Abstract. Conductive plastics have become an alternative to traditional metallic cabinets to shield boxes from electromagnetic interferences. The wide range of available conductivities with these materials can satisfy any particular design. A design with an outer metallic layer and an inner layer of conductive dielectric can obtain advantages from both materials. In this paper the design by means of genetic algorithms of electronic systems cabinets made of new plastic materials to reduce electromagnetic radiated interferences in enclosures with an aperture is described. This optimization procedure requires the use of electromagnetic simulators with a high computational cost. A 2D simulation tool is used in this work for evaluating 3D structures, reducing drastically the computation time. The relationship between obtained solutions and skin depth parameter is evaluated to help in design procedures. A commercial 3D full wave electromagnetic tool has been used to validate the obtained results.

1 Introduction

Electronic systems are enclosed in a cabinet with many functions: protection against physical aggressions, good looking of the product, electromagnetic shielding; this last feature becomes important in environments where the electromagnetic spectrum is polluted by many radiating sources.

Enclosures have been traditionally manufactured with metals. However, there has been a rising interest of the plastic industry on the use of conductive polymers in electromagnetic shielding tasks in the last years due to different reasons: they are lighter, have no corrosion problems and designs are aesthetically more interesting. An optimized design for every application is possible due to the wide range of available conductivities in these materials.

Genetic algorithms (GA) have become a general purpose optimization technique widely used in the last years for electromagnetic design [1]. Shielding electromagnetic properties of a multilayer cylinder for low frequencies have been

J. Mira and J.R. Álvarez (Eds.): IWINAC 2007, Part I, LNCS 4527, pp. 578–586, 2007.

studied through genetic algorithms in [2] by means of its analytical expressions. On the other hand shielding properties of metallic enclosures against electromagnetic interferences have been studied through many techniques. Analytical tools provide fast results. In [3] an analytical and fast formulation is used to obtain the shielding effectiveness of empty rectangular enclosures with apertures. However, numerical methods are necessary when specific problems cannot be solved by analytical tools although the computation time and the memory requirements are high. In [4] the Method of Moments (MoM) is used to study the shielding properties of a metallic enclosure, both empty and with inner elements as printed circuit boards (PCBs) or absorbers. The Finite-Difference Time-Domain (FDTD) method has also been applied to study the shielding effectiveness (SE) of an empty metallic box with apertures [5]. In [6] a work to determine the shielding effectiveness of a double layered spherical shell with no apertures in its surface showed that there is an optimum conductivity for resonance suppression. In [7] a simple approximation to study the behaviour of conductive dielectrics inside metallic enclosures with an aperture is presented. In this paper a study of electromagnetic interference suppression by means of GA with the aid of a 2D simulation tool to evaluate 3D structures shows how these materials can be used jointly with metallic enclosures to improve the shielding behaviour of a cabinet.

2 Theory

2.1 Shielding Effectiveness

Shielding effectiveness for a particular shielding configuration is defined as the ratio between the field in the selected placement without enclosure and the field with the enclosure. In this study the shielding effectiveness has been obtained for the electric field

$$SE(dB) = 20 \log_{10} \left| \frac{E_i}{E_t} \right| . \tag{1}$$

For the two dimensional study the effect of the width of the aperture w is evaluated through the following factor as reported in [7]

$$SE(dB) = SE_{2-D}(dB) + 10 \log_{10} \left(\frac{b}{w} \right) . \tag{2}$$

where b is the height of the cavity, w is the width of the aperture and $\mathrm{SE}_{2-\mathrm{D}}$ is the shielding effectiveness obtained for the 2-D configuration through any numerical method. As long as the frequency increases, new enclosure resonances reduce the shielding properties of the cabinet. A metallic shield provides high shielding effectiveness levels but apertures in the surface of the shield allow the coupling of energy from the outer part to the inner one. Resonances associated to the enclosure dimensions produce high field levels that can affect the normal operation of inner electronic equipment. In the present study it is shown that

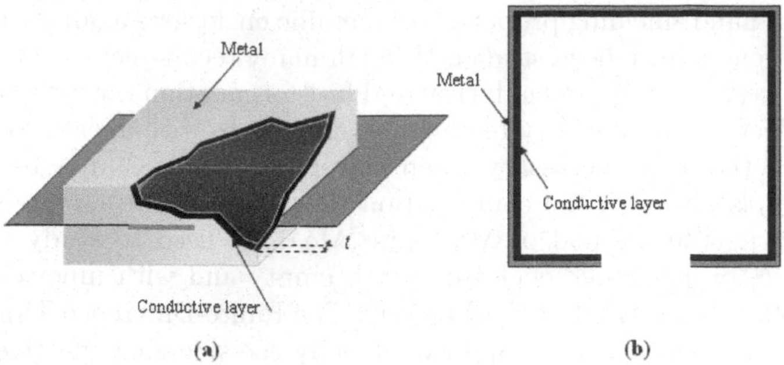

Fig. 1. Studied cabinet with an external metallic layer and an inner conductive layer (b) 2-D equivalent studied enclosure

the metallic surface can provide the required SE levels and an inner lossy layer can be designed to absorb high field levels associated to the enclosure resonances (Fig. 1).

The skin depth, obtained by (3), depends on frequency f and conductivity σ and is a penetration parameter of a wave in a medium. This parameter can help us to design the inner lossy layer.

$$\delta = \frac{1}{\sqrt{\pi f \mu_0 \sigma}} \ . \tag{3}$$

2.2 Genetic Algorithms

Genetic algorithms are optimization procedures based on the principles of natural selection and evolution. Global optimization avoiding local minima is provided and designs with multiple parameters can be achieved. In Fig. 2 the flow chart of the evolutionary procedure is depicted.

An initial population formed by individuals or possible solutions (dielectric constant and conductivity σ) is randomly selected. After evaluating the initial population, operations of selection, mutation and recombination are performed. The best individuals have more possibilities to survive in the next generation. Results obtained using genetic algorithms are compared to the skin depth parameter taking into account the resonance frequency of the enclosure with the aperture, the thickness of the conductive dielectric sample and the optimum conductivity obtained for the sample. Genetic algorithms have been implemented by using a Matlab^{TM} code [8].

In Table 1 the parameters of the GA have been listed. The fitness function evaluates the value of E_i/E_t where $E_i = 1V/m$ and E_t is the field obtained in the location under study for the chosen resonance frequency. The number of generations and individuals has been limited to 30 and 20, respectively, due to the high computation times required by a two dimensional full wave simulator. An optimal solution is obtained by GA when the number of generations and

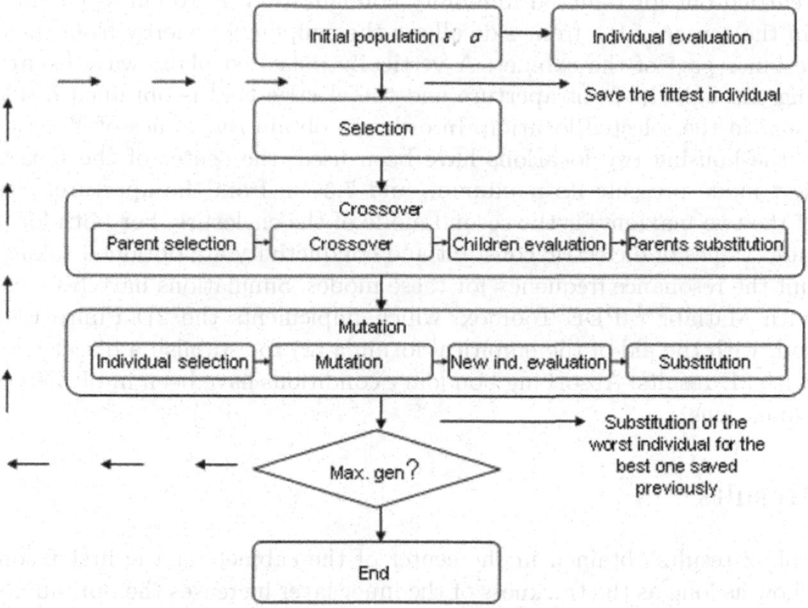

Fig. 2. Genetic algorithm flow chart

Table 1. Genetic algorithm parameters

Design parameters	ϵ_r' and σ
Population	20
Generations	30
Selection type	Geometric normalized
Crossover prob.	10 crossover/generation
Crossover type	Arithmetic
Mutation prob.	2 mutations/generation
Mutation type	Non-uniform
ϵ_r' limits	1-10
σ limits	0-600 S/m

individuals is high enough. However, the high computational cost required to evaluate each individual with the help of an electromagnetic simulator leads to limit the number of generations and individuals per generation. Results as the skin depth parameter or two dimensional approaches to the solution can provide an initial orientation to solve a more complex problem.

3 Set Up

The cabinet used for all the simulations has as inner dimensions 30 cm x 12 cm x 30 cm. The inner layer thickness varies from 0.1 cm to 0.3 cm. The study has been carried out for radiated immunity configuration. A 10 cm x 0.5 cm aperture in the center of the front side allows the coupling of energy from the outer to the inner part of the cabinet. A vertically polarized plane wave excites the housing through the front aperture and the electric field is obtained inside the enclosure in the selected location. In order to obtain the values of electric field inside the housing two locations have been used: the center of the box where the first mode presents its maximum, and 7.5 cm from the aperture, close to one of the two maxima for the second mode of the enclosure. For both locations optimum values of dielectric constant and conductivity are obtained taking into account the resonance frequency for these modes. Simulations have been carried out with Matlab^{TM} PDE Toolbox, which implements the 2D Finite Element Method, with the aid of the empirical formula (2) to establish a direct relationship with 3D results. Absorbing boundary conditions have been implemented for the simulations.

4 Results

In Table 2 results obtained in the center of the cabinet for the first resonance show how as long as the thickness of the inner layer increases the optimum value for the conductivity decrease. This behavior is directly related to the values of the skin depth as reported in [6]. In that case an approximated value of $t/\delta \approx 1.15$ was obtained. Concerning to $\epsilon_r^{'}$ variation it tends to the higher values of the studied limits in this case. However obtained results show that its influence is negligible against the conductivity of the studied material as the skin depth does not depend on the $\epsilon_r^{'}$. This conclusion is very interesting since both values are usually related in fabrication processes and only the conductivity will be interesting in electromagnetic interference reduction.

Table 2. Best values for the first resonance

Thickness t(cm)	Frequency (MHz)	$\epsilon_r^{'}$	$\sigma(S/m)$	t/δ
0.1	692	7.64	490.31	1.15
0.2	694	8.69	126.27	1.17
0.3	696	6.08	52.97	1.14

In Table 3 the second resonance of the cabinet has been studied. Once again results show a dependence with the skin depth parameter, pointing out the conductivity as the main design parameter for this application. Best values obtained in Table 2 have been simulated, represented and compared with the empty cabinet simulation from 500 MHz to 1200 MHz in Fig. 3. As the thickness of the inner layer increases, higher values of the minimum for the resonances are obtained as expected due to the inclusion of more absorbing material.

In Fig. 4 best values have been depicted for the second resonance. The measurement position has been taken 7.5 cm from the side of the aperture. Similar

Table 3. Best values for the second resonance

Thickness t(cm)	Frequency (MHz)	ϵ_r'	$\sigma(S/m)$	t/δ
0.1	1073	7.56	302.22	1.13
0.2	1076	6.48	78.42	1.15
0.3	1079	9.11	38.14	1.20

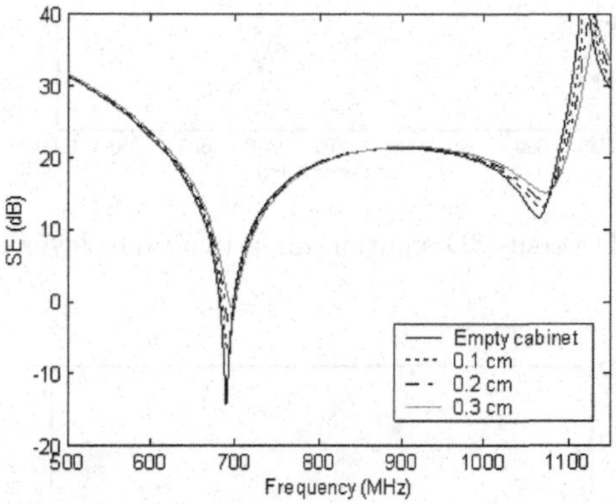

Fig. 3. Simulation of the best values obtained for the first resonance study

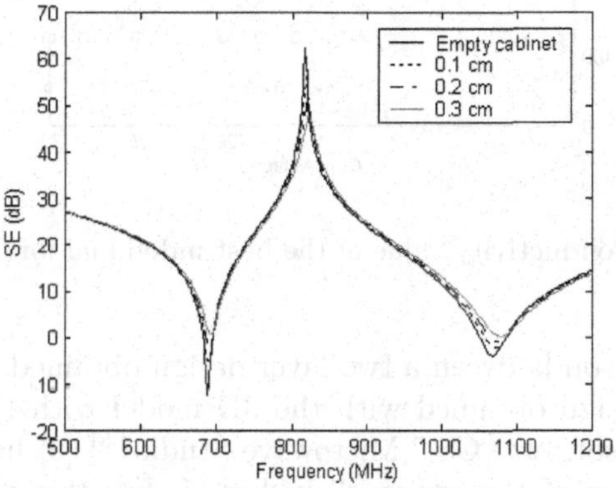

Fig. 4. Simulation of the best values obtained for the second resonance study

conclusions can be extracted for this figure. For this problem the 2D solution becomes a good approach due to the vertical invariance of the first appearing modes although in the 2D approach the upper and bottom sides of the cavity are not taken into account. For both preceding figures although the resonance suppression is optimized for a specific frequency, the rest of the frequencies are also affected in a minor way. The inclusion of a second inner layer can provide a new degree of freedom to the solution. In order to validate the described

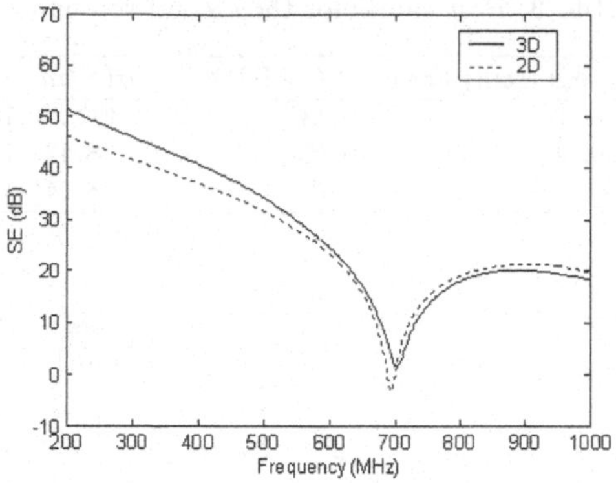

Fig. 5. 2D versus 3D solution for a two inner layer design

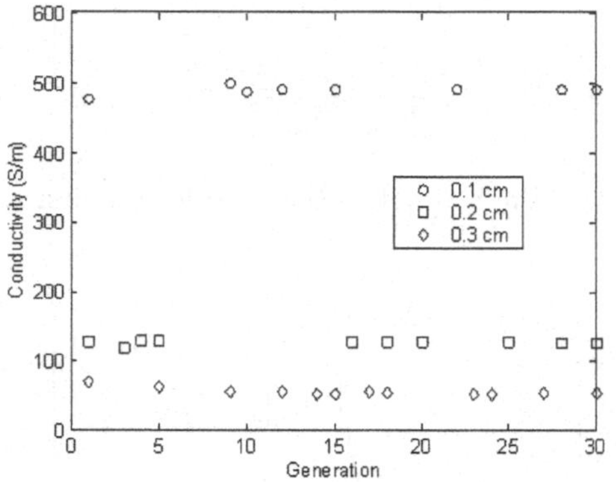

Fig. 6. Conductivity value of the best individual for Table 2

technique a comparison between a two layer design obtained with the proposed 2D model and a design obtained with the 3D model with the aid of the simulation commercial software CST Microwave Studio^{TM} [9] has been performed showing the goodness of the approach in Fig. 5. For this scenario the design parameters are the conductivity of the two inner layers with a 0.1 cm thickness: σ_1 and σ_2. Dielectric constant values have been fixed to 2 and $\sigma \in [0\ 300]$ S/m. Solutions obtained are for the 2D case were $\sigma_1 = 14.27$ S/m and $\sigma_2 = 135.43$ S/m and for the 3D case $\sigma_1 = 7.92$ S/m and $\sigma_2 = 151.6$ S/m. Good agreement has been found between both curves and solutions show similar values. The result obtained for this two layers case (0.1 cm + 0.1 cm) is slightly better than the equivalent one layer solution (0.2 cm) due to the increase of the degrees of freedom.

In Fig. 6 the evolution of the best individuals of each generation for the conductivity value is obtained for the 2D technique in the case of the Table

2 values. As it converges very quickly this information can be used to reduce computation times in more complex problems.

5 Conclusion

In this study the design of electronic systems cabinets to reduce electromagnetic radiated interferences in enclosures with an aperture has been carried out. A 2D simulation tool has been used to evaluate 3D structures jointly with GA.

The proposed technique is applied to optimize a multilayer structure in which the design parameters are the material permittivity and conductivity. Relationship between the thickness and the conductivity of an inner conductive layer has been evaluated with the help of the skin depth parameter showing interesting easy design rules. An example with two layers has been included to validate the proposed technique with the commercial software CST Microwave Studio^{TM} showing good agreement.

Plastic materials are very attractive to manufacture shielding cabinets for electronic equipment. Designs with the help of a full wave electromagnetic simulator jointly with techniques as GA may improve the shielding capabilities of a protecting cabinet. On the other hand, computational times with three dimensional CAD tools become very high. Two dimensional approaches or easy rules may be useful to initiate three dimensional optimization processes reducing drastically computation times in the study of problems with arbitrary shapes and contents.

Acknowledgments. This work was supported in part by Fundación Séneca under project reference 00700/PPC/04.

References

1. Ramat-Samii, Y., Michielssen, E.: Electromagnetic Optimization by Genetic Algorithms. John Wiley & Sons, Inc. Canada (1999)
2. Öktem, M. H., Saka, B.: Design of Multilayered Cylindrical Shields Using a Genetic Algorithm. IEEE Transactions on Electromagnetic Compatibility, **43** (2001) 170–176
3. Robinson, M. P.,Benson, T. M.,Christopoulos, C.,Dawson, J. F.,Ganley, M. C.,Marvin, A. C.,Porter, S. J.,Thomas, D. W. P.: Analytical Formulation for the Shielding Effectiveness of Enclosures with Apertures. IEEE Transactions on Electromagnetic Compatibility, **40** (1998) 240–247
4. Olyslager, F., Laermans, E., Zutter, D., Criel, S., Smedt, R. D., Nietaert N., Clercq, A. D., : Numerical and Experimental Study of the Shielding Effectiveness of a Metallic Enclosure. IEEE Transactions on Electromagnetic Compatibility, **41** (1999) 282–294
5. Georgakopoulos, S. V., Birtcher, C. R., Balanis, C. A.: HIRF Penetration Through Apertures: FDTD Versus Measurements. IEEE Transactions on Electromagnetic Compatibility, **43** (2001) 282–294
6. Yamane, T., Nishikata, A., Shimizu, Y.: Resonance Suppression of a Spherical Electromagnetic Shielding Enclosure by Using Conductive Dielectrics. IEEE Transactions on Electromagnetic Compatibility, **42** (2000) 441–448

7. Lozano, A. J., Díaz, A. Balbastre, J. V., Calvo, A. B., Nuño, L.: Damping of Resonances in a Metallic Enclosure Through Conductive Polymers. European Microwave Conference, (2005) 1399–1402
8. Houck, C.R., Joines, J.A., Kay, M.G.: A Genetic Algorithm for Function Optimization: a MATLAB Implementation. The Mathworks, Natick, MA. NCSU-IE TR (1995) 95–09
9. CST Microwave Studio Manual.

Evolutionary Tool for the Incremental Design of Controllers for Collective Behaviors

Pilar Caamaño, Abraham Prieto, Jose Antonio Becerra, Richard Duro, and Francisco Bellas

Integrated Group for Engineering Research, Universidade da Coruña, 15403, Ferrol, Spain
`http://www.gii.udc.es`

Abstract. In this paper we present a software tool for the automatic design of collective behaviors in animated feature films. The most successful existing commercial solutions used in animation studios require an explicit knowledge by the designer of the AI or other techniques and involve the hand design of many parameters. Our main motivation consists in developing a design tool that permits creating the behaviors of the characters from a high level perspective, using general concepts related to the final desired objectives, and to judge these behaviors from a visual point of view, thus abstracting the designer from the computational techniques in the system core. In this case, a bioinspired approach has been followed consisting in the incremental generation of controllers for simulated agents using evolution. An example of flocking activity is created with the system.

Keywords: Computer Animation, Automatic Design, Collective Behavior, Evolutionary Techniques.

1 Introduction

The creation of animated films with scenes containing hundreds or thousands of characters is a highly time consuming task requiring very repetitive and tedious manual work. In fact, classical solutions consisted simply on the replication of the behavior of a simple character in order to create a collective behavior. In order to automate the design of these kinds of scenes with multiple interacting characters, the animation studios started to develop simulation tools based on artificial intelligence techniques, where the creation of collective behaviors in simulated agents have been widely studied.

This research has led to the conclusion that obtaining controllers for collective behaviors implies two simultaneous tasks. On one hand, the controller for each individual participating in the collective behavior must be designed and, at the same time, the interactions between participants so as to obtain the desired collective goal behavior have to be programmed. Many authors have created an environment with its physics (or operational rules) and introduced sets of agents and some type of fixed or adaptive control strategy for each one of the

J. Mira and J.R. Álvarez (Eds.): IWINAC 2007, Part I, LNCS 4527, pp. 587–596, 2007.

participating individuals and tried to explain what came out. If the result was not satisfactory, they started to fidget with environmental parameters and tried to establish some type of relationship. This is what we would call the *biological approach*: the researchers set some conditions and try to observe what comes out. In this line, some authors take inspiration from natural systems [1],[2],[3] or even try to reproduce complete natural closed systems [4]. Introducing evolution in these types of systems has also been considered by many authors [1],[2],[5]. Some paradigmatic examples of this approach are Collins' [1] AntFarm in which cooperation among individuals is studied. Ray's Tierra [3], which is a simulation of an artificial ecosystem consisting of computer virtual memory (space), virtual machine code (matter) and CPU time (energy). Sims [5] evolved virtual creatures simultaneously considering their morphology and their controllers through artificial neural networks.

Here we are more interested in what could be considered an *engineering approach* rather than a biological or observational approach. We want to obtain systems that produce specific behaviors required for particular tasks or processes. We are thus contemplating a design process with the objective of generating the appropriate controllers for the individual entities participating in the collective process so that the whole does what it is supposed to do. Examples of these approaches can be found in the field of nanotechnology and nanorobotics where thousands of individual entities must be controlled so that their collective behavior is the desired one [6]. One of the classical references in the design of collective behaviors is Reynolds [7] who successfully animated a virtual flock ecosystem, by simulating the local behavior of the boids response to their local surroundings. Reynolds' boids are purely stimulus-driven, that is, they base their action only on direct perception input and their behavior follows a set of predefined local rules. Tu [8] simulated a complex marine environment where fish have the capacity to base their reactions on desires, habit strings and perceptions.

Returning to the field of film animation, the first intelligent techniques applied where based on Particle Swarm Optimization. In these systems, each particle represents a separate animated individual [9]. Examples of the use of this technique are found in the films *Gladiator* or *Star Wars Episode II: Attack of the Clones.* In the film *Lord of the Rings: The Two Towers* a scene of 50,000 battling characters was animated using a software system called MASSIVE [9]. In this case, instead of particle systems, the animators made use of an agent based environment where the agents had complex controllers (using fuzzy logic to design their characters' responses) to produce particular behaviors according to their environments. Another commercial software that has been widely used in animation studios is *Softimage Behavior* that permits the designer to implement different behavioral algorithms.

These commercial tools require an explicit knowledge of the computational techniques underlying them by the designer and a complex design of many parameters by-hand (scripting problem). Our main motivation is related to developing a design tool that permit creating the collective behaviors from a high level perspective, using general concepts having to do with the final desired

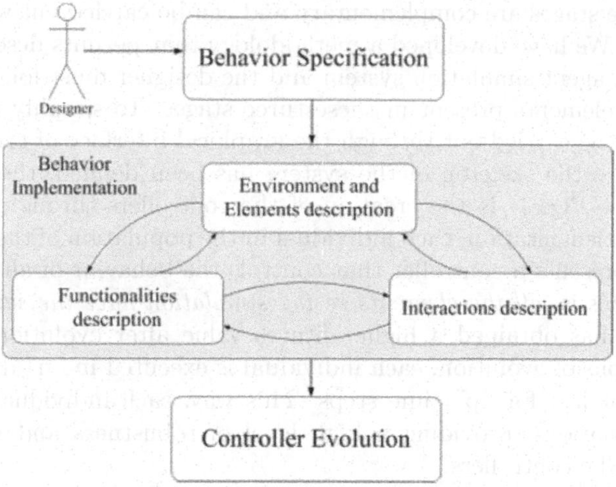

Fig. 1. Simplified flow diagram of BDesigner

objective, and to judge these behaviors from a visual point of view, thus abstracting the designer from the computational techniques in the system core. To achieve these requirements, we decided to use *Evolutionary Techniques* as search algorithms to obtain the controllers of the behaviors. These techniques provide us three main advantages: it allowed using different encodings for the controllers (they could be Artificial Neural Networks, rules, parametric functions, etc), using non-supervised learning and selecting the controller using an indirect measure obtained after observing the behavior through the use of a subjective fitness function. The tool has been called BDesigner (Behavior Designer) and will be presented in the next section.

2 The BDesigner Tool

Fig. 1 shows a simplified flow diagram of the BDesigner tool, where the three main stages of a behavior creation are displayed: the *Behavior Specification*, the *System Implementation* and the *Controller Evolution*. The first thing a designer has to do is to specify the behavior that "he wants to see" from his subjective point of view through the establishment of the *private and global utility functions*, which are described in the next section.

After this initial stage, the designer must *implement the system* by describing:

- *The Environment and the Elements*: the dimensions and the elements present in the environment must be defined to create the simulation skeleton.
- *The Interactions*: the designer must define sensors, effectors and possible interactions among elements and between the elements and the environment.
- *The Functionalities*: to simplify the creation of a complex behavior, it will be divided into simple parts that must be defined here.

These three stages are complementary and can be carried out without a particular order. We have developed a methodology that permits describing all the features of an agent-simulation system and the designer must follow it in order to define the elements present in these three stages. To simplify this task, the definition can be carried out through the graphical interface of the BDesigner.

Finally, once the skeleton of the system has been defined, the last stage of BDesigner (see Fig. 1) is the creation of the controllers through evolution. In this initial implementation, each individual in the population of the evolutionary algorithm represents a controller that controls the behavior of all the agents in the system, this is, *all the elements in the simulation have the same controller*, the one that has obtained a higher fitness value after evolution. During the "m" generations of evolution, each individual is executed in "n" different initial situations that last for "p" time steps. This way, each individual is evaluated in n*m*p situations, providing a high level of robustness and generalization capability to the controllers.

There are two main features in this evolution process that must be considered in detail: the fitness function and the incremental creation of controllers.

2.1 Fitness Function

As commented before, one of the main reasons behind the application of evolutionary algorithms in this tool is that they permit an abstract classification of the controllers through the establishment of a subjective fitness function, this is, a quantitative comparison that depends on what the designer sees in the simulator, if he likes it or not. This quantitative measure is typically denoted in the references as the *utility function* [10]. The utility can be divided into private utility and global utility. The first one is a measure that depends on each individual's behavior and that must be maximized. The second one depends on the global behavior of all the agents in the system, this is, it depends on the degree of collaboration between the elements. The maximization of the global utility is the main objective of the collaborative systems we are dealing with, but to define this global utility is quite complex because in most cases it cannot be derived as a simple sum of the private utilities of all the agents.

Wolpert [10] defines the conditions the private utility function must verify so that an improvement of its value leads to the improvement of the global utility function. In the BDesigner tool we have used these conditions in order to create the private utility function automatically from the global utility function, which is the *only quantity* the designer must define to establish the fitness function.

2.2 Incremental Creation of Controllers

The controllers in the BDesigner tool can be represented by Artificial Neural Networks (as we will see in the examples), rules, parametric functions, etc. The system does not impose any restriction in this sense. As in the general case, these controllers have as inputs the sensorial information corresponding to each particular agent and must provide as output the action that must be executed through the effectors. The problem with the controllers arises from the way they are evaluated (globally) as it is quite hard to determine a progressive construction of complex behaviors. To solve, in part, this problem we have decided to *define a complex behavior by dividing it into simpler behaviors* and using an incremental architecture developed in our group [11] to efficiently combine them.

The approach followed to progressively construct complex behaviors and, what is more important, reuse behaviors that have been previously obtained is described as follows: First, simple low level behaviors, which take the control of effectors, are identified and obtained separately or, they can be reused from previously obtained libraries. The next step is to obtain the higher level behavior controllers. These do not take the control of effectors but modify the way lower level behaviors act. Obviously, in many cases, the available low level behaviors are not appropriate for the higher level modules to be able to implement the desired global behaviors and thus it is necessary to simultaneously obtain another low level behavior that complements those already present and performs every task not covered by them. By iterating this process as required a tree-like control architecture is obtained. Higher level modules modify the operation of lower level modules by acting on their inputs or their outputs through product operations. We have called these process input / sensor modulation and output / effector modulation. A more formal description of the components of the architecture can be found in [11].

The use of effector modulators leads to a continuous range of behaviors for the transitions between those determined by the individual controllers. This is due to the fact that effector values can now be linear combinations of those produced by low level modules. The sensor modulators permit changing how a module reacts under a given input pattern transforming it to a different one. This way, it is very easy to make changes in that reaction for already learnt input patterns. Even if the construction of the control architecture produces an apparently hierarchical structure, this is due to the incremental nature of this process, taking into account the way modulators act.

3 Application Example

To show the basic operation of BDesigner, a flocking behavior, similar to those created by Reynolds [7], has been designed. In our case, we have increased the complexity of the behavior by adding more elements to the system. The behavior we want to simulate is the following: *there is a bird flock where each bird has an internal hunger sensor and an internal fear sensor. If the bird is hungry, it wants to reach food and if it is not hungry, depending on the fear level, it wants to fly close to the rest of the birds or it wants to fly alone.* Two internal sensors (hunger and fear) and an external element (food) have been added to the original boids system. This final behavior has been divided into simpler ones according to the philosophy previously presented.

The initial step the designer must accomplish consists in implementing the system where the behavior will be simulated. First, the designer describes the environment and the elements belonging to this environment. In this case, the environment is a wall-limited world with the height and width established by the designer (700x1000 pixels). The necessary elements are *bird elements* and *food sources*. Once the designer has described the elements and the environment, he will define the sensors and effectors that allow the interaction between them. In this example we need:

- *Center of mass sensor*: it measures the distance and angle to the element neighborhood center of mass.
- *Food source sensor*: it measures the distance and angle to the food source.
- *Hunger and Fear sensors*: internal sensors that measure the hunger and fear level of an element.
- *Movement effector*: effector that allows to move an element through the environment.

To implement these sensors and effectors the designer has to set the values of their characteristics: range, vision distance or direction on the sensors, and maximun speed on the motion effector. As commented before, the designer divides the global behavior into parts called funcionalities. The funcionalities in this example are: *Approach a food source*, *Escape from a food source*, *Approach a group of birds* and *Escape from a group of birds*. According to the incremental architecture presented in section 2.2, to create the final behavior controller, we have designed a hierarchical structure shown in Fig. 2. It contains *four low level modules* acting directly over the effectors, a *continuous selector* that combines the output of two low level modules depending on the fear input and an *effector modulator* that combines the output of the two remaining low level modules and the continuous selector depending on the hunger input.

Once the general structure of the controllers required to implement the final behavior have been defined, they must be obtained through evolution. As a general case, most of the low level behaviors will be present in the behavior library of the system as they will have been previously obtained for other cases. Anyway, we will describe the process assuming that none are present.

The first controller that must be implemented is the one corresponding to *approaching a food source* for the case when the hunger level is high. This controller has been represented by a delay based multilayer perceptron network [12] with 2 inputs (distance and angle to the food source), one hidden layer of 4 neurons and 2 outputs (linear and angular speed). To obtain this controller we have used a classical genetic algorithm with 300 individuals, 80% multipoint crossover, 5% random mutation and a tournament selection with a pool size of 2.

The global utility function establishes the fitness function of the genetic algorithm and, in this case, measures the goodness of the bird motion to reach the food source. This global utility is automatically obtained, in this simple case, as the sum of the private utilities of all the birds in the flock. The private utility function measures the absolute and relative distance from the bird to the food source. With these two values, absolute and relative distance, we prevent the private utility function value depending on the initial position of the bird. The

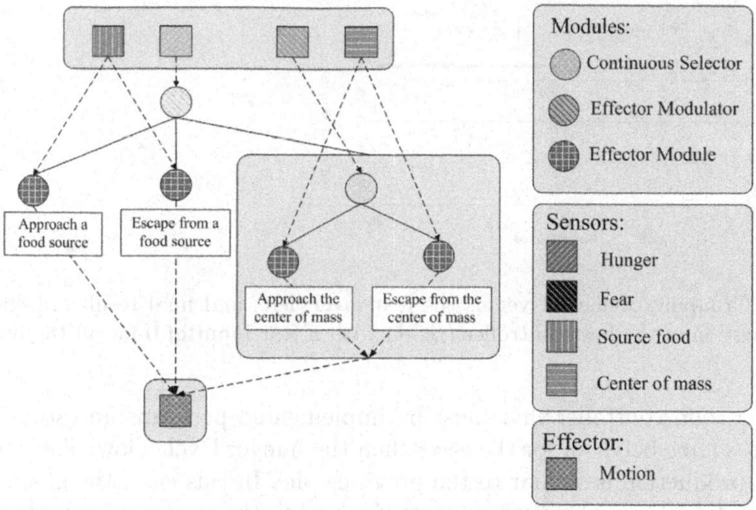

Fig. 2. Final Behavior controller structure

genetic algorithm ran 450 generations containing 5 simulations with different initial positions of 100 steps each.

Fig. 3 left shows the evolution of the fitness function (global utility) in this case. As we can see, in about 200 generations, it reaches the maximum that is maintained until the final generation 450. In the first generations, the controller has already a high fitness value as a consequence to the simplicity of the desired behavior. Anyway, if we simulate the behavior of an early controller (for example, generation 10) we obtain a more inefficient motion than the final one. To illustrate the behavior provided by the best individual after evolution, in Fig. 3 right we have represented the motion of the bird as is seen by the designer in the BDesigner interface.

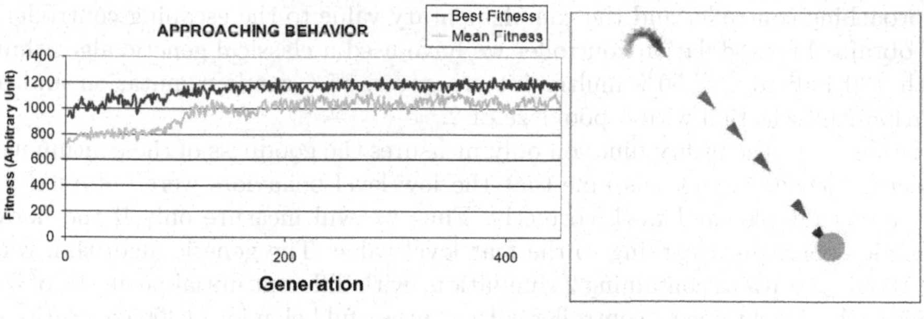

Fig. 3. Evolution of the fitness value for the controller of the behavior for approaching the food (left) and final result obtained with the execution of the best controller (right)

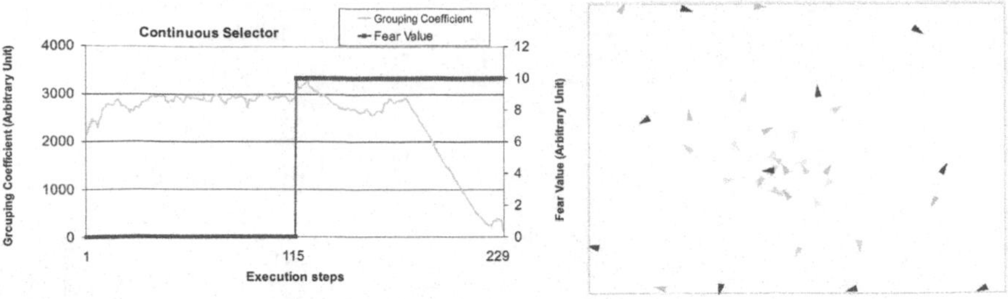

Fig. 4. Grouping coefficient versus the fear level (left) and final result obtained with the execution of the best controller (right) with a fear input of 0 for all the birds

The second controller that must be implemented performs an *escaping from the food source* behavior for the case when the hunger level is low. The controller and its production is similar to the previous one. In this case, the global utility function depends on the distance from the bird to the food source, with a higher utility corresponding to a higher distance.

At this point, we realized that the previously obtained controllers could be generalized to behaviors corresponding to "Approaching a point" and "Escaping from a point". This way, the behaviors of *approaching and escaping from a group of birds* could be obtained directly by just changing the sensor "distance and angle to food source" to "distance and angle to center of mass of a bird neighbourhood". This approach provided us with a successful result so we obtained two new low level controllers without the need for evolving them.

Once the four low level controllers were obtained we decided to create a higher level (second level) controller that, depending on the fear input, creates thighter or more disperse groups of birds. This controller corresponds to a continuous selector module that acts over the low level modules. To do that, this controller has been represented by a multilayer perceptron network with 1 input (fear level), two hidden layers of 2 neurons and 1 output (the modulation level). This modulation is performed assigning the output value provided by the network to the approaching controller and the complementary value to the escaping controller. To obtain this modulation controller we have used a classical genetic algorithm with 130 individuals, 50% multipoint crossover, 15% random mutation and a tournament selection with a pool size of 2.

In this case, the utility function only measures the goodness of the continuous selector, because we can assume that the low level behaviors were obtained in previous examples and work correctly. Thus we will measure only if the flock behavior performs according to the fear level value. The genetic algorithm ran for 1000 generations containing 2 simulations with different initial positions of 50 steps each. We obtained a controller with a successful behavior in 300 generations of evolution.

To show the operation of this module, in Fig. 4 left we have represented the grouping coefficient, that measures the distance between the birds, against the fear value of the flock. As the figure shows, when the fear level reaches a value

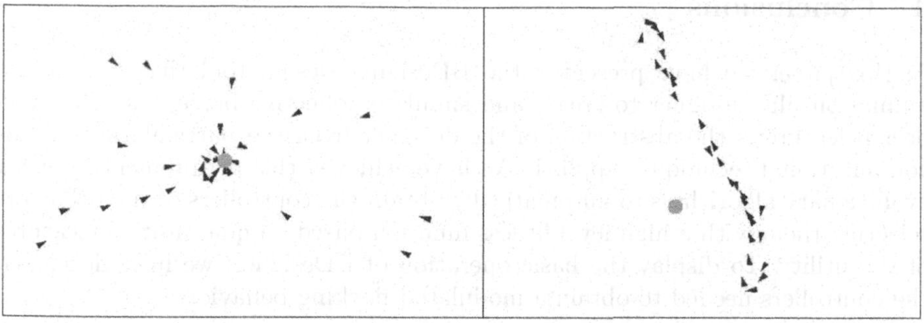

Fig. 5. Final behavior in the case of hunger 10 (left image) and in the case of hunger 0 and fear 10 (right image) when we execute the best controller obtained

of 0, the grouping coefficient takes high values, what means that the birds are separated from each others. In Fig. 4 right we show this behavior with fear 0 as the designer can see it in the BDesigner graphical user interface. If we change the fear level, for example, to 10, the grouping coefficient decreases, what means that the birds form a tighter group when flying together. This behavior with a high fear value is the same as the one shown by the original Reynolds boids.

The final behavior is obtained by creating a higher level controller (third level) that, with a high hunger level, makes the birds approach the food source without paying attention to the fear level. In the case of a low hunger level the birds must pay attention to the fear level flying grouped or dispersed. To achieve this behavior, we decided to create an effector modulator controller, that selects if the bird motion is controlled by the *Approach a food source* controller, by the *Escape from a food source* controller or by the *continuous selector*. This controller has been represented by a multilayer perceptron network with 1 input (hunger level), one hidden layer with 3 neurons and 3 outputs. These outputs correspond to the contribution of each module to the final values to the effectors (linear and angular speed). To obtain this controller we have used a classical genetic algorithm with 180 individuals, 50% multipoint crossover, 15% random mutation and a tournament selection with a pool size of 2.

The global utility is created simply by rewarding the degree of satisfaction of the desired behavior by the whole flock, this is, with a high hunger value, the birds must approach the food and with a low value they must act according to the fear level. The genetic algorithm ran for 1000 generations containing 2 simulations with different initial positions of 50 steps each. In this case, we obtained a controller with a successful behavior in 600 generations of evolution.

In Fig. 5 left we have represented the final behavior the designer can see in the BDesigner interface in the case of hunger 10 (left image) and in the case of hunger 0 and fear 10 (right image) when we execute the best controller obtained. These are just two illustrative behaviors, but we could obtain each intermediate response of the flock by just adjusting the hunger and fear values. Obviously, richer behaviors are obtained than in the case of using just a couple or rules.

4 Conclusions

In this paper we have presented the BDesigner design tool that permits an animation film designer to create and simulate collective behaviors. The main design feature is the abstraction of the designer from the particularities of the computational techniques applied. We have achieved this requirement by using evolutionary algorithms to automatically obtain the controllers of the behaviors selecting them with a high level fitness function based on quantitative measures of the utility. To display the basic operation of BDesigner, we have developed the controllers needed to obtain a modulated flocking behavior.

Acknowledgments. This work was partially funded by the MEC through projects DEP2006-56158-C03-02 and DPI2006-15346-C03-01.

References

1. Collins, R.J., Jefferson, D.: Antfarm: Towards simulated evolution. Artificial Life II. Addison-Wesley, (1992) 579-601
2. Yaeger, L: Computational genetics, physiology, metabolism, neural systems, learning, vision, and behavior or PolyWorld: Life in a new context. Proc. Workshop on Artificial Life (ALIFE92), volume 17. Addison-Wesley, (1994) 263-298
3. Ray, T. S.: An approach to the synthesis of life. Artificial Life II. Addison-Wesley, (1992) 371-408
4. Terzopoulos, D., Tu, X., Grzeszczuk, R.: Artificial fishes with autonomous locomotion, perception, behavior, and learning in a simulated physical world. Proc. ArtificialLifeIV. MIT Press, (1994) 17-27
5. Sims, K.: Evolving virtual creatures. Comp. Graphics, 28 (1994) 15-22
6. Adriano Cavalcanti, A., Freitas Jr. R., Kretly, L.: Nanorobotics Control Design: A Practical Approach Tutorial. Robotics Today, Dearborn, Mich.: SME Society of Manufacturing Engineers, 4th Quarter, Vol. 18, no. 4, (2005)
7. Reynolds, C.W.: Flocks, herds, and schools: A distributed behavioral model. Computer Graphics, volume 21, (1987) 25-34
8. Tu, X: Artificial animals for computer animation: biomechanics, locomotion, perception, and behavior. LNCS Vol. 1635. Springer-Verlag (1999)
9. Jones, A. R.: Animating interactions between agents who change their minds. Honours Thesis, Monash University, (2003)
10. Wolpert, D., Tume, K.: Optimal payoff functions for members of collectives. Advances in Complex Systems, 4(2/3), (2001) 265-279
11. Becerra, J.A., Bellas, F., Santos, J., Duro,R.J.: Complex Behaviours Through Modulation in Autonomous Robot Control. LNCS 3512, (2005) 717-724
12. Duro, R.J., Santos J.: Discrete Time Backpropagation for training Synaptic Delay Based Artificial Neural Networks. IEEE Trans. on Artificial Neural Networks, v. 10, (1999) 779-789

A Possibilistic Approach for Mining Uncertain Temporal Relations from Diagnostic Evolution Databases

Francisco Guil[1], Jose M. Juarez[2], and Roque Marin[2]

[1] Departamento de Lenguajes y Computacion
Universidad de Almeria
04120 Almeria
francisco.guil@ual.es

[2] Dept. Ingenieria de la Informacion y las Comunicaciones
Universidad de Murcia
30071 Espinardo Murcia
{jmjuarez,roque}@dif.um.es

Abstract. In this paper we propose a method for building possibilistic temporal constraint networks that better summarizes the huge set of mined timed-stamped sequences from a temporal data mining process. It belongs to the well-known second-order data mining problem, where the vast amount of simple sequences or patterns needs to be summarized further. It is a very important topic because the huge number of temporal associations extracted in the temporal data mining step makes the knowledge discovery process practically unmanageable for human experts. The method is based on the Theory of Evidence of Shafer as a mathematical tool for obtaining the fuzzy measures involved in the temporal network. This work also presents briefly a practical example describing an application of this proposal in the Intensive Care domain.

1 Introduction

Temporal data mining can be defined as the activity of looking for interesting correlations (or patterns) in large sets of temporal data accumulated for other purposes. It has the capability of mining activity, inferring associations of contextual and temporal proximity, that could also indicate a cause-effect association. This important kind of knowledge can be overlooked when the temporal component is ignored or treated as a simple numeric attribute [10].

In [4] we presented an algorithm, named $TSET$, based on the inter-transactional framework for mining frequent sequences from several kind of datasets, mainly transactional and relational datasets. The improvement of the proposed solution was the use of a unique structure to store all frequent sequences. The data structure used is the well-known set-enumeration tree, widely used in the data mining area, in which the temporal semantic is incorporated. The result is a set of frequent sequences describing partially the dataset. This set forms a potential base of temporal information that, after the experts analysis, can be

J. Mira and J.R. Álvarez (Eds.): IWINAC 2007, Part I, LNCS 4527, pp. 597–606, 2007.

very useful to obtain valuable knowledge. However, the overwhelming number of discovered frequent sequences may make such task absolutely impossible in practice. This problem can be viewed as a second-order data mining problem, which consists in the necessity of obtaining a more understandable and useful sort of knowledge from a huge volume of temporal associations resulting after the data mining process. In this paper, we propose an extension of a previous work [6], which consists on the description of the building of a special model of temporal network formed by a set of uncertain relations amongst temporal points. The temporal model, proposed by HadjAli, Dubois and Prade in [7], is based on the Possibility Theory as expressive tool for the representation and management of uncertainty in point-based temporal relations. The uncertainty is represented by a vector describing three possibility values, expressing the relative plausibility of the three basic relations between two temporal points, that is, "before", "at the same time" and "after". Thus, the authors define the basic operations (inversion, composition, combination and negation) that allow to infer new temporal information and to propagate uncertainty in a possibilistic way. Once the sequences base is obtained (characterized by a frequency distribution), we propose a Shafer Theory-based technique which: firstly divides the sequence base into a set of nested subsets and then it normalizes the frequencies of each nested subset so they add to 1. Secondly, for each nested subset, it builds a temporal constraint network calculating, for each pair of temporal points or events, the possibility degrees of the three basic temporal relations. The result is an enumeration of temporal constraint networks that better summarizes the temporal information existing in the dataset. In other words, they permit the qualitative representation of uncertain temporal relations and they are based on formal sound theory for reasoning with uncertainty.

The remainder of the paper is organized as follows. Section 2 describes briefly the *TSET* algorithm and gives a formal description of the problem of mining frequent sequences from datasets. Section 3 describes briefly the representation aspects of the possibilistic temporal model. In Section 4 we describe the approach for obtaining the uncertain vectors associated with the basic temporal relations from the divided sequences base. Section 5 presents a practical experience at Intensive Care Unit (hereinafter ICU) that illustrates the proposed approach. Conclusions and future work are finally drawn in Section 6.

2 The *TSET* Algorithm

TSET is an algorithm designed for mining frequent sequences (or frequent temporal pattern) from large relational datasets. It is based on the 1-dimensional inter-transactional framework [8], and therefore, the aim is to find associations of events amongst different records (or transactions), and not only the associations of events within records. The main improvement of *TSET* is that it uses a unique tree-based structure to store all frequent sequences. The data structure used is the well known set-enumeration tree, in which the temporal semantic is incorporated.

The algorithm follows the same basic principles as most apriori-based algorithms [1]. Frequent sequence mining is an iterative process, and the focus is on a *level-wise* pattern generation. Firstly, all frequent 1-sequences (frequent events) are found, these are used to generate frequent 2-sequences, then 3-sequences are found using frequent 2-sequences, and so on. In other words, (k+1)-sequences are generated only after all k-sequences have been generated. On each cycle, the *downward closure* property is used to prune the search space. This property, also called anti-monotonicity property, indicates that if a sequence is infrequent, then all super-sequence must also be infrequent.

In the sequel, we will introduce the terminologies and the definitions necessary to establish the problem of mining frequent sequences from large datasets.

2.1 Concepts and Terminologies

Definition 1. *A* dataset D *is an ordered sequence of records* $D[0]$, $D[1]$,...,$D[r-1]$ *where each* $D[i]$ *can have c columns or attributes,* $A[0], ..., A[c-1]$. *The 0-attribute will be the dimensional attribute, the temporal data associated with the record, expressed in temporal units. The rest of attributes can be quantitative or categorical.*

We assume that the domain of each attribute is a finite subset of non-negative integers, and we also assume that the structure of time is discrete and linear. Due to every event registered has its absolute date identified, we represent the time for events with an absolute dating system [9].

Definition 2. *An* event e *is a 3-tuple* $(A[i], v, t)$, *where* $0 < i < c$, $v \in dom\{A[i]\}$, *and* $t \in dom\{A[0]\}$, *that is,* $t \in \mathbb{N}$. *Events are "things that happen", and they usually represent the dynamic aspect of the world [9].*

In our case, an event is related to the fact that a value v is assigned to a certain attribute A[i] with the occurrence time t. The set of all distinct pairs (A[i],v) can be also called event types. *We will use the notation e.a, e.v, and e.t to set and get the attribute, value, and time variables related to the event e, and e.type to get the event type associated with it.*

Definition 3. *Given two events* e_1 *and* e_2, *we define the* $\leq$ *relation as follows:*

1. $e_1 = e_2$ *iff* $(e_1.t = e_2.t) \wedge (e_1.a = e_2.a) \wedge (e_1.v = e_2.v)$
2. $e_1 < e_2$ *iff* $(e_1.t < e_2.t) \vee ((e_1.t = e_2.t) \wedge (e_1.a < e_2.a))$
3. $e_1 \leq e_2$ *iff* $(e_1 < e_2) \vee (e1 = e2)$

We assume that a lexicographic ordering exists among the pairs (attribute, value), the events types, in the dataset.

Definition 4. *A* sequence (or event sequence) *is an ordered set of events* $S = \{e_0, e_1, ..., e_{k-1}\}$, *where for all* $i < j$, $e_i < e_j$. *Obviously,* $|S| = k$.

Definition 5. *Let* U_{tmin} *be the minimal dimensional value associated to the sequence S. In other words,* $U_{tmin} = min\{e_i.t\}$, *for* $e_i \in S$. *If* $U_{tmin} = 0$, *we say* *that S is a* normalized sequence. *Note that any non-normalized sequence can be transformed into a normalized one through a normalization function.*

Let U_{tmax} be the maximal dimensional value associated to the sequence S. This value indicates the maximum distance amongst the events belonging to the normalized sequence S. In other words, $U_{tmax} = e_k.t$, where $|S| = k$. From both, confidence and complexity points of view [8], this value will be always less than or equal to a user-defined parameter called maxspan, denoted by ω.

Definition 6. *The* support *(frequency) of a sequence is defined as:*

$$support(S) = \frac{f_r(S)}{|D|},$$

where $f_r(S)$ denotes the number of occurrences of the sequence S in the dataset, and $|D|$ is the number of records in the dataset D, in other words, r.

Definition 7. *A* frequent sequence *is a normalized sequence whose support is greater than or equal to a user-specified threshold called minimum support. We denote this user-defined parameter as minsup, or simply σ.*

Definition 8. *A sequence is a* frequent maximal sequence *if and only if it is frequent and no proper super-sequence (superset) of it is frequent.*

Given a dataset D and the user-defined parameters ω and σ, the goal of sequence mining is to determine in the dataset the set $\mathcal{S}_f^{D,\sigma,\omega}$, formed by all the frequent sequences whose support are greater than or equal to σ, that is,

$$\mathcal{S}_f^{D,\sigma,\omega} = \{S_i | support(S_i) \geq \sigma\}.$$

This set, formed by a large number of time-stamped sequences, is the goal of the temporal data mining algorithm and the input of the method proposed in this paper for obtaining a temporal constraint network. Basically, the idea is to divide it into a set of nested subsets and, for each subset, obtain a temporal constraint model which summarize better the existing temporal information in the sequences.

3 Representation of Uncertain Temporal Relations

In literature can be found a large amount of work trying to handle uncertainty in temporal reasoning. However, very few work deal with time points as ontological primitives for expressing temporal elements. Basically, two temporal point-based approaches have been recently proposed for representing and managing uncertain relations between events, the probabilistic model done by Ryabov and Puuronen [11], and the possibilistic model proposed by HadjAli, Dubois, and Prade [7]. In this paper, the authors argued the main differences between these two approaches. Mainly, there are two main differences. First, the possibilistic modeling can be purely qualitative, avoiding the necessity of quantifying uncertainty if information is poor. Second, their proposal is capable of modeling ignorance in a non-biased way. In our case, the selection of the possibilistic model is reinforced by the fact that we need a model which make the fusion of mined and expert knowledge easier [3].

The selected model is based on possibility theory [2] for the representation and management of uncertainty in temporal relations between two point-based events. Uncertainty is represented as a vector involving three *possibility values* expressing the relative plausibility of the three basic relations (" $<$ "," $=$ ", and " $>$ ") that can hold between these points. Also, they describe the inference rules (that form the basis of the reasoning method) defining a set of operations: inversion, composition, combination, and negation, the operations that govern the uncertainty propagation in the inference process. The authors show that the whole reasoning process can actually be handled in possibilistic logic.

Three basic relations can hold between two temporal points, "before ($<$)", "at the same time ($=$)", and "after ($>$)". An uncertain relation between temporal points is expressed as any possible disjunction of basic relations:

$$\begin{array}{c} \hline \leq \iff < \mathit{or} = \\ \geq \iff > \mathit{or} = \\ \neq \iff < \mathit{or} > \\ ? \iff <, =, \mathit{or} > \end{array}$$

The last case represents *total ignorance*, that is, any of the three basic relations is possible. The representation is extended using the Possibility Theory for modeling the plausibility degree of each basic relation. Given two temporal points, a and b, an uncertain relation r_{ab} between them is represented by a *normalized vector* $\Pi_{ab} = (\Pi_{ab}^{<}, \Pi_{ab}^{=}, \Pi_{ab}^{>})$, such that $max(\Pi_{ab}^{<}, \Pi_{ab}^{=}, \Pi_{ab}^{>}) = 1$, where $\Pi_{ab}^{<}$ (respectively, $\Pi_{ab}^{=}$, $\Pi_{ab}^{>}$) is the possibility of $a < b$ (respectively $a = b$, $a > b$).

From the uncertain vector $(\Pi_{ab}^{<}, \Pi_{ab}^{=}, \Pi_{ab}^{>})$, and using the duality between possibility and necessity, namely

$$N(A) = 1 - \Pi(A^c), \quad \text{where } A^c \text{ is the complement of A}$$

we can derive the possibility and necessity degree of each basic relation and their disjunctions.

As,

$$\Pi_{ab}^{\leq} = max(\Pi_{ab}^{<}, \Pi_{ab}^{=})$$

$$\Pi_{ab}^{\geq} = max(\Pi_{ab}^{=}, \Pi_{ab}^{>})$$

$$\Pi_{ab}^{\neq} = max(\Pi_{ab}^{<}, \Pi_{ab}^{>}),$$

we can obtain the necessity degrees of the basic relations,

$$N_{ab}^{<} = N(a < b) = 1 - \Pi_{ab}^{\geq}$$

$$N_{ab}^{=} = N(a = b) = 1 - \Pi_{ab}^{\neq}$$
$$N_{ab}^{>} = N(a > b) = 1 - \Pi_{ab}^{\leq}.$$

Moreover, the authors defined the rules that enable us to infer new temporal information and to propagate uncertainty in a possibilistic way. The reasoning tool relies on four operations expressing:

inversion	$\iff \widetilde{r}_{ab} = r_{ba}$
composition	$\iff r_{ac} = r_{ab} \otimes r_{bc}$
combination	$\iff r_{ab} = r1_{ab} \oplus r2_{ab}$
negation	$\iff \neg$

These rules complete the definition of a model for representing and reasoning with uncertain temporal relations that uses the Possibility Theory as an expressive tool for dealing with uncertainty in temporal reasoning.

4 Extracting Uncertain Temporal Relations

In this section, we propose a technique for extract the uncertain temporal relation between each pair of event types from the sequences base. The uncertain temporal relation is represented by an uncertain vector formed by three possibility values, expressing the plausibility degree for each basic temporal relation. We propose the use of Shafer Theory of Evidence [12] to obtain the plausibility degrees from the frequencies values associated with the set of sequences. The result will be a set of temporal constraint networks, which belong to a a suitable model for representing and reasoning with temporal information where uncertainty is presented.

4.1 Shafer's Theory of Evidence

The Shafer Theory of Evidence, also known as Dempster-Shafer Theory, is a theory of uncertainty developed specially for modelling complex systems. It is based on a special fuzzy measure called *belief measure*. Beliefs can be assigned to propositions to express the uncertainty associated to them being discerned. Given a finite universal set $\mathcal{U}$, the *frame of discernment*, the beliefs are usually computed based on a density function $m : 2^{\mathcal{U}} \to [0, 1]$ called *basic probability assignment* (bpa):

$$m(\emptyset) = 0, \text{ and } \sum_{A \subseteq \mathcal{U}} m(A) = 1.$$

$m(A)$ represents the belief exactly committed to the set A. If $m(A) > 0$, then A is called a *focal element*. The set of focal elements constitute a core:

$$\mathcal{F} = \{A \subseteq \mathcal{U} : m(A) > 0\}$$

The core and its associated bpa define a *body of evidence*, from where a belief function $Bel : 2^{\mathcal{U}} \rightarrow [0,1]$, and its dual measure (the *plausibility measure*), $Pl : 2^{\mathcal{U}} \rightarrow [0,1]$ are defined:

$$Bel(A) = \sum_{B|B \subseteq A} m(B) \qquad Pl(A) = \sum_{B|B \cap A \neq \emptyset} m(B).$$

It can be verified [12] that the functions *Bel* and *Pl* are, respectively, a possibility (or necessity) measure if and only if the focal elements form a nested or consonant set, that is, if it can be ordered in such a way that each is contained within the next. In that case, the associated *belief* and *plausibility* measures posses the following properties: For all $A, B \in 2^{\mathcal{U}}$,

$$Bel(A \cap B) = N(A \cap B) = min[Bel(A), Bel(B)]$$
$$Pl(A \cup B) = \Pi(A \cup B) = max[Pl(A), Pl(B)]$$

4.2 Calculating the Possibility Measures of Temporal Relations

In our proposal, the sequences base is formed by a set of linked nested set, each one corresponding to a frequent maximal sequence and its subsequences. From an algorithm point of view, each nested set corresponds with a branch of the tree. So the proposed method build the temporal constraint networks in a linear time, just with a depth-first traversal of the tree.

Following the notation of Shafer's Theory, our core is each set of nested sequences $\mathcal{NS} \subseteq \mathcal{BS}^{D,\sigma,\omega}$ which is formed by a set of focal elements or sequences. We normalize the frequencies of each nested subset so they add to 1.

Let Ω be the set of event types presented in the dataset, that is,

$$\Omega = \{(A[i], v) | v \in dom(A[i])\}.$$

Taking into account the *maxspan* constraint, the set of events is defined as an extension of the Ω set in this way:

$$\Omega^{\omega} = \{(A[i], v, t) | v \in dom(A[i]) \wedge 0 \leq t \leq w)\}$$

This set is our frame of discernment, that is, $\Omega^{\omega} = \mathcal{U}$. So, the set of focal elements, the nested sequences base, is defined:

$$\mathcal{NS} = \{S_i \subseteq \Omega^{\omega} | m(S_i) > 0\},$$

where m is the bpa function derived from the frequencies of the sequences, such that $m : 2^{\Omega^{\omega}} \rightarrow [0,1]$,

$$m(\emptyset) = 0, \sum_i m(S_i) = 1$$

We will denote a temporal relation between two events e_1, e_2 as $e_1 \Theta e_2$. Since we are only interested in the basic temporal relations,

$$\Theta \in \{<, =, >\}.$$

For each pair of event types presented in the nested set, we need to obtain the possibility degree of each basic temporal relation between them. In order to compute the possibility of a temporal relation, it is necessary to consider all focal elements, that is, all sequences which make the temporal relation possible. However, from complexity point of view, we will obtain the possibility degrees from the necessity ones, calculated over the complement of the basic temporal relation, that is,

$$\Theta^c \in \{>=, <>, <=\}.$$

Proposition 1. *Let suppose the qualitative temporal relation $e_1\Theta^c e_2$. This relation induces a parameterized set:*

$$\mathcal{X}_{e_1\Theta^c e_2} = \{(e_i e_j)\},$$

where $e_i, e_j \in \Omega^\omega, e_i.type = e_1,\ e_j.type = e_2,$ and $e_i.t\Theta^c e_j.t$.

Proposition 2. *In order to obtain the set of sequences involved in the temporal relation, we introduce the assessment operator Γ, defined as:*

$$\Gamma(\mathcal{X}_{e_1\Theta^c e_2}) = \{S_i | S_i \subseteq \mathcal{X}_{e_1\Theta^c e_2}\},$$

where $S_i \in \mathcal{NS}$.

Proposition 3. *The possibility degree of the temporal relation $e_1\Theta e_2$ is defined as:*

$$\Pi(e_1\Theta e_2) = 1 - N(e_1\Theta^c e_2) = 1 - \sum_{S_i \in \Gamma(\mathcal{X}_{e_1\Theta e_2})} m(S_i)$$

5 A Practical Experience at Intensive Care Unit

The Intensive Care Unit (ICU) is a medical service to provide critical attention of medically recoverable patients. One of the fundamental characteristics of this domain is that patients require a permanent availability of monitoring equipment and specialist care. Thus, clinicians work in shifts in order to provide a 24 hours service. In this sense, the temporal evolution of patients is permanently recorded. Physicians at ICU are daily required to provide reports, describing the different diagnosis hypotheses that they assume and the posterior actions (tests, treatments, or requiring new laboratory analysis). In our particular case, the ICU service has a Health Information System (HIS) that stores this information and generates the reports.

Due to the amount of information (different medical areas implied), and the importance of the temporal dimension (implicitly and explicitly analysed in patients' evolution), we consider that the ICU is a suitable domain to apply our second-order temporal data mining proposal.

In ICU domains, as well as the final diagnosis (like other hospital services), there are *evolutive diagnoses* that state the diagnostic hypotheses. These hypotheses are daily made by physicians during patient's stay at the ICU service.

Furthermore, they can be considered high-level medical information since it is obtained from physician's knowledge and medical observations (like EKGs, tests, or nursing care data).

Despite the importance of other clinical information within the health record, such as treatments or demographic data, we consider in our experiment that the evolution of these diagnosis are a good representation of patient problems and the discovery of temporal pattern diagnosis could be useful in many AI systems for temporal diagnosis or prognosis.

In our experiment, each patient is represented in the database by a temporal sequence of diagnoses (temporal points) and the data mining process results are frequent temporal patterns (or frequent sequences) of diagnosis evolution. In the analysis of this data, different parameters have been empirically stated ($maxspan = 24$, and *support* value = 3, 5, 9) depending of the dataset of 144 patients. In Table 1 is shown a summary of some of the results obtained from the proposed data mining process.

Table 1. Practical experiments considering independent patients and complete data. Supp = data mining parameter of minimum support. N = number of sequences obtained. Max = maximum size of the sequences.

Supp	Patient Patt		Tot Patt	
3	N= 936	Max=5	N=379374	Max=12
5	N=122	Max=3	N=115810	Max=11
9	N= 49	Max=1	N=20837	Max=9

In [5] we can see the complete example. This paper shows how a very representative pattern is obtained from a nested set of frequent sequences.

6 Conclusions and Future Work

In this paper, we propose an initial approach for building qualitative temporal constraint networks from a set of mined frequent sequences with the aim of obtaining a more understandable, useful, and manageable sort of knowledge. The selected temporal model is the proposed by HadjAli, Dubois, and Prade, which uses the Possibility Theory as an expressive tool for representing and reasoning with uncertain temporal relations between point-based events. We propose a Shafer's Theory-based technique to obtain these possibility degrees involved in the network from the frequencies of the sequences.

In order to demonstrate the viability of this proposal we have applied it to the temporal evolution of diagnosis hypotheses at a ICU service. Despite that the clinical validation is not yet performed, the presented results points out the simplicity of representation and the advantage for expert's comprehension.

In future work, we intend to analyze in depth the networks obtained from the set of mined frequent sequences. We also propose to extend the model of temporal network in order to represent not only qualitative but also quantitative temporal relations, taking advantage of the temporal information presented in the time-stamped sequences extracted by $TSET$.

Acknowledgments

This work is supported in part by MEC grant TIN2006-15460-C04-01 and MEC grant TIN2004-05694.

References

1. R. Agrawal, T. Imielinski, and A. N. Swami. Mining association rules between sets of items in large databases. In P. Buneman and S. Jajodia, editors, *Proc. of the ACM SIGMOD Int. Conf. on Management of Data, Washington, D.C., May 26-28, 1993*, pages 207–216. ACM Press, 1993.
2. D. Dubois and H. Prade. *Possibility Theory.* Plenum Press, 1988.
3. D. Dubois, H. Prade, and G. Yager. Merging fuzzy information. In *Fuzzy Sets in Approximate Reasoning and Information Systems*, pages 335–401. Kluwer Academic Publishers, 1999.
4. F. Guil, A. Bosch, and R. Marín. TSET: An algorithm for mining frequent temporal patterns. In *Proc. of the First Int. Workshop on Knowledge Discovery in Data Streams, in conjunction with ECML/PKDD 2004*, pages 65–74, 2004.
5. F. Guil and J. M. Juárez R. Marín. Mining possibilistic temporal constraint networks: A case study in diagnostic evolution at intensive care units. In *Intelligen Data Anlisis in Biomedicine and Pharmacology (IDAMAP 2006)*, 2006.
6. F. Guil and R. Marín. Extracting uncertain temporal relations from mined frequent sequences. In *Proc. of the 13th Int. Symposium on Temporal Representation and Reasoning (TIME 2006)*, pages 152–159, 2006.
7. A. HadjAli, D. Dubois, and H. Prade. A possibility theory-based approach for handling of uncertain relations between temporal points. In *11th International Symposium on Temporal Representation and Reasoning (TIME 2004)*, pages 36–43. IEEE Computer Society, 2004.
8. H. Lu, L. Feng, and J. Han. Beyond intra-transaction association analysis: Mining multi-dimensional inter-transaction association rules. *ACM Transactions on Information Systems (TOIS)*, 18(4):423–454, 2000.
9. A. K. Pani. Temporal representation and reasoning in artificial intelligence: A review. *Mathematical and Computer Modelling*, 34:55–80, 2001.
10. J. F. Roddick and M. Spiliopoulou. A survey of temporal knowledge discovery paradigms and methods. *IEEE Transactions on Knowledge and Data Engineering*, 14(4):750–767, 2002.
11. V. Ryabov and S. Puuronen. Probabilistic reasoning about uncertain relations between temporal points. In *8th International Symposium on Temporal Representation and Reasoning (TIME 2001)*, pages 1530–1511. IEEE Computer Society, 2001.
12. G. Shafer. *A Mathematical Theory of Evidence.* Princenton University Press, Princenton, NJ, 1976.

Temporal Abstraction of States Through Fuzzy Temporal Constraint Networks

M. Campos[1], J.M. Juárez[2], J. Salort[2], J. Palma[2], and R. Marín[2]

[1] Informatics and Systems Dept. Computer Science faculty. University of Murcia
mcampos@dif.um.es

[2] Information and Communications Engineering Dept. Computer Science faculty. University of Murcia*

Abstract. Temporal abstraction methods produce high level descriptions of a parameter evolution from collections of temporal data. As the level of abstraction of the data is increased, it becomes easier to use them in a reasoning process based on high-level explicit knowledge. Furthermore, the volume of data to be treated is reduced and, subsequently, the reasoning becomes more efficient. Besides, there exist domains, such as medicine, in which there is some imprecision when describing the temporal location of data, especially when they are based on subjective observations. In this work, we describe how the use of fuzzy temporal constraint networks enables temporal imprecision to be considered in temporal abstraction.

1 Introduction

Medical decision support systems (MDSS) apply explicit medical knowledge to the patient's clinical data in order to support task such as diagnosis, deciding upon a therapy, monitoring the effects of a therapy, etc. This explicit knowledge is commonly defined using generalizations that can be structured as association rules, causal models or behavioural models. In other words, MDSSs deal with knowledge expressed at a high level of abstraction. In contrast, patient data to be used are obtained at a low abstraction level, e.g. results of laboratory analysis. Thus, it is necessary to abstract these specific data to bring them towards the generalizations used to formalize the knowledge and thus get a matching from which to draw conclusions.

Time is a reference framework to describe evolutionary clinical processes and supports the concept of change. The dynamic implicit in the evolution of data (for example pathologies in the case of medical domains) needs to be managed explicitly by temporal models and methods [9,7]. Data abstraction processes in which time plays a fundamental role are known as Temporal Abstraction (TA) processes. TA is supported on temporal reasoning methods which are basically focused on establishing temporal relations between the instances of the concepts

* This work was supported by the Spanish Ministry of Education and Science(MEC) and the European Regional Development Fund of the European Commission(FEDER)under grant TIN2006-15460-C04-01.

J. Mira and J.R. Álvarez (Eds.): IWINAC 2007, Part I, LNCS 4527, pp. 607–616, 2007.

which make up our knowledge domain, and on obtaining different types of generalizations from the raw data. The aim of TA techniques is to abstract high level concepts and patterns from sets of temporal data, i.e. from data which contain a time mark [9].

In this paper we present a temporal abstraction model which focuses on the search for a possible temporal explanation of the observations collected. In our model, an abstract explanation of the data consists of a *temporal sequence of interval states* which has to account for all the data. It must be consistent with the temporal dynamic expected and it should be as simple as possible. The application of an abductive method formally guarantees a temporal explanation that fulfills these conditions. We will model a part of the temporal abstraction process as a temporal constraints satisfaction problem.

A final aspect to be considered is the possible existence of imprecision in the temporal data as well as in the medical knowledge itself. Physicians therefore use ambiguous expressions like "a few minutes later", or "some 30 or 40 minutes" instead of precise time constraints between manifestations and diagnostic hypothesis. Consequently, it is necessary to apply suitable techniques to represent and manage imprecision within the time component. Our proposal is based on Fuzzy Temporal Constraint Networks (FTCN) [6], which allows us to capture the temporal imprecision associated to the temporal relations between the sequences abstracted and the sequences of the measurements taken.

The rest of this paper is organized as follows. Section 2 presents the temporal reasoning model on which our proposal is based. Section 3 provides a definition of all the elements which make up the temporal abstraction process. Section 4 describes the proposed temporal abstraction method in detail. Finally, we include some related works, the conclusions and some future research.

2 Temporal Framework

In our model, temporal concepts can be represented as time points or time intervals, and they can be related by means of quantitative relations (between points, referred to as MPP) or qualitative relations (between points, between points and intervals and between intervals, referred to as QPP, QPI or QIP, and QII respectively). Since reasoning with the full algebra for temporal relations is a NP-complete problem, we have chosen one of the tractable subalgebras: the set of convex relations implemented in *FTCN* formalism [6].

An *FTCN* is a pair $\mathcal{N} = \langle \mathcal{T}, \mathcal{L} \rangle$ consisting of a finite set of temporal variables, $\mathcal{T} = \{T_0, T_1, ..., T_n\}$, and a finite set of binary temporal constraints, $\mathcal{L} = \{L_{ij},\ 0 \leq i, j \leq n\}$ defined on the variables of $\mathcal{T}$. An *FTCN* can be represented by means of a directed constraint graph, where nodes represent temporal variables and arcs represent binary temporal constraints.

Each binary constraint L_{ij} on two temporal variables T_i and T_j is defined by means of a fuzzy number, that is a convex possibility distribution $\pi_{L_{ij}}$, which restricts the possible values of the time elapsed between both temporal variables.

An unknown relation between two variables corresponds to a universal constraint given by $\pi_U(t) = 1, \forall t \in \mathbb{Z}$.[1]

An *FTCN* network $\mathcal{N}$ is *consistent* if and only if there exits a non-empty σ-*possible solution* given a previously established threshold Π_{th} for σ, being that threshold the minimum degree of possibility allowed for all the constraints. The inference of unknown relations is carried out by applying a constraint propagation algorithm. By means of constraint propagation, a new *FTCN*, called minimal, that is equivalent to the original one is obtained. The minimal network makes all the implicit constraints in the network explicit, and always corresponds to a complete graph with the most precise temporal information consistent with the temporal information provided. This operation has an affordable computational cost, $O(n^3)$, by a trade off between representation capacity or expressivity and efficiency.

Figure 1 shows an example of a temporal constraint network corresponding to the temporal distribution of an episode of subarachnoid hemorrhage (SAH) in a patient at ICU. In the figure, squares represent temporal intervals and circles represent time points. Every interval can be translated into a point representation, and each qualitative relation can be translated into a quantitative one in order to obtain a FTCN. By means of the constraint propagation, we can know any implicit temporal relation, as for example the constraint between the vasospasm complication and the loss of consciousness symtomp.

To fill the gap between the *FTCN* and the high-level temporal language, a temporal reasoner call FuzzyTIME (*Fuzzy Temporal Information Management Engine*) [2] has been implemented. FuzzyTIME provides procedures for maintaining and querying temporal information (with both points or intervals, and

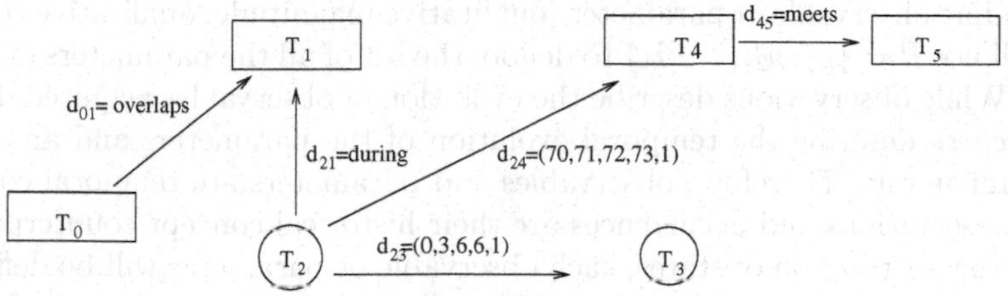

d01 = (Headache ,T_0, *overlaps* with loss of consciousness, T_1);
d21 = (Admission, T_2, *during* loss of consciousness, T_1);
d32 = (CT-Scan, T_3, *in less than 6 hours after* Admission, T_2);
d52 = (Low blood pressure, T_4, *approximately 72 hours after* Admission, T_2);
d45 = (Low blood pressure, T_4, *meets* vasospasm, T_5);

Fig. 1. Possible FTCN example of events of a patient with subarachnoid hemorrhage (SAH)

[1] A convex possibility distribution can be represented by means of a trapezoid defined by a 5-tuple fuzzy number (a, b, c, d, h), which indicates that the event associated to it necessarily occurs in interval $[a - c, b + d]$ (referred to as support), but possibly occurs in interval $[a, b]$ (referred to as kernel), with a possibility of h.

quantitative or qualitative relations) at $FTCN$ level. Within FuzzyTIME, it is possible to formulate queries that give a necessity measure about the occurrence of an event in the temporal network; this ability can be used for complex abstractions.

3 Temporal Abstraction Process

3.1 Temporal Ontology

Our temporal ontology is composed of three types of concepts domain, temporal and historical[3]. Domain concepts provide the organizational structure of the variables handled in the domain. Temporal concepts are used define the temporal organization of those domain concepts which have a temporal component. Historical concepts are used to describe the temporal evolution of the various temporal concepts. From the temporal abstraction perspective, the temporal concepts can be grouped together under ***observables*** and ***parameters***. An ***observable*** corresponds to a primitive variable of the domain whose value can be directly measured. An ***observation*** is the result of applying a measuring action to an observable at a given moment, and it describes the value of the observable at the discrete instant of time at which the measurement was made. An observation $m = (o, v, T)$ is made up of an observable $o \in O$, a value belonging to the set of values of the observable $v \in V(o)$ and a temporal variable which indicates the moment at which the measurement was made, $T \in \mathcal{T}$.

A *parameter* represents a variable whose value cannot be directly measured but which can be derived from the values of a given observable or, even, another parameter. In other words, a parameter abstracts some characteristic of the corresponding observable or parameter (qualitative magnitude, qualitative trend,..). We will use $P = \{p_1, p_2, \ldots, p_n\}$ to denote the set of all the parameters of the domain. While observations describe the evolution of observables acquired directly, *occurrences* describe the temporal evolution of the parameters and are defined in a similar way. Therefore, observables and parameters are temporal concepts, while observations and occurrences are their historical concept counterpart.

For the abstraction of states, each observable or parameter will be defined by the following attributes:

- ***Persistence*** ***(***δ***):*** The period of maximum validity of an observation. In other words, the maximum period during which we can suppose that an observation does not change its value.
- ***Granularity(***g***):*** This establishes the minimum duration of a state. This constraint allows to control that state do not change too fast.

Figure 2 shows an example of an observable, a real value of temperature, along with an associated parameter -the presence of fever, whose possible values are true and false. In the example, the persistence of the temperature is 1 hour and the granularity of the fever state is 90 minutes.

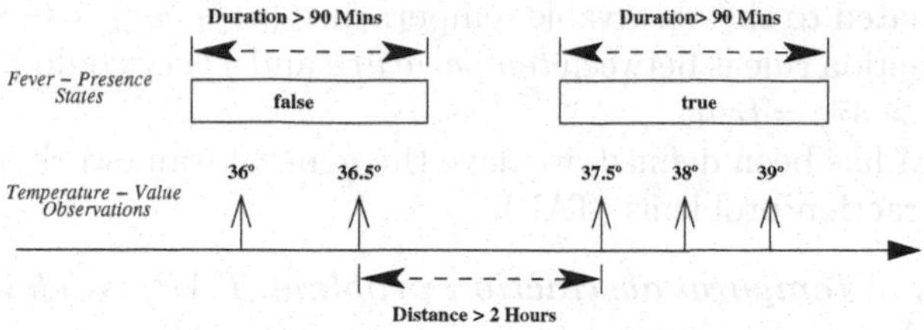

Fig. 2. Example of an observable and a parameter

As a result of the repeated application of a measurement to an observable, a *sequence of observations* is obtained which describes the evolution of the observable. The abstracted sequence of a temporal concept is defined as a sequence of states (which do not have to be consecutive, since gaps may exist in the explanation). A state defines a time interval in which a parameter maintains a constant qualitative value. Each state is represented by the tuple $s = (p, v, T^b, T^e)$, where $p \in P$ is a parameter , $v \in V(p)$ is a qualitative value belonging to the set of values of the parameter domain, $T^b \in \mathcal{T}$ is a temporal variable representing the start of the maximum time interval in which the parameter maintain a qualitative value, and $T^e \in \mathcal{T}$ is a temporal variable representing the end of the said interval. It is always assumed that $T^b \prec T^e$ since states of zero duration are not admitted.

3.2 Temporal Abstraction Problem

The basic knowledge used to solve a temporal abstraction problem comes under what is known as the *abstraction model* (AM). An AM defines the values of a parameter which are inferred from observables or from other parameters through a set of abstraction rules (functions f_p^o) which establish a correspondence between the possible parameter values and the possible values of the observables.

Definition 1. *An* ***abstraction model,*** *AM, is defined by the tuple* $D =< O, P, a, F_v >$*where:*

- O is a set of observables.
- P is a set of parameters.
- a is a suprajective application $a : P \cup O \to P$ defined as $\forall p' \in P, \exists x \in O \cup P \mid a(x) = p'$, i.e. given a parameter p', it tells us which observable or parameter is used for the abstraction.
- F_v is a set of abstraction rules, $F_v = \{f_p^x\}$, with $f_p^x(F_x) = v_p$ where F_x is a logical formula on $V(x)$ and $v_p \in V(p)$.[2]

For example, the series of temperature measurements would correspond to the observables ($temperature \in O$), while the states that describe the presence or absence of fever would correspond to the abstracted parameter, fever ($fever \in P$)

[2] We have defined in [3] a series of basic abstractions (qualitative, generalizations, etc) which can be expressed in this way.

which is associated to the observable temperature ($a(fever) = temperature$). A possible abstraction rule is between $temperature$ and $fever$ could be $f_{fever}^{temperature}$ $(temperature > 37) = true$.

Once an AM has been defined, we have the general framework in which to set temporal abstraction problems (TAP).

Definition 2. *A* ***temporal abstraction problem****, TAP, is defined by the tuple s $TAP = \langle D, \mathcal{N}_A, H_A \rangle$, where:*

- D *is an* AM.
- $\mathcal{N}_A = \langle \mathcal{T}_A, \mathcal{L}_A \rangle$ is a FCTN in which $\mathcal{T}_A$ is the set of all the temporal variables of the available observations, and $\mathcal{L}_A$ is a set of temporal constraints among the variables of $\mathcal{T}_A$ and the time origin T_0, $\mathcal{L}_A = \{L_{i0}\}$.
- H_A is the set formed by all the available observations, $H_A = \{(o_i, v_i, T_i) | o_i \in O \wedge v_i \in V(o_i) \wedge T_i \in \mathcal{T}_A\}$.

In our case we will consider that $\mathcal{L}$ is a set of constraints between the time origin and each observation, i.e. the observations are going to be associated to absolute dates. This will allow us to work with an input sequence of data that can have a precise time mark, as is the case of the majority of information gathering systems (e.g. a clinical information system).

Our approach is based on the consideration that the temporal abstraction process can be seen as an abductive process, i.e. the solution to a temporal abstraction problem is a possible explanation of the sequence of observations defining the TAP. We call such an explanation an *abstraction hypothesis.*

Definition 3. *Given a $TAP = \langle D, \mathcal{N}_A, H_A \rangle$, an* ***abstraction hypothesis****, AH for that TAP can be formally defined as the tuple $AH = \langle \mathcal{N}_{AH}, S_{AH}, \mathcal{L}_{AB} \rangle$, where:*

- $\mathcal{N}_{AH} = \langle \mathcal{T}_{AH}, \mathcal{L}_{AH} \rangle$ is a minimal FCTN in which $\mathcal{T}_{AH}$ is a set of temporal variables associated to the events that define the states, and $\mathcal{L}_{AH}$ is a set of temporal constraints between the temporal variables, $\mathcal{L}_{AH} = \{\mathcal{L}(T_i, T_j) \mid T_i \in \mathcal{T}_{AH} \wedge T_j \in \mathcal{T}_{AH}\}$.
- $S_{AH} = \{S_{AH_I} = (p, v_i, T^b_{AH_i}, T^e_{AH_i}) \mid p \in P \wedge v_i \in V(p) \wedge T^b_{AH_i} \in \mathcal{T}_{AH} \wedge T^e_{AH_i} \in \mathcal{T}_{AH}\}$ is the set of states representing $T^b_{AH_i} \in \mathcal{T}_{AH}$ and $T^e_{AH_i} \in \mathcal{T}_{AH}$ the events at the beginning and end of the state, respectively.
- $\mathcal{L}_{AB}$ is a set of temporal constraints between the temporal variables $\mathcal{T}_{AH}$ and those of $\mathcal{T}_A$: $\mathcal{L}_{AB} = \{\mathcal{L}(T_i, T_j) \mid T_i \in \mathcal{T}_A \wedge T_j \in \mathcal{T}_{AH}\}$

As with any abductive problem, different explanations can be obtained for the same set of input data according to how we understand the concept of explanation. Thus, to obtain a single hypothesis abstraction it is necessary to establish some criteria that will allow us to select the best hypothesis. The criteria that are going to be required of an AH in our case for it to be selected as a solution are:

- *Covering*: Each observation has to be explained by some state, i.e. the states included in the hypothesis must imply the observations available.

- *Temporal Consistency*: The global FTCN obtained as the union of the TAP and the AH must be consistent. (For the definition of consistency, see Section 2).
- *Exclusivity*: Each observation must be included at most in a state.
- *Parsimony*: Every conjectured state must be supported by at least one observation. There is no need to create unnecessary states since gaps can exist in the explanation.
- *Dynamic Compatibility*: The duration of each state must necessarily be greater than the granularity and all the observations that form part of the same state must have a distance between them of, at most, twice that of the persistence of the associated observable. Otherwise, gaps will be created.

4 Temporal Abstraction of States Algorithm

The abstraction process we explain below rests on three pillars. In the first place, the observations are processed increasingly. Hence, all the data processed which have not produced any type of inconsistency will be correct, i.e. any temporal inconsistency in the set of states is produced by the observation being processed and it will therefore be discarded. In the second place, we consider time to be discrete. Finally, we assume that from a single observable, various sequences of states describing the evolution of different concept characteristics can be obtained, i.e. different states: qualitative value, trend, gradient, etc.

Taking into account the above considerations, the temporal abstraction process can be described considering each one of the different cases that we can find when processing a new observation.

Case 1: Generation of a new non consecutive state. This base case deals with the situation in which an observation has to generate a new state (the first observation to be processed by the system or one which does not meet the persistence constraints). Let us suppose that the following observation arrives $m_i = (o_i, v_i, t_i)$ and that we generate a new state $S_j \leftarrow (p, f_p^{o_i}(v_i), T_{S_j}^b, T_{S_j}^e, g_p, \delta_{o_i})$, where g_p is the granularity of the parameter P and δ_{o_i} is the persistence of the observable o_i. The creation of a new state gives rise to the temporal relationships in Figure 3.

C1 and C2 constraints establish that the temporal distance between the starting event $T_{S_j}^b$ and the end event $T_{S_j}^b$ of the state S_j and the time instant t_i of the observation m_i is located within the time interval $[0, \delta_{o_i}]$. Note that theses constraints establish the maximum period of validity for the generated state. C3 constraint establishes that the temporal distance between the starting event $T_{S_j}^b$ and the end event $T_{S_j}^e$ of the state is greater than the granularity g_p.

Case 2: Subsumption of an observation in a state. In order to process a new observation m_i, we need to analyze the last state $S_j = (p, f_p^{o_i}(v_{i-1}), T_{S_j}^b, T_{S_j}^e, g_p, \delta_{o_i})$ associated with the observation m_{i-1}. There are two aspects to look at: abstract value and fulfillment of the persistence condition. The persistence

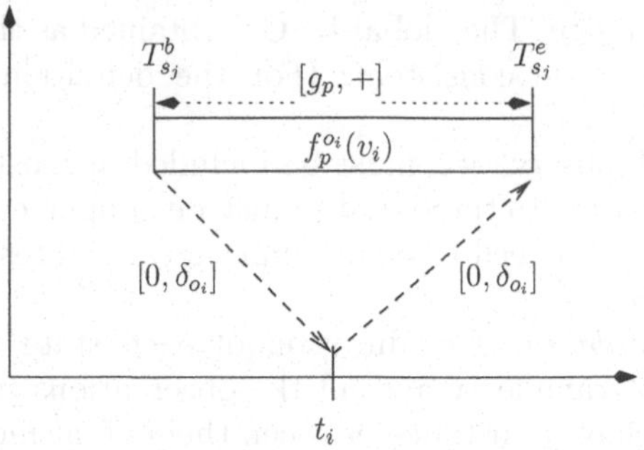

C1: $MPP(T^b_{S_j}, t_i, LESSEQ\_THAN(\delta_{o_i}))$ and $QPP(T^b_{S_j}, t_i, BEFORE)$
C2: $MPP(t_i, T^e_{S_j}, LESSEQ\_THAN(\delta_{o_i}))$ and $QPP(t_i, T^e_{S_j}, BEFORE)$
C3: $MPP(T^b_{S_j}, T^e_{S_j}, MORE\_THAN(g_p))$

Fig. 3. Creating a new state

condition indicate us whether two states necessarily do not overlap or if the time mark of the observation is consistent with the maximum period of validity for the previous state. Given the previous assumptions, an intuitive way of checking this condition is by using the following temporal query: $IS(MPP(t_i, t_{i-1}, MORE\_THAN(2*\delta_{o_i}+1)))$. A positive answer leads us to the creation of a new state S_{j+1}, analogously to the base case; a negative answer leads us to Case 2 or Case 3.

In Case 2, we check if the values of the previous and the current observations coincide, $f_p^{o_i}(v_{i-1}) = f_p^{o_i}(v_i)$. If these values are equal, then the new observation m_i is subsumed in the previous state S_j. A subsumption process will consist of retracting the temporal constraints which existed between the end of the state and the last observation m_{i-1} (constraint C2 in Figure 3) and of asserting the new temporal constraints for the end of the state $T^e_{S_j}$ with m_i (see C5 in Figure 4). The rest of constraints are still fulfilled.

Case 3: Generation of a consecutive case. In the case where the observations have a different abstract value, i.e. $f_p^{o_i}(v_{i-1}) \neq f_p^{o_i}(v_i)$, a new state, S_{j+1}, is created with the same constraints as those in Case 1. It only remains to make it explicit that the new state is consecutive to the previous one by establishing that distance between the extremity of the end of the previous state and the start of the new one is a time unit. This case can be seen in Figure 5.

When the observations has been processed it is necessary to check the consistency of the resulting FTCN. If the FTCN is not consistent, the observation will be discarded, since it will be considered as wrong, and we will proceed to the following observation. It is only necessary to verify the consistency of the states when inserting an observation with an abstract value which differs from that of the previous observation, since the subsumption of an observation in a state will never cause inconsistencies within the network (see Figure 4). The same TA process can be applied recursively to the sequence of the generated states making the concatenation of basic abstractions possible.

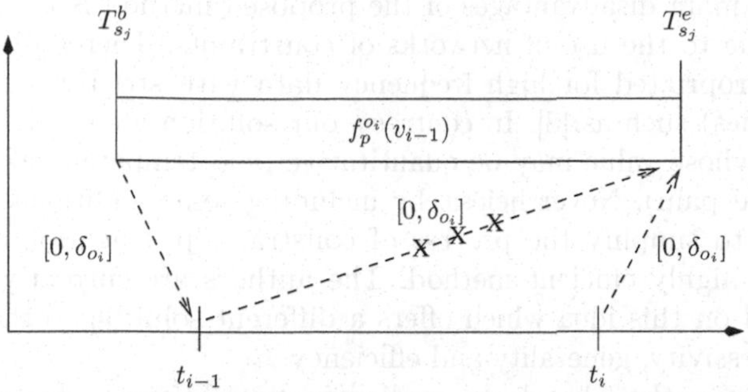

C4: Retract $QPP(t_{i-1}, T^e_{S_j}, (BEFORE))$ and $MPP(t_{i-1}, T^e_{S_j}, LESS\_THAN(\delta_{o_i}))$
C5: Assert $QPP(t_i, T^e_{S_j}, (BEFORE))$ and $MPP(t_i, T^e_{S_j}, LESSEQ\_THAN(\delta_{o_i}))$

Fig. 4. Include the observation of the state

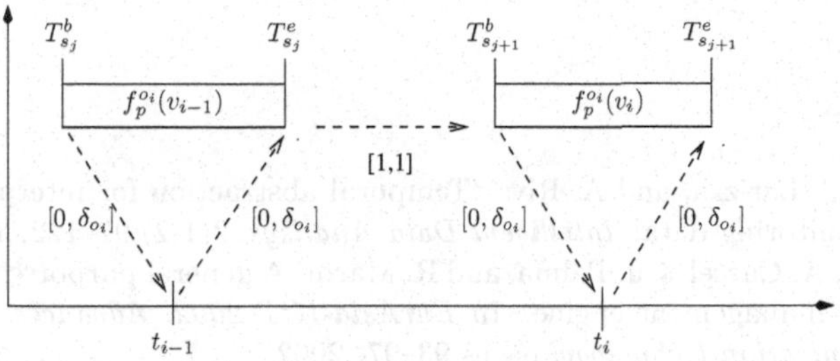

C6: $MPP(T^e_{S_j}, T^b_{S_{j+1}}, 1)$

Fig. 5. Creation of consecutive states

5 Conclusions

This paper presents a method of temporal abstraction of states which allows the incorporation of the concept of fuzzy time in a high level description of a collection of temporal data. This method has been implemented in a general framework for temporal abstraction described in [3]. The proposed method is easily extensible since new abstractions only require the definition of the semantic properties of the concepts implied and the abstractions functions for those properties. Furthermore, the same method can be recursively applied to generate abstract explanations from sequences of states.

The main contribution with respect to other works is the capacity to treat temporal imprecision in the input data and in the states generated, by the use of FTCNs. Another advantage is that the proposed method can deal homogeneously with disperse data or with continuous data. Similar approaches have been proposed [9,1] but don't consider temporal imprecision. In [5] a TA framework that is also based in fuzzy time management is proposed but is limited to the abstraction of trends through the use of temporal patterns.

One of the main disadvantages of the proposed method is its computational complexity due to the use of networks of constraints. Therefore, other models are more appropriated for high frequency data with strictly numerical values (temporal series) such as [8]. In contrast, our solution allows the processing of observations whose value may be quantitative (e.g. temperature) or qualitative (e.g.abdominal pain). Nevertheless, by including some additional suppositions, it is possible to simplify the process of constraint propagation and to obtain a similar but highly efficient method. The authors are currently developing a method based on this idea which offers a different solution to the compromise between expressivity, generality and efficiency.

Finally, this method has been applied in two different scenarios. The first consists of a pre-processing step in a model for discovering temporal knowledge, not only for a significant reduction of the volume of the data but also coping with the problem of obtaining a more concise and complete representation. The second case is a patient visual display unit in which a series of data abstractions is generated from the patient's data.

References

1. R. Bellazi, C. Larizza, and A. Riva. Temporal abstraction for interpreting diabetic patients monitoring data. *Intelligent Data Analisys*, 2(1-2):97–122, 1998.
2. M. Campos, A. Cárceles, J. Palma, and R. Marín. A general purporse fuzzy temporal information management engine. In *EurAsia-ICT 2002. Advances in information and communication technology*, pages 93–97, 2002.
3. M. Campos, A. Martínez, J. Palma, and R. Marín. Modelo genérico de abstracción temporal de datos. In *Proceedings of the XI Conferencia de la Asociación Española para la inteligencia artificial.CAEPIA05*, volume 2, pages 51–60, 2005.
4. P. Felix, S. Barro, and R. Marín. Fuzzy constraint networks for signal pattern recognition. *Artificial Intelligence*, 148(1-2):103–140, August 2003. Special issue: Fuzzy set and possibility theory-based methods in artificial intelligence.
5. I. J. Haimowitz and I. S. Kohane. Managing temporal worlds for medical trend diagnosis. *Artificial Intelligence in Medicine*, 8:299–321, 1996.
6. R. Marín, J. Mira, R. Patón, and S. Barro. A model and a language for the fuzzy representation and handling of time. *Fuzzy Sets and Systems*, 61:153–165, 1994.
7. J. Palma, J.M. Juarez, M. Campos, and R. Marin. Fuzzy theory approach for temporal model-based diagnosis: An application to medical domains. *Artificial Intelligence in Medicine*, 38(2):197–218, 2006.
8. A. Seyfang, S. Miksch, W. Horn, M.S. Urschitz, C. Popow, and C.F. Poets. Using Time-Oriented Data Abstraction Methods to Optimize Oxygen Supply for Neonates. In *Proceedings of European Conference on Artificial Intelligence in Medicine (AIME 2001)*, pages 217–226, Cascais, Portugal, 2001.
9. Y. Shahar. A framework for knowledge-based temporal abstraction. *Artificial Intelligence*, 90(1-2):79–133, 1997.

Spieldose: An Interactive Genetic Software for Assisting to Music Composition Tasks

Ángel Sánchez, Juan José Pantrigo, Jesús Virseda, and Gabriela Pérez

Departamento de Ciencias de la Computación
Universidad Rey Juan Carlos, C/Tulipán, s/n,
28933 Móstoles, Madrid, Spain
{angel.sanchez,juanjose.pantrigo,j.virseda,gabriela.perez}@urjc.es

Abstract. We describe a new software tool, called Spieldose (in English, musical box), suited to the automatic music composition task. Our system is based on the paradigm of Interactive Genetic Algorithms (abbreviated as Interactive GA) where the parent selection stage in a typical GA is made by the user according to his/her subjective criteria. The tool permits to integrate the interaction between the system and the potential users when they create their melodies. One important contribution of this work is the proposal of specific musical genetic operators (different types of crossover, mutation and improvement operators) which ensure that the generated melodies are in concordance with Music Theory and they are also nice to listening. Moreover, our software tool can be customized to a particular musical style by including the specific musical knowledge domain in the system. For validation purposes, we used Spieldose to compose different pieces corresponding to the classicism.

1 Introduction

In general, Computer Music is related to the theory and application of different techniques to the musical generation (or composition) with the aid of computers. Musical information analysis from different sources (like digital audio, digital partitures or metadata) is also a component area in Computer Music. Therefore, it is a multidisciplinary field related to disciplines like Digital Signal Processing, Artificial Intelligence, Acoustics, Mathematics or Image Processing, among others.

A good survey on automatic music composition is presented by López de Mantaras and Arcos [1]. This paper describes a set of representative Computer Music systems (related to compositional, improvisation and performance aspects) that use AI techniques. The pioneering work in automatic music composition is due to Hiller and Isaacson in 1958. Many fundamental compositional processes in music may be described as taken an existing musical idea and changing it to produce a new related piece [2]. Modern musical composers follow many times some simple rules during their creative process. These rules refer to the intervals between notes, to the notes used in each tone and in each musical style and to the types of rhythmic articulations [3].

J. Mira and J.R. Álvarez (Eds.): IWINAC 2007, Part I, LNCS 4527, pp. 617–626, 2007.

In general, music composition can be a very complex task for the people without musical knowledge or skill. To overcome this difficulty, different automatic music composition systems have been proposed since the appearance of computers [4].

Several computational intelligence methods have been applied to music synthesis and analysis tasks. Neural Networks (NN), Genetic Algorithms (GA), and Genetic Programming (GP) are the main ones. A survey of the application of Artificial Intelligence (AI) methods to music generation can be found in [5]. Genetic and evolutionary algorithms have demonstrated to be an effective method to find solutions in complex search spaces. Marques et al [3] have applied GA to the generation of musical sequences using melodic and musical theory concepts. Khalifa and Foster [6] proposed a composition system in two stages: first, musical sound patterns are identified and then they are combined in a suitable way. The MusicBlox project [4] is based on small music fragments or blocks which are combined using GA producing musical successful results. The combination of GA and NN (in particular, Multilayer Perceptrons) for music composition has also been exploited in the work by Göksu et al [7]. Some other AI techniques like constraint programming provide a suitable tool for automatic music generation. Henz et al [8] developed an experimental platform called COMPOzE for intention-based composition. Jewell et al [9] have described the architecture of an agent-based distributed system which is oriented to musical composition.

This paper presents an interactive music composition system based on the application of Genetic Algorithms (GA) to assist the user in this complex task. Our work takes into account the fact that when several people listen to the same musical piece, their impressions are not necessarily the same. In a related work, Unehara et al [10][11] have also remarked that: "any music generation system must reflect composers subjectivity towards music". To achieve this goal, these authors have proposed the application of Interactive GA to music composition. In these algorithms the parent selection stage is performed the human. In this way, our adapted GA is used as an optimization method to generate and evolve a population of musical pieces (the individuals in a standard GA) by the application of some specific genetic operators which are introduced to hold the principles of Musical Theory. At each iteration during the genetic evolution, the user selects several melodies or pieces considered as "good" candidates according to musical subjective criteria [10]. In this way, an initial population of automatically generated compositions is evolved until a termination condition is met. These musical works, which can comply with a given musical style or author, are generated by considering some specific compositional criteria. The developed software tool, called Spieldose, aims to include criteria of musical composition into the Interactive GA. This integration is the core of the proposed work and it will be detailed in successive sections.

The rest od the paper is organized as follows. Section 2 offers a global description of the presented musical composition system. This description is focused on the detailed presentation of our specific Interactive GA and on the application Graphical User Interface (GUI). The different components of our Interactive GA

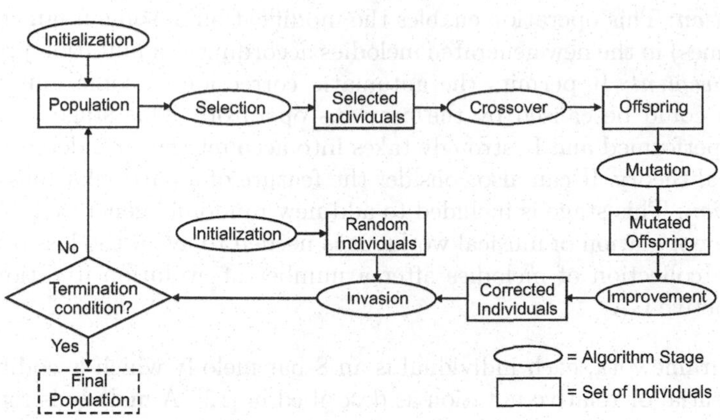

Fig. 1. Overview of the proposed Interactive Genetic Algorithm

are detailed in Section 3. Experimental results are resumed in section 4. Section 5 outlines the conclusions and provides future research lines.

2 System Overview

Our approach is focused on an specific Interactive GA for the automatic music composition task. These kind of GAs were initially applied to fields like the industrial design where the fitness functions were difficult to be defined. We have followed a similar approach as presented in [10][11] by Unehara and Osinawa. Spieldose is a prototype tool which extends the work of these authors. We created new types of genetic operators, in particular, new mechanisms for crossover, mutation, improvement and invasion operations. A complete tool GUI aimed to assist both unskilled and expert user in the music composition stages has also been developed. Our proposed Interactive GA includes the stages represented by the Figure 1:

- *Initialization*: The set of individuals that form the initial GA population are generated in this stage. This operation is defined to produce an initial set of musical works or melodies that are created using some specific knowledge from music theory. In our application, we have only produced 8-bar length musical melodies.
- *Selection*: The best melodies (individuals) in the population are selected by a human expert at each iteration of the GA. In this stage, a variable-size subset of melodies is chosen by the user (by listening them one by one) considering his/her musical preferences or guided by the characteristics of a predetermined musical style.
- *Crossover*: Those individuals selected are combined using three possible crossover mechanisms (according to crossover probabilities), to produce a new generation of child melodies (offspring).

- *Mutation*: This operation enables the modification of some fragments (chromosomes) in the new generated melodies according to a mutation probability.
- *Improvement*: It permits the automatic correction of some musical errors which could be caused by the previous operators. This stage is automatically performed and It strongly takes into account the considered criteria of musical theory. It can also consider the feature of a particular musical style.
- *Invasion*: This stage is included to add new randomly generated individuals to the population of musical works. It is needed to avoid the loss of diversity in the collection of melodies after a number of evolution iterations in the Interactive GA.

In our framework, each individual is an 8-bar melody which is codified using the differential or relative notation as described in [12]. A melody is represented by a sequence of notes, where each note has its pitch, length and type attributes (as shown in Figure 2)). The corresponding data structure to represent a melody is a vector where each position has two fields: the first one describes the note pitch expressed in half tones (where a zero value represents a ligature of the same pitch), and the second one implicitly codifies the note (or silence) length using three possible capital letters: *'N'* represents a note, *'S'* represents a silence and *'L'* stands for a note (or silence) ligature. Some complementary remarks are now pointed out:

- Note types are also implicitly represented and each one holds a determined number of vector positions equals to its length using as minimal reference unit a semiquaver note (for example a black note requires four vector positions, a quaver note requires two vector positions, and so on).
- Each note pitch is expressed with respect to the previous note except for the first note of the melody (represented by an absolute number of half tones with respect to a reference octave). The pitch is given by a positive or negative number of half tones required to obtain the note sound from the previous note in the piece.
- In order to simplify the implementation of GA operators, we set the size of all individuals (melodies) of the population to 128. This value corresponds to 8-bar pieces at 4/4 measures, where the minimal considered note length is a semiquaver. Therefore, length of a note is determined by adding one to the number of consecutive *'L'* vector positions preceded by a *'N'* vector position.

One important contribution of this works is the variety and its effective implementation of the operators in the Interactive GA. We have properly combined in this work the knowledge of experts in Musical Theory and experts in Combinatorial Optimization methods. The following section describes in detail each of the involved GA stages and their components. Another contribution of this work is the complete graphical user interface (GUI) of Spieldose that offers the user the appropriate functionality for the musical composition task and also the

4	0	0	0	1	0	2	0	0	0	0	0	4	0	0	0
N	L	L	L	N	L	N	L	S	L	L	L	N	L	L	L

Fig. 2. Example of the codification of the first bar in a given melody

GUI hides the implementation details of the interactive GA. Figure 3 shows the main GUI window of Spieldose (left) and the initialization stage window (right). This GUI offers the user the following main options:

- Initialize a population of melodies (both initial and invader populations).
- Interactively listen to the created musical works in order to choose those ones that are considered the best ones according to the user preferences.
- Select the best subset of melodies using a tournament algorithm.
- Save the favorite melodies at each iteration in Waveform Audio Format (wav) and/or in text format (tex).
- Interactively edit a musical piece in text format such that the corresponding audio is also modified at the same time (there exists a complete updated equivalence between both formats for each melody).
- Modify different features of the proposed Interactive GA, such as the crossover and mutation mechanisms, the population size, etc.

3 Main Components of the Interactive GA

This section describes the main components and the GA operators of our algorithm, as represented in Figure 1.

3.1 Algorithm Initialization

For the generation of the initial set of melodies (individuals), we used the knowledge of the experts in Musical Theory. Two main substages have been considered for this goal: (a) initialization of the common features corresponding to all individuals in the population and (b) initialization of particular aspects from each specific individual. All the individuals share the same time signature or rhythms and also the harmonic structure in order to simplify the application of the GA operators on the melodies. Of course, individuals differ in the melody they represent (their chromosome structure). The common information to all the individuals is generated as follows.

- The permitted rhythms are 2/4, 3/4 and 4/4, and they can be set manually or random.
- The considered harmonic structure assigns a fundamental chord to each of the melody bars in the musical piece. This structure determines that

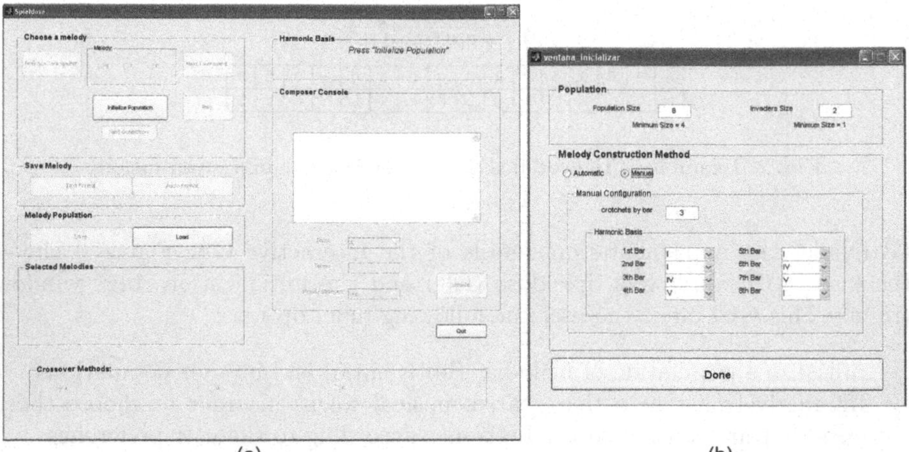

Fig. 3. A view of Spieldose application GUI

a melody is correctly created and also that it is human listening. Figure 4(a) represents a octave of three-note chords. For example, if tonic (or first) note is C in major-chord, then this chord I is composed by the notes: *C*, *E* and *G*, respectively. Figure 4(b) describes the considered tonal possibilities for each of the 8 bars in the harmonic basis of the melodies. The first bar always starts with a chord I (and it also represents the major/minor tonality of the melody). Note that the bar positions in the second and third rows with respect to the corresponding ones in the first row of Figure 4(b) means that these chords in a given bar can be exchanged.

(a)

I	I	IV	V	I	IV	V	I
	III	II	VII	III	II	VII	
	IV	VI		IV	VI		

(b)

Fig. 4. Solution initialization: the harmonic structure

As pointed out, each individual in the GA stores an 8 bars melody. Next, ee describe the stages followed by the system to create a musical piece.

– *Rhythmic structure*: First, the melody rhythm is constructed. It consists in a vector of notes as represented by the Figure 2. For this purpose, three possible values can be assigned to the second component in each vector position: *'N'*, *'S'* and *'L'*. The first position in a bar of this 2-tuple vector is assigned a value *'N'* to ensure the independence of each bar in the melody

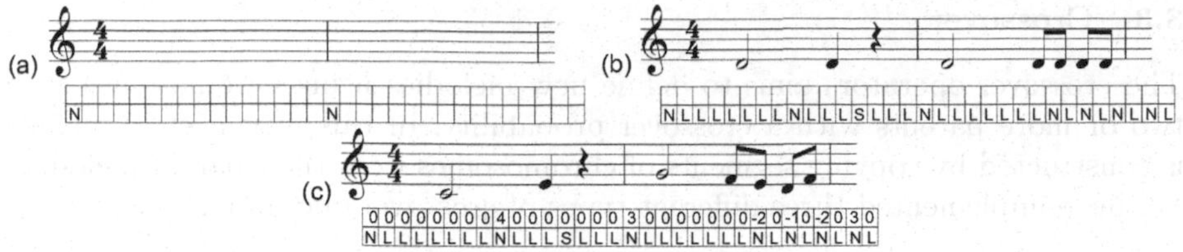

Fig. 5. Sketch of an example of the automatic creation of a melody using the Spieldose tool (for shortness, the first two bars are only represented)

thus discarding the ligatures between notes of different bars (Figure 5(a)). Next, the rest of vector positions are completed by raffling the three possible values (*'N'*, *'S'* and *'L'*) as shown in Figure 5(b). These values can have different probabilities of occurrence.

– *Pitch sequence*: The corresponding pitch values are assigned to the notes in the rhythmic sequence (see Figure 5(c)). These values are subsequently added to the melody such that they are chosen with a probability proportional to an assigned pitch weight. This weight is computed according to the considered tonality and following the compositional rules of a concrete musical style (in our framework we have focussed on the classicism). Therefore, those notes belonging to a considered triad chord have a higher probability to be chosen, and later those notes with pleasant dissonances (in particular, the seventh and ninth notes of the chord). The probabilities associated to each tonality jump are predefined and they have been suggested by the experts in Music Theory.

3.2 Parent Selection Scheme

This stage in our Interactive GA is performed by a tournament method and it is the only operation than necessarily requires from the human participation. A number of N different melodies automatically generated by the system are presented in groups of M (where $M < N$). Then, after listening the M melodies, the user chooses one in each group according to his/her musical preferences. These selected pieces will be the parents for the next generation of individuals, and the rest of the melodies are discarded. The tournament is a very appropriate mechanism for implementing the selection stage in an Interactive GA since the user is only responsible to select those individuals are the best ones in the actual population. This task, as performed manually, requires that user listen to all the generated musical pieces. However, it results much easier than assigning a fitness value to each individual. While performing this stage, the user can save those melodies considered as "good". Moreover, there also exists the possibility to edit and modify on-line the created melodies. This way, the user can interactively transform some parts or aspects of the musical pieces.

3.3 Crossover

The crossover operators aims to define new melodies using chromosomes from two or more parents with a crossover probability. In this way, a child melody is constructed by copying segments of chromosomes from their parent melodies. We have implemented three different types of crossover operators:

- Type A: Two new melodies are created from two parents. For each of the two selected 8-bar musical works, we determine two crossing points (one in each parent), and all the bars for the two child individuals are generated by exchanging the corresponding chromosomes of both parents.
- Type B: In this case, two new melodies are also created from two parents. However, several (in general, more than two) crossing positions are selected at different places in the parent melodies, and the two children are formed by the alternate selection of chromosomes from both parents.
- Type C: This is a generalization of crossover type B that is extended to a large number N ($N > 2$) of parents. As a result, N child melodies are created by selecting several crossing points in the parents and, in a similar way, we alternate the selection of parent chromosomes to build the N children.

It is important to remark that when using all types of crossover operators, the new resulting melodies are automatically repaired to preserve their corresponding tonality.

3.4 Mutation

It is possible to apply different types of mutation operators to the chromosomes of the individuals. We have implemented the following ones: (a) rotation of the notes in a segment (sequence of notes) of a melody, (b) melodic inversion of a segment and (c) the variation of a note (in its length and/or pitch). A given melody can mutate its chromosomes according to one of these three possibilities with a given mutation probability.

3.5 Improvement

This is one of the most interesting contributions of our Interactive GA. In the improvement stage, it is possible to use different heuristics to model the expert's musical knowledge domain with the aim to create musical works that are correct respect to musical theory and also pleasant to be listened. As an example of these improvements, we correct the large pitch jumps between notes of different bars to reduce these differences to less than one octave.

3.6 Invasion

This operator also generates new melodies by a similar procedure of the initialization stage. These new created individuals are added to the actual population.

The aim of this operation is to prevent the loss of diversity and also to avoid a premature convergence of the proposed Interactive GA after a small number of iterations.

4 Experimental Results

Due to the subjectivity of each particular human user when composing musical pieces with Spieldose, it is not very appropriate to show quantitative results in this work. A way to demonstrate the validity of our proposal, is creating a repository of musical pieces generated by Spieldose. This can be found in WAV format at the following URL: `http://gavab.escet.urjc.es/recursos.html`. All the pieces correspond to the classicism musical style. These 8-bar melodies have been produced by applying the different crossover and mutation operators considered by our Interactive GA. Spieldose source code in Matlab and the application tool can also be downloaded from the previous URL.

5 Conclusion and Future Work

This paper has presented a software tool called Spieldose for assisting the human user in the automatic music composition task. Our approach is based on the paradigm of Interactive GA where the parent selection stage is directly performed by the user. The application permits to integrate the interaction between the system and users when they compose their own musical pieces. This principle is useful both for experts in Music Theory and also for musically unskilled people. The aim when developing this software was also to show the natural integration of automatic music composition tasks into The GA. One main contribution is the proposal of new genetic operators (in special, different types of crossover, mutation and improvement operators) in the music context, to guarantee that composed melodies respect the principles of Music Theory and they are also pleasant to listening. This software can be customized to a particular musical style by including in it the specific musical knowledge domain. In particular, in order to validate our approach, we have used Spieldose to create different 8-bar pieces emulating the classicism style. The developed GUI of Spieldose offers to an unskilled music user the appropriate functionality for the compositional tasks. It also enables an immediate equivalence between the textual and audio formats of the musical work being composed.

As future work, we propose two types of improvements for Spieldose: those referred to the Interactive GA itself, and those related to the application interface. With respect to the first type, we intend to develop a system of modular GA components which enable the user to configure his/her own Interactive GA. As an example of this feature, new types of fitness functions could be included for both unskilled and expert users in the musical field. These functions would provide different weighted fitness criteria combining the user subjectivity and the objective musical quality for melody being composed. With respect to the application GUI, we plan to embed into the tool a musical editor which would display in musical notation a composed piece.

Acknowledgements

This research has been partially supported the grant URJC-TIC-C51-1, from Universidad Rey Juan Carlos (URJC) and Comunidad de Madrid (CAM), Spain.

References

1. López de Mantaras R., Arcos, J.L. AI and Music: From composition to expressive performance. AI Magazine, vol. 23, no. 3, pp. 43-57, 2002.
2. Gartland-Jones, A., Copley, P. The Suitability of Genetic Algorithms for Musical Composition. Contemporary Music Review vol. 22:3, 43-55, 2003.
3. Marques, M., Oliveira, V., Vieira, S., Rosa, A.C. Music composition using genetic evolutionary algorithms. Proceedings of the 2000 Congress on Evolutionary Computation, Vol 1, pp. 714-719, 2000.
4. Gartland-Jones, A. MusicBlox: A Real-Time Algorithmic Composition System Incorporating a Distributed Interactive Genetic Algorithm. LNCS 2611, pp: 490-501, 2003.
5. Burton, A.R., Vladimirova, T. Generation of Musical Sequences with Genetic Techniques. Comput. Music J., vol 23 pp:59-73, 1999.
6. Khalifa, Y., Foster, R. A Two-Stage Autonomous Evolutionary Music Composer. LNCS 3907 pp: 717-721, 2006
7. Hüseyin Göksu, Paul Pigg and Vikas Dixit. Music Composition Using Genetic Algorithms (GA) and Multilayer Perceptrons (MLP). LNCS 3612, pp:1242-1250, 2005.
8. Henz, M., Lauer, S., Zimmermann, D. COMPOzE-intention-based music composition through constraint programming. Proceedings Eighth IEEE International Conference on Tools with Artificial Intelligence. pp:118-121, 1996.
9. Michael O. Jewell, Lee Middleton, Mark S. Nixon, Adam Prgel-Bennett and Sylvia C. Wong. A Distributed Approach to Musical Composition. LNCS 3683, pp: 642-648, 2005
10. Unehara, M. Onisawa, T. Music composition system based on subjective evaluation. IEEE International Conference on Systems, Man and Cybernetics Vol. 1 pp. 980-986, 2003.
11. Unehara, M., Onisawa, T. Construction of Music Composition System with Interactive Genetic Algorithm. Proc. of 6th Asian Design Int. Conf. 2003.
12. Cruz-Alcázar, P.P., Vidal-Ruiz, E. Learning Regular Grammars to Model Musical Style: Comparing Different Coding Schemes. Lecture Notes in Computer Science, Vol. 1433/1998 pp. 211-222,1889.

Author Index

Lecture Notes in Computer Science

For information about Vols. 1–4445

please contact your bookseller or Springer

Vol. 4543: A.K. Bandara, M. Burgess (Eds.), Inter-Domain Management. XII, 237 pages. 2007.

Vol. 4542: P. Sawyer, B. Paech, P. Heymans (Eds.), Requirements Engineering: Foundation for Software Quality. IX, 384 pages. 2007.

Vol. 4541: T. Okadome, T. Yamazaki, M. Makhtari (Eds.), Pervasive Computing for Quality of Life Enhancemanet. IX, 248 pages. 2007.

Vol. 4539: N.H. Bshouty, C. Gentile (Eds.), Learning Theory. XII, 634 pages. 2007. (Sublibrary LNAI).

Vol. 4538: F. Escolano, M. Vento (Eds.), Graph-Based Representations in Pattern Recognition. XII, 416 pages. 2007.

Vol. 4537: K.C.-C. Chang, W. Wang, L. Chen, C.A. Ellis, C.-H. Hsu, A.C. Tsoi, H. Wang (Eds.), Advances in Web and Network Technologies, and Information Management. XXIII, 707 pages. 2007.

Vol. 4534: I. Tomkos, F. Neri, J. Solé Pareta, X. Masip Bruin, S. Sánchez Lopez (Eds.), Optical Network Design and Modeling. XI, 460 pages. 2007.

Vol. 4531: J. Indulska, K. Raymond (Eds.), Distributed Applications and Interoperable Systems. XI, 337 pages. 2007.

Vol. 4530: D.H. Akehurst, R. Vogel, R.F. Paige (Eds.), Model Driven Architecture- Foundations and Applications. X, 219 pages. 2007.

Vol. 4529: P. Melin, O. Castillo, L.T. Aguilar, J. Kacprzyk, W. Pedrycz (Eds.), Foundations of Fuzzy Logic and Soft Computing. XIX, 830 pages. 2007. (Sublibrary LNAI).

Vol. 4528: J. Mira, J.R. Álvarez (Eds.), Nature Inspired Problem-Solving Methods in Knowledge Engineering, Part II. XXII, 650 pages. 2007.

Vol. 4527: J. Mira, J.R. Álvarez (Eds.), Bio-inspired Modeling of Cognitive Tasks, Part I. XXII, 630 pages. 2007.

Vol. 4526: M. Malek, M. Reitenspieß, A. van Moorsel (Eds.), Service Availability. X, 155 pages. 2007.

Vol. 4525: C. Demetrescu (Ed.), Experimental Algorithms. XIII, 448 pages. 2007.

Vol. 4524: M. Marchiori, J.Z. Pan, C.d.S. Marie (Eds.), Web Reasoning and Rule Systems. XI, 382 pages. 2007.

Vol. 4523: Y.-H. Lee, H.-N. Kim, J. Kim, Y. Park, L.T. Yang, S.W. Kim (Eds.), Embedded Software and Systems. XIX, 829 pages. 2007.

Vol. 4522: B.K. Ersbøll, K.S. Pedersen (Eds.), Image Analysis. XVIII, 989 pages. 2007.

Vol. 4521: J. Katz, M. Yung (Eds.), Applied Cryptography and Network Security. XIII, 498 pages. 2007.

Vol. 4519: E. Franconi, M. Kifer, W. May (Eds.), The Semantic Web: Research and Applications. XVIII, 830 pages. 2007.

Vol. 4517: F. Boavida, E. Monteiro, S. Mascolo, Y. Koucheryavy (Eds.), Wired/Wireless Internet Communications. XIV, 382 pages. 2007.

Vol. 4516: L. Mason, T. Drwiega, J. Yan (Eds.), Managing Traffic Performance in Converged Networks. XXIII, 1191 pages. 2007.

Vol. 4515: M. Naor (Ed.), Advances in Cryptology - EUROCRYPT 2007. XIII, 591 pages. 2007.

Vol. 4514: S.N. Artemov, A. Nerode (Eds.), Logical Foundations of Computer Science. XI, 513 pages. 2007.

Vol. 4513: M. Fischetti, D.P. Williamson (Eds.), Integer Programming and Combinatorial Optimization. IX, 500 pages. 2007.

Vol. 4510: P. Van Hentenryck, L. Wolsey (Eds.), Integration of AI and OR Techniques in Constraint Programming for Combinatorial Optimization Problems. X, 391 pages. 2007.

Vol. 4509: Z. Kobti, D. Wu (Eds.), Advances in Artificial Intelligence. XII, 552 pages. 2007. (Sublibrary LNAI).

Vol. 4508: M.-Y. Kao, X.-Y. Li (Eds.), Algorithmic Aspects in Information and Management. VIII, 428 pages. 2007.

Vol. 4507: F. Sandoval, A. Prieto, J. Cabestany, M. Graña (Eds.), Computational and Ambient Intelligence. XXVI, 1167 pages. 2007.

Vol. 4506: D. Zeng, I. Gotham, K. Komatsu, C. Lynch, M. Thurmond, D. Madigan, B. Lober, J. Kvach, H. Chen (Eds.), Intelligence and Security Informatics: Biosurveillance. XI, 234 pages. 2007.

Vol. 4505: G. Dong, X. Lin, W. Wang, Y. Yang, J.X. Yu (Eds.), Advances in Data and Web Management. XXII, 896 pages. 2007.

Vol. 4504: J. Huang, R. Kowalczyk, Z. Maamar, D. Martin, I. Müller, S. Stoutenburg, K.P. Sycara (Eds.), Service-Oriented Computing: Agents, Semantics, and Engineering. X, 175 pages. 2007.

Vol. 4501: J. Marques-Silva, K.A. Sakallah (Eds.), Theory and Applications of Satisfiability Testing – SAT 2007. XI, 384 pages. 2007.

Vol. 4500: N. Streitz, A. Kameas, I. Mavrommati (Eds.), The Disappearing Computer. XVIII, 304 pages. 2007.

Vol. 4497: S.B. Cooper, B. Löwe, A. Sorbi (Eds.), Computation and Logic in the Real World. XVIII, 826 pages. 2007.

Vol. 4496: N.T. Nguyen, A. Grzech, R.J. Howlett, L.C. Jain (Eds.), Agent and Multi-Agent Systems: Technologies and Applications. XXI, 1046 pages. 2007. (Sublibrary LNAI).

Vol. 4495: J. Krogstie, A. Opdahl, G. Sindre (Eds.), Advanced Information Systems Engineering. XVI, 606 pages. 2007.

Vol. 4494: H. Jin, O.F. Rana, Y. Pan, V.K. Prasanna (Eds.), Algorithms and Architectures for Parallel Processing. XIV, 508 pages. 2007.

Vol. 4493: D. Liu, S. Fei, Z. Hou, H. Zhang, C. Sun (Eds.), Advances in Neural Networks – ISNN 2007, Part III. XXVI, 1215 pages. 2007.

Vol. 4492: D. Liu, S. Fei, Z. Hou, H. Zhang, C. Sun (Eds.), Advances in Neural Networks – ISNN 2007, Part II. XXVII, 1321 pages. 2007.

Vol. 4491: D. Liu, S. Fei, Z.-G. Hou, H. Zhang, C. Sun (Eds.), Advances in Neural Networks – ISNN 2007, Part I. LIV, 1365 pages. 2007.

Vol. 4490: Y. Shi, G.D. van Albada, J. Dongarra, P.M.A. Sloot (Eds.), Computational Science – ICCS 2007, Part IV. XXXVII, 1211 pages. 2007.

Vol. 4489: Y. Shi, G.D. van Albada, J. Dongarra, P.M.A. Sloot (Eds.), Computational Science – ICCS 2007, Part III. XXXVII, 1257 pages. 2007.

Vol. 4488: Y. Shi, G.D. van Albada, J. Dongarra, P.M.A. Sloot (Eds.), Computational Science – ICCS 2007, Part II. XXXV, 1251 pages. 2007.

Vol. 4487: Y. Shi, G.D. van Albada, J. Dongarra, P.M.A. Sloot (Eds.), Computational Science – ICCS 2007, Part I. LXXXI, 1275 pages. 2007.

Vol. 4486: M. Bernardo, J. Hillston (Eds.), Formal Methods for Performance Evaluation. VII, 469 pages. 2007.

Vol. 4485: F. Sgallari, A. Murli, N. Paragios (Eds.), Scale Space and Variational Methods in Computer Vision. XV, 931 pages. 2007.

Vol. 4484: J.-Y. Cai, S.B. Cooper, H. Zhu (Eds.), Theory and Applications of Models of Computation. XIII, 772 pages. 2007.

Vol. 4483: C. Baral, G. Brewka, J. Schlipf (Eds.), Logic Programming and Nonmonotonic Reasoning. IX, 327 pages. 2007. (Sublibrary LNAI).

Vol. 4482: A. An, J. Stefanowski, S. Ramanna, C.J. Butz, W. Pedrycz, G. Wang (Eds.), Rough Sets, Fuzzy Sets, Data Mining and Granular Computing. XIV, 585 pages. 2007. (Sublibrary LNAI).

Vol. 4481: J. Yao, P. Lingras, W.-Z. Wu, M. Szczuka, N.J. Cercone, D. Ślęzak (Eds.), Rough Sets and Knowledge Technology. XIV, 576 pages. 2007. (Sublibrary LNAI).

Vol. 4480: A. LaMarca, M. Langheinrich, K.N. Truong (Eds.), Pervasive Computing. XIII, 369 pages. 2007.

Vol. 4479: I.F. Akyildiz, R. Sivakumar, E. Ekici, J.C.d. Oliveira, J. McNair (Eds.), NETWORKING 2007. Ad Hoc and Sensor Networks, Wireless Networks, Next Generation Internet. XXVII, 1252 pages. 2007.

Vol. 4478: J. Martí, J.M. Benedí, A.M. Mendonça, J. Serrat (Eds.), Pattern Recognition and Image Analysis, Part II. XXVII, 657 pages. 2007.

Vol. 4477: J. Martí, J.M. Benedí, A.M. Mendonça, J. Serrat (Eds.), Pattern Recognition and Image Analysis, Part I. XXVII, 625 pages. 2007.

Vol. 4476: V. Gorodetsky, C. Zhang, V.A. Skormin, L. Cao (Eds.), Autonomous Intelligent Systems: Multi-Agents and Data Mining. XIII, 323 pages. 2007. (Sublibrary LNAI).

Vol. 4475: P. Crescenzi, G. Prencipe, G. Pucci (Eds.), Fun with Algorithms. X, 273 pages. 2007.

Vol. 4474: G. Prencipe, S. Zaks (Eds.), Structural Information and Communication Complexity. XI, 342 pages. 2007.

Vol. 4472: M. Haindl, J. Kittler, F. Roli (Eds.), Multiple Classifier Systems. XI, 524 pages. 2007.

Vol. 4471: P. Cesar, K. Chorianopoulos, J.F. Jensen (Eds.), Interactive TV: a Shared Experience. XIII, 236 pages. 2007.

Vol. 4470: Q. Wang, D. Pfahl, D.M. Raffo (Eds.), Software Process Dynamics and Agility. XI, 346 pages. 2007.

Vol. 4468: M.M. Bonsangue, E.B. Johnsen (Eds.), Formal Methods for Open Object-Based Distributed Systems. X, 317 pages. 2007.

Vol. 4467: A.L. Murphy, J. Vitek (Eds.), Coordination Models and Languages. X, 325 pages. 2007.

Vol. 4466: F.B. Sachse, G. Seemann (Eds.), Functional Imaging and Modeling of the Heart. XV, 486 pages. 2007.

Vol. 4465: T. Chahed, B. Tuffin (Eds.), Network Control and Optimization. XIII, 305 pages. 2007.

Vol. 4464: E. Dawson, D.S. Wong (Eds.), Information Security Practice and Experience. XIII, 361 pages. 2007.

Vol. 4463: I. Măndoiu, A. Zelikovsky (Eds.), Bioinformatics Research and Applications. XV, 653 pages. 2007. (Sublibrary LNBI).

Vol. 4462: D. Sauveron, K. Markantonakis, A. Bilas, J.-J. Quisquater (Eds.), Information Security Theory and Practices. XII, 255 pages. 2007.

Vol. 4459: C. Cérin, K.-C. Li (Eds.), Advances in Grid and Pervasive Computing. XVI, 759 pages. 2007.

Vol. 4453: T. Speed, H. Huang (Eds.), Research in Computational Molecular Biology. XVI, 550 pages. 2007. (Sublibrary LNBI).

Vol. 4452: M. Fasli, O. Shehory (Eds.), Agent-Mediated Electronic Commerce. VIII, 249 pages. 2007. (Sublibrary LNAI).

Vol. 4451: T.S. Huang, A. Nijholt, M. Pantic, A. Pentland (Eds.), Artifical Intelligence for Human Computing. XVI, 359 pages. 2007. (Sublibrary LNAI).

Vol. 4450: T. Okamoto, X. Wang (Eds.), Public Key Cryptography – PKC 2007. XIII, 491 pages. 2007.

Vol. 4448: M. Giacobini et al. (Ed.), Applications of Evolutionary Computing. XXIII, 755 pages. 2007.

Vol. 4447: E. Marchiori, J.H. Moore, J.C. Rajapakse (Eds.), Evolutionary Computation,Machine Learning and Data Mining in Bioinformatics. XI, 302 pages. 2007.

Vol. 4446: C. Cotta, J. van Hemert (Eds.), Evolutionary Computation in Combinatorial Optimization. XII, 241 pages. 2007.